			7A (17)	**8A** (18)

Partial periodic table:

			3A (13)	**4A** (14)	**5A** (15)	**6A** (1‿)	**7A** (17)	**8A** (18)

	1B (11)	**2B** (12)	Boron 5 **B** 10.811	Carbon 6 **C** 12.011	Nitrogen			Helium 2 **He**
(10)			Aluminum 13 **Al** 26.9815	Silicon 14 **Si** 28.0855	Phosph. 15 **P** 30.9738	32.066	Chlorine 17 **Cl** 35.4527	**Ne** 20.1797 Argon 18 **Ar** 39.948
Nickel 28 **Ni** 58.693	Copper 29 **Cu** 63.546	Zinc 30 **Zn** 65.39	Gallium 31 **Ga** 69.723	Germanium 32 **Ge** 72.61	Arsenic 33 **As** 74.9216	Selenium 34 **Se** 78.96	Bromine 35 **Br** 79.904	Krypton 36 **Kr** 83.80
Palladium 46 **Pd** 106.42	Silver 47 **Ag** 107.8682	Cadmium 48 **Cd** 112.411	Indium 49 **In** 114.82	Tin 50 **Sn** 118.710	Antimony 51 **Sb** 121.757	Tellurium 52 **Te** 127.60	Iodine 53 **I** 126.9045	Xenon 54 **Xe** 131.29
Platinum 78 **Pt** 195.08	Gold 79 **Au** 196.9665	Mercury 80 **Hg** 200.59	Thallium 81 **Tl** 204.3833	Lead 82 **Pb** 207.2	Bismuth 83 **Bi** 208.9804	Polonium 84 **Po** (209)	Astatine 85 **At** (210)	Radon 86 **Rn** (222)
110 Discovered Nov. 1994	111 Discovered Dec. 1994							

Europium 63 **Eu** 151.965	Gadolinium 64 **Gd** 157.25	Terbium 65 **Tb** 158.9253	Dysprosium 66 **Dy** 162.50	Holmium 67 **Ho** 164.9303	Erbium 68 **Er** 167.26	Thulium 69 **Tm** 168.9342	Ytterbium 70 **Yb** 173.04	Lutetium 71 **Lu** 174.967

Americium 95 **Am** (243)	Curium 96 **Cm** (247)	Berkelium 97 **Bk** (247)	Californium 98 **Cf** (251)	Einsteinium 99 **Es** (252)	Fermium 100 **Fm** (257)	Mendelevium 101 **Md** (258)	Nobelium 102 **No** (259)	Lawrencium 103 **Lr** (260)

Organic Laboratory Techniques

SMALL-SCALE APPROACH

INTRODUCTION TO

Organic Laboratory Techniques

SMALL-SCALE APPROACH

FIRST EDITION

Donald L. Pavia
Gary M. Lampman
George S. Kriz
Western Washington University, Bellingham, Washington

Randall G. Engel
Edmonds Community College, Lynnwood, Washington

Saunders Golden Sunburst Series
SAUNDERS COLLEGE PUBLISHING
Harcourt Brace College Publishers

Fort Worth   Philadelphia   San Diego   New York   Orlando   Austin
San Antonio   Toronto   Montreal   London   Sydney   Tokyo

Publisher: John Vondeling
Developmental Editor: Sandra Kiselica
Cover Designer: Kathleen Flanagan
Cover Photo Credit: Reginald Wickham Photography
Production Services: York Production Services
Product Manager: Angus McDonald

Printed in the United States of America
Introduction to Organic Laboratory Techniques: Small Scale Approach
ISBN: 0-03-024519-2
Library of Congress Catalog Card Number: 94-066238
4567890123   039   10987654321

"WARNINGS ABOUT SAFETY PRECAUTIONS
Some of the experiments contained in this Laboratory Manual involve a degree of risk on the part of the instructor and student. Although performing the experiments is generally safe for the college laboratory, unanticipated and potentially dangerous reactions are possible for a number of reasons, such as improper measurement or handling of chemicals, improper use of laboratory equipment, failure to follow laboratory safety procedures, and other causes. Neither the Publisher nor the Authors can accept any responsibility for personal injury or property damage resulting from the use of this publication."

This book is dedicated to the many students who have developed and tested experiments, contributed ideas, and otherwise helped us to make our book successful.

PREFACE

As we set out to write a new edition to our organic chemistry laboratory textbook, we initially saw it as a "fourth edition" to *Introduction to Organic Laboratory Techniques: A Contemporary Approach*. In the meantime, we had gathered experience with microscale techniques in the organic laboratory through the development of experiments for the microscale version of our laboratory textbook. That experience taught us that students *can* learn to do careful work in the organic laboratory on a small scale. They do not have to consume large quantities of chemicals, and they do not have to work in very large flasks to learn the standard techniques. At that same time, we recognized that many instructors do not wish to abandon the traditional-scale approach to their course, and many colleges and universities cannot afford to convert all of their glassware to microscale. Some instructors prefer to use equipment that more closely resembles research-style equipment. As a result, we have re-cast this textbook as a **small-scale** approach to the laboratory.

In the traditional approach to teaching this subject (called **standard-scale** or **"macroscale"**), the quantities of chemicals used were on the order of 5–100 grams. The approach used in this textbook, a **small-scale** approach, differs from the traditional laboratory in that nearly all of the experiments use smaller amounts of chemicals (1–10 grams). However, the glassware and methods used in small-scale experiments are identical to the glassware and methods used in standard-scale experiments.

The advantages of the small-scale approach include improved safety in the laboratory, reduced risk of fire and explosion, and reduced exposure to hazardous vapors. This approach decreases the need for hazardous waste disposal, leading to reduced contamination of the environment.

Another approach, a **microscale** approach, is used at many colleges and universities and differs from the traditional laboratory in that the experiments use *very* small amounts of chemicals (0.050–1.000 gram). Some of the microscale glassware is very different from standard-scale glassware, and there are a few special techniques that are unique to the microscale laboratory. Because of the widespread use of microscale methods, some reference to microscale techniques will be made in the technique chapters of this book. In fact, a few experiments in this textbook will feature microscale methods. However, these experiments have been designed to use ordinary glassware; they do not require specialized microscale equipment.

In preparing this new edition, therefore, we focused on converting our earlier experiments to a small scale. At the same time, we also devoted considerable effort toward improving the safety of all of the experiments. The introductory chapter, "Laboratory Safety," places strong emphasis on the safe use and disposal of hazardous chemicals. Information describing Material Safety Data Sheets and Right-to-Know laws is included. We have continued to update and improve instructions for the handling of waste products that are produced in the experiments. In this textbook we recommend that virtually all waste, including aqueous solutions, be placed into appropriate waste containers.

Besides modifying virtually all the experiments from the original edition, we have added some new experiments. These include:

Experiment 5C Isolation of Caffeine from "Mountain Dew" Syrup
Experiment 12   Analysis of Fats and Oils by Gas Chromatography
Experiment 25B Hydrolysis of Alkyl Chlorides (Computer-Interfaced Method)
Experiment 26B Chromic Acid Oxidation of Alcohols—UV-VIS Spectrophotometer Method
Experiment 26C Chromic Acid Oxidation of Alcohols—Computer-Interfaced Method
Experiment 31   Carbonation of an Unknown Aromatic Halide
Experiment 32   The Preparation of Pinacol and the Pinacol Rearrangement
Experiment 49   Preparation of an α,β-Unsaturated Ketone *via* Michael and Aldol Condensation Reactions
Experiment 50   An Introduction to Molecular Modeling
Experiment 53B Synthesis of β-Benzopinacolone: The Acid-Catalyzed Rearrangement of Benzpinacol
Experiment 65   The Fatty Acids: An Exercise in Molecular Modeling

We have also revised a number of experiments which have appeared in our microscale textbook, converting them for use in this edition. Included among these are:

Experiment 1   Isolation of the Active Ingredient in an Analgesic Drug
Experiment 9   Preparation of a C-4 or C-5 Acetate Ester
Experiment 18   Isolation of Essential Oils from Allspice, Cloves, Cumin, Caraway, Cinnamon, or Fennel
Experiment 27   4-Methylcyclohexene
Experiment 34   Relative Reactivities of Several Aromatic Compounds
Experiment 51   The Diels–Alder Reaction of Cyclopentadiene with Maleic Anhydride
Experiment 56   Preparation and Properties of Polymers: Polyester, Nylon, and Polystyrene
Experiment 59   Analysis of a Diet Soft Drink by HPLC
Experiment 63   Chiral Reduction of Ethyl Acetoacetate: Optical Purity Determination Using a Chiral Shift Reagent
Experiment 64   Esterification Reactions of Vanillin: The Use of NMR to Solve a Structure Proof Problem

The essays have been updated, and new discussions have been included in many of them. Among the new essays are:

Molecular Modeling and Molecular Mechanics
The Chemistry of Sweeteners

Part Two of this book contains the chapters on techniques. These chapters have been reorganized from the previous edition of the "macroscale" book. The technique chapters have been rewritten to focus on traditional-scale laboratory techniques, but since the use of microscale methods is increasing in popularity, we have included a discussion of the corresponding microscale methods as well. A new chapter on high-performance liquid chromatography has also been added.

In a departure from most laboratory textbooks, we have included methods that bring the computer and the laboratory together. We have provided experiments that introduce students to computer-interfaced experiments and to molecular modeling. If you would like to see more of these types of experiments in future editions, let us know!

This book also includes several experiments that require the students to develop their own laboratory procedure or to solve an interesting chemical problem through the application of appropriate experimental methods. By adding "open-ended" experiments such as these, we hope to demonstrate to the student something of the intellectual approach that is part of scientific research or the practice of chemistry in industry.

We would also like to point out that an instructor's manual accompanies our textbook and is available from Saunders College Publishing or from your Saunders representative. The Instructor's Manual contains complete instructions for the preparation of reagents and equipment for each experiment, as well as answers to each of the questions. Other comments that should prove helpful to the instructor are also included in the instructor's manual. We strongly recommend that instructors obtain a copy of this manual.

We owe our sincere thanks to the many colleagues who have used our textbooks and who have offered their suggestions for changes and improvements in our laboratory procedures. Although we cannot mention everyone who has made valuable contributions, we must make special mention of Richard Bosman (Washington State Patrol), Rosemary Fowler (Cottey College), Caroline Mann (Edmonds Community College), Vicky Minderhout (Seattle University), Mary O'Brien (Edmonds Community College), Karalee Stokes (St. Joseph Hospital), and Mary Syre (University of Washington). We have also received a great deal of important assistance from persons who work on our own campuses. We especially thank Dennis Fitch, Armando Herbelin, Denice Hougen, and Ruth McCrea. Production of this textbook was capably handled by Saunders College Publishing and York Production Services. We thank all who contributed, with special thanks to our Developmental Editor, Sandi Kiselica and to Dolores Wolfe at York Production Services.

We are especially grateful to the students and friends who have volunteered to participate in the development of experiments or who offered help and criticism. We thank Robert Ashby, Brian Bales, Jeffery Bowles, David Buck, Annie Chapman, Brian Colclazier, David Germack, Ned Ingalls, Derick Krutsinger, John Logan, Godfrey Miles, Heath Miller, Nadia Morgen, Katharine Popejoy, Andrea Powers-Probstfeld, Amber Ratcliffe, Robert Rienstra, Brenda Smits, and Alex Walters.

If you wish to contact us with comments, questions, or suggestions, we have a special electronic mail address for this purpose (plke@chem.wwu.edu). We encourage you to visit our home page at http://atom.chem.wwu.edu/dept/plkhome.html. You may also wish to visit the Saunders College Publishing web site at http://www.saunderscollege.com.

Finally, we must thank our families and special friends, especially Neva-Jean Pavia, Marian Lampman, Carolyn Kriz, and Marianne LaBarre, for their encouragement, support, and patience.

Donald L. Pavia
Gary M. Lampman
George S. Kriz
Randall G. Engel
April 1997

Saunders College Publishing may provide complimentary instructional aids and supplements or supplement packages to those adopters qualified under our adoption policy. Please contact your sales representative for more information. If as an adopter or potential user you receive supplements you do not need, please return them to your sales representative or send them to:

Attn: Returns Department
Troy Warehouse
465 South Lincoln Drive
Troy, MO 63379

TABLE OF CONTENTS

Part Two   THE TECHNIQUES

APPENDICES

Introduction

WELCOME TO ORGANIC CHEMISTRY!

Organic chemistry can be fun, and we hope to prove it to you. The work in this laboratory course will teach you a lot. The personal satisfaction that comes with performing a sophisticated experiment skillfully and successfully will be great.

To get the most out of the laboratory course, you should strive to do several things. First, you need to understand the organization of this laboratory manual and how to use it effectively. The manual is your guide to learning. Second, you must try to understand both the purpose and the principles behind each experiment you do. Third, you must try to organize your time effectively *before* each laboratory period.

ORGANIZATION OF THE TEXTBOOK

Consider briefly how this textbook is organized. There are four introductory chapters, of which this Welcome is the first; a chapter on laboratory safety is second; advance preparation and laboratory records make up the third; and laboratory glassware is the fourth. Beyond these introductory chapters, the textbook is divided into three parts. Part One consists of 64 experiments, which may be assigned as part of your laboratory course. Besides the experiments, you will also find 26 essays. These essays provide background information related to the experiments, and they place the experiments into a larger overall context, showing how the experiments and compounds can be applied to areas of everyday concern and interest. Your instructor will choose a set of these experiments.

Part Two is composed of a series of detailed instructions and explanations dealing with the techniques of organic chemistry. The techniques are extensively developed and used, and you will become familiar with them in the context of the experiments. Within each experiment, you will find a section, "Required Reading," that indicates which techniques you should study to do that experiment. Extensive cross-referencing to the techniques chapters in Part Two is included in each experiment. Each experiment also contains a section called "Special Instructions," which lists special safety precautions and specific instructions to you the student. Finally, each experiment contains a section entitled "Waste Disposal," which provides instruction on the correct means of disposing of the reagents and materials used during the experiment.

The Appendices to this textbook contain sections dealing with infrared spectroscopy, nuclear magnetic resonance, and ^{13}C nuclear magnetic resonance. Many of the experiments included in Part One utilize these spectroscopic techniques, and your instructor may choose to add them to other experiments.

ADVANCE PREPARATION

It is essential to plan carefully for each laboratory period so that you will be able to keep abreast of the material you will learn in your organic chemistry laboratory course. You should not treat these experiments as a novice cook would treat *The Good Housekeeping Cookbook*. You should come to the laboratory with a plan for the use of your time and some understanding of what you are about to do. A really good cook does not follow the recipe line-by-line with a finger, nor does a good mechanic fix your car with the instruction manual in one hand and a wrench in the other. In addition, it is unlikely that you will learn much if you try to follow the instructions blindly, without understanding them. We can't emphasize strongly enough that you should come to the lab *prepared.*

If there are items or techniques that you do not understand, you should not hesitate to ask questions. You will learn more, however, if you figure things out on your own. Don't rely on others to do your thinking for you.

You should read the chapter entitled "Advance Preparation and Laboratory Records" right away. Although your instructor will undoubtedly have a preferred format for keeping records, much of the material here will help you learn to think constructively about laboratory experiments in advance. It would also save time if, as soon as possible, you read the first six techniques chapters in Part Two. These techniques are basic to all experiments in this textbook. The laboratory class will begin with experiments almost immediately, and a thorough familiarity with this particular material will save you much valuable laboratory time.

It is also very important to read the chapter called "Laboratory Safety." It is your responsibility to know how to perform the experiments safely and how to understand and evaluate the risks that are associated with laboratory experiments. Knowing what to do and what not to do in the laboratory is of paramount importance, since the laboratory has many potential hazards associated with it.

BUDGETING TIME

As just mentioned in "Advance Preparation," you should read several chapters of this book even before your first laboratory class meeting. You should also read the assigned experiment carefully before every class meeting. Having read the experiment will allow you to schedule your time wisely. Often you will be doing more than one experiment at a time. Experiments like the fermentation of sugar or the chiral reduction of ethyl acetoacetate require a few minutes of advance preparation *one week* ahead of the actual experiment. At other times you will have to catch up on some unfinished details of a previous experiment. For instance, usually it is not possible to determine a yield accurately or a melting point of a product immediately after you first obtain the product. Products must be free of solvent to give an accurate weight or melting point range; they have to be "dried." Usually, this drying is done by leaving the product in an open container on your desk or in your locker. Then, when you have a pause in your schedule during the subsequent experiment, you can determine these missing data using a sample that is dry. Through careful planning you can set aside the time required to perform these miscellaneous experimental details.

PURPOSE

The main purpose of an organic laboratory course is to teach you the techniques necessary for a person dealing with organic chemicals. You will also learn the techniques needed for separating and purifying organic compounds. If the appropriate experiments are included in your course, you may also learn how to identify unknown compounds. The experiments themselves are only the vehicle for learning these techniques. The technique chapters in Part Two are the heart of this textbook, and you should learn these techniques thoroughly. Your instructor may provide laboratory lectures and demonstrations explaining the techniques, but the burden is on you to master them by familiarizing yourself with the chapters in Part Two.

Besides good laboratory technique and the methods of carrying out basic laboratory procedures, other things you will learn from this laboratory course are

1. How to take data carefully.
2. How to record relevant observations.
3. How to use your time effectively.
4. How to assess the efficiency of your experimental method.
5. How to plan for the isolation and purification of the substance you prepare.
6. How to work safely.
7. How to solve problems and think like a chemist.

In choosing experiments, we have tried whenever possible to make them relevant, and more important, interesting. To that end, we have tried to make them a learning experiment of a different kind. Most experiments are prefaced by a background essay to place things in context and to provide you with some new information. We hope to show you that organic chemistry pervades your lives (drugs, foods, plastics, perfumes, and so on). Furthermore, you should leave your course well trained in organic laboratory techniques. We are enthusiastic about our subject and hope you will receive it with the same spirit.

This textbook discusses the important laboratory techniques of organic chemistry and illustrates many important reactions and concepts. In the traditional approach to teaching this subject (called **standard scale** or **macroscale**), the quantities of chemicals used were on the order of 5–100 grams. The approach used in this textbook, a **small-scale** approach, differs from the traditional laboratory in that nearly all of the experiments use smaller amounts of chemicals (1–10 grams). However, the glassware and methods used in small-scale experiments are identical to the glassware and methods used in standard-scale experiments.

The advantages of the small-scale approach include improved safety in the laboratory, reduced risk of fire and explosion, and reduced exposure to hazardous vapors. This approach decreases the need for hazardous waste disposal, leading to reduced contamination of the environment.

Another approach, a **microscale** approach, differs from the traditional laboratory in that the experiments use *very* small amounts of chemicals (0.050–1.000 g). Some microscale glassware is very different from standard-scale glassware, and there are a few techniques that are unique to the microscale laboratory. Because of the widespread use of microscale methods, some reference to microscale techniques will be made in the technique chapters. A few experiments in this textbook feature microscale methods. These experiments have been designed to use ordinary glassware; they do not require specialized microscale equipment.

LABORATORY SAFETY

In any laboratory course, familiarity with the fundamentals of laboratory safety is critical. Any chemistry laboratory, particularly an organic chemistry laboratory, can be a dangerous place in which to work. Understanding potential hazards will serve well in minimizing that danger for you. We have attempted to point out specific hazards in the experiments found in this textbook. However, it is ultimately your responsibility, along with the laboratory instructor, to make sure that all laboratory work is carried out in a safe manner.

SAFETY GUIDELINES

In the organic chemistry laboratory, it is vital that you take necessary precautions. Your laboratory instructor will advise you of the specific rules for your laboratory. The following list of safety guidelines should be observed in all organic laboratories.

1. Eye Safety

Always Wear Approved Safety Glasses or Goggles. This sort of eye protection must be worn whenever you are in the laboratory. Even if you are not actually carrying out an experiment, a person near you might have an accident that could endanger your eyes, so eye protection is essential. Even dish washing may be hazardous. We know of cases in which a person has been cleaning glassware only to have an undetected piece of reactive material explode, sending fragments into the person's eyes. To avoid this sort of accident, it is necessary to wear your safety glasses at all times. We recommend that contact lenses not be worn in the laboratory, even when safety goggles are worn over them. Contact lenses can trap chemicals against the eye and make it difficult to wash the eye quickly. Soft contact lenses present an additional hazard because they can absorb harmful chemical vapors.

Learn the Location of Eyewash Facilities. If there are eyewash fountains in your laboratory, you should determine which one is nearest to you before you start to work. In case any chemical enters your eyes, go immediately to the eyewash fountain and flush your eyes and face with large amounts of water. If an eyewash fountain is not available, the laboratory will usually have at least one sink fitted with a piece of flexible hose. When the water is turned on, this hose can be aimed upward and directly into the face, thus working much like an eyewash fountain. The water flow rate should not be set too high and the temperature should be slightly warm in order to avoid damage to the eyes.

2. Fires

Use Care with Open Flames in the Laboratory. Because an organic chemistry laboratory course deals with flammable organic solvents, the danger of fire is frequently present. Because of this danger, DO NOT SMOKE IN THE LABORATORY. Furthermore, exercise supreme caution when you light matches or use any open flame. Always check to see whether your neighbors on either side, across the bench, and behind you are using flammable solvents. If so, either delay your use of a flame or move to a safe location, such as a fume hood, to use your open flame. Many flammable organic substances are the source of dense vapors that can travel for some distance down a bench. These vapors present a fire danger, and you should be careful, since the source of those vapors may be far away from you. Do not use the bench sinks to dispose of flammable solvents. If your bench has a trough running along it, pour only *water* (No flammable solvents!) into it. The troughs and sinks are designed to carry water from the condenser hoses and aspirators—not flammable materials.

Learn the Location of Fire Extinguishers, Fire Showers, and Fire Blankets. For your own protection in case of a fire, you should learn immediately where the nearest fire extinguisher, fire shower, and fire blanket are. You should learn how these safety devices are operated, particularly the fire extinguisher. Your instructor can demonstrate how to operate it.

If you have a fire, the best advice is to get away from it and let the instructor or laboratory assistant take care of it. DON'T PANIC! Time spent in thought before action is never wasted. If it is a small fire in a container, it usually can be extinguished quickly by placing a wire gauze screen with a ceramic fiber center or, possibly, a watch glass, over the mouth of the container. It is good practice to have a wire screen or watch glass handy whenever you are using a flame. If this method does not take care of the fire and if help from an experienced person is not readily available, then extinguish the fire yourself with a fire extinguisher.

Should your clothing catch on fire, DO NOT RUN. Walk *purposefully* toward the fire shower station or the nearest fire blanket. Running will fan the flames and intensify them.

3. Organic Solvents: Their Hazards

Avoid Contact with Organic Solvents. It is essential to remember that most organic solvents are **flammable** and will burn if they are exposed to an open flame or a match. Remember also that upon repeated or excessive exposure some may be toxic or carcinogenic (cancer causing) or both. For example, many chlorocarbon solvents, when accumulated in the body, cause liver deterioration similar to the cirrhosis caused by the excessive use of ethanol. The body does not rid itself easily of chlorocarbons nor does it **detoxify** them; thus, they build up over time and may cause illness in the future. Some chlorocarbons are also suspected to be carcinogens. MINIMIZE YOUR EXPOSURE. Long-term exposure to benzene may cause a form of leukemia. Don't sniff benzene, and avoid spilling it on yourself. Many other solvents, such as chloroform and ether, are good anesthetics and will put you to sleep if you breathe too much of them. They subsequently cause nausea. Many of these solvents have a synergistic effect with ethanol, meaning that they enhance its effect. Pyridine causes temporary impotence. In other words, organic solvents are just as dangerous as corrosive chemicals, such as sulfuric acid, but manifest their hazardous nature in other, more subtle ways.

If you are pregnant, you may want to consider taking this course at a later time. Some exposure to organic fumes is inevitable, and any possible risk to an unborn baby should be avoided.

In this textbook, carcinogenic chemicals or highly toxic solvents are called for only in a few experiments and when they are absolutely necessary to perform a particular procedure. Special precautions will be given when the use of these chemicals is required.

Minimize any direct exposure to solvents and treat them with respect. The laboratory room should be well ventilated. Normal cautious handling of solvents should not cause any health problem. If you are trying to evaporate a solution in an open container, you must do the evaporation in the hood. Excess solvents should be discarded in a container specifically intended for waste solvents, rather than down the drain at the laboratory bench.

A sensible precaution is to wear gloves when working with solvents. Gloves made from polyethylene are inexpensive and provide good protection. The disadvantage of polyethylene gloves is that they are slippery. Disposable surgical gloves provide a better grip on glassware and other equipment, but they do not offer as much protection as polyethylene gloves. Nitrile gloves offer better protection (see p. 10).

Do Not Breathe Solvent Vapors. If you want to check the odor of a substance, you should be careful not to inhale very much of the material. The technique for smelling flowers is not advisable here; you could inhale dangerous amounts of the compound. Rather, a technique for smelling minute amounts of a substance is used. You should pass a stopper or spatula moistened with the substance (if it is a liquid) under your nose. Alternatively, you may hold the substance away from you and waft the vapors toward you with your hand. But you should never hold your nose over the container and inhale deeply!

The hazards associated with organic solvents you are likely to encounter in the organic laboratory are discussed in detail beginning on page 13. By using proper safety precautions, your exposure to harmful organic vapors will be minimized and should present no health risk.

4. Waste Disposal

Do Not Place Any Liquid or Solid Waste into Sinks; Use Appropriate Waste Containers. Many substances are toxic, flammable, and difficult to degrade; it is neither legal nor acceptable to dispose of organic solvents or other liquid or solid reagents by pouring them down the sink. Municipal sewage-treatment plants are not equipped to remove these materials from sewage. Furthermore, with volatile and flammable materials, a spark or an open flame can cause an explosion in the sink or further down the drains.

The appropriate disposal method for wastes is to place them into appropriately labeled waste containers. These containers should be placed in the hoods in the laboratory. The waste containers will be disposed of safely by qualified persons using approved protocols.

Specific guidelines for disposing of waste will be determined by the people in charge of your particular laboratory and by local regulations. One system for handling waste disposal is presented here. For each experiment in this textbook, you will be instructed to dispose of all wastes in one of the following ways:

Nonhazardous Solids. Nonhazardous solids such as paper and corks can be placed into an ordinary wastebasket.

Broken Glassware. Broken glassware should be put into a container specifically designated for broken glassware.

Organic Solids. Solid products that are not turned in or any other organic solids should be disposed of in the container designated for organic solids.

Inorganic Solids. Solids such as alumina, silica gel, and activated charcoal should be placed into a container specifically designated for them.

Nonhalogenated Organic Solvents. Organic solvents such as diethyl ether, hexane, toluene, or any solvent that does not contain a halogen atom, should be disposed of in the container designated for nonhalogenated organic solvents.

Halogenated Solvents. Methylene chloride (dichloromethane), chloroform, and carbon tetrachloride are examples of common halogenated organic solvents. Dispose of all halogenated solvents into the container designated for them.

Strong Inorganic Acids and Bases. Strong acids such as hydrochloric, sulfuric, and nitric acid will be collected in a specially marked container. Strong bases such as sodium hydroxide and potassium hydroxide will also be collected in a specially designated container.

Aqueous Solutions. Aqueous solutions will be collected in a specially marked waste container. It is not necessary to separate each type of aqueous solution (unless it contains heavy metals); rather, unless otherwise instructed, you may combine all aqueous solutions into the same waste container. While many types of solutions (aqueous sodium bicarbonate, aqueous sodium chloride, etc.) may seem innocuous, and one may believe that their disposal down the sink drain is not likely to cause harm, nevertheless we must appreciate that many communities are becoming increasingly restrictive about what they will permit to enter municipal sewage-treatment systems. In light of this trend toward greater caution, it is important to develop good laboratory habits regarding the disposal of *all* chemicals.

Heavy Metals. Many heavy metal ions such as mercury and chromium are highly toxic and should be disposed of into specifically designated waste containers.

5. Use of Flames

Even though organic solvents are frequently flammable (for example, hexane, diethyl ether, methanol, acetone, and petroleum ether), there are certain laboratory procedures for which a flame may be used. Most often these procedures involve an aqueous solution. In fact, as a general rule, use a flame to heat only aqueous solutions. Heating methods that do not use a flame are discussed in detail in Technique 2, starting on page 600. Most organic solvents boil below 100°C, and a steam bath or a water bath may be used to heat these solvents safely. A listing of common organic solvents is given in Table 3.1, page 612, of Technique 3. Solvents marked in that table with boldface type will burn. Diethyl ether, pentane, and hexane are especially dangerous because, in combination with the correct amount of air, they may explode.

Some commonsense rules apply to using a flame in the presence of flammable solvents. Again, we stress that you should check to see whether anyone in your vicinity is using flammable solvents before you ignite any open flame. If someone is using a flammable solvent, move to a safer location before you light your flame. Your laboratory should have an area set aside for using a burner to prepare micropipets or other pieces of glassware.

The drainage troughs or sinks should never be used to dispose of flammable organic solvents. They will vaporize if they are low boiling and may encounter a flame further down the bench on their way to the sink.

6. Inadvertently Mixed Chemicals

To avoid unnecessary hazards of fire and explosion, **never pour any reagent back into a stock bottle.** There is always the chance that you may accidentally pour back some foreign substance that will react explosively with the chemical in the stock bottle. Of course, by pouring reagents back into stock bottles you may introduce impurities that could spoil the experiment for the person using the stock reagent after you. Thus, pouring things back into bottles is not only a dangerous practice, but it is also inconsiderate. This also means that you should not take more chemicals than you need.

7. Unauthorized Experiments

You should never undertake any unauthorized experiments. The risk of an accident is high, particularly with an experiment that has not been completely checked to reduce the hazard. Also, you should never work alone in the laboratory. The laboratory instructor or supervisor must always be present.

8. Food in the Laboratory

Because all chemicals are potentially toxic, you should avoid accidentally ingesting any toxic substance; therefore, never eat or drink any food in the laboratory. There is always the possibility that whatever you are eating or drinking may become contaminated with a potentially hazardous material.

9. Clothing

You should always wear shoes in the laboratory. Note that open-toed shoes or sandals offer inadequate protection against spilled chemicals or broken glass. Do not wear your best clothing in the laboratory because some chemicals can make holes or permanent stains on your clothing. To protect yourself and your clothing, it is advisable to wear a full-length laboratory apron or coat.

When working with chemicals that are very toxic, you should wear some type of gloves. Disposable gloves are inexpensive, offer good protection, and provide acceptable "feel." Many departmental stockrooms or college bookstores offer disposable gloves for sale. Disposable latex surgical or polyethylene gloves are the least expensive type of glove; they are satisfactory when working with inorganic reagents and solutions. Better protection is afforded by disposable "nitrile" gloves. This type of glove provides good protection against organic chemicals and solvents. Heavier nitrile gloves are also available.

On a final note, you should tie back hair that is shoulder length or longer. This is especially important if you are working with a burner.

Note: Any injury, no matter how small, must be reported to your instructor immediately.

10. First Aid: Cuts, Minor Burns, and Acid or Base Burns

If any chemical enters your eyes, as instructed in "Eye Safety," immediately irrigate the eyes with copious quantities of water. Tempered (slightly warm) water, if it is available, is preferable. Be sure that the eyelids are kept open. Continue flushing the eyes in this way for 15 minutes.

In case of a cut, wash the wound well with water, unless you are specifically instructed to do otherwise. If necessary, apply pressure to the wound to stop the flow of blood.

Minor burns caused by flames or contact with hot objects may be soothed by immediately immersing the burned area in cold water or cracked ice for about 5 minutes. Applying salves to burns is discouraged. Severe burns must be examined and treated by a physician.

For chemical acid or base burns, rinse the burned area with copious quantities of water for at least 10 minutes.

If you accidentally ingest a chemical, immediately begin drinking large quantities of water while proceeding immediately to the nearest medical assistance. It is important that the examining physician be informed of the exact nature of the substance ingested.

RIGHT-TO-KNOW LAWS

The federal government and most state governments now require that employers provide their employees with complete information about hazards in the work place. These regulations are often referred to as **Right-to-Know Laws.** At the federal level, the

Occupational Safety and Health Administration (OSHA) is charged with enforcing these regulations.

In 1990, the federal government extended the Hazard Communication Act, which established the Right-to-Know Laws, to include a provision that requires the establishment of a Chemical Hygiene Plan at all academic laboratories. Every college and university chemistry department should have a Chemical Hygiene Plan. Having this plan means that all the safety regulations and laboratory procedures will be written in a manual. The plan also provides for the training of all employees in laboratory safety. Your laboratory instructor and assistants should have this training.

One of the components of Right-to-Know Laws is that employees and students have access to information about the hazards of any chemicals with which they are working. In this textbook, we alert you to dangers to which you need to pay particular attention. However, you may want to seek additional information. Two excellent sources of information are labels on the bottles that come from a chemical manufacturer and Material Safety Data Sheets (MSDSs). The MSDSs are also provided by the manufacturer and must be kept available for all chemicals used at educational institutions.

Material Safety Data Sheets

Reading an MSDS for a chemical can be a daunting experience, even for an experienced chemist. MSDSs contain a wealth of information, some of which must be decoded to be understood. A partial MSDS for methanol is shown on pages (14–18). Only the information that might be of interest to you is shown there. You may wish to refer to those pages while reading the paragraphs that follow.

Section I. The first part of Section I identifies the substance by name, formula, and various numbers and codes. Most organic compounds have more than one name. In this case, the systematic (or IUPAC) name is methanol, and the other names are common names or are from an older system of nomenclature. The CAS No. (Chemical Abstract Service Number) is often used to identify a substance, and it may be used to access extensive information about a substance found in many computer databases or in the library.

The Baker SAF-T-DATA System is found on all MSDSs and bottle labels for chemicals supplied by J. T. Baker, Inc. For each category listed, the number indicates the degree of hazard. The lowest number is 0 (very low hazard) and the highest number is 4 (extreme hazard). The Health category refers to the danger involved when a substance is inhaled, ingested, or absorbed. Flammability indicates the tendency of a substance to burn. Reactivity refers to how reactive a substance is with air, water, or other substances. The last category, Contact, refers to how hazardous a substance is when it comes in contact with external parts of the body. Note that this rating scale is applicable only to Baker MSDSs and labels; other rating scales with different meanings are also in common use.

Section II. The information contained in Section II refers to hazards associated with mixing of chemicals. In the organic laboratory, you will be dispensing pure substances; therefore, the information contained in Section II is not likely to be useful.

Section III. The odor threshold in Section III can sometimes be of use in determining whether a substance can be detected by odor before it has reached a dangerous level. If the odor threshold for a substance is lower than the threshold limit value (TLV, discussed in Section V), then most people will be able to smell a substance before the concentration has reached a toxic level. For example, chlorine gas has a TLV of 1 ppm (parts per million), while the odor threshold is 0.314 ppm. Because these levels are similar, the concentration level is near the toxic level if you can smell chlorine gas.

Section IV.   In Section IV is found the NFPA (National Fire Protection Association) rating. This is similar to the Baker SAF-T-DATA System (discussed in Section I), except that the number represents the hazards when a fire is present. The order here is Health, Flammability, and Reactivity. Often this is presented in graphic form on a label (see figure). The small diamonds are often color coded: blue for health, red for flammability, and yellow for reactivity. The bottom diamond (white) is sometimes used to display graphic symbols denoting unusual reactivity, hazards, or special precautions to be taken.

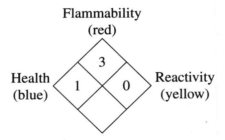

Section V.   Much valuable information is found in Section V. To help you understand this material, some of the more important terms used here are defined:

Threshold Limit Value (TLV).   The American Conference of Governmental Industrial Hygienists (ACGIH) developed the TLV. This is the maximum concentration of a substance in air that a person should be exposed to on a regular basis. It is usually expressed in ppm (parts per million) or mg/m^3. Note that this value assumes that a person is exposed to the substance 40 hours per week, on a long-term basis. This value may not be particularly applicable in the case of a student performing an experiment in a single laboratory period.

Permissible Exposure Limit (PEL).   This has the same meaning as TLV; however, PELs were developed by OSHA. Note that for methanol the TLV and PEL are both 200 ppm.

Lethal Dose, 50% Mortality (LD_{50}).   This is the dose of a substance that will kill 50% of the animals administered a single dose of this amount. Different means of administration are used, such as oral, intraperitoneal (injected into the lining of the abdominal cavity), subcutaneous (injected under the skin), and applied to the surface of the skin. The LD_{50} is usually expressed in milligrams (mg) of substance per kilogram (kg) of animal weight. The lower the value of LD_{50} the more toxic the substance. It is assumed that the toxicity in humans will be similar.

Without considerably more knowledge about chemical toxicity, this information is most useful for comparing the toxicity of one substance to another. For example, TLV for methanol is 200 ppm, whereas the TLV for benzene is 10 ppm. Clearly, performing an experiment involving benzene would require much more stringent precautions than an experiment involving methanol. One of the LD_{50} values for methanol is 5628 mg/kg. The comparable LD_{50} value of aniline is 250 mg/kg. Clearly, aniline is much more toxic, and since it is easily absorbed through the skin it presents a significant hazard.

It should also be mentioned that both TLV and PEL ratings assume that the worker comes in contact with a substance on a repeated and long-term basis. Thus, even if a chemical has a relatively low TLV or PEL, this does not mean that using it for one experiment will present a danger to you. Furthermore, by performing experiments at the small-scale level with proper safety precautions, your exposure to organic chemicals in this course will be minimal.

Section V also provides helpful information for emergency and first aid procedures. Specific information about the reactivity of a substance is given in Section VI. This information could be important to consider before carrying out an experiment not previously done. The last part of this MSDS, Section VII, deals with procedures for handling spills and disposal. The information about spills could be very helpful, particularly if a large amount of a chemical were spilled.

Bottle Labels

Reading the label on a bottle can be a very helpful way of learning about the hazards of a chemical. The amount of information varies greatly, depending on which company supplied the chemical.

Apply some common sense when you read MSDSs and bottle labels. Using these chemicals does not mean you will suffer the consequences described for each chemical. For example, an MSDS for sodium chloride states: "Exposure to this product may have serious adverse health effects." In spite of the apparent severity of this cautionary statement, it would not be reasonable to expect people to stop using sodium chloride in a chemistry experiment or to stop sprinkling a small amount of it (as table salt) on eggs to enhance their flavor. In many cases, the consequences described in MSDSs from exposure to chemicals are somewhat overstated, particularly for students using these chemicals to perform a small-scale laboratory experiment.

COMMON SOLVENTS

Most organic experiments involve an organic solvent at some step in the procedure. A list of common organic solvents follows, with a discussion of toxicity, possible carcinogenic properties, and precautions that you should use when handling these solvents. A tabulation of the compounds currently suspected of being carcinogens appears at the end of this chapter.

Acetic Acid.   Glacial acetic acid is corrosive enough to cause serious acid burns on the skin. Its vapors can irritate the eyes and nasal passages. Care should be exercised not to breathe the vapors and not to allow them to escape into the laboratory.

Acetone.   Relative to other organic solvents, acetone is not very toxic. It is flammable, however. Do not use near open flames.

J. T. BAKER CHEMICAL CO. 222 RED SCHOOL LANE, PHILLIPSBURG, NJ 08865
M A T E R I A L S A F E T Y D A T A S H E E T
24-HOUR EMERGENCY TELEPHONE — (908) 859-2151
CHEMTREC # (800) 424-9300 — NATIONAL RESPONSE CENTER # (800) 424-8802

M2015 M06   METHANOL
EFFECTIVE: 03/09/92   ISSUED: 03/28/92

J. T. BAKER INC., 222 RED SCHOOL LANE, PHILLIPSBURG, NJ 08865

SECTION I - PRODUCT IDENTIFICATION

PRODUCT NAME: METHANOL
COMMON SYNONYMS: METHYL ALCOHOL; WOOD ALCOHOL; CARBINOL;
                     METHYLOL; WOOD SPIRIT
CHEMICAL FAMILY: ALCOHOLS
FORMULA: CH_3OH
FORMULA WT.:   32.04
CAS NO.:   67-56-1
NIOSH/RTECS NO.:   PC 1400000
PRODUCT USE: LABORATORY REAGENT
PRODUCT CODES: 9049,9063,9076,9091,5370,9074,6808,9127,9098,5807,9073,
                     9071,5811,5217,9075,9090,9093,P704,9069,5536,9263,9070,
                     9077,9072,9068

PRECAUTIONARY LABELING

BAKER SAF-T-DATA* SYSTEM

HEALTH	-	3	SEVERE (POISON)
FLAMMABILITY	-	3	SEVERE (FLAMMABLE)
REACTIVITY	-	1	SLIGHT
CONTACT	-	1	SLIGHT

LABORATORY PROTECTIVE EQUIPMENT

GOGGLES & SHIELD; LAB COAT & APRON; VENT HOOD; PROPER GLOVES; CLASS B EXTINGUISHER

U. S. PRECAUTIONARY LABELING

POISON DANGER

FLAMMABLE. HARMFUL IF INHALED. CANNOT BE MADE NON-POISONOUS. MAY BE FATAL OR CAUSE BLINDNESS IF SWALLOWED.
KEEP AWAY FROM HEAT, SPARKS, FLAME. DO NOT GET IN EYES, ON SKIN, ON CLOTHING.
AVOID BREATHING VAPOR. KEEP IN TIGHTLY CLOSED CONTAINER. USE WITH ADEQUATE VENTILATION. WASH THOROUGHLY AFTER HANDLING. IN CASE OF FIRE, USE ALCOHOL FOAM, DRY CHEMICAL, CARBON DIOXIDE - WATER MAY BE INEFFECTIVE. FLUSH SPILL AREA WITH WATER SPRAY.

INTERNATIONAL LABELING

HIGHLY FLAMMABLE. TOXIC BY INHALATION AND IF SWALLOWED.
KEEP OUT OF REACH OF CHILDREN. KEEP CONTAINER TIGHTLY CLOSED. KEEP
AWAY FROM SOURCES OF IGNITION - NO SMOKING. AVOID CONTACT WITH SKIN.

SAF-T-DATA* STORAGE COLOR CODE: RED (FLAMMABLE)

SECTION II - HAZARDOUS COMPONENTS

...

SECTION III - PHYSICAL DATA

BOILING POINT: 65 C (149 F)
   (AT 760 MMHG)

VAPOR PRESSURE (MMHG): 96
   (20 C)

MELTING POINT: −98 C (−144 F)
   (AT 760 MMHG)

VAPOR DENSITY (AIR = 1): 1.11

SPECIFIC GRAVITY: 0.79
   (H_2O = 1)

EVAPORATION RATE: 4.6
   (BUTYL ACETATE = 1)

SOLUBILITY(H_2O): COMPLETE (100%)

% VOLATILES BY VOLUME: 100
   (21 C)

PH: N/A

ODOR THRESHOLD (P.P.M.): N/A

PHYSICAL STATE: LIQUID

COEFFICIENT WATER/OIL DISTRIBUTION: N/A

APPEARANCE & ODOR: CLEAR, COLORLESS LIQUID. PUNGENT ODOR.

SECTION IV - FIRE AND EXPLOSION HAZARD DATA

FLASH POINT (CLOSED CUP): 12 C (54 F)   NFPA 704M RATING: 1-3-0

AUTOIGNITION TEMPERATURE: 463 C (867 F)

FLAMMABLE LIMITS: UPPER - 36.0%   LOWER - 6.0%

FIRE EXTINGUISHING MEDIA
   USE ALCOHOL FOAM, DRY CHEMICAL OR CARBON DIOXIDE. (WATER MAY BE
   INEFFECTIVE.)

SPECIAL FIRE-FIGHTING PROCEDURES
FIREFIGHTERS SHOULD WEAR PROPER PROTECTIVE EQUIPMENT AND
SELF-CONTAINED BREATHING APPARATUS WITH FULL FACEPIECE OPERATED
IN POSITIVE PRESSURE MODE. MOVE CONTAINERS FROM FIRE AREA IF IT
CAN BE DONE WITHOUT RISK. USE WATER TO KEEP FIRE-EXPOSED
CONTAINERS COOL.

UNUSUAL FIRE & EXPLOSION HAZARDS
VAPORS MAY FLOW ALONG SURFACES TO DISTANT IGNITION SOURCES AND
FLASH BACK. CLOSED CONTAINERS EXPOSED TO HEAT MAY EXPLODE.
CONTACT WITH STRONG OXIDIZERS MAY CAUSE FIRE. BURNS WITH A
CLEAR, ALMOST INVISIBLE FLAME.

TOXIC GASES PRODUCED
CARBON MONOXIDE, CARBON DIOXIDE, FORMALDEHYDE

EXPLOSION DATA-SENSITIVITY TO MECHANICAL IMPACT
NONE IDENTIFIED

EXPLOSION DATA-SENSITIVITY TO STATIC DISCHARGE
NONE IDENTIFIED

SECTION V - HEALTH HAZARD DATA

THRESHOLD LIMIT VALUE (TLV/TWA): 260 MG/M^3 (200 PPM)

THE TLV LISTED DENOTES TLV (SKIN).

SHORT-TERM EXPOSURE LIMIT (STEL): 310 MG/M^3 (250 PPM)

PERMISSIBLE EXPOSURE LIMIT (PEL): 260 MG/M^3 (200 PPM)

TOXICITY OF COMPONENTS

ORAL RAT LD$_{50}$ FOR METHANOL   5628 MG/KG
INTRAPERITONEAL RAT LD$_{50}$ FOR METHANOL   9540 MG/KG
SUBCUTANEOUS MOUSE LD$_{50}$ FOR METHANOL   9800 MG/KG
SKIN RABBIT LD$_{50}$ FOR METHANOL   20 G/KG
CARCINOGENICITY: NTP: NO   IARC: NO   Z LIST: NO   OSHA REG: NO

CARCINOGENICITY
NONE IDENTIFIED.

REPRODUCTIVE EFFECTS
NONE IDENTIFIED.

EFFECTS OF OVEREXPOSURE

INHALATION: IS HARMFUL AND MAY BE FATAL, HEADACHE, NAUSEA, VOMITING, DIZZINESS, NARCOSIS, RESPIRATORY FAILURE, LOW BLOOD PRESSURE, CENTRAL NERVOUS SYSTEM DEPRESSION

SKIN CONTACT: IRRITATION, PROLONGED CONTACT MAY CAUSE DERMATITIS

EYE CONTACT: IRRITATION, MAY CAUSE TEMPORARY CORNEAL DAMAGE

SKIN ABSORPTION: NONE IDENTIFIED

INGESTION: IS HARMFUL AND MAY BE FATAL, BLINDNESS, HEADACHE, NAUSEA, VOMITING, DIZZINESS, GASTROINTESTINAL IRRITATION, CENTRAL NERVOUS SYSTEM DEPRESSION, HEARING LOSS

CHRONIC EFFECTS: KIDNEY DAMAGE, LIVER DAMAGE

TARGET ORGANS
EYES, SKIN, CENTRAL NERVOUS SYSTEM, GI TRACT, RESPIRATORY SYSTEM, LUNGS

MEDICAL CONDITIONS GENERALLY AGGRAVATED BY EXPOSURE
EYE DISORDERS, SKIN DISORDERS, LIVER OR KIDNEY DISORDERS

PRIMARY ROUTES OF ENTRY
INHALATION, INGESTION, EYE CONTACT, SKIN CONTACT, ABSORPTION

EMERGENCY AND FIRST AID PROCEDURES

INGESTION: CALL A PHYSICIAN. IF SWALLOWED, IF CONSCIOUS, GIVE LARGE AMOUNTS OF WATER. INDUCE VOMITING.

INHALATION: IF INHALED, REMOVE TO FRESH AIR. IF NOT BREATHING, GIVE ARTIFICIAL RESPIRATION. IF BREATHING IS DIFFICULT, GIVE OXYGEN.

SKIN CONTACT: IN CASE OF CONTACT, IMMEDIATELY FLUSH SKIN WITH PLENTY OF WATER FOR AT LEAST 15 MINUTES WHILE REMOVING CONTAMINATED CLOTHING AND SHOES. WASH CLOTHING BEFORE RE-USE.

EYE CONTACT: IN CASE OF EYE CONTACT, IMMEDIATELY FLUSH WITH PLENTY OF WATER FOR AT LEAST 15 MINUTES.

SARA/TITLE III HAZARD CATEGORIES AND LISTS

ACUTE: YES   CHRONIC: YES   FLAMMABILITY: YES   PRESSURE: NO
REACTIVITY: NO

EXTREMELY HAZARDOUS SUBSTANCE: NO
CERCLA HAZARDOUS SUBSTANCE: YES   CONTAINS METHANOL (RQ = 5000 LB)
SARA 313 TOXIC CHEMICALS: YES   CONTAINS METHANOL
    GENERIC CLASS: CO5
TSCA INVENTORY: YES

SECTION VI - REACTIVITY DATA

STABILITY: STABLE   HAZARDOUS POLYMERIZATION: WILL NOT OCCUR

CONDITIONS TO AVOID: HEAT, FLAME, OTHER SOURCES OF IGNITION

INCOMPATIBLES: STRONG OXIDIZING AGENTS, STRONG ACIDS, ZINC,
                           ALUMINUM, MAGNESIUM

DECOMPOSITION PRODUCTS: CARBON MONOXIDE, CARBON DIOXIDE,
                           FORMALDEHYDE

SECTION VII - SPILL & DISPOSAL PROCEDURES

STEPS TO BE TAKEN IN THE EVENT OF A SPILL OR DISCHARGE
     WEAR SELF-CONTAINED BREATHING APPARATUS AND FULL PROTECTIVE
     CLOTHING. SHUT OFF IGNITION SOURCES; NO FLARES, SMOKING OR
     FLAMES IN AREA. STOP LEAK IF YOU CAN DO SO WITHOUT RISK. USE
     WATER SPRAY TO REDUCE VAPORS. TAKE UP WITH SAND OR OTHER NON-
     COMBUSTIBLE ABSORBENT MATERIAL AND PLACE INTO CONTAINER FOR
     LATER DISPOSAL. FLUSH AREA WITH WATER.

J. T. BAKER SOLUSORB(R) SOLVENT ABSORBENT IS RECOMMENDED FOR SPILLS
OF THIS PRODUCT.

DISPOSAL PROCEDURE
     DISPOSE IN ACCORDANCE WITH ALL APPLICABLE FEDERAL, STATE, AND
     LOCAL ENVIRONMENTAL REGULATIONS.

EPA HAZARDOUS WASTE NUMBER: U154 (TOXIC WASTE)

SECTION VIII - INDUSTRIAL PROTECTIVE EQUIPMENT

...

SECTION IX - STORAGE AND HANDLING PRECAUTIONS

...

SECTION X - TRANSPORTATION DATA AND ADDITIONAL INFORMATION

...

Benzene. Benzene can cause damage to bone marrow; it is a cause of various blood disorders, and its effects may lead to leukemia. Benzene is considered a serious carcinogenic hazard. It is absorbed rapidly through the skin. It also poisons the liver and kidneys. In addition, benzene is flammable. Because of its toxicity and its carcinogenic properties, benzene should not be used in the laboratory; you should use some less dangerous solvent instead. In this textbook, *no experiments call for benzene.* Toluene is considered a safe alternative solvent in procedures that specify benzene.

Carbon Tetrachloride. Carbon tetrachloride can cause serious liver and kidney damage as well as skin irritation and other problems. It is absorbed rapidly through the skin. In high concentrations, it can cause death, owing to respiratory failure. Moreover, carbon tetrachloride is suspected of being a carcinogenic material. Although this solvent has the advantage of being nonflammable (in the past, it was used on occasion as a fire extinguisher), it should not be used routinely in the laboratory since it causes health problems. If no reasonable substitute exists, however, it must be used in small quantities, as in preparing samples for infrared (IR) and nuclear magnetic resonance (NMR) spectroscopy. In such cases, you must use it in a hood.

Chloroform. Chloroform is like carbon tetrachloride in its toxicity. It has been used as an anesthetic. However, chloroform is currently on the list of suspect carcinogens. Because of this, do not use chloroform routinely as a solvent in the laboratory. Occasionally, it may be necessary to use chloroform as a solvent for special samples. Then, you must use it in a hood. Methylene chloride is usually found to be a safer substitute in procedures that specify chloroform as a solvent. Deuterochloroform $CDCl_3$ is a common solvent for NMR spectroscopy. Caution dictates that you should treat it with the same respect as chloroform.

1,2-Dimethoxyethane (Ethylene Glycol Dimethyl Ether or Monoglyme). Because it is miscible with water, 1,2-dimethoxyethane is a useful alternative to solvents such as dioxane and tetrahydrofuran, which may be more hazardous. It is flammable and should not be handled near open flames. On long exposure to light and oxygen, explosive peroxides may form. 1,2-Dimethoxyethane is a possible reproductive toxin.

Dioxane. Dioxane has been used widely because it is a convenient, water-miscible solvent. It is now suspected, however, of being carcinogenic. Additionally, it is toxic, affecting the central nervous system, liver, kidneys, skin, lungs, and mucous membranes. Dioxane is also flammable and tends to form explosive peroxides when it is exposed to light and air. Because of its carcinogenic properties, it is no longer used in the laboratory unless absolutely necessary. Either 1,2-dimethoxyethane or tetrahydrofuran is a suitable, water-miscible alternative solvent.

Ethanol. Ethanol has well-known properties as an intoxicant. In the laboratory, the principal danger arises from fires, since ethanol is a flammable solvent. When using ethanol, take care to work where there are no open flames.

Diethyl Ether (Ether). The principal hazard associated with diethyl ether is fire or explosion. Ether is probably the most flammable solvent one is likely to find in the laboratory. Because the vapors are much more dense than air, they may travel along a laboratory bench for a considerable distance from their source before being ignited. Before you begin to use ether, it is very important to be sure that no one is working with matches or any open flame. Ether is not a particularly toxic solvent, although in high enough concentrations it can cause drowsiness and perhaps nausea. It has been used as a general anesthetic. Ether can form highly explosive peroxides when exposed to air. Consequently, you should never distill it to dryness.

Hexane.   Hexane may be irritating to the respiratory tract. It can also act as an intoxicant and a depressant of the central nervous system. It can cause skin irritation because it is an excellent solvent for skin oils. The most serious hazard, however, comes from its flammable nature. The precautions recommended for the use of diethyl ether in the presence of open flames apply equally to hexane.

Ligroin.   See Hexane.

Methanol.   Much of the material outlining the hazards of ethanol applies to methanol. Methanol is more toxic than ethanol; ingestion can cause blindness and even death. Because methanol is more volatile, the danger of fires is more acute.

Methylene Chloride (Dichloromethane).   Methylene chloride is not flammable. Unlike other members of the class of chlorocarbons, it is not currently considered a serious carcinogenic hazard. Recently, however, it has been the subject of much serious investigation, and there have been proposals to regulate it in industrial situations where workers have high levels of exposure on a day-to-day basis. Methylene chloride is less toxic than chloroform and carbon tetrachloride. It can cause liver damage when ingested, however, and its vapors may cause drowsiness or nausea.

Pentane.   See Hexane.

Petroleum Ether.   See Hexane.

Pyridine.   There is some fire hazard associated with pyridine. The most serious hazard arises from its toxicity, however. Pyridine may cause depression of the central nervous system; irritation of the skin and respiratory tract; damage to the liver, kidneys, and gastrointestinal system; and even temporary sterility. You should treat pyridine as a highly toxic solvent and handle it only in the fume hood.

Tetrahydrofuran.   Tetrahydrofuran may cause irritation of the skin, eyes, and respiratory tract. It should never be distilled to dryness, since it tends to form potentially explosive peroxides on exposure to air. Tetrahydrofuran does present a fire hazard.

Toluene.   Unlike benzene, toluene is not considered a carcinogen. However, it is at least as toxic as benzene. It can act as an anesthetic and cause damage to the central nervous system. If benzene is present as an impurity in toluene, then one must expect the hazards associated with benzene to manifest themselves. Toluene is also a flammable solvent, and the usual precautions about working near open flames should be applied.

You should not use certain solvents in the laboratory because of their carcinogenic properties. Benzene, carbon tetrachloride, chloroform, and dioxane are among these solvents. For certain applications, however, notably as solvents for infrared or NMR spectroscopy, there may be no suitable alternative solvent. When it is necessary to use one of these solvents, safety precautions are recommended, or you will be referred to the discussion in Technique 19.

Because relatively large amounts of solvents may be used in a large organic laboratory class, your laboratory supervisor must take care to store these substances safely. Only the amount of solvent that is needed for a particular experiment being conducted should be kept in the laboratory room. The preferred location for bottles of solvents being used during a class period is in a hood. When the solvents are not being used, they should be stored in a fireproof storage cabinet for solvents. If possible, this cabinet should be ventilated into the fume hood system.

CARCINOGENIC SUBSTANCES

A **carcinogenic substance** is one that causes cancer in living tissue. It should be pointed out that, in determining whether a substance is carcinogenic, the normal procedure is to expose laboratory animals to high dosages over a long period. It is not clear whether short-term exposure to these chemicals carries with it a comparable risk, but it is prudent to use these substances with special precautions. There are only a few experiments in this book in which a procedure calls for the use of a carcinogenic substance. We clearly indicate when this occurs and give special precautions. If you follow all safety precautions given in this textbook for handling such substances, your exposure will be very brief and minimal.

Many regulatory agencies have compiled lists of carcinogenic substances or substances suspected of being carcinogenic. Because there are inconsistencies in these lists, compiling a definitive list of carcinogenic substances is difficult. The accompanying table includes common substances that are found in many of these lists.

Acetamide	4-Methyl-2-oxetanone (β-butyrolactone)
Acrylonitrile	1-Naphthylamine
Asbestos	2-Naphthylamine
Benzene	*N*-Nitroso compounds
Benzidine	2-Oxetanone (β-propiolactone)
Carbon tetrachloride	Phenacetin
Chloroform	Phenylhydrazine and its salts
Chromic oxide	Polychlorinated biphenyl (PCB)
Coumarin	Progesterone
Diazomethane	Styrene oxide
1,2-Dibromoethane	Tannins
Dimethyl sulfate	Testosterone
p-Dioxane	Thioacetamide
Ethylene oxide	Thiourea
Formaldehyde	*o*-Toluidine
Hydrazine and its salts	Trichloroethylene
Lead(II) acetate	Vinyl chloride

REFERENCES

Aldrich Catalog and Handbook of Fine Chemicals. Milwaukee, WI: Aldrich Chemical Co., current edition.

Armour, M. A. *Hazardous Laboratory Chemicals: Disposal Guide.* Boca Raton, FL: CRC Press, 1991.

Fire Protection Guide on Hazardous Materials, 10th ed. Quincy, MA: National Fire Protection Association, 1991.

Flinn Chemical Catalog Reference Manual. Batavia, IL: Flinn Scientific, Inc., current edition.

Gosselin, R. E., Smith, R. P., and Hodge, H. C. *Clinical Toxicology of Commercial Products,* 5th ed. Baltimore: Williams and Wilkins, 1984.

Lenga, R. E., ed. *The Sigma-Aldrich Library of Chemical Safety Data.* Milwaukee, WI: Sigma-Aldrich Corp., 1985.

Lewis, R. J. *Carcinogenically Active Chemicals: A Reference Guide.* New York: Van Nostrand Reinhold, 1990.

Merck Index, 12th ed. Rahway, NJ: Merck and Co., 1996.

Prudent Practices for Disposal of Chemicals from Laboratories. Washington, DC: Committee on Hazardous Substances in the Laboratory, National Research Council, National Academy Press, 1983.

Prudent Practices for Handling Hazardous Chemicals in Laboratories. Washington, DC: Committee on Hazardous Substances in the Laboratory, National Research Council, National Academy Press, 1981.

Renfrew, M. M., ed. *Safety in the Chemical Laboratory.* Easton, PA: Division of Chemical Education, American Chemical Society, 1967–1991.

Safety in Academic Chemistry Laboratories, 4th ed. Washington, DC: Committee on Chemical Safety, American Chemical Society, 1985.

Sax, N. I., and Lewis, R. J. *Dangerous Properties of Industrial Materials,* 7th ed. New York: Van Nostrand Reinhold, 1988.

Sax, N. I., and Lewis, R. J., eds. *Rapid Guide to Hazardous Chemicals in the Work Place,* 2nd ed. New York: Van Nostrand Reinhold, 1990.

Useful Safety-Related Internet Addresses

Enviro-Net site: http://www.enviro-net.com/technical/msds (This site mirrors the University of Utah site (see below), and it has a user-friendly interface.)

Fisher Scientific, Inc. site: http://www.fisher1.com (This site provides information about each chemical in the Fisher Scientific catalogue, including MSDS sheets for each chemical.)

Northwest Fisheries site: http://research.nwfsc.noaa.gov/msds.html

Text-Trieve Internet Services site: http://halcyon.com/ttrieve/msdshome.html (This site provides useful links to other internet sites that specialize in chemical safety information.)

University of Utah MSDS site: http://atlas.chem.utah.edu:70/11/MSDS

Oregon State University gopher site: gopher://gaia.ucs.orst.edu:70/11/osu-i+s/osu-d+o/ehs (Note: Using this site may be slightly cumbersome, but it provides lots of information.)

ADVANCE PREPARATION AND LABORATORY RECORDS ■

In the welcoming chapter of this book, we mentioned the importance of advance preparation for laboratory work. Presented here are some suggestions about what specific information you should try to obtain in your advance studying. Because much of this information must be obtained while preparing your laboratory notebook, the two subjects, advance study and notebook preparation, are developed simultaneously.

An important part of any laboratory experience is learning to maintain very complete records of every experiment undertaken and every item of data obtained. Far too often, careless recording of data and observations has resulted in mistakes, frustration, and lost time due to needless repetition of experiments. If reports are required, you will find that proper collection and recording of data can make your report writing much easier.

Because organic reactions are seldom quantitative, special problems result. Frequently, reagents have to be used in large excess to increase the amount of product. Some reagents are expensive, and, therefore, care in the amounts of these substances used is necessary. Very often, many more reactions take place than you desire. These extra reactions, or **side reactions,** may form other products besides the desired product. These are called **side products.** For all these reasons, you must plan your experimental procedure carefully before undertaking the actual experiment.

THE NOTEBOOK

For recording data and observations during experiments, use a *bound notebook.* The notebook should have consecutively numbered pages. If it does not, you should number the pages immediately. A spiral-bound notebook or any other type from which the pages can be removed easily is not acceptable, since the possibility of losing the pages is great.

All data and observations must be recorded in the notebook. Paper towels, napkins, toilet tissue, or scratch paper have a tendency to become lost or destroyed. It is bad laboratory practice to record information on such random and perishable pieces of paper. All entries in your notebook must be recorded in *permanent ink.* It can be frustrating to have important information disappear from the notebook because it was recorded in washable ink and could not survive a flood caused by the student at the next position on the bench. Because you will be using your notebook in the laboratory, it is quite likely that the book will become soiled or stained by chemicals, filled with scratched-out entries, or even slightly burned.

Your instructor may check your notebook at any time, so you should always have it up to date. If your instructor requires reports, you can prepare them expeditiously from the material recorded in the laboratory notebook.

NOTEBOOK FORMAT

Advance Preparation

Individual instructors vary greatly in the type of notebook format they prefer; such variations stem from differences in philosophies and experience. You must obtain specific directions from your own instructor for preparing a notebook. Certain features, however, are common to most notebook formats. The following discussion presents what might be included in a typical notebook.

You can save much time in the laboratory if you understand fully the procedure of the experiment and the theory underlying it. It will be very helpful if you know the main reactions, the potential side reactions, the mechanism, the stoichiometry, and the procedure before you come to the laboratory. Understanding the procedure by which the desired product is to be separated from undesired materials is also very important. If you examine each of these topics before coming to class, you will be prepared to do the experiment efficiently. You will have your equipment and reagents already prepared when they are to be used. Your reference material will be at hand when you need it. Finally, with your time efficiently organized, you will be able to take advantage of long reaction or reflux periods to perform other tasks, such as doing shorter experiments or finishing previous ones.

For experiments in which a compound is synthesized from other reagents, that is, with **preparative experiments,** it is essential to know the main reaction. In order to perform stoichiometric calculations, the equation for the main reaction should be balanced. Therefore, before you begin the experiment, your notebook should contain the balanced equation for the pertinent reaction. Using the preparation of isopentyl acetate, or banana oil (Experiment 8), as an example, you should write:

$$\underset{\text{Acetic acid}}{CH_3-\overset{\overset{\displaystyle O}{\|}}{C}-OH} + \underset{\text{Isopentyl alcohol}}{CH_3-\overset{\overset{\displaystyle CH_3}{|}}{CH}-CH_2-CH_2-OH} \xrightarrow{\ H^+\ }$$

$$\underset{\text{Isopentyl acetate}}{CH_3-\overset{\overset{\displaystyle O}{\|}}{C}-O-CH_2-CH_2-\overset{\overset{\displaystyle CH_3}{|}}{CH}-CH_3} + H_2O$$

You should also enter in the notebook the possible side reactions that divert reagents into contaminants (side products), before beginning the experiment. You will have to separate these side products from the major product during purification.

You should list physical constants such as melting points, boiling points, densities, and molecular weights in the notebook when this information is needed to perform an experiment or to do calculations. These data are located in such sources as the *Handbook of Chemistry and Physics,* the *Merck Index,* or Lange's *Handbook of Chemistry.* In many of the experiments in this textbook, some of this information is given within the experimental procedure. Physical constants required for an experiment should be written in your notebook before you come to class.

Advance preparation may also include examination of some subjects—information not necessarily recorded in the notebook—that should prove useful in understanding the experiment. Included among these subjects are an understanding of the mechanism of the reaction, an examination of other methods by which the same compound might be prepared, and a detailed study of the experimental procedure. Many students find that an outline of the procedure, prepared *before* they come to class, helps them use their time more efficiently once they begin the experiment. Such an outline could very well be prepared on some loose sheet of paper, rather than in the notebook itself.

Once the reaction has been completed, the desired product does not magically appear as purified material; it must be isolated from a frequently complex mixture of side products, unreacted starting materials, solvents, and catalysts. You should try to outline a **separation scheme** in your notebook for isolating the product from its contaminants. At each stage you should try to understand the reason for the particular instruction given in the experimental procedure. This not only will familiarize you with the basic separation and purification techniques used in organic chemistry but also will help you understand when to use these techniques. Such an outline might take the form of a flowchart. For example, see the separation scheme for isopentyl acetate (see p. 26). Careful attention to understanding the separation may, besides familiarizing you with the procedure by which the desired product is separated from impurities in your particular experiments, also prepare you for original research, where no experimental procedure exists.

In designing a separation scheme, you should note that the scheme outlines those steps undertaken once the reaction period has been concluded. For this reason, the represented scheme did not include such steps as the addition of the reactants (isopentyl alcohol and acetic acid) and the catalyst (sulfuric acid) or the heating of the reaction mixture.

For experiments in which a compound is isolated from a particular source and not prepared from other reagents, some of the information described in this section will not be applicable. Such experiments are called **isolation experiments.** A typical isolation experiment involves isolating a pure compound from a natural source. Some examples include the isolation of caffeine from tea or the isolation of cinnamaldehyde from cinnamon. Although isolation experiments require somewhat different advance preparation, this advance study may include looking up physical constants for the compound isolated and outlining the isolation procedure. A detailed examination of the separation scheme is very important here because it is the heart of such an experiment.

Laboratory Records

When you begin the actual experiment, your notebook should be kept nearby so that you will be able to record in it those operations you perform. When you are working in the laboratory, the notebook serves as a place where a rough transcript of your experimental method is recorded. Data from actual weighings, volume measurements, and determinations of physical constants are also noted. This section of your notebook should *not* be prepared in advance. The purpose here is not to write a recipe, but rather to provide a record of what you *did* and what you *observed.* These observations will help you to write reports without resorting to memory. They will also help you or other workers to

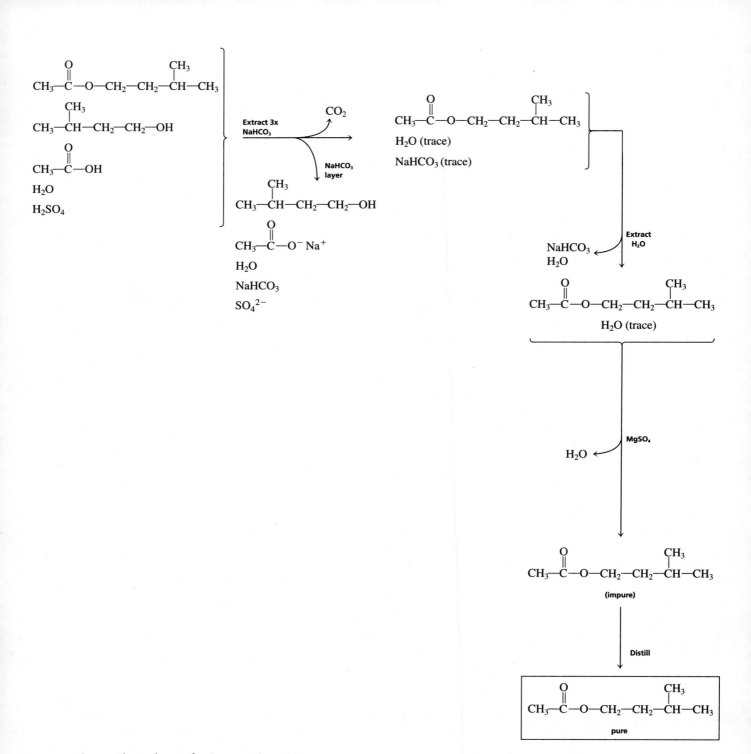

Separation scheme for isopentyl acetate.

repeat the experiment in as nearly as possible the same way. The sample notebook pages found in this section (see p. 29) illustrate the type of data and observations that should be written in your notebook.

When your product has been prepared and purified, or isolated if it is an isolation experiment, you should record such pertinent data as the melting point or boiling point of the substance, its density, its index of refraction, and the conditions under which spectra were determined.

Calculations

A chemical equation for the overall conversion of the starting materials to products is written on the assumption of simple ideal stoichiometry. Actually, this assumption is seldom realized. Side reactions or competing reactions will also occur, giving other products. For some synthetic reactions, an equilibrium state will be reached in which an appreciable amount of starting material is still present and can be recovered. Some of the reactant may also remain if it is present in excess or if the reaction was incomplete. A reaction involving an expensive reagent illustrates another need for knowing how far a particular type of reaction converts reactants to products. In such a case, it is preferable to use the most efficient method for this conversion. Thus, information about the efficiency of conversion for various reactions is of interest to the person contemplating the use of these reactions.

The quantitative expression for the efficiency of a reaction is given by a calculation of the **yield** for the reaction. The **theoretical yield** is the number of grams of the product expected from the reaction on the basis of ideal stoichiometry, with side reactions, reversibility, and losses ignored. In order to calculate the theoretical yield, it is first necessary to determine the **limiting reagent.** The limiting reagent is the reagent that is not present in excess and on which the overall yield of product depends. The method for determining the limiting reagent in the isopentyl acetate experiment is illustrated in the sample notebook on pages 29–30. You should consult your general chemistry textbook for more complicated examples. The theoretical yield is then calculated from the expression

$$\text{Theoretical yield} = (\text{moles of limiting reagent})(\text{ratio})(\text{MW of product})$$

The ratio here is the stoichiometric ratio of product to limiting reagent. In the preparation of isopentyl acetate, that ratio is $1:1$. One mole of isopentyl alcohol, under ideal circumstances, should yield one mole of isopentyl acetate.

The **actual yield** is simply the number of grams of desired product obtained. The **percentage yield** describes the efficiency of the reaction and is determined by

$$\text{Percentage yield} = \frac{\text{Actual yield}}{\text{Theoretical yield}} \times 100$$

Calculation of the theoretical yield and percentage yield can be illustrated using hypothetical data for the isopentyl acetate preparation:

$$\text{Theoretical yield} = (6.94 \times 10^{-2} \text{ mol isopentyl alcohol})\left(\frac{1 \text{ mol isopentyl acetate}}{1 \text{ mol isopentyl alcohol}}\right)$$

$$\times \left(\frac{130.2 \text{ g isopentyl acetate}}{1 \text{ mol isopentyl acetate}}\right) = 9.03 \text{ g isopentyl acetate}$$

$$\text{Actual yield} = 3.81 \text{ g isopentyl acetate}$$

$$\text{Percentage yield} = \frac{3.81 \text{ g}}{9.03 \text{ g}} \times 100 = 42.2\%$$

For experiments that have the principal objective of isolating a substance such as a natural product, rather than preparing and purifying some reaction product, the **weight percentage recovery** and not the percentage yield is calculated. This value is determined by

$$\text{Weight percentage recovery} = \frac{\text{Weight of substance isolated}}{\text{Weight of original material}} \times 100$$

Thus, for instance, if 0.014 g of caffeine were obtained from 2.3 g of tea, the weight percentage recovery of caffeine would be

$$\text{Weight percentage recovery} = \frac{0.014 \text{ g caffeine}}{2.3 \text{ g tea}} \times 100 = 0.61\%$$

LABORATORY REPORTS

Various formats for reporting the results of the laboratory experiments may be used. You may write the report directly into your notebook in a format similar to the sample notebook pages included in this section. Alternatively, your instructor may require a more formal report that you write separately from your notebook. When original research is performed, these reports should include a detailed description of all the experimental steps undertaken. Frequently, the style used in scientific periodicals such as *Journal of the American Chemical Society* is applied to writing laboratory reports. Your instructor is likely to have his or her own requirements for laboratory reports and should describe the requirements to you.

THE PREPARATION OF ISOPENTYL ACETATE (BANANA OIL)

Main Reaction

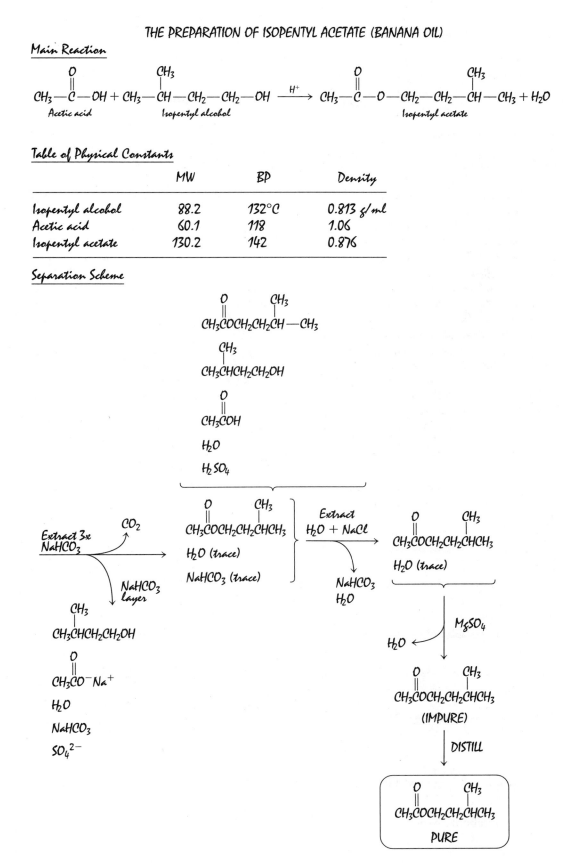

Table of Physical Constants

	MW	BP	Density
Isopentyl alcohol	88.2	132°C	0.813 g/ml
Acetic acid	60.1	118	1.06
Isopentyl acetate	130.2	142	0.876

Separation Scheme

<u>Data and Observations</u>

7.5 mL of isopentyl alcohol was added to a pre-weighed 50-mL round-bottomed flask:

Flask + alcohol	139.75 g
Flask	133.63 g
	6.12 g isopentyl alcohol

     Glacial acetic acid (10 mL) and 2 mL of concentrated sulfuric acid were also added to the flask, with swirling, along with several boiling stones. A water-cooled condenser was attached to the flask. The reaction was allowed to boil, using a heating mantle, for about one hour. The color of the reaction mixture was brownish-yellow.

     After the reaction mixture had cooled to room temperature, the boiling stones were removed, and the reaction mixture was poured into a separatory funnel. About 30 mL of cold water was added to the separatory funnel. The reaction flask was rinsed with 5 mL of cold water, and the water was also added to the separatory funnel. The separatory funnel was shaken, and the lower aqueous layer was removed and discarded. The organic layer was extracted twice with two 10–15-mL portions of 5% aqueous sodium bicarbonate. During the first extraction, much CO_2 was given off, but the amount of gas evolved was markedly diminished during the second extraction. The organic layer was a light yellow in color. After the second extraction, the aqueous layer turned red litmus blue. The bicarbonate layers were discarded, and the organic layer was extracted with a 10–15-mL portion of water. A 2–3 mL portion of saturated sodium chloride solution was added during this extraction. When the aqueous layer had been removed, the upper, organic phase was transferred to a 15-mL Erlenmeyer flask. 2 g of anhydrous magnesium sulfate was added. The flask was stoppered, swirled gently, and allowed to stand for 15 mins.

     The product was transferred to a 25-mL round-bottomed flask, and it was distilled by simple distillation. The distillation continued until no liquid could be observed dripping into the collection flask. After the distillation, the ester was transferred to a pre-weighed sample vial.

Sample vial + product	9.92 g
Sample vial	6.11 g
	3.81 g isopentyl acetate

     The product was colorless and clear. The observed boiling point obtained during the distillation, was 140°C. An IR spectrum was obtained of the product.

<u>Calculations</u>

     Determine limiting reagent:

$$\text{isopentyl alcohol } 6.12 \text{ g} \left(\frac{1 \text{ mol isopentyl alcohol}}{88.2 \text{ g}} \right) = 6.94 \times 10^{-2} \text{ mol}$$

$$\text{acetic acid: } (10 \text{ mL}) \left(\frac{1.06 \text{ g}}{\text{mL}} \right) \left(\frac{1 \text{ mol acetic acid}}{60.1 \text{ g}} \right) = 1.76 \times 10^{-1} \text{ mol}$$

Since they react in a 1:1 ratio, isopentyl alcohol is the limiting reagent. Theoretical yield:

$$(6.94 \times 10^{-2} \text{ mol isopentyl alcohol}) \left(\frac{1 \text{ mol isopentyl acetate}}{1 \text{ mol isopentyl alcohol}} \right) \left(\frac{130.2 \text{ g isopentyl acetate}}{1 \text{ mol isopentyl acetate}} \right)$$

$$= 9.04 \text{ g isopentyl acetate}$$

$$\text{Percentage yield} = \frac{3.81 \text{ g}}{9.03 \text{ g}} \times 100 = 42.2\%$$

A sample notebook, page 2.

SUBMISSION OF SAMPLES

In all preparative experiments, and in some of the isolation experiments, you will be required to submit to your instructor the sample of the substance you prepared or isolated. How this sample is labeled is very important. Again, learning a correct method of labeling bottles and vials can save time in the laboratory, because fewer mistakes will be made. More importantly, learning to label properly can decrease the danger inherent in having samples of material that cannot be identified correctly at a later date.

Solid materials should be stored and submitted in containers that permit the substance to be removed easily. For this reason, narrow-mouthed bottles or vials are not used for solid substances. Liquids should be stored in containers that will not let them escape through leakage. You should be careful not to store volatile liquids in containers that have plastic caps, unless the cap is lined with an inert material such as Teflon. Otherwise, the vapors from the liquid are likely to come in contact with the plastic and dissolve some of it, thus contaminating the substance being stored.

On the label, print the name of the substance, its melting or boiling point, the actual and percentage yields, and your name. An illustration of a properly prepared label follows:

Isopentyl Acetate
BP 140°C
Yield 3.81 g (42.2%)
Joe Schmedlock

LABORATORY GLASSWARE ▪

Since your glassware is expensive and you are responsible for it, you will want to give it proper care and respect. If you read this section carefully and follow the procedures presented here, you may be able to avoid some unnecessary expense. You may also save time, since cleaning problems and replacing broken glassware are time-consuming.

For those of you who are unfamiliar with the equipment found in an organic laboratory or who are uncertain about how such equipment should be treated, this section will provide some useful information. Topics such as cleaning glassware, caring for glassware when using corrosive or caustic reagents, and assembling components from your organic laboratory kit are included. At the end of this chapter are illustrations and names of most of the equipment you are likely to find in your drawer or locker.

CLEANING GLASSWARE

Glassware can be cleaned easily if you clean it immediately. It is good practice to do your "dish washing" right away. With time, the organic tarry materials left in a container begin to attack the surface of the glass. The longer you wait to clean glassware, the more extensively this interaction will have progressed. Cleaning is then more difficult, because water will no longer wet the surface of the glass as effectively. If you are not able to wash your glassware immediately after use, you should soak the dirty pieces in soapy water. A half-gallon plastic container provides a convenient vessel in which to soak and wash your glassware. The use of a plastic container also helps to prevent the loss of small pieces of equipment.

Various soaps and detergents are available for washing glassware. They should be tried first when washing dirty glassware. Organic solvents can also be used, since the residue remaining in dirty glassware is likely to be soluble in some organic solvent. After the solvent has been used, the flask probably will have to be washed with soap and water to remove the residual solvent. When you use solvents in cleaning glassware, use caution since the solvents are hazardous (see the chapter entitled "Laboratory Safety"). You should try to use fairly small amounts of a solvent for cleaning purposes. Usually less than 5 mL will be sufficient. Acetone is commonly used, but it is expensive. Your **wash acetone** can be used effectively several times before it is "spent." Once your acetone is spent, dispose of it as directed by your instructor. If acetone does not work, other organic solvents such as methylene chloride or toluene can be used in the same way as acetone.

> **Caution:** Acetone is very flammable. Do not use it around flames.

For troublesome stains and residues that insist on adhering to the glass in spite of your best efforts, a mixture of sulfuric acid and nitric acid can be used. Cautiously add about 20 drops of concentrated sulfuric acid and five drops of concentrated nitric acid to the flask.

> **Caution:** You must wear safety glasses when you are using this cleaning solution. Do not allow the solution to come into contact with your skin or your clothing. It will cause severe burns and create holes in your clothing. It is also possible that the acids will react with the residue in the container.

Swirl the acid mixture in the container for a few minutes. If necessary, place the glassware in a warm water bath and heat cautiously to accelerate the cleaning process. Continue heating until any sign of a reaction ceases. When the cleaning procedure is completed, decant the mixture into an appropriate waste container.

Caution: Do not pour the acid solution into a waste container that is intended for organic wastes.

Rinse the piece of glassware thoroughly with water and then wash with soap and water. For most common organic chemistry applications, any stains that survive this treatment are not likely to cause difficulty in subsequent laboratory procedures.

If the glassware is contaminated with stopcock grease, rinse the glassware with a small amount (1–2 mL) of methylene chloride. Discard the rinse solution into an appropriate waste container. Once the grease is removed, wash the glassware with soap or detergent and water.

DRYING GLASSWARE

The easiest way to dry glassware is to allow it to stand overnight. Vials, flasks, and beakers should be stored upside down on a piece of paper towel to permit the water to drain from them. Drying ovens can be used to dry glassware if they are available, and if they are not being used for other purposes. Rapid drying can be achieved by rinsing the glassware with acetone and air-drying it or placing it in an oven. First, thoroughly drain the glassware of water. Then rinse it with one or two *small* portions (1–2 mL) of acetone. Do not use any more acetone than is suggested here. Return the used acetone to a waste acetone container for recycling. After you rinse the glassware with acetone, dry it by placing it in a drying oven for a few minutes or allow it to air-dry at room temperature. The acetone can also be removed by aspirator suction. In some laboratories, it may be possible to dry the glassware by blowing a *gentle* stream of dry air into the container. (Your laboratory instructor will indicate if you should do this.) Before drying the glassware with air, you should make sure that the air line is not filled with oil. Otherwise, the oil will be blown into the container, and you will have to clean it again. It is not necessary to blast the acetone out of the glassware with a wide-open stream of air; a gentle stream of air is just as effective and will not startle other people in the room.

You should not dry your glassware with a paper towel unless the towel is lint-free. Most paper will leave lint on the glass that can interfere with subsequent procedures performed in the equipment. Sometimes it is not necessary to dry a piece of equipment thoroughly. For example, if you are going to place water or an aqueous solution in a container, it does not need to be completely dry.

GROUND-GLASS JOINTS

It is likely that the glassware in your organic kit has **standard-taper ground-glass joints.** Joints may be either inner (male) or outer (female). Each joint is ground to a precise size, which is designated by the symbol $\mathbf{T}$ followed by two numbers. A common joint size in most organic glassware kits is ⚒ 19/22. The first number indicates the diameter (in millimeters) of the joint at its widest point, and the second number refers to its length. One advantage of standard-taper joints is that the pieces fit together snugly and form a good seal. In addition, standard-taper joints allow all glassware components with the same joint size to be connected, thus permitting the assembly of a wide variety of apparatus. One disadvantage of glassware with ground-glass joints, however, is that it is very expensive.

It is important to make sure no solid or liquid is on the joint surfaces. Such material will lessen the efficiency of the seal, and the joints may leak. Also, if the apparatus is to be heated, material caught between the joint surfaces will increase the tendency for the joints to stick. If the joint surfaces are coated with liquid or adhering solid, you should wipe them with a cloth or lint-free paper towel before assembling.

Another disadvantage of glassware with ground-glass joints is that the joints have some tendency to stick together. To prevent this sticking, you can use a lubricant. Typical lubricants include a variety of hydrocarbon-based stopcock greases, such as Lubriseal and Apiezon. Silicone greases are also used. To apply the grease, coat the inner (male) joint with a *very thin* film of lubricant. You should be careful not to apply too much grease, since excess grease may contaminate the materials contained within the apparatus. A properly lubricated joint appears clear, with no striations. Contamination due to excess grease is a particular problem with silicone grease, which is much harder to remove than hydrocarbon-based greases. Because of this possible contamination, many chemists prefer to use grease only when necessary. The joints usually seal well without grease. In some situations, grease is necessary, including distillations under reduced pressure—when joints must be airtight—and in reactions involving strong caustic solutions (see below, "Etching Glassware"). Whenever there is doubt, it never hurts to use grease. Before lubricating the joints, always make sure that they are free of any adhering liquid or solid.

SEPARATING GROUND-GLASS JOINTS

The most important thing you can do to prevent ground-glass joints from becoming "frozen" or stuck together is to disassemble the glassware as soon as possible after a procedure is completed. Even when this precaution is followed, ground-glass joints may become stuck tightly together. The same is true of glass stoppers in bottles. Since smaller items of glassware may be small and very fragile, it is relatively easy to break a piece of glassware when trying to pull two pieces apart. If the pieces do not separate easily, you must be careful when you try to pull them apart. The best way is to hold the two pieces, with both hands touching, as close as possible to the joint. With a firm grasp, try to loosen the joint with a slight twisting motion (do not twist very hard). If this does not work, try to pull your hands apart without pushing sideways on the glassware.

If it is not possible to pull the pieces apart, the following methods may help. A frozen joint can sometimes be loosened if you tap it *gently* with the wooden handle of a spatula. Then, try to pull it apart as already described. If this procedure fails, you may try heating the joint in hot water or a steam bath. If this heating fails, the instructor may be able to advise you. As a last resort, you may try heating the joint in a flame. You should not try this unless the apparatus is hopelessly stuck, because heating by flame often causes the joint to expand rapidly and crack or break. If you use a flame, make sure the joint is clean and dry. Heat the outer part of the joint slowly, in the yellow portion of a low flame, until it expands and breaks away from the inner section. Heat the joint very slowly and carefully, or it may break.

ETCHING GLASSWARE

Glassware that has been used for reactions involving strong bases such as sodium hydroxide or sodium alkoxides must be cleaned thoroughly *immediately* after use. If these caustic materials are allowed to remain in contact with the glass, they will etch the glass permanently. The etching makes later cleaning more difficult, since dirt particles may become trapped within the microscopic surface irregularities of the etched glass. Furthermore, the glass is weakened, so the lifetime of the glassware is decreased. If caustic materials are allowed to come into contact with ground-glass joints without being removed promptly, the joints will become fused or "frozen." It is extremely difficult to separate fused joints without breaking them.

ASSEMBLING THE APPARATUS

Care must be taken when assembling the glass components into the desired apparatus. You should always remember that Newtonian physics applies to chemical apparatus, and unsecured pieces of glassware are certain to respond to gravity. You should always clamp the glassware securely to a ring stand. Throughout this textbook, the illustrations of the various glassware arrangements include the clamps that attach the apparatus to a ring stand. You should assemble your apparatus using the clamps as shown in the illustrations.

ATTACHING RUBBER TUBING TO EQUIPMENT

When you attach rubber tubing to the glass apparatus or when you insert glass tubing into rubber stoppers, you should lubricate the rubber tubing or the rubber stopper with either water or glycerin beforehand. Without such lubrication, it can be difficult to attach rubber tubing to the sidearms of items of glassware such as condensers and filter flasks. Furthermore, glass tubing may break when it is inserted into rubber stoppers. Water is a good lubricant for most purposes. Do not use water as a lubricant when it might contaminate the reaction. Glycerin is a better lubricant than water and should be used when there is considerable friction between the glass and rubber. If glycerin is the lubricant, be careful not to use too much.

DESCRIPTION OF EQUIPMENT

The figures that follow on pages 37–39 include glassware and equipment that are commonly used in the organic laboratory. Your glassware and equipment may vary slightly from the pieces shown.

The components of the organic kit recommended for use in this textbook are shown in the figure on page 37. Notice that most of the joints in these pieces of glassware are ⌖ 19/22. The organic kits used in your laboratory may have different joint sizes, but these kits will work as well with the experiments in this book as the glassware recommended in the figure.

25-mL Round-bottom
boiling flask

50-mL Round-bottom
boiling flask

100-mL Round-bottom
boiling flask

250-mL Round-bottom
boiling flask

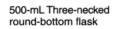

500-mL Three-necked
round-bottom flask

Vacuum
adapter

Distillation
head

Stopper

Claisen head

Thermometer
adapter (with
rubber fitting)

Ebulliator
tube

Condenser
(West)

125-mL
Separatory funnel

Fractionating
column

Components of the organic laboratory kit.

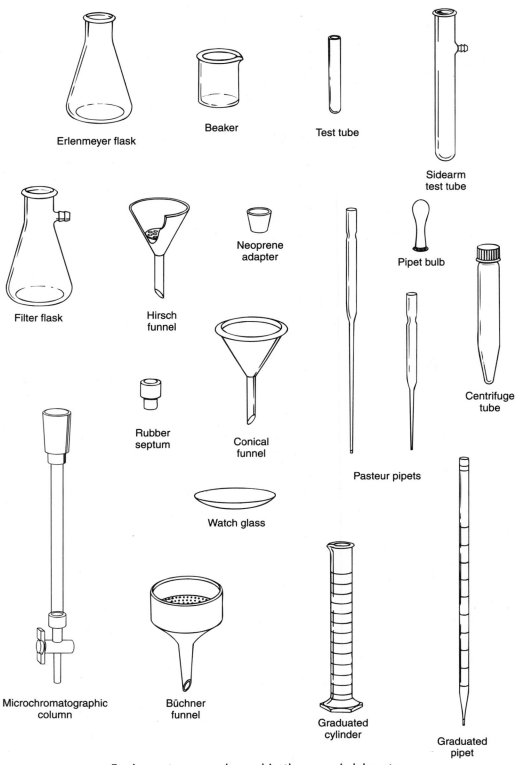

Erlenmeyer flask

Beaker

Test tube

Sidearm test tube

Filter flask

Hirsch funnel

Neoprene adapter

Pipet bulb

Rubber septum

Conical funnel

Centrifuge tube

Pasteur pipets

Watch glass

Microchromatographic column

Büchner funnel

Graduated cylinder

Graduated pipet

Equipment commonly used in the organic laboratory.

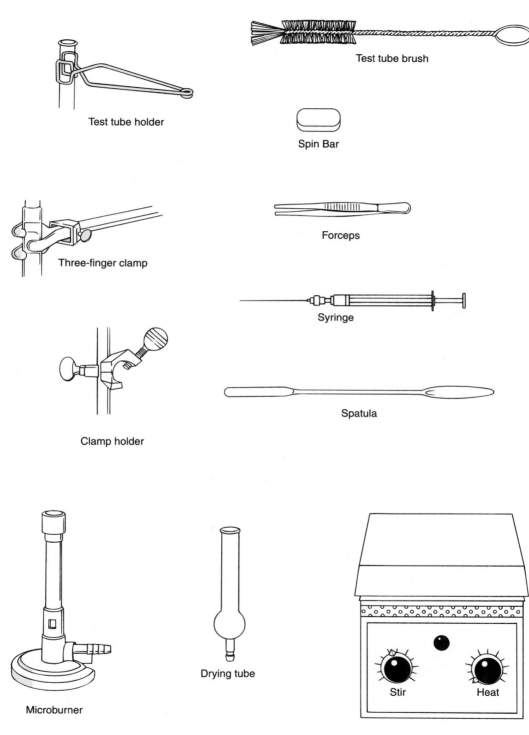

Test tube brush

Test tube holder

Spin Bar

Three-finger clamp

Forceps

Syringe

Clamp holder

Spatula

Microburner

Drying tube

Stir   Heat

Hot plate / Stirrer

Equipment commonly used in the organic laboratory.

Preparations and Reactions of Organic Compounds

EXPERIMENT 1

Isolation of the Active Ingredient in an Analgesic Drug

Extraction
Filtration
Melting point

The major analgesic (pain-relieving) drugs found on the shelves of any drug or grocery store generally fall into one of four categories. These drugs may contain **acetylsalicyclic acid, acetaminophen,** or **ibuprofen** as the active ingredient, or some **combination** of these compounds may be used in a single preparation. All tablets, regardless of type, contain a large amount of starch or other inert substance. This material acts as a binder to keep the tablet from falling apart and to make it large enough to handle. Some analgesic drugs also contain caffeine or buffering agents. In addition, many tablets are coated to make them easier to swallow and to prevent users from experiencing the unpleasant taste of the drugs.

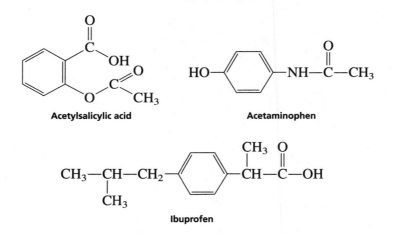

The three drugs, along with their melting points and common brand names, follow:

Drug	MP	Brand Names
Acetylsalicylic acid	135–136°C	Aspirin, ASA, Acetylsalicyclic acid, Generic aspirin, Empirin
Acetaminophen	169–170.5°C	Tylenol, Datril, Panadol, Non-aspirin pain reliever (various brands)
Ibuprofen	75–77°C	Advil, Brufen, Motrin, Nuprin

The purpose of this experiment is to demonstrate some important techniques that are applied throughout this textbook and to allow you to become accustomed to working in the laboratory. More specifically, you will extract (dissolve) the active ingredient of an analgesic drug by mixing the powdered tablet with a solvent, methanol. Two steps are required to remove the fine particles of binder, which remain suspended in the solvent. First, you will use centrifugation to remove most of the binder. The second step will be a filtration technique using a Pasteur pipet packed with alumina (finely ground aluminum oxide). The solvent will then be evaporated to yield the solid analgesic, which will be collected by filtration on a Hirsch funnel. Finally, you will test the purity of the drug by doing a melting point determination.

Required Reading

New: Technique 3   Reaction Methods, Section 3.11
        Technique 4   Filtration, Sections 4.1–4.6
        Technique 6   Physical Constants, Part A, Melting Points

Special Instructions

You will be allowed to select an analgesic that is a member of one of the categories described previously. You should use an uncoated tablet that contains only a single ingredient analgesic and binder. If it is necessary to use a coated tablet, try to remove the coating when the tablet is crushed. To avoid decomposition of aspirin, it is essential to minimize the length of time that it remains dissolved in methanol. Do not stop this experiment until after the drug is dried on the Hirsch funnel.

Waste Disposal

Dispose of any remaining methanol in the waste container for nonhalogenated organic solvents. Place the alumina in the container designated for wet alumina.

Procedure

Extraction of Active Ingredient.   If you are isolating aspirin or acetaminophen, use *one* tablet in this procedure. If you are isolating ibuprofen, use *two* tablets. Crush the tablet (or tablets) between two pieces of weighing paper using a pestle. If the tablet is coated, try to remove fragments of the coating material with forceps after the tablet is first crushed. Add all of the powdered material to a centrifuge tube with a tight-fitting cap. Add about 2 mL of methanol to the centrifuge tube, cap the tube, and mix thoroughly by shaking. Loosen the cap at least once during the mixing process in order to release any pressure that may build up in the tube.

Allow the undissolved portion of the powder to settle in the centrifuge tube. A cloudy suspension may remain even after 5 minutes or more. You should wait only until it is obvious that the larger particles have settled completely. Using a Pasteur pipet, transfer the liquid phase to another centrifuge tube. Add a second 2-mL portion of methanol to the original centrifuge tube and repeat the shaking process described previously. After the solid has settled, transfer the liquid phase to the centrifuge tube containing the first extract.

Place the tube containing the combined extracts in a centrifuge along with another centrifuge tube of equal weight on the opposite side. Centrifuge the mixture for 2–3 minutes. The suspended solids should collect on the bottom of the tube, leaving a clear or nearly clear **supernatant liquid,** the liquid above the solid. If the liquid is still quite cloudy, repeat the centrifugation for a longer period of time or at a higher speed. Being careful not to disturb the solid at the bottom of the tube, transfer the supernatant liquid with a Pasteur pipet to a test tube or small beaker.

Column Chromatography.   Prepare an alumina column using a Pasteur pipet, as shown in the figure. Insert a small ball of cotton into the top of the column. Using a long thin object such as a glass stirring rod or a wooden applicator stick, push the cotton down so that it fits into the Pasteur pipet where the constriction begins. Add about 0.5 g of alumina to the pipet and tap the column with your finger to pack the alumina. Clamp the pipet in a vertical position so that the liquid can drain from the column into a small beaker or a test tube (16 × 100-mm). Place a small beaker under the column. With a Pasteur pipet, add about 2 mL of methanol to the column and allow the liquid to drain until the level of the methanol just reaches the top of the alumina. Once methanol has been added to the alumnia, the top of the alumina in the column should not be allowed to run dry. If necessary, add more methanol.

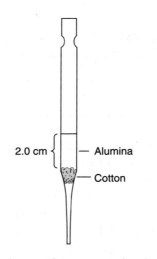

Column for purifying an analgesic drug.

Note: It is essential that the methanol not be allowed to drain below the surface of the alumina.

When the level of the methanol reaches the surface of the alumina, transfer the solution containing the drug from the beaker or test tube to the column, using a Pasteur pipet. Collect the liquid that passes through the column into a test tube (16 × 100-mm). When all the liquid from the beaker has been added to the column and has penetrated the alumina, add an additional 1 mL of methanol to the column and allow it to drain. This ensures that all of the analgesic drug has been eluted from the column.

Evaporation of Solvent.   *If you are isolating aspirin,* it is essential that the following evaporation procedure be completed in 10–15 minutes. Otherwise, the aspirin may partially decompose. Evaporate the methanol in the test tube using a water bath at about 50°C (Technique 3, Section 3.11, p. 630). Clamp the test tube to hold it in a vertical position. To speed evaporation, direct a gentle stream of dry air or nitrogen into the test tube containing the liquid.

When the solvent has completely evaporated or it is apparent that the remaining liquid is no longer evaporating, remove the test tube from the water bath and allow it to cool to room temperature. (The volume of liquid should be less than 0.5 mL when you discontinue evaporation.) If liquid remains, which is likely with the ibuprofen- or acetaminophen-containing analgesics, place the cool test tube in an ice-water bath for 10–15 minutes. Prepare the ice-water bath in a small beaker using both ice and water. Crystallization of the product may occur more readily if you scrape the inside of the tube with a small spatula or a glass rod (not fire polished). If the solid is very hard and clumped together, you should use a small spatula to break up the solid as much as possible before going on to the next step.

Vacuum Filtration.   Set up a Hirsch funnel for vacuum filtration (see Technique 4, Section 4.3, and Figure 4.5, p. 641). Moisten the filter paper with a few drops of methanol, and turn on the vacuum (or aspirator) to the fullest extent. Use a small spatula to transfer the material in the test tube to the Hirsch funnel. The vacuum will draw any remaining solvent from the crystals. Allow the crystals to dry for 5–10 minutes while air is drawn through the crystals in the Hirsch funnel.

Carefully scrape the dried crystals from the filter paper onto a tared (previously weighed) watch glass. If necessary, use a spatula to break up any remaining large pieces of solid. Allow the crystals to air-dry on the watch glass. To determine when the crystals are dry, move them around with a dry spatula. When the crystals no longer clump together or cling to the spatula, they should be dry. If you are working with ibuprofen, the solid will be slightly sticky even when it is completely dried. Weigh the watch glass with the crystals to determine the weight of analgesic drug that you have isolated. Use the weight of the active ingredient specified on the label of the container as a basis for calculating the weight percentage recovery.

Use a small sample of the crystals to determine the melting point (see Technique 6, Sections 6.5–6.8, pp. 669–674). Crush the crystals into a powder, using a stirring rod, in order to determine their melting point. You may observe some "sweating" or shrinkage (see Section 6.8, p. 672) before the substance actually begins to melt. The beginning of the melting point range is when actual melting is observed, not when the solid takes on a slightly wet or shiny appearance or when shrinkage occurs. If you have isolated ibuprofen, the melting point may be somewhat lower than that given on page 42.

At the instructor's option, place your product in a small vial, label it properly (p. 31), and submit it to your instructor.

QUESTIONS

1. Why was the percentage recovery less than 100%? Give several reasons.
2. Why was the tablet crushed?
3. What was the purpose of the centrifugation step?
4. What was the purpose of the alumina column?
5. If 185 mg of acetaminophen were obtained from a tablet containing 350 mg of acetaminophen, what would be the weight percentage recovery?
6. A student who was isolating aspirin stopped the experiment after the filtration step with alumina. One week later, the methanol was evaporated and the experiment was completed. The melting point of the aspirin was found to be 110–115°C. Explain why the melting point was very low and why the melting range was so wide.

E S S A Y

Aspirin

Aspirin is one of the most popular cure-alls of modern life. Even though its curious history began over 200 years ago, we still have much to learn about this enigmatic remedy. No one knows exactly how or why it works, yet more than 20 million pounds of aspirin are consumed each year in the United States.

The history of aspirin began on June 2, 1763, when Edward Stone, a clergyman, read a paper to the Royal Society of London entitled "An Account of the Success of the Bark of the Willow in the Cure of Agues." By *ague*, Stone was referring to what we now call malaria, but his use of the word *cure* was optimistic; what his extract of willow bark actually did was to reduce the feverish symptoms of the disease. Almost a century later, a Scottish physician was to find that extracts of willow bark would also alleviate the symptoms of acute rheumatism. This extract was ultimately found to be a powerful **analgesic** (pain reliever), **antipyretic** (fever reducer), and **anti-inflammatory** (reduces swelling) drug.

Soon thereafter, organic chemists working with willow bark extract and flowers of the meadowsweet plant (which gave a similar principle) isolated and identified the active ingredient as salicylic acid (from *salix*, the Latin name for the willow tree). The substance could then be chemically produced in large quantities for medical use. It soon became apparent that using salicylic acid as a remedy was severely limited by its acidic properties. The substance caused severe irritation of the mucous membranes lining the mouth, gullet, and stomach. The first attempts at circumventing this problem by using the less acidic sodium salt (sodium salicylate) were only partially successful. This substance gave less

Salicylic acid   **Sodium salicylate**   **Acetylsalicylic acid**
                                                          (Aspirin)

irritation but had such an objectionable sweetish taste that most people could not be induced to take it. The breakthrough came at the turn of the century (1893) when Felix Hofmann, a chemist for the German firm of Bayer, devised a practical route for synthesizing acetylsalicylic acid, which was found to have all the same medicinal properties without the highly objectionable taste or the high degree of mucosal-membrane irritation. Bayer called its new product "aspirin," a name derived from *a* for acetyl, and the root *-spir*, from the Latin name for the meadowsweet plant, *spirea*.

The history of aspirin is typical of many of the medicinal substances in current use. Many began as crude plant extracts or folk remedies whose active ingredients were isolated and structures determined by chemists, who then improved on the original.

In the last few years, the mode of action of aspirin has just begun to unfold. A whole new class of compounds, called **prostaglandins,** has been found to be involved in the

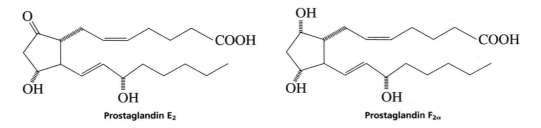

Prostaglandin E$_2$ **Prostaglandin F$_{2\alpha}$**

body's immune responses. Their synthesis is provoked by interference with the body's normal functioning by foreign substances or unaccustomed stimuli.

These substances are involved in a wide variety of physiological processes and are thought to be responsible for evoking pain, fever, and local inflammation. Aspirin has recently been shown to prevent bodily synthesis of prostaglandins and thus to alleviate the symptomatic portion (fever, pain, inflammation, menstrual cramps) of the body's immune responses (that is, the ones that let you know something is wrong). A recent report suggests that the role of aspirin may be to inactivate one of the enzymes responsible for the synthesis of prostaglandins. The natural precursor for prostaglandin synthesis is **arachi-**

donic acid. This substance is converted to a peroxide intermediate by an enzyme called **cyclo-oxygenase,** or prostaglandin synthase. This intermediate is converted further to

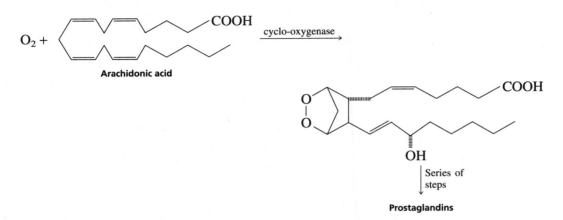

prostaglandin. The apparent role of aspirin is to attach an acetyl group to the active site of cyclo-oxygenase, thus rendering it unable to convert arachidonic acid to the peroxide intermediate. In this way, prostaglandin synthesis is blocked.

Aspirin tablets (5-grain) are usually compounded of about 0.32 g of acetylsalicylic acid pressed together with a small amount of starch, which binds the ingredients. Buffered aspirin usually contains a basic buffering agent to reduce the acidic irritation of mucous membranes in the stomach, since the acetylated product is not totally free of this irritating effect. Bufferin contains 0.325 g of aspirin together with calcium carbonate, magnesium oxide, and magnesium carbonate as buffering agents. Combination pain relievers usually contain aspirin, acetaminophen, and caffeine. Extra Strength Excedrin, for instance, contains 0.250 g of aspirin, 0.250 g of acetaminophen, and 0.065 g of caffeine.

REFERENCES

"Aspirin Cuts Deaths After Heart Attacks." *New Scientist, 118* (April 7, 1988): 22.

Collier, H. O. J. "Aspirin," *Scientific American, 209* (November 1963): 96.

Collier, H. O. J. "Prostaglandins and Aspirin." *Nature, 232* (July 2, 1971): 17.

Disla, E., Rhim, H. R., Reddy, A., and Taranta, A. "Aspirin on Trial as HIV Treatment." *Nature, 366* (November 18, 1993): 198.

Kingman, S. "Will an Aspirin a Day Keep the Doctor Away?" *New Scientist, 117* (February 11, 1988): 26.

Kolata, G. "Study of Reye's—Aspirin Link Raises Concerns." *Science, 227* (January 25, 1985): 391.

Macilwain, C. "Aspirin on Trial as HIV Treatment." *Nature, 364* (July 29, 1993): 369.

Nelson, N. A., Kelly, R. C., and Johnson, R. A. "Prostaglandins and the Arachidonic Acid Cascade." *Chemical and Engineering News* (August 16, 1982): 30.

Pike, J. E. "Prostaglandins." *Scientific American, 225* (November 1971): 84.

Roth, G. J., Stanford, N., and Majerus, P. W. "Acetylation of Prostaglandin Synthase by Aspirin." *Proceedings of the National Academy of Science of the U.S.A., 72* (1975): 3073.

Street, K. W. "Method Development for Analysis of Aspirin Tablets." *Journal of Chemical Education, 65* (October 1988): 914.

Vane, J. R. "Inhibition of Prostaglandin Synthesis as a Mechanism of Action for Aspirin-like Drugs." *Nature-New Biology, 231* (June 23, 1971): 232.

Weissmann, G. "Aspirin." *Scientific American, 264* (January 1991): 84.

E X P E R I M E N T   2

Acetylsalicylic Acid

Crystallization
Vacuum filtration
Melting point
Esterification

Aspirin (acetylsalicylic acid) can be prepared by the reaction between salicylic acid and acetic anhydride:

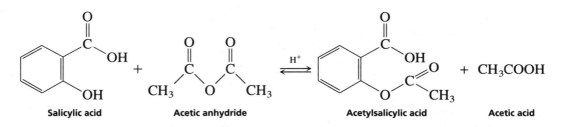

Salicylic acid   Acetic anhydride   Acetylsalicylic acid   Acetic acid

In this reaction, the **hydroxyl group** (—OH) on the benzene ring in salicylic acid reacts with acetic anhydride to form an **ester** functional group. Thus, the formation of acetylsalicylic acid is referred to as an **esterification** reaction. This reaction requires the presence of an acid catalyst, indicated by the H^+ above the equilibrium arrows.

When the reaction is complete, some unreacted salicylic acid and acetic anhydride will be present along with acetylsalicylic acid, acetic acid, and the catalyst. The technique used to purify the acetylsalicylic acid from the other substances is called **crystallization.** The basic principle is quite simple. At the end of this reaction, the reaction mixture will be hot, and all substances will be in solution. As the solution is allowed to cool, the solubility of acetylsalicylic acid will decrease, and it will gradually come out of solution, or crystallize. Since the other substances are either liquids at room temperature or are present in much smaller amounts, the crystals formed will be composed mainly of acetylsalicylic acid. Thus, a separation of acetylsalicylic acid from the other materials will have been accomplished. The purification process is facilitated by the addition of water after the crystals have formed. The water decreases the solubility of acetylsalicylic acid and dissolves some of the impurities.

To purify the product even more, a recrystallization procedure will also be performed. In order to prevent the decomposition of acetylsalicylic acid, ethyl acetate, rather than water, will be used as the solvent for recrystallization.

The most likely impurity in the product after purification is salicylic acid itself, which can arise from incomplete reaction of the starting materials or from **hydrolysis** (reaction with water) of the product during the isolation steps. The hydrolysis reaction of acetylsalicylic acid produces salicylic acid. Salicylic acid and other compounds that contain a hydroxyl group on the benzene ring are referred to as **phenols.** Phenols form a highly colored complex with ferric chloride (Fe^{3+} ion). Aspirin is not a phenol, because it does not possess a hydroxyl group directly attached to the ring. Since aspirin will not give the color reaction with ferric chloride, the presence of salicylic acid in the final product is easily detected. The purity of your product will also be determined by obtaining the melting point.

Required Reading

Review: Technique 4   Filtration, Sections 4.1–4.6
           Technique 6   Physical Constants, Part A, Melting Points

New: Technique 1   Measurement of Volume and Weight
           Technique 2   Heating and Cooling Methods
           Technique 3   Reaction Methods, Sections 3.1, 3.4–3.6
           Technique 5   Crystallization
           Essay   Aspirin

Special Instructions

This experiment involves concentrated sulfuric acid, which is highly corrosive. It will cause burns if it is spilled on the skin. Exercise care in handling it.

Waste Disposal

Dispose of the aqueous filtrate in the container for aqueous waste. The filtrate from the recrystallization in ethyl acetate should be disposed of in the container for nonhalogenated organic waste.

Procedure

Preparation of Acetylsalicylic Acid (Aspirin).   Weigh 2.0 g of salicyclic acid (*MW* = 138.1) and place this in a 125-mL Erlenmeyer flask. Add 5.0 mL of acetic anhydride ($MW = 102.1$, $d = 1.08$ g/mL), followed by 5 drops of concentrated sulfuric acid, and

> **Caution:** Concentrated sulfuric acid is highly corrosive. You must handle it with great care.

swirl the flask gently until the salicylic acid dissolves. Heat the flask gently on the steam bath or in a hot water bath at about 50°C (see Technique 2, Figure 2.4, p. 604) for at least 10 minutes. Allow the flask to cool to room temperature, during which time the acetylsalicylic acid should begin to crystallize from the reaction mixture. If it does not, scratch the walls of the flask with a glass rod and cool the mixture slightly in an ice bath until crystallization has occurred. After crystal formation is complete (usually when the product appears as a solid mass), add 50 mL of water and cool the mixture in an ice bath.

Vacuum Filtration.   Collect the product by vacuum filtration on a Büchner funnel (see Technique 4, Section 4.3, p. 640 and Figure 4.5, p. 641). The filtrate can be used to rinse the Erlenmeyer flask repeatedly until all crystals have been collected. Rinse the crystals several times with small portions of *cold* water. Continue drawing air through the crystals on the Büchner funnel by suction until the crystals are free of solvent (5–10 minutes). Remove the crystals for air drying. Weigh the crude product, which may contain some unreacted salicylic acid, and calculate the percentage yield of crude acetylsalicylic acid ($MW = 180.2$).

Ferric Chloride Test for Purity.   You can perform this test on a sample of your product that is not completely dry. To determine if there is any salicylic acid remaining in your product, carry out the following procedure. Obtain three small test tubes. Add 0.5 mL of water to each test tube. Dissolve a small amount of salicylic acid in the first tube. Add a similar amount of your product to the second tube. The third test tube, which contains only solvent, will serve as the control. Add one drop of 1% ferric chloride solution to each tube and note the color after shaking. Formation of an iron-phenol complex with Fe(III) gives a definite color ranging from red to violet, depending on the particular phenol present.

Recrystallization.   Water is not a suitable solvent for crystallization because aspirin will partially decompose when heated in water. Follow the general instructions described in Technique 5, Section 5.3, p. 651, and Figure 5.3, p. 652. Dissolve the product in a *minimum* of hot ethyl acetate (no more than 2–3 mL) in a 25-mL Erlenmeyer flask, while gently and continuously heating the mixture on a steam bath or a hot-plate.[1]

---

[1]It will usually not be necessary to filter the hot mixture. If an appreciable amount of solid material remains, add 5 mL of additional ethyl acetate, heat the solution to boiling, and filter the hot solution by gravity into an Erlenmeyer flask through a fluted filter. Be sure to preheat the short-stemmed funnel by pouring hot ethyl acetate through it (see Technique 4, Section 4.1, p. 634, and Technique 5, Section 5.3, p. 651). Reduce the volume until crystals appear. Add a minimum additional amount of hot ethyl acetate until the crystals dissolve. Let the filtered solution stand.

When the mixture cools to room temperature, the aspirin should crystallize. If it does not, evaporate some of the ethyl acetate solvent to concentrate the solution and cool the solution in ice water while scratching the inside of the flask with a glass rod (not a fire-polished one). Collect the product by vacuum filtration, using a Büchner funnel. Any remaining material can be rinsed out of the flask with a few milliliters of cold petroleum ether. Dispose of the residual solvents in the waste container for nonhalogenated organic waste. Test the aspirin for purity with ferric chloride as described on page 51. Determine the melting point of your product (see Technique 6, Sections 6.5–6.8, pp. 669–674). The melting point must be obtained with a completely dried sample. Pure aspirin has a melting point of 135–136°C.

Place your product in a small vial, label it properly (p. 31), and submit it to your instructor.

ASPIRIN TABLETS

Aspirin tablets are acetylsalicylic acid pressed together with a small amount of inert binding material. Common binding substances include starch, methylcellulose, and microcrystalline cellulose. You can test for the presence of starch by boiling approximately one-fourth of an aspirin tablet with 2 mL of water. Cool the liquid and add a drop of iodine solution. If starch is present, it will form a complex with the iodine. The starch–iodine complex is deep blue violet. Repeat this test with a commercial aspirin tablet and with the acetylsalicylic acid prepared in this experiment.

QUESTIONS

1. What is the purpose of the concentrated sulfuric acid used in the first step?

2. What would happen if the sulfuric acid were left out?

3. If you used 5.0 g of salicylic acid and excess acetic anhydride in the preceding synthesis of aspirin, what would be the theoretical yield of acetylsalicylic acid in moles? In grams?

4. What is the equation for the decomposition reaction that can occur with aspirin?

5. Most aspirin tablets contain five grains of acetylsalicylic acid. How many milligrams is this?

6. A student performed the reaction in this experiment using a water bath at 90°C instead of 50°C. The final product was tested for the presence of phenols with ferric chloride. This test was negative (no color observed); however, the melting point of the dry product was 122–125°C. Explain these results as completely as possible.

7. If the aspirin crystals were not completely dried before the melting point was determined, what effect would this have on the observed melting point?

ESSAY

Analgesics

Acylated aromatic amines (those having an acyl group, $R—\overset{\overset{\displaystyle O}{\|}}{C}—$, substituted on nitrogen) are important in over-the-counter headache remedies. Over-the-counter drugs are those you may buy without a prescription. Acetanilide, phenacetin, and acetaminophen are mild analgesics (relieve pain) and antipyretics (reduce fever) and are important, along with aspirin, in many nonprescription drugs.

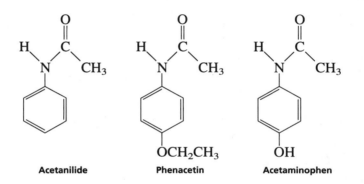

Acetanilide   **Phenacetin**   **Acetaminophen**

The discovery that acetanilide was an effective antipyretic came about by accident in 1886. Two doctors, Cahn and Hepp, had been testing naphthalene as a possible **vermifuge** (an agent that expels worms). Their early results on simple worm cases were very discouraging, so Dr. Hepp decided to test the compound on a patient with a larger variety of complaints, including worms—a sort of shotgun approach. A short time later, Dr. Hepp excitedly reported to his colleague, Dr. Cahn, that naphthalene had miraculous fever-reducing properties.

In trying to verify this observation, the doctors discovered that the bottle they thought contained naphthalene had apparently been mislabeled. In fact, the bottle brought to them by their assistant had a label so faint as to be illegible. They were sure that the sample was not naphthalene since it had no odor. Naphthalene has a strong odor reminiscent of mothballs. So close to an important discovery, the doctors were nevertheless stymied; they appealed to a cousin of Hepp, who was a chemist in a nearby dye factory, to help them identify the unknown compound. This compound turned out to be acetanilide, a compound with a structure not at all like that of naphthalene. Certainly, Hepp's unscientific and risky approach would be frowned on by doctors today, and to be sure, the Food and Drug

Naphthalene

Administration (FDA) would never allow human testing before extensive animal testing (consumer protection has progressed). Nevertheless, Cahn and Hepp made an important discovery.

In another instance of serendipity, the publication of Cahn and Hepp, describing their experiments with acetanilide, caught the attention of Carl Duisberg, director of research at the Bayer Company in Germany. Duisberg was confronted with the problem of profitably getting rid of nearly 50 tons of *p*-aminophenol, a by-product of the synthesis of one of Bayer's other commercial products. He immediately saw the possibility of converting *p*-aminophenol to a compound similar in structure to acetanilide, by putting an acyl group on the nitrogen. It was then believed, however, that all compounds having a hydroxyl group on a benzene ring (that is, phenols) were toxic. Duisberg devised a scheme of structural modification of *p*-aminophenol to get the compound phenacetin. The reaction scheme is shown here.

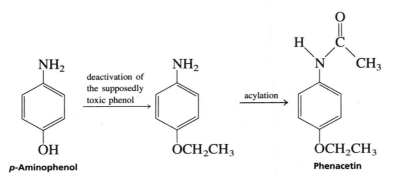

Phenacetin turned out to be a highly effective analgesic and antipyretic. A common form of combination pain reliever, called an APC tablet, was once available. An APC tablet contained **A**spirin, **P**henacetin, and **C**affeine (hence, **APC**). Phenacetin is no longer used in commercial pain-relief preparations. It was later found that not all aromatic hydroxyl groups lead to toxic compounds, and today the compound acetaminophen is very widely used as an analgesic in place of phenacetin.

Another analgesic, structurally similar to aspirin, that has found some application is **salicylamide.** Salicylamide is found as an ingredient in some pain-relief preparations, although its use is declining.

Salicylamide

On continued or excessive use, acetanilide can cause a serious blood disorder called **methemoglobinemia.** In this disorder, the central iron atom in hemoglobin is converted from Fe(II) to Fe(III) to give methemoglobin. Methemoglobin will not function as an oxygen carrier in the bloodstream. The result is a type of anemia (deficiency of hemoglobin or lack of red blood cells). Phenacetin and acetaminophen cause the same disorder, but to a much lesser degree. Since they are also more effective as antipyretic and analgesic drugs than acetanilide, they are preferred remedies. Acetaminophen is marketed under the trade names Tylenol and Datril and is often used successfully by persons who are allergic to aspirin.

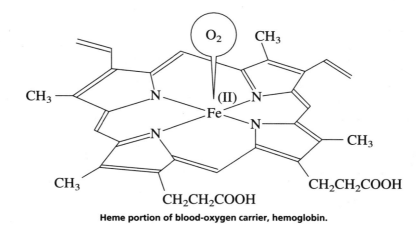

Heme portion of blood-oxygen carrier, hemoglobin.

More recently, a new drug has appeared in over-the-counter preparations. This drug is **ibuprofen,** which is marketed as a prescription drug in the United States under the name Motrin. Ibuprofen was first developed in England in 1964. United States marketing rights were obtained in 1974. Ibuprofen is now sold without prescription under brand names that include Advil and Nuprin. Ibuprofen is principally an anti-inflammatory drug, but it is also effective as an analgesic and an antipyretic. It is particularly effective in the treatment of the symptoms of rheumatoid arthritis and menstrual cramps. Ibuprofen appears to control the production of prostaglandins, which parallels the mode of action of aspirin. An important advantage of ibuprofen is that it is a very powerful pain reliever. One 200-mg tablet is as effective as two tablets (650 mg) of aspirin. Furthermore, ibuprofen has a more advantageous dose-response curve, which means that taking two tablets of this drug is approximately twice as effective as one tablet for certain types of pain. Aspirin and acetaminophen reach their maximum effective dose at two tablets. Little additional relief is gained at doses above that level. Ibuprofen, however, continues to increase its effectiveness up to the 400-mg level (the equivalent of four tablets of aspirin or acetaminophen). Ibuprofen is a relatively safe drug, but its use should be avoided in cases of aspirin allergy, kidney problems, ulcers, asthma, hypertension, or heart disease.

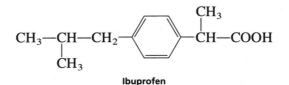

Ibuprofen

The Food and Drug Administration has also approved two other drugs with similar structures to ibuprofen for over-the-counter use as pain relievers. These new drugs are known by their generic names, **naproxen** and **ketoprofen.** Naproxen is often administered in the form of its sodium salt. Naproxen and ketoprofen can be used to alleviate the pain of headaches, toothaches, muscle aches, backaches, arthritis, and menstrual cramps, and they can also be used to reduce fever. They appear to have a longer duration of action than the older analgesics.

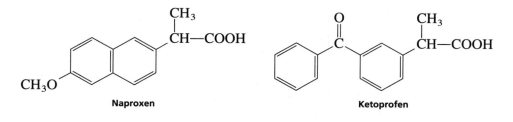

Naproxen   Ketoprofen

Analgesics and Caffeine in Some Common Preparations

	ASPIRIN	ACETAMINOPHEN	CAFFINE	SALICYLAMIDE
Aspirin*	0.325 g	—	—	—
Anacin	0.400 g	—	0.032 g	—
Bufferin	0.325 g	—	—	—
Cope	0.421 g	—	0.032 g	—
Excedrin (Extra-Strength)	0.250 g	0.250 g	0.065 g	—
Tylenol	—	0.325 g	—	—
B. C. Tablets	0.325 g	—	0.016 g	0.095 g

NOTE: Nonanalgesic ingredients (e.g., buffers) are not listed.
*5-grain tablet (1 grain = 0.0648 g).

REFERENCES

Barr, W. H., and Penna, R. P. "O-T-C Internal Analgesics." In G. B. Griffenhagen, ed., *Handbook of Non-Prescription Drugs,* 7th ed., Washington, DC: American Pharmaceutical Association, 1982.

Bugg, C. E., Carson, W. M., and Montgomery, J. A. "Drugs by Design." *Scientific American, 269* (December 1993): 92.

Flower, R. J., Moncada, S., and Vane, J. R. "Analgesic-Antipyretics and Anti-inflammatory Agents; Drugs Employed in the Treatment of Gout." In A. G. Gilman, L. S. Goodman, T. W. Rall, and F. Murad. *The Pharmacological Basis of Therapeutics,* 7th ed. New York: Macmillan, 1985.

Hansch, C. "Drug Research or the Luck of the Draw." *Journal of Chemical Education, 51* (1974): 360.

"The New Pain Relievers." *Consumer Reports, 49* (November 1984): 636–638.

Ray, O. S. "Internal Analgesics." *Drugs, Society, and Human Behavior,* 2nd ed. St. Louis: C. V. Mosby, 1978.

Senozan, N. M. "Methemoglobinemia: An Illness Caused by the Ferric State." *Journal of Chemical Education, 62* (March 1985): 181.

EXPERIMENT 3

Acetanilide

Crystallization
Gravity filtration
Vacuum filtration
Decolorization
Preparation of an amide

An amine can be treated with an acid anhydride to form an amide. In this experiment, aniline, the amine, is reacted with acetic anhydride to form acetanilide, the amide, and acetic acid. Crude acetanilide contains a colored impurity that is removed by decolorization with charcoal. Acetanilide is purified further by crystallization from hot water.

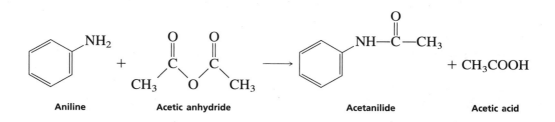

| Aniline | Acetic anhydride | Acetanilide | Acetic acid |

Amines can be acylated in several ways. Among these are the use of acetic anhydride, acetyl chloride, or glacial acetic acid. The procedure with glacial acetic acid is of commercial interest since it is economical. It requires long heating, however. Acetyl chloride is unsatisfactory for several reasons. Principally, it reacts vigorously, liberating HCl; this converts half of the amine to its hydrochloride salt, rendering it incapable of participating in the reaction. Acetic anhydride is preferred for a laboratory synthesis, and is used in this experiment. Its rate of hydrolysis (reaction with water) is low enough to allow the acetylation of amines to be carried out in aqueous solutions. The procedure gives a product of high purity and in good yield, but it is not suitable for use with deactivated amines (weak bases) such as *ortho*- and *para*-nitroanilines.

Acetylation is often used to "protect" a primary or a secondary amine functional group. Acylated amines are less susceptible to oxidation (Experiment 44), less reactive in aromatic substitution reactions (Experiment 34 and 43), and less prone to participate in many of the typical reactions of free amines, since they are less basic. The amino group can be regenerated readily by hydrolysis in acid or base (Experiment 43).

Required Reading

Review:	Techniques 1, 2	
	Technique 3	Reaction Methods, Sections 3.1 and 3.4–3.6
	Technique 4	Filtration, Sections 4.1–4.5
	Technique 6	Physical Constants, Part A, Melting Points
New:	Technique 5	Crystallization: Purification of Solids
	Essay	Analgesics

Special Instructions

Acetic anhydride can cause irritation of tissue, especially in nasal passages. Avoid breathing the vapor and avoid contact with skin and eyes. Aniline is a toxic substance, and it can be absorbed through the skin. Care should be exercised in handling it. You should be careful not to stop the experiment at any place where solid has not been filtered from the solution.

Waste Disposal

Aqueous solutions obtained from filtration operations should be poured into the container designated for aqueous wastes.

Notes to the Instructor

Aniline acquires a black color upon standing due to air oxidation. Since charcoal is very effective in decolorizing the crude acetanilide, it is recommended to use *impure* aniline in this experiment. When very dark aniline is used, students can see just how effective charcoal can be in purifying organic compounds.

Procedure

Reaction Mixture. Weigh 2.0 g of aniline into a 125-mL Erlenmeyer flask. To avoid spills or contact with the skin, use a Pasteur pipet to transfer the aniline to the flask. Add 15 mL of water to the flask. Next, swirl the flask gently while adding 2.5 mL of acetic anhydride (density 1.08 g/mL) slowly. Note in your laboratory notebook any changes that might occur during this addition.

Once the crude acetanilide precipitates during this reaction, it must be crystallized. This crystallization will be accomplished in the same flask as the original reaction. Add 50 mL of water, along with a boiling stone, to the flask. Heat the mixture on a hotplate until all the solid and oily materials have dissolved. After removing the flask from the hotplate, pour about 1 mL of the hot solution into a small beaker and allow this material to cool.

> **Caution:** The Erlenmeyer flask will be hot, so it is advisable to handle it with a paper towel. Do not use a test tube clamp or crucible tongs to remove the flask.

Decolorization.   Add a small amount of activated charcoal, or Norite, to the Erlenmeyer flask, and bring the mixture to a boil again. About one spatulaful should be enough charcoal. It is important not to add the charcoal to the solution while it is boiling vigorously, or violent frothing is likely. Swirl the mixture and allow it to boil gently for a few minutes.

Gravity Filtration.   While the solution is boiling, assemble an apparatus for gravity filtration, equipped with a funnel (stemless, if possible) and fluted filter paper in a 125- or a 250-mL Erlenmeyer flask (see Technique 4, Section 4.1, p. 634). Also prepare about 25 mL of boiling water, which will be needed as a washing solvent in the steps that follow.

Warm the funnel by pouring about 10 mL of the boiling water through it. Discard this warming water, put in the fluted filter paper (Figure 4.3, page 637) and then filter the acetanilide-charcoal mixture as quickly as possible, but in small portions, through the fluted filter. Keep the solution warm on the hotplate until it is poured into the funnel. While pouring the solution through the filter, keep the collection flask warmed on the hotplate. The purpose of keeping these flasks on the hotplate is to allow the water vapor to warm the funnel stem and to reduce the likelihood that crystals will form in the funnel, clogging it. If all goes well, the charcoal remains behind on the fluted filter paper and the aqueous solution containing acetanilide passes through the filter. If crystallization should begin in the funnel, add some hot water to the funnel to dissolve the crystals. Rinse the flask and the solid materials on the fluted filter paper with a little hot water and set the flask containing the filtrate aside to cool slowly to room temperature.

Crystallization and Vacuum Filtration.   To complete the crystallization, place the flask in an ice bath for about 15 minutes. Using the instructions given in Section 4.3, p. 640, prepare a Büchner funnel with filter paper. Collect the crystals by vacuum filtration and dry them as completely as possible by allowing air to be drawn through them while they remain on the Büchner funnel. Complete the drying by spreading the crystals on a watchglass, covering them with an inverted beaker, and allowing them to dry until the next laboratory period.

Collect the crystals that were not treated with charcoal in a Hirsch funnel or small Büchner funnel and compare the color of these crystals with the color of the crystals that were treated with charcoal. Record the results of the comparison in your laboratory notebook.

Record the weight of the pure crystallized product and calculate the percentage yield. Determine the melting point of both the pure and the impure samples of dried acetanilide. Place each sample of crystals in a separate, labeled vial, and submit both samples to the instructor with your laboratory report.

QUESTIONS

1. Aniline is basic and acetanilide is not basic. Explain this difference.
2. If 10 g of aniline is allowed to react with excess acetic anhydride, what is the theoretical yield of acetanilide in moles? In grams?
3. Give equations for the reactions of aniline with acetyl chloride and with acetic acid, to give acetanilide.
4. In the introduction to this experiment, the hydrolysis of acetic anhydride is mentioned as a competing reaction. Write an equation for this reaction.
5. During the crystallization of acetanilide, why was the mixture cooled in an ice bath?
6. Give two reasons why the crude product in most reactions is not pure.

EXPERIMENT 4

Acetaminophen

Vacuum filtration
Decolorization
Crystallization
Preparation of an amide

Preparation of acetaminophen involves treating an amine with an acid anhydride to form an amide. In this case, *p*-aminophenol, the amine, is treated with acetic anhydride to form acetaminophen (*p*-acetamidophenol), the amide.

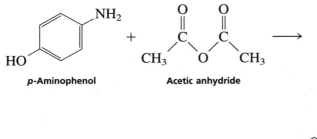

p-Aminophenol   Acetic anhydride

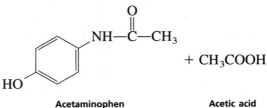

Acetaminophen   Acetic acid

The crude solid acetaminophen contains dark impurities carried along with the *p*-aminophenol starting material. These impurities, which are dyes of unknown structure, are formed from oxidation of the starting phenol. While the amount of the dye impurity is small, it is intense enough to impart color to the crude acetaminophen. Most of the colored impurity is destroyed by heating the crude product with sodium dithionite (sodium hydrosulfite $Na_2S_2O_4$). The dithionite reduces double bonds in the colored dye to produce colorless substances.

The decolorized acetaminophen is collected on a Büchner funnel. It is further purified by crystallization from a mixture of methanol and water.

Required Reading

Review:	Techniques 1, 2	
	Technique 3	Reaction Methods, Sections 3.1, 3.4–3.6
	Technique 4	Filtration, Sections 4.1–4.5
	Technique 6	Physical Constants, Part A, Melting Points
New:	Technique 5	Crystallization: Purification of Solids
	Essay	Analgesics

Special Instructions

Acetic anhydride can cause irritation of tissue, especially in nasal passages. Avoid breathing the vapor and avoid contact with skin and eyes. *p*-Aminophenol is a skin irritant and is toxic.

Waste Disposal

Aqueous solutions obtained from filtration operations should be poured into the container designated for aqueous wastes. This includes the filtrate from the methanol and water crystallization steps.

Notes to the Instructor

The *p*-aminophenol acquires a black color upon standing due to air oxidation. It is best to use a recently purchased sample, which usually has a gray color. If necessary, black material can be decolorized by heating it in a 10% aqueous solution of sodium dithionite (sodium hydrosulfite) prior to starting the experiment.

Procedure

Reaction Mixture.   Weigh about 1.5 g of *p*-aminophenol (*MW* = 109.1), and place this in a 50-mL Erlenmeyer flask. Using a graduated cylinder, add 4.5 mL of water and 1.7 mL of acetic anhydride (*MW* = 102.1, *d* = 1.08 g/mL). Place a magnetic stir bar in the flask.

Heating.   Heat the reaction mixture, with stirring, directly on a hotplate using a thermometer to monitor the internal temperature (about 100°C). After the solid has dissolved (it may dissolve, precipitate, and redissolve), heat the mixture for an additional 10 minutes at about 100°C to complete the reaction.

Isolation of Crude Acetominophen.   Remove the flask from the hotplate and allow the flask to cool to room temperature. If crystallization has not occurred, scratch the inside of the flask with a glass stirring rod to initiate crystallization (Technique 5, Section 5.7, p. 661). Cool the mixture thoroughly in an ice bath for 15–20 minutes and collect the crystals by vacuum filtration on a small Büchner funnel (Technique 4, Section 4.3, p. 640). Rinse the flask with about 5 mL of ice water and transfer this mixture to the Büchner funnel. Wash the crystals on the funnel with two additional 5 mL portions of ice water. Dry the crystals for 5–10 minutes by allowing air to be drawn through them while they remain on the Büchner funnel. During this drying period, break up any large clumps of crystals with a spatula. Transfer the product to a watch glass and allow the crystals to dry in air. It may take several hours for the crystals to dry completely, but you may go on to the next step before they are totally dry. Weigh the crude product and set aside a small sample for a melting point determination and a color comparison after the next step. Calculate the percentage yield of crude acetaminophen (*MW* = 151.2). Record the appearance of the crystals in your notebook.

Decolorization of Crude Acetaminophen.   Dissolve 2.0 g of sodium dithionite (sodium hydrosulfite) in 15 mL of water in a 50-mL Erlenmeyer flask. Add your crude acetaminophen to the flask. Heat the mixture to about 100°C for 15 minutes, with occasional stirring with a spatula. Some of the acetaminophen will dissolve during the decolorization process. Cool the mixture thoroughly in an ice bath for about 10 minutes to reprecipitate the decolorized acetaminophen (scratch the inside of the flask if necessary to induce crystallization). Collect the purified material by vacuum filtration on a small Büchner funnel using small portions (about 5 mL total) of ice water to aid the transfer. Dry the crystals for 5–10 minutes by allowing air to be drawn through them while they remain on the Büchner funnel. You may go on to the next step before the material is totally dry. Weigh the purified acetaminophen and compare the color of the purified material to that obtained above.

Crystallization of Acetaminophen.   Place the purified acetaminophen in a 50-mL Erlenmeyer flask. Crystallize the material from a solvent mixture composed of 50% water and 50% methanol by volume. Follow the crystallization procedure described in Technique 5, Section 5.3, p. 651. The solubility of acetaminophen in this hot (nearly boiling) solvent is about 1 g/5 mL. Although you can use this as a rough indication of how much solvent is required to dissolve the solid, you should still use the technique shown in Figure 5.3, p. 652 to determine how much solvent to add. Add small portions of hot solvent until the solid dissolves. Step 2 in Figure 5.3 (removal of insoluble impurities) should not be required in this crystallization. When dissolved, allow the mixture to cool slowly to room temperature.

When the mixture has cooled to room temperature, place the flask in an ice bath for at least 10 minutes. If necessary, induce crystallization by scratching the inside of the flask with a glass stirring rod. Because acetaminophen may crystallize *slowly* from the solvent, it is necessary to cool the flask in an ice bath for the 10-minute period. Collect the crystals using a Büchner funnel as shown in Technique 4, Figure 4.5 on page 641. Dry the crystals for 5–10 minutes by allowing air to be drawn through them while they remain on the Büchner funnel. Alternatively, you may allow the crystals to dry until the next laboratory period.

Yield Calculation and Melting Point Determination. Weigh the crystallized acetaminophen (*MW* = 151.2) and calculate the percentage yield. This calculation should be based on the original amount of *p*-aminophenol used at the beginning of this procedure. Determine the melting point of the product. Compare the melting point of this final product with that of the crude acetaminophen. Also compare the colors of the crude, decolorized, and pure acetaminophen. Pure acetaminophen melts at 169.5–171°C. Place your product in a properly labeled vial and submit it to your instructor.

QUESTIONS

1. During the crystallization of acetaminophen, why was the mixture cooled in an ice bath?
2. In the reaction between *p*-aminophenol and acetic anhydride to form acetaminophen, 4.5 mL of water was added. What was the purpose of the water?
3. Why should you use a minimum amount of water to rinse the flask while transferring the purified acetaminophen to the Büchner funnel?
4. If 1.30 g of *p*-aminophenol is allowed to react with excess acetic anhydride, what is the theoretical yield of acetaminophen in moles? In grams?
5. Give two reasons why the crude product in most reactions is not pure.
6. Phenacetin has the structure shown. Write an equation for its preparation starting from 4-ethoxyaniline.

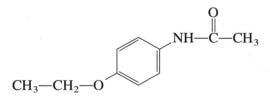

E S S A Y

Caffeine

The origins of coffee and tea as beverages are so old that they are lost in legend. Coffee is said to have been discovered by an Abyssinian goatherd who noticed an unusual friskiness in his goats when they consumed a certain little plant with red berries. He decided to try the berries himself and discovered coffee. The Arabs soon cultivated the coffee plant, and one of the earliest descriptions of its use is found in an Arabian medical book circa A.D. 900. The great systematic botanist, Linnaeus, named the tree *Coffea arabica.*

One legend of the discovery of tea—from the Orient, as you might expect—attributes the discovery to Daruma, the founder of Zen. Legend has it that he inadvertently fell asleep one day during his customary mediations. To be assured that this indiscretion would not recur, he cut off both eyelids. Where they fell to the ground, a new plant took root that had the power to keep a person awake. Although some experts assert that the medical use of tea was reported as early as 2737 B.C. in the pharmacopeia of Shen Nung, an emperor of China, the first indisputable reference is from the Chinese dictionary of Kuo P'o, which appeared in A.D. 350. The nonmedical, or popular, use of tea appears to have spread slowly. Not until about A.D. 700 was tea widely cultivated in China. Since tea is native to upper Indochina and upper India, it must have been cultivated in these places before its introduction to China. Linnaeus named the tea shrub *Thea sinensis;* however, tea is more properly a relative of the camellia, and botanists have renamed it *Camellia thea.*

The active ingredient that makes tea and coffee valuable to humans is **caffeine.** Caffeine is an **alkaloid,** a class of naturally occurring compounds containing nitrogen and having the properties of an organic amine base (alkaline, hence, *alkaloid*). Tea and coffee are not the only plant sources of caffeine. Others include kola nuts, maté leaves, guarana seeds, and in small amount, cocoa beans. The pure alkaloid was first isolated from coffee in 1821 by the French chemist Pierre Jean Robiquet.

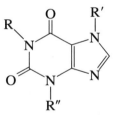

XANTHINES
Xanthine R = R′ = R″ = H
Caffeine R = R′ = R″ = CH$_3$
Theophylline R = R″ = CH$_3$, R′ = H
Theobromine R = H, R′ = R″ = CH$_3$

Caffeine belongs to a family of naturally occurring compounds called **xanthines.** The xanthines, in the form of their plant progenitors, are possibly the oldest known stimulants. They all, to varying extents, stimulate the central nervous system and the skeletal muscles. This stimulation results in an increased alertness, the ability to put off sleep, and an increased capacity for thinking. Caffeine is the most powerful xanthine in this respect. It is the main ingredient of the popular No-Doz keep-alert tablets. While caffeine has a powerful effect on the central nervous system, not all xanthines are as effective. Thus theobromine, the xanthine found in cocoa, has fewer central nervous system effects. It is, however, a strong **diuretic** (induces urination) and is useful to doctors in treating patients with severe water-retention problems. Theophylline, a second xanthine found in tea, also has fewer central nervous system effects but is a strong **myocardial** (heart muscle) stimulant; it **dilates** (relaxes) the coronary artery that supplies blood to the heart. Its most important use is in the treatment of bronchial asthma, since it has the properties of a **bronchodilator** (relaxes the bronchioles of the lungs). Because it is also a **vasodilator** (relaxes blood vessels), it is often used in treating hypertensive headaches. It is also used to alleviate and to reduce the frequency of attacks of **angina pectoris** (severe chest pain). In addition, it is a more powerful diuretic than theobromine.

One can develop both a tolerance for the xanthines and a dependence on them, particularly caffeine. The dependence is real, and a heavy user (>5 cups of coffee per day) will experience lethargy, headache, and perhaps nausea after about 18 hours of abstinence. An excessive intake of caffeine may lead to restlessness, irritability, insomnia, and muscular tremor. Caffeine can be toxic, but to achieve a lethal dose of caffeine, one would have to drink about 100 cups of coffee over a relatively short period.

Caffeine is a natural constituent of coffee, tea, and kola nuts (*Kola nitida*). Theophylline is found as a minor constituent of tea. The chief constituent of cocoa is theobromine. The amount of caffeine in tea varies from 2 to 5%. In one analysis of black tea, the following compounds were found: caffeine, 2.5%; theobromine, 0.17%; theophylline, 0.013%; adenine, 0.014%; and guanine and xanthine, traces. Coffee beans can contain up to 5% by weight of caffeine, and cocoa contains around 5% theobromine. Commercial cola is a beverage based on a kola nut extract. We cannot easily get kola nuts in this country, but we can get the ubiquitous commercial extract as a syrup. The syrup can be converted into "cola." The syrup contains caffeine, tannins, pigments, and sugar. Phosphoric acid is added, and caramel is added to give the syrup a deep color. The final drink is prepared by adding water and carbon dioxide under pressure, to give the bubbly mixture. The Food and Drug Administration currently requires that a "cola" contain *some* caffeine but limits this amount to a maximum of 5 milligrams per ounce. To achieve a regulated level of caffeine, manufacturers must remove all caffeine from the kola extract and then re-add the correct amount to the syrup. The caffeine content of various beverages is listed in the accompanying table.

Amount of Caffeine (mg/oz) Found in Beverages

Brewed coffee	12–30	Tea	4–20
Instant coffee	8–20	Cocoa (but 20 mg/oz theobromine)	0.5–2
Espresso (1 serving = 1.5–2 oz)	50–70	Coca-Cola	3.75
Decaffeinated coffee	0.4–1.0		

NOTE: The average cup of coffee or tea contains about 5–7 oz of liquid. The average bottle of cola contains about 12 oz of liquid.

With the recent popularity of gourmet coffee beans and espresso stands, it is interesting to consider the caffeine content of these specialty beverages. Certainly, gourmet coffee has more flavor than the typical ground coffee that you may find on any grocery store shelf, and the concentration of brewed gourmet coffee tends to be higher than ordinary drip grind coffee. Brewed gourmet coffee probably contains something on the order of 20–25 mg of caffeine per ounce of liquid. Espresso coffee is a very concentrated, dark-brewed coffee. While the darker roasted beans used for espresso actually contain less caffeine per gram than regularly roasted beans, the method of preparing espresso (extraction using pressurized steam) is more efficient, and a higher percentage of the total caffeine in the beans is extracted. The caffeine content per ounce of liquid, therefore, is substantially higher than in most brewed coffees. The serving size for espresso coffee, however, is much smaller than for ordinary coffee (about 1.5–2 oz per serving), so that the total caffeine available in a serving of espresso turns out to be about the same as in a serving of ordinary coffee.

Because of the central nervous system effects from caffeine, many persons prefer to use **decaffeinated** coffee. In the most commonly used process, the caffeine is removed from coffee by extracting the whole beans with an organic solvent. Then, the solvent is drained off, and the beans are steamed to remove any residual solvent. The beans are dried and roasted to bring out the flavor. Decaffeination reduces the caffeine content of coffee to the range of 0.03 to 1.2% caffeine. The extracted caffeine is used in various pharmaceutical products, such as APC tablets.

Among coffee lovers there is some controversy about the best method to remove the caffeine from coffee beans. **Direct contact** decaffeination uses an organic solvent (usually methylene chloride) to remove the caffeine from the beans. When the beans are subsequently roasted at 200°C, virtually all traces of the solvent are removed, because methylene chloride boils at 40°C. The advantage of direct contact decaffeination is that the method removes only the caffeine (and some waxes), but leaves the substances responsible for the flavor of the coffee intact in the bean.

Water process decaffeination is favored among many drinkers of decaffeinated coffee, because it does not use organic solvents. In this method, hot water and steam are used to remove caffeine and other soluble substances from the coffee. The resulting solution is then passed through activated charcoal filters to remove the caffeine. While this method does not use organic solvents, the disadvantage is that water is not a very selective decaffeinating agent. Many of the flavor oils in the coffee are removed at the same time, resulting in a coffee with a somewhat bland flavor.

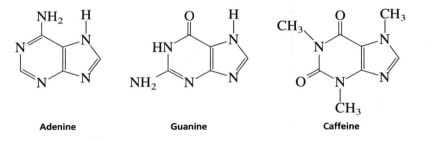

Adenine   **Guanine**   **Caffeine**

Caffeine has always been a controversial compound. Some religions forbid the use of beverages containing caffeine, which they consider an addictive drug. In fact, many people consider caffeine an addictive drug. Recently, there has been concern because caffeine is structurally similar to the purine bases adenine and guanine, which are two of the five principal bases that organisms use to form the nucleic acids DNA and RNA. It is feared that the substitution of caffeine for adenine or guanine in either of these genetically important substances could lead to chromosome defects.

A portion of the structure of a DNA molecule is shown on p. 67. The typical mode of incorporation of both adenine and guanine is specifically shown. If caffeine were substituted for either of these, the hydrogen bonding necessary to link the two chains would be disturbed. Although caffeine is most similar to guanine, it could not form the central hydrogen bond, since it has a methyl group instead of a hydrogen in the necessary position. Hence, the genetic information would be garbled, and there would be a break in the chain. Fortunately, little evidence exists of chromosome breaks due to caffeine. Many cultures have been using tea and coffee for centuries without any apparent genetic problems.

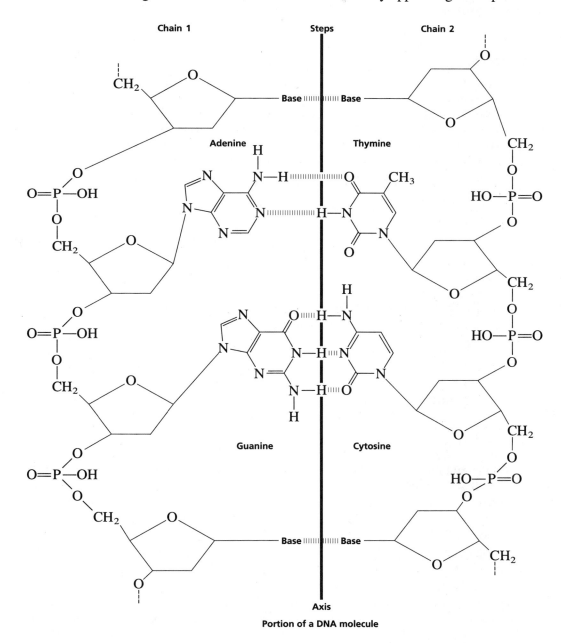

Portion of a DNA molecule

Another problem, not related to caffeine but rather to the beverage tea, is that in some cases persons who consume high quantities of tea may show symptoms of Vitamin B_1 (thiamine) deficiency. It is suggested that the tannins in the tea may complex with the thiamine, rendering it unavailable for use. An alternative suggestion is that caffeine may reduce the levels of the enzyme transketolase, which depends on the presence of thiamine for its activity. Lowered levels of transketolase would produce the same symptoms as lowered levels of thiamine.

REFERENCES

"Caffeine May Damage Genes by Inhibiting DNA Polymerase." *Chemical and Engineering News,* *45* (November 20, 1967): 19.

Emboden, W. "The Stimulants." *Narcotic Plants,* rev. ed. New York: Macmillan, 1979.

Ray, O., and Ksir, C. "Caffeine." *Drugs, Society and Human Behavior,* 7th ed. St. Louis: C. V. Mosby, 1996.

Ritchie, J. M. "Central Nervous System Stimulants. II: The Xanthines." In L. S. Goodman and A. Gilman, eds. *The Pharmacological Basis of Therapeutics,* 8th ed. New York: Macmillan, 1990.

Taylor, N. *Plant Drugs That Changed the World.* New York: Dodd, Mead, 1965. Pp. 54–56.

Taylor, N. "Three Habit-Forming Nondangerous Beverages." In *Narcotics—Nature's Dangerous Gifts.* New York: Dell, 1970. (Paperbound revision of *Flight from Reality.*)

E X P E R I M E N T   5

Isolation of Caffeine

Isolation of a natural product
Extraction
Sublimation

In this experiment, caffeine is isolated from tea leaves. The chief problem with the isolation is that caffeine does not exist alone in tea leaves but is accompanied by other natural substances from which it must be separated. The main component of tea leaves is cellulose, which is the principal structural material of all plant cells. Cellulose is a polymer of glucose. Since cellulose is virtually insoluble in water, it presents no problems in the isolation procedure. Caffeine, on the other hand, is water soluble and is one of the main substances extracted into the solution called tea. Caffeine constitutes as much as 5% by weight of the leaf material in tea plants.

Tannins also dissolve in the hot water used to extract tea leaves. The term **tannin** does not refer to a single homogeneous compound, or even to substances that have similar chemical structure. It refers to a class of compounds that have certain properties in common. Tannins are phenolic compounds having molecular weights between 500 and 3000. They are widely used to tan leather. They precipitate alkaloids and proteins from aqueous solutions. Tannins are usually divided into two classes: those that can be **hydrolyzed** (react with water) and those that cannot. Tannins of the first type that are found in tea generally yield glucose and gallic acid when they are hydrolyzed. These tannins are esters of gallic acid and glucose. They represent structures in which some of the hydroxyl groups in glucose have been esterified by digalloyl groups. The nonhydrolyzable tannins found in tea are condensation polymers of catechin. These polymers are not uniform in structure; catechin molecules are usually linked at ring positions 4 and 8.

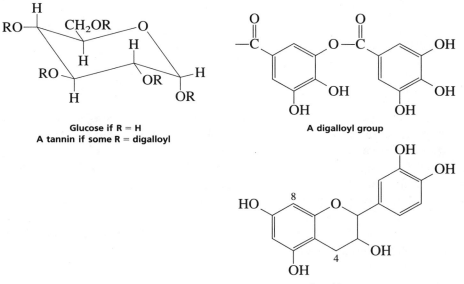

Glucose if R = H
A tannin if some R = digalloyl

A digalloyl group

Catechin

When tannins are extracted into hot water, some of these compounds are partially hydrolyzed to form free gallic acid. The tannins, because of their phenolic groups, and gallic acid, because of its carboxyl groups, are both acidic. If sodium carbonate, a base, is added to tea water, these acids are converted to their sodium salts that are highly soluble in water.

Although caffeine is soluble in water, it is much more soluble in the organic solvent methylene chloride. Caffeine can be extracted from the basic tea solution with methylene chloride, but the sodium salts of gallic acid and the tannins remain in the aqueous layer.

The brown color of a tea solution is due to flavonoid pigments and chlorophylls and to their respective oxidation products. Although chlorophylls are soluble in methylene chloride, most other substances in tea are not. Thus, the methylene chloride extraction of the basic tea solution removes nearly pure caffeine. The methylene chloride is easily removed by evaporation (bp 40°C) to leave the crude caffeine. The caffeine is then purified by sublimation.

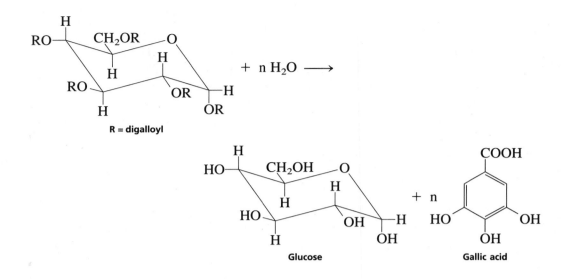

R = digalloyl

Glucose

Gallic acid

Experiment 5A outlines the isolation of caffeine from tea using standard-scale techniques. An optional procedure in Experiment 5A allows the student to convert caffeine to a **derivative.** A derivative of a compound is a second compound, of known melting point, formed from the original compound by a simple chemical reaction. In trying to make a positive identification of an organic compound, it is often necessary to convert it to a derivative. If the first compound, caffeine in this case, and its derivative have melting points that match those reported in the chemical literature (a handbook, for instance), it is assumed that there is no coincidence and that the identity of the first compound, caffeine, has been established conclusively.

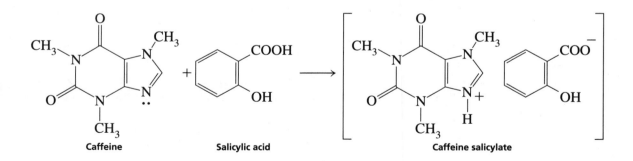

Caffeine   Salicylic acid   Caffeine salicylate

Caffeine is a base and will react with an acid to give a salt. With salicylic acid, a derivative **salt** of caffeine, caffeine salicylate, can be made to establish the identity of the caffeine isolated from tea leaves.

In Experiment 5B, the isolation of caffeine is accomplished using microscale methods. In this experiment, you will be asked to isolate the caffeine from the tea contained in a single tea bag. Finally, Experiment 5C describes the isolation of caffeine from the syrup concentrate used to prepare "Mountain Dew" soft drinks.

Required Reading

Review:	Techniques 1, 2	
	Technique 3	Reaction Methods, Section 3.11
	Technique 6	Physical Constants, Part A, Melting Points
New:	Technique 7	Extractions, Separations, and Drying Agents, Sections 7.1–7.5, and 7.7–7.9
	Technique 16	Sublimation
	Essay	Caffeine

Special Instructions

Be careful when handling methylene chloride. It is a toxic solvent, and you should not breathe it excessively or spill it on yourself. In Experiment 5B, the extraction procedure with methylene chloride calls for two centrifuge tubes with screwcaps. Corks can also be used to seal the tubes; however, the corks will absorb a small amount of the liquid. Rather than shaking the centrifuge tube, agitation can be accomplished conveniently with a vortex mixer.

Waste Disposal

You must dispose of methylene chloride in a waste container marked for the disposal of halogenated organic waste. When you are discarding tea leaves, do not put them in the sink; they will clog the drain. Dispose of them in a waste container. Dispose of the tea bags in a waste container, not in the sink. The aqueous solutions obtained after the extraction steps must be disposed of in a waste container labeled for aqueous waste.

EXPERIMENT 5A

Isolation of Caffeine from Tea Leaves

Procedure

Preparing the Tea Solution. Place 5 g of tea leaves, 2 g of calcium carbonate powder, and 50 mL of water in a 100-mL round-bottomed flask equipped with a condenser for reflux (Technique 3, Figure 3.8, p. 619). Heat the mixture under gentle reflux, being careful to prevent any bumping, for about 20 minutes. Use a heating mantle to heat the mixture. Shake the flask occasionally during this heating period. While the solution is still hot, vacuum-filter it through a fast filter paper such as E&D No. 617 or S&S No. 595 (Technique 4, Section 4.3, p. 640). A 125-mL filter flask is appropriate for this step.

Extraction and Drying. Cool the filtrate (filtered liquid) to room temperature, and using a 125-mL separatory funnel, extract it (Technique 7, Section 7.4, page 690) with a 10-mL portion of methylene chloride (dichloromethane). Shake the mixture vigorously for 1 minute. The layers should separate after standing for several minutes, although some emulsion will be present in the lower organic layer (Technique 7, Section 7.9, p. 699). The emulsion can be broken and the organic layer dried at the same time by passing the lower layer *slowly* through anhydrous magnesium sulfate, according to the following method. Place a small piece of cotton (not glass wool) in the neck of a conical funnel and add a 1-cm layer of anhydrous magnesium sulfate on top of the cotton. Pass the organic layer directly from the separatory funnel into the drying agent, and collect the filtrate in a dry Erlenmeyer flask. Rinse the magnesium sulfate with 1 or 2 mL of fresh methylene chloride solvent. Repeat the extraction with another 10-mL portion of methylene chloride on the aqueous layer remaining in the separatory funnel, and repeat the drying, as described above, with a *fresh* portion of anhydrous magnesium sulfate. Collect the organic layer in the flask containing the first methylene chloride extract. These extracts should now be clear, showing no visible signs of water contamination. If some water should pass through the filter, repeat the drying, as described above, with a fresh portion of magnesium sulfate. Collect the dried extracts in a dry Erlenmeyer flask.

Distillation. Pour the dry organic extracts into a 50-mL round-bottomed flask. Assemble an apparatus for simple distillation (Technique 8, Figure 8.4, p. 711), add a boiling stone, and remove the methylene chloride by distillation on a steam bath or heating mantle. The residue in the distillation flask contains the caffeine and is purified by crystallization and sublimation. Save the methylene chloride that was distilled; you may use some of it in the next step. The remaining methylene chloride must be placed in a waste container marked for halogenated waste; it must *not* be discarded in the sink.

Crystallization (Purification). Dissolve the residue from the methylene chloride extraction of the tea solution in about 5 mL of the methylene chloride that you saved from the distillation. You may have to heat the mixture on a steam bath or heating mantle to dissolve the solid. Transfer the solution to a 25-mL Erlenmeyer flask. Rinse the distillation flask with an additional 2–3 mL of methylene chloride and combine this solution with the contents of the Erlenmeyer flask. Add a boiling stone and evaporate the now light green solution to dryness by heating it on a steam bath or a hotplate *in the hood.*

The residue obtained on evaporation of the methylene chloride is next crystallized by the mixed-solvent method (Technique 5, Section 5.9, p. 663). Using a steam bath or hotplate, dissolve the residue in a small quantity (about 2 mL) of hot acetone[1] and add dropwise just enough low-boiling (bp 30–60°C) petroleum ether to turn the solution faintly cloudy. Cool the solution and collect the crystalline product by vacuum filtration, using a small Büchner funnel. A small amount of petroleum

---

[1]If the residue does not dissolve in this quantity of acetone, magnesium sulfate may be present as an impurity (drying agent). Add additional acetone (up to about 5 mL), gravity-filter the mixture to remove the solid impurity, and reduce the volume of the filtrate to about 2 mL. Now add petroleum ether as indicated in the procedure.

ether can be used to help in transferring the crystals to the Büchner funnel. A second crop of crystals can be obtained by concentrating the filtrate. Weigh the product (an analytical balance may be necessary). Calculate the weight percentage yield based on the 5 g of tea originally used and determine the melting point. The melting point of pure caffeine is 236°C. Note the color of the solid for comparison with the material obtained after sublimation.

Sublimation of Caffeine.   Caffeine can be purified by sublimation (Technique 16, p. 824). Assemble a sublimation apparatus as shown in Figure 16.2C, p. 828. Insert a 15 × 125-mm test tube into a No. 2 neoprene adapter, using a *little* water as a lubricant, until the tube is fully inserted. Place the crude caffeine into a 20 × 150-mm sidearm test tube. Next place the 15 × 120-mm test tube into the sidearm test tube, making sure they fit together tightly. Turn on the aspirator or house vacuum and make sure a good seal is obtained. At the point at which a good seal has been achieved, you should hear or observe a change in the water velocity in the aspirator. At this time, also make sure that the central tube is centered in the sidearm test tube; this will allow for optimal collection of the purified caffeine. Once the vacuum has been established, place small chips of ice in the test tube to fill it.[2] When a good vacuum seal has been obtained and ice has been added to the inner test tube, heat the sample gently and carefully with a microburner to sublime the caffeine. Hold the burner in your hand (hold it at the *base,* not by the hot barrel) and apply heat by moving the flame back and forth under the outer tube and up the sides. If the sample begins to melt, remove the flame for a few seconds before you resume heating. When sublimation is complete, remove the burner and allow the apparatus to cool. As the apparatus is cooling, and before you disconnect the vacuum, remove the water and ice from the inner tube using a Pasteur pipet.

When the apparatus has cooled and the water has been removed from the tube, you may disconnect the vacuum. The vacuum should be removed carefully to avoid dislodging the crystals from the inner tube by the sudden rush of air into the apparatus. *Carefully* remove the inner tube of the sublimation apparatus. If this operation is done carelessly, the sublimed crystals may be dislodged from the inner tube and fall back into the residue. Scrape the sublimed caffeine onto weighing paper, using a small spatula. Determine the melting point of this purified caffeine and compare it in melting point and color with the caffeine obtained following crystallization. Submit the sample to the instructor in a labeled vial, or if the instructor directs, prepare the caffeine salicylate derivative.

THE DERIVATIVE (OPTIONAL)

The amounts given in this part, including solvents, should be adjusted to fit the quantity of caffeine you obtained. Use an analytical balance. Dissolve 25 mg of caffeine and 18 mg of salicylic acid in 2 mL of toluene in a small Erlenmeyer flask by

---

[2]It is very important that ice not be added to the inner test tube until the vacuum has been established. If the ice is added before the vacuum is turned on, condensation on the outer walls of the inner tube will contaminate the sublimed caffeine.

warming the mixture on a steam bath or hotplate. Add about 0.5 mL (10 drops) of high-boiling (bp 60–90°C) petroleum ether or ligroin and allow the mixture to cool and crystallize. It may be necessary to cool the flask in an ice-water bath or to add a small amount of extra petroleum ether to induce crystallization. Collect the crystalline product by vacuum filtration, using a Hirsch funnel or a small Büchner funnel. Dry the product by allowing it to stand in the air and determine its melting point. Pure caffeine salicylate melts at 137°C. Submit the sample to the instructor in a labeled vial.

E X P E R I M E N T   5 B

Isolation of Caffeine from a Tea Bag

Procedure

Preparing the Tea Solution. Place 20 mL of water in a 50-mL beaker. Cover the beaker with a watch glass and heat the water on a hotplate until it is almost boiling. Place a tea bag[3] into the hot water so that it lies flat on the bottom of the beaker and is covered as completely as possible with water. Replace the watch glass and continue heating for about 15 minutes. During this heating period, it is important to push down *gently* on the tea bag with a test tube, so that all the tea leaves are in constant contact with water. As the water evaporates during this heating step, replace it by adding water from a Pasteur pipet.

Using a Pasteur pipet, transfer the concentrated tea solution to two centrifuge tubes fitted with screwcaps. Try to keep the liquid volume in each centrifuge tube approximately equal. To squeeze additional liquid out of the tea bag, hold the tea bag on the inside wall of the beaker and roll a test tube back and forth while exerting *gentle* pressure on the tea bag. Press out as much liquid as possible without breaking the bag. Combine this liquid with the solution in the centrifuge tubes. Place the tea bag on the bottom of the beaker again, and pour 2 mL of hot water over the bag. Squeeze the liquid out, as just described, and transfer this liquid to the centrifuge tubes. Add 0.5 g of sodium carbonate to the hot liquid in each centrifuge tube. Cap the tubes and shake the mixture until the solid dissolves.

Extraction and Drying. Cool the tea solution to room temperature. Using a calibrated Pasteur pipet (p. 596), add 3 mL of methylene chloride to each centrifuge tube to extract the caffeine (Technique 7, Section 7.7, p. 696). Cap the centrifuge tubes and gently shake the mixture for several seconds. Vent the tubes to release the pressure, being careful that the liquid does not squirt out toward you. Shake the

---

[3]The weight of tea in the bag will be given to you by your instructor. This can be determined by opening several bags of tea and determining the average weight. If this is done carefully, the tea can be returned to the bags, which can be restapled.

mixture for an additional 30 seconds with occasional venting. To separate the layers and break the emulsion (see Technique 7, Section 7.9, p. 699), centrifuge the mixture for several minutes (be sure to balance the centrifuge by placing the two centrifuge tubes on opposite sides). If an emulsion still remains (indicated by a green-brown layer between the clear methylene chloride layer and the top aqueous layer), centrifuge the mixture again.

Remove the lower organic layer with a Pasteur pipet and transfer it to a test tube. Be sure to squeeze the bulb before placing the tip of the Pasteur pipet into the liquid, and try not to transfer any of the dark aqueous solution along with the methylene chloride layer. Add a fresh 3-mL portion of methylene chloride to the aqueous layer remaining in each centrifuge tube, cap the centrifuge tubes, and shake the mixture in order to carry out a second extraction. Separate the layers by centrifugation, as described previously. Combine the organic layers from each extraction into one test tube. If there are visible drops of the dark aqueous solution in the test tube, transfer the methylene chloride solution to another test tube using a clean, dry Pasteur pipet. If necessary, leave a small amount of the methylene chloride solution behind in order to avoid transferring any of the aqueous mixture. Add a small amount of granular anhydrous sodium sulfate to dry the organic layer (Technique 7, Section 7.8, p. 696). If all the sodium sulfate clumps together when the mixture is stirred with a spatula, add some additional drying agent. Allow the mixture to stand for 10–15 minutes. Stir occasionally with a spatula.

Evaporation.   Transfer the dry methylene chloride solution with a Pasteur pipet to a dry, pre-weighed 25-mL Erlenmeyer flask, while leaving the drying agent behind. Evaporate the methylene chloride by heating the flask in a hot water bath (Technique 3, Section 3.11, p. 630). This should be done in a hood and can be accomplished more rapidly if a stream of dry air or nitrogen gas is directed at the surface of the liquid. When the solvent is evaporated, the crude caffeine will coat the bottom of the flask. Do not heat the flask after the solvent has evaporated, or you may sublime some of the caffeine. Weigh the flask and determine the weight of crude caffeine. Calculate the weight percentage recovery (see p. 28) of caffeine from tea leaves, using the weight of tea given to you by your instructor. You may store the caffeine by simply placing a stopper firmly into the flask.

Sublimation of Caffeine.   Caffeine can be purified by sublimation (Technique 16, Section 16.4, p. 827). Follow the method described in Experiment 5A. Add approximately 0.5 mL of methylene chloride to the Erlenmeyer flask and transfer the solution to the sublimation apparatus using a clean and dry Pasteur pipet. Add a few more drops of methylene chloride to the flask in order to rinse the caffeine out completely. Transfer this liquid to the sublimation apparatus. Evaporate the methylene chloride from the outer tube of the sublimation apparatus by gentle heating in a warm water bath under a stream of dry air or nitrogen.

Assemble the apparatus as described in Experiment 5A. Be sure that the inside of the assembled apparatus is clean and dry. If you are using an aspirator, install a trap between the aspirator and the sublimation apparatus. Turn on the vacuum and check to make sure that all joints in the apparatus are sealed tightly. Place *ice cold* water in the inner tube of the apparatus. Heat the sample gently and carefully with a microburner to sublime the caffeine. Hold the burner in your hand (hold it at its base, *not* by the hot barrel) and apply the heat by moving the flame back and forth under the outer test tube and up the sides. If the sample begins to melt, remove the flame for a few seconds before you resume heating. When sublimation is complete, discontinue heating. Remove the cold water and remaining ice from the inner tube and allow the apparatus to cool while continuing to apply the vacuum.

When the apparatus is at room temperature, remove the vacuum and *carefully* remove the inner tube. If this operation is done carelessly, the sublimed crystals may be dislodged from the inner tube and fall back into the residue at the bottom of the outer test tube. Scrape the sublimed caffeine onto a tared piece of smooth paper and determine the weight of caffeine recovered. Calculate the weight percentage recovery (see p. 28) of caffeine after the sublimation. Compare this value to the percentage recovery determined after the evaporation step. Determine the melting point of the purified caffeine. The melting point of pure caffeine is 236°C; however, the observed melting point will be lower. Submit the sample to the instructor in a labeled vial.

EXPERIMENT 5C

Isolation of Caffeine from "Mountain Dew" Syrup

Procedure

Dilute 50 mL of "Mountain Dew" syrup[4] with 50 mL of water in a 250-mL Erlenmeyer flask. Add 10 mL of concentrated ammonium hydroxide and 50 mL of methylene chloride (dichloromethane) to the solution. Pour the solution into a 250-mL separatory funnel. Shake the mixture vigorously for 1 minute. The layers should separate after standing for several minutes, although some emulsion will be present in the lower organic layer (Technique 7, Section 7.9, p. 699). The emulsion can be broken and the organic layer dried at the same time by passing the lower layer *slowly* through anhydrous magnesium sulfate, according to the following method. Place a small piece of cotton (not glass wool) in the neck of a conical funnel and add a 1-cm layer of anhydrous magnesium sulfate on top of the cotton. Pass the organic layer directly from the separatory funnel into the drying agent, and collect the filtrate in a dry Erlenmeyer flask. Rinse the magnesium sulfate with 1 or 2 mL of fresh methylene chloride solvent. Repeat the extraction with another 10-mL portion of methylene chloride on the aqueous layer remaining in the separatory funnel, and repeat the drying, as described above, with a *fresh* portion of anhydrous magnesium sulfate. Collect all organic layers in the flask containing the first methylene chloride extract. These extracts should now be clear, showing no visible signs

---

[4]"Mountain Dew" syrup can be obtained from local soft drink distributors.

of water contamination. If some water should pass through the filter, repeat the drying, as described above, with a fresh portion of magnesium sulfate. Collect the dried extracts in a dry Erlenmeyer flask.

Remove the methylene chloride by distillation as described in Experiment 5A (page 72). You will not need to save the methylene chloride; pour it into a container designated for halogenated organic waste.

Caffeine can be purified by sublimation (Technique 16, Section 16.4, p. 827). Follow the method described in Experiment 5A. Add approximately 0.5 mL of methylene chloride to the round-bottom flask and transfer the solution to the sublimation apparatus using a clean and dry Pasteur pipet. Add a few more drops of methylene chloride to the flask in order to rinse the caffeine out completely. Transfer this liquid to the sublimation apparatus. Evaporate the methylene chloride from the outer tube of the sublimation apparatus by gentle heating in a warm water bath under a stream of dry air or nitrogen.

Proceed with the sublimation following the instructions given in Experiment 5B. When the sublimation is complete, scrape the sublimed caffeine onto a tared piece of smooth paper and determine the weight of caffeine recovered. Calculate the weight percentage recovery (see p. 28) of caffeine after the sublimation. Determine the melting point of the purified caffeine. The melting point of pure caffeine is 236°C; however, the observed melting point will be lower. Submit the sample to the instructor in a labeled vial.

QUESTIONS

1. Outline a separation scheme for isolating caffeine from tea (Exp. 5A or Exp. 5B). Use a flowchart similar in format to that shown in the "Advance Preparation and Laboratory Records" chapter (see p. 26).
2. Why was the sodium carbonate added in Exp. 5B? Why was calcium carbonate added in Exp. 5A?
3. The crude caffeine isolated from tea has a green tinge. Why?
4. What are some possible explanations for why the melting point of your isolated caffeine was lower than the literature value (236°C)?
5. What would happen to the caffeine if the sublimation step were performed at atmospheric pressure?

E S S A Y

Identification of Drugs

Frequently, a chemist is called on to identify a particular unknown substance. If there is no prior information to work from, this can be a formidable task. There are several million known compounds, both inorganic and organic. For a completely unknown substance, the chemist must often use every available method for identification. If the unknown substance is a mixture, then the mixture must be separated into its components and each component identified separately. A pure compound can often be identified from its physical properties (melting point, boiling point, density, refractive index, and so on) and a knowledge of its functional groups. These can be identified by the reactions that the compound is observed to undergo or by spectroscopy (infrared, ultraviolet, NMR, and mass spectroscopy). The techniques necessary for this type of identification are introduced in a later section.

A somewhat simpler situation often arises in drug identification. The scope of drug identification is more limited, and the chemist working in a hospital trying to identify the source of a drug overdose or the law enforcement officer trying to identify a suspected illicit drug or a poison usually has some prior clues to work from. So does the medicinal chemist working for a pharmaceutical manufacturer who might be trying to discover why a competitor's product is better than that of his or her company.

Consider a drug overdose case as an example. The patient is brought into the emergency ward of a hospital. This person may be in a coma or a hyperexcited state, have an allergic rash, or clearly be hallucinating. These physiological symptoms are themselves a clue to the nature of the drug. Samples of the drug may be found in the patient's possession. Correct medical treatment may require a rapid and accurate identification of a drug powder or capsule. If the patient is conscious, the necessary information can be elicited orally; if not, the drug must be examined. If the drug is a tablet or a capsule, the process is often simple, since many drugs are coded by a manufacturer's trademark or logo, by shape (round, oval, bulletshape), by formulation (tablet, gelatin capsule, time-release microcapsules), and by color.

It is more difficult to identify a powder, but under some circumstances such identification may be easy. Plant drugs are often easily identified since they contain microscopic bits and pieces of the plant from which they are obtained. This cellular debris is often characteristic for certain types of drugs, and the drug can be identified on this basis alone. A microscope is all that is needed. Sometimes, chemical color tests can be used as confirmation. Certain drugs give rise to characteristic colors when treated with special reagents. Other drugs form crystalline precipitates of characteristic color and crystal structure when treated with appropriate reagents.

If the drug itself is not available and the patient is unconscious (or dead), identification may be more difficult. It may be necessary to pump the stomach or bladder contents of the patient (or corpse), or to obtain a blood sample, and work on these. These samples of stomach fluid, urine, or blood would be extracted with an appropriate organic solvent, and the extract would be analyzed.

Often the final identification of a drug, as an extract of urine, serum, or stomach fluid, hinges on some type of chromatography. Thin-layer chromatography (TLC) is often used. Under specified conditions, many drug substances can be identified by their R_f values and by the colors that their TLC spots turn when treated with various reagents or when they are observed under certain visualization methods. In the experiment that follows, TLC is applied to the analysis of an unknown analgesic drug.

REFERENCES

Keller, E. "Forensic Toxicology: Poison Detection and Homicide." *Chemistry, 43* (1970): 14

Keller, E. "Origin of Modern Criminology." *Chemistry, 42* (1969): 8.

Lieu, V. T. "Analysis of APC Tablets." *Journal of Chemical Education, 48* (1971): 478.

Neman, R. L. "Thin Layer Chromatography of Drugs." *Journal of Chemical Education, 49* (1972): 834.

Rodgers, S. S. "Some Analytical Methods Used in Crime Laboratories." *Chemistry, 42* (1969): 29.

Tietz, N. W. *Fundamentals of Clinical Chemistry.* Philadelphia: W. B. Saunders, 1970.

Walls, H. J. *Forensic Science.* New York: Praeger, 1968.

A collection of articles on forensic chemistry can be found in:

Berry, K., and Outlaw, H. E., eds. "Forensic Chemistry—A Symposium Collection." *Journal of Chemical Education, 62* (1985): 1043–1065.

E X P E R I M E N T   6

TLC Analysis of Analgesic Drugs

Thin-layer chromatography

In this experiment, thin-layer chromatography (TLC) will be used to determine the composition of various over-the-counter analgesics. If the instructor chooses, you may also be required to identify the components and actual identity (trade name) of an unknown analgesic. You will be given either two or three commercially prepared TLC plates with a flexible backing and a silica gel coating with a fluorescent indicator. On the first TLC plate, a reference plate, you will spot five standard compounds often used in analgesic formulations. In addition, a standard reference mixture containing four of these same compounds will also be spotted. Ibuprofen is omitted from this standard mixture because it would overlap with salicylamide after the plate is developed. If your instructor wishes, you will spot five additional reference substances on a second reference plate, including the newest analgesic drugs. On the final plate (the sample plate) you will spot several commercial analgesic preparations in order to determine their composition. At your instructor's option, one or more of these may be an unknown.

Reference Plate One		Reference Plate Two (optional)		Sample Plate
Acetaminophen	(Ac)	Aspirin	(Asp)	Five commercial preparations (or unknowns) plus the reference mixture
Aspirin	(Asp)	Ibuprofen	(Ibu)	
Caffeine	(Cf)	Ketoprofen	(Kpf)	
Ibuprofen	(Ibu)	Naproxen sodium	(Nap)	
Salicylamide	(Sal)	Salicylamide	(Sal)	
Reference mixture 1	(Ref-1)	Reference mixture 2	(Ref-2)	

The standard compounds will be all available as solutions of 1 g of each dissolved in 20 mL of a 50:50 mixture of methylene chloride and ethanol. The purpose of the first reference plate is to determine the order of elution (R_f values) of the known substances and to index the standard reference mixture. On the second reference plate (optional), several of the substances have similar R_f values, but you will note a different behavior for each spot with the visualization methods. On the sample plate, the standard reference mixture will be spotted, along with several solutions that you will prepare from commercial analgesic tablets. These tablets will each be crushed and dissolved in a 50:50 methylene chloride–ethanol mixture for spotting.

Two methods of visualization will be used to observe the positions of the spots on the developed TLC plates. First, the plates will be observed while under illumination from a short-wavelength ultraviolet (UV) lamp. This is done best in a darkened room or in a fume hood that has been darkened by taping butcher paper or aluminum foil over the lowered glass cover. Under these conditions, some of the spots will appear as dark areas on the plate, while others will fluoresce brightly. This difference in appearance under UV illumination will help to distinguish the substances from one another. You will find it convenient to outline very lightly in *pencil* the spots observed and to place a small **x** inside those spots that fluoresce. For a second means of visualization, iodine vapor will be used. Not all the spots will become visible when treated with iodine, but some will develop yellow, tan, or deep brown colors. The differences in the behaviors of the various spots with iodine can be used to further differentiate among them.

It is possible to use several developing solvents for this experiment, but ethyl acetate with 0.5% glacial acetic acid added is preferred. The small amount of glacial acetic acid supplies protons and suppresses ionization of aspirin, ibuprofen, naproxen sodium and ketoprofen, allowing them to travel upward on the plates in their protonated form. Without the acid, these compounds do not move.

In some analgesics, you may find ingredients besides the five mentioned previously. Some include an antihistamine and some a mild sedative. For instance, Midol contains *N*-cinnamylephedrine (cinnamedrine), an antihistamine, while Excedrin PM contains the sedative methapyrilene hydrochloride. Cope contains the related sedative methapyrilene fumarate. Some tablets may be colored with a chemical dye.

Required Reading

Review:	Essay	Analgesics
New:	Technique 12	Column Chromatography, Sections 12.1–12.3
	Technique 14	Thin-Layer Chromatography
	Essay	Identification of Drugs

Special Instructions

You must examine the developed plates under ultraviolet light first. After comparisons of *all* plates have been made with UV light, iodine vapor can be used. The iodine permanently affects some of the spots, making it impossible to go back and repeat the UV visualization. Take special care to notice those substances that have similar R_f values; these spots each have a different appearance when viewed under UV illumination, or a different staining color with iodine, allowing you to distinguish among them.

Aspirin presents some special problems since it is present in a large amount in many of the analgesics and since it hydrolyzes easily. For these reasons, the aspirin spots often show excessive tailing.

Waste Disposal

Dispose of all development solvent in the container for nonhalogenated organic solvents. Dispose of the ethanol–methylene chloride mixture in the container for halogenated organic solvents. The micropipets used for spotting the solution should be placed in a special container labeled for that purpose. The TLC plates should be stapled in your lab notebook.

Notes to the Instructor

If you wish, students may work in pairs on this experiment, each one preparing one of the two reference plates, and then cooperating on the third plate. Alternatively, especially if students are to work alone, you may wish to omit Plate Two and forgo identifying the two newest analgesics (ketoprofen and naproxen sodium).

Perform the thin-layer chromatography with flexible Silica Gel 60 F-254 plates (EM Science 5554-7). If the TLC plates have not been purchased recently, you should place them in an oven at 100°C for 30 minutes and store them in a desiccator until used. If you use different thin-layer plates, try out the experiment before using them with a class. Other plates may not resolve all five substances.

Ibuprofen and salicylamide have approximately the same R_f value, but they show up differently under the detection methods. For reasons that are not yet clear, ibuprofen sometimes gives two or even three spots. Naproxen sodium and ketoprofen have approximately the same R_f as aspirin. Once again, however, they show up differently under the detection methods. Fortunately, none of the new drugs appears in combination, either together, or with aspirin or ibuprofen, in any current commercial product.

Procedure

Initial Preparations. You will need at least 12 capillary micropipets (18 if both reference plates are prepared) to spot the plates. The preparation of these pipets is described and illustrated in Technique 14, Section 14.4, page 797. A common error is to pull the center section out too far when making these pipets, with the result that too little sample is applied to the plate. If this happens, you won't see *any* spots. Follow the directions carefully.

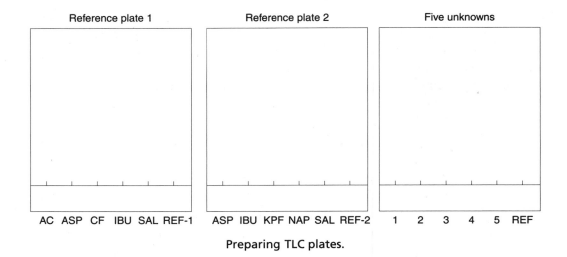

Reference plate 1   Reference plate 2   Five unknowns

AC ASP CF IBU SAL REF-1   ASP IBU KPF NAP SAL REF-2   1   2   3   4   5   REF

Preparing TLC plates.

After preparing the micropipets, obtain two (or three) 10-cm × 6.6-cm TLC plates (EM Science Silica Gel 60 F-254, No. 5554-7) from your instructor. These plates have a flexible backing, but they should not be bent excessively. Handle them carefully or the adsorbent may flake off. Also, you should handle them only by the edges; the surface should not be touched. Using a lead pencil (not a pen), *lightly* draw a line across the plates (short dimension) about I cm from the bottom. Using a centimeter ruler, move its index about 0.6 cm in from the edge of the plate and lightly mark off six 1-cm intervals on the line (see figure). These are the points at which the samples will be spotted. If you are preparing two reference plates, it would be a good idea to mark a small number **1** or **2** in the upper right hand corner of each plate to allow easy identification.

Spotting the First Reference Plate.   On the first plate (marked 1), starting from left to right, spot acetaminophen, then aspirin, caffeine, ibuprofen, and salicylamide. This order is alphabetic and will avoid any further memory problems or confusion. Solutions of these compounds will be found in small bottles on the side shelf. The standard reference mixture (Ref-1), also found on the side shelf, is spotted in the last position. The correct method of spotting a TLC plate is described in Technique 14, Section 14.4, page 797. It is important that the spots be made as small as possible, but not too small. With too much sample, the spots will tail and will overlap one another after development. With too little sample, no spots will be observed after development. The optimum applied spot should be about 1–2 mm (1/16 in.) in diameter. If scrap pieces of the TLC plates are available, it would be a good idea to practice spotting on these before preparing the actual sample plates.

Spotting the Second Reference Plate (Optional).   On the second plate (marked 2), starting from left to right, spot aspirin, ibuprofen, ketoprofen, naproxen sodium, salicylamide and the reference mixture (Ref-2). Follow the same procedure and take the precautions noted above for the first plate.

Preparing the Development Chamber.   When the reference plate (or plates) have been spotted, obtain a 16-oz wide-mouthed screwcap jar (or other suitable container) for use as a development chamber. The preparation of a development chamber is described in Technique 14, Section 14.5, page 799. Since the backing on the TLC plates is very thin, if they touch the filter paper liner of the development chamber *at any point,* solvent will begin to diffuse onto the absorbent surface at that point. To avoid this, you may either omit the liner or make the folllowing modification.

If you wish to use a liner, use a very narrow strip of filter paper (approximately 5 cm wide). Fold it into an L shape that is long enough to traverse the bottom of the jar and extend up the side to the top of the jar. TLC plates placed in the jar for development should *straddle* this liner strip but not touch it.

When the development chamber has been prepared, obtain a small amount of the development solvent (0.5% glacial acetic acid in ethyl acetate). Your instructor should prepare this mixture; it contains such a small amount of acetic acid that small individual portions are difficult to prepare. Fill the chamber with the development solvent to a depth of about 0.5–0.7 cm. If you are using a liner, be sure it is saturated with the solvent. Recall that the solvent level must not be above the spots on the plate or the samples will dissolve off the plate into the reservoir instead of developing.

Development of the Reference TLC Plates.   Place the spotted plate (or plates) in the chamber (straddling the liner if one is present) and allow the spots to develop. If you are doing two reference plates, both plates may be placed in the same development jar. Be sure the plates are placed in the developing jar so that their bottom edge is parallel to the bottom of the jar (straight, not tilted); if not, the solvent front will not advance evenly, increasing the difficulty of making good comparisons. The plates should face each other and slant or lean back in opposite directions. When the solvent has risen to a level about 0.5 cm from the top of the plate, remove each plate from the chamber (in the hood) and, using a lead pencil, mark the position of the solvent front. Set the plate on a piece of paper towel to dry. It may be helpful to place a small object under one end to allow optimum air flow around the drying plate.

UV Visualization of the Reference Plates.   When the plates are dry, observe them under a short-wavelength UV lamp, preferably in a darkened hood or a darkened room. Lightly outline all of the observed spots with a pencil. Carefully notice any differences in behavior between the spotted substances, especially those on Plate Two. Several compounds have similar R_f values, but the spots have a different appearance under UV illumination or iodine staining. Currently, there are no commercial analgesic preparations containing any compounds that have the same R_f values, but you will need to be able to distinguish them from one another to identify which one is present. Before proceeding, make a sketch of the plates in your notebook and note the differences in appearance that you observed. Using a ruler marked in millimeters, measure the distance that each spot has traveled relative to the solvent front. Calculate R_f values for each spot (Technique 14, Section 14.9, p. 803).

Analysis of Commercial Analgesics or Unknowns (Sample Plate).   Next, obtain half a tablet of each of the analgesics to be analyzed on the final TLC plate. If you were issued an unknown, you may analyze four other analgesics of your choice; if not, you may analyze five. The experiment will be most interesting if you make your choices in a way that gives a wide spectrum of results. Try to pick at least one analgesic each containing aspirin, acetaminophen, ibuprofen, a newer analgesic and, if available, salicylamide. If you have a favorite analgesic, you may wish to include it among your samples. Take each analgesic half-tablet, place it on a smooth piece of notebook paper, and crush it well with a spatula. Transfer each crushed half-tablet to a small, labeled test tube or Erlenmeyer flask. Using a graduated cylinder, mix 15 mL of absolute ethanol and 15 mL of methylene chloride. Mix the solution well. Add 5 mL of this solvent to each of the crushed half-tablets and then heat each of them *gently* for a few minutes on a steam bath or sand bath at about 100°C. Not all the tablet will dissolve, since the analgesics usually contain an insoluble binder. In addition, many of them contain inorganic buffering agents or coatings that are insoluble in this solvent mixture. After heating the samples, allow them to settle and then spot the clear liquid extracts on the sample plate. At the sixth position, spot the standard reference solution (Ref-1 or Ref-2). Develop the plate in 0.5% glacial acetic acid–ethyl acetate as before. Observe the plate under UV illumination and mark the visible spots as you did for the first plate. Sketch the plate in your notebook and record your conclusions about the contents of each tablet. This can be done by directly comparing your plate to the reference plate(s)—they can all be placed under the UV light at the same time. If you were issued an unknown, try to determine its identity (trade name).

Iodine Analysis.   Do not perform this step until UV comparisons of all the plates are complete. When ready, place the plates in a jar containing a few iodine crystals, cap the jar, and warm it gently on a steam bath or warm hotplate until the spots begin to appear. Notice which spots become visible and note their relative colors. You can directly compare colors of the reference spots to those on the unknown plate(s). Remove the plates from the jar and record your observations in your notebook.

QUESTIONS

1. What happens if the spots are made too large when preparing a TLC plate for development?

2. What happens if the spots are made too small when preparing a TLC plate for development?

3. Why must the spots be above the level of the development solvent in the developing chamber?

4. What would happen if the spotting line and positions were marked on the plate with a ballpoint pen?

5. Is it possible to distinguish two spots that have the same R_f value, but represent different compounds? Give two different methods.

6. Name some advantages of using acetaminophen (Tylenol) instead of aspirin as an analgesic.

ESSAY

Ethanol and Fermentation Chemistry

The fermentation processes involved in making bread, making wine, and brewing are among the oldest chemical arts. Even though fermentation had been known as an art for centuries, not until the 19th century did chemists begin to understand this process from the point of view of science. In 1810 Gay-Lussac discovered the general chemical equation for the breakdown of sugar into ethanol and carbon dioxide. The manner in which the process took place was the subject of much conjecture until Louis Pasteur began his thorough examination of fermentation. Pasteur demonstrated that yeast was required in the fermentation. He was also able to identify other factors that controlled the action of the yeast cells. His results were published in 1857 and 1866.

For many years, scientists believed that the transformation of sugar into ethanol and carbon dioxide by yeasts was inseparably connected with the life process of the yeast cell. This view was abandoned in 1897, when Büchner demonstrated that yeast extract would bring about alcoholic fermentation in the absence of any yeast cells. The fermenting activity of yeast is due to a remarkably active catalyst of biochemical origin, the enzyme zymase. It is now recognized that most of the chemical transformations that go on in living cells of plants and animals are brought about by enzymes. The enzymes are organic compounds, generally proteins, and establishment of structures and reaction mechanisms of these compounds is an active field of present-day research. Zymase is now known to be a complex of at least 22 separate enzymes, each of which catalyzes a specific step in the fermentation reaction sequence.

Enzymes show an extraordinary specificity—a given enzyme acts on a specific compound or a closely related group of compounds. Thus, zymase acts on only a few select sugars and not on all carbohydrates; the digestive enzymes of the alimentary tract are equally specific in their activity.

The chief sources of sugars for fermentation are the various starches and the molasses residue obtained from refining sugar. Corn (maize) is the chief source of starch in the United States, and ethyl alcohol made from corn is commonly known as **grain alcohol.** In preparing alcohol from corn, the grain, with or without the germ, is ground and cooked to give the **mash.** The enzyme diastase is added in the form of **malt** (sprouted barley that has been dried in air at 40°C and ground to a powder) or of a mold such as *Aspergillus oryzae.* The mixture is kept at 40°C until all the starch has been converted to the sugar **maltose** by hydrolysis of ether and acetal bonds. This solution is known as the **wort.**

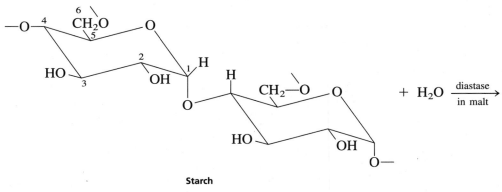

Starch

This is a glucose polymer with 1,4- and
1,6- glycosidic linkages. The linkages
at C-1 are α.

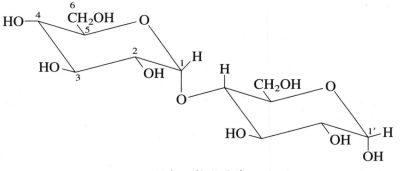

Maltose ($C_{12}H_{22}O_{11}$)

The α linkage still exists at C-1.
The —OH is shown α at the 1′ position
(axial), but it can also be β (equatorial).

The wort is cooled to 20°C and diluted with water to 10% maltose, and a pure yeast culture is added. The yeast culture is usually a strain of *Saccharomyces cerevisiae* (or *ellipsoidus*). The yeast cells secrete two enzyme systems: maltase, which converts the maltose into glucose, and zymase, which converts the glucose into carbon dioxide and alcohol. Heat is liberated, and the temperature must be kept below 35°C by cooling to prevent destruction of the enzymes. Oxygen in large amounts is initially necessary for the optimum reproduction of yeast cells, but the actual production of alcohol is anaerobic. During fermentation, the evolution of carbon dioxide soon establishes anaerobic conditions. If oxygen were freely available, only carbon dioxide and water would be produced.

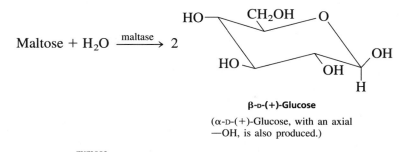

$$\text{Maltose} + H_2O \xrightarrow{\text{maltase}} 2$$

β-D-(+)-Glucose

(α-D-(+)-Glucose, with an axial
—OH, is also produced.)

$$\text{Glucose} \xrightarrow{\text{zymase}} 2\ CO_2 + 2\ CH_3CH_2OH + 26\ \text{kcal}$$

$C_6H_{12}O_6$

After 40–60 hours, fermentation is complete, and the product is distilled to remove the alcohol from solid matter. The distillate is fractionated by means of an efficient column. A small amount of acetaldehyde (bp 21°C) distills first and is followed by 95% alcohol. Fusel oil is contained in the higher boiling fractions. The fusel oil consists of a mixture of higher alcohols, chiefly 1-propanol, 2-methyl-1-propanol, 3-methyl-1-butanol, and 2-methyl-1-butanol. The exact composition of fusel oil varies considerably; it particularly depends on the type of raw material that is fermented. These higher alcohols are not formed by fermentation of glucose. They arise from certain amino acids derived from the proteins present in the raw material and the yeast. These fusel oils cause the headaches associated with drinking alcoholic beverages.

Industrial alcohol is ethyl alcohol used for nonbeverage purposes. Most commercial alcohol is denatured to avoid payment of taxes, the biggest cost in the price of liquor. The denaturants render the alcohol unfit for drinking. Methanol, aviation fuel, and other substances are used for this purpose. The difference in price between taxed and nontaxed alcohol is more than $20 a gallon. Before efficient synthetic processes were developed, the chief source of industrial alcohol was fermented blackstrap molasses, the noncrystallizable residue from refining cane sugar (sucrose). Most industrial ethanol in the United States is now manufactured from ethylene, a product of the "cracking" of petroleum hydrocarbons. By reaction with concentrated sulfuric acid, ethylene becomes ethyl hydrogen sulfate, which is hydrolyzed to ethanol by dilution with water. The alcohols 2-propanol, 2-butanol, 2-methyl-2-propanol, and higher secondary and tertiary alcohols also are produced on a large scale from alkenes derived from cracking.

Yeasts, molds, and bacteria are used commercially for the large-scale production of various organic compounds. An important example, in addition to ethanol production, is the anaerobic fermentation of starch by certain bacteria to yield 1-butanol, acetone, ethanol, carbon dioxide, and hydrogen.

REFERENCES

Amerine, M. A. "Wine." *Scientific American, 211* (August 1964): 46.

Hallberg, D. E. "Fermentation Ethanol." *ChemTech, 14* (May 1984): 308.

Ough, C. S. "Chemicals Used in Making Wine." *Chemical and Engineering News, 65* (January 5, 1987): 19.

Van Koevering, T. E., Morgan, M. D., and Younk, T. J. "The Energy Relationships of Corn Production and Alcohol Fermentation." *Journal of Chemical Education, 64* (January 1987): 11.

Webb, A. D. "The Science of Making Wine." *American Scientist, 72* (July–August 1984): 360.

Students wanting to investigate alcoholism and possible chemical explanations for alcohol addiction may consult the following references:

Cohen, G., and Collins, M. "Alkaloids from Catecholamines in Adrenal Tissue: Possible Role in Alcoholism." *Science, 167* (1970): 1749.

Davis, V. E., and Walsh, M. J. "Alcohol Addiction and Tetrahydropapaveroline." *Science, 169* (1970): 1105.

Davis, V. E., and Walsh, M. J. "Alcohols, Amines, and Alkaloids: A Possible Biochemical Basis for Alcohol Addiction." *Science, 167* (1970): 1005.

Seevers, M. H., Davis, V. E., and Walsh, M. J. "Morphine and Ethanol Physical Dependence: A Critique of a Hypothesis." *Science, 170* (1970): 1113.

Yamanaka, Y., Walsh, M. J., and Davis, V. E. "Salsolinol, An Alkaloid Derivative of Dopamine Formed in Vitro during Alcohol Metabolism." *Nature, 227* (1970): 1143.

E X P E R I M E N T   7

Ethanol from Sucrose

Fermentation
Fractional distillation
Azeotropes

Either sucrose or maltose can be used as the starting material for making ethanol. Sucrose is a disaccharide with the formula $C_{12}H_{22}O_{11}$. It has one glucose molecule combined with fructose. Maltose consists of two glucose molecules. The enzyme **invertase** is used to catalyze the hydrolysis of sucrose. **Maltase** is more effective in catalyzing the hydrolysis of maltose. The hydrolysis of maltose is discussed in the essay on ethanol and fermentation (p. 85). **Zymase** is used to convert the hydrolyzed sugars to alcohol and carbon dioxide. Pasteur observed that growth and fermentation were promoted by adding small amounts of mineral salts to the nutrient medium. Later it was found that before fermentation actually begins, the hexose sugars combine with phosphoric acid, and the resulting hexose-phosphoric acid combination is then degraded into carbon dioxide and ethanol. The carbon dioxide is not wasted in the commercial process because it is converted to dry ice.

The fermentation is inhibited by its end-product ethanol; it is not possible to prepare solutions containing more than 10–15% ethanol by this method. More concentrated ethanol can be isolated by fractional distillation. Ethanol and water form an azeotropic mixture consisting of 95% ethanol and 5% water by weight, which is the most concentrated ethanol that can be obtained by fractionation of dilute ethanol–water mixtures.

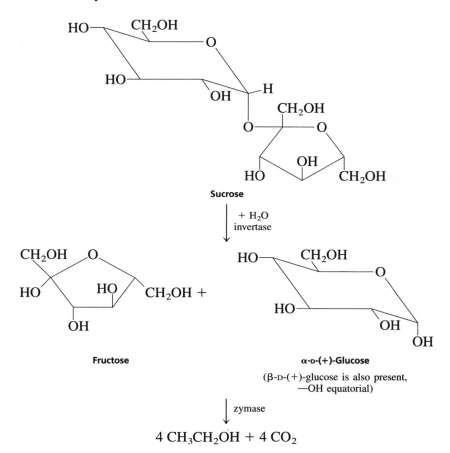

Sucrose

+ H_2O
invertase

Fructose

α-D-(+)-Glucose

(β-D-(+)-glucose is also present,
—OH equatorial)

zymase

$4\ CH_3CH_2OH + 4\ CO_2$

Required Reading

Review: Technique 4   Sections 4.3 and 4.4
          Technique 6   Physical Constants, Part B, Boiling Points

New: Technique 6   Physical Constants, Part D, Density
          Technique 10 Fractional Distillation, Azeotropes
          Essay   Ethanol and Fermentation Chemistry

Special Instructions

Start the fermentation at least 1 week before the period in which the ethanol will be isolated. When the aqueous ethanol solution is to be separated from the yeast cells, it is important to transfer carefully as much of the clear, supernatant liquid as possible, without agitating the mixture.

Waste Disposal

Discard all aqueous solutions in the waste container marked for the disposal of aqueous waste. Filter Aid may be discarded in the trash containers.

Notes to the Instructor

It may be necessary to use an external heat source to maintain a temperature of 30–35°C. Place a lamp in the hood to act as a heat source.

This experiment can also be performed without doing the fermentation. Provide each student with 20 mL of a 10% ethanol solution. This solution is used in place of the fermentation mixture in the Fractional Distillation section of the Procedure.

Procedure

Fermentation.   Place 20.0 g of sucrose in a 250-mL Erlenmeyer flask. Add 175 mL of water warmed to 25–30°C; 20 mL of Pasteur's salts;[1] and 2.0 g of *dried* baker's yeast. Shake the contents vigorously to mix them, then fit the flask with a one-hole rubber stopper with a glass tube leading to a beaker or test tube containing a solution of barium hydroxide. Protect the barium hydroxide from air by adding some mineral oil or xylene to form a layer above the barium hydroxide (see p. 91). A precipitate of barium carbonate will form, indicating that CO_2 is being evolved. Alternatively, a balloon may be substituted for the barium hydroxide trap. Oxygen from the atmosphere is excluded from the chemical reaction by these techniques. If oxygen were allowed to continue in contact with the fermenting solution, the ethanol could be further oxidized to acetic acid or even all the way to carbon dioxide and water. As long as carbon dioxide continues to be liberated, ethanol is being formed.

Allow the mixture to stand at about 30–35°C until fermentation is complete, as indicated by the cessation of gas evolution. Usually about 1 week is required. After

---

[1]A solution of Pasteur's salts consists of potassium phosphate, 2.0 g; calcium phosphate, 0.20 g; magnesium sulfate, 0.20 g; and ammonium tartrate, 10.0 g, dissolved in 860 ml water.

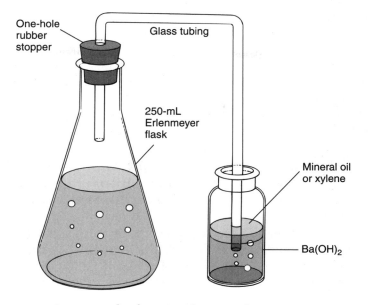

One-hole rubber stopper

Glass tubing

250-mL Erlenmeyer flask

Mineral oil or xylene

Ba(OH)$_2$

Apparatus for fermentation experiment.

this time, *carefully* move the flask away from the heat source and remove the stopper. Without disturbing the sediment, transfer the clear, supernatant liquid solution to another container by decanting.

If the liquid is not clear, clarify it by the following method. Place about 1 tablespoon of Filter Aid (Johns-Manville Celite) in a beaker with about 100 mL of water. Stir the mixture vigorously and then pour the contents into a Büchner funnel (with filter paper) while applying a vacuum, as in a vacuum filtration (Technique 4, Section 4.3, p. 640). This procedure will cause a thin layer of Filter Aid to be deposited on the filter paper (Technique 4, Section 4.4, p. 642). Discard the water that passes through this filter. The pipeted liquid containing the ethanol is then passed through this filter under gentle suction. The extremely tiny yeast particles are trapped in the pores of the Filter Aid. The liquid contains ethanol in water, plus smaller amounts of dissolved metabolites (fusel oils) from the yeast.

Fractional Distillation.   Assemble the apparatus shown in Technique 10, Figure 10.2. p. 734; select a round-bottom flask that will be filled between half and two-thirds full by the liquid to be distilled. Use a heating mantle for the heat source. Pack the condenser *uniformly* with stainless steel cleaning pad material (no soap!) (Technique 10, Figure 10.7, p. 741, and Section 10.6, p. 743).

Caution: You should wear heavy cotton gloves when handling the stainless steel cleaning pad. The edges are very sharp and can easily cut into the skin.

Add about 10 g of potassium carbonate to the filtered solution for each 20 mL of liquid. After the solution has become saturated with potassium carbonate, transfer it to the round-bottom flask of the distillation apparatus. Distill the liquid slowly through the fractionating column to get the best possible separation. Once distillation begins, the temperature in the distillation head will increase to about 78°C and then rise gradually until the ethanol fraction is distilled. Collect the fraction boiling between 78 and 88°C and discard the residue in the distillation flask. You should collect about 4–5 mL of distillate. The distillation should then be interrupted by removing the apparatus from the heat source.

Analysis of Distillate.   Determine the total weight of the distillate. Determine the approximate density of the distillate by transferring a known volume of the liquid with an automatic pipet or graduated pipet to a tared vial. Reweigh the vial and calculate the density. This method is good to two significant figures. Using the following table, determine the percentage composition by weight of ethanol in your distillate from the density of your sample. The extent of purification of the ethanol is limited since ethanol and water form a constant-boiling mixture, an azeotrope, with a composition of 95% ethanol and 5% water.

Percentage Ethanol by Weight	Density at 20°C (g/mL)	Density at 25°C (g/mL)
75	0.856	0.851
80	0.843	0.839
85	0.831	0.827
90	0.818	0.814
95	0.804	0.800
100	0.789	0.785

Calculate the percentage yield of the alcohol, and submit the ethanol to the instructor in a labeled vial.[2]

---

[2]A careful analysis by flame-ionization gas chromatography on a typical student-prepared ethanol sample provided the following results:

Acetaldehyde	0.060%
Diethylacetal of acetaldehyde	0.005%
Ethanol	88.3% (by hydrometer)
1-Propanol	0.032%
2-Methyl-1-propanol	0.092%
5-Carbon and higher alcohols	0.140%
Methanol	0.040%
Water	11.3% (by difference)

QUESTIONS

1. Write a balanced equation for the conversion of sucrose into ethanol.

2. By doing some library research, see whether you can find the commercial method or methods used to produce **absolute ethanol.**

3. Why is the air trap necessary in the fermentation?

4. How does acetaldehyde impurity arise in the fermentation?

5. The diethylacetal of acetaldehyde can be detected by gas chromatography. How does this impurity arise in fermentation?

6. Calculate how many milliliters of carbon dioxide would be produced theoretically from 20 g of sucrose of 25°C and 1 atmosphere pressure.

E S S A Y

Esters—Flavors and Fragrances

Esters are a class of compounds widely distributed in nature. They have the general formula

The simple esters tend to have pleasant odors. In many cases, although not exclusively so, the characteristic flavors and fragrances of flowers and fruits are due to compounds with the ester functional group. An exception is the case of the essential oils. The **organoleptic** qualities (odors and flavors) of fruits and flowers may often be due to a single ester, but more often, the flavor or the aroma is due to a complex mixture in which a single ester predominates. Some common flavor principles are listed in Table One.

Food and beverage manufacturers are thoroughly familiar with these esters and often use them as additives to spruce up the flavor or odor of a dessert or beverage. Many times, such flavors or odors do not even have a natural basis, as is the case with the "juicy fruit" principle, isopentenyl acetate. An instant pudding that has the flavor of rum may never have seen its alcoholic namesake: This flavor can be duplicated by the proper admixture, along with other minor components, of ethyl formate and isobutyl propionate. The natural flavor and odor are not exactly duplicated, but most people can be fooled. Often, only a trained person with a high degree of gustatory perception—a professional taster—can tell the difference.

A single compound is rarely used in good-quality imitation flavoring agents. A formula for an imitation pineapple flavor that might fool an expert is listed in Table Two. The formula includes 10 esters and carboxylic acids that can easily be synthesized in the laboratory. The remaining seven oils are isolated from natural sources.

TABLE ONE   Ester Flavors and Fragrances

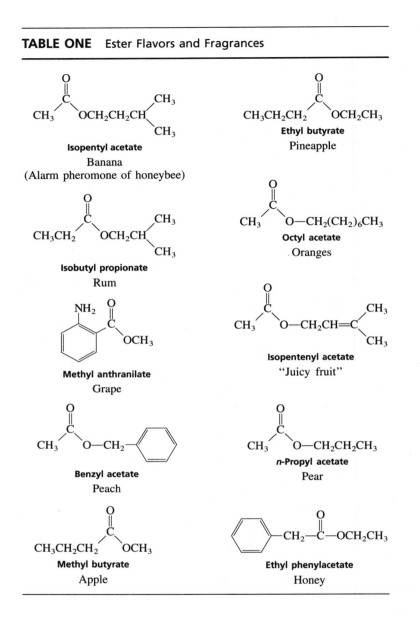

Isopentyl acetate
Banana
(Alarm pheromone of honeybee)

Ethyl butyrate
Pineapple

Isobutyl propionate
Rum

Octyl acetate
Oranges

Methyl anthranilate
Grape

Isopentenyl acetate
"Juicy fruit"

Benzyl acetate
Peach

n-Propyl acetate
Pear

Methyl butyrate
Apple

Ethyl phenylacetate
Honey

Flavor is a combination of taste, sensation, and odor transmitted by receptors in the mouth (taste buds) and nose (olfactory receptors). The stereochemical theory of odor is discussed in the essay that precedes Experiment 21. The four basic tastes (sweet, sour, salty, and bitter) are perceived in specific areas of the tongue. The sides of the tongue perceive sour and salty tastes, the tip is most sensitive to sweet tastes, and the back of the tongue detects bitter tastes. The perception of flavor, however, is not so simple. If it were, it would require only the formulation of various combinations of four basic substances: a bitter substance (a base), a sour substance (an acid), a salty substance (sodium chloride), and a sweet substance (sugar), to duplicate any flavor! In fact, we cannot duplicate fla-

---

TABLE TWO Artificial Pineapple Flavor

Pure Compounds	%	Essential Oils	%
Allyl caproate	5	Oil of sweet birch	1
Isopentyl acetate	3	Oil of spruce	2
Isopentyl isovalerate	3	Balsam Peru	4
Ethyl acetate	15	Volatile mustard oil	1
Ethyl butyrate	22	Oil cognac	5
Terpinyl propionate	3	Concentrated orange oil	4
Ethyl crotonate	5	Distilled oil of lime	2
Caproic acid	8		19
Butyric acid	12		
Acetic acid	5		
	81		

---

vors in this way. The human possesses 9000 taste buds. The combined response of these taste buds is what allows perception of a particular flavor.

Although the "fruity" tastes and odors of esters are pleasant, they are seldom used in perfumes or scents that are applied to the body. The reason for this is chemical. The ester group is not as stable to perspiration as the ingredients of the more expensive essential-oil perfumes. The latter are usually hydrocarbons (terpenes), ketones, and ethers extracted from natural sources. Esters, however, are used only for the cheapest toilet waters, since on contact with sweat they undergo hydrolysis, giving organic acids. These acids, unlike their precursor esters, generally do not have a pleasant odor.

$$
\underset{R}{\overset{O}{\underset{\|}{C}}}\text{OR}' + H_2O \longrightarrow \underset{R}{\overset{O}{\underset{\|}{C}}}\text{OH} + R'OH
$$

Butyric acid, for instance, has a strong odor like that of rancid butter (of which it is an ingredient) and is a component of what we normally call body odor. It is this substance that makes foul-smelling humans so easy for an animal to detect when the animal is downwind of them. It is also of great help to the bloodhound, which is trained to follow small traces of this odor. Ethyl butyrate and methyl butyrate, however, which are the esters of butyric acid, smell like pineapple and apple, respectively.

A sweet fruity odor also has the disadvantage of possibly attracting fruit flies and other insects in search of food. Isopentyl acetate, the familiar solvent called banana oil, is particularly interesting. It is identical to the alarm **pheromone** of the honeybee. Pheromone is the name applied to a chemical secreted by an organism that evokes a specific response in another member of the same species. This kind of communication is common between insects who otherwise lack means of intercourse. When a honeybee worker stings an intruder, an alarm pheromone, composed partly of isopentyl acetate, is secreted along with the sting venom. This chemical causes aggressive attack on the intruder by other bees, who swarm after the intruder. Obviously, it wouldn't be wise to wear a perfume compounded of isopentyl acetate near a beehive. Pheromones are discussed in more detail in the essay preceding Experiment 16.

REFERENCES

Bauer, K., and Garbe, D. *Common Fragrance and Flavor Materials.* New York: VCH Publishers, 1985.

The Givaudan Index. New York: Givaudan-Delawanna, 1949. (Gives specifications of synthetics and isolates for perfumery.)

Gould, R. F., ed. *Flavor Chemistry, Advances in Chemistry,* No. 56. Washington, DC: American Chemical Society, 1966.

Layman, P. L. "Flavors and Fragrances Industry Taking on New Look." *Chemical and Engineering News* (July 20, 1987): 35.

Moyler, D. "Natural Ingredients for Flavours and Fragrances." *Chemistry and Industry* (January 7, 1991): 11.

Rasmussen, P. W. "Qualitative Analysis by Gas Chromatography—G. C. versus the Nose in Formulation of Artificial Fruit Flavors." *Journal of Chemical Education, 61* (January 1984): 62.

Shreve, R. N., and Brink, J. *Chemical Process Industries,* 4th ed. New York: McGraw-Hill, 1977.

Welsh, F. W., and Williams, R. E. "Lipase Mediated Production of Flavor and Fragrance Esters from Fusel Oil." *Journal of Food Science, 54* (November/December 1989): 1565.

EXPERIMENT 8

Isopentyl Acetate (Banana Oil)

Esterification
Heating under reflux
Separatory funnel
Extraction
Simple distillation

In this experiment you will prepare an ester, isopentyl acetate. This ester is often referred to as banana oil, since it has the familiar odor of this fruit.

$$
\underset{\substack{\text{Acetic acid}\\\text{(excess)}}}{CH_3{-}\overset{\displaystyle O}{\overset{\|}{C}}{-}OH} + \underset{\text{Isopentyl alcohol}}{CH_3{-}\overset{\displaystyle CH_3}{\overset{|}{CH}}{-}CH_2{-}CH_2{-}OH} \underset{}{\overset{H^+}{\rightleftharpoons}}
$$

$$
\underset{\text{Isopentyl acetate}}{CH_3{-}\overset{\displaystyle O}{\overset{\|}{C}}{-}O{-}CH_2{-}CH_2{-}\overset{\displaystyle CH_3}{\overset{|}{CH}}{-}CH_3} + H_2O
$$

Isopentyl acetate is prepared by the direct esterification of acetic acid with isopentyl alcohol. Since the equilibrium does not favor the formation of the ester, it must be shifted to the right, in favor of the product, by using an excess of one of the starting materials. Acetic acid is used in excess because it is less expensive than isopentyl alcohol and more easily removed from the reaction mixture.

In the isolation procedure, much of the excess acetic acid and the remaining isopentyl alcohol are removed by extraction with sodium bicarbonate and water. After drying with anhydrous sodium sulfate, the ester is purified by distillation. The purity of the liquid product is analyzed by determining the infrared spectrum.

Required Reading

Review: Techniques 1 and 2
New: Technique 3
           Technique 6   Physical Constants, Part B, Boiling Points
           Technique 7   Extractions, Separations, and Drying Agents
           Technique 8   Simple Distillation
           Essay   Esters—Flavors and Fragrances

If performing the optional infrared spectroscopy, also read:

Technique 19 Preparation of Samples for Spectroscopy, Part A

Special Instructions

Be careful when dispensing sulfuric and glacial acetic acids. They are very corrosive and will attack your skin if you make contact with them. If you get one of these acids on your skin, wash the affected area with copious quantities of running water for 10–15 minutes.

Since a 1-hour reflux is required, you should start the experiment at the very beginning of the laboratory period. During the reflux period, you may perform other experimental work.

Waste Disposal

Any aqueous solutions should be placed in a container specially designated for dilute aqueous waste. Place any excess ester in the nonhalogenated organic waste container.

Notes to the Instructor

This experiment has been carried out successfully using Dowex 50X2-100 ion exchange resin instead of the sulfuric acid.

Procedure

Apparatus. Assemble a reflux apparatus, using a 25-mL round-bottom flask and a water-cooled condenser (refer to Technique 3, Fig. 3.8, p. 619). Use a heating mantle to heat. In order to control vapors, place a drying tube packed with calcium chloride on top of the condenser.

Reaction Mixture. Weigh (tare) an empty 10-mL graduated cylinder and record its weight. Place approximately 5.0 mL of isopentyl alcohol in the graduated cylinder and reweigh it to determine the weight of alcohol. Disconnect the round-bottom flask from the reflux apparatus and transfer the alcohol into it. Do not clean or wash the graduated cylinder. Using the same graduated cylinder, measure approximately 7.0 mL of glacial acetic acid ($MW = 60.1$, $d = 1.06$ g/mL) and add it to the alcohol already in the flask. Using a calibrated Pasteur pipet, add 1 mL of concentrated sulfuric acid, mixing *immediately* (swirl), to the reaction mixture contained in the flask. Add a corrundum boiling stone and reconnect the flask. Do not use a calcium carbonate (marble) boiling stone because it will dissolve in the acidic medium.

Reflux. Start water circulating in the condenser and bring the mixture to a boil. Continue heating under reflux for 60–75 minutes. Then, disconnect or remove the heating source and allow the mixture to cool to room temperature.

Extractions. Disassemble the apparatus and transfer the reaction mixture to a separatory funnel (125-mL) placed in a ring that is attached to a ring stand. Be sure that the stopcock is closed and, using a funnel, pour the mixture into the top of the separatory funnel. Also be careful to avoid transferring the boiling stone, or you will need to remove it after the transfer. Add 10 mL of water, stopper the funnel, and mix the phases by careful shaking and venting (Technique 7, Section 7.4 and Fig. 7.6, pp. 689 and 692). Allow the phases to separate and then unstopper the funnel and drain the lower aqueous layer through the stopcock into a beaker or other suitable container. Next, extract the organic layer with 5 mL of 5% aqueous sodium bicarbonate just as you did previously with water. Extract the organic layer once again, this time with 5 mL of saturated aqueous sodium chloride.

Drying. Transfer the crude ester to a clean, dry 25-mL Erlenmeyer flask and add approximately 1.0 g of anhydrous sodium sulfate. Cork the mixture and allow it to stand for 10–15 minutes while you prepare the apparatus for distillation. If the mixture does not appear dry (the drying agent clumps and does not "flow," the solution is cloudy, or drops of water are obvious), transfer the ester to a new clean, dry 25-mL Erlenmeyer flask and add a new 0.5-g portion of anhydrous sodium sulfate to complete the drying.

Distillation. Assemble a distillation apparatus using your smallest round-bottom flask to distill from (Technique 8, Fig. 8.4, p. 711). Use a heating mantle to heat. Pre-weigh (tare) and use another small round-bottom flask, or an Erlenmeyer flask, to collect the product. Immerse the collection flask in a beaker of ice to ensure condensation and to reduce odors. You should look up the boiling point of your expected product in a handbook so you will know what to expect. Continue distilla-

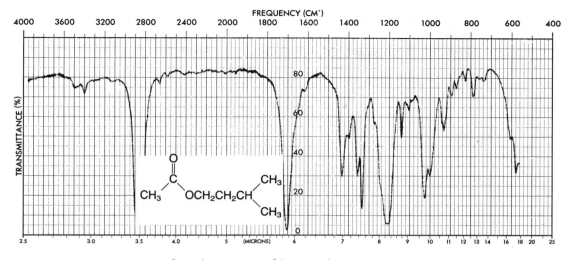

Infrared spectrum of isopentyl acetate, neat.

tion until only one or two drops of liquid remain in the distilling flask. Record the observed boiling point *range* in your notebook.

Yield Determination. Weigh the product and calculate the percentage yield of the ester. At the option of your instructor, determine the boiling point using one of the methods described in Technique 6, Sections 6.11 and 6.12, pages 677 and 680.

Spectroscopy. At your instructor's option, obtain an infrared spectrum using salt plates (Technique 19, Section 19.2, p. 843). Compare your spectrum with the one reproduced in the text. Interpret the spectrum and include it in your report to the instructor. You may also be required to determine and interpret the proton and carbon-13 NMR spectra (Technique 19, Sections 19.9 and 19.10, pp. 855 and 859). Submit your sample in a properly labeled vial with your report.

QUESTIONS

1. One method of favoring the formation of an ester is to add excess acetic acid. Suggest another method, involving the right-hand side of the equation, that will favor the formation of the ester.

2. Why is the mixture extracted with sodium bicarbonate? Give an equation and explain its relevance.

3. Why are gas bubbles observed?

4. Which starting material is the limiting reagent in this procedure? Which reagent is used in excess? How great is the molar excess (how many times greater)?

5. Outline a separation scheme for isolating pure isopentyl acetate from the reaction mixture.

6. Interpret the principal absorption bands in the infrared spectrum of isopentyl acetate or, if you did not determine the infrared spectrum of your ester, do this for the spectrum of isopentyl acetate shown above. (Appendix 3 may be of some help.)

7. Write a mechanism for the acid-catalyzed esterification of acetic acid with isopentyl alcohol.

8. Why is glacial acetic acid designated as "glacial"? (*Hint:* Consult a handbook of physical properties.)

EXPERIMENT 9

Preparation of a C-4 or C-5 Acetate Ester

Esterification
Separatory funnel
Simple distillation

In this experiment, we prepare an ester from acetic acid and a C-4 or a C-5 alcohol. This experiment is similar to the preparation of isopentyl acetate, which is described in Experiment 8. However, for this experiment, either your instructor will assign, or you will pick, one of the following C-4 or C-5 alcohols to react with acetic acid:

1-butanol (*n*-butyl alcohol)   1-pentanol (*n*-pentyl alcohol)

2-butanol (*sec*-butyl alcohol)   2-pentanol

2-methyl-1-propanol (isobutyl alcohol)   3-pentanol

3-methyl-l-butanol (isopentyl alcohol)   cyclopentanol

If an NMR spectrometer is available, your instructor may wish to give you one of these alcohols as an unknown, leaving it to you to determine which alcohol was issued. For this purpose, you could use the infrared and NMR spectra as well as the boiling points of the alcohol and its ester.

Required Reading

Review: Essay   Esters—Flavors and Fragrances
           Experiment 8
           Technique 6   Part B, Boiling Points
           Techniques 7, 8

Special Instructions

Be careful when dispensing sulfuric and glacial acetic acids. They are very corrosive and will attack your skin if you make contact with them. If you get one of these acids on your skin, wash the affected area with copious quantities of running water for 10–15 minutes.

If you select 2-butanol, reduce the amount of concentrated sulfuric acid to 0.5 mL. Also reduce the heating time to 50 minutes or less. Secondary alcohols have a tendency to give a significant percentage of elimination in strongly acidic solutions.

Waste Disposal

Any aqueous solutions should be placed in the appropriate container for dilute aqueous waste. Place any excess ester in the nonhalogenated organic waste container.

Procedure

Apparatus.   Assemble a reflux apparatus using a 25-mL round-bottom flask and a water-cooled condenser (refer to Technique 3, Fig. 3.8, p. 619). In order to control vapors, place a drying tube packed with calcium chloride on top of the condenser. Use a heating mantle to heat the reaction.

Reaction Mixture.   Weigh (tare) an empty 10-mL graduated cylinder and record its weight. Place approximately 5.0 mL of your chosen alcohol in the graduated cylinder and reweigh it to determine the weight of alcohol. Disconnect the round-bottom flask from the reflux apparatus and transfer the alcohol into it. Do not clean or wash the graduated cylinder. Using the same graduated cylinder, measure approximately 7.0 mL of glacial acetic acid ($MW = 60.1$, $d = 1.06$ g/mL) and add it to the alcohol already in the flask. Using a calibrated Pasteur pipet, add 1 mL of concentrated sulfuric acid (0.5 mL if you have chosen 2-butanol), mixing *immediately* (swirl), to the reaction mixture contained in the flask. Add a corrundum boiling stone and reconnect the flask. Do not use a calcium carbonate (marble) boiling stone because it will dissolve in the acidic medium.

Reflux.   Start water circulating in the condenser and bring the mixture to a boil. Continue heating under reflux for 60–75 minutes. Then, disconnect or remove the heating source and allow the mixture to cool to room temperature.

Extractions.   Disassemble the apparatus and transfer the reaction mixture to a separatory funnel (125-mL) placed in a ring attached to a ring stand. Be sure that the stopcock is closed and, using a funnel, pour the mixture into the top of the separatory funnel. Also be careful to avoid transferring the boiling stone, or you will need to remove it after the transfer. Add 10 mL of water, stopper the funnel, mix the phases by careful shaking and venting (Technique 7, Section 7.4 and Fig. 7.6, pp. 690 and 692). Allow the phases to separate and then unstopper the funnel and drain the lower aqueous layer through the stopcock into a beaker or other suitable container. Next, extract the organic layer with 5 mL of 5% aqueous sodium bicarbonate just as you did previously with water. Extract the organic layer once again, this time with 5 mL of saturated aqueous sodium chloride.

Drying.   Transfer the crude ester to a clean, dry 25-mL Erlenmeyer flask and add approximately 1.0 g of anhydrous sodium sulfate. Cork the mixture and allow it to stand for 10–15 minutes while you prepare the apparatus for distillation. If the mixture does not appear dry (the drying agent clumps and does not "flow," the solution is cloudy, or drops of water are obvious), transfer the ester to a new clean, dry 25-mL Erlenmeyer flask and add a new 0.5-g portion of anhydrous sodium sulfate to complete the drying.

Distillation.   Assemble a distillation apparatus using your smallest round-bottom flask to distill from (Technique 8, Figure 8.4, p. 711). Use a heating mantle to heat. Pre-weigh (tare) and use a 50-mL round-bottom flask or a small Erlenmeyer flask to collect the product. Immerse the collection flask in a beaker of ice to ensure condensation and to reduce odors. If your alcohol is not an unknown, you can look up its boiling point in a handbook; otherwise, you can expect your ester to have a boiling point between 95 and 150°C. Continue distillation until only one or two drops of liquid remain in the distilling flask. Record the observed boiling point *range* in your notebook.

Yield Determination.   Weigh the product and calculate the percentage yield of the ester. At the option of your instructor, determine the boiling point using one of the methods described in Technique 6, Sections 6.11 and 6.12, pages 677 and 680.

Spectroscopy.   At your instructor's option, obtain an infrared spectrum using salt plates (Technique 19, Section 19.2, p. 843). Compare the spectrum with the one reproduced in Experiment 8 (p. 99). The spectrum of your ester should have similar features to the one shown. Interpret the spectrum and include it in your report to the instructor. You may also be required to determine and interpret the proton and carbon-13 NMR spectra (Technique 19, Sections 19.9 and 19.10, p. 859). Submit your sample in a properly labeled vial with your report.

QUESTIONS

1. One method of favoring the formation of an ester is to add excess acetic acid. Suggest another method, involving the right-hand side of the equation, that will favor the formation of the ester.

2. Why is the mixture extracted with sodium bicarbonate? Give an equation and explain its relevance.

3. Why are gas bubbles observed?

4. Using your alcohol, determine which starting material is the limiting reagent in this procedure. Which reagent is used in excess? How great is the molar excess (how many times greater)?

5. Outline a separation scheme for isolating your pure ester from the reaction mixture.

6. Interpret the principal absorption bands in the infrared spectrum of your ester or, if you did not determine the infrared spectrum of your ester, do this for the spectrum of isopentyl acetate on page 99. (Appendix 3 may be of some help.)

7. Write a mechanism for the acid-catalyzed esterification that uses your alcohol and acetic acid. You may need to consult the chapter on carboxylic acids in your lecture textbook.

8. Tertiary alcohols do not work well in the procedure outlined for this experiment; they give a different product than you might expect. Explain this and draw the expected product from *t*-butyl alcohol (2-methyl-2-propanol).

9. Why is glacial acetic acid designated as "glacial"? (*Hint:* Consult a handbook of physical properties.)

EXPERIMENT 10

Methyl Salicylate (Oil of Wintergreen)

Synthesis of an ester
Separatory funnel
Extraction
Vacuum distillation

In this experiment, we prepare a familiar-smelling organic ester—oil of wintergreen. Methyl salicylate was first isolated in 1843 by extraction from the wintergreen plant (*Gaultheria*). It was soon found that this compound had analgesic and antipyretic character almost identical to that of salicylic acid (see the essay "Aspirin," p. 46) when taken internally. This medicinal character probably derives from the ease with which methyl salicylate is hydrolyzed to salicylic acid under the alkaline conditions found in the intestinal tract. Salicylic acid is known to have such analgesic and antipyretic properties. Methyl salicylate can be taken internally or absorbed through the skin; thus, it finds much use in liniment preparations. Applied to the skin, it produces a mild burning or soothing sensation, which probably comes from its phenolic hydroxyl group. This ester also has a pleasant odor, and is used to a small extent as a flavoring principle.

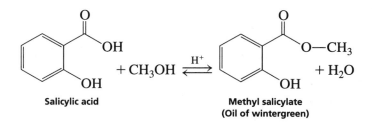

Methyl salicylate will be prepared from salicylic acid, which is esterified at the carboxyl group with methanol. You should recall from your organic chemistry lecture that esterification is an acid-catalyzed equilibrium reaction. When equimolar amounts of the reactants are used, the equilibrium does not lie far enough to the right to favor the formation of the ester in high yield. More product can be formed by increasing the concentration of one of the reactants. In this experiment, a large excess of methanol will shift the equilibrium to favor more completely the formation of the ester.

The experiment also illustrates the use of distillation under reduced pressure (vacuum distillation) for purifying high-boiling liquids. Distillation of high-boiling liquids at atmospheric pressure is often unsatisfactory. At the high temperatures required, the material being distilled (the ester, in this case) may partially or even completely decompose and cause loss of product and contamination of the distillate. When the total pressure inside the distillation apparatus is reduced, however, the boiling point of the substance is lowered. In this way, the substance can be distilled conveniently without being decomposed.

Required Reading

Review:	Techniques 1, 2	
	Experiment 8	
	Technique 6	Physical Constants Part B, Boiling Points
New:	Essay	Esters—Flavors and Fragrances
	Techniques 7, 8, 9	

Special Instructions

Be careful when dispensing sulfuric acid; it is very corrosive and will attack your skin if you make contact with it. If you get sulfuric acid on your skin, wash the affected area with copious quantities of running water for 10–15 minutes.

It is recommended that two students combine their crude ester and work as a pair to perform the vacuum distillation. The distillation will work better with a larger quantity of material.

The experiment should be started at the very beginning of the period, since a long reflux time is needed to esterify salicylic acid and obtain a respectable yield. Enough time should remain at the end of the period to perform the extractions and to place the product over the drying agent. Since vacuum distillation is sometimes a bit tricky, the distillation should be performed during a second period if possible.

Waste Disposal

Any aqueous solutions should be placed in the appropriate container for dilute aqueous waste. Place any excess ester in the nonhalogenated organic waste container.

Procedure

Apparatus. Assemble a reflux apparatus using a 100-mL round-bottom flask and a water-cooled condenser (refer to Technique 3, Fig. 3.8, p. 619). Assemble the entire apparatus on a single ring stand so that the reaction mixture can be swirled if necessary. Use a heating mantle to heat the reaction.

Reaction Mixture. Place 9.7 g of salicylic acid (0.07 mole) and 25 mL of methanol (density 0.8 g/mL) in the flask. Swirl the flask and heat the contents slightly to help dissolve the salicylic acid. *Carefully* add 10 mL of concentrated sulfuric acid *in small portions* to the mixture in the flask. After each addition, immediately swirl the flask. A white precipitate may form, but it will dissolve after the mixture is heated. Add a corrundum boiling stone and reconnect the flask. Do not use a calcium carbonate (marble) boiling stone because it will dissolve in the acidic medium.

Reflux. Heat the mixture under gentle reflux for 1 hour. Occasionally swirl the mixture to assure that the reactants are well mixed. During the heating period, the mixture will turn cloudy, and a layer of product will form on top of the mixture. When the reflux period is finished, disconnect or remove the heating source and allow the reaction to cool to room temperature.

Extractions. Disassemble the apparatus and transfer the reaction mixture to a separatory funnel (125-mL) placed in a ring attached to a ring stand. Be sure that the stopcock is closed and, using a funnel, pour the mixture into the top of the separatory funnel. Also be careful to avoid transferring the boiling stone, or you will need to remove it after the transfer. Rinse the reaction flask with 25 mL of methylene chloride and add it to the separatory funnel. Stopper the funnel, mix the phases by careful shaking and venting (Technique 7, Section 7.4 and Fig. 7.6, pp. 690 and 693). Allow the phases to separate and then unstopper the funnel and drain the lower methylene chloride layer through the stopcock into a beaker or other suitable container. Next, extract the contents with a second 25-mL portion of methylene chloride and combine it in the beaker with the first extraction. Discard the remaining contents of the separatory funnel into the halogenated organic waste since they may contain methylene chloride. Return the saved methylene chloride layers (in the beaker) to the separatory funnel. Extract the combined methylene chloride layers with 25 mL of water. Drain the lower organic layer, discard the top aqueous layer into the aqueous waste container, and return the organic layer to the separatory funnel. Now extract the organic layer with 25 mL of 5% aqueous sodium bicarbonate just as you did previously with water. When finished, repeat the extraction once again with another 25 mL of 5% sodium bicarbonate.

Drying. Transfer the crude ester to a clean, dry 125-mL Erlenmeyer flask and add approximately 2.0 g of anhydrous sodium sulfate. Cork the mixture, swirl it, and allow it to stand for 10–15 minutes. If the mixture does not appear dry (the drying agent clumps and does not "flow," the solution is cloudy, or drops of water are obvious), transfer the ester to a new clean, dry 125-mL Erlenmeyer flask and add a new 1.0-g portion of anhydrous sodium sulfate to complete the drying. If you are going to wait until the next period to distill your product, you may stopper it here (over the sodium sulfate) with a cork (do not use a rubber stopper) and leave it in your locker.

Distillation. Assemble an apparatus for distillation under reduced pressure using your smallest round-bottom flask to distill from (Technique 9, Fig. 9.1, p. 718). Use a heating mantle to heat and install a monometer in the system if one is available (Technique 9, Fig. 9.9, p. 729). Fit the distilling flask with a capillary ebulliator tube that reaches to the bottom of the flask (Technique 9, Fig. 9.1, p. 718). The bleeder tube supplied with most organic kits will serve as an ebulliator if it is equipped with a piece of rubber tubing. The tubing is closed by a screw clamp until a gentle stream of bubbles is passed into the solution when the vacuum is applied to the system. A trap must be used between the aspirator (or vacuum line) and the distillation

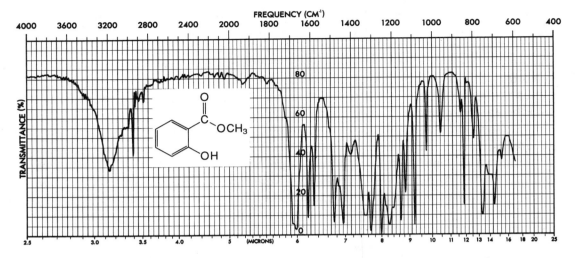

Infrared spectrum of methyl salicylate, neat.

apparatus (Technique 9, Fig. 9.3 or 9.4, p. 721). All joints should be lightly greased. Pre-weigh (tare) and use another round-bottom flask to collect the product. At the option of your instructor, you may be asked to combine your crude ester with that from another student to perform the vacuum distillation.

Using a funnel, carefully decant your product away from the drying agent into your distilling flask. (A funnel avoids getting grease mixed into your product.) Reconnect the apparatus and test the vacuum before you begin distillation. It is important that all vacuum tubing connections and ground glass joints fit tightly. If you do not have a measurable vacuum, test each connection carefully. Begin distillation by applying a vacuum and adjusting the screw clamp on the ebulliator until a small stream of bubbles issues from the tube. When the vacuum is satisfactory, apply heat from a heating mantle to the distilling flask and begin the distillation. Follow the directions in Technique 9 (Section 9.2, p. 722) when you begin the distillation. Collect all the liquid that distills at a temperature of 100°C or higher.[1] If a manometer was used, record the pressure. Also record the observed boiling point *range* of your product in your notebook. When almost all of the liquid has distilled, stop the heating, remove the hose from the aspirator or open the trap valve, and turn off the water. Always disconnect the vacuum hose before turning off the water flow through the aspirator.

Yield Determination.   Weigh the product and calculate the percentage yield of the ester. At the option of your instructor, determine the boiling point using one of the methods described in Technique 6, Sections 6.11 and 6.12, pages 677–680.

---

[1]The boiling point of methyl salicylate at atmospheric pressure is 222°C, but the ester will distill at 105°C at a pressure of 14 mm. A pressure of 14 mm is the lowest that can be obtained with a good aspirator and with a water temperature of 16.5°C. A manometer should be connected to the system to measure the pressure precisely (Technique 9, Section 9.4, p. 726).

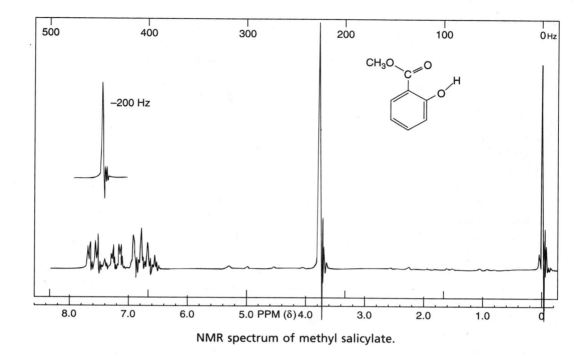

NMR spectrum of methyl salicylate.

Spectroscopy. At your instructor's option, obtain an infrared spectrum using salt plates (Technique 19, Section 19.2, p. 843). Compare your spectrum with the one reproduced on page 106. Interpret the spectrum and include it in your report to the instructor. You may also be required to determine and interpret the proton and carbon-13 NMR spectra (Technique 19, Section 19.9, p. 855). Submit your sample in a properly labeled vial with your report.

QUESTIONS

1. Write a mechanism for the acid-catalyzed esterification of salicylic acid with methanol.

2. What is the function of the sulfuric acid in this reaction? Is it consumed?

3. In this experiment, excess methanol was used to shift the equilibrium toward the formation of more product. Describe other methods for achieving the same result.

4. How are excess sulfuric acid and the excess methanol removed from the crude ester after the reaction has been completed?

5. Why was 5% $NaHCO_3$ used in the extraction? What would happen if 5% NaOH had been used?

6. Interpret the principal absorption bands in the infrared and NMR spectra of methyl salicylate.

EXPERIMENT 11   ■

Hydrolysis of Methyl Salicylate

Hydrolysis of an ester
Heating under reflux
Filtration, crystallization
Melting point determination

Esters can be hydrolyzed to their constituent carboxylic acid and alcohol parts under either acidic or basic conditions. In this experiment, methyl salicylate, an ester known as oil of wintergreen because of its natural source, is treated with aqueous base. The immediate product of this hydrolysis, besides methanol and water, is the sodium salt of salicylic acid. The reaction mixture is acidified with sulfuric acid, which converts the sodium salt to the free acid. The overall organic products of the reaction, therefore, are salicylic acid and methanol. The salicylic acid is a solid, which can be isolated and purified by crystallization. The chemical equations that describe this experiment are

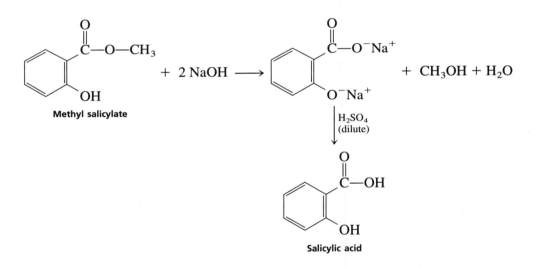

Because the phenolic hydroxyl group is acidic, it is also converted to the corresponding sodium salt during the basic hydrolysis. In the subsequent acidification, this group also becomes reprotonated.

Required Reading

Review: Techniques 1, 2, 3, and 4

Special Instructions

This experiment can be stopped at nearly any point in the procedure.

Waste Disposal

All aqueous solutions may be placed in the dilute aqueous waste.

Procedure

Apparatus. Assemble a reflux apparatus, using a 250-mL round-bottom flask and a water-cooled condenser (refer to Technique 3, Fig. 3.8, p. 619). Use a heating mantle to heat.

Reaction Mixture. Dissolve 10 g of sodium hydroxide in 50 mL of water in a 250-mL Erlenmeyer flask. When the solution has cooled, place it, along with 5.0 g of methyl salicylate, in the 250-mL round-bottom flask and add one or two boiling stones. Reassemble the apparatus for reflux.

Reflux. Start water circulating in the condenser and bring the mixture to a boil. Continue heating under vigorous reflux for 30–40 minutes. Then, disconnect or remove the heating source and allow the mixture to cool to room temperature.

Crystallization and Filtration. When the solution is cool, transfer it to a 250-mL beaker and carefully add enough 1M sulfuric acid to make the solution acidic to litmus paper (blue litmus turns pink). It may be necessary to add as much as 150 mL of 1M sulfuric acid. When the litmus turns pink, add an additional 15 mL of 1M sulfuric acid. This should cause the salicylic acid to precipitate from the solution. Cool the mixture in an ice-water bath to about 0°C. Allow the mixture to settle, and then collect the product by vacuum filtration using a Büchner funnel with filter paper. See Technique 4, Section 4.3, p. 640 for the details of this method. The filtration can be conducted most easily by decanting most of the supernatant liquid through the Büchner funnel before swirling the final mixture and filtering the mass of crystals.

Recrystallization. Recrystallize the crude salicylic acid from water in a 250-mL Erlenmeyer flask. Add 100 mL of hot water and a boiling stone, and heat the mixture to boiling to dissolve the solid. If the solid does not dissolve on boiling, add enough extra water to dissolve it. Do not be concerned if a few small pieces of solid remain that refuse to dissolve; these might be insoluble impurities. Filter the hot solution by gravity filtration through a fluted filter paper (Technique 4, Section 4.1, p. 634, and Figure 4.3, p. 637). This gravity filtration must be carried out carefully, keeping the solution hot, and filtering only a small quantity at a time. Use a short-stemmed funnel for the filtration to reduce the probability that crystals might form in the stem, clogging the funnel. The filtration assembly should be kept warm. If salicylic acid begins to crystallize in the funnel, add to the filter the minimum amount of boiling water needed to redissolve the crystals.

After the filtered solution has cooled and begun to crystallize, place the flask in an ice-water bath to aid crystallization. Scratch the inside of the flask with an un-polished glass rod to induce crystallization if necessary. When the crystals of salicylic acid have formed, collect them by vacuum filtration (Technique 4, Section 4.3, p. 640). Allow the crystals to dry overnight on a watch glass covered by a large inverted beaker (to prevent dust from settling), or in a small beaker covered by a larger inverted one.

Yield Determination. When the crystals are thoroughly dry, weigh the product and calculate the percentage yield of the ester.

Melting Point Determination. Determine the melting point of the pure mate-rial (Technique 6, Sections 6.5 and 6.7, pp. 669 and 670). Look up the melting point of pure salicylic acid in a handbook and compare it to your result. Submit your sam-ple in a properly labeled vial with your report.

Spectroscopy. At your instructor's option, obtain an infrared spectrum using a KBr pellet (Technique 19, Section 19.4, p. 847). Interpret the spectrum and include it in your report to the instructor. You may also be required to determine and inter-pret the proton and carbon-13 NMR spectra (Technique 19, Section 19.9 and 19.10, pp. 855–859).

QUESTIONS

1. What is the white solid that forms when methyl salicylate is added to aqueous solutions of sodium hydroxide?

2. Give a mechanism for the hydrolysis of methyl salicylate.

3. Is sodium hydroxide consumed in the reaction or is it a catalyst? Explain.

4. Interpret the principal absorption bands in the infrared spectrum of salicylic acid.

ESSAY

Fats and Oils

In the normal human diet, about 25 to 50% of the caloric intake consists of fats and oils. These substances are the most concentrated form of food energy in our diet. When metabolized, fats produce about 9.5 kcal of energy per gram. Carbohydrates and proteins produce less than half this amount. For this reason, animals tend to build up fat deposits as a reserve source of energy. They do this, of course, only when their food intake ex-ceeds their energy requirements. In times of starvation, the body metabolizes these stored fats. Even so, some fats are required by animals for bodily insulation and as a protective sheath around some vital organs.

The constitution of fats and oils was first investigated by the French chemist Chevreul during the years 1810 to 1820. He found that when fats and oils were hydrolyzed, they gave rise to several "fatty acids" and the trihydroxylic alcohol glycerol. Thus, fats and oils are **esters** of glycerol, called **glycerides** or **acylglycerols.** Because glycerol has three hydroxyl groups, it is possible to have mono-, di-, and triglycerides. Fats and oils are predominantly triglycerides (triacylglycerols), constituted as follows:

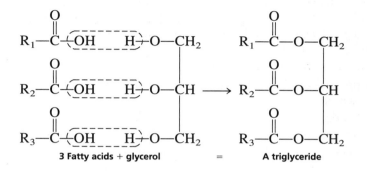

3 Fatty acids + glycerol   =   A triglyceride

Thus, most fats and oils are esters of glycerol, and their differences result from the differences in the fatty acids with which glycerol may be combined. The most common fatty acids have 12, 14, 16, or 18 carbons, although acids with both lesser and greater numbers of carbons are found in several fats and oils. These common fatty acids are listed in Table One along with their structures. As you can see, these acids are both saturated and unsaturated. The saturated acids tend to be solids, while the unsaturated acids are usually liquids. This circumstance also extends to fats and oils. Fats are made up of fatty acids that are mostly saturated, while oils are primarily composed of fatty acid portions that have greater numbers of double bonds. In other words, unsaturation lowers the melting point. Fats (solids) are usually obtained from animal sources, while oils (liquids) are commonly obtained from vegetable sources. Therefore, vegetable oils usually have a higher degree of unsaturation.

About 20 to 30 different fatty acids are found in fats and oils, and it is not uncommon for a given fat or oil to be composed of as many as 10 to 12 (or more) different fatty

TABLE ONE   Common Fatty Acids

C_{12} acids	Lauric	$CH_3(CH_2)_{10}COOH$
C_{14} acids	Myristic	$CH_3(CH_2)_{12}COOH$
C_{16} acids	Palmitic	$CH_3(CH_2)_{14}COOH$
	Palmitoleic	$CH_3(CH_2)_5CH{=}CH{-}CH_2(CH_2)_6COOH$
C_{18} acids	Stearic	$CH_3(CH_2)_{16}COOH$
	Oleic	$CH_3(CH_2)_7CH{=}CH{-}CH_2(CH_2)_6COOH$
	Linoleic	$CH_3(CH_2)_4(CH{=}CH{-}CH_2)_2(CH_2)_6COOH$
	Linolenic	$CH_3CH_2(CH{=}CH{-}CH_2)_3(CH_2)_6COOH$
	Ricinoleic	$CH_3(CH_2)_5CH(OH)CH_2CH{=}CH(CH_2)_7COOH$

acids. Typically, these fatty acids are randomly distributed among the triglyceride molecules, and the chemist cannot identify anything more than an average composition for a given fat or oil. The average fatty acid composition of some selected fats and oils is given in Table Two on page 113. As indicated, all the values in the table may vary in percentage, depending, for instance, on the locale in which the plant was grown or on the particular diet on which the animal subsisted. Thus, perhaps there is a basis for the claims that corn-fed hogs or cattle taste better than animals maintained on other diets.

Vegetable fats and oils are usually found in fruits and seeds and are recovered by three principal methods. In the first method, **cold pressing,** the appropriate part of the dried plant is pressed in a hydraulic press to squeeze out the oil. The second method is **hot pressing,** which is the same as the first method but done at a higher temperature. Of the two methods, cold pressing usually gives a better grade of product (more bland); the hot pressing method gives a higher yield, but with more undesirable constituents (stronger odor and flavor). The third method is **solvent extraction.** Solvent extraction gives the highest recovery of all and can now be regulated to give bland, high-grade food oils.

Animal fats are usually recovered by **rendering,** which involves cooking the fat out of the tissue by heating it to a high temperature. An alternative method involves placing the fatty tissue in boiling water. The fat floats to the surface and is easily recovered. The most common animal fats, lard (from hogs) and tallow (from cattle), can be prepared in either way.

Many triglyceride fats and oils are used for cooking. We use them to fry meats and other foods and to make sandwich spreads. Almost all commercial cooking fats and oils, except lard, are prepared from vegetable sources. Vegetable oils are liquids at room temperature. If the double bonds in a vegetable oil are hydrogenated, the resultant product becomes solid. Manufacturers, in making commercial cooking fats (Crisco, Spry, Fluffo, etc.), hydrogenate a liquid vegetable oil until the desired degree of consistency is achieved. This makes a product that still has a high degree of unsaturation (double bonds) left. The same technique is used for margarine. "Polyunsaturated" oleomargarine is produced by the partial hydrogenation of oils from corn, cottonseed, peanut, and soybean sources. The final product has a yellow dye (β-carotene) added to make it look like butter; milk, about 15% by volume, is mixed into it to form the final emulsion. Vitamins A and D are also commonly added. Because the final product is tasteless (try Crisco), salt, acetoin, and biacetyl are often added. The latter two additives mimic the characteristic flavor of butter.

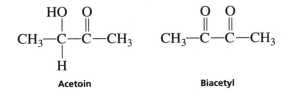

Acetoin   Biacetyl

TABLE TWO   Average Fatty Acid Composition (by Percentage) of Selected Fats and Oils

	Saturated Fatty Acids (No Double Bonds)						Unsaturated (1 Double Bond)			Unsaturated (>1 Double Bond)			
	C_{10} C_8 C_6 C_4	C_{12} Lauric	C_{14} Myristic	C_{16} Palmitic	C_{18} Stearic	C_{20} C_{22} C_{24}	C_{16} Palmitoleic	C_{18} Oleic	C_{18} Ricinoleic	C_{18} Linoleic (2)	C_{18} Linolenic (3)	C_{18} Eleostearic (3)	C_{20} C_{22} C_{24} Unsaturated
Animal fats													
Tallow			2–3	24–32	14–32		1–3	35–48		2–4			
Butter	7–10	2–3	7–9	23–26	10–13		5	30–40		4–5			2
Lard			1–2	28–30	12–18		1–3	41–48		6–7			2
Animal oils													
Neat's foot				17–18	2–3			74–77					
Whale			4–5	11–18	2–4		13–18	33–38					17–31
Sardine			6–8	10–16	1–2		6–15			24–30			12–19
Vegetable oils													
Corn			0–2	7–11	3–4		0–2	43–49		34–42			
Olive			0–1	5–15	1–4		0–1	69–84		4–12			
Peanut				6–9	2–6	3–10	0–1	50–70		13–26			
Soybean			0–1	6–10	2–6			21–29		50–59	4–8		
Safflower				6–10	1–4			8–18		70–80	2–4		
Castor bean				0–1				0–9	80–92	3–7			
Cottonseed			0–2	19–24	1–2		0–2	23–33		40–48			
Linseed				4–7	2–5			9–38		3–43	25–58		
Coconut	10–22	45–51	17–20	4–10	1–5			2–10		0–2			
Palm			1–3	34–43	3–6			38–40		5–11			
Tung								4–16		0–1		74–91	

Many producers of margarine claim it to be more beneficial to health because it is "high in polyunsaturates." Animal fats are low in unsaturated fatty acid content and are generally excluded from the diets of persons who have high cholesterol levels in the blood. Such people have difficulty in metabolizing saturated fats correctly and should avoid them because they encourage cholesterol deposits to form in the arteries. This ultimately leads to high blood pressure and heart trouble. People who pay close attention to their intake of fats tend to avoid consuming large quantities of saturated fats, knowing that eating these fats increases the risk of heart disease. Diet-conscious people try to limit their fat consumption to unsaturated fats, and they make use of the current mandatory food labeling to obtain information on the fat content of the food they eat.

Unfortunately, not all of the unsaturated fats appear to be equally safe. When we eat partially hydrogenated fats, we increase our consumption of ***trans*-fatty acids.** These acids, which are isomers of the naturally occurring *cis*-fatty acids, have been implicated in a variety of conditions, including heart disease, cancer, and diabetes. The strongest evidence that *trans*-fatty acids may be harmful comes in studies of the incidence of coronary heart disease. Ingestion of *trans*-fatty acids appears to increase blood cholesterol levels, in particular the ratio of low-density lipoproteins (LDL, or "bad" cholesterol) to high-density lipoproteins (HDL, or "good" cholesterol). The *trans*-fatty acids appear to exhibit harmful effects on the heart that are similar to those shown by saturated fatty acids.

The *trans*-fatty acids do not occur naturally to any significant extent. Rather, they are formed during the partial hydrogenation of vegetable oils to make margarine and solid forms of shortening. For a small percentage of *cis*-fatty acids subjected to hydrogenation, only one hydrogen atom is added to the carbon chain. This process forms an intermediate free radical, which is able to rotate its conformation by 180° before it releases the extra hydrogen atom back to the reaction medium. The result is an isomerization of the double bond.

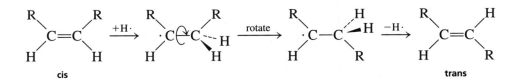

Concern over the health and nutrition of the public, particularly over the average fat intake of most Americans, has prompted food chemists and technologists to develop a variety of **fat replacers.** The objective has been to discover substances that have the taste and mouth-feel of a real fat but do not have deleterious effects on the cardiovascular system. One product that has recently appeared in certain snack foods is **olestra** (marketed under the trade name **Olean,** by The Procter and Gamble Company). Olestra is not an acylglycerol; rather it is composed of a molecule of **sucrose** that has been substituted by long-chain fatty acid residues. It is a **polyester,** and the body's enzyme systems are not capable of attacking it and catalyzing its breakdown into smaller molecules.

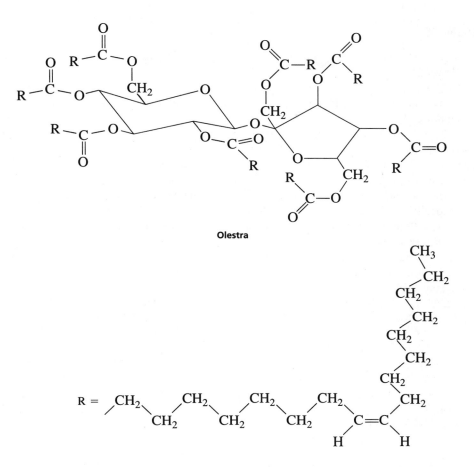

Olestra

Because the body's enzyme systems are unable to break this molecule down, it does not contain any usable dietary calories. Furthermore, it is heat stable, which makes it ideal for frying and other cooking. Unfortunately, for some individuals there may be harmful or unpleasant side effects. The use of olestra has been reported to cause the depletion of certain fat-soluble vitamins, particularly Vitamins A, D, E, and K. For this reason, products prepared with olestra have these vitamins added to offset this effect. Also, for some people, incidences of diarrhea and abdominal cramps have been reported.

Is the development of fat replacers such as olestra part of the wave of the future? As the average American's appetite for snack foods continues to grow and as health problems arising from obesity also increase, a demand for satisfying foods that are less fattening will always be strong. In the long run, however, it would probably be better if we all learned to curtail our appetite for fatty foods and, instead, tried to increase our intake of fruits, vegetables, and other healthful foods. At the same time, a change from a sedentary lifestyle to one that included regular exercise would be much more beneficial to our health than a search for nonfattening food additives.

REFERENCES

Dawkins, M. J. R., and Hull, D. "The Production of Heat by Fat." *Scientific American, 213* (August 1965): 62.

Dolye, E. "Olestra? The Jury's Still Out." *Journal of Chemical Education*, 74, (April 1997): 370.

Eckey, E. W., and Miller, L. P. *Vegetable Fats and Oils.* ACS Monograph No. 123. New York: Reinhold, 1954.

Farines, M., Soulier, F., and Soulier, J. "Analysis of the Triglycerides of Some Vegetable Oils." *Journal of Chemical Education, 65* (May 1988): 464.

Gunstone, F. D. "The Composition of Hydrogenated Fats Determined by High Resolution ^{13}C NMR Spectroscopy." *Chemistry and Industry* (November 4, 1991): 802.

Heinzen, H., Moyna, P., and Grompone, A. "Gas Chromatographic Determination of Fatty Acid Compositions." *Journal of Chemical Education, 62* (May 1985): 449.

Jandacek, R. J. "The Development of Olestra, a Noncaloric Substitute for Dietary Fat." *Journal of Chemical Education, 68* (June 1991): 476.

Kalbus, G. E., and Lieu, V. T. "Dietary Fat and Health: An Experiment on the Determination of Iodine Number of Fats and Oils by Coulometric Titration." *Journal of Chemical Education, 68* (January 1991): 64.

Lemonick, M. D. "Are We Ready for Fat-Free Fat?" *Time, 147* (January 8, 1996): 52.

Martin, C. "TFA's—A Fat Lot of Good?" *Chemistry in Britain, 32* (October 1996): 34.

Nawar, W. W. "Chemical Changes in Lipids Produced by Thermal Processing." *Journal of Chemical Education, 61* (April 1984): 299.

Shreve, R. N., and Brink, J. "Oils, Fats, and Waxes." *The Chemical Process Industries,* 4th ed. New York: McGraw-Hill, 1977.

Thayer, A. M. "Food Additives." *Chemical and Engineering News, 70* (June 15, 1992): 26.

Wootan, M., Liebman, B., and Rosofsky, W. "*Trans*: The Phantom Fat." *Nutrition Action Health Letter, 23* (September 1996): 10.

E X P E R I M E N T   1 2

Analysis of Fats and Oils by Gas Chromatography

Transesterification
Gas chromatography

The purpose of this experiment is to determine the fatty acid composition of various common fats and oils. As seen in the preceding essay, "Fats and Oils," oils vary considerably in their fatty acid content. Triglycerides are triesters of glycerol, but they rarely have the same acid groups attached to glycerol. You may find, for example, that a typical oil such as soybean oil may have some molecules of a triglyceride with all of the same fatty acids attached. You will also find molecules in soybean oil that have different fatty acids attached to glycerol. Since there are as many as ten different fatty acids represented in fats and oils, you could imagine that there are many different combinations of triglyceride molecules present in the soybean oil (see Question 3). To solve the problems inherent in analyzing a complex mixture of different triglyceride molecules present in soybean oil, it is necessary to break apart the triglycerides into their individual fatty acids. After this is done, the *average composition* of the soybean oil can be determined.

The classical procedure for obtaining the fatty acids from a fat is **saponification** (soap-making), or hydrolysis with alkali to give the sodium salts of the fatty carboxylic acids mixed with glycerol. Such a mixture constitutes a soap (see the essay on soaps preceding Experiment 14). Acidifying the soap mixture gives the fatty acids:

$$RCOO^-Na^+ + HCl \rightarrow RCOOH + NaCl$$

Unfortunately, saponification is slow at room temperature; and at higher temperatures, the unsaturated acids show a tendency to isomerize the positions of their double bonds. This clearly interferes with any analysis. An additional complication is that although some of the acids are solids, others are liquids, and all are high-boiling.

Instead of saponification, we shall use the **transesterification** reaction, in which one alcohol group of an ester is exchanged for another. If we use a monohydroxylic alcohol like methanol, three different methyl esters and one molecule of glycerol will be produced. The methyl esters have lower boiling points than the acids, and can easily be analyzed by gas chromatography. If sodium methoxide in methanol is used, the triglycerides are converted cleanly into the methyl esters. Acetic acid is added after the reaction to acidify the mixture, and the methyl esters are extracted into hexane. Following evaporation of the hexane, the mixture of esters is dissolved in methylene chloride and analyzed by gas chromatography.

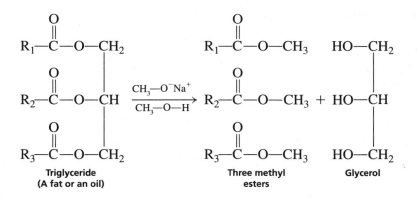

The gas chromatograms of methyl esters prepared from samples of olive oil and of soybean oil are shown in the figure. You will obtain a similar gas chromatogram, starting with a fat or oil of your choice, and from the chromatogram you will calculate the percentage of each of the methyl esters of fatty acids formed from the fat or oil. In most cases, you should be able to see up to five methyl esters derived from palmitic, stearic, oleic, linoleic, and linolenic acids. These methyl esters can be identified by comparing your chromatogram to the reference chromatogram made available by your instructor.

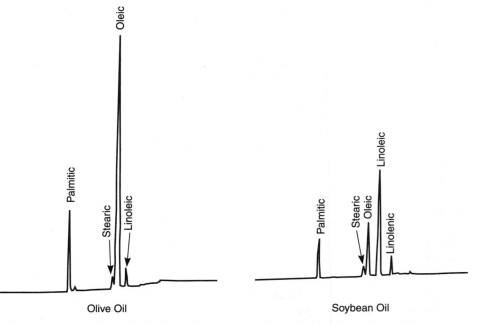

Gas chromatograms of methyl esters prepared from samples of olive oil and soybean oil.

Required Reading

New: Technique 15   Gas Chromatography
        Essay   Fats and Oils

Special Instructions

Although some oils will be made available for this experiment in the laboratory, you may choose to bring an oil or fat of your own. It is recommended that you use an oil rather than a solid (shortening or lard) because the oils lead to more satisfying chromatographic results. Good choices of oils include corn oil, olive oil, soybean oil, and safflower oil.

Waste Disposal

Pour aqueous wastes obtained in this experiment into the waste container provided for you.

Notes to the Instructor

You can modify this experiment by issuing three or four oils as unknowns. After gas chromatography, students would then calculate the percentages and determine the type of oil issued to them by using the table in the Fats and Oils essay. Students can also compare their results with others in the class.

It is recommended that a capillary gas chromatography column be used for this experiment in order to obtain better resolution of the methyl esters. The conditions listed here are for a Hewlett-Packard 5890, flame ionization gas chromatograph: J&W DB-wax column (polyethylene glycol, J&W No. 125-7012); 15 m long with a 0.53-mm ID; helium carrier gas flow of 103 ml/min; injector temperature 250°C; detector temperature 270°C; temperature programming: initial temperature at 50°C for 2 min, then ramp at the rate of 8 degrees/min to a final temperature of 230°C, temperature held at this final value for 10 min; total analysis time, 34.5 min.

It is recommended that a set of standards be analyzed on your instrument and column using the conditions listed above, to check for resolution. At a minimum you should have a set of standards that includes the following five methyl esters (approximate retention times in parentheses): methyl palmitate (19.7 min), methyl stearate (22.2 min), methyl oleate (22.5 min), methyl linoleate (23.0 min), and methyl linolenate (23.6 min). You may also want to include methyl laurate (14.6 min) and methyl myristate (17.3 min), although these esters will only be minor components with common oils such as corn, olive, and soybean.

It is recommended that the instructor or assistant dispense the 0.5 M sodium methoxide in methanol solution to students in order to ensure that the reagent remains anhydrous. The reagent as purchased from Aldrich Chemical Company (No. 40,306-7) is packaged under nitrogen in Sure/Seal bottles. There is a liner in the cap that can be punctured with syringe needles. In order to keep the reagent anhydrous, we recommend that you transfer material from the bottle using a syringe while simultaneously replacing the volume removed with dry nitrogen gas. To do this without removing the cap on the reagent bottle, you will need to insert one short syringe needle into the cap as a source of nitrogen. Another, longer syringe needle is inserted into the bottle to remove the reagent. The longer needle is attached to a syringe that is used to remove the sodium methoxide. Remove a 2-mL sample of sodium methoxide with the syringe and transfer it directly into the student's test tube.

The instructor or assistant will need to dry some tetrahydrofuran over molecular sieves (type 4A, 1.6-mm pellets) well in advance of the laboratory class. The hexane does not need to be dried.

Procedure

Running the Reaction. In the procedure that follows, it is important to maintain anhydrous conditions until the acetic acid is added. Place one or two drops of your oil into a *dry* 25 × 150-mm test tube and add 1 mL of *dry*[1] tetrahydrofuran (THF). With the help of your instructor or assistant, add 2 mL of 0.5 M sodium methox-

---

[1]The tetrahydrofuran (THF) needs to be dried well in advance of the laboratory, as described in the Notes to the Instructor.

ide in anhydrous methanol. Swirl the mixture gently to mix the reagents. Insert a cork loosely into the test tube containing the homogeneous mixture. Place the test tube into a beaker of warm water (40°C) and clamp it into position. Use a hotplate to heat the water. Allow the mixture to react for 10 minutes at 40°C in the beaker of warm water.

Extracting the Methyl Esters.   Remove the test tube from the water bath, and add one drop of glacial acetic acid while gently swirling. Add 5 mL of distilled water, and mix with a stirring rod. Next, add 5 mL of hexane to extract the methyl esters out of the mixture. Stir the mixture for a few minutes in order to extract the mixture thoroughly. There should now be two separate layers. If an emulsion forms, it should dissipate on standing after a few minutes. Remove the top hexane layer with a Pasteur pipet and place it in a 25-mL beaker or Erlenmeyer flask. Thoroughly extract the aqueous mixture remaining in the test tube with a second 5-mL portion of hexane. Again, remove the upper hexane layer with a Pasteur pipet and add it to the first hexane extract in the beaker or flask.

Dry the hexane extracts over anhydrous sodium sulfate that contains 10% by weight of potassium bicarbonate. After the sample is dried, remove the solution from the drying agent with a Pasteur pipet and pass it through a filtering pipet (Technique 4, Section 4.1, p. 634). The filtering pipet should remove any remaining drying agent from the solution. Collect the filtrate in a 25-mL round-bottom flask. Make sure that no particles of drying agent are present in the final filtered solution.

Remove the hexane by evaporating it in a warm water bath. Use a gentle stream of nitrogen gas to aid the evaporation process. When the hexane has been removed, one or two drops of a mixture of methyl esters will remain. Add 1 mL of methylene chloride to the esters and gently swirl the flask to dissolve the mixture of esters.

Gas Chromatography.   Analyze the mixture of methyl esters using gas chromatography. If you need to store the product until the next laboratory period, stopper the flask (no cork, use ground-glass stopper), and, if possible, store it in a refrigerator (properly labeled). Your instructor or assistant will help you operate the gas chromatograph. You will need to vary the amount of sample you inject depending on the type of chromatograph available to you. From the chromatogram obtained, identify the various peaks by comparing the retention times with those of the known substances in the reference mixture (you may observe minor differences in retention times). A reference chromatogram will be made available to you by your instructor. Some minor components may also be present in the chromatogram. Try to identify all of the components by referring to the table in the essay, "Fats and Oils," and to the chromatogram of the reference mixture. Using the method explained in Technique 15, Section 15.11, p. 820, calculate the percentage composition of the methyl esters found in the mixture (some instruments have integrators attached to the chromatograph, and percentages can be calculated from the numbers that are printed with the chromatogram). Compare your percentages obtained from the oil or fat with the values given in the essay, "Fats and Oils."

REFERENCES

Christie, W. W. *Lipid Analysis.* Oxford: Pergamon Press, 1982. Pp. 53–54.

QUESTIONS

1. Draw a mechanism for the transesterification reaction that converts a triglyceride to methyl esters of fatty acids using sodium methoxide in methanol.
2. Why do you not observe methanol or glycerol in the gas chromatogram?
3. If an oil were composed of a mixture of triglycerides formed from only *two* different fatty acids, how many different possible triglycerides could be found in the oil? Draw them.

EXPERIMENT 13

Methyl Stearate from Methyl Oleate

Catalytic hydrogenation
Filtration (Pasteur pipet)
Recrystallization
Unsaturation tests

In this experiment, you will convert the liquid methyl oleate, an "unsaturated" fatty acid ester, to solid methyl stearate, a "saturated" fatty acid ester, by catalytic hydrogenation.

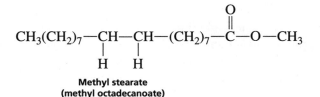

$$CH_3(CH_2)_7-CH=CH-(CH_2)_7-\overset{\overset{\textstyle O}{\|}}{C}-O-CH_3 \xrightarrow[H_2]{Pd/C}$$

Methyl oleate
(methyl *cis*-9-octadecenoate)

$$CH_3(CH_2)_7-\underset{\underset{\textstyle H}{|}}{CH}-\underset{\underset{\textstyle H}{|}}{CH}-(CH_2)_7-\overset{\overset{\textstyle O}{\|}}{C}-O-CH_3$$

Methyl stearate
(methyl octadecanoate)

By commercial methods similar to those described in this experiment, the unsaturated fatty acids of vegetable oils are converted to margarine (see the essay, "Fats and Oils"). However, rather than using the mixture of triglycerides that would be present in a cooking oil such as Mazola (corn oil), we use as a model the pure chemical methyl oleate.

For this procedure, a chemist would usually use a cylinder of hydrogen gas. Because many students will be following the procedure simultaneously, however, we use the simpler expedient of causing zinc metal to react with dilute sulfuric acid:

$$Zn + H_2SO_4 \xrightarrow{\text{H}_2\text{O}} H_2(g) + ZnSO_4$$

The hydrogen so generated will be passed into a solution containing methyl oleate and the palladium on carbon catalyst (10% Pd/C).

Required Reading

Review: Techniques 1, 2 and 4

New: Technique 4   Filtration, Sections 4.3–4.5
         Technique 6   Physical Constants, Part A, Melting Points
         Essay   Fats and Oils

You should also read those sections in your lecture textbook that deal with catalytic hydrogenation. If the instructor indicates that you should perform the optional unsaturation tests on your starting material and product, read the descriptions of the Br_2/CH_2Cl_2 test on page 126 and in Experiment 27 on page 248.

Special Instructions

Because this experiment calls for generating hydrogen gas, no flames will be allowed in the laboratory.

Caution: No flames allowed.

Because a buildup of hydrogen is possible within the apparatus, it is especially important to remember to wear your safety goggles; you can thus protect yourself against the possibility of minor "explosions" from joints popping open, from fires, or from any glassware accidentally cracking under pressure.

Caution: Wear safety goggles.

When you operate the hydrogen generator, be sure to add sulfuric acid at a rate that does not cause hydrogen gas to evolve too rapidly. The hydrogen pressure in the flask should not rise much above atmospheric pressure. Neither should the hydrogen evolution be allowed to stop. If this happens, your reaction mixture may be "sucked back" into your hydrogen generator.

Waste Disposal

Carefully dilute the sulfuric acid (from the hydrogen generator) with water and place it in a container provided for this purpose. Place any leftover zinc in the solids container designated for unreacted zinc. After centrifugation, transfer the Pd/C catalyst to a specially designated container for later recycling. After collecting the methyl stearate by filtration, place the methanol filtrate in the nonhalogenated organic waste. Discard the solutions that remain after the bromine test for unsaturation into a waste container designated for the disposal of halogenated organic solvents.

Notes to the Instructor

Use methyl oleate that is 100% (or nearly 100%) pure. Avoid the practical grades, which may be only 70–80% of methyl oleate. We use Aldrich Chemical Co., No. 31,111-1. A commercial cooking oil could be substituted for methyl oleate in this experiment, but the results would not be as clear-cut.

Procedure

Apparatus.   Assemble the apparatus as illustrated in the figure on page 124. The apparatus consists of basically three parts:

1. hydrogen generator
2. reaction flask
3. mineral oil bubbler trap

The **hydrogen generator** is a 20 × 150-mm sidearm test tube, fitted with a No. 3 rubber stopper. The **reaction flask** is a 50-mL round-bottom flask with a Claisen head attached. The hydrogen enters the reaction flask through a bubbler tube (ebulliator) attached to the top of the Claisen head by using the thermometer adapter. A small magnetic stirring bar is placed in the round-bottom flask, and the bubbler tube is adjusted to be just high enough to avoid contact, but to allow hydrogen to bubble *through* the solution. A second thermometer adapter, fitted with a short piece of glass tubing, allows connection to the mineral oil bubbler. (No. 2 one-hole rubber stoppers could be substituted for the thermometer adapters if your joint size is $19/22.) A 150-mL beaker, filled with water and placed on a stirring hotplate, provides the heating bath. The **mineral oil bubbler,** a 20 × 150-mm sidearm test tube, has two functions. First, it allows you to keep a pressure of hydrogen within the sys-

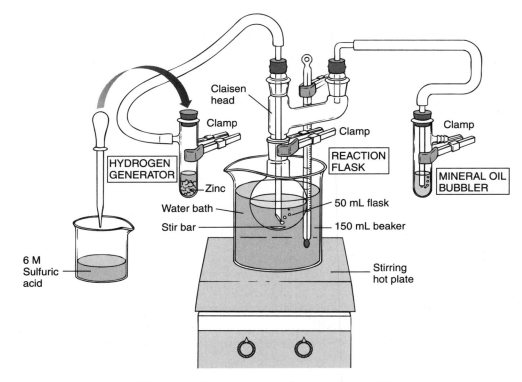

Hydrogenation apparatus for Experiment 13.

tem that is slightly above atmospheric. Second, it prevents back-diffusion of air into the system. The functions of the other two units are self-explanatory.

So that hydrogen leakage is prevented, the tubing used to connect the various subunits of the apparatus should be either relatively new rubber tubing, without cracks or breaks, or Tygon tubing. The tubing can be checked for cracks or breaks simply by stretching and bending it before use. It should be of such size that it will fit onto all connections tightly. Similarly, if any rubber stoppers are used, they should be fitted with a size of glass tubing that fits firmly through the holes in their centers. If the seal is tight, it will not be easy to slide the glass tubing up and down in the hole.

Preparing for the Reaction. Fill the bubbler trap (second side-arm test tube) about one-fourth full with mineral oil. The end of the glass tube should be submerged below the surface of the oil.

To charge the hydrogen generator, weigh out about 3 g of mossy zinc and place it in the side-arm test tube. Seal the large opening at its top using a rubber stopper. Obtain about 10 mL of 6M sulfuric acid and place it in a small Erlenmeyer flask or beaker, but do not add it yet.

Weigh a 10-mL graduated cylinder and record its weight. Place 2.5 mL of methyl oleate into it. Reweigh the cylinder in order to obtain the exact amount of methyl oleate used. Detach the 50-mL round-bottom flask, place it in a small beaker to keep it upright, and transfer the methyl oleate. Do not clean the graduated cylinder. Instead, pour two consecutive 8-mL portions of methanol solvent (16 mL total) into the cylinder to rinse it, and pour each of them into the reaction flask. Also remember to place a magnetic stirring bar into the flask. Using smooth weighing paper, weigh about 0.050 g (50 mg) of 10% Pd/C. Carefully place about one-third of the catalyst into the flask and gently swirl the liquid until the solid catalyst has sunk into the liquid. Repeat this with the rest of the catalyst, adding one-third of the original amount each time.

> **Caution:** Be careful when adding the catalyst; sometimes it will cause a flame. Do not hold onto the flask; it should be in a small beaker on the lab bench. Have a watch glass handy to cover the opening and smother the flame should this occur.

Running the Reaction.   Complete the assembly of the apparatus, making sure that all the seals are gas tight. Place the round-bottom flask in a warm water bath maintained at 40°C. This will help to keep the product dissolved in the solution throughout the course of the reaction. If the temperature rises above 40°C, you will lose a significant amount of the methanol solvent (bp 65°C). If this occurs, do not hesitate to add more methanol to the reaction flask through the side arm of the Claisen head. Begin stirring the reaction mixture with the magnetic stirring bar. Avoid stirring too fast or a vortex will form, leaving the bubbler tube out of the solution. Start the evolution of hydrogen by removing the rubber stopper and adding a portion of the 6 M sulfuric acid solution (about 6 mL) to the hydrogen generator (use a small disposable Pasteur pipet). Replace the rubber stopper. A good rate of bubbling in the reaction flask is about three to four bubbles a second. Continue the evolution of hydrogen for at least 60 minutes. If necessary, open the generator, *empty it,* and refresh the zinc and sulfuric acid. (Keep in mind that the acid is used up as hydrogen is produced and becomes more dilute as the zinc reacts. As the acid solution becomes more dilute, the rate of hydrogen evolution will slow down.)

Stopping the Reaction.   After the reaction is complete, stop the reaction by disconnecting the generator from the reaction flask. Decant the acid in the side-arm test tube into a designated waste container, being careful not to transfer any zinc metal. Rinse the zinc in the test tube several times with water and then place any unreacted zinc in a waste container provided for this purpose.

Keep the temperature of the reaction mixture at about 40°C until you perform the centrifugation; otherwise, the methyl stearate may crystallize and interfere with removal of the catalyst. There should not be any white solid (product) in the round-bottom flask. If there is a white solid, add more methanol and stir until the solid dissolves.

Removal of the Catalyst.   Pour the reaction mixture into a centrifuge tube. Place the centrifuge tube into the water bath at 40°C until just before you are ready to centrifuge the mixture. (If the solution won't fit in a single centrifuge tube, divide it between two tubes and place them opposite each other in the centrifuge.) Centrifuge the mixture for several minutes. After centrifugation, the black catalyst should be at the bottom of the tube. If some of the catalyst is still suspended in the liquid, heat the mixture to 40°C and centrifuge the mixture again. Carefully pour (or remove with a Pasteur pipet) the supernatant liquid (leaving the black catalyst in the centrifuge tube) into a small beaker and cool to room temperature.

Crystallization and Isolation of Product.   Place the beaker in an ice bath to induce crystallization. If crystals do not form or if only a few crystals form, you may need to reduce the volume of solvent. Do this by heating the beaker in a water bath and directing a slow stream of air into the beaker, using a Pasteur pipet for a nozzle (Technique 3, Fig. 3.19A, p. 632). If crystals begin to form while you are evaporating the solvent, remove the beaker from the water bath. If crystals do not form, reduce the volume of the solvent by about one-third. Allow the solution to cool and then place it in an ice bath.

Collect the crystals by vacuum filtration, using a small Büchner funnel (Technique 4, Section 4.3, p. 640). Save both the crystals and the filtrate for the tests below. After the crystals are dry, weigh them and determine their melting point (literature, 39°C). Calculate the percentage yield. Submit your remaining sample to your instructor in a properly labeled container along with your report.

Unsaturation Tests (OPTIONAL).   Using a solution of bromine in methylene chloride, test for the number of drops of this solution decolorized by:

1. about 0.1 mL of methyl oleate dissolved in a small amount of methylene chloride.
2. a small spatulaful of your methyl stearate product dissolved in a small amount of methylene chloride.
3. about 0.1 mL of the filtrate that you saved as directed above.

Use small test tubes and Pasteur pipets to make these tests. Include the results of the tests and your conclusions in your report.

QUESTIONS

1. Using the information in the essay on fats and oils, draw the structure of the triacylglycerol (triglyceride) formed from oleic acid, linoleic acid, and stearic acid. Give a balanced equation and show how much hydrogen would be needed to reduce the triacylglycerol completely; show the product.

2. A 0.150-g sample of a pure compound subjected to catalytic hydrogenation takes up 25.0 mL of H_2 at 25°C and 1 atm pressure. Calculate the molecular weight of the compound, assuming that it has only one double bond.

3. A compound with the formula C_5H_6 takes up 2 moles of H_2 on catalytic hydrogenation. Give one possible structure that would fit the information given.

4. A compound of formula C_6H_{10} takes up 1 mole of H_2 on reduction. Give one possible structure that would fit the information.

5. How would this experiment differ in outcome if you used a commercial cooking oil instead of methyl oleate?

ESSAY

Soaps and Detergents

Soaps as we know them today were virtually unknown before the first century A.D. Clothes were cleaned primarily by the abrasive action of rubbing them on rocks in water. Somewhat later, it was discovered that certain types of leaves, roots, nuts, berries, and barks formed soapy lathers that solubilized and removed dirt from clothes. We now refer to these natural materials that lather as **saponins.** Many saponins contain pentacyclic triterpene carboxylic acids, such as oleanolic acid or ursolic acid, chemically combined with a sugar molecule. These acids also appear in the uncombined state. Saponins were probably the first known "soaps." They may have also been an early source of pollution in that they are known to be toxic to fish. The pollution problem associated with the development of soap and detergents has been long and controversial.

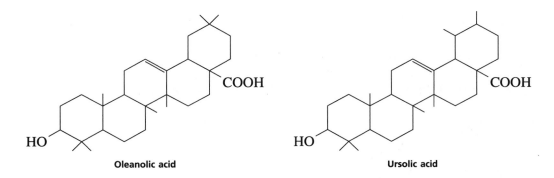

Oleanolic acid   **Ursolic acid**

Soap as we know it today has evolved over many centuries from experimentation with crude mixtures of alkaline and fatty materials. Pliny the Elder described the manufacture of soap during the first century A.D. A modest soap factory was even built in Pompeii. During the Middle Ages, cleanliness of the body or clothing was not considered important. Those who could afford perfumes used them to hide their body odor. Perfumes, like fancy clothes, were status symbols for the rich. An interest in cleanliness again emerged during the eighteenth century, when disease-causing microorganisms were discovered.

SOAPS

The process of making soap has remained practically unchanged for 2000 years. The procedure involves the basic hydrolysis or **saponification** of an animal fat or a vegetable oil. Chemically, fats and oils are referred to as **triglycerides** or **triacylglycerols.** They contain ester functional groups. Saponification involves heating a fat or oil with an alkaline solution. This alkaline solution was originally obtained by leaching wood ashes or from the evaporation of natural alkaline waters. Today, lye (sodium hydroxide) is used as the source of the alkali. The alkaline solution hydrolyzes the fat or oil into its component parts, the sodium salt of a long-chain carboxylic acid (soap) and an alcohol (glycerol). When common salt is added, the soap precipitates. The soap is washed free of unreacted sodium hydroxide and molded into bars. The following equation shows how soap is produced from a fat or oil.

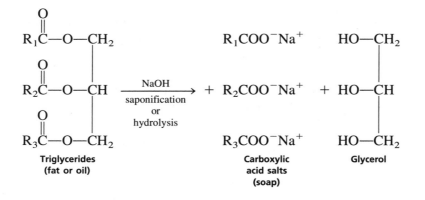

Triglycerides (fat or oil) → **Carboxylic acid salts (soap)** + **Glycerol**

The carboxylic acids represented in soap are rarely of a single type in any given fat or oil. In fact, a single triglyceride (triacylglycerol) molecule in a fat may contain three different acid residues (R_1COOH, R_2COOH, R_3COOH), and not every triglyceride in the substance will be identical. Each fat or oil, however, has a characteristic *statistical distribution* of the various types of acids possible. The carboxylic acid salts of soap usually contain 12–18 carbons arranged in a straight chain. The carboxylic acids containing even numbers of carbon atoms predominate, and the chains may contain unsaturation. The composition of the common fats and oils is given in the essay "Fats and Oils" (p. 110).

The fats and oils that are most common in soap preparations are lard and tallow from animal sources and coconut, palm, and olive oils from vegetable sources. The length of the hydrocarbon chain and the number of double bonds in the carboxylic acid portion of the fat or oil determine the properties of the resulting soap. For example, a salt of a saturated long-chain acid makes a harder, more insoluble soap. Chain length also affects solubility.

Tallow is the principal fatty material used in making soap. The solid fats of cattle are melted with steam, and the tallow layer formed at the top is removed. Soap-makers usually blend tallow with coconut oil and saponify this mixture. The resulting soap contains mainly the salts of palmitic, stearic, and oleic acids from the tallow and the salts of lauric and myristic acids from the coconut oil. The coconut oil is added to produce a softer, more soluble soap. Lard (from hogs) differs from tallow (from cattle or sheep) in that lard contains more oleic acid.

Pure coconut oil yields a soap that is very soluble in water. The soap contains essentially the salt of lauric acid, with some myristic acid. It is so soft (soluble) that it will lather even in seawater. Palm oil contains mainly two acids, palmitic acid and oleic acid,

Tallow $CH_3(CH_2)_{14}COOH$ $CH_3(CH_2)_{16}COOH$
Palmitic acid **Stearic acid**

$CH_3(CH_2)_7CH{=}CH(CH_2)_7COOH$
Oleic acid

Coconut oil $CH_3(CH_2)_{10}COOH$ $CH_3(CH_2)_{12}COOH$
Lauric acid **Myristic acid**

in about equal amounts. Saponification of this oil yields a soap that is an important constituent of toilet soaps. Olive oil contains mainly oleic acid. It is used to prepare Castile soap, named after the region in Spain in which it was first made.

Toilet soaps generally have been carefully washed free of any alkali remaining from the saponification. As much glycerol as possible is usually left in the soap, and perfumes and medicinal agents are sometimes added. Floating soaps are produced by blowing air into the soap as it solidifies. Soft soaps are made by using potassium hydroxide, yielding potassium salts rather than the sodium salts of the acids. They are used in shaving creams and liquid soaps. Scouring soaps have abrasives added, such as fine sand or pumice.

A disadvantage of soap is that it is an ineffective cleanser in hard water. Hard water contains salts of magnesium, calcium, and iron in solution. When soap is used in hard water, "calcium soap," the insoluble calcium salts of the fatty acids, and other precipitates are deposited as **curds.** This precipitate, or curd, is referred to as bathtub ring. Although soap is a poor cleanser in hard water, it is an excellent cleanser in soft water.

$$2 \, R{-}\underset{\text{Soap}}{\underset{\|}{\overset{\overset{\text{O}}{\|}}{C}}}{-}O^- Na^+ + Ca^{2+} \longrightarrow \left(R{-}\underset{\text{Curd}}{\overset{\overset{\text{O}}{\|}}{C}}{-}\bar{O} \right)_2 Ca^{2+}$$

Water softeners are added to soaps to help remove the troublesome hard-water ions so that the soap will remain effective in hard water. Sodium carbonate or trisodium phosphate will precipitate the ions as the carbonate or phosphate. Unfortunately, the precipitate may become lodged in the fabric of items being laundered, causing a grayish or streaked appearance.

$$Ca^{2+} + CO_3^{2-} \longrightarrow CaCO_3\downarrow$$

$$3 \, Ca^{2+} + 2 \, PO_4^{3-} \longrightarrow Ca_3(PO_4)_2\downarrow$$

An important advantage of soap is that it is **biodegradable.** Microorganisms can consume the linear soap molecules and convert them to carbon dioxide and water. The soap is thus eliminated from the environment.

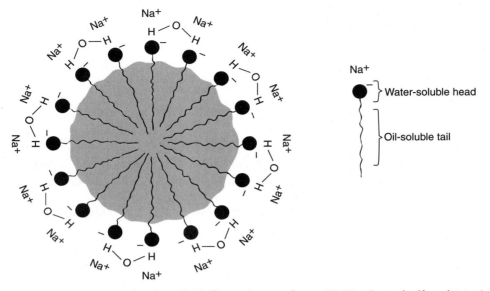

A soap micelle solvating a droplet of oil (from Linstromberg, W. W., *Organic Chemistry: A Brief Course*, D. C. Heath, 1978).

ACTION OF SOAP IN CLEANING

Dirty clothes, skin, or other surfaces have particles of dirt suspended in a layer of oil or grease. Polar water molecules cannot remove the dirt embedded in nonpolar oil or grease. One can remove the dirt with soap, however, because of its dual nature. The soap molecule has a polar, *water-soluble* head (carboxylate salt) and a long, *oil-soluble* tail (the hydrocarbon chain). The hydrocarbon tail of soap dissolves in the oily substance, but the ionic end remains outside the oily surface. When enough soap molecules have oriented themselves around an oil droplet with their hydrocarbon ends dissolved in the oil, the oil droplet, together with the suspended dirt particles, is removed from the surface of the cloth or skin. The oil droplet is removed because the heavily negatively charged oil droplet is now strongly attracted to water and solvated by the water. The solvated oil droplet is called a **micelle.**

DETERGENTS

Detergents are synthetic cleaning compounds, often referred to as **syndets.** They were developed as an alternative to soaps because they are effective in *both* soft and hard water. No precipitates form when calcium, magnesium, or iron ions are present in a detergent solution. One of the earliest detergents developed was sodium lauryl sulfate. It is prepared by the action of sulfuric acid or chlorosulfonic acid on lauryl alcohol (1-dodecanol). This detergent is relatively expensive, however. The following reactions show one industrial method of preparation:

$$CH_3(CH_2)_{10}CH_2OH + H_2SO_4 \longrightarrow CH_3(CH_2)_{10}CH_2OSO_3H + H_2O$$

Lauryl alcohol

$$CH_3(CH_2)_{10}CH_2OSO_3H + Na_2CO_3 \longrightarrow CH_3(CH_2)_{10}CH_2OSO_3^- Na^+ + NaHCO_3$$

Sodium lauryl sulfate

The first of the inexpensive detergents appeared about 1950. These detergents, called alkylbenzenesulfonates (ABS), can be prepared from inexpensive petroleum sources by the following set of reactions:

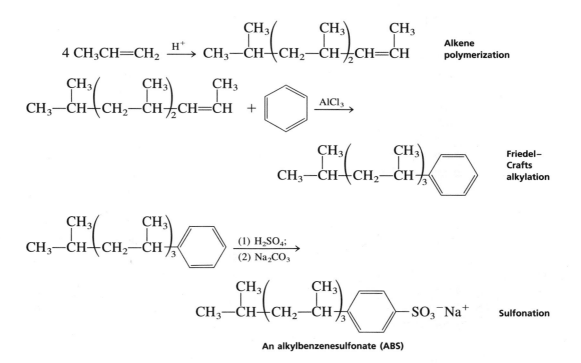

An alkylbenzenesulfonate (ABS)

Detergents became very popular because they could be used effectively in all types of water and were cheap. They rapidly displaced soap as the most popular cleaning agent. A problem with the detergents was that they passed through sewage-treatment plants without being degraded by the microorganisms present, a process necessary for full sewage treatment. Rivers and streams in many sections of the country became polluted with detergent foam. The detergents even found their way into the drinking water supplies of numerous cities. The reason for the persistence of the detergents was that bacterial enzymes, which could degrade straight-chain soaps and sodium lauryl sulfate, could not destroy the highly branched detergents such as ABS.

It was soon found that the bacterial enzymes could degrade only a chain of carbons that contained, at the most, one branch. Numerous cities and states banned the sale of the nonbiodegradable detergents, and by 1966, they were replaced by the new biodegradable detergents called linear alkylsulfonates (LAS). One example of an LAS detergent is shown here. Notice that there is one branch next to the aromatic ring.

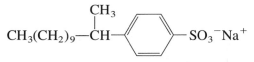

A linear alkylsulfonate detergent (LAS)

NEW PROBLEMS WITH DETERGENTS

Detergents (also soaps) are not sold as pure compounds. A typical heavy-duty, controlled **sudser** may contain only 8–20% of the linear alkylsulfonate. A large quantity (30–50%) of a **builder** such as sodium tripolyphosphate $Na_5P_3O_{10}$ may be present. Other additives include corrosion inhibitors, antideposition agents, and perfumes. Optical brighteners are also added. Brighteners absorb invisible ultraviolet light and reemit it as visible light, so laundry appears white and thus "clean." The phosphate builder is added to complex the hard-water ions, calcium and magnesium, and keep them in solution. Builders seem to enhance the washing ability of the LAS and also act as a cheap filler.

Unfortunately, phosphates speed the **eutrophication** of lakes and other bodies of water. The phosphates, along with other substances, are nutrients for algae. When algae begin to die and decompose, they consume so much dissolved oxygen from the water that no other life can exist in that water. The lake rapidly "dies." This is eutrophication.

Because phosphates have this undesirable effect, a search was initiated for a replacement for the phosphate builders. Some replacements have been made, but most also have problems associated with them. Two replacement builders, sodium metasilicate and sodium perborate, are highly basic substances, and they have caused injuries to children. In addition, they appear to destroy bacteria in sewage-treatment plants and may have other unknown environmental effects.

Many people have suggested a return to soap. The main problem is that we probably cannot produce enough soap to meet the demand because of the limited amount of animal fat available. Where do we go from here?

REFERENCES

Ainsworth, S. J. "Soaps and Detergents." *Chemical and Engineering News, 72* (January 24, 1994): 34.

Ainsworth, S. J. "Soaps and Detergents." *Chemical and Engineering News, 73* (January 23, 1995): 30.

Ainsworth, S. J. "Soaps and Detergents." *Chemical and Engineering News, 74* (January 22, 1996): 32.

Davidson, A., and Milwidsky, B. M. *Synthetic Detergents,* 6th ed. New York: Wiley, 1978.

Greek, B. F. "Detergent Components Become Increasingly Diverse, Complex." *Chemical and Engineering News, 66* (January 25, 1988): 21.

Greek, B. F., and Layman, P. L. "Higher Costs Spur New Detergent Formulations." *Chemical and Engineering News, 67* (January 23, 1989): 29.

"LAS Detergents End Stream Foam." *Chemical and Engineering News, 45* (1967): 29.

Levey, M. "The Early History of Detergent Substances." *Journal of Chemical Education, 31* (1954): 521.

Meloan, C. E. "Detergents—Soaps and Syndets." *Chemistry, 49* (September 1976): 6.

Rosen, M. J. "Geminis: A New Generation of Surfactants." *Chemtech, 23* (March 1993): 30.

Rosen, M. J. *Surfactants and Interfacial Phenomena,* 2nd ed. New York: Wiley, 1989.

Rosen, M. J. "Surfactants: Designing Structure for Performance." *Chemtech, 15* (May 1985): 292.

Snell, F. D. "Soap and Glycerol." *Journal of Chemical Education, 19* (1942): 172.

Snell, F. D., and Snell, C. T. "Syndets and Surfactants." *Journal of Chemical Education, 35* (1958): 271.

Stinson, S. C. "Consumer Preferences Spur Innovation in Detergents." *Chemical and Engineering News, 65* (January 26, 1987): 21.

Thayer, A. M. "Soaps and Detergents." *Chemical and Engineering News, 71* (January 25, 1993): 26.

EXPERIMENT 14

Preparation of Soap

Hydrolysis of a fat (ester)
Filtration

In this experiment we prepare soap from animal fat (lard). Animal fats and vegetable oils are esters of carboxylic acids; they have a high molecular weight and contain the alcohol glycerol. Chemically, these fats and oils are called **triglycerides.** The principal acids in animal fats and vegetable oils can be prepared from the natural triglycerides by alkaline hydrolysis (saponification). See the preceding essay for a complete discussion of soaps and detergents.

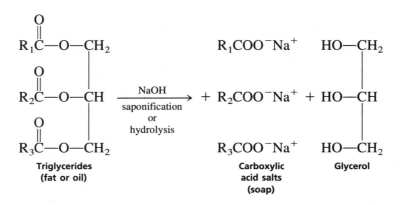

Triglycerides (fat or oil) → Carboxylic acid salts (soap) + Glycerol

Required Reading

Review: Techniques 2, 4

New: Essay Soaps and Detergents

Special Instructions

This experiment is short and can easily be scheduled with another experiment. Avoid contact with sodium hydroxide. It is very caustic.

Waste Disposal

Dispose of all filtrates obtained in this experiment in the waste container provided for aqueous solutions.

Procedure

Reaction Mixture. Prepare a solution of 2.5 g of sodium hydroxide dissolved in a mixture of 10 mL of water and 10 mL of 95% ethanol. Place 5 g of shortening, cooking oil, fat, or lard (a commercial solid shortening works best) in a 250-mL beaker and add the basic solution to it.

> **Caution:** This solution is caustic. Avoid skin contact.

Heat this mixture in a boiling water bath. The boiling water bath consists of a larger beaker that is partly filled with boiling water and heated on a hotplate. Heat the mixture for at least 45 minutes in the boiling water bath. Prepare 20 mL of a 50:50 solution of ethanol-water and add it in small portions to the mixture over the 45-minute period. Stir the mixture occasionally with a stirring rod.

Isolation of Soap. Prepare a solution of 25 g of sodium chloride in 75 mL of water in a 400-mL beaker. If the solution must be heated to dissolve the salt, it should be cooled before proceeding. Pour the hot mixture into the cooled salt solution. Stir the mixture thoroughly for several minutes and then cool it to room temperature in an ice bath.

Collect the precipitated soap by vacuum filtration, using a Büchner funnel equipped with fast filter paper (Technique 4, Sections 4.2 and 4.3, pp. 639–642). Rinse the soap in the funnel with portions of ice cold water. Continue to draw air through the soap to dry the product partially. Allow the soap to dry until the next laboratory period. Weigh the soap.

If you are preparing the detergent in Experiment 15, save a small sample of the soap for the short comparison tests. Submit the remainder of the soap to your laboratory instructor in a labeled vial.

QUESTIONS

1. Why should the potassium salts of fatty acids yield soft soaps?
2. Why is the soap derived from coconut oil so soluble?
3. Why does adding a salt solution cause soap to precipitate?
4. Why do you suppose a mixture of ethanol and water instead of simply water itself is used for saponification?
5. Sodium acetate and sodium propionate are poor soaps. Why?

E X P E R I M E N T   1 5

Preparation of a Detergent

Preparation of a sulfonate ester
Properties of soaps and detergents

In this experiment, you will prepare the detergent sodium lauryl sulfate. A detergent is usually defined as a synthetic cleaning agent, whereas a soap is derived from a natural source—a fat or an oil.

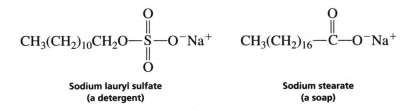

Sodium lauryl sulfate
(a detergent)

Sodium stearate
(a soap)

The differences between the two basic types of cleaning agents are discussed in the essay "Soaps and Detergents," which precedes Experiment 14. Following the preparation of sodium lauryl sulfate, you will compare the properties of soap with the properties of the prepared detergent.

In the first step of the synthesis, lauryl alcohol is allowed to react with chlorosulfonic acid to give the lauryl ester of sulfuric acid. In the second step, aqueous sodium carbonate is added to produce the sodium salt (detergent).

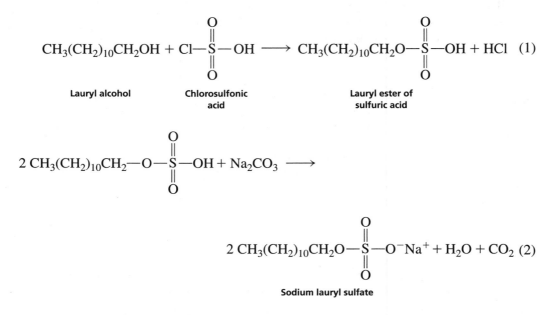

The aqueous mixture is saturated with solid sodium carbonate and extracted with 1-butanol. Sodium carbonate must be added to give phase separation; otherwise, 1-butanol would be soluble in water. The sodium salt (detergent) is more soluble in 1-butanol than in the aqueous layer because of the long hydrocarbon chain, which gives the salt considerable organic (nonpolar) character.

Required Reading

Review: Techniques 2, 3, 7

New: Essay   Soaps and Detergents

Special Instructions

Chlorosulfonic acid must be handled with care because it is a corrosive liquid and reacts violently with water. Be certain to use dry glassware.

Waste Disposal

You may dispose of the aqueous layers by placing them into the aqueous waste bottle.

Notes to the Instructor

The 1-dodecanol (lauryl alcohol) is best handled as a liquid. If necessary, melt the alcohol (mp 24–27°C) and pour the liquid into a small container. Keep the alcohol in the liquid state by placing the container on a hotplate. In this way, the alcohol will be available to the class as a liquid.

Procedure

PART A.   SODIUM LAURYL SULFATE

Preparation of Chlorosulfonic Acid/Acetic Acid Solution.   Transfer 1.0 mL of concentrated (glacial) acetic acid into a *dry* 25-mL Erlenmeyer flask. Cork the flask and cool it in an ice bath for about 5 minutes. In a hood, remove 0.350 mL of chlorosulfonic acid ($d = 1.77$ g/mL) using a carefully calibrated pipet provided by the instructor. Add the chlorosulfonic acid *dropwise* to the flask containing the acetic acid in the ice bath. (Use safety glasses!)

> **Caution:** Use chlorosulfonic acid with extreme care. Avoid getting water or ice in the flask. Chlorosulfonic acid reacts violently with water to form hydrochloric acid. Transfer the material directly into your flask without dripping the liquid. Chlorosulfonic acid is an extremely strong acid similar to concentrated sulfuric acid. It will cause immediate burns on the skin.

Reaction of Chlorosulfonic Acid with 1-Dodecanol.   In a hood, remove the flask from the ice bath and add *slowly* 1.2 mL 1-dodecanol (lauryl alcohol, $d = 0.831$ g/mL) into the flask. Let this mixture sit at room temperature for 15 minutes, with occasional swirling of the flask. After this time, *carefully add dropwise* with a Pasteur pipet 5 mL of ice cold water to the flask over a period of 5 minutes. Swirl the flask occasionally while adding the water.

> **Caution:** Excess chlorosulfonic acid will react violently with water. Work in a good fume hood and add the water dropwise, at least at the beginning.

Extraction of the Detergent with 1-Butanol.   Add 3 mL of 1-butanol to the flask and swirl the mixture occasionally for 5 minutes. While stirring with a stirring rod, add slowly 1.5 g of sodium carbonate (anhydrous) to neutralize the acids and to aid in the separation of layers. The sodium carbonate will dissolve. Transfer the mixture to a screw-cap centrifuge tube (Technique 7, Section 7.7, p. 694). Gently invert the centrifuge tube over several minutes so that the 1-butanol will extract the detergent from the aqueous layer. Avoid vigorous shaking so as to avoid a difficult emulsion (Technique 7, Section 7.9, p. 699). Allow the layers to separate for 5–10 minutes, or until a complete separation has been achieved (you may need to add some water to break the emulsion). The 1-butanol layer will be on top. Remove the lower aqueous layer from the centrifuge tube and transfer it to another centrifuge tube (save it). Save the remaining organic layer in the centrifuge tube since it contains the detergent product.

Reextract the aqueous layer. To do this, add 3 mL of 1-butanol to the aqueous layer, cap the centrifuge tube, and shake it gently. Allow the tube to stand for about 10 minutes or until a complete separation has been achieved. Remove the lower aqueous layer with a Pasteur pipet and discard it.

Combine the *two* 1-butanol organic layers in a *dry* centrifuge tube. Allow these combined layers to stand for a few minutes to see if any further separation of layers occurs. If some further separation is observed, remove the lower aqueous layer and discard it. Otherwise, transfer the 1-butanol extracts into a pre-weighed beaker.

Evaporation of 1-Butanol. At the option of your instructor, either store the 1-butanol solution of your detergent in your locker or place it in a hood until the next laboratory period. In either case, do not cover the beaker. During this time, the 1-butanol should evaporate to yield the detergent. If the 1-butanol has not evaporated after this time, place the beaker in an 80°C water bath and direct a gentle stream of air into the beaker until a solid is obtained. If an odor of 1-butanol still remains, place the beaker in an oven maintained at about 130°C until the solid is thoroughly dry and odor-free.

Use your spatula to break up the solid. Weigh the product and calculate the percentage yield (*MW* = 288.4). If the detergent is not totally free of 1-butanol, the apparent yield may exceed 100%. If necessary, continue to dry the sample. After doing the tests that follow, submit the remaining detergent to the instructor in a labeled vial.

PART B. TESTS ON SOAPS AND DETERGENTS

Soap. Pour 3 mL of soap solution[1] into a 10-mL graduated cylinder. Place your thumb over the opening of the cylinder and shake it vigorously for about 15 seconds. Allow the solution to stand for 30 seconds and observe the level of the foam. Add 2 drops of 4% calcium chloride solution from a Pasteur pipet. Shake the mixture for 15 seconds and allow it to stand for about 30 seconds. Observe the effect

---

[1]A large batch of soap solution should be prepared by the instructor, as follows: Add one bar of Ivory soap to 1 L of distilled water. Stir the solution occasionally and allow the mixture to stand overnight. Remove the remainder of the bar. The mixture can be used directly. Alternatively, a 0.05-g sample of soap prepared in Experiment 14 can be added to 10 mL of distilled water.

of the calcium chloride on the foam. Do you observe anything else? Add 0.3 g of trisodium phosphate and shake the mixture again for about 15 seconds. Allow the solution to stand for 30 seconds. What do you observe? Explain these tests in your laboratory report.

Detergent. Place a 0.05-g sample of your prepared detergent in a 10-mL graduated cylinder and add 3 mL of distilled water. Hold your thumb over the opening and shake the graduated cylinder vigorously for 30 seconds. Allow the solution to stand for about 30 seconds and observe the level of the foam. Add 2 drops of 4% calcium chloride solution. Shake the mixture for 15 seconds and allow it to stand for 30 seconds. What do you observe? Explain the results of these tests in your laboratory report.

QUESTIONS

1. Draw a mechanism for the reaction of lauryl alcohol with chlorosulfornic acid.
2. Why do you suppose sodium carbonate, instead of some other base, is used for neutralization?
3. Propose a model to explain how a cationic detergent works. A cationic detergent has its polar end positively charged.
4. Sodium methyl sulfate $CH_3OSO_2{}^-Na^+$ is a poor detergent. Why?
5. Sodium lauryl sulfate can be prepared by replacing chlorosulfonic acid with another reagent. What could be used? Show the equations.
6. Suggest a method for synthesizing the linear alkyl sulfonate detergent shown on page 132, starting with lauryl alcohol, benzene, and any needed inorganic compounds.

E S S A Y

Pheromones: Insect Attractants and Repellents

It is difficult for humans, who are accustomed to heavy reliance on visual and verbal forms of communication, to imagine that there are forms of life that depend primarily on the release and perception of **odors** to communicate with one another. Among insects, however, this is perhaps the chief form of communication. Many species of insects have developed a virtual "language" based on the exchange of odors. These insects have well-developed scent glands, often of several different types, which have as their sole purpose the synthesis and release of chemical substances. When these chemical substances, known as **pheromones,** are secreted by insects and detected by other members of the same species, they induce a specific and characteristic response. Pheromones are usually of two distinct types: releaser pheromones and primer pheromones. **Releaser pheromones** produce an immediate **behavioral** response in the recipient insect; **primer pheromones** trigger a series of **physiological** changes in the recipient. Some pheromones, however, combine both releaser and primer effects.

SEX ATTRACTANTS

Among the most important types of releaser pheromones are the sex attractants. **Sex attractants** are pheromones secreted by either the female or, less commonly, the male of the species to attract the opposite member for the purpose of mating. In large concentrations, sex pheromones also induce a physiological response in the recipient (for example, the changes necessary to the mating act), and thus have a primer effect and so are misnamed.

Anyone who has owned a female cat or dog knows that sex pheromones are not limited to insects. Female cats or dogs widely advertise, by odor, their sexual availability when they are "in heat." This type of pheromone is not uncommon to mammals. Some persons even believe that there are human pheromones responsible for attracting certain sensitive males and females to one another. This idea is, of course, responsible for many of the perfumes and fragrances now widely available. Whether or not the idea is correct cannot yet be established, but there are proven sexual differences in the ability of humans to smell certain substances. For instance, Exaltolide, a synthetic lactone of 14-hydroxy-tetradecanoic acid, can be perceived only by females or by males after they have been injected with an estrogen. Exaltolide is very similar in overall structure to civetone (civet cat) and muskone (musk deer), which are two naturally occurring compounds believed to be mammalian sex pheromones. Whether human males emit pheromones has never been established. Curiously, Exaltolide is used in fragrances intended for female as well as male use! But while the odor may lead a woman to believe that she smells pleasant, it cannot possibly have any effect on the male. The "musk oils," civetone and muskone, have also been long used in expensive perfumes.

One of the first identified insect attractants belongs to the gypsy moth, *Lymantria dispar.* This moth is a common agricultural pest, and it was hoped that the sex attractant that females emitted could be used to lure and trap males. Such a method of insect control would be preferable to inundating large areas with DDT and would be species-specific. Nearly 50 years of work were expended in identifying the chemical substance responsible. Early in this period, researchers found that an extract from the tail sections of female gypsy moths would attract males, even from a great distance. In experiments with the isolated gypsy moth pheromone, it was found that the male gypsy moth has an almost unbelievable ability to detect extremely small amounts of the substance. He can detect it in concentrations lower than a few hundred *molecules* per cubic centimeter (about 10^{-19}–10^{-20} g/cc)! When a male moth encounters a small concentration of pheromone, he immediately turns into the wind and flies upward in search of higher concentrations and the female. In only a mild breeze, a continuously emitting female can activate a space 300 ft high, 700 ft wide, and almost 14,000 ft (nearly 3 miles) long!

In subsequent work, 20 mg of a pure chemical substance was isolated from solvent extracts of the two extreme tail segments collected from each of 500,000 female gypsy moths (about 0.1 μg/moth). This emphasizes that pheromones are effective in very minute amounts and that chemists must work with very small amounts to isolate them and prove their structures. It is not unusual to process thousands of insects in order to get even a very small sample of these substances. Very sophisticated analytical and instrumental methods, like spectroscopy, must be used to determine the structure of a pheromone.

In spite of these techniques, the original researchers assigned an incorrect structure to the gypsy moth pheromone and proposed for it the name gyplure. Because of its great

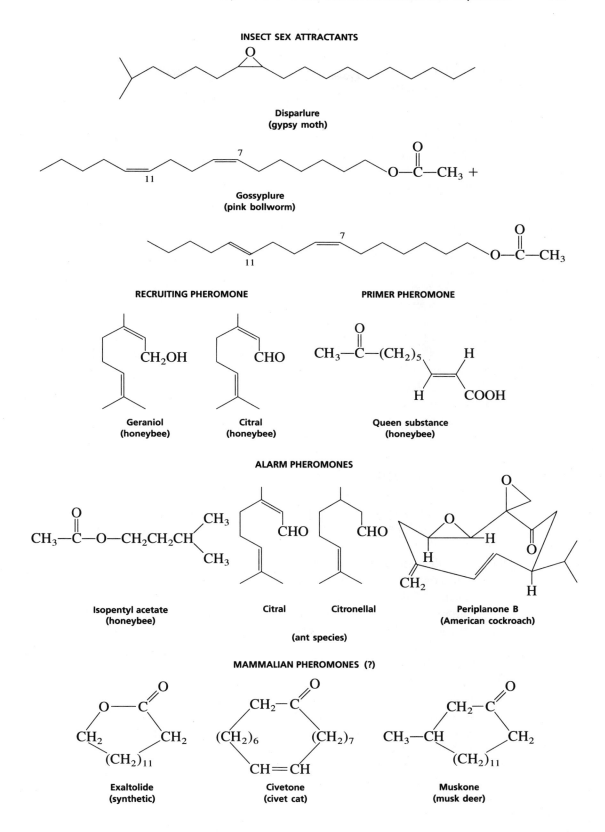

INSECT SEX ATTRACTANTS

Disparlure
(gypsy moth)

Gossyplure
(pink bollworm)

RECRUITING PHEROMONE

Geraniol
(honeybee)

Citral
(honeybee)

PRIMER PHEROMONE

Queen substance
(honeybee)

ALARM PHEROMONES

Isopentyl acetate
(honeybee)

Citral   Citronellal

(ant species)

Periplanone B
(American cockroach)

MAMMALIAN PHEROMONES (?)

Exaltolide
(synthetic)

Civetone
(civet cat)

Muskone
(musk deer)

promise as a method of insect control, gyplure was soon synthesized. The synthetic material turned out to be totally inactive. After some controversy about why the synthetic material was incapable of luring male gypsy moths (see the references for the complete story), it was finally shown that the proposed structure for the pheromone (that is, the gyplure structure) was incorrect. The actual pheromone was found to be *cis*-7,8-epoxy-2-methyloctadecane, also named (7R,8S)-epoxy-2-methyloctadecane. This material was soon synthesized, found to be active, and given the name disparlure. In recent years, disparlure traps have been found to be a convenient and economical method for controlling the gypsy moth.

A similar story of mistaken identity can be related for the structure of the pheromone of the pink bollworm, *Pectinophora gossypiella.* The originally proposed structure was called propylure. Synthetic propylure turned out to be inactive. Subsequently the pheromone was shown to be a mixture of two isomers of 7,11-hexadecadien-1-yl acetate, the *cis,cis* (7Z,11Z) isomer and the *cis,trans* (7Z,11E) isomer. It turned out to be quite easy to synthesize a 1:1 mixture of these two isomers, and the 1:1 mixture was named gossyplure. Curiously, adding as little as 10% of either of the other two possible isomers, *trans,cis* (7E,11Z) or *trans,trans* (7E,11E), to the 1:1 mixture greatly diminishes its activity, apparently masking it. Stereoisomerism can be important! The details of the gossyplure story can also be found in the references.

Both these stories have been partly repeated here to point out the difficulties of research on pheromones. The usual method is to propose a structure determined by work on *very tiny* amounts of the natural material. The margin for error is great. Such proposals are usually not considered "proved" until synthetic material is shown to be as biologically effective as the natural pheromone.

OTHER PHEROMONES

The most important example of a **primer pheromone** is found in honeybees. A bee colony consists of one queen bee, several hundred male drones, and thousands of worker bees, or undeveloped females. It has recently been found that the queen, the only female that has achieved full development and reproductive capacity, secretes a primer pheromone called the **queen substance.** The worker females, while tending the queen bee, continuously ingest quantities of the queen substance. This pheromone, which is a mixture of compounds, prevents the workers from rearing any competitive queens and prevents the development of ovaries in all other females in the hive. The substance is also active as a sex attractant; it attracts drones to the queen during her "nuptial flight." The major component of queen substance is shown in the figure.

Honeybees also produce several other important types of pheromones. It has long been known that bees will swarm after an intruder. It has also been known that isopentyl acetate induces a similar behavior in bees. Isopentyl acetate (Experiment 8) is an **alarm pheromone.** When an angry worker bee stings an intruder, she discharges, along with the sting venom, a mixture of pheromones that incites the other bees to swarm upon and attack the intruder. Isopentyl acetate is an important component of the alarm pheromone mixture. Alarm pheromones have also been identified in many other insects. In insects less aggressive than bees or ants, the alarm pheromone may take the form of a **repellent,** which induces the insects to go into hiding or leave the immediate vicinity.

Honeybees also release **recruiting** or **trail pheromones.** These pheromones attract others to a source of food. Honeybees secrete recruiting pheromones when they locate flowers in which large amounts of sugar syrup are available. Although the recruiting pheromone is a complex mixture, both geraniol and citral have been identified as components. In a similar fashion, when ants locate a source of food, they drag their tails along the ground on their way back to the nest, continuously secreting a trail pheromone. Other ants follow the trail to the source of food.

In some species of insects, **recognition pheromones** have been identified. In carpenter ants, a caste-specific secretion has been found in the mandibular glands of the males of five different species. These secretions have several functions, one of which is to allow members of the same species to recognize one another. Insects not having the correct recognition odor are immediately attacked and expelled from the nest. In one species of carpenter ant, the recognition pheromone has been shown to have methyl anthranilate as an important component.

We do not yet know all the types of pheromones that any given species of insect may use, but it seems that as few as 10 or 12 pheromones could constitute a "language" that could adequately regulate the entire life cycle of a colony of social insects.

INSECT REPELLENTS

Currently, the most widely used **insect repellent** is the synthetic substance *N,N*-diethyl-*m*-toluamide (Experiment 16), also called Deet. It is effective against fleas, mosquitoes, chiggers, ticks, deerflies, sandflies, and biting gnats. A specific repellent is known for each of these types of insects, but none has the wide spectrum of activity that this repellent has. Exactly why these substances repel insects is not yet fully understood. The most extensive investigations have been carried out on the mosquito.

Originally, many investigators thought that repellents might simply be compounds that provided unpleasant or distasteful odors to a wide variety of insects. Others thought that they might be alarm pheromones for the species affected, or that they might be the alarm pheromones of a hostile species. Early research with the mosquito indicates that at least for several varieties of mosquitoes, none of these is the correct answer.

Mosquitoes seem to have hairs on their antennae that are receptors enabling them to find a warm-blooded host. These receptors detect the convection currents arising from a warm and moist living animal. When a mosquito encounters a warm and moist convection current, it moves steadily forward. If it passes out of the current into dry air, it turns until it finds the current again. Eventually it finds the host and lands. Repellents cause a mosquito to turn in flight and become confused. Even if it should land, it becomes confused and flies away again.

Researchers have found that the repellent prevents the moisture receptors of the mosquito from responding normally to the raised humidity of the subject. At least two sensors are involved, one responsive to carbon dioxide and the other responsive to water vapor. The carbon dioxide sensor is activated by the repellent, but if exposure to the chemical continues, adaptation occurs, and the sensor returns to its usual low output of signal. The moisture sensor, on the other hand, simply seems to be deadened, or turned off, by the repellent. Therefore, mosquitoes have great difficulty in finding and interpreting a host when they are in an environment saturated with repellent. They fly right through warm and humid convection currents as if the currents did not exist. Only time will tell if other biting insects respond likewise.

REFERENCES

Agosta, W. C. "Using Chemicals to Communicate." *Journal of Chemical Education, 71* (March 1994): 242.

Batra, S. W. T. "Polyester-Making Bees and Other Innovative Insect Chemists." *Journal of Chemical Education, 62* (February 1985): 121.

Katzenellenbogen, J. A. "Insect Pheromone Synthesis: New Methodology." *Science, 194* (October 8, 1976): 139.

Leonhardt, B. A. "Pheromones." *ChemTech, 15* (June 1985): 368.

Prestwick, G. D. "The Chemical Defenses of Termites." *Scientific American, 249* (August 1983): 78.

Silverstein, R. M. "Pheromones: Background and Potential Use for Insect Control." *Science, 213* (September 18, 1981): 1326.

Stine, W. R. "Pheromones: Chemical Communication by Insects." *Journal of Chemical Education, 63* (July 1986): 603.

Villemin, D. "Olefin Oxidation: A Synthesis of Queen Bee Pheromone." *Chemistry and Industry* (January 20, 1986): 69.

Wilson, E. O. "Pheromones." *Scientific American, 208* (May 1963): 100.

Winston, M. L., and Slessor, K. N. "The Essence of Royalty: Honey Bee Queen Pheromone." *American Scientist, 80* (July–August 1992): 374.

Wood, W. F. "Chemical Ecology: Chemical Communication in Nature." *Journal of Chemical Education, 60* (July 1983): 531.

Wright, R. H. "Why Mosquito Repellents Repel." *Scientific American, 233* (July 1975): 105.

Yu, H., Becker, H., and Mangold, H. K. "Preparation of Some Pheromone Bouquets." *Chemistry and Industry* (January 16, 1989): 39.

Gypsy Moth

Beroza, M., and Knipling, E. F. "Gypsy Moth Control with the Sex Attractant Pheromone." *Science, 177* (1972): 19.

Bierl, B. A., Beroza, M., and Collier, C. W. "Potent Sex Attractant of the Gypsy Moth: Its Isolation, Identification, and Synthesis." *Science, 170* (1970): 87.

Pink Bollworm

Anderson, R. J., and Henrick, C. A. "Preparation of the Pink Bollworm Sex Pheromone Mixture, Gossyplure." *Journal of the American Chemical Society, 97* (1975): 4327.

Hummel, H. E., Gaston, L. K., Shorey, H. H., Kaae, R. S., Byrne, K. J., and Silverstein, R. M. "Clarification of the Chemical Status of the Pink Bollworm Sex Pheromone." *Science, 181* (1973): 873.

American Cockroach

Adams, M. A., Nakanishi, K., Still, W. C., Arnold, E. V., Clardy, J., and Persoon, C. J. "Sex Pheromone of the American Cockroach: Absolute Configuration of Periplanone-B." *Journal of the American Chemical Society, 101* (1979): 2495.

Still, W. C. "(±)-Periplanone-B: Total Synthesis and Structure of the Sex Excitant Pheromone of the American Cockroach." *Journal of the American Chemical Society,* 101 (1979): 2493.

Stinson, S. C. "Scientists Synthesize Roach Sex Excitant." *Chemical and Engineering News, 57* (April 30, 1979): 24.

Spider

Schulz, S., and Toft, S. "Identification of a Sex Pheromone from a Spider." *Science, 260* (June 11, 1993): 1635.

Silkworm

Emsley, J. "Sex and the Discerning Silkworm." *New Scientist, 135* (July 11, 1992): 18.

Aphids

Coghlan, A. "Aphids Fall for Siren Scent of Pheromones." *New Scientist, 127* (July 21, 1990): 32.

Snakes

Mason, R. T., Fales, H. M., Jones, T. H., Pannell, L. K., Chinn, J. W., and Crews, D. "Sex Pheromones in Snakes." *Science, 245* (July 21, 1989): 290.

Oriental Fruit Moth

Mithran, S., and Mamdapur, V. R. "A Facile Synthesis of the Oriental Fruit Moth Sex Pheromone." *Chemistry and Industry* (October 20, 1986): 711.

EXPERIMENT 16

N,N-Diethyl-*m*-Toluamide: The Insect Repellent "OFF"

Preparation of an amide
Extraction
Column chromatography

In this experiment, you will synthesize the active ingredient of the insect repellent "OFF," *N,N*-diethyl-*m*-toluamide. This substance belongs to the class of compounds called **amides.** Amides have the generalized structure

The amide to be prepared in this experiment is a disubstituted amide. That is, two of the hydrogens on the amide —NH_2 group have been replaced with ethyl groups. Amides cannot be prepared directly by mixing a carboxylic acid with an amine. If an acid and an amine are mixed, an acid–base reaction occurs, giving the conjugate base of the acid, which will not react further while in solution:

$$RCOOH + R_2NH \longrightarrow [RCOO^-R_2NH_2{}^+]$$

However, if the amine salt is isolated as a crystalline solid and strongly heated, the amide can be prepared:

$$[RCOO^-R_2NH_2{}^+] \xrightarrow{\text{heat}} RCONR_2 + H_2O$$

This is not a convenient laboratory method because of the high temperature required for this reaction.

Amides are usually prepared via the acid chloride, as in this experiment. In Step 1, *m*-toluic acid is converted to its acid chloride derivative using thionyl chloride ($SOCl_2$).

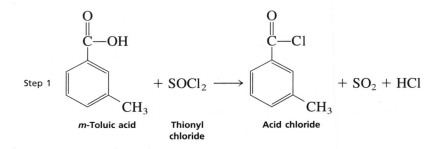

The acid chloride is not isolated or purified, and it is allowed to react directly with diethylamine in Step 2. An excess of diethylamine is used in this experiment to react with the hydrogen chloride produced in Step 2.

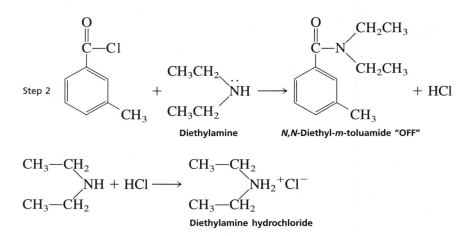

Required Reading

Review: Techniques 1, 2, 3, 4
          Technique 7   Sections 7.4, 7.5, 7.8, 7.9, 7.10
          Technique 12   Sections 12.6–12.9

New: Essay   Pheromones: Insect Attractants and Repellents

Special Instructions

All equipment used in this experiment should be dry, because thionyl chloride reacts with water to liberate HCl and SO_2. Likewise, *anhydrous* ether should be used, because water reacts with both thionyl chloride and the intermediate acid chloride.

Thionyl chloride is a noxious and corrosive chemical and should be handled with care. If it is spilled on the skin, serious burns will result. Thionyl chloride and diethylamine must be dispensed **in the hood** from bottles that should be kept tightly closed when not in use. Diethylamine is also noxious and corrosive. In addition, it is quite volatile (bp 56°C) and must be cooled in a hood prior to use.

Waste Disposal

All aqueous extracts should be poured into the waste bottle designated for aqueous waste.

Procedure

Apparatus Assembly. Assemble the apparatus shown in the figure in a good fume hood. Omit the syringe for now. The gas trap shown in the figure will trap the hydrogen chloride and sulfur dioxide gases that are evolved in the reaction (Technique 3, Section 3.9, p. 625). Moisten the cotton with a little water, but avoid getting water into the condenser. If the apparatus is set up in a good fume hood, you may not need the gas trap (see your instructor). You can save time by warming a beaker of water on your hotplate to about 90°C prior to measuring reagents (alternatively, you may use a heating mantle to avoid any possibility of water vapor coming in contact with thionyl chloride).

Preparation of the Acid Chloride. Place 1.81 g of *m*-toluic acid (3-methylbenzoic acid, *MW* = 136.1) into a dry 100-mL round-bottom flask. In a hood, transfer 2.0 mL of thionyl chloride (*MW* = 118.9, *d* = 1.64 g/mL) into a test tube with the dry graduated pipet provided with the reagent. Stopper the test tube while bringing it to your hood.

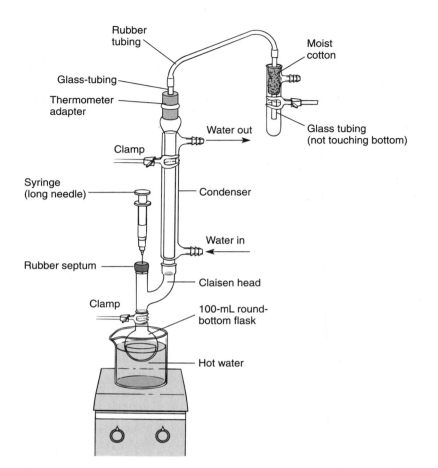

Apparatus for Experiment 16. Note: A long syringe needle is recommended.

Caution: The thionyl chloride is kept in a hood. Do not breathe the vapors of this noxious and corrosive chemical. Use dry equipment when handling this material as it reacts violently with water. Do not get it on your skin. If you are using a hot water bath, be sure to remove it from the vicinity of your apparatus so that there is no danger of pouring thionyl chloride into the water.

Remove the rubber septum and transfer the thionyl chloride into the flask with a Pasteur pipet. Add a boiling stone, reattach the septum, start the circulation of water in the reflux condenser, and heat the mixture to reflux. Boil the mixture gently for 15 minutes.

Preparation of the Amide.   Raise the apparatus from the water bath and allow the flask to cool to *room temperature.* Remove the hotplate and hot water bath, as the next part of this reaction sequence is conducted at room temperature.

After the mixture has cooled to room temperature, remove the rubber septum and pour 25.0 mL of *anhydrous* ether into the reaction flask. Reattach the septum. While holding the ring stand, swirl the mixture until a homogeneous solution is obtained. In a hood, remove 4.5 mL of ice cold diethylamine (*MW* = 73.1, *d* = 0.71 g/mL) and place it in a small Erlenmeyer flask for storage. Add 8 mL of *anhydrous* ether to the amine in the flask.

Draw up some of the diethylamine solution into a *dry* syringe and carefully insert the needle through the rubber septum of your apparatus. Add the solution of diethylamine and ether *dropwise* over a 10–15 minute period to the round-bottom flask, refilling the syringe when necessary with the diethylamine/ether solution in the storage flask. As the solution is added, a voluminous white cloud of diethylamine hydrochloride will form in the flask. Holding the ring stand, swirl the reaction mixture occasionally.

After adding the diethylamine, swirl the mixture occasionally over a 10 minute period. After this time, remove the rubber septum and pour 14 mL of a 10% aqueous sodium hydroxide solution into the flask in small portions. Swirl the contents of the flask occasionally over a 15 minute period. During this time, the sodium hydroxide converts most of the remaining acid chloride to the sodium salt of *m*-toluic acid. This salt is soluble in the aqueous layer. Diethylamine hydrochloride is also water soluble. Any remaining thionyl chloride is destroyed by water present in the aqueous sodium hydroxide. The desired amide is soluble in ether. If any solid remains, you can add water to dissolve it. It may also be helpful to add some ether.

Extraction of Product.   Remove the gas trap assembly, the condenser, and the Claisen head. Transfer all the liquid to a separatory funnel. Shake the funnel vigorously for 2–3 minutes. The shaking operation helps to complete the conversion of any remaining acid chloride to the sodium salt. Allow the layers to separate and then remove the lower aqueous layer, leaving the desired ether layer behind in the separatory funnel. Discard the aqueous layer. Add another 14-mL portion of 10% sodium hydroxide to the remaining ether layer and again shake the funnel vigorously for 2–3 minutes. Allow the layers to separate, and again remove the lower aqueous layer and discard it.

Now shake the ether layer remaining in the separatory funnel with a 14-mL portion of 10% hydrochloric acid to remove any remaining diethylamine as its hydrochloride salt. Allow the layers to separate and then drain off the lower aqueous layer and discard it. Finally, shake the ether layer with a 14-mL portion of water and drain off the lower aqueous layer remaining after the layers have separated. Keep the upper ether layer.

Transfer the ether layer containing the amide product to a dry Erlenmeyer flask and dry the ether phase with granular anhydrous sodium sulfate. Decant the ether phase away from the drying agent into another *dry* preweighed flask. Use a small amount of additional ether to rinse the drying agent. Evaporate the ether by placing the flask into a hot water bath at about 50°C, in a good fume hood. Use a stream of air or nitrogen to aid in the evaporation process (Technique 3, Section 3.11, p. 630). A crude dark brown liquid amide will remain. Reweigh the flask to determine the amount of crude product obtained. Column chromatography is used to remove much of the dark color from the product.

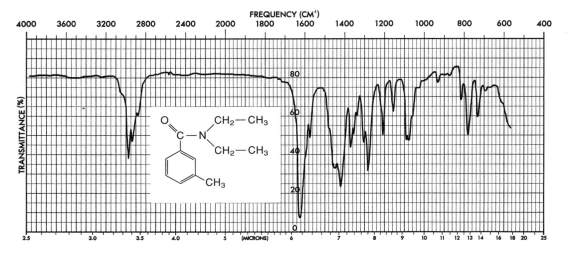

Infrared spectrum of *N,N*-diethyl-*m*-toluamide, neat.

Column Chromatography.   Preweigh a 100-mL beaker for use in collecting the material eluted from the column. Prepare a column for column chromatography using a commercially available chromatography column such as that shown in Technique 12, Figure 12.8, p. 773. Alternatively, prepare a column like that shown in Figure 12.9, p. 776 using a 20-cm length of 10-mm glass tubing with a constriction at one end. Place a small piece of cotton into the column and gently push it into the constriction. Pour 7.3 g of alumina[1] into the column while gently tapping the column with a pencil or finger. Obtain about 30 mL of hexane. The hexane will be used to prepare the column, dissolve the crude product, and elute the purified product as described in the next paragraph.

Clamp the column above the preweighed beaker. Then add about 7 mL of hexane to the column and allow it to percolate through the alumina. Allow the solvent to drain until the solvent level just begins to enter the alumina. Dissolve the crude product in about 2 mL of hexane before adding it to the column. Add the solution of crude product and hexane to the top of the column using a Pasteur pipet and allow the mixture to pass onto the column. Use about 4 mL hexane to rinse the flask that contained the crude product. When the first batch of crude product has passed fully into the alumina so that the surface of the liquid just begins to enter the top of the alumina, add the hexane rinse on the column using a pipet.

When the solvent level has again reached the top of the alumina, add more hexane with a pipet to elute the product into the beaker. You should add about 14 mL of hexane, in portions, to the column to elute the product. Collect all of the liquid that passes through the column as one fraction (yellow material). Place the beaker in a warm water bath (about 50°C) and evaporate the hexane with a light stream of air or nitrogen in a hood to give the *N,N*-diethyl-*m*-toluamide as a light

---

[1]EM Science (No. AX0612-1). The particle sizes are 80–200 mesh and the material is Type F-20.

tan liquid. If necessary, use a few drops of hexane to rinse the product from the side of the beaker into the bottom. Evaporate this solvent.

   Analysis of Product.   Reweigh the beaker to determine the weight of product. Calculate the percentage yield (*MW* = 193.1) of the crude product as well as the column purified product. Determine the infrared spectrum of your product. Submit the remaining sample to the instructor. The infrared spectrum can be compared to the one reproduced on page 150.

REFERENCE

Wang, B. J-S. "An Interesting and Successful Organic Experiment." *Journal of Chemical Education, 51* (October 1974): 631. (The synthesis of *N,N*-diethyl-*m*-toluamide.)

QUESTIONS

1. Write an equation that describes the reaction of thionyl chloride with water.
2. What reaction would take place if the acid chloride of *m*-toluic acid were mixed with water?
3. Why is the reaction mixture extracted with 10% aqueous sodium hydroxide? Write an equation.
4. Write a mechanism for each step in the preparation of *N,N*-diethyl-*m*-toluamide.
5. Interpret each of the principal peaks in the infrared spectrum of *N,N*-diethyl-*m*-toluamide.
6. A student determined the infrared spectrum of the product and found an absorption at 1785 cm^{-1}. The rest of the spectrum resembled the one given in this experiment. Assign this peak and provide an explanation for this unexpected result.

E S S A Y

Petroleum and Fossil Fuels

   Crude petroleum is a liquid that consists of hydrocarbons as well as some related sulfur, oxygen, and nitrogen compounds. Other elements, including metals, may be present in trace amounts. Crude oil is formed by the decay of marine animal and plant organisms that lived millions of years ago. Over many millions of years, under the influence of temperature, pressure, catalysts, radioactivity, and bacteria, the decayed matter was converted into what we now know as crude oil. The crude oil is trapped in pools beneath the ground by various geological formations.

   Most crude oils have a specific gravity between 0.78 and 1.00 g/mL. As a liquid, crude oil may be as thick and black as melted tar or as thin and colorless as water. Its characteristics depend on the particular oil field from which it comes. Pennsylvania crude oils are high in straight-chain alkane compounds (called **paraffins** in the petroleum industry); those crude oils are therefore useful in the manufacture of lubricating oils. Oil fields in California and Texas produce crude oil with a higher percentage of cycloalkanes (also called **naphthenes** by the petroleum industry). Some Middle East fields produce crude oil containing up to 90% cyclic hydrocarbons. Petroleum contains molecules in which the number of carbons ranges from 1 to 60.

TABLE ONE   Fractions Obtained from the Distillation of Crude Oil

Petroleum Fraction	Composition	Commercial Use
Natural gas	C_1 to C_4	Fuel for heating
Gasoline	C_5 to C_{10}	Motor fuel
Kerosene	C_{11} to C_{12}	Jet fuel and heating
Light gas oil	C_{13} to C_{17}	Furnaces, diesel engines
Heavy gas oil	C_{18} to C_{25}	Motor oil, paraffin wax, petroleum jelly
Residuum	C_{26} to C_{60}	Asphalt, residual oils, waxes

When petroleum is refined to convert it into a variety of usable products, it is initially subjected to a fractional distillation. Table One lists the various fractions obtained from fractional distillation. Each of these fractions has its own particular uses. Each fraction may be subjected to further purification, depending on the desired application.

The gasoline fraction obtained directly from the distillation of crude oil is called **straight-run gasoline.** An average barrel of crude oil will yield about 19% straight-run gasoline. This yield presents two immediate problems. First, there is not enough gasoline contained in crude oil to satisfy current needs for fuel to power automobile engines. Second, the straight-run gasoline obtained from crude oil is a poor fuel for modern engines. It must be "refined" at a chemical refinery.

The initial problem of the small quantity of gasoline available from crude oil can be solved by **cracking** and **polymerization.** Cracking is a refinery process by which large hydrocarbon molecules are broken down into smaller molecules. Heat and pressure are required for cracking, and a catalyst must be used. Silica–alumina and silica–magnesia are among the most effective cracking catalysts. A mixture of saturated and unsaturated hydrocarbons is produced in the cracking process. If gaseous hydrogen is also present during the cracking, only saturated hydrocarbons are produced. The hydrocarbon mixtures produced by these cracking processes tend to have a fairly high proportion of branched-chain isomers. These branched isomers improve the quality of the fuel.

$$C_{16}H_{34} + H_2 \xrightarrow[\text{heat}]{\text{catalyst}} 2\ C_8H_{18} \qquad \text{Cracking}$$

In the polymerization process, also carried out at a refinery, small molecules of alkenes are caused to react with one another to form larger molecules, which are also alkenes.

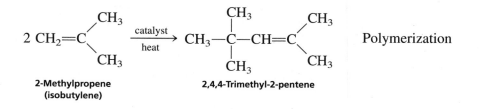

2-Methylpropene
(isobutylene)   2,4,4-Trimethyl-2-pentene

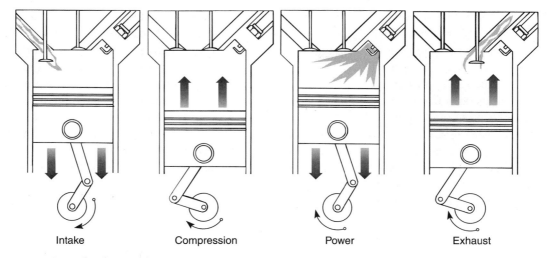

Intake   Compression   Power   Exhaust

Operation of a four-cycle engine.

The newly formed alkenes may be hydrogenated to form alkanes. The reaction sequence shown here is a very common and important one in petroleum refining because the product, 2,2,4-trimethylpentane (or "isooctane"), forms the basis for determining the quality of gasoline. By these refining methods, the percentage of gasoline that can be obtained from a barrel of crude oil may rise to as much as 45 or 50%.

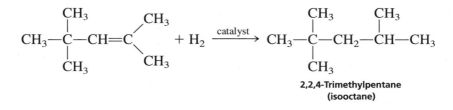

2,2,4-Trimethylpentane
(isooctane)

The internal combustion engine, as it is found in most automobiles, operates in four cycles or **strokes.** They are illustrated in the figure. The power stroke is of greatest interest from the chemical point of view since combustion occurs during this stroke.

When the air–fuel mixture is ignited, it does not explode. Rather, it burns at a controlled, uniform rate. The gases closest to the spark are ignited first; then they in turn ignite the molecules farther from the spark; and so on. The combustion proceeds in a wave of flame or a **flame front,** which starts at the spark plug and proceeds uniformly outward from that point until all the gases in the cylinder have been ignited. Because a certain time is required for this burning, the initial spark is timed to ignite just before the piston has reached the top of its travel. In this way, the piston will be at the very top of its travel at the precise instant that the flame front and the increased pressure that accompanies it reach the piston. The result is a smoothly applied force to the piston, driving it downward.

If heat and compression should cause some of the air–fuel mixture to ignite before the flame front has reached it or to burn faster than expected, the timing of the combustion sequence is disturbed. The flame front arrives at the piston before the piston has reached the very top of its travel. When the combustion is not perfectly coordinated with the motion of the piston, we observe **knocking** or **detonation** (sometimes called "pinging"). The transfer of power to the piston under these conditions is much less effective than in normal combustion. The wasted energy is merely transferred to the engine block in the form of additional heat. The opposing forces that occur in knocking may eventually damage the engine.

It has been found that the tendency of a fuel to knock is a function of the structures of the molecules composing the fuel. Normal hydrocarbons, those with straight carbon chains, have a greater tendency to lead to knocking than do those alkanes with highly branched chains. The quality of a gasoline, then, is a measure of its resistance to knocking, and this quality is improved by increasing the proportion of branched-chain alkanes in the mixture. Such chemical refining processes as **reforming** and **isomerization** are used to convert normal alkanes to branched-chain alkanes, thus improving the knock-resistance of gasoline.

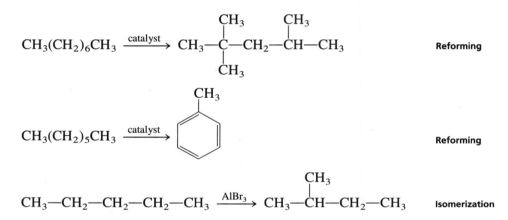

None of these processes converts all the normal hydrocarbons into branched-chain isomers; consequently, additives are also put into gasoline to improve the knock-resistance of the fuel. Aromatic hydrocarbons can be considered additives that are effective in improving the knock-resistance of gasoline, and they are used extensively in unleaded as well as leaded gasolines. The most common additive used to reduce knocking has been **tetraethyllead.** Gasoline that contains tetraethyllead is called **leaded** gasoline, while gasoline produced without tetraethyllead is called **unleaded** gasoline. In recent years, because of concern over the possible health hazard associated with emission of lead into the atmosphere and because the presence of lead will inactivate the catalytic converters found on new cars, the Environmental Protection Agency has issued regulations that are leading to the gradual elimination of tetraethyllead in gasoline. As a consequence, oil companies are testing other additives that will improve the antiknock properties of gasoline without producing harmful emissions.

$$CH_3-CH_2-\underset{\underset{CH_2-CH_3}{|}}{\overset{\overset{CH_2-CH_3}{|}}{Pb}}-CH_2-CH_3$$

Tetraethyllead

New cars are designed to operate on unleaded gasoline, which contains no lead compounds. The quality of the gasoline is maintained by adding increased quantities of hydrocarbons that have very high antiknock properties themselves. Typical are the aromatic hydrocarbons, including benzene, toluene, and xylene.

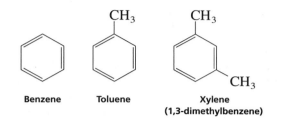

Benzene   **Toluene**   **Xylene**
                            (1,3-dimethylbenzene)

More expensive refining processes such as **hydrocracking** (cracking in the presence of hydrogen gas) and **reforming** produce mixtures of hydrocarbons that are more knock-resistant than typical gasoline components. Adding the products of hydrocracking and reforming to gasoline improves its performance. Increasing the proportion of aromatic hydrocarbons brings with it certain hazards, however. These substances are toxic, and benzene is considered a serious carcinogenic hazard. The risk that illness will be contracted by workers in refineries, and especially by persons who work in service stations, is increased.

Considerable research is also being directed toward development of non-hydrocarbon compounds that can improve the quality of unleaded gasoline. To this end, compounds such as methyl *tert*-butyl ether (MTBE), ethanol, and other oxygenates (oxygen-containing compounds) are added to improve the antiknock properties of fuels. In particular, ethanol is attractive because it is formed by fermentation of living material, a renewable resource (see essay, "Ethanol and Fermentation Chemistry," p. 85). Thus, ethanol not only would improve the antiknock properties of gasolines but it would also potentially help the country to reduce its dependence on imported petroleum. Substituting ethanol for hydrocarbons in petroleum would have the effect of increasing the "yield" of fuel produced from a barrel of crude oil. Like many stories that are too good to be true, it is not clear that the energy needed to produce the ethanol by fermentation and distillation is significantly smaller than the amount of energy that is produced when the ethanol is burned in an engine!

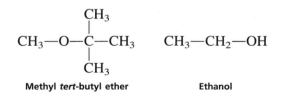

Methyl *tert*-butyl ether   **Ethanol**

In an effort to improve air quality in urban areas, the Clean Air Act mandated the addition of oxygen-containing compounds in many urban areas during the winter months (November to February). These compounds are expected to reduce carbon monoxide emissions produced when the gasoline burns in cold engines by helping to oxidize carbon monoxide to carbon dioxide. Refineries add "oxygenates," such as ethanol or methyl *tert*-

butyl ether, to the gasoline sold in the carbon monoxide-containment areas. By law, gasoline must contain at least 2.7% oxygen by weight, and the areas must use it for a minimum of the 4 winter months. The 2.7% oxygen content corresponds to a 15% by volume content of MTBE in gasoline.

Although methyl *tert*-butyl ether is still the most widely used oxygenate additive, the use of ethanol is becoming more common. There are several reasons for an increasing preference for ethanol. First, ethanol is cheaper than MTBE because of special tax breaks and subsidies that have been granted to producers of ethanol formed by fermentation. Second, there has been some concern that MTBE may cause some health problems. It is known, for example, that people notice the odor of gasoline more easily when MTBE is present in the fuel. The volatility problem may be partly solved by the replacement of methyl *tert*-butyl ether with ethyl *tert*-butyl ether. The ethyl ether is less volatile than the methyl ether derivative. There is no current hard evidence, however, to suggest that MTBE is a health hazard. It has been used as an antiknock additive since 1979 without major concerns.

The use of ethanol and methyl *tert*-butyl ether in summer months is very controversial. There is evidence that the inclusion of oxygenates increases the volatility of fuels. This would have the effect of increasing the emissions of volatile organic compounds (VOCs) in the air and increasing air pollution during the summer smog season. There is even some evidence that the presence of oxygenates in fuel may not significantly reduce carbon monoxide emissions at all, even during winter months. One statistical study showed that the reduction of carbon monoxide was considerably less than predicted. In fact, it has been suggested that the replacement of old cars with new ones may have a much more significant effect on carbon monoxide reduction because of the efficiency of modern engines. In addition, studies have suggested that oxygenated fuel increases the formation of atmospheric aldehydes, such as acetaldehyde, formed from ethanol. Because acetaldehyde is a precursor to peroxyacetyl nitrate (see p. 157), it may be possible that *increased* air pollution may result from use of oxygenated fuel.

A fuel can be classified according to its antiknock characteristics. The most important rating system is the **octane rating** of gasoline. In this method of classification, the antiknock properties of a fuel are compared in a test engine with the antiknock properties of a standard mixture of heptane and 2,2,4-trimethylpentane. This latter compound is called "isooctane," hence the name "octane rating." A fuel that has the same antiknock properties as a given mixture of heptane and "isooctane" has an octane rating numerically equal to the percentage of "isooctane" in that reference mixture. Today's 87 octane unleaded gasoline is a mixture of compounds that have, taken together, the same antiknock characteristics as a test fuel composed of 13% heptane and 87% "isooctane." Other substances besides hydrocarbons may also have high resistance to knocking. Table Two presents a list of organic compounds with their octane ratings.

The number of grams of air required for the complete combustion of one mole of gasoline (assuming the formula C_8H_{18}) is 1735 g. This gives rise to a theoretical air–fuel ratio of 15.1:1 for complete combustion. For several reasons, however, it is neither easy nor advisable to supply each cylinder with a theoretically correct air–fuel mixture. The power and performance of an engine improve with a slightly richer mixture (lower air–fuel ratio). Maximum power is obtained from an engine when the air–fuel ratio is near 12.5:1,

TABLE TWO   Octane Ratings of Organic Compounds*

Compound	Octane Number	Compound	Octane Number
Octane	−19	1-Butene	97
Heptane	0	2,2,4-Trimethylpentane	100
Hexane	25	Cyclopentane	101
Pentane	62	Ethanol	105
Cyclohexane	83	Benzene	106
1-Pentene	91	Methanol	106
2-Hexene	93	*m*-Xylene	118
Butane	94	Toluene	120
Propane	97		

*The octane values in this table are determined by the **research method.**

while maximum economy is obtained when the air–fuel ratio is near 16:1. Under conditions of idling or full load (that is, acceleration), the air–fuel ratio is lower than what would be theoretically correct. As a result, complete combustion does not take place in an internal combustion engine, and carbon monoxide (CO) is produced in the exhaust gases. Other types of nonideal combustion behavior give rise to the presence of unburned hydrocarbons in the exhaust. The high combustion temperatures cause the nitrogen and oxygen of the air to react, forming a variety of nitrogen oxides in the exhaust. Each of these materials contributes to air pollution. Under the influence of sunlight, which has enough energy to break covalent bonds, these materials may react with each other and with air to produce **smog.** Smog consists of **ozone,** which deteriorates rubber and damages plant life; **particulate matter,** which produces haze; **oxides of nitrogen,** which produce a brownish color in the atmosphere; and a variety of eye irritants, such as **peroxyacetyl nitrate** (PAN). Lead particles from tetraethyllead may also cause problems because they are toxic. Sulfur compounds in the gasoline may lead to the production of noxious gases in the exhaust.

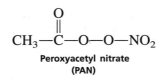

**Peroxyacetyl nitrate
(PAN)**

Current efforts to reverse the trend of deteriorating air quality caused by automotive exhaust have taken many forms. Initial efforts at modifying the air–fuel mixture of engines produced some improvements in emissions of carbon monoxide, but at the cost of increased nitrogen oxide emissions and poor engine performance. With the more stringent air-quality standards imposed by the Environmental Protection Agency, attention has been turned to alternative sources of power. Recently, there has been much interest in the **diesel**

engine as a power plant for passenger cars. The diesel engine has the advantage of producing only very small quantities of carbon monoxide and unburned hydrocarbons. It does, however, produce significant amounts of nitrogen oxides, soot (containing polynuclear aromatic hydrocarbons), and odor-causing compounds. At present, there are no legally established standards for the emission of soot or odor by motor vehicles. This does not mean that these substances are harmless; it means merely that there is no reliable method of analyzing exhaust gases quantitatively for the presence of these materials. Soot and odor may well prove to be harmful, but the emission of these substances remains unregulated. An additional advantage of diesel engines, important in these times of high crude oil prices, is that they tend to yield higher fuel mileage than gasoline engines of a similar size. Research has also been directed at developing internal combustion engines that operate using propane, methane, or even hydrogen as fuels. These engines are not likely to appear in commercial use in the near future, however, because significant technical problems remain to be solved.

In the meantime, because the standard gasoline engine remains the most attractive power plant as a result of its great flexibility and reliability, efforts at improving its emissions continue. The advent of **catalytic converters,** which are muffler-like devices containing catalysts that can convert carbon monoxide, unburned hydrocarbons, and nitrogen oxides into harmless gases, has resulted from such efforts. Unfortunately, the catalysts are rendered inactive by the lead additives in leaded gasoline. Unleaded gasoline must be used, but it takes more crude oil to make a gallon of unleaded gasoline than it does to make leaded gasoline. Other hydrocarbons must be added as antiknock agents to replace tetraethyllead. The active metals in the catalytic converters, principally platinum, palladium, and rhodium, are scarce and extremely expensive. Also, there has been concern that traces of other harmful substances may be produced in the exhaust gases by reactions catalyzed by these metals.

Some success in reducing exhaust emissions has been attained by modifying the design of combustion chambers of internal combustion engines. Additionally, the use of computerized control of ignition systems shows promise. Efforts have also been directed at developing alternative fuels that would give greater mileage, lower emissions, better performance, and a lower demand on crude oil supplies. Methanol has been proposed as an alternative to gasoline as a fuel. Some preliminary tests have indicated that the amount of the principal air pollutants in automobile exhaust is greatly lowered when methanol is used instead of gasoline in a typical automobile. Experiments with methane have also been promising. Methane has a very high octane number, and the proportion of carbon monoxide and unburned hydrocarbons in the exhaust of a methane-powered engine is very small. The production of methane does not require the expensive and inefficient refining processes that are needed to produce gasoline. Experiments are even under way toward developing hydrogen gas as a future fuel. Although the technology for using these alternative fuels remains to be developed fully, the future should bring some interesting advances in engine design in an effort to solve our transportation needs while improving the quality of our air.

REFERENCES

Anderson, E. V. "Brazil's Program to Use Ethanol as Transportation Fuel Loses Steam." *Chemical and Engineering News, 71* (October 18, 1993): 13.

Anderson, E. V. "Health Studies Indicate MTBE Is Safe Gasoline Additive." *Chemical and Engineering News, 71* (September 20, 1993): 9.

Calvin, M. "High-Energy Fuels and Materials from Plants." *Journal of Chemical Education, 64* (April 1987): 335.

Chen, C. T. "Understanding the Fate of Petroleum Hydrocarbons in the Subsurface Environment." *Journal of Chemical Education, 69* (May 1992): 357.

Haggin, J. "Interest in Coal Chemistry Intensifies." *Chemical and Engineering News, 60* (August 9, 1982): 17.

Illman, D. "Oxygenated Fuel Cost May Outweigh Effectiveness." *Chemical and Engineering News, 71* (April 12, 1993): 28.

Kimmel, H. S., and Tomkins, R. P. T. "A Course on Synthetic Fuels." *Journal of Chemical Education, 62* (March 1985): 249.

Schriescheim, A., and Kirschenbaum, I. "The Chemistry and Technology of Synthetic Fuels." *American Scientist, 69* (September–October 1981): 536.

Shreve, R. N., and Brink, J. "Petroleum Refining." *The Chemical Process Industries,* 4th ed. New York: McGraw-Hill, 1977.

Shreve, R. N., and Brink, J. "Petrochemicals." *The Chemical Process Industries,* 4th ed. New York: McGraw-Hill, 1977.

Vartanian, P. F. "The Chemistry of Modern Petroleum Product Additives." *Journal of Chemical Education, 68* (December 1991): 1015.

E X P E R I M E N T   1 7

Gas Chromatographic Analysis of Gasolines

Gasoline
Gas chromatography

In this experiment, you will analyze samples of gasoline by gas chromatography. From your analysis, you should learn something about the composition of these fuels. Although all gasolines are compounded from the same basic hydrocarbon components, each company blends these components in different proportions in order to obtain a gasoline with properties similar to those of competing brands.

Sometimes the composition of the gasoline may vary depending on the composition of the crude petroleum from which the gasoline was derived. Frequently, refineries vary the composition of gasoline in response to differences in climate or seasonal changes or environmental concerns. In the winter or in cold climates, the relative proportion of butane and pentane isomers is increased to improve the volatility of the fuel. This increased volatility permits easier starting. In the summer or in warm climates, the relative proportion of these volatile hydrocarbons is reduced. The decreased volatility reduces the possibility of vapor-lock formation. Occasionally, differences in composition can be detected by examining the gas chromatograms of a particular gasoline over several months. In this experiment, we do not try to detect such small differences.

There are different octane rating requirements for "regular" and "premium" gasolines. You may be able to observe differences in the composition of these two types of fuels. You should pay particular attention to increases in the proportions of those hydrocarbons that raise octane ratings in the premium fuels.

In some areas of the country, manufacturers are required from November to February to control the amounts of carbon monoxide produced when the gasoline burns. To do this, they add oxygenates, such as ethanol or methyl *tert*-butyl ether (MTBE), to the gasoline. You should try to observe the presence of these oxygenates, which may be observed in gasolines produced in carbon monoxide-containment areas.

The class will analyze samples of regular unleaded, premium unleaded, and regular leaded gasolines. If available, the class will analyze oxygenated fuels. If different brands are analyzed, equivalent grades from the different companies should be compared.

Discount service stations usually buy their gasoline from one of the large petroleum-refining companies. If you analyze gasoline from a discount service station, you may find it interesting to compare that gasoline with an equivalent grade from a major supplier, noting particularly the similarities.

Required Reading

New: Technique 15 Gas Chromatography
Essay Petroleum and Fossil Fuels

Special Instructions

Your instructor may want each student in the class to obtain a sample of gasoline from a service station. The instructor will compile a list of the different gasoline companies represented in the nearby area. Each student then will be assigned to collect a sample from a different company. You should collect the gasoline sample in a labeled screw-cap jar. An easy way to collect a gasoline sample for this experiment is to drain the excess gasoline from the nozzle and hose of the pump into the jar after the gasoline tank of a car has been filled. The collection of gasoline in this manner must be done *immediately after* the gas pump has been used. If not, the volatile components of the gasoline may evaporate, thus changing the composition of the gasoline. Only a very small sample (a few *milli*liters) is required, because the gas chromatographic analysis requires no more than a few *micro*liters (μL) of material. Be certain to close the cap of the sample jar tightly to prevent the selective evaporation of the most volatile components. The label on the jar should list the brand of gasoline and the grade (unleaded regular, unleaded premium, oxygenated unleaded, etc.). Alternatively, your instructor may supply samples for you.

> **Caution:** Gasoline contains many highly volatile and flammable components. Do not breathe the vapors and do not use open flames near gasoline. Also, leaded gasoline contains tetraethyllead and is therefore toxic.

This experiment may be assigned along with another short one, because it requires only a few minutes of each student's time to carry out the actual gas chromatography. For this experiment to work as efficiently as possible, you may be asked to schedule an appointment for using the gas chromatograph.

Waste Disposal

Dispose of all gasoline samples in the container designated for nonhalogenated wastes.

Notes to the Instructor

You need to adjust your particular gas chromatograph to the proper conditions for the analysis. We recommend that you prepare and analyze the reference mixture listed in the Procedure section. Most chromatographs will be able to separate this mixture cleanly, with the possible exception of the xylenes. One possible set of conditions for a Gow-Mac model 69-350 chromatograph is the following: column temperature, 110–115°C; injection port temperature, 110–115°C; carrier gas flow rate, 40–50 mL/min; column length, approximately 12 ft long. The column should be packed with a nonpolar stationary phase similar to silicone oil (SE-30) on Chromosorb W or with some other stationary phase that separates components principally according to boiling point.

The chromatograms shown in this experiment were obtained on a Hewlett Packard model 5890 gas chromatograph. A 30-meter, DB 5 capillary column (0.32 mm, with 0.25 micron film) was used. A temperature program was run starting at 5°C and ramping to 150°C. Each run took about 8 minutes. A flame ionization detector was used. The conditions are given in the Instructor's Manual. Superior separations are obtained using capillary columns and they are recommended. Even better results are obtained with longer columns.

Procedure

Reference Mixture.   First, analyze a standard mixture that includes pentane, hexane (or hexanes), benzene, heptane, toluene, and xylenes (a mixture of *meta, para,* and *ortho* isomers). Inject a 0.5-μL sample into the gas chromatograph, or use an alternative sample size as indicated by your instructor. Measure the retention time of each component in the reference mixture on your chromatogram (Technique 15, Section 15.7, p. 817). The previously listed compounds elute in the order given (pentane first and xylenes last). Compare your chromatogram to the one posted near the gas chromatograph or the one reproduced in this experiment.

Your instructor or a laboratory assistant may prefer to perform the sample injections. The special microliter syringes used in the experiment are very delicate and expensive. If you are performing the injections yourself, be sure to obtain instruction beforehand.

Oxygenated Fuel Reference Mixture.   Oxygenated compounds are added to gasolines in carbon monoxide-containment areas during the months of November through February. Currently, ethanol and methyl *tert*-butyl ether are in most common use. Your instructor may have available a reference mixture that includes all the previously listed compounds and either ethanol or methyl *tert*-butyl ether. Again, you need to inject a sample of this mixture and analyze the chromatogram to obtain the retention times for each component in this mixture.

Gasoline Samples.   Inject a sample of a regular unleaded, premium unleaded, or oxygenated gasoline into the gas chromatograph and wait for the gas chromatogram to be recorded. Compare the chromatogram to the reference mixture. Determine the retention times for the major components and identify as many of the components as possible. For comparison, gas chromatograms of a premium unleaded gasoline and the reference mixture are shown on page 163. A list of the major components in gasolines is shown on page 164. Notice that the oxygenate methyl *tert*-butyl ether appears in the C_6 region. Does your oxygenated fuel show this component? See if you can notice a difference between regular and premium unleaded gasolines.

Analysis.   Be certain to compare very carefully the retention times of the components in each fuel sample with the standards in the reference mixture. Retention times of compounds vary with the conditions under which they are determined. It is best to analyze the reference mixture and each of the gasoline samples in succession to reduce the variations in retention times that may occur over time. Compare the gas chromatograms with those of students who have analyzed gasolines from other dealers.

Report.   The report to the instructor should include the actual gas chromatograms as well as an identification of as many of the components in each chromatogram as possible.

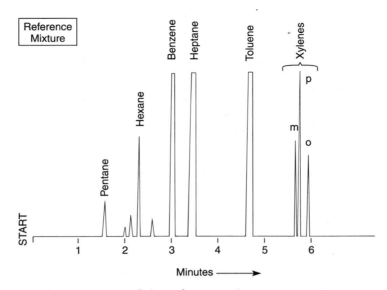

Gas chromatogram of the reference mixture.

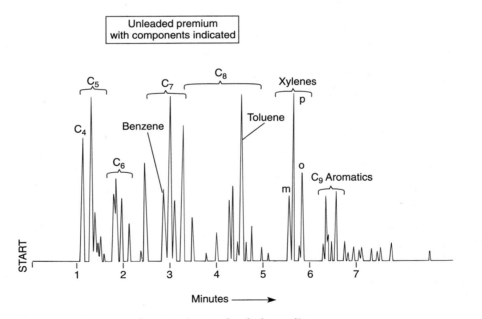

Gas chromatogram of a premium unleaded gasoline.

Major Components in Gasolines*

C_4 compounds	Isobutane
	Butane
C_5 compounds	Isopentane
	Pentane
C_6 compounds and oxygenates	2,3-Dimethylbutane
	2-Methylpentane
	3-Methylpentane
	Hexane
	Methyl *tert*-butyl ether (oxygenate)
C_7 compounds and aromatics (benzene)	2,4-Dimethylpentane
	Benzene (C_6H_6)
	2-Methylhexane
	3-Methylhexane
	Heptane
C_8 compounds and aromatics (toluene, ethylbenzene, and xylenes)	2,2,4-Trimethylpentane (isooctane)
	2,5-Dimethylhexane
	2,4-Dimethylhexane
	2,3,4-Trimethylpentane
	2,3-Dimethylhexane
	Toluene (C_7H_8)
	Ethylbenzene (C_8H_{10})
	m-, *p*-, *o*-Xylenes (C_8H_{10})
C_9 aromatic compounds	1-Ethyl-3-methylbenzene
	1,3,5-Trimethylbenzene
	1,2,4-Trimethylbenzene
	1,2,3-Trimethylbenzene

*Approximate order of elution.

QUESTIONS

1. If you had a mixture of benzene, toluene, and *m*-xylene, what would be the expected order of retention times? Explain.

2. If you were a forensic chemist working for the police department, and the fire marshal brought you a sample of gasoline found at the scene of an arson attempt, could you identify the service station at which the arsonist purchased the gasoline? Explain.

3. How could you use infrared spectroscopy to detect the presence of ethanol in an oxygenated fuel?

ESSAY

Terpenes and Phenylpropanoids

Anyone who has walked through a pine or cedar forest, or anyone who loves flowers and spices, knows that many plants and trees have distinctively pleasant odors. The essences or aromas of plants are due to volatile or **essential oils,** many of which have been valued since antiquity for their characteristic odors (frankincense and myrrh, for example). A list of the commercially important essential oils would run to over 200 entries. Allspice, almond, anise, basil, bayberry, caraway, cinnamon, clove, cumin, dill, eucalyptus, garlic, jasmine, juniper, orange, peppermint, rose, sandalwood, sassafras, spearmint, thyme, violet, and wintergreen are but a few familiar examples of such valuable essential oils. Essential oils are used for their pleasant odors in perfumes and incense. They are also used for their taste appeal as spices and flavoring agents in foods. A few are valued for antibacterial and antifungal action. Some are used medicinally (camphor and eucalyptus) and others as insect repellents (citronella). Chaulmoogra oil represents one of the few known curative agents for leprosy. Turpentine is used as a solvent for many paint products.

Essential oil components are often found in the glands or intercellular spaces in plant tissue. They may exist in all parts of the plant but are often concentrated in the seeds or flowers. Many components of essential oils are steam-volatile and can be isolated by steam distillation. Other methods of isolating essential oils include solvent extraction and pressing (expression) methods. Esters (see essay, p. 93) are frequently responsible for the characteristic odors and flavors of fruits and flowers, but other types of substances may also be important components of odor or flavor principles. Besides the esters, the ingredients of essential oils may be complex mixtures of hydrocarbons, alcohols, and carbonyl compounds. These other components usually belong to one of the two groups of natural products called **terpenes** or **phenylpropanoids.**

TERPENES

Chemical investigations of essential oils in the nineteenth century found that many of the compounds responsible for the pleasant odors contained exactly ten carbon atoms. These ten-carbon compounds came to be known as terpenes if they were hydrocarbons and as **terpenoids** if they contained oxygen and were alcohols, ketones, or aldehydes. Eventually, it was found that there are also minor and less volatile plant constituents containing 15, 20, 30, and 40 carbon atoms. Because compounds of ten carbons were originally called terpenes, they came to be called **monoterpenes.** The other terpenes were classified in the following way.

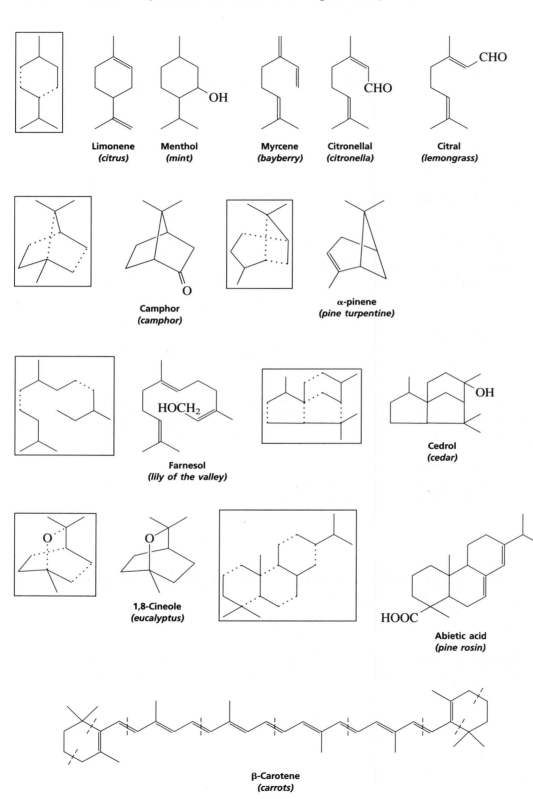

Limonene
(citrus)

Menthol
(mint)

Myrcene
(bayberry)

Citronellal
(citronella)

Citral
(lemongrass)

Camphor
(camphor)

α-pinene
(pine turpentine)

Farnesol
(lily of the valley)

Cedrol
(cedar)

1,8-Cineole
(eucalyptus)

Abietic acid
(pine rosin)

β-Carotene
(carrots)

Selected terpenes.

Class	No. of Carbons	Class	No. of Carbons
Hemiterpenes	5	Diterpenes	20
Monoterpenes	10	Triterpenes	30
Sesquiterpenes	15	Tetraterpenes	40

Further chemical investigations of the terpenes, all of which contain multiples of five carbons, showed them to have a repeating structural unit based on a five-carbon pattern. This structural pattern corresponds to the arrangement of atoms in the simple five-carbon compound isoprene. Isoprene was first obtained by the thermal "cracking" of natural rubber.

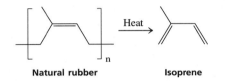

Natural rubber   **Isoprene**

As a result of this structural similarity, a diagnostic rule for terpenes, called the **isoprene rule,** was formulated. This rule states that a terpene should be divisible, at least formally, into **isoprene units.** The structures of a number of terpenes, along with a diagrammatic division of their structures into isoprene units, are shown in the figure that accompanies this essay. Many of these compounds represent odors or flavors that should be very familiar to you.

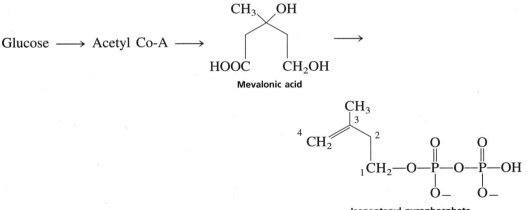

Modern research has shown that terpenes do not arise from isoprene; it has never been detected as a natural product. Instead, the terpenes arise from an important biochemical precursor compound called **mevalonic acid.** This compound arises from acetyl coenzyme A, a product of the biological degradation of glucose (glycolysis), and is converted to a compound called isopentenyl pyrophosphate. Isopentenyl pyrophosphate and its isomer 3,3-dimethylallyl pyrophosphate (double bond moved to the second position) are the five-carbon building blocks used by nature to construct all the terpene compounds.

PHENYLPROPANOIDS

Aromatic compounds, those containing a benzene ring, are also a major type of compound found in essential oils. Some of these compounds, like *p*-cymene, are actually cyclic terpenes that have been aromatized (had their ring converted to a benzene ring), but most are of a different origin.

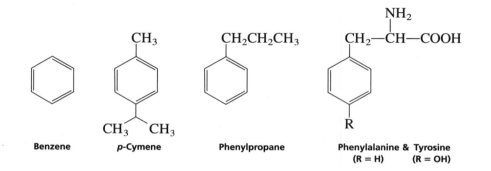

Benzene   *p*-Cymene   Phenylpropane   Phenylalanine & Tyrosine
                                                             (R = H)   (R = OH)

Many of these aromatic compounds are **phenylpropanoids,** compounds based on a phenyl-propane skeleton. Phenylpropanoids are related in structure to the common amino acids phenylalanine and tyrosine, and many are derived from a biochemical pathway called the **shikimic acid pathway.**

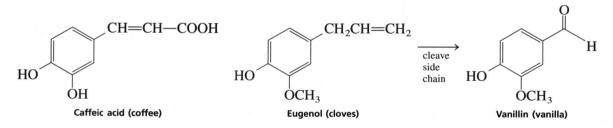

Caffeic acid (coffee)   Eugenol (cloves)   Vanillin (vanilla)

It is also common to find compounds of phenylpropanoid origin that have had the three-carbon side chain cleaved. As a result, phenylmethane derivatives, such as vanillin, are also quite common in plants.

REFERENCES

Cornforth, J. W. "Terpene Biosynthesis." *Chemistry in Britain, 4* (1968): 102.
Geissman, T. A., and Crout, D. H. G. *Organic Chemistry of Secondary Plant Metabolism.* San Francisco: Freeman, Cooper and Co., 1969.

Hendrickson, J. B. *The Molecules of Nature.* New York: W. A. Benjamin, 1965.
Pinder, A. R. *The Terpenes.* New York: John Wiley and Sons, 1960.
Ruzicka, L. "History of the Isoprene Rule." *Proceedings of the Chemical Society (London)* (1959): 341.
Sterret, F. S. "The Nature of Essential Oils. Part I. Production." *Journal of Chemical Education, 39* (1962): 203.
Sterret, F. S. "The Nature of Essential Oils. Part II. Chemical Constituents. Analysis." *Journal of Chemical Education, 39* (1962): 246.

EXPERIMENT 18

Isolation of Essential Oils from Allspice, Cloves, Cumin, Caraway, Cinnamon, or Fennel

Steam distillation
Extraction
Infrared spectroscopy

In this experiment, you will steam distill the essential oil from a spice. Either you will choose, or the instructor will assign you, a spice from the following list: allspice, cloves, cumin, caraway, cinnamon, or fennel. Each spice produces a relatively pure essential oil. The structures for the major essential oil components of the spices are shown here. Your spice will yield one of these compounds. You are to determine which structure represents the essential oil that was distilled from your spice.

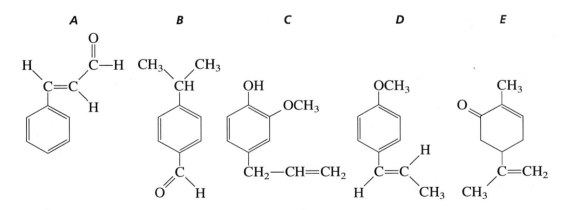

In trying to determine your structure, be sure to look for the following features (stretching frequencies) in the infrared spectrum: C=O (ketone or aldehyde), C—H (aldehyde), O—H (phenol), C—O (ether), benzene ring, and C=C (alkene). Also be sure to look for the aromatic-ring, out-of-plane bending frequencies, which may help you determine the substitution patterns of the benzene rings (see p. 902). The out-of-plane bending region may also be of help in determining the degree of substitution on the alkene double bond where it exists (see p. 901). There are enough differences in the infrared spectra of the five possible compounds that you should be able to identify your essential oil.

If NMR spectroscopy is available, it will provide a nice confirmation of your conclusions. Carbon-13 NMR would be even more informative than proton magnetic resonance. However, neither of these techniques is required for a solution.

Required Reading

Review: Technique 7

New: Essay   Terpenes and Phenylpropanoids
          Technique 11   Steam Distillation

Special Instructions

Foaming can be a serious problem if you use finely ground spices. It is recommended that you use clove buds, whole allspice, or cinnamon sticks in place of the ground spices. However, be sure to cut or break up the large pieces or crush them with a mortar and pestle.

Waste Disposal

Any aqueous solutions should be placed in the container specially designated for aqueous wastes. Be sure to place any solid spice residues in the garbage can because they will plug the sink.

Notes to the Instructor

If ground spices are used (not recommended), you may want to have the students insert a Claisen head between the round-bottom flask and the distillation head to allow extra volume in case the mixture foams.

Procedure

Apparatus. Using a 100-mL round-bottom flask to distill and a 50-mL round-bottom flask to collect, assemble a distillation apparatus similar to that shown in Figure 8.4, page 711. Use a heating mantle to heat. The collection flask may be immersed in ice to ensure condensation of the distillate.

Preparing the Spice.   Weigh approximately 3.0 g of your spice onto a weighing paper and record the exact weight. If your spice is already ground, you may proceed without grinding it; otherwise, break up the seeds using a mortar and pestle, or cut larger pieces into smaller ones using a scissors. Mix the spice with 35–40 mL of water in the 100-mL round-bottom flask, add a boiling stone, and reattach it to your distillation apparatus. Allow the spice to soak in the water for about 15 minutes before beginning the heating. Be sure that all the spice gets thoroughly wetted. Swirl the flask gently, if necessary.

Steam Distillation.   Turn on the cooling water in the condenser and begin heating the mixture to provide a steady rate of distillation. If you approach the boiling point too quickly, you may have difficulty with frothing or bump-over. You will need to find the amount of heating that provides a steady rate of distillation but avoids frothing and/or bumping. A good rate of distillation would be to have one drop of liquid collected every 2–5 seconds. Continue distillation until at least 15 mL of distillate has been collected.

Normally, in a steam distillation the distillate will be somewhat cloudy due to separation of the essential oil as the vapors cool. However, you may not notice this and still obtain satisfactory results.

Extraction of the Essential Oil.   Transfer the distillate to a separatory funnel and add 5.0 mL of methylene chloride (dichloromethane) to extract the distillate. Shake the funnel vigorously, venting frequently. Allow the layers to separate.

The mixture may be spun in a centrifuge if the layers do not separate well. Stirring gently with a spatula sometimes helps to resolve an emulsion. It may also help to add about 1 mL of a saturated sodium chloride solution. For the following directions, however, be aware that the saturated salt solution is quite dense, and the aqueous layer may change places with the methylene chloride layer, which is normally on the bottom.

Transfer the lower methylene chloride layer to a clean, dry Erlenmeyer flask. Repeat this extraction procedure with a fresh 5.0-mL portion of methylene chloride and place it in the same Erlenmeyer flask as you placed the first extraction. If there are visible drops of water, you need to transfer the methylene chloride solution carefully to a clean, dry flask, leaving the drops of water behind.

Drying.   Dry the methylene chloride solution by adding about 1 g of granular anhydrous sodium sulfate to the Erlenmeyer flask (see Technique 7, Section 7.8, p. 696). Let the solution stand for 10–15 minutes and swirl it occasionally.

Evaporation.   While the organic solution is being dried, obtain a clean and dry medium-size test tube and weigh (tare) it accurately. Decant a portion (about one-third) of the dried organic layer to this tared test tube, leaving the drying agent behind. Add a boiling stone and, working in a hood, evaporate the methylene chloride from the solution by using a gentle stream of air or nitrogen and heating to about 40°C with a water bath (see Technique 3, Section 3.11, p. 630). When the first portion is reduced to a small volume of liquid, add a second portion of the methylene chloride solution and evaporate as before. When you add the final portion, use small amounts of clean methylene chloride to rinse the drying agent, allowing you to transfer all of the remaining solution into the tared test tube. Be careful to keep any of the sodium sulfate from being transferred.

> **Caution:** The stream of air or nitrogen must be very gentle or you will blast your solution out of the test tube. In addition, do not overheat the sample or your sample may "bump" out of the tube. Do not continue the evaporation beyond the point where all the methylene chloride has evaporated. Your product is a volatile oil (i.e., liquid). If you continue to heat and evaporate, you will lose it. It would be better to leave some methylene chloride than to lose your sample.

Yield Determination.   When the solvent has been removed, reweigh the test tube. Calculate the weight percentage recovery of the oil from the original amount of spice used.

SPECTROSCOPY

Infrared.   Obtain the infrared spectrum of the oil as a pure liquid sample (Technique 19, Section 19.2, p. 843). It may be necessary to use a Pasteur pipet with a narrow tip to transfer a sufficient amount to the salt plates. If even this fails, you may add one or two drops of carbon tetrachloride (tetrachloromethane) to aid in the transfer. This solvent will not interfere with the infrared spectrum. Include the infrared spectrum in your laboratory report, along with an interpretation of the principal peaks.

Nuclear Magnetic Resonance.   At the instructor's option, determine the nuclear magnetic resonance spectrum of the oil (Technique 19, Section 19.9, p. 855).

REPORT

From the infrared spectrum (and any other data you have used) you should determine the structure (A-E) that best matches the essential oil you isolated from your spice. Label the major peaks in the infrared spectrum, and give an argument that supports your choice of structure. Also be sure to include your weight percentage recovery calculation.

QUESTIONS

1. Take a sheet of paper and build a matrix by drawing each of the five possible essential oil compounds given previously down the left side of the sheet and by listing each of the possible infrared spectral features given previously along the top of the sheet. Draw lines to form boxes, and inside the boxes opposite each compound, note the expected infrared observation. Is the peak expected to be present or absent? If not absent, give the expected number of peaks and the probable frequencies. A good set of correlation charts and tables will help you with this (see Appendix 3).

2. Why does the newly condensed steam distillate appear cloudy?

3. After the drying step, what observations will help you to determine if the extracted solution is "dry" (i.e., free of water)?

EXPERIMENT 19

Thin-Layer Chromatography for Monitoring the Oxidation of Borneol to Camphor

Chromic acid oxidation
Monitoring reactions
Thin-Layer chromatography

A typical reaction of secondary alcohols is their oxidation to ketones by such reagents as chromic acid, sodium dichromate, or potassium permanganate. A specific example is shown for the oxidation of borneol to camphor.

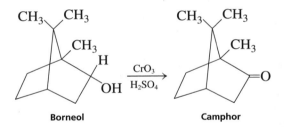

Borneol   Camphor

For a more detailed discussion of this reaction, read the introductory material in Experiment 20.

In this experiment, borneol is oxidized to camphor with chromic acid. This reaction is virtually identical to the oxidation step of Experiment 20. The object is to determine the time needed for the reaction to reach completion. The means used to follow the reaction will be thin-layer chromatography (TLC). Although both the starting material and the product are colorless, they can be visualized by iodine vapor.

The purpose of this experiment is to demonstrate the basic principles and techniques of thin-layer chromatography. This experiment also illustrates one method of monitoring the progress of a reaction, and it demonstrates that organic reactions often proceed slowly.

Required Reading

New: Techniques 12, 14
       Experiment 20 (introductory material)

Special Instructions

This experiment involves diethyl ether, which is a highly flammable solvent. Avoid the use of flames in the laboratory. Avoid contact with sulfuric acid, it is extremely corrosive. Chromic oxide is a possible carcinogen (see p. 21) and teratogen (birth defects); skin contact should be avoided. If you prepare hand-dipped TLC slides, it is recommended that you use gloves and that you work in a hood while preparing your plates. Although the quantities of hazardous materials are quite small, safety precautions should still be exercised.

If UV visualization is used, it should be done in a darkened room or in a hood that has had its sash covered with butcher paper, or aluminum foil, to darken it.

Waste Disposal

Any leftover solutions containing chromium, a heavy metal, should be placed in a container specially designated for that purpose. Unused developing solvent contains methylene chloride and should be placed in the halogenated organic waste container. Any unused portions of solutions A or C may be placed in the nonhalogenated organic waste container. Special containers should be provided for used capillary pipets (glass waste) and glass microscope slides, if they were used. At the option of your instructor, glass microscope slides can be cleaned and reused. If you used commercial TLC plates, these should also be collected in a special container for proper disposal.

Procedure

Initial Preparations. About ten TLC plates will be needed. You can make these by hand-dipping glass microscope slides into a slurry of silica gel G in a methylene chloride–methanol (2:1) solution (Technique 14, Section 14.3, p. 794) or by cutting sheets of the commercially prepared TLC plates with an aluminum backing (recommended) to the same approximate size with a pair of scissors.

Prepare a developing chamber from a 4-oz wide-mouthed screw-cap jar as described in Technique 14, Section 14.5, p. 799. Add some of the methylene chloride–methanol solvent to the developing chamber. When you are finished, prepare about 12 capillary micropipets as described in Technique 14, Section 14.4, p. 797. Be sure to do this in an area set aside for the use of flames.

Reagent Solutions. For this experiment you use three solutions: Solution A is a 2% solution of borneol in diethyl ether; solution B is a solution of 10% chromic oxide and 5% sulfuric acid in water; solution C is a 2% solution of camphor in diethyl ether. Solutions A and C provide reference spots on each TLC slide, and solution B provides the oxidizing medium to oxidize borneol to camphor.

Monitoring the Oxidation Reaction. Mix about 1 mL of solution A with 1 mL of solution B in a small test tube. Write down the time at which this mixture is prepared. Shake the mixture briefly. Spot a TLC plate with solution A, solution C, and the upper (ether) layer of the reaction mixture. Place the plate in the developing chamber and develop it as described in Technique 14, Section 14.5, p. 799. After the reaction has been allowed to take place for 5 minutes, spot a second TLC plate with solution A, solution C, and the upper layer of the reaction mixture. Develop this plate as before. Continue spotting and developing plates at 5-minute intervals for a total reaction time of 40 minutes. During this period, shake the reaction test tube periodically. Be sure to mark each plate (in pencil at the top) with a number after removing it from the development chamber. Place the plates face-up on a paper towel in the hood until all the solvents evaporate.

Iodine Visualization of the Results.   After each plate has become free of the solvents, it can be placed in a jar containing a few crystals of iodine. Cap the jar and warm it gently until spots begin to appear (Technique 14, Section 14.7, p. 801).

UV Visualization (Alternate Method).   If commercial plates with a fluorescent indicator were used, it should be possible to observe the spots on the TLC plates using short-wavelength UV (ultraviolet) light for visualization. This method may not be successful, however, if the iodine visualization has been performed first. Iodine will permanently affect some of the spots. If both methods are to be used, visualization by UV light should be performed first. Wait until each plate is free of the developing solvent before observation under UV.

Analysis of Results.   You will notice that the R_f values for borneol and camphor differ, with camphor more mobile than borneol. By comparing these R_f values with the R_f values for the spots formed in the reaction mixture, you should be able to identify the presence of borneol and camphor in the mixture. Compare the intensities of the borneol and camphor spots in the reaction mixture on each of the TLC plates you have developed. Using this information, determine the time required for the oxidation reaction to reach completion. Record the reaction time and the R_f values for borneol and camphor in your notebook and sketch a typical plate showing the position of the spots and the associated R_f values.

REFERENCE

Davis, M. "Using TLC to Follow the Oxidation of a Secondary Alcohol to a Ketone." *Journal of Chemical Education, 45* (March 1968): 192.

QUESTIONS

1. What would you expect to happen to the R_f values for borneol and camphor if pure methanol were used to develop the plates? What would you expect if pure methylene chloride were used?

2. Why do you suppose a mixture of methylene chloride and methanol is used to develop the TLC plates in this experiment rather than either pure solvent?

3. Why is camphor more mobile (higher R_f) than borneol? Base your answer on structural considerations.

4. Suppose you had used a primary alcohol, piperonal alcohol, for this experiment and observed *three* spots instead of the expected two spots. How could you explain this?

Piperonal alcohol

EXPERIMENT 20

An Oxidation–Reduction Scheme: Borneol, Camphor, Isoborneol

Sodium dichromate oxidation
Sodium borohydride reduction
Stereochemistry
Spectroscopy (infrared, proton NMR, carbon-13 NMR)
Gas chromatography (optional)

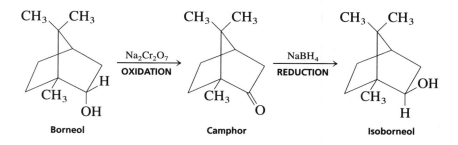

This experiment will illustrate the use of an oxidizing agent (sodium dichromate) for converting a secondary alcohol (borneol) to a ketone (camphor). The camphor is then reduced by sodium borohydride to give the **isomeric** alcohol isoborneol. The spectra of borneol, camphor, and isoborneol will be compared to detect structural differences and to determine the extent to which the final step produces a pure alcohol isomeric with the starting material.

OXIDATION OF BORNEOL WITH SODIUM DICHROMATE

Secondary alcohols are easily oxidized to ketones with sodium dichromate or other chromium (VI) compounds, such as chromium trioxide or sodium chromate. The half-reactions and the overall reaction used in the oxidation of borneol with dichromate are as follows:

HALF-REACTION $Cr_2O_7^{2-} + 14H^+ + 6e^- \longrightarrow 2Cr^{3+} + 7H_2O$

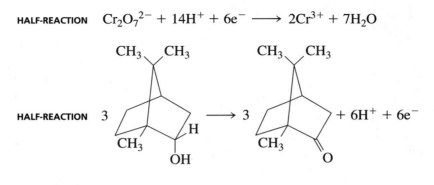

NET REACTION

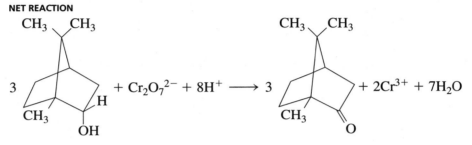

The mechanism for the oxidation is as follows:

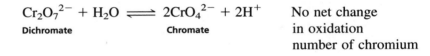

$Cr_2O_7^{2-} + H_2O \rightleftharpoons 2CrO_4^{2-} + 2H^+$ No net change
in oxidation
number of chromium

Dichromate   **Chromate**

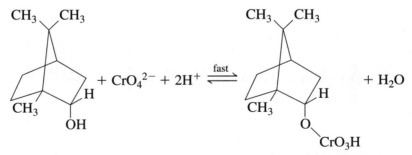

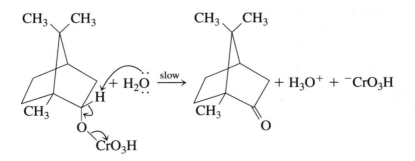

REDUCTION OF CAMPHOR WITH SODIUM BOROHYDRIDE

Metal hydrides of the Group III elements (sources of $H:^-$), such as lithium aluminum hydride $LiAlH_4$ and sodium borohydride $NaBH_4$, are widely used in reducing carbonyl groups. Lithium aluminum hydride, for example, reduces many compounds containing carbonyl groups, such as aldehydes, ketones, carboxylic acids, esters, or amides, whereas sodium borohydride reduces only aldehydes and ketones. The reduced reactivity of borohydride allows it to be used even in alcohol and water solvents, whereas lithium aluminum hydride reacts violently with these solvents to produce hydrogen gas and, therefore, must be used in nonhydroxylic solvents. In the present experiment, sodium borohydride is used because it is easily handled, and the results of reductions using either of the two reagents are essentially the same. The same care need not be taken in keeping sodium borohydride away from water as is required with lithium aluminum hydride.

The mechanism of action of sodium borohydride in reducing a ketone is as follows:

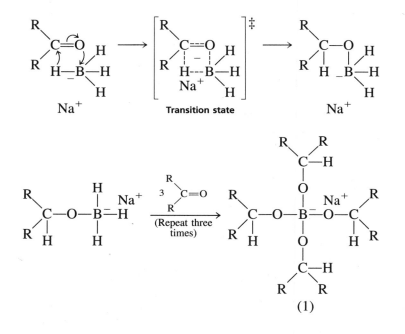

(1)

Note in this mechanism that all four hydrogen atoms are available as hydrides ($H:^-$), and thus one mole of borohydride can reduce four moles of a ketone. All the steps are irreversible. Usually, *excess* borohydride is used because there is often uncertainty regarding its purity and because of the fact that some of it may react with the solvent.

Once the final tetraalkoxyboron compound (1) is produced, it can be decomposed (along with excess borohydride) by methanol at elevated temperatures as shown:

$$(R_2CH-O)_4B^-Na^+ + 4\ R'OH \longrightarrow 4\ R_2CHOH + (R'O)_4B^-Na^+$$

(1)

The stereochemistry of the reduction is very interesting. The hydride can approach the camphor molecule more easily from the bottom side (**endo** approach) than from the top side (**exo** approach). If attack occurs at the top, a large steric repulsion is created by one of the two **geminal** methyl groups. (Geminal methyl groups are groups that are attached to the same carbon.) Attack at the bottom avoids this steric interaction.

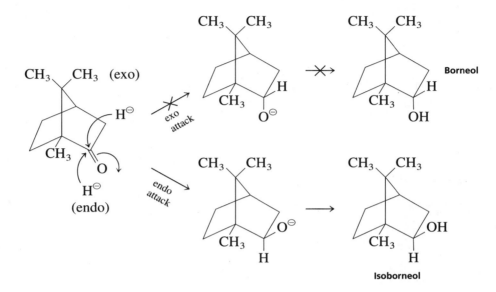

It is expected, therefore, that **isoborneol,** the alcohol produced from the attack at the *least*-hindered position, will *predominate but will not be the exclusive product* in the final reaction mixture: the reaction is stereoselective, not stereospecific. The percentage composition of the mixture can be determined by spectroscopy or by gas chromatography.

It is interesting to note that when the methyl groups are absent (as in 2-norbornanone), the top side (**exo** approach) is favored, and the opposite stereochemical result is obtained. Again, the reaction is stereoselective, but does not give exclusively one product.

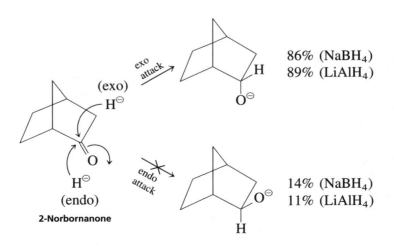

Bicyclic systems such as camphor and 2-norbornanone react predictably according to steric influences. This effect has been termed **steric approach control.** Stereoselectivity controls the reaction. In the reduction of simple acyclic and monocyclic ketones, however, the reaction seems to be influenced primarily by thermodynamic factors: the product of lowest energy is favored. This effect has been termed **product development control.** In the reduction of 4-*t*-butylcyclohexanone, for instance, the thermodynamically more stable product is produced by product development control.

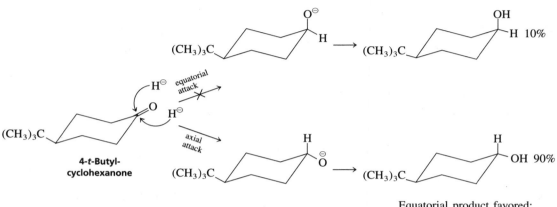

Equatorial product favored;
"product development control"

Required Reading

Review: Technique 3   Section 3.11
          Technique 4   Section 4.1
          Technique 7
          Technique 16
          Technique 19   Sections 19.2, 19.9, 19.10
          Appendices 3, 4, and 5

Special Instructions

The reactants and products are all highly volatile and must be stored in tightly closed containers. Sodium dichromate and other chromium compounds are suspect carcinogens (p. 21). Since volatile chromium compounds may be present above acidic solutions of dichromate, you should work in a hood during the first part of the oxidation step (first paragraph of Procedure). If the instructor chooses, you may be asked to follow the progress of the oxidation of borneol to camphor by TLC (Experiment 19). The reduction of camphor to isoborneol involves diethyl ether, which is extremely flammable. Be certain that no open flames of any sort are in your vicinity when you are using ether.

Waste Disposal

Any aqueous solutions containing chromium should be placed in a container designated for aqueous chromium wastes (this includes any leftover oxidizing reagent and the extractions performed prior to the one using sodium bicarbonate). Other aqueous wastes (such as the bicarbonate and water washes) should be placed in the usual container designated for aqueous solutions. Any leftover ether solutions may be placed in the nonhalogenated organic liquid waste container.

Notes to the Instructor

You can have students monitor the progress of the oxidation by following the procedures in Experiment 19. If the oxidation step is not complete (as best judged by the infrared spectrum), the quantitative results will be affected, since the unreacted borneol will be a part of the final reduction mixture.

In performing the reduction step, the sodium borohydride should be checked to see whether it is active. Place a small amount of powdered material in some methanol and heat it gently. The solution should bubble vigorously if the hydride is active.

Procedure

PART A.   OXIDATION OF BORNEOL TO CAMPHOR

Preparations.   Dissolve 2.0 g of sodium dichromate dihydrate in 8 mL of water, and *carefully* add 1.6 mL of concentrated sulfuric acid with a disposable Pasteur pipet (or an eyedropper). Place the oxidizing solution in an ice bath.

> **Caution:** Sodium dichromate should be handled with caution. See page 21. Work in a hood.

While this solution is cooling, dissolve 1.0 g of racemic borneol in 4 mL of ether in a 25-mL Erlenmeyer flask, and also cool it in an ice bath.

Oxidation by Sodium Dichromate.   Remove 6 mL of the cooled sodium dichromate oxidizing mixture you have prepared, and slowly add it with a Pasteur pipet to the *cold* ether solution of borneol over a period of 10 minutes. Swirl the reaction mixture in the ice bath between additions, and continue swirling for an additional 5 minutes following the final addition of oxidant.

Extraction of Camphor.   Pour the final mixture into a separatory funnel and rinse the Erlenmeyer flask, first with a 10-mL portion of ether, and then with 10 mL of water. Add both of the rinsings to the separatory funnel. Mix the contents of the separatory funnel by shaking with occasional venting. Then place the separatory funnel in a ring attached to a ring stand and allow the phases to separate.

The entire mixture will be very dark, and it may be difficult to see the interface between the aqueous and the organic phase. In this case, use a light source such as a lamp or a flashlight to detect the interface.

Drain the aqueous phase from the bottom of the funnel, using a beaker to catch it, and then transfer the organic phase to an Erlenmeyer flask for storage. Return the aqueous phase to the separatory funnel and extract it, as before, with two successive 10-mL portions of ether. Each time, drain the aqueous phase first, and then add the ether phase to the original ether phase stored in the Erlenmeyer flask.

When finished, return the combined ether extracts to the separatory funnel and extract them with 10 mL of 5% sodium bicarbonate. A small amount of solid material may be produced at the interface. Carefully remove the lower aqueous layer and as much of this solid as possible without losing any of the ether layer. Finally, wash the ether layer with 10 mL of water and remove the lower aqueous layer. The upper ether layer contains the desired camphor. Pour the ether layer out the top of the separatory funnel, into an Erlenmeyer flask, decanting it away from any remaining solids. Dry the ether thoroughly by adding about 1 g of anhydrous sodium sulfate. Stopper the flask, swirling it gently from time to time, until the ether solution is completely clear.[1]

Isolation of Product. Preweigh an Erlenmeyer flask, decant the dry ether phase into it, add a boiling stone, and evaporate the solvent in the hood using a warm water bath (or a steam bath) and a stream of dry air. When the ether has evaporated and a solid has appeared, remove the flask from the heat source *immediately,* otherwise the product may sublime prematurely and be lost. Weigh the flask to determine the weight of product. At least 0.5 g of crude camphor should be obtained. If you choose to stop here, the camphor (which is volatile) should be stored in a tightly sealed container until the next period. Calculate a *crude yield* of camphor, based on your original sample of borneol.

Infrared Spectrum. Before proceeding to the next step, you should verify that the oxidation was successful. This can be done by determining the infrared spectrum of your crude material. For the infrared spectrum, dissolve a small amount of camphor in carbon tetrachloride, place the solution between salt plates (Technique 19, Section 19.2, page 843), and then determine the spectrum. By examination of the infrared peaks, determine if the borneol (an alcohol) is absent and has been oxidized to camphor (a ketone). For comparison, an infrared spectrum of camphor is shown at the end of the experiment. If your oxidation was not totally successful (pure camphor), consult your instructor for your options.

Sublimation. Purify your camphor by vacuum sublimation using an aspirator (or your house vacuum) and the apparatus shown in Technique 16, Figure 16-2C, page

---

[1]Some solutions may not be clear. If you do not have a clear solution, gravity-filter the solution to remove as much solid as possible, then continue the procedure as indicated. The colored inorganic materials will be left behind when the camphor is sublimed.

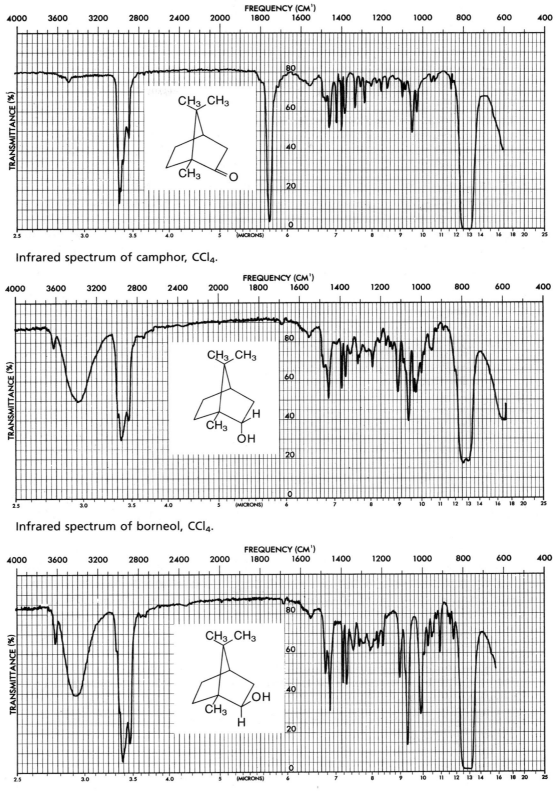

Infrared spectrum of camphor, CCl₄.

Infrared spectrum of borneol, CCl₄.

Infrared spectrum of isoborneol, CCl₄.

828. Refer to the sublimation procedure given in Experiment 5, page 68, and follow the details given there. A microburner is a convenient heating source for the sublimation, but great care must be taken to avoid fires. You must be certain that no one is using ether near your desk. Check with your instructor for clearance. You should sublime your camphor in portions. Be certain that the apparatus is *vacuum tight* before the solid is sublimed. Scrape the purified material from the cold finger onto a preweighed smooth piece of paper with a spatula, reweigh the paper, and determine the amount of material recovered from the sublimation. Calculate a percentage yield of purified camphor, based on the original amount of borneol that you used.

Analysis of Camphor.   Determine the melting point of your camphor; the melting point of pure racemic camphor is 174°C. The melting point of your camphor, however, will often be low due to impurities.[2] The remainder of your camphor is reduced in the next step to isoborneol. Store the camphor in a tightly sealed container until needed.

PART B.   REDUCTION OF CAMPHOR TO ISOBORNEOL (AND BORNEOL)

Reduction.   The procedure below is for 0.5 g of camphor. However, you may use whatever amount of camphor you obtained in the first step unless it is less than 0.4 g. If it is less than 0.4 g, obtain some camphor from the supply shelf (or your instructor) to supplement your yield. If you obtained more than 0.5 g of camphor, scale up the reagents from the amounts given below.

In a 25-mL Erlenmeyer flask, dissolve 0.5 g of the racemic camphor (obtained above) in 2 mL of methanol. In portions, cautiously and intermittently add 0.3 g of sodium borohydride to the solution. If necessary, cool the flask in an ice bath to keep the reaction mixture at room temperature and prevent boiling. Remember that a small amount of bubbling is normal since some of the borohydride will react with the methanol. When all the borohydride is added, gently boil the contents of the flask on a warm hotplate or a steam bath for about 2 minutes.

Isolation.   Pour the hot reaction mixture into about 15 g of chipped ice, using small portions of methanol to aid the transfer. When the ice melts, collect the white solid by vacuum filtration using a Hirsch funnel. Transfer the solid to a small flask and add about 20 mL of ether to dissolve it. Add a spatula of anhydrous sodium sulfate to dry the solution. Decant the solution from the drying agent into a preweighed Erlenmeyer flask and evaporate the solvent in the hood. Store the product in a tightly sealed container until analysis is to be performed.

Analysis of Product.   Reweigh the flask to determine the weight of the product and calculate the percentage yield. Determine the melting point; pure racemic isoborneol melts at 212°C. Determine the infrared spectrum of the product by the method used above for camphor. Compare it with the spectra for borneol and isoborneol shown in the figures.

---

[2]See Question 4 at the end of the experiment. A small amount of impurity drastically reduces the melting point and increases the range.

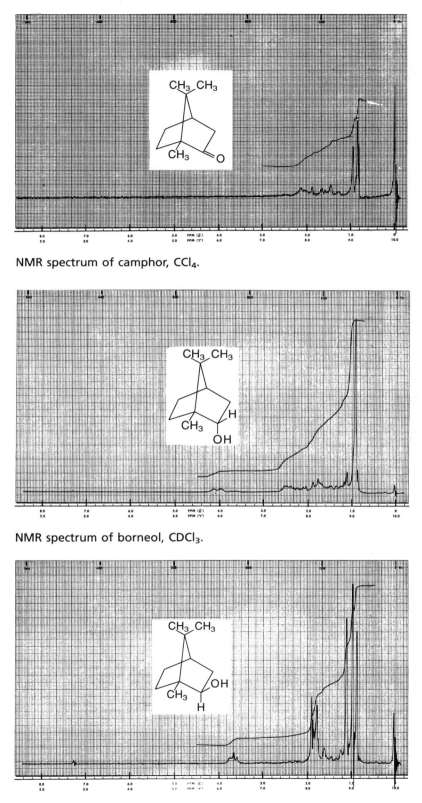

NMR spectrum of camphor, CCl$_4$.

NMR spectrum of borneol, CDCl$_3$.

NMR spectrum of isoborneol, CDCl$_3$.

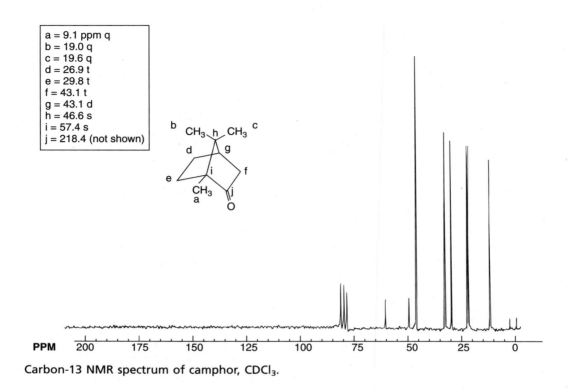

Carbon-13 NMR spectrum of camphor, CDCl₃.

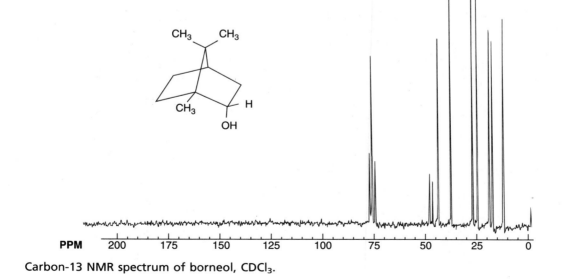

Carbon-13 NMR spectrum of borneol, CDCl₃.

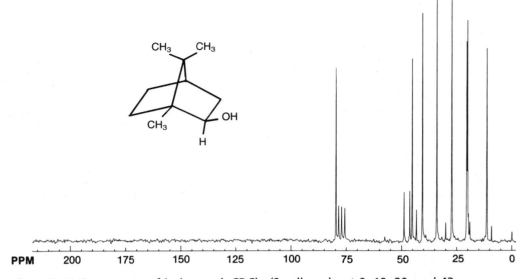

Carbon-13 NMR spectrum of isoborneol, $CDCl_3$. (Small peaks at 9, 19, 30, and 43 ppm are due to impurities.)

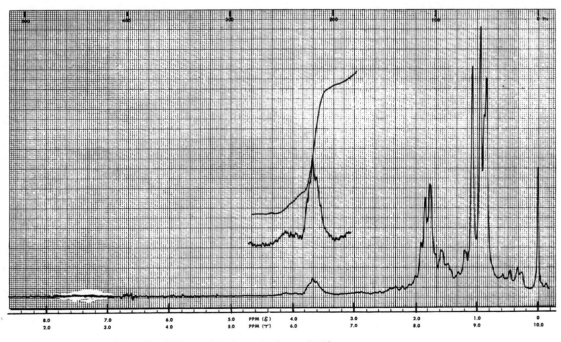

NMR spectrum of borohydride reduction product, $CDCl_3$.

PART C.   PERCENTAGES OF ISOBORNEOL AND BORNEOL OBTAINED FROM THE REDUCTION OF CAMPHOR

NMR Determination. The percentage of each of the isomeric alcohols in the borohydride reduction mixture can be determined from the NMR spectrum. (See Technique 19, Section 19.9, p. 855, and Appendix 4.) The NMR spectra of the pure alcohols are shown on page 185. The hydrogen on the carbon bearing the hydroxyl group appears at 4.0 ppm for borneol and 3.6 ppm for isoborneol. To obtain the product ratio, integrate these peaks (using an expanded presentation) in the NMR spectrum of isoborneol obtained after the borohydride reduction. In the spectrum shown on page 187, the isoborneol–borneol ratio 5:1 was obtained. The percentages obtained are 83% isoborneol and 17% borneol.

Gas Chromatography. Approximate percentages of borneol and isoborneol can also be determined by gas chromatography. We use a Gow-Mac 69-930 instrument with an 8-foot column of 10% Carbowax 20M, and operate the column at 180°C with a 40 mL/min flow rate. The compounds are dissolved in methylene chloride for analysis. Under these conditions, the retention times for camphor, isoborneol and borneol are 8, 10, and 11 minutes, respectively.

REFERENCES

Brown, H. C., and Muzzio, J. "Rates of Reaction of Sodium Borohydride with Bicyclic Ketones." *Journal of the American Chemical Society, 88* (1966): 2811.

Dauben, W. G., Fonken, G. J., and Noyce, D. S. "Stereochemistry of Hydride Reductions." *Journal of the American Chemical Society, 78* (1956): 2579.

Flautt, T. J., and Erman, W. F. "The Nuclear Magnetic Resonance Spectra and Stereochemistry of Substituted Boranes." *Journal of the American Chemical Society, 85* (1963): 3212.

Markgraf, J. H. "Stereochemical Correlations in the Camphor Series." *Journal of Chemical Education, 44* (1967): 36.

Mohrig, J. R., Nienhuis, C. F., Van Zoeren, C., Fox, B. G., and Mahaffy, P. G. "The Design of Laboratory Experiments in the 1980's." *Journal of Chemical Education, 62* (1985): 519.

QUESTIONS

1. Interpret the major absorption bands in the infrared spectra of camphor, borneol, and isoborneol.

2. Explain why the *gem*-dimethyl groups appear as separate peaks in the proton NMR spectrum of isoborneol although they are not resolved in borneol.

3. A sample of isoborneol prepared by reduction of camphor was analyzed by infrared spectroscopy and showed a band at 1760 cm^{-1}. This result was unexpected. Why?

4. The observed melting point of camphor is often low. Look up the molal freezing-point-depression constant K for camphor and calculate the expected depression of the melting point of a quantity of camphor that contains 0.5 molal impurity. *Hint:* Look in a general chemistry book under freezing point depression or colligative properties of solutions.

5. Why was the ether layer washed with sodium bicarbonate in the procedure for the preparation of camphor?

6. The peak assignments are shown on the carbon-13 NMR spectrum of camphor. Using these assignments as a guide, assign as many peaks as possible in the carbon-13 spectra of borneol and isoborneol.

ESSAY

Stereochemical Theory of Odor

The human nose has an almost unbelievable ability to distinguish odors. Just consider for a few moments the different substances you are able to recognize by odor alone. Your list should be a very long one. A person with a trained nose—a perfumer, for instance—can often recognize even individual components in a mixture. Who has not met at least one chef who could sniff almost any culinary dish and identify the seasonings and spices that were used? The olfactory centers in the nose can identify odorous substances even in very small amounts. With some substances, studies have shown that as little as one ten-millionth of a gram (10^{-7} g) can be perceived. Many animals—for example, dogs and insects—have an even lower threshold of smell than humans (see essay on pheromones that precedes Experiment 16).

There have been many theories of odor, but few have persisted very long. Strangely enough, one of the oldest theories, although in modern dress, is still the most current theory. Lucretius, one of the early Greek atomists, suggested that substances having odor gave off a vapor of tiny "atoms," all of the same shape and size, and that they gave rise to the perception of odor when they entered pores in the nose. The pores would have to be of various shapes and the odor perceived would depend on which pores the atoms were able to enter. We now have many similar theories regarding the action of drugs (receptor-site theory) and the interaction of enzymes with their substrates (the lock-and-key hypothesis).

A substance must have certain physical characteristics to have the property of odor. First, it must be volatile enough to give off a vapor that can reach the nostrils. Second, once it reaches the nostrils, it must be somewhat water-soluble, even if only to a small degree, so that it can pass through the layer of moisture (mucus) that covers the nerve endings in the olfactory area. Third, it must have lipid solubility to allow it to penetrate the lipid (fat) layers that form the surface membranes of the nerve cell endings.

Once we pass these criteria, we come to the heart of the question. Why do substances have different odors? In 1949, R. W. Moncrieff, a Scotsman, resurrected Lucretius' hypothesis. He proposed that in the olfactory area of the nose there is a system of receptor cells of several different types and shapes. He further suggested that each receptor site corresponded to a different type of primary odor. Molecules that would fit these receptor sites would display the characteristics of that primary odor. It would not be necessary for the entire molecule to fit into the receptor, so that for larger molecules, any portion might fit into the receptor and activate it. Molecules having complex odors would presumably be able to activate several different types of receptors.

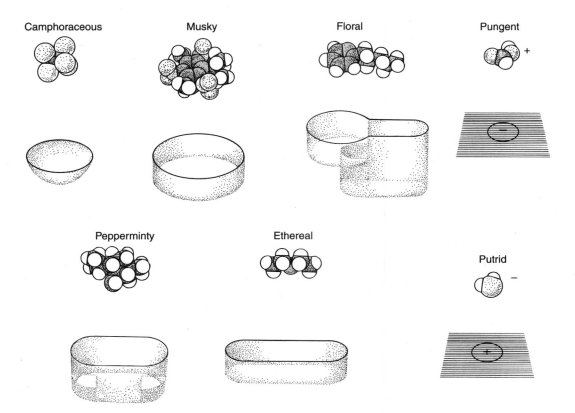

Moncrieff's hypothesis has been strengthened substantially by the work of J. E. Amoore, who began studying the subject as an undergraduate at Oxford in 1952. After an extensive search of the chemical literature, Amoore concluded that there were only seven basic primary odors. By sorting molecules with similar odor types, he even formulated possible shapes for the seven necessary receptors. For instance, from the literature, he culled more than 100 compounds that were described as having a "camphoraceous" odor. Comparing the sizes and shapes of all these molecules, he postulated a three-dimensional shape for a camphoraceous receptor site. Similarly, he derived shapes for the other six receptor sites. The seven primary receptor sites he formulated are shown in the figure, along with a typical prototype molecule of the appropriate shape to fit the receptor. The shapes of the sites are shown in perspective. Pungent and putrid odors are not thought to require a particular shape in the odorous molecules but rather to need a particular type of charge distribution.

You can verify quickly that compounds with molecules of roughly similar shape have similar odors if you compare nitrobenzene and acetophenone with benzaldehyde or *d*-camphor and hexachloroethane with cyclooctane. Each group of substances has the same basic odor **type** (primary), but the individual molecules differ in the **quality** of the odor. Some of the odors are sharp, some pungent, others sweet, and so on. The second group of substances all have a camphoraceous odor, and the molecules of these substances all have approximately the same shape.

An interesting corollary to the Amoore theory is the postulate that if the receptor sites are chiral, then optical isomers (enantiomers) of a given substance might have *different* odors. This circumstance proves true in several cases. It is true for (+)- and (−)-carvone; we investigate the idea in Experiment 21 in this textbook.

Several researchers have tested Amoore's hypothesis by experiment. The results of these studies are generally favorable to the hypothesis—so favorable that some chemists now elevate the hypothesis to the level of theory. In several cases, researchers have been able to "synthesize" odors that are almost indistinguishable from the real thing by properly blending primary odor substances. The primary odor substances used are unrelated to the chemical substances composing the natural odor. These experiments, and others, are described in the articles listed at the end of this essay.

REFERENCES

Amoore, J. E. *The Molecular Basis of Odor.* American Lecture Series Publication, No. 773. Springfield, IL: Thomas, 1970.

Amoore, J. E., Johnston, J. W., Jr., and Rubin, M. "The Stereochemical Theory of Odor." *Scientific American, 210* (February 1964): 1.

Amoore, J. E., Rubin, M., and Johnston, J. W., Jr. "The Stereochemical Theory of Olfaction." *Proceedings of the Scientific Section of the Toilet Goods Association* (Special Supplement to No. 37) (October 1962): 1–47.

Buckenmayer, R. H. "Odor Modification." *Encyclopedia of Chemical Technology,* 3rd ed., Vol. 16. New York: Wiley-Interscience, 1978.

Burton, R. *The Language of Smell.* London: Routledge & Kegan Paul, 1976.

Christian, R. J. *Sensory Experience,* 2nd ed. New York: Harper and Row, 1979.

Lipkowitz, K. B. "Molecular Modeling in Organic Chemistry: Correlating Odors with Molecular Structure." *Journal of Chemical Education, 66* (April 1989): 275.

Moncrieff, R. W. *The Chemical Senses.* London: Leonard Hill, 1951.

Sbrollini, M. C. "Olfactory Delights." *Journal of Chemical Education, 64* (September 1987): 799.

Sell, C. "On the Right Scent." *Chemistry in Britain, 33* (March 1997): 39.

Theimer, E. T., ed. *Fragrance Chemistry.* New York: Academic Press, 1982.

EXPERIMENT 21

Spearmint and Caraway Oil: (+)- and (−)-Carvones

Stereochemistry
Gas chromatography
Polarimetry
Spectroscopy
Refractometry

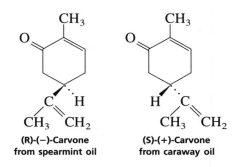

(R)-(−)-Carvone
from spearmint oil

(S)-(+)-Carvone
from caraway oil

In this experiment you will compare (+)-carvone from caraway oil to (−)-carvone from spearmint oil, using gas chromatography. If you have the proper preparative-scale gas chromatographic equipment, it should be possible to prepare pure samples of each of the carvones from their respective oils. If this equipment is not available, the instructor will provide pure samples of the two carvones obtained from a commercial source, and any gas chromatographic work will be strictly analytical.

The odors of the two enantiomeric carvones are distinctly different from each other. The presence of one or the other isomer is responsible for the characteristic odors of each oil. The difference in the odors is to be expected, because the odor receptors in the nose are chiral (see essay, "Stereochemical Theory of Odor," p. 189). This phenomenon, in which a chiral receptor interacts differently with each of the enantiomers of a chiral compound, is called **chiral recognition.**

Although we should expect the optical rotations of the isomers (enantiomers) to be of opposite sign, the other physical properties should be identical. Thus, for both (+)- and (−)-carvone, we predict that the infrared and nuclear magnetic resonance spectra, the gas-chromatographic retention times, the refractive indices, and the boiling points will be identical. Hence, the only differences in properties you will observe for the two carvones are the odors and the signs of rotation in a polarimeter.

Caraway oil contains mainly limonene and (+)-carvone. The gas chromatogram for this oil is shown in the figure. The (+)-carvone (bp 230°C) can easily be separated from the lower-boiling limonene (bp 177°C) by gas chromatography, as shown in the figure on page 193. If one has a preparative gas chromatograph, the (+)-carvone and limonene can be collected separately as they elute from the gas chromatography column.

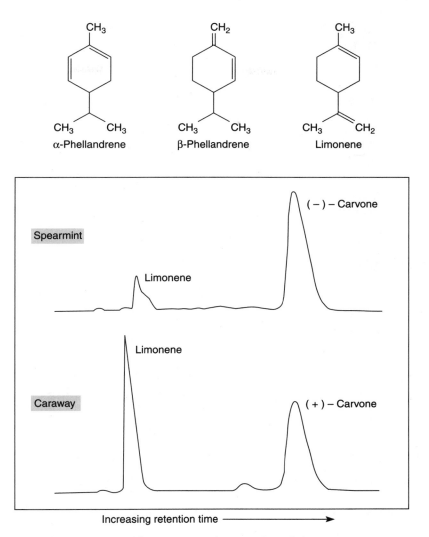

Gas chromatograms of caraway and spearmint oil.

Spearmint oil contains mainly (−)-carvone with a smaller amount of limonene and very small amounts of the lower-boiling terpenes, α- and β-phellandrene. The gas chromatogram for this oil is also shown in the figure. With preparative equipment, you can easily collect the (−)-carvone as it exits the column. It is more difficult, however, to collect limonene in a pure form. It is likely to be contaminated with the other terpenes, since they all have similar boiling points.

Required Reading

Review: Technique 19   Preparation of Samples for Spectroscopy, Part A, Infrared
               Appendix 3   Infrared Spectroscopy

New: Technique 15   Gas Chromatography
        Technique 17   Polarimetry
        Essay   Stereochemical Theory of Odor

If performing any of the optional procedures, read, as appropriate:

Technique 6   Physical Constants, Part B, Boiling Points
Technique 18   Refractometry
Technique 19   Preparation of Samples for Spectroscopy, Part B, Nuclear
               Magnetic Resonance
Appendix 4   Nuclear Magnetic Resonance Spectroscopy
Appendix 5   Carbon-13 Nuclear Magnetic Resonance Spectroscopy

Special Instructions

Your instructor will either assign you spearmint or caraway oil or have you choose one. You will also be given instructions on which procedures from Part A you are to perform. You should compare your data with someone who has studied the other enantiomer.

> **Note:** If a gas chromatograph is not available, this experiment can be performed with spearmint and caraway oils and pure commercial samples of the (+)- and (−)-carvones.

If the proper equipment is available, your instructor may require you to perform a gas chromatographic analysis. If preparative gas chromatography is available, you will be asked to isolate the carvone from your oil (Part B). Otherwise, if you are using analytical equipment, you will be able to compare only the retention times and integrals from your oil to those of the other essential oil.

While preparative gas chromatography will yield enough sample to do spectra, it will not yield enough material to do the polarimetry. Therefore, if you are required to determine the optical rotation of the pure samples, whether or not you perform preparative gas chromatography, your instructor will provide a prefilled polarimeter tube for each sample.

Notes to the Instructor

This experiment may be scheduled along with another experiment. It is best if students work in pairs, each student using a different oil. An appointment schedule for using the gas chromatograph should be arranged so that students are able to make efficient use of their time. You should prepare chromatograms using both carvone isomers and limonene as reference standards. Appropriate reference standards include a mixture of

(+)-carvone and limonene and a second mixture of (−)-carvone and limonene. The chromatograms should be posted with retention times, or each student should be provided with a copy of the appropriate chromatogram.

The gas chromatograph should be prepared as follows: column temperature, 200°C; injection and detector temperature, 210°C; carrier gas flow rate, 20 mL/min. The recommended column is 8 feet long with a stationary phase such as Carbowax 20M. It is convenient to use a Gow-Mac 69-350 instrument with the preparative accessory system for this experiment.

You should fill polarimeter cells (0.5 dm) in advance with the undiluted (+)- and (−)-carvone. There should also be four bottles containing spearmint and caraway oils and (+)- and (−)-carvone. Both enantiomers of carvone are commercially available.

Procedure

PART A.   ANALYSIS OF THE CARVONES

The samples (either those obtained from gas chromatography, Part B, or commercial samples) should be analyzed by the following methods. The instructor will indicate which methods to use. Compare your results with those obtained by someone who used a different oil. In addition, measure the observed rotation of the commercial samples of (+)-carvone and (−)-carvone. The instructor will supply prefilled polarimeter tubes.

Analyses to be Performed on Spearmint and Caraway Oils:

Odor.   Carefully smell the containers of spearmint and caraway oil and of the two carvones. About 8–10% of the population cannot detect the difference in the odors of the optical isomers. Most people, however, find the difference quite obvious. Record your impressions.

Analytical Gas Chromatography.   If you separated your sample by preparative gas chromatography in Part B, you should already have your chromatogram. In this case, you should compare it to one done by someone using the other oil. Be sure to obtain retention times and integrals, or obtain a copy of the other person's chromatogram.

If you did not perform Part B, obtain the analytical gas chromatograms of your assigned oil—spearmint or caraway—and obtain the result from the other oil from someone else. The instructor may prefer to perform the sample injections or have a laboratory assistant perform them. The sample injection procedure requires careful technique, and the special microliter syringes that are required are very delicate and expensive. If you are to perform the injections yourself, your instructor will give you adequate instruction beforehand.

For both oils, determine the retention times of the components (see Technique 15, Section 15.11, p. 820). Calculate the percentage composition of the two essential oils by one of the methods explained in the same section.

Analyses to be Performed on the Purified Carvones:

Polarimetry. With the help of the instructor or assistant, obtain the observed optical rotation α of the pure (+)-carvone and (−)-carvone samples. These are provided in prefilled polarimeter tubes. The specific rotation $[\alpha]_D$ is calculated from the relationship given on page 832 of Technique 17. The concentration c will equal the density of the substances analyzed at 20°C. The values, obtained from actual com-

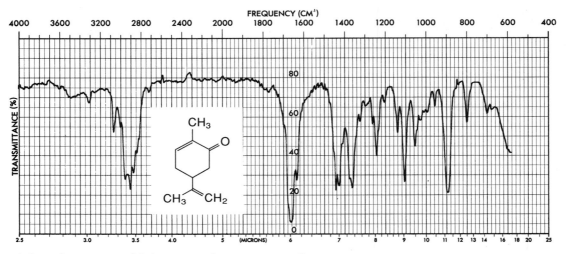

Infrared spectrum of (+)-carvone from caraway oil, neat.

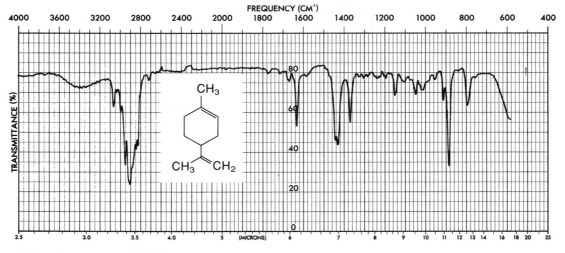

Infrared spectrum of (+)-limonene, neat.

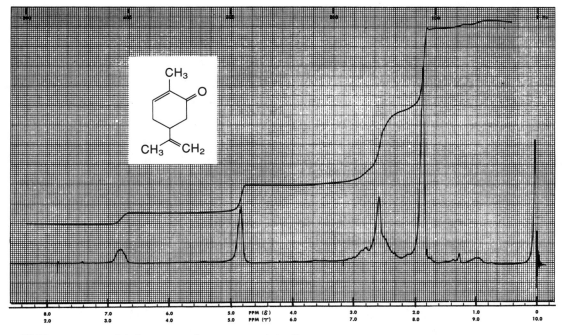

NMR spectrum of (−)-carvone from spearmint oil.

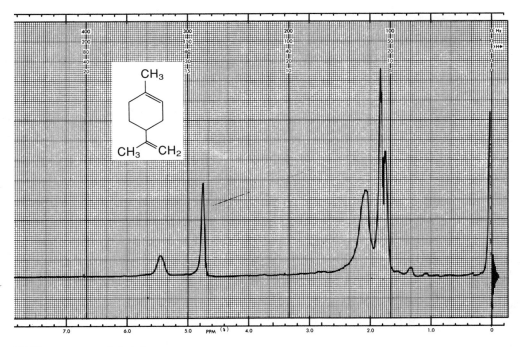

NMR spectrum of (+)-limonene.

mercial samples, are 0.9608 g/mL for (+)-carvone and 0.9593 g/mL for (−)-carvone. The literature values for the specific rotations are as follows: $[\alpha]_D^{20} = +61.7°$ for (+)-carvone and −62.5° for (−)-carvone. These values are not identical because trace amounts of impurities are present.

Polarimetry does not work well on the crude spearmint and caraway oils because of the presence of large amounts of limonene and other impurities.

Infrared Spectroscopy.   Obtain the infrared spectrum of the (−)-carvone sample from spearmint or of the (+)-carvone sample from caraway (see Technique 19, Section 19.2, p. 843). Compare your result with that of a person working with the other isomer. At the option of the instructor, obtain the infrared spectrum of the (+)-limonene, which is found in both oils. If possible, determine all spectra using neat samples. If you isolated the sample by preparative gas chromatography, it may be necessary to add one to two drops of carbon tetrachloride to the sample. Thoroughly mix the liquids by drawing the mixture into a Pasteur pipet and expelling several times. It may be helpful to draw the end of the pipet to a narrow tip in order to withdraw all the liquid in the conical vial. As an alternative, use a microsyringe. Obtain a spectrum of this solution, as described in Technique 19, Section 19.5, page 850.

Nuclear Magnetic Resonance Spectroscopy.   Using an NMR instrument, obtain a proton NMR spectrum of your carvone. Compare your spectrum with the NMR spectra for (−)-carvone and (+)-limonene shown in this experiment. Attempt to assign as many peaks as you can. If your NMR instrument is capable of obtaining a carbon-13 NMR spectrum, determine a carbon-13 spectrum. Compare your spectrum of carvone with the carbon-13 NMR spectrum shown in this experiment. Once again, attempt to assign the peaks.

Boiling Point.   Determine the boiling point of the carvone you were assigned. Use the micro boiling point technique (Technique 6, Section 6.11, p. 677). The boiling points for both carvones are 230°C at atmospheric pressure. Compare your result to that of someone using the other carvone.

Refractive Index.   Use the technique for obtaining the refractive index on a small volume of liquid, as described in Technique 18, Section 18.2, page 838. Obtain the refractive index for the carvone you separated (Part B) or for the one assigned. Compare your value to that obtained by someone using the other isomer. At 20°C, the (+)- and (−)-carvones have the same refractive index, equal to 1.4989.

PART B.   SEPARATION BY GAS CHROMATOGRAPHY (OPTIONAL)

The instructor may prefer to perform the sample injections or have a laboratory assistant perform them. The sample injection procedure requires careful technique, and the special microliter syringes that are required are very delicate and expensive. If you are to perform the sample injections, your instructor will give you adequate instruction beforehand.

Inject 50 μL of caraway or spearmint oil onto the gas chromatography column. Just before a component of the oil (limonene or carvone) elutes from the column, install a gas collection tube at the exit port, as described in Technique 15, Section

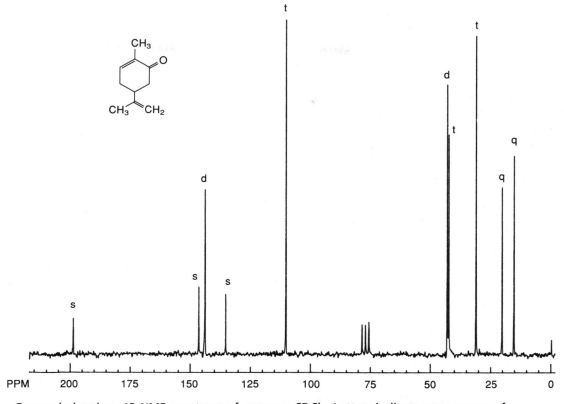

Decoupled carbon-13 NMR spectrum of carvone, $CDCl_3$. Letters indicate appearance of spectrum when carbons are coupled to protons (s = singlet, d = doublet, t = triplet, q = quartet).

15.10, page 818. To determine when to connect the gas collection tube, refer to the chromatograms prepared by your instructor. These chromatograms have been run on the same instrument you are using and under the same conditions. Ideally, you should connect the gas collection tube just before the limonene or carvone elutes from the column and remove the tube as soon as all the component has been collected, but before any other compound begins to elute from the column. You can accomplish this most easily by watching the recorder as your sample passes through the column. The collection tube is connected (if possible) just before a peak is produced, or as soon as a deflection is observed. When the pen returns to the baseline, remove the gas collection tube.

This procedure is relatively easy for collecting the carvone component of both oils and for collecting the limonene in caraway oil. Because of the presence of several terpenes in spearmint oil, it is somewhat more difficult to isolate a pure sample of limonene from spearmint oil (see chromatogram in the figure, p. 193). In this case, you must try to collect only the limonene component and not any other compounds, such as the terpene, which produces a shoulder on the limonene peak in the chromatogram for spearmint oil.

After collecting the samples, insert the ground joint of the collection tube into a 0.1-mL conical vial, using an O-ring and screwcap to fasten the two pieces together securely. Place this assembly into a centrifuge tube, as shown in Technique 15, Figure 15.10, page 819. Put cotton on the bottom of the tube and use a rubber septum cap at the top to hold the assembly in place and to prevent breakage. Balance the centrifuge by placing a tube of equal weight on the opposite side (this could be your other sample or someone else's sample). During centrifugation, the sample is forced into the bottom of the conical vial. Disassemble the apparatus, cap the vial, and perform the analyses described in Part A. You should have enough sample to perform the infrared and NMR spectroscopy, but your instructor may need to provide additional sample to perform the other procedures.

REFERENCES

Friedman, L., and Miller, J. G. "Odor, Incongruity and Chirality." *Science, 172* (1971): 1044.

Murov, S. L., and Pickering, M. "The Odor of Optical Isomers." *Journal of Chemical Education, 50* (1973): 74.

Russell, G. F., and Hills, J. I. "Odor Differences between Enantiomeric Isomers," *Science, 172* (1971): 1043.

QUESTIONS

1. Interpret the infrared spectra for carvone and limonene and the proton and carbon-13 NMR spectra of carvone.

2. Identify the chiral centers in α-phellandrene, β-phellandrene, and limonene.

3. Explain how carvone fits the isoprene rule (see essay, "Terpenes and Phenylpropanoids," p. 165).

4. Using the Cahn–Ingold–Prelog sequence rules, assign priorities to the groups around the chiral carbon in carvone. Draw the structural formulas for (+)- and (−)-carvone with the molecules oriented in the correct position to show the R and S configurations.

5. Explain why limonene elutes from the column before either (+)- or (−)-carvone.

6. Explain why the retention times for both carvone isomers are the same.

7. The toxicity of (+)-carvone in rats is about 400 times greater than that of (−)-carvone. How do you account for this?

E X P E R I M E N T   2 2

Reactivities of Some Alkyl Halides

S_N1/S_N2 reactions
Relative rates
Reactivities

The reactivities of alkyl halides in nucleophilic substitution reactions depend on two important factors: reaction conditions and substrate structure. The reactivities of several substrate types will be examined under both S_N1 and S_N2 reaction conditions in this experiment.

SODIUM IODIDE OR POTASSIUM IODIDE IN ACETONE

A reagent composed of sodium iodide or potassium iodide dissolved in acetone is useful in classifying alkyl halides according to their reactivity in an S_N2 reaction. Iodide ion is an excellent nucleophile, and acetone is a nonpolar solvent. The tendency to form a precipitate increases the completeness of the reaction. Sodium iodide and potassium iodide are soluble in acetone, but the corresponding bromides and chlorides are not soluble. Consequently, as bromide ion or chloride ion is produced, the ion is precipitated from the solution. According to LeChâtelier's Principle, the precipitation of a product from the reaction solution drives the equilibrium toward the right; such is the case in the reaction described here:

$$R—Cl + Na^+I^- \longrightarrow RI + NaCl\downarrow$$

$$R—Br + Na^+I^- \longrightarrow RI + NaBr\downarrow$$

SILVER NITRATE IN ETHANOL

A reagent composed of silver nitrate dissolved in ethanol is useful in classifying alkyl halides according to their reactivity in an S_N1 reaction. Nitrate ion is a poor nucleophile, and ethanol is a moderately powerful ionizing solvent. The silver ion, because of its ability to coordinate the leaving halide ion to form a silver halide precipitate, greatly assists the ionization of the alkyl halide. Again, a precipitate as one of the reaction products also enhances the reaction.

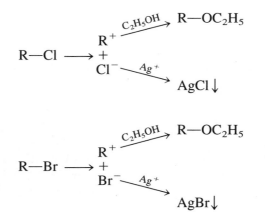

Required Reading

Before this experiment, review the chapters dealing with nucleophilic substitution in your lecture textbook.

Special Instructions

Some compounds used in this experiment, particularly crotyl chloride and benzyl chloride, are powerful lachrymators. **Lachrymators** cause eye irritation and the formation of tears.

> **Caution:** Because some of these compounds are lachrymators, perform these tests in a hood. Be careful to dispose of the test solutions in a waste container marked for halogenated organic waste. After testing, rinse the test tubes with acetone and pour the contents into the same waste container.

Waste Disposal

Dispose of all the halide wastes into the container marked for halogenated waste. Any acetone washings should also be placed in the same container. Any leftover sodium iodide-in-acetone or silver nitrate-in-ethanol solutions should be discarded in specially marked waste containers.

Notes to the Instructor

Each of the halides should be checked with NaI/acetone and $AgNO_3$/ethanol to test for their purity before the class performs this experiment.

Procedure

PART A.   SODIUM IODIDE IN ACETONE

The Experiment.   Label a series of ten clean and dry test tubes (10 × 75-mm test tubes may be used) from 1 to 10. In each test tube add 2 mL of a 15% NaI-in-acetone solution. Now add 4 drops of one of the following halides to the appropriate test tube: (1) 2-chlorobutane, (2) 2-bromobutane, (3) 1-chlorobutane, (4) 1-bromobutane, (5) 2-chloro-2-methylpropane (*t*-butyl chloride), (6) crotyl chloride $CH_3CH{=}CHCH_2Cl$ (see Special Instructions), (7) benzyl chloride (α-chlorotoluene) (see Special Instructions), (8) bromobenzene, (9) bromocyclohexane, and (10) bromocyclopentane. Make certain that you return the dropper to the proper container so as to avoid cross contamination of these halides.

Reaction at Room Temperature.   After adding the halide, shake the test tube well to ensure adequate mixing of the alkyl halide and the solvent. Record the times needed for any precipitate or cloudiness to form.

Reaction at Elevated Temperature.   After about 5 minutes, place any test tubes that do not contain a precipitate in a 50°C water bath. Be careful not to allow the temperature of the water bath to exceed 50°C, since the acetone will evaporate or boil out of the test tube. After about 1 minute of heating, cool the test tubes to room temperature and note whether a reaction has occurred. Record the results.

Observations.   Generally, reactive halides give a precipitate within 3 minutes at room temperature, moderately reactive halides give a precipitate when heated, and unreactive halides do not give a precipitate even after being heated. Ignore any color changes.

Report.   Record your results in tabular form in your notebook. Explain why each compound has the reactivity that you observed. Explain the reactivities in terms of structure.

PART B.   SILVER NITRATE IN ETHANOL

The Experiment.   Label a series of ten clean and dry test tubes from 1 to 10, as described in the previous section. Add 2 mL of a 1% ethanolic silver nitrate solution to each test tube. Now add 4 drops of the appropriate halide to each test tube, using the same numbering scheme indicated for the sodium iodide test. Return the dropper to the proper container so as to avoid cross contamination of these halides.

Reaction at Room Temperature.   After adding the halide, shake the test tube well to ensure adequate mixing of the alkyl halide and the solvent. After thoroughly mixing the samples, record the times needed for any precipitate or cloudiness to form. Record your results as dense precipitate, cloudiness, or no precipitate/cloudiness.

Reaction at Elevated Temperature.   After about 5 minutes, place any test tubes that do not contain a precipitate or cloudiness in a hot water bath at about 100°C. After about 1 minute of heating, cool the test tubes to room temperature and note whether a reaction has occurred. Record your results as dense precipitate, cloudiness, or no precipitate/cloudiness.

Observations.   Reactive halides give a precipitate (or cloudiness) within 3 minutes at room temperature, moderately reactive halides give a precipitate (or cloudiness) when heated, and unreactive halides do not give a precipitate even after being heated. Ignore any color changes.

Report.   Record your results in tabular form in your notebook. Explain why each compound has the reactivity that you observed. Explain the reactivities in terms of structure. Compare relative reactivities for compounds with similar structures.

QUESTIONS

1. In the tests with sodium iodide in acetone and silver nitrate in ethanol, why should 2-bromobutane react faster than 2-chlorobutane?

2. Why is benzyl chloride reactive in both tests, while bromobenzene is unreactive?

3. When benzyl chloride is treated with sodium iodide in acetone, it reacts much faster than 1-chlorobutane, even though both compounds are primary alkyl chlorides. Explain this rate difference.

4. 2-Chlorobutane reacts much more slowly than 2-chloro-2-methylpropane in the silver nitrate test. Explain this difference in reactivity.

5. Bromocyclopentane is more reactive than bromocyclohexane when heated with sodium iodide in acetone. Explain this difference in reactivity.

6. How do you expect the following series of compounds to compare in behavior in the two tests?

$$CH_3—CH{=}CH—CH_2—Br \qquad CH_3—\underset{\underset{Br}{|}}{C}{=}CH—CH_3 \qquad CH_3—CH_2—CH_2—CH_2—Br$$

E X P E R I M E N T   2 3

Synthesis of *n*-Butyl Bromide and *t*-Pentyl Chloride

Synthesis of alkyl halides
Extraction
Simple distillation

The synthesis of two alkyl halides from alcohols is the basis for these experiments. In the first experiment, a primary alkyl halide *n*-butyl bromide is prepared as shown in Equation 1.

$$CH_3CH_2CH_2CH_2OH + NaBr + H_2SO_4 \longrightarrow CH_3CH_2CH_2CH_2Br + NaHSO_4 + H_2O \quad (1)$$

   n-Butyl alcohol   *n*-Butyl bromide

In the second experiment, a tertiary alkyl halide *t*-pentyl chloride (*t*-amyl chloride) is prepared as shown in Equation 2.

$$CH_3CH_2—\underset{\underset{OH}{|}}{\overset{\overset{CH_3}{|}}{C}}—CH_3 + HCl \longrightarrow CH_3CH_2—\underset{\underset{Cl}{|}}{\overset{\overset{CH_3}{|}}{C}}—CH_3 + H_2O \quad (2)$$

           t-Pentyl alcohol   *t*-Pentyl chloride

These reactions provide an interesting contrast in mechanisms. The *n*-butyl bromide synthesis proceeds by an S_N2 mechanism, while *t*-pentyl chloride is prepared by an S_N1 reaction.

n-BUTYL BROMIDE

The primary alkyl halide *n*-butyl bromide can be prepared easily by allowing *n*-butyl alcohol to react with sodium bromide and sulfuric acid by Equation 1. The sodium bromide reacts with sulfuric acid to produce hydrobromic acid.

$$2 \text{ NaBr} + \text{H}_2\text{SO}_4 \longrightarrow 2 \text{ HBr} + \text{Na}_2\text{SO}_4$$

Excess sulfuric acid serves to shift the equilibrium and thus to speed the reaction by producing a higher concentration of hydrobromic acid. The sulfuric acid also protonates the hydroxyl group of *n*-butyl alcohol so that water is displaced rather than the hydroxide ion OH^-. The acid also protonates the water as it is produced in the reaction and deactivates it as a nucleophile. Deactivation of water keeps the alkyl halide from being converted back to the alcohol by nucleophilic attack of water. The reaction of the primary substrate proceeds via an S_N2 mechanism.

$$\text{CH}_3\text{CH}_2\text{CH}_2\text{CH}_2\text{—O—H} + \text{H}^+ \xrightarrow{\text{fast}} \text{CH}_3\text{CH}_2\text{CH}_2\text{CH}_2\text{—}\overset{+}{\text{O}}\text{—H}$$
$$\underset{\text{H}}{|}$$

$$\text{CH}_3\text{CH}_2\text{CH}_2\text{CH}_2\text{—}\overset{+}{\underset{\overset{|}{\text{H}}}{\text{O}}}\text{—H} + \text{Br}^- \xrightarrow[S_N2]{\text{slow}} \text{CH}_3\text{CH}_2\text{CH}_2\text{CH}_2\text{—Br} + \text{H}_2\text{O}$$

During the isolation of the *n*-butyl bromide, the crude product is washed with sulfuric acid, water, and sodium bicarbonate to remove any remaining acid or *n*-butyl alcohol.

t-PENTYL CHLORIDE

The tertiary alkyl halide can be prepared by allowing *t*-pentyl alcohol to react with concentrated hydrochloric acid according to Equation 2. The reaction is accomplished simply by shaking the two reagents in a separatory funnel. As the reaction proceeds, the insoluble alkyl halide product forms an upper phase. The reaction of the tertiary substrate occurs via an S_N1 mechanism.

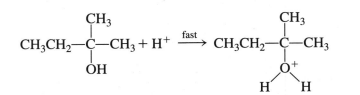

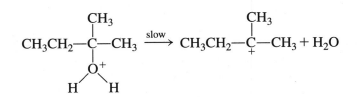

A small amount of alkene, 2-methyl-2-butene, is produced as a by-product in this reaction. If sulfuric acid had been used as it was for *n*-butyl bromide, a much larger amount of this alkene would have been produced.

Required Reading

Review: Techniques 1, 2, 3, 7, and 8

Special Instructions

Caution: Take special care with concentrated sulfuric acid; it causes severe burns.

As your instructor indicates, perform either the *n*-butyl bromide or the *t*-pentyl chloride procedure, or both.

Waste Disposal

Dispose of all aqueous solutions produced in this experiment in the container for aqueous waste.

EXPERIMENT 23A

n-Butyl Bromide

Procedure

Preparation of *n*-Butyl Bromide.   Place 17.0 g of sodium bromide in a 100-mL round-bottom flask and add 17 mL of water and 10.0 mL of *n*-butyl alcohol (1-butanol, *MW* = 74.1, *d* = 0.81 g/mL). Cool the mixture in an ice bath and slowly add 14 mL of concentrated sulfuric acid with continuous swirling in the ice bath. Add several boiling stones to the mixture and assemble the reflux apparatus and trap shown in the figure. The trap absorbs the hydrogen bromide gas evolved during the reaction period. Heat the mixture to a gentle boil for 60–75 minutes.

Extraction.   Remove the heat source and allow the apparatus to cool until you can disconnect the round-bottom flask without burning your fingers.

Note: Do not allow the reaction mixture to cool to room temperature. Complete the operations in this paragraph as quickly as possible. Otherwise, salts may precipitate, making this procedure more difficult to perform.

Disconnect the round-bottom flask and carefully pour the reaction mixture into a 125-mL separatory funnel. The *n*-butyl bromide layer should be on top. If the reaction is not yet complete, the remaining *n*-butyl alcohol will sometimes form a *second organic layer* on top of the *n*-butyl bromide layer. Treat both organic layers as if they were one. Drain the lower aqueous layer from the funnel.

The organic and aqueous layers should separate as described in the following instructions. However, to make sure that you do not discard the wrong layer, it would be a good idea to add a drop of water to any aqueous layer you plan to discard. If a drop of water dissolves in the liquid, you can be confident that it is an aqueous layer. Add 14 mL of 9*M* H_2SO_4 to the separatory funnel and shake the mixture (Technique 7, Section 7.4, p. 690). Allow the layers to separate. Because any remaining *n*-butyl alcohol is extracted by the H_2SO_4 solution, there should now be only one organic layer. The organic layer should be the top layer. Drain and discard the lower aqueous layer.

Add 14 mL H_2O to the separatory funnel. Stopper the funnel and shake it, venting occasionally. Allow the layers to separate. Drain the lower layer, which contains *n*-butyl bromide (*d* = 1.27 g/mL), into a small beaker. Discard the aqueous layer after making certain the correct layer has been saved. Return the alkyl halide to the funnel. Add 14 mL of saturated aqueous sodium bicarbonate, a little at a time, while

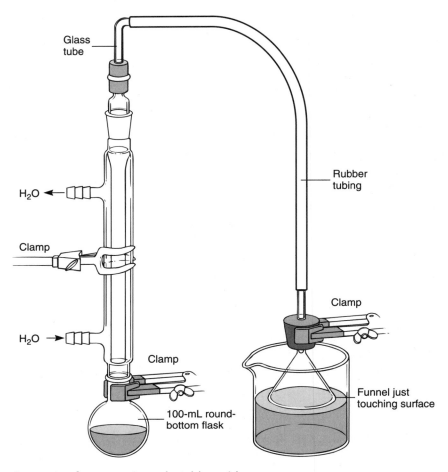

Glass
tube

Rubber
tubing

H₂O ←

Clamp

H₂O →

Clamp

Clamp

Funnel just
touching surface

100-mL round-
bottom flask

Apparatus for preparing *n*-butyl bromide.

swirling. Stopper the funnel and shake it for 1 minute, venting frequently to relieve
any pressure that is produced. Drain the lower alkyl halide layer into a dry Erlenmeyer
flask. Add 1.0 g of anhydrous calcium chloride to dry the solution (Technique 7,
Section 7.8, p. 696). Stopper the flask and swirl the contents until the liquid is *clear.*
The drying process can be accelerated by *gently* warming the mixture on a steam
bath.

   Distillation.   Transfer the clear liquid to a *dry* 25-mL round-bottom flask using
a Pasteur pipet. Add a boiling stone and distill the crude *n*-butyl bromide in a *dry*
apparatus (Technique 8, Section 8.3, Figure 8.4, p. 711). Collect the material that boils
between 94 and 102°C. Weigh the product, calculate the percentage yield and determine a microscale boiling point. Determine the infrared spectrum of the product
using salt plates (Technique 19, Section 19.2, p. 843). Submit the remainder of your
sample in a properly labeled vial, along with the infrared spectrum, when you submit your report to the instructor.

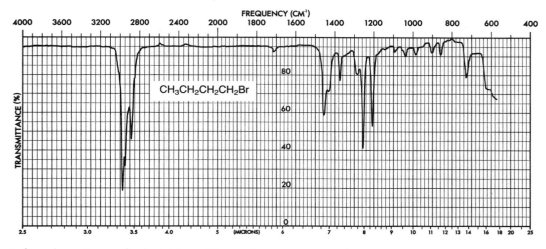

Infrared spectrum of *n*-butyl bromide, neat.

EXPERIMENT 23B

t-Pentyl Chloride

Procedure

Preparation of *t*-Pentyl Chloride. In a 125-mL separatory funnel, place 10.0 mL of *t*-pentyl alcohol (2-methyl-2-butanol, *MW* = 88.2, *d* = 0.805 g/mL) and 25 mL of concentrated hydrochloric acid (*d* = 1.18 g/mL). Do not stopper the funnel. Gently swirl the mixture in the separatory funnel for about 1 minute. After this period of swirling, stopper the separatory funnel and carefully invert it. Without shaking the separatory funnel, immediately open the stopcock to release the pressure. Close the stopcock, shake the funnel several times, and again release the pressure through the stopcock (Technique 7, Section 7.4, p. 690). Shake the funnel for 2 to 3 minutes, with occasional venting. Allow the mixture to stand in the separatory funnel until the two layers have completely separated. The *t*-pentyl chloride (*d* = 0.865 g/mL) should be the top layer, but be sure to verify this by adding a few drops of water. The water should dissolve in the lower (aqueous) layer. Drain and discard the lower layer.

Extraction. The operations in this paragraph should be done as rapidly as possible since the *t*-pentyl chloride is unstable in water and sodium bicarbonate solution. It is easily hydrolyzed back to the alcohol. In each of the following steps, the organic layer should be on top; however, you should add a few drops of water to make sure. Wash (swirl and shake) the organic layer with 10 mL of water. Separate the layers and discard the aqueous phase after making certain that the proper layer has been saved. Add a 10-mL portion of 5% aqueous sodium bicarbonate to the sep-

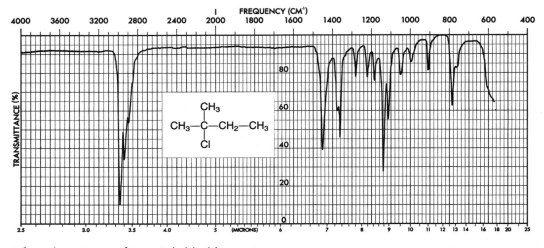

Infrared spectrum of *t*-pentyl chloride, neat.

aratory funnel. Gently swirl the funnel (unstoppered) until the contents are thoroughly mixed. Stopper the funnel, and carefully invert it. Release the excess pressure through the stopcock. Gently shake the separatory funnel, with frequent release of pressure. Following this, vigorously shake the funnel, again with release of pressure, for about 1 minute. Allow the layers to separate, and drain the lower aqueous layer. Wash (swirl and shake) the organic layer with one 10-mL portion of water, and again drain the lower aqueous layer.

Transfer the organic layer to a small dry Erlenmeyer flask by pouring it from the top of the separatory funnel. Dry the crude *t*-pentyl chloride over 1.0 g of anhydrous calcium chloride until it is clear (Technique 7, Section 7.8, p. 696). Swirl the alkyl halide with the drying agent to aid the drying.

Distillation.   Transfer the clear liquid to a *dry* 25-mL round-bottom flask using a Pasteur pipet. Add a boiling stone and distill the crude *t*-pentyl chloride in a *dry* apparatus (Technique 8, Section 8.3, Figure 8.4, p. 711). Collect the pure *t*-pentyl chloride in a receiver cooled in ice. Collect the material that boils between 78 and 84°C. Weigh the product, calculate the percentage yield, and determine the microscale boiling point. Determine the infrared spectrum of the product using salt plates (Technique 19, Section 19.2, p. 843). Submit the remainder of your sample in a properly labeled vial, along with the infrared spectrum, when you submit your report to the instructor.

QUESTIONS

n-Butyl Bromide

1. What are the formulas of the salts that may precipitate when the reaction mixture is cooled?
2. Why does the alkyl halide layer switch from the top layer to the bottom layer at the point where water is used to extract the organic layer?
3. An ether and an alkene are formed as by-products in this reaction. Draw the structures of these by-products and give mechanisms for their formation.

4. Aqueous sodium bicarbonate was used to wash the crude *n*-butyl bromide.

(**a**) What was the purpose of this wash? Give equations.

(**b**) Why would it be undesirable to wash the crude halide with aqueous sodium hydroxide?

5. Look up the density of *n*-butyl chloride (1-chlorobutane). Assume that this alkyl halide was prepared instead of the bromide. Decide whether the alkyl chloride would appear as the upper or the lower phase at each stage of the separation procedure: after the reflux, after the addition of water, and after the addition of sodium bicarbonate.

6. Why must the alkyl halide product be dried carefully with anhydrous calcium chloride before the distillation? (*Hint:* See Technique 10, Section 10.7).

t-Pentyl Chloride

1. Aqueous sodium bicarbonate was used to wash the crude *t*-pentyl chloride.

(**a**) What was the purpose of this wash? Give equations.

(**b**) Why would it be undesirable to wash the crude halide with aqueous sodium hydroxide?

2. Some 2-methyl-2-butene may be produced in the reaction as a by-product. Give a mechanism for its production.

3. How is unreacted *t*-pentyl alcohol removed in this experiment? Look up the solubility of the alcohol and the alkyl halide in water.

4. Why must the alkyl halide product be dried carefully with anhydrous calcium chloride before the distillation? (*Hint:* See Technique 10, Section 10.7).

5. Will *t*-pentyl chloride (2-chloro-2-methylbutane) float on the surface of water? Look up its density in a handbook.

E X P E R I M E N T   2 4

Nucleophilic Substitution Reactions: Competing Nucleophiles

Nucleophilic substitution
Heating under reflux
Extraction
Refractometry
Gas chromatography
NMR spectroscopy

The purpose of this experiment is to compare the relative nucleophilicities of chloride ions and bromide ions toward each of the following alcohols: 1-butanol (*n*-butyl alcohol), 2-butanol (*sec*-butyl alcohol), and 2-methyl-2-propanol (*t*-butyl alcohol). The two nucleophiles will be present at the same time in each reaction, in equimolar concentrations, and they will be competing for substrate.

In general, alcohols do not react readily in simple nucleophilic displacement reactions. If they are attacked by nucleophiles directly, hydroxide ion, a strong base, must be displaced. Such a displacement is not energetically favorable, and it cannot occur to any reasonable extent:

$$X^- + ROH \not\longrightarrow R-X + OH^-$$

To avoid this problem, you must carry out nucleophilic displacement reactions on alcohols in acidic media. In a rapid initial step, the alcohol is protonated; then water, a very stable molecule, is displaced. This displacement is energetically very favorable, and the reaction proceeds in high yield:

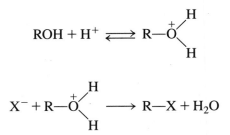

Once the alcohol is protonated, it reacts by either the S_N1 or the S_N2 mechanism, depending on the structure of the alkyl group of the alcohol. For a brief review of these mechanisms, you should consult the chapters on nucleophilic substitution in your lecture textbook.

You will analyze the products of the three reactions in this experiment by a variety of techniques to determine the relative amounts of alkyl chloride and alkyl bromide formed in each reaction. That is, using equimolar concentrations of chloride ions and bromide ions reacting with 1-butanol, 2-butanol, and 2-methyl-2-propanol, you will try to determine which ion is the better nucleophile. In addition, you will try to determine for which of the three substrates (reactions) this difference is important and whether an S_N1 or S_N2 mechanism predominates in each case.

Required Reading

Review:	Techniques 1, 2	
	Technique 3	Reaction Methods, Sections 3.4, 3.6, 3.7, and 3.9
	Technique 7	Extractions, Separations, and Drying Agents, Sections 7.5, 7.7, 7.8, and 7.10
	Technique 15	Gas Chromatography
	Technique 18	Refractometry
	Technique 19	Preparation of Samples for Spectroscopy, Part B, Nuclear Magnetic Resonance
	Appendix 4	Nuclear Magnetic Resonance Spectroscopy

Before beginning this experiment, review the appropriate chapters on nucleophilic substitution in your lecture textbook.

Special Instructions

Each student will carry out the reaction with 2-methyl-2-propanol. Your instructor will also assign you either 1-butanol or 2-butanol. By sharing your results with other students, you will be able to collect data for all three alcohols. You should begin this experiment by preparing the solvent–nucleophile medium and then starting Experiment 24A. During the lengthy reflux period, you will be instructed to go on to Experiment 24B. When you have prepared the product of that experiment, you will return to complete Experiment 24A. In order to analyze the results of both experiments, your instructor will assign specific analysis procedures in Experiment 24C that the class will accomplish.

The solvent–nucleophile medium contains a high concentration of sulfuric acid. Sulfuric acid is very corrosive; be careful when handling it.

In each experiment, the longer your product remains in contact with water or aqueous sodium bicarbonate, the greater the risk that your product will decompose, leading to errors in your analytical results. Before coming to class, prepare ahead so that you know exactly what you are supposed to do during the purification stage of the experiment.

Waste Disposal

When you have completed the three experiments and all the analyses have been completed, discard any remaining alkyl halide mixture in the organic waste container marked for the disposal of halogenated substances. All aqueous solutions produced in this experiment should be disposed of in the container for aqueous waste.

Notes to the Instructor

Be certain that the *tert*-butyl alcohol has been melted before the beginning of the laboratory period.

The gas chromatograph should be prepared as follows: column temperature, 100°C; injection and detector temperature, 130°C; carrier gas flow rate, 50 mL/min. The recommended column is 8 feet long with a stationary phase such as Carbowax 20M. If you wish to analyze the products from the reaction of *tert*-butyl alcohol (Experiment 24B) by gas chromatography, be sure that the *tert*-butyl halides do not undergo decomposition under the conditions set for the gas chromatograph. *tert*-Butyl bromide is very susceptible to elimination.

Procedure

Caution: Take special care with concentrated sulfuric acid; it causes severe burns.

Place 10 g of ice in a 100-mL beaker and carefully add 7.6 mL of concentrated sulfuric acid. Carefully weigh 1.90 g of ammonium chloride and 3.50 g of ammonium bromide into a beaker. Crush any lumps of the reagents to powder and then, using a powder funnel, transfer the halides to a 125-mL Erlenmeyer flask. Exercising caution, add the sulfuric acid mixture to the ammonium salts. Swirl the mixture vigorously to dissolve the salts. It will probably be necessary to heat the mixture on a steam bath or a hotplate to achieve total solution. If necessary, you may add as much as 1 mL of water at this stage. Do not worry if a few small granules do not dissolve.

When solution has been achieved, allow the liquid to cool slightly. However, if you allow it to cool too much and the salts begin to precipitate, you will need to heat the mixture again to achieve solution. Place 6.0 mL of water into a 15-mL screw-cap centrifuge tube with a tight-fitting cap. Make a mark on the tube at the 6.0 mL level and pour out the water. While holding the centrifuge tube close to the Erlenmeyer flask containing the solvent–nucleophile mixture, carefully transfer 6.0 mL of the mixture to the centrifuge tube using a Pasteur pipet. Cap the centrifuge tube. Pour the remainder of the mixture into a 25-mL round-bottom flask. A small portion of the salts in the centrifuge tube or the flask or both may precipitate as the solution cools. Do not worry about these at this point; they will redissolve during the reactions.

E X P E R I M E N T   2 4 A

Competitive Nucleophiles with 1-Butanol or 2-Butanol

Procedure

Apparatus.   Assemble an apparatus for reflux using the 25-mL round-bottom flask containing the solvent–nucleophile mixture, a reflux condenser, and a trap, as shown in the figure. Use a heating mantle as the heat source. The beaker of water will trap the hydrogen chloride and hydrogen bromide gases produced during the reaction.

Using the following procedure, add 1.0 mL of 1-butanol (n-butyl alcohol) or 1.0 mL of 2-butanol (sec-butyl alcohol), depending on which alcohol you were assigned, to the solvent–nucleophile mixture contained in the reflux apparatus. Remove the condenser, and with a Pasteur pipet dispense the alcohol into the round-bottom flask. Also add an inert boiling stone.[1] Replace the condenser and start circulating the cooling water. Adjust the heat from the heating mantle so that this mixture maintains a

---

[1]Do not use calcium carbonate–based stones or Boileezers, because they will partially dissolve in the highly acidic reaction mixture.

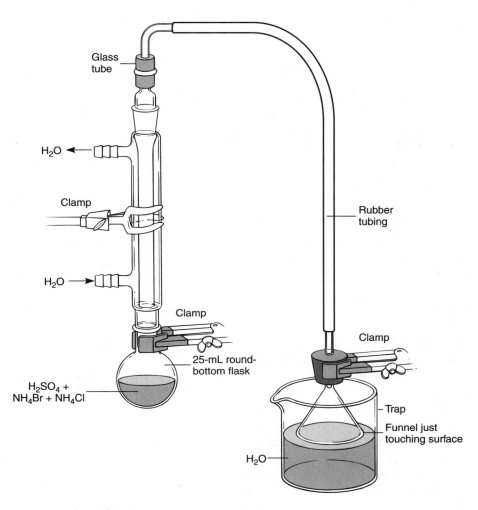

Glass tube

H₂O ←

Clamp

H₂O →

Clamp

Rubber tubing

Clamp

25-mL round-bottom flask

$H_2SO_4 +$
$NH_4Br + NH_4Cl$

Trap

Funnel just touching surface

H₂O

Apparatus for reflux.

gentle boiling action. Be very careful to adjust the reflux ring, if one is visible, so that it remains in the lower fourth of the condenser. Violent boiling will cause loss of product. Continue heating the reaction mixture containing 1-butanol for 75 minutes. Heat the mixture containing 2-butanol for 60 minutes. During this heating period, go on to Experiment 24B and complete as much of it as possible before returning to this procedure.

Purification.   When the period of reflux has been completed, discontinue heating, remove the heating mantle, and allow the reaction mixture to cool. Do not remove the condenser until the flask is cool. Be careful not to shake the hot solution as you lift it from the heating block, or a violent boiling and bubbling action will result; this could allow material to be lost out the top of the condenser. After the mixture has cooled for about 10 minutes, immerse the round-bottom flask (with condenser attached) in a beaker of cold tap water (no ice) and wait for this mixture to cool down to room temperature.

There should be an organic layer present at the top of the reaction mixture. Add 1.0 mL of pentane to the mixture and *gently* swirl the flask. The purpose of the pentane is to increase the volume of the organic layer so that the following operations are easier to accomplish. Using a Pasteur pipet, transfer most (about 10 mL) of the bottom (aqueous) layer to another container. Be very careful that all of the top organic layer remains in the boiling flask. Transfer the remaining aqueous layer and the organic layer to a centrifuge tube with a tight-fitting cap, taking care to leave behind any solids that may have precipitated. Allow the phases to separate and remove the bottom (aqueous) layer using a Pasteur pipet.

Note: For the following sequence of steps, be certain to be well-prepared in advance. If you find that you are taking longer than 5 minutes to complete the entire extraction sequence, you probably have affected your results adversely!

Add 1.5 mL of water to the centrifuge tube, cap the tube, and *gently* shake this mixture. Allow the layers to separate and remove the aqueous layer, which is still on the bottom. Extract the organic layer with 2 mL of saturated sodium bicarbonate solution and remove the bottom aqueous layer.

Drying. Using a clean dry Pasteur pipet, transfer the remaining organic layer into a small test tube (10 × 75 mm) containing about 0.4 g of anhydrous granular sodium sulfate. Stir the mixture with a small spatula, put a stopper on the tube, and set it aside for 10–15 minutes or until the solution is clear. If the mixture does not turn clear, add more anhydrous sodium sulfate. Transfer the halide solution with a clean, dry Pasteur pipet to a small, dry vial, taking care not to transfer any solid. *Be sure the cap is screwed on tightly.* It is a good idea to cover the cap with Parafilm. Do not store the liquid in a container with a cork or a rubber stopper, because these will absorb the halides. If it is necessary to store the sample overnight, store it in a refrigerator. This sample can now be analyzed by as many of the methods in Experiment 24C as your instructor indicates. However, it cannot be analyzed by refractometry because of the presence of pentane.

EXPERIMENT 24B

Competitive Nucleophiles with 2-Methyl-2-Propanol

Procedure

Transfer 1.0 mL of 2-methyl-2-propanol (*tert*-butyl alcohol, mp 25°C) to the 15-mL centrifuge tube containing the other portion of the solvent–nucleophile mixture, which should have cooled to room temperature by this time. Replace the cap and make sure that it doesn't leak.

> **Caution:** The solvent-nucleophile mixture contains concentrated sulfuric acid.

Shake the tube vigorously, venting occasionally, for 5 minutes. Any solids that were originally present in the centrifuge tube should dissolve during this period. After shaking, allow the layer of alkyl halides to separate (10–15 minutes at most). A fairly distinct top layer containing the products should have formed by this time.

> **Caution:** *tert*-Butyl halides are very volatile and should not be left in an open container any longer than necessary.

Slowly remove most of the bottom aqueous layer with a Pasteur pipet and transfer it to a beaker. After waiting 10–15 seconds, remove the remaining lower layer in the centrifuge tube, including a small amount of the upper organic layer, so as to be certain that the organic layer is not contaminated by any water.

> **Note:** For the following purification sequence, be certain to be well-prepared in advance. If you find that you are taking longer than 5 minutes to complete the entire sequence, you probably have affected your results adversely!

Using a dry Pasteur pipet, transfer the remainder of the alkyl halide layer into a small test tube (10 × 75 mm) containing about 0.05 g of solid sodium bicarbonate. As soon as the bubbling stops and a clear liquid is obtained, transfer it with a Pasteur pipet into a small, dry vial, taking care not to transfer any solid. *Be sure the cap is screwed on tightly.* Again, it is a good idea to cover the cap with Parafilm. Do not store the liquid in a container with a cork or a rubber stopper, because these will absorb the halides. If it is necessary to store the sample overnight, store it in a refrigerator. This sample can now be analyzed by as many of the methods in Experiment 24C as your instructor indicates. When you have finished this procedure, return to Experiment 24A.

EXPERIMENT 24C

Analysis

Procedure

The ratio of 1-chlorobutane to 1-bromobutane, 2-chlorobutane to 2-bromobutane, or *tert*-butyl chloride to *tert*-butyl bromide must be determined. At your instructor's option, you may do this by one of three methods: gas chromatography, refractive index, or NMR spectroscopy. The products obtained from the reactions of 1-butanol and 2-butanol, however, cannot be analyzed by the refractive index method (they contain pentane). The products obtained from the reaction of *tert*-butyl alcohol may be difficult to analyze by gas chromatography, because the *tert*-butyl halides sometimes undergo elimination in the gas chromatograph.[2]

GAS CHROMATOGRAPHY[3]

The instructor or a laboratory assistant may either make the sample injections or allow you to make them. In the latter case, your instructor will give you adequate instruction beforehand. A reasonable sample size is 2.5 μL. Inject the sample into the gas chromatograph and record the gas chromatogram. The alkyl chloride, because of its greater volatility, has a shorter retention time than the alkyl bromide.

Once the gas chromatogram has been obtained, determine the relative areas of the two peaks (Technique 15, Section 15.11, p. 820). While the peaks may be cut out and weighed on an analytical balance as a method of determining areas, triangulation is the preferred method. Record the percentages of alkyl chloride and alkyl bromide in the reaction mixture.

REFRACTIVE INDEX

Measure the refractive index of the product mixture (Technique 18). To determine the composition of the mixture, assume a linear relation between the refractive index and the molar composition of the mixture. At 20°C the refractive indices of the alkyl halides are

---

[2]Note to the Instructor: To obtain reasonable results for the gas chromatographic analysis of the *tert*-butyl halides, it may be necessary to supply the students with response factor corrections (Technique 15, Section 15.11, p. 820)

[3]Note to the Instructor: If pure samples of each product are available, check the assumption used here that the gas chromatograph responds equally to each substance. Response factors (relative sensitivities) are easily determined by injecting an equimolar mixture of products and comparing the peak areas.

tert-butyl chloride   1.3877

tert-butyl bromide   1.4280

If the temperature of the laboratory room is not 20°C, the refractive index must be corrected. Add 0.00045 refractive index unit to the observed reading for each degree above 20°C and subtract the same amount for each degree below this temperature. Record the percentages of alkyl chloride and alkyl bromide in the reaction mixture.

NUCLEAR MAGNETIC RESONANCE SPECTROSCOPY

The instructor or a laboratory assistant will record the NMR spectrum of the reaction mixture.[4] Submit a sample vial containing the mixture for this spectral determination. The spectrum will also contain integration of the important peaks (Appendix 4, Nuclear Magnetic Resonance Spectroscopy).

If the substrate alcohol was 1-butanol, the resulting halide and pentane mixture will give rise to a complicated spectrum. Each alkyl halide will show a downfield triplet caused by the CH_2 group nearest the halogen. This triplet will appear further downfield for the alkyl chloride than for the alkyl bromide. In a 60-MHz spectrum, these triplets will overlap, but one branch of each triplet will be available for comparison. Compare the integral of the *downfield* branch of the triplet for 1-chlorobutane with the *upfield* branch of the triplet for 1-bromobutane. The upper spectrum on page 220 provides an example. The relative heights of these integrals correspond to the relative amounts of each halide in the mixture.

If the substrate alcohol was 2-methyl-2-propanol, the resulting halide mixture will show two peaks in the NMR spectrum. Each halide will show a singlet, because all the CH_3 groups are equivalent and are not coupled. In the reaction mixture the upfield peak is due to *tert*-butyl chloride, and the downfield peak is caused by *tert*-butyl bromide. Compare the integrals of these peaks. The lower spectrum on page 220 provides an example. The relative heights of these integrals correspond to the relative amounts of each halide in the mixture.

REPORT

Record the percentages of alkyl chloride and alkyl bromide in the reaction mixture for each of the three alcohols. You need to share your data from the reaction with 1-butanol or 2-butanol with other students in order to do this. The report must include the percentages of each alkyl halide determined by each method used in this experiment for the two alcohols you studied. On the basis of product distribution, develop

---

[4]It is difficult to determine the ratio of 2-chlorobutane to 2-bromobutane using nuclear magnetic resonance. This method requires at least a 90-MHz instrument. At 300 MH₃, the peaks are completely resolved.

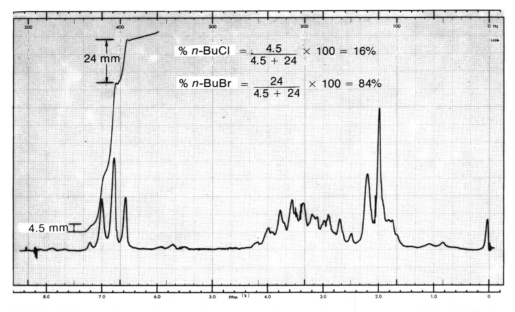

$$\% \text{ } n\text{-BuCl} = \frac{4.5}{4.5 + 24} \times 100 = 16\%$$

$$\% \text{ } n\text{-BuBr} = \frac{24}{4.5 + 24} \times 100 = 84\%$$

A 60-MHz NMR spectrum of 1-chlorobutane and 1-bromobutane, sweep width 250 Hz (no pentane in sample).

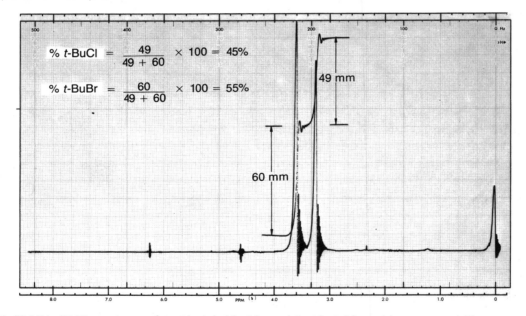

$$\% \text{ } t\text{-BuCl} = \frac{49}{49 + 60} \times 100 = 45\%$$

$$\% \text{ } t\text{-BuBr} = \frac{60}{49 + 60} \times 100 = 55\%$$

A 60-MHz NMR spectrum of *tert*-butyl chloride and *tert*-butyl bromide, sweep width 250 Hz.

an argument for which mechanism (S_N1 or S_N2) predominated for each of the three alcohols studied. The report should also include a discussion of which is the better nucleophile, chloride ion, or bromide ion, based on the experimental results. All gas chromatograms, refractive index data, and spectra should be attached to the report.

QUESTIONS

1. Draw complete mechanisms that explain the resultant product distributions observed for the reactions of *tert*-butyl alcohol and 1-butanol under the reaction conditions of this experiment.

2. Which is the better nucleophile, chloride ion or bromide ion? Try to explain this in terms of the nature of the chloride ion and the bromide ion.

3. What is the principal organic by-product of these reactions?

4. A student left some alkyl halides (RCl and RBr) in an open container for several minutes. What happened to the composition of the halide mixture during that time? Assume that some liquid remains in the container.

5. What would happen if all the solids in the nucleophile medium were not dissolved? How might this affect the outcome of the experiment?

6. What might have been the product ratios observed in this experiment if an aprotic solvent like dimethyl sulfoxide had been used instead of water?

7. Explain the order of elution you observed while performing the gas chromatography for this experiment. What property of the product molecules seems to be the most important in determining relative retention times?

8. Does it seem reasonable to you that the refractive index should be a temperature-dependent parameter? Try to explain.

9. When you calculate the percentage composition of the product mixture, exactly what kind of "percentage" (i.e., volume percent, weight percent, mole percent) are you dealing with?

10. When a pure sample of *tert*-butyl bromide is analyzed by gas chromatography, two components are usually observed. One of them is *tert*-butyl bromide and the other one is a decomposition product. As the temperature of the injector is increased the amount of the decomposition product increases and the amount of *tert*-butyl bromide decreases.

(a) What is the structure of the decomposition product?

(b) Why does the amount of the decomposition product increase with increasing temperature?

(c) Why does *tert*-butyl bromide decompose much more easily than *tert*-butyl chloride?

EXPERIMENT 25

Hydrolysis of Some Alkyl Chlorides

Synthesis of an alkyl halide
Use of a separatory funnel
Titration
Kinetics
Computer-interfaced experiment

Two chemical reactions are of interest in this experiment. The first is the preparation of the alkyl chlorides whose hydrolysis rates are to be measured. The chloride formation is a simple nucleophilic substitution reaction carried out in a separatory funnel.

Because the concentration of the initial alkyl chloride does not need to be determined for the kinetic experiment, isolation and purification of the alkyl chloride are not required.

$$ROH + HCl \longrightarrow RCl + H_2O$$

The second reaction is the actual hydrolysis, and the rate of this reaction will be measured. Under the conditions of this experiment, the reaction proceeds by an S_N1 pathway. The reaction rate is monitored by measuring the rate of appearance of hydrochloric acid. In this experiment, the concentration of hydrochloric acid is determined by titration with aqueous sodium hydroxide (Experiment 25A) or by measuring the pH of the solution using a pH probe attached to the computer interface (Experiment 25B).

$$R\!-\!Cl + H_2O \xrightarrow{\text{solvent}} R\!-\!OH + HCl$$

The rate equation for the S_N1 hydrolysis of an alkyl chloride is

$$+\frac{d[HCl]}{dt} = k[RCl]$$

Let c equal the initial concentration of RCl. At some time, t, x moles per liter of alkyl chloride will have decomposed and x moles per liter of HCl will have been produced. The remaining concentration of alkyl chloride at that value of time equals $c - x$. The rate equation becomes

$$+\frac{dx}{dt} = k(c - x)$$

On integration, this becomes

$$\ln\left(\frac{c}{c - x}\right) = kt$$

which, converted to base 10 logarithms, is

$$2.303 \log\left(\frac{c}{c - x}\right) = kt$$

This equation is of the form appropriate for a straight line $y = mx + b$ with slope m and with intercept b equal to zero. If the reaction is indeed first-order, a plot of log $[c/(c - x)]$ versus t will provide a straight line whose slope is $k/2.303$.

Evaluation of the term $c/(c - x)$ remains a problem, because it is experimentally difficult to determine the concentration of alkyl chloride. We can, however, determine the concentration of hydrochloric acid produced by titrating with base or by measuring it with a pH probe. Because the stoichiometry of the reaction indicates that the number of moles of alkyl chloride consumed equals the number of moles of hydrochloric acid produced, c

must also equal the number of moles of HCl produced when the reaction has gone to completion (the so-called infinity concentration of HCl), and x equals the number of moles of HCl produced at some particular value at time t. For Experiment 25A, we can rewrite the integrated rate expression in terms of volume of base used in the titration. At the end point of the titration,

$$\text{number of moles HCl} = \text{number of moles NaOH}$$

or

$$x = \text{number of moles NaOH at time } t$$

and

$$c = \text{number of moles NaOH at time } \infty$$

$$\text{number of moles NaOH} = [\text{NaOH}]V$$

where V is the volume. Substituting and cancelling gives

$$\left(\frac{c}{c-x}\right) = \frac{\cancel{[\text{NaOH}]}V_\infty}{\cancel{[\text{NaOH}]}(V_\infty - Vt)}$$

where V_∞ is the volume of NaOH used when the reaction is complete and V_t is the volume of NaOH used at time t. This integrated rate equation becomes

$$2.303 \log\left(\frac{V_\infty}{V_\infty - V_t}\right) = kt$$

The concentration of base used in the titration cancels out of this equation, so that it is necessary to know neither the concentration of base nor the amount of alkyl chloride used in the experiment.

A plot of $\log V_\infty/(V_\infty - V_t)$ versus t will provide a straight line whose slope equals $k/2.303$. The slope is determined according to the figure on page 224. If the time is measured in minutes, the units of k are min^{-1}. The experimental points plotted on the graph may contain a certain amount of scatter, but the line drawn is the best *straight* line. The line should pass through the origin of the graph. With some reactions, competing processes may cause the line to contain a certain amount of curvature. In these cases, the slope of the initial portion of the line is used before the curvature becomes too important.

For Experiment 25B, the signal produced by the pH probe can be used to provide a measure of c. (Remember that

$$pH = -\log[H^+]$$

and therefore

$$[H^+] = c = 10^{-pH})$$

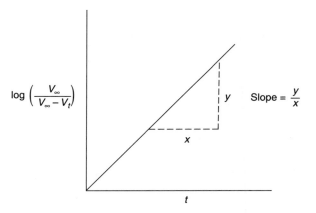

A plot of log $[V_\infty/(V_\infty - V_t)]$ vs. t provides a straight line with a slope equal to $k/2.303$.

A plot of ln $[c/(c - x)]$ versus t will provide a straight line whose slope equals k. The slope of the line is determined directly by the Labworks software. If the time is measured in seconds, the units of k are $\sec^{-1}$. The experimental points plotted on the graph may contain a certain amount of scatter, but the line drawn is the best *straight* line. Again, a best straight (or *regression*) line is calculated using the Labworks software. The line should pass through the origin of the graph. With some reactions, competing processes may cause the line to contain a certain amount of curvature. In these cases, the slope of the initial portion of the line is used before the curvature becomes too important.

Strictly speaking the value of c is determined by taking the final value of the concentration of HCl (c_∞) and correcting it by subtracting the *initial* value of the concentration, before the reaction has begun ($c_\infty - c_0$). For this experiment, we will assume that the initial value of the concentration is so low as to be negligible. Therefore c in the above equation will be taken as c_∞.

One other value often cited in kinetic studies is the **half-life** of the reaction τ. The half-life is the time required for one-half of the reactant to undergo conversion to products. During the first half-life, 50% of the available reactant is consumed. At the end of the second half-life, 75% of the reactant has been consumed. For a first-order reaction, the half-life is calculated by

$$\tau = \frac{\ln 2}{k} = \frac{0.69315}{k}$$

Two alkyl chlorides will be studied by the class in a variety of solvents. The class data will be compared, so that relative reactivities of the alkyl chlorides can be determined.

Required Reading

Review: Before beginning this experiment, you should read the material dealing with the methods of kinetics in your lecture textbook.

New: Technique 7 Extractions, Separations, and Drying Agents, especially Section 7.4, page 690.

Special Instructions

You must work in pairs in this experiment in order to make the measurements rapidly. You should alternate jobs on each run: For Experiment 25A, one partner should do the titrations while the other reads the stopwatch and records the data. For Experiment 25B, one partner should prepare the solutions while the other performs the actual kinetic run.

> **Caution:** Concentrated hydrochloric acid is corrosive; avoid any direct contact and avoid breathing the vapors.

Waste Disposal

Dispose of any unused acetone solutions of alkyl chlorides in the waste container marked for the disposal of halogenated organic waste. All aqueous solutions produced in this experiment should be collected and discarded into a waste container designated for aqueous waste. The concentration of unreacted alkyl chloride that may be present in the reaction mixture (after titration) is too low to warrant any special disposal procedure.

Note to the Instructor

The procedure for Experiment 25B is written to be used with an SCI Technologies interface and Labworks software. If your laboratory is equipped with another type of computer interface system, it will be necessary to prepare a modified laboratory procedure that can be used with your system.

EXPERIMENT 25A

Hydrolysis of Alkyl Chlorides (Titration Method)

Procedure

PART A.   PREPARATION OF THE ALKYL CHLORIDES

Select an alcohol. The choices include *tert*-butyl alcohol (2-methyl-2-propanol) and α-phenylethyl alcohol (1-phenylethanol). Place the alcohol (11 mL) in a 125-mL separatory funnel along with 25 mL of cold, concentrated hydrochloric acid (specific gravity 1.18; 37.3% hydrogen chloride). Shake the separatory funnel vigorously, venting frequently to relieve any excess pressure, over 30 minutes. Remove the aqueous layer. Wash the organic phase quickly with three 5-mL portions of cold water, followed by a washing with 5 mL of 5% sodium bicarbonate solution. Place the organic product in a small Erlenmeyer flask over 3–4 g of anhydrous calcium chloride. Shake the flask occasionally over 5 minutes. Carefully decant the alkyl chloride from the drying agent into a small Erlenmeyer flask, which can then be stoppered tightly. The alkyl chloride is used in this experiment without prior distillation. Because the true concentration of alkyl chloride introduced into the hydrolysis reaction is determined by titration, it is not necessary to purify the product prepared in this part of the experiment.

PART B.   KINETIC STUDY OF THE HYDROLYSIS OF AN ALKYL CHLORIDE

Because the chlorides hydrolyze rapidly under the conditions used in this experiment, work in pairs to perform the kinetic studies. One partner performs the titrations, while the other measures the time and records the data.

Prepare a stock solution of alkyl chloride by dissolving about 0.6 g of alkyl chloride in 50 mL of dry, reagent-grade acetone. Store this solution in a stoppered container to protect it from moisture. Use a 125-mL Erlenmeyer flask to carry out the hydrolysis. The flask should contain a magnetic stirring bar, 50 mL of solvent (see Table One, for the appropriate solvent), and two to three drops of bromthymol blue indicator. Use absolute ethanol in preparing the aqueous ethanol solvent. Do not use denatured ethanol, as the denaturing agents may interfere with the reactions being studied. Bromthymol blue has a yellow color in acid solution and a blue color in alkaline solution.

Place a 50-mL buret filled with approximately 0.01 N sodium hydroxide above the flask. The exact concentration of sodium hydroxide does not need to be known. Record the initial volume of sodium hydroxide at time t equal to 0.0 minutes. Add about 2 mL of sodium hydroxide from the buret to the Erlenmeyer flask and precisely record the new volume in the buret. Start the stirrer. At time 0.0 minutes, *rapidly* add 1.0 mL of the acetone solution of the alkyl chloride from a pipet. Start the timer when the pipet is about half empty. The indicator will undergo a color change, passing from blue through green to yellow when enough hydrogen chloride has been formed in the reaction to neutralize the sodium hydroxide in the flask. Record the time at which

---

TABLE ONE   Experimental Conditions

Compound	Solvent Mixtures (volume percentage of organic phase in water)
tert-Butyl Chloride	40% Ethanol
	25% Acetone
	10% Acetone
α-Phenylethyl Chloride	50% Ethanol
	40% Ethanol
	35% Ethanol

the color changed. This color change may not be rapid. Try to use the same color as the end point each time. Add another 2 mL of sodium hydroxide from the buret, and precisely record the volume and the time at which this second volume of sodium hydroxide is consumed. Repeat the sodium hydroxide addition twice more (four total). Finally, allow the reaction to go to completion for an hour without excess sodium hydroxide present. Stopper the Erlenmeyer flask during this period.

After the reaction has gone to completion, *accurately* titrate the amount of hydrogen chloride in solution to the end point. The end point is reached when the color of the solution remains constant for at least 30 seconds. The time corresponding to this final volume is infinity ($t = \infty$). Repeat this process in the other two solvent mixtures indicated in Table One. These experiments can be carried out while you are waiting for the infinity titration of the previous experiments, provided that a separate buret is used for each run, so that the infinity concentrations of hydrogen chloride produced can be accurately determined.

EXPERIMENT 25B   ■

Hydrolysis of Alkyl Chlorides (Computer-Interfaced Method)

Procedure

Select an alcohol. Following the procedure given in Experiment 25A, page 225, prepare the alkyl chloride from the corresponding alcohol.

In order to share the workload and to prevent decomposition of the alkyl halides or loss of the halides through evaporation, work in pairs to perform the kinetic studies. One partner performs the kinetic measurements, while the other prepares the alkyl halides and the solutions.

From the Labworks main menu, select **Build/Edit Experiment**. Select **Build/Edit** and use the editor to enter the program shown in the figure, on page 228.

Select **Save Expt.** You will have to enter a file name for your program. Be sure to use the .**EXP** extension on your file name.

Prepare a stock solution of alkyl chloride by dissolving about 0.6 g of alkyl chloride in 50 mL of dry, reagent-grade acetone. Store this solution in a stoppered container to protect it from moisture. Use a 125-mL Erlenmeyer flask to carry out the hydrolysis. The flask should contain a magnetic stirring bar and 50 mL of solvent (see Table One for the appropriate solvent). Use absolute ethanol in preparing the aqueous ethanol solvent. Do not use denatured ethanol, as the denaturing agents may interfere with the reactions being studied.

From the Labworks main menu, select **Perform Experiment.** You will have to enter the file name for the program that you just created. The program will ask you to select the screen mode in which to run the experiment. Select **Graph Right.** You

```
 1  CLEAR WHOLE SCREEN
 2  PRINT "In this experiment the rate constant" ON LINE 1
 3  PRINT "of a reaction will be determined by"
 4  PRINT "monitoring the change in pH as a"
 5  PRINT "function of time"
 6  CLEAR TEXT SCREEN
 7  PRINT "You will be adding 1 mL of RCl to" ON LINE 1
 8  PRINT "50 mL of solvent."
 9  PRINT "Before starting experiment, be sure"
10  PRINT "interface is ON and switch "X" is up."
11  PRINT "Press ENTER with the addition of the RCl."
12  PRINT "Times and corresponding pH readings will"
13  PRINT "be stored in a spreadsheet. When the"
14  PRINT "reaction goes to completion (i.e., the"
15  PRINT "graph flattens out) push "X" down."
16  PRINT "This will end the experiment."
17  PAUSE
18  START Timer1 FOR sec
19  SEND INPUT FROM Timer1 TO COL-A
20  SEND INPUT FROM Timer1 TO GraphX
21  SEND INPUT FROM pH1 TO COL-B
22  SEND INPUT FROM pH1 TO GraphY
23  IF INPUT FROM SwX = Off GOTO 26
24  DELAY 10 sec
25  IF INPUT FROM SwX = On GOTO 19
26  SAVE DATA TO FILE
27  CLEAR TEXT SCREEN
28  PRINT "You can now exit and go to spreadsheet."
29  PAUSE
30  STOP
31  CLEAR TEXT SCREEN
```

Computer program for kinetics of the hydrolysis of an alkyl chloride.

will then be prompted to state the source of your data. Select **Interface.** You will then have to enter the file name for the data that will be created during the experiment. Be sure to use the **.DAT** extension for this file name.

Start the stirrer. Make sure that the computer interface is switched on and that the "X" switch is in the up position. At time 0.0 minutes, *rapidly* add 1.0 mL of the acetone solution of the alkyl chloride from a pipet. Press the RETURN key when the pipet is about half empty. At this point the reaction will proceed, and the computer will begin to collect and display the data.

After the reaction has gone to completion, that is, when there is no apparent change in the observed pH value, move switch "X" to the down position. At this point, the display should prompt you that the experiment has gone to completion and you should calculate your results using the spreadsheet. Repeat this experiment using the other two solvent mixtures indicated in Table One.

From the Labworks main menu, select **Spreadsheet.** Locate the final pH reading, determine the value of 10^{-pH}, and write this value down for future reference. From the editor, enter the following relationship into column E of the spreadsheet:

$10^\wedge - B$.

Select **Analysis** and then select **Recompute Column** to fill column E with the calculated values.

In column F, enter

(*xxx*/(*xxx* − B))

where *xxx* is the final value of the concentration of HCl that you determined and saved. Again, select **Analysis** and **Recompute Column**. Finally, in column G, enter

LN(F).

As before, it is necessary to allow the spreadsheet to compute the values in column G, using the method described previously.

Select **Graph** from the spreadsheet menu. For the value of the X-axis, use the times that are recorded in column A of the spreadsheet. For the value of the Y-axis, use the values that have been computed for column G. Enter the title that you want displayed at the top of the graph. Finally, select **Plot** and then select **Linear Regression**. The display on the screen will show a plot of *ln(F)* on the Y-axis, and the time on the X-axis. A calculated linear regression line will be displayed as drawn through the experimental points. At the top of the graph, the value of the slope (which is equal to the rate constant, *k*) will be displayed, along with the value of the intercept of the linear regression line. In addition, a correlation value will also be shown; this value provides a measure of how well the calculated line fits the experimental points (a correlation value of 1.000 would mean that there was a perfect fit between the line and the data points).

You may print the graph by selecting **Print** from the spreadsheet menu. When you are finished with the calculations and other utilities, select **Quit** from the spreadsheet menu. Be sure that your data and experimental method have been saved. Select **Exit** from the main menu.

Report

Prepare your report to the instructor. Be sure to include the rate constant (*k*) and the value of the half-life (τ) for each kinetic run and for each solvent that you tested. Your report to the instructor should include the plots of your data as well as a table of data (for Experiment 25A). A sample table of data is shown in Table Two.

TABLE TWO   Hydrolysis of α-Phenylethyl Chloride in 50% Ethanol

Time (min)	Vol. NaOH Recorded	Vol. NaOH Used	$V_\infty - V_t$	$\dfrac{V_\infty}{V_\infty - V_t}$	$\ln\left(\dfrac{V_\infty}{V_\infty - V_t}\right)$
0.00	0.2	0.0	6.9	1.00	0.000
8.46	2.2	2.0	4.9	1.41	0.343
18.25	4.2	4.0	2.9	2.37	0.863
31.80	5.9	5.7	1.2	5.75	1.750
47.72	6.8	6.6	0.3	23.00	3.136
100 (∞)	7.1	6.9	0.0	. . .	. . .

Explain your results, especially the effect of changing the water content of the solvent on the rate of the reaction. If the instructor so desires, the results from the entire class may be compared.

QUESTIONS

1. Plot the data given in Table Two. Determine the rate constant and the half-life for this example.
2. What are the principal by-products of these reactions? Give the rate equations for these competing reactions. Should the production of these by-products go on at the same rate as the hydrolysis reactions? Explain.
3. Compare the energy diagrams for an S_N1 reaction in solvents with two different percentages of water. Explain any differences in the diagrams and their effect on the reaction rate.
4. Compare the expected rates of hydrolysis of *tert*-cumyl chloride (2-c　　-2-phenylpropane) and α-phenylethyl chloride (1-chloro-1-phenylethane) in the same solvent. 　　any differences that might be expected.

E S S A Y

Detection of Alcohol: The Breathalyzer

If one places organic compounds on a scale ranking their extent of oxidation, a general order such as

$$R—CH_3 < R—CH_2OH < R—CHO \text{ (or } R_2CO) < R—COOH < CO_2$$

is obtained. According to this scale, you can see that alcohols represent a relatively reduced form of organic compound, while carbonyl compounds and carboxylic acid derivatives represent highly oxidized structures. Using appropriate oxidizing agents, it should be possible to oxidize an alcohol to an aldehyde, a ketone, or a carboxylic acid depending on the substrate and the oxidation conditions.

Primary alcohols can be oxidized to aldehydes by various oxidizing agents, including potassium permanganate, potassium dichromate, and nitric acid:

$$R—CH_2OH \xrightarrow{[O]} \left[R—\overset{\displaystyle O}{\overset{\displaystyle \|}{C}}—H \right] \xrightarrow{[O]} R—\overset{\displaystyle O}{\overset{\displaystyle \|}{C}}—OH$$

The aldehyde formed in this oxidation is unstable relative to further oxidation, and consequently the aldehyde is usually oxidized further to the corresponding carboxylic acid. The aldehyde is seldom isolated from such an oxidation, unless the oxidizing agent is relatively mild.

Chromium(VI) is a very useful oxidizing agent. It appears in various chemical forms, including chromium trioxide CrO_3, chromate ion CrO_4^{2-}, and dichromate ion $Cr_2O_7^{2-}$. The chromium(VI) compounds are typically red to yellow. During the oxidation, they are reduced to Cr^{3+}, which is green. As a result, an oxidation reaction can be monitored by the color change. A typical chromium(VI) oxidation, to illustrate the role of both the oxidizing and the reducing species, is the dichromate oxidation of ethanol to acetaldehyde:

$$3\ CH_3CH_2OH + Cr_2O_7^{2-} + 8\ H^+ \longrightarrow 3\ CH_3-\overset{\overset{\displaystyle O}{\|}}{C}-H + 2\ Cr^{3+} + 7\ H_2O$$

Because the aldehyde is also susceptible to oxidation, a second oxidation step of acetaldehyde to acetic acid can also take place:

$$3\ CH_3-\overset{\overset{\displaystyle O}{\|}}{C}-H + Cr_2O_7^{2-} + 8\ H^+ \longrightarrow 3\ CH_3-\overset{\overset{\displaystyle O}{\|}}{C}-OH + 2\ Cr^{3+} + 4\ H_2O$$

This oxidation reaction of alcohols by dichromate ion leads to a standard method of analysis for alcohols. The material to be tested is treated with acidic potassium dichromate solution, and the green chromic ion formed in the oxidation of the alcohol is measured spectrophotometrically by measuring the amount of light absorbed at 600 nm. By this method, it is possible indirectly to determine from 1 to 10 mg of ethanol per liter of blood with an accuracy of 5%. The alcohol content of beer can be determined within 1.4% accuracy.

THE BREATHALYZER

An interesting application of the oxidation of alcohols appears in a quantitative method of determining the amount of ethanol in the blood of a person who has been drinking. The ethanol contained in alcoholic beverages may be oxidized by dichromate according to the equation shown above. During this oxidation, the color of the chromium-containing reagent changes from reddish orange ($Cr_2O_7^{2-}$) to green (Cr^{3+}). Law-enforcement officials use the color change in this reaction to estimate the alcohol content of the breath of suspected drunken drivers. This value can be converted to an alcohol content of the blood.

In most states, the usual legal definition of being under the influence of alcohol is based on a 0.10% alcohol content in the blood. Because the air deep within the lungs is in equilibrium with the blood passing through the pulmonary arteries, the amount of alcohol in the blood can be determined by measuring the alcohol content of the breath. The proper breath–blood ratio can be determined by simultaneous blood and breath tests. As a result of this equilibration, police officers do not need to be trained to administer blood tests. Instead, a simple instrument, a breath analyzer, which does not require any particular sophistication for its operation, can be used in the field.

In the simplest form, a breath analyzer contains a potassium dichromate–sulfuric acid reagent impregnated on particles of silica gel in a sealed glass ampoule. Before the instrument is to be used, the ends of the ampoule are broken off, and one end is fitted with a mouthpiece while the other is attached to the neck of an empty plastic bag. The person being tested blows into the tube to inflate the plastic bag. As air containing ethanol passes through the tube, a chemical reaction takes place, and the reddish-orange dichromate reagent is reduced to the green chromium sulfate (Cr^{3+}). When the green color extends beyond a certain point along the tube (the halfway point), it is determined that there is a relatively high alcohol concentration in the motorist's breath; the motorist is usually taken to the police station for more precise tests. The device described here is simple, and its precision is not high. It is used primarily as a **screening device** for suspected drunken drivers. An example of this simple device is shown in Figure 1. The use of this type of simple screening device has been supplanted in recent years by hand-held instruments, which are simpler and more accurate.

A more precise instrument, the "Breathalyzer," is shown in Figure 2. Air is blown into a cylinder *A,* whereupon a piston is raised. When the cylinder is full, the piston is allowed to fall and pump the measured volume of breath through a reaction ampoule *B* containing the potassium dichromate solution in sulfuric acid. As the alcohol-laden air is bubbled through this solution, the alcohol is oxidized to acetaldehyde and further to acetic acid, while the dichromate ion is reduced to Cr^{3+}. The instrument contains a light source *C*. Filters are used to select light in the blue region of the spectrum. This blue light passes through the reaction ampoule and is detected by a photocell *D*. The light also passes through a sealed standard reference ampoule *E,* which contains exactly the same concentration of potassium dichromate in sulfuric acid as the reaction ampoule *B* had originally. No alcohol is allowed to enter this reference ampoule. The light passing through the reference ampoule is detected by another photocell. A meter *F,* calibrated in milligrams of ethanol per 100 mL of blood, or in percentage of blood alcohol, registers the difference in voltages between the two photocells. Before the test, both ampoules transmit blue light to the same extent, so the meter reads zero. After the test, the reaction ampoule transmits more blue light than the reference ampoule, and a voltage is registered on the meter.

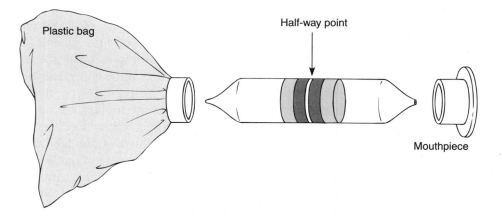

Figure 1.   Breath alcohol screening device.

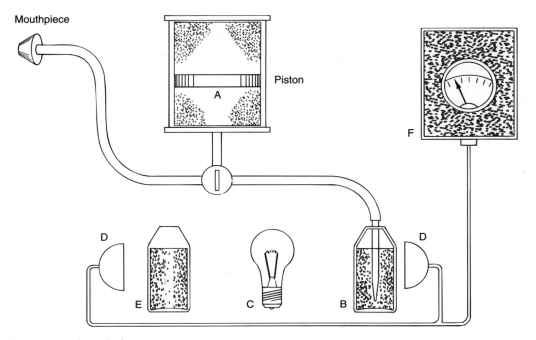

Mouthpiece

Figure 2.   A breathalyzer.

Such an instrument, although more complicated and more delicate than the simple device shown in Figure 1, can be used in the field without a support laboratory. The instrument is portable, permitting it to be easily transported in the trunk of a patrol car.

A similar method is used in Experiment 26 to follow the rate of oxidation of several alcohols by dichromate ion. The color change that accompanies the oxidation is monitored by a spectrophotometer.

MODERN BREATH-TESTING METHODS

The most current methods for determining the blood alcohol concentration are instrumental techniques based on infrared spectroscopy. The typical breath-testing instrument contains a simple infrared spectrophotometer (see Appendix 3) set to measure absorbance at two frequencies. Absorbance at 2910 cm^{-1} and at 2880 cm^{-1} is measured. The bands correspond to the C—H stretching motions of the methyl and methylene groups of ethanol. The instrument determines the total absorbance in the breath sample in order to determine the percentage of ethanol present. At the same time, the instrument also measures the ratio of the absorbances at 2910 and 2880 cm^{-1} to confirm that the specific compound being analyzed is indeed ethanol. While the majority of organic compounds absorb at these two frequencies, no other compound that might be found in human breath, except ethanol, is likely to have the same absorbance *ratio* at these two frequencies. The microprocessor in the instrument has several built-in routines that ensure that calibration is maintained, that a valid breath sample has been obtained, and that interference from other substances is minimized. A built-in printer produces a printed ticket with the relevant blood-alcohol percentages marked on the ticket, along with information that provides identification of the suspect.

This type of instrument generally is not carried in a patrol car. Rather, it is located at a police station. As previously indicated, the arresting officer in the field uses a hand-held breath-screening device to determine whether there is a likelihood that the suspect is intoxicated. If the screening device indicates that additional analysis is warranted, the suspect is taken to the police station, where the complete breath analysis, using the infrared breath scanner, is accomplished.

REFERENCES

Anderson, J. M. "Oxidation–Reduction in Blood Analysis: Demonstrating the Reaction in a Breathalyzer." *Journal of Chemical Education, 67* (March 1990): 263.

Denney, R. C. "Analysing for Alcohol." *Chemistry in Britain, 6* (1970): 533.

Labianca, D. A. "The Chemical Basis of the Breathalyzer." *Journal of Chemical Education, 67* (March 1990): 259.

Labianca, D. A. "Estimation of Blood-Alcohol Concentration." *Journal of Chemical Education, 69* (August 1992): 628.

Lovell, W. S. "Breath Tests for Determining Alcohol in the Blood." *Science, 178* (1972): 264.

Timmer, W. C. "An Experiment in Forensic Chemistry—The Breathalyzer." *Journal of Chemical Education, 63* (October 1986): 897.

Treptow, R. S. "Determination of Alcohol in Breath for Law Enforcement." *Journal of Chemical Education, 51* (1974): 651.

EXPERIMENT 26

Chromic Acid Oxidation of Alcohols

Chromic acid oxidation of an alcohol
Kinetics
Ultraviolet-visible spectrophotometry
Computer-interfaced experiment

The chemical reaction of interest in this experiment is the oxidation of an alcohol to the corresponding aldehyde by an acidic solution of potassium dichromate:

$$3\ RCH_2OH + Cr_2O_7^{2-} + 8\ H^+ \longrightarrow 3\ R\overset{\displaystyle O}{\overset{\displaystyle \|}{-C}}-H + 2\ Cr^{3+} + 7\ H_2O$$

Normally, the aldehyde formed is also susceptible to oxidation by the dichromate ion, yielding the corresponding acid:

$$3 \; \underset{\displaystyle R-\overset{\textstyle O}{\overset{\|}{C}}-H}{} + Cr_2O_7{}^{2-} + 8 \; H^+ \longrightarrow 3 \; \underset{\displaystyle R-\overset{\textstyle O}{\overset{\|}{C}}-OH}{} + 2 \; Cr^{3+} + 4 \; H_2O$$

In this experiment, however, the alcohol is present in large excess, and the likelihood that the second reaction will take place is thereby greatly reduced. A secondary alcohol is oxidized to a ketone by a similar process. A ketone is not easily oxidized further by the dichromate reagent.

$$3 \; \underset{\displaystyle \underset{R'}{|}}{R-CH-OH} + Cr_2O_7{}^{2-} + 8 \; H^+ \longrightarrow 3 \; \underset{\displaystyle R-\overset{\textstyle O}{\overset{\|}{C}}-R'}{} + 2 \; Cr^{3+} + 7 \; H_2O$$

Although various mechanisms have been proposed to explain how dichromate ion oxidizes alcohols, the most commonly accepted mechanism is the one F. H. Westheimer first proposed in 1949. In acid solution, dichromate ion forms two molecules of chromic acid H_2CrO_4:

$$2 \; H_3O^+ + Cr_2O_7{}^{2-} \underset{}{\overset{fast}{\rightleftharpoons}} 2 \; H_2CrO_4 + H_2O \tag{1}$$

The chromic acid, in a rapid, reversible step, forms a chromate ester with the alcohol:

$$RCH_2OH + H_2CrO_4 \overset{fast}{\rightleftharpoons} R-CH_2-O-\overset{\textstyle O}{\underset{\textstyle O}{\overset{\|}{\underset{\|}{Cr}}}}-OH + H_2O \tag{2}$$

The chromate ester undergoes a rate-determining decomposition by a two-electron transfer with cleavage of the α-carbon-hydrogen bond, as seen in Step 3.

$$R-\underset{\displaystyle \underset{H}{|}}{\overset{\displaystyle \overset{H}{|}}{C}}-O-\overset{\textstyle O}{\underset{\textstyle O}{\overset{\|}{\underset{\|}{Cr}}}}-OH \overset{slow}{\longrightarrow} R-\underset{\displaystyle \underset{H}{|}}{C}=O + H_2CrO_3 \tag{3}$$

Cr in +6   Cr in +4
oxidation state   oxidation state

The H_2CrO_3 is further reduced to Cr^{3+} by interaction with chromium in various oxidation states and by further interaction with the alcohol. All these subsequent steps are very rapid relative to Step 3. Consequently, they are not involved in the rate-determining step of the mechanism and need not be considered further.

The rate-determining step, Step 3, involves only one molecule of the chromate ester, which in turn arises from a prior equilibrium involving the combination of one molecule of alcohol and one molecule of chromic acid (Step 2). As a result, this reaction, which is first-order in chromate ester, turns out to be *second-order* for the reacting alcohol and the dichromate reagent. The kinetic equation therefore is

$$-\frac{d[\text{Cr}_2\text{O}_7{}^{2-}]}{dt} = k[\text{RCH}_2\text{OH}][\text{Cr}_2\text{O}_7{}^{2-}]$$

The presence of the chromium atom strongly affects the distribution of electrons in the remainder of the chromate ester molecule. Electrons need to be transferred to the chromium atom during the cleavage step. If the R group includes an electron-withdrawing group, it diminishes the necessary electron density needed for reaction. Consequently, the reaction proceeds more slowly. An electron-releasing group would be expected to have the opposite effect.

THE EXPERIMENTAL METHOD

The rate of the reaction is measured by following the rate of disappearance of the dichromate ion as a function of time. The dichromate ion $\text{Cr}_2\text{O}_7{}^{2-}$ is yellow-orange, absorbing light at 350 and 440 nm. The chromium is reduced to the green Cr^{3+} during the reaction. The ion Cr^{3+} does not absorb light significantly at 350 or 440 nm, but rather at 406, 574, and 666 nm. Therefore, if we measure the amount of light absorbed at a single wavelength, such as 440 nm, we can follow the rate of disappearance of dichromate ion without any interfering absorption of light by the ion Cr^{3+}.

The instrument used to measure the amount of light absorbed at a particular wavelength, when that light lies within the visible region of the electromagnetic spectrum, is a **spectrophotometer.** This type of instrument can be described simply. Ordinary visible light is passed through the sample and then through a prism, where the light of the particular wavelength being studied is selected. This selected light is directed against a photocell, where its intensity is measured. A meter provides a visible display of the intensity of the light of the desired wavelength. In a computer-interfaced experiment, the output of the spectrophotometer is connected to an interface module, which converts the analog signal from the spectrophotometer into a digital signal that can be directed to a computer.

The true rate equation for this reaction is second-order. However, because a large excess of alcohol will be used, its concentration will change imperceptibly during the reaction. The rate equation, under these conditions, will simplify to that of a pseudo first-order reaction. The mathematics involved will become much simpler as a result.

The rate equation for a first-order (or a pseudo first-order) reaction is

$$-\frac{d[\text{A}]}{dt} = k[\text{A}]$$

In this experiment, the rate equation becomes

$$-\frac{d[Cr_2O_7{}^{2-}]}{dt} = k[Cr_2O_7{}^{2-}]$$

Let a equal the initial concentration of dichromate ion. At some time t, an amount x moles/L of dichromate will have undergone reaction, and x moles/L of aldehyde will have been produced. The remaining concentration of dichromate at that value of time equals $a - x$. The rate equation becomes

$$-\frac{dx}{dt} = k(a - x)$$

Integration provides

$$\ln\left(\frac{a}{a - x}\right) = kt$$

Converting to base 10 logarithms gives

$$2.303 \log\left(\frac{a}{a - x}\right) = kt$$

This equation is of the form appropriate for a straight line with intercept equal to zero. If the reaction is indeed first-order, a plot of log $[a/(a - x)]$ versus t will provide a straight line whose slope is $k/2.303$.

Because it is experimentally difficult to measure directly how much dichromate ion is consumed during this reaction, evaluating the term $[a/(a - x)]$ requires an indirect approach. What is needed is some measurable quantity from which the concentration of dichromate can be derived. Such a quantity is the amount of light absorbed by the solution at wavelength 440 nm.

The Beer-Lambert law relates the amount of light absorbed by a molecule or an ion to its concentration, according to the equation

$$A = \epsilon c l$$

where A is the absorbance of the solution, ϵ is the molar absorptivity (a measure of the efficiency with which the sample absorbs the light), c is the concentration of the solution, and l is the path length of the cell in which the solution is contained. The absorbance is read by the spectrophotometer.

At the initial concentration a of dichromate ion, we may write

$$A_0 = \epsilon a l$$

or

$$a = \frac{A_0}{\epsilon l}$$

The amount of dichromate ion remaining unreacted at any particular time t, which equals $a - x$ becomes

$$A_t = \epsilon(a - x)l$$

or

$$a - x = \frac{A_t}{\epsilon l}$$

Substituting absorbance values for concentrations and cancelling provides

$$\frac{a}{a - x} = \frac{A_0}{\cancel{\epsilon l}} \frac{\cancel{\epsilon l}}{A_t}$$

or

$$\frac{a}{a - x} = \frac{A_0}{A_t}$$

At this point, a correction must be introduced. When the reaction reaches completion, at "infinite" time, a certain degree of absorption of 440-nm light remains. In other words, at time $t = \infty$, the value A_∞ does not equal zero. Therefore, this residual absorbance must be subtracted from each of the absorbance terms written above. The difference $A_0 - A_\infty$ gives the actual amount of dichromate ion present initially, and the difference $A_t - A_\infty$ gives the actual amount of dichromate ion remaining unreacted at a value t of time. Introducing these corrections, we have

$$\frac{a}{a - x} = \frac{A_0 - A_\infty}{A_t - A_\infty}$$

The integrated rate equation becomes

$$2.303 \log \left(\frac{A_0 - A_\infty}{A_t - A_\infty} \right) = kt \qquad (4)$$

Because the dimensions of the cell and the molar absorptivity cancel out of this equation, it is not necessary to have any particular knowledge of these parameters.

A plot of $\ln [(A_0 - A_\infty)/(A_t - A_\infty)]$ versus time (see figure) will provide a straight line whose slope equals k. The slope is determined as shown on the graph. For Experiment 26C, the slope of the line is determined directly by the Labworks software that is included in the computer interfacing instrument package. If time is measured in seconds, the units

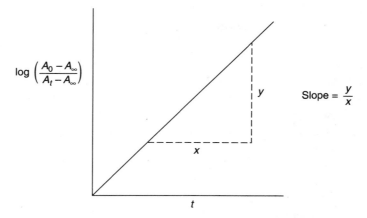

A plot of log $[(A_0 - A_\infty)/(A_t - A_\infty)]$ versus time provides a straight line with slope equal to $k/2.303$.

of k are $\sec^{-1}$. The experimental points plotted on the graph may contain a certain amount of scatter, but the line drawn is the best **straight line** (use some mathematical method such as the method of averages or of least squares). In Experiment 26C, a best straight (or *regression*) line is calculated using the Labworks software.

One other value often cited in kinetic studies is the **half-life** τ of the reaction. The half-life is the time required for half of the reactant to undergo conversion to products. During the first half-life, 50% of the available reactant is consumed. At the end of the second half-life, 75% of the reactant has been consumed. For a first-order reaction, the half-life is calculated by

$$\tau = \frac{\ln 2}{k} = \frac{0.69315}{k}$$

The experimental method used in Experiment 26C utilizes a Bausch & Lomb Spectronic 21 spectrophotometer which is connected to the SCI Technologies interface by means of the positive and negative millivolt outputs of the spectrophotometer. The data which are collected by the interface, therefore, are not absorbance values; rather they are *millivolt* readings. For this experiment, the millivolt readings, which are directly proportional to *percent transmittance,* must be converted into absorbance. This conversion requires two steps. First, the output in millivolts, which has a range of 0 to 1000 millivolts, must be converted to percent transmittance, which ranges from 0 to 100. To accomplish this step, we must divide the millivolt value by 10. Second, another division by 100 must be included to convert *percent transmittance* to *transmittance.* We may simply combine these two steps by dividing the millivolt value by 1000 in order to obtain the corresponding transmittance value (Equation 5). The transmittance is then converted to absorbance using Equation 6.

$$\text{Transmittance} = T = \text{millivolts}/1000 \tag{5}$$

$$\text{Absorbance} = A = \log(1/T) \tag{6}$$

The conversion from millivolts to absorbance will be accomplished using the spreadsheet, during the data treatment phase of this experiment.

The class will study several alcohols in this experiment. The class data will be compared in determining the relative reactivities of the alcohols. Two particular alcohols, 2-methoxyethanol and 2-chloroethanol, react more slowly than the other alcohols used in this experiment. In spite of this lower reactivity, the reactions will not be followed for more than a few minutes, because in these compounds, the second reaction—the oxidation of the aldehyde product to the corresponding carboxylic acid—becomes more important over longer periods. As a result of this second reaction, dichromate ion becomes consumed more rapidly than the calculations would suggest, and the graph of log $[(A_0 - A_\infty)/(A_t - A_\infty)]$ versus time becomes curved. So that this complication is avoided, only the first few minutes of the reaction are used to calculate an initial reaction rate, which corresponds to the reaction being studied in this experiment. The other alcohols are sufficiently reactive that the second reaction does not introduce any significant error.

Required Reading

Review: Read the sections on kinetics in your lecture textbook.

New: Essay   Detection of Alcohol: The Breathalyzer

Special Instructions

Primary and secondary alcohols are oxidized in this experiment to aldehydes and ketones, respectively. The experimental procedure is identical for both types of alcohols. This experiment should be done by pairs of students, in order to distribute the workload of preparing solutions and making the measurements. There are three alternative methods that are available in this experiment. Experiment 26A uses a Bausch & Lomb Spectronic 21 spectrophotometer to collect data. Experiment 26B is similar to Experiment 26A, except that it has been written for use with a Hewlett-Packard 8452 Diode-Array UV-VIS Spectrophotometer. Finally, Experiment 26C makes use of a Bausch & Lomb Spectronic 21 spectrophotometer which is connected to the SCI Technologies computer interface. Each kinetic run requires about 1.5 to 2 hours.

> **Caution:** Potassium dichromate solutions have been determined to be potential carcinogens.

This experiment involves using an acidic solution of potassium dichromate. The dichromate solution will be prepared as a stock solution for the entire class to use. This stock solution should be stored in a hood. Students should wear gloves and use pipet bulbs when using this stock solution.

The Occupational Safety and Health Administration is considering regulating the use of 2-methoxyethanol because there is mounting evidence that it is a reproductive toxin. When using this substance, use gloves and avoid breathing the vapors. All the alcohols should be stored in a hood for use by the class.

Waste Disposal

Dispose of all aqueous solutions containing chromium in a container designated for chromium wastes. Use gloves when handling this waste.

EXPERIMENT 26A

Chromic Acid Oxidation of Alcohols—Visible Spectrophotometer Method

Procedure

Preparation of Reagents. Select an alcohol from one of the following: ethanol, 1-propanol, 2-propanol, 2-methoxyethanol, 2-chloroethanol, ethylene glycol, and 1-phenylethanol (methylbenzyl alcohol). A stock solution of 3.9M sulfuric acid and a carefully prepared solution of 0.0196M potassium dichromate solution (prepared using distilled water in a volumetric flask) should be available for the entire class to use.

Instrument Preparation. Turn on the instrument and allow it to warm up. Select the tungsten lamp as the light source. Select an operating mode that allows the instrument to operate at a fixed wavelength of 440 nm and to record data as absorbance.

Running the Experiment. Using a small flask, prepare the test solution by transferring 1 mL of the stock dichromate solution and 10 mL of the stock sulfuric acid solution into the flask with volumetric pipets supplied by the instructor. (**Caution:** *Use pipet bulbs.*) Shake the solution well. Rinse a sample cuvette three times with this acidic dichromate solution and then fill the cell. Wipe the cuvette clean and dry. Place the cuvette into the sample cell compartment and place a cuvette filled with distilled water into the reference cell compartment. Close the cell compartment lid and allow the chromic acid solution to reach temperature equilibrium by allowing it to remain in the instrument (with the instrument running) for 20 minutes. This preheating minimizes the problem of the solution being slowly heated during the experiment by the tungsten lamp, which is the light source in the spectrophotometer. Such heating would tend to accelerate the reaction as time passed.

At the end of the preheating, record the absorbance A_0 of the chromic acid solution, along with a time value of 0.0 minutes. Withdraw a 10.0-μL sample of the alcohol being studied into a hypodermic syringe and transfer it rapidly to the chromic acid solution. As the transfer is made, start the timer. Withdraw the sample cuvette from the cell compartment, shake it vigorously for 20 to 30 seconds, and return it to the cell compartment. Be sure to wipe off the cell again. Close the compartment lid. The measurements can now be taken.

Take readings of the absorbance A_t and the corresponding time at 1-minute intervals over a 6-minute period (8 minutes for 2-propanol). At the end of this time, remove the cuvette from the cell compartment of the spectrophotometer and allow the solution in the cuvette to stand undisturbed for at least 1 hour (another student may use the instrument during this time). After this period, return the cuvette to the cell compartment (warm up the instrument first, if it was turned off) and read the absorbance value. This final value corresponds to "infinite" time A_∞.

The instructor may require each student to perform a duplicate run. If so, repeat the experiment under precisely the same conditions used for the first run.

Analysis of Data.   Plot the data according to the method described in the introductory section of the experiment. A table of sample data is shown on page 247. Report the value of each rate constant determined in this experiment (and the average of the rate constants, if duplicate determinations were made). Also, report the value of the half-life τ. Include all data and graphs in the report. The results from the entire class may be compared, at the option of the instructor.

E X P E R I M E N T   2 6 B

Chromic Acid Oxidation of Alcohols—UV-VIS Spectrophotometer Method

Procedure

Preparation of Reagents.   Prepare the reagents according to the procedure given in Experiment 26A, page 241.

Instrument Setup.   Turn on the spectrophotometer and the computer connected to it. The Hewlett Packard Program should appear on the screen. Using the arrow keys, move the highlight bar down to **OPERATOR NAME** and press **ENTER**. Input an operator name, press **ENTER** and move the highlight bar up to **GENERAL SCANNING** and press **ENTER**. Make sure that the flow cell has been removed from the beam path and that the cuvette holder has been installed. The spectrometer setup screen should be displayed at this time, with a listing of function key (F1 . . . F10) commands listed below. Select **ACQUISITION** (F4), and move the highlighted bar down to **STANDARD DEVIATION**. If an **X** does not appear in the brackets to the left of **STANDARD DEVIATION** then press **ENTER**. Press **ESC** to return to the setup screen. Select **SAMP. INPUT** (F8), highlight **MANUAL INPUT** and press **ENTER**. Press **ESC** to exit. Move to the **OPTIONS** selection (F5) and choose **OVERLAY SPECTRA**. While still in the setup mode, insert a cuvette filled with deionized water (DI-H_2O) into the sample cell holder and select **MEAS. BLANK** by pressing **F2**.

Sample Scans.   A spectrum of DI-H_2O should appear on the screen in what is known as the Graphics mode. Select **REGISTERS** (F5), and then press **F3** to clear all registers. The blank scan is automatically stored to another file and does not need to be included in the sample spectrum. Press **ESC** to return to the Graphics mode. The spectrometer is now ready for sample scans to take place. The same cuvette used for the blank scan should also be used for the experimental portion to maintain consistency.

Using a small flask, prepare the test solution by transferring 1 mL of the stock dichromate solution and 10 mL of the stock sulfuric acid solution into the flask with the volumetric pipettes supplied. Shake the solution well. Rinse a sample cuvette three times with this acidic dichromate solution and then fill the cell. Wipe the cuvette clean and dry. Place the cuvette into the sample cell compartment.

Withdraw a 10.0-μL sample of ethanol into a hypodermic syringe and transfer it rapidly to the chromic acid solution. Simultaneously, as you transfer the sample, press **F1**. This starts the timer and records the value of absorbance, A_0. Withdraw the sample cuvette from the cell compartment, shake it vigorously for 20–30 seconds, and return it to the cell compartment. Be sure to wipe off the cell again. Close the compartment lid.

Once the experiment is initiated, all that is required is to press **F1 (Meas. Sample)** at each time interval for the time dependent part of the experiment. Only the time need be recorded, the absorbance and other data can be obtained later when time is not a constraint. Each time **F1** is pressed, a spectrum will be displayed, overlayed on top of the last scan(s) taken.

Take readings of the absorbance A_t and the corresponding time at 1-minute intervals over a 6-minute period. At the end of this time, it will be necessary to wait for approximately one hour to allow the reaction to go to completion. After the waiting period, read the final absorbance value. This final value corresponds to "infinite" time, A_∞.

Data Workup.   The peak of interest for this experiment is the absorbance value found at 440 nm. Select the **RESCALE** (F3) function and highlight **ZOOM IN**. Press **ENTER** and a cursor will appear on the screen. Move the cursor, using the arrow keys, to a point below and to the left of the area you wish to zoom in on. Press **ENTER** and move the cursor above and to the right of the area to be isolated, drawing a box around the area. Pressing **ENTER** will bring the portion of the spectrum into view. If a mistake is made, select **RESCALE** (F3) again, and choose **ZOOM OUT**. This will return the full spectrum to the screen, and the zooming process can be repeated.

Each peak in this portion of the spectrum can be marked using the cursor feature. Select **CURSOR** (F2) and an arrow should appear on Register A. (See just below the spectrum for a description of the arrow placement and on which spectrum it is located.) The left and right arrow keys will move the arrow along the spectrum, while the up and down arrow keys will move the arrow to a different register/spectrum. Place the arrow in Register A, which should correspond to the spectrum at time = 0.0 minutes. Move the arrow to the peak at 440 nm and press **MARK** (F1). Move down to Register B, mark the peak at 440 nm and repeat for all of the remaining spectra. Once all of the peak(s) are marked, press **LIST MARKS** (F3) to display the absorbances for each peak at 440 nm. Exit the cursor mode and return to the Graphics mode by selecting **RETURN** (F10).

To obtain a printout of the spectrum and the absorbances, select **HARDCOPY** (F9), highlight **PRINT SPECTRA,** and press **RETURN. (NOTE:** *Do not* select **TABULATE SPECTRA** or you will receive about thirty seconds of useless material!!)

Analysis of Data. Plot the data according to the method described in Experiment 26A, page 242. Report the value of the rate constant determined in this experiment. Also, report the value of the half-life, τ. Include all data and graphs in the report. The results from the entire class may be compared, at the option of the instructor.

EXPERIMENT 26C

Chromic Acid Oxidation of Alcohols—Computer-Interfaced Method

Procedure

Preparation of Reagents. Prepare the reagents according to the procedure given in Experiment 26A, page 241.

From the Labworks main menu, select **Build/Edit Experiment.** Select **Build/Edit** and use the editor to enter the program shown in the figure on page 245.

Select **Save Expt.** You will have to enter a file name for your program. Be sure to use the **.EXP** extension on your file name.

From the Labworks main menu, select **Perform Experiment.** You will have to enter the file name for the program that you just created. The program will ask you to select the screen mode in which to run the experiment. Select **Graph Right.** You will then be prompted to state the source of your data. Select **Interface.** You will then have to enter the file name for the data that will be created during the experiment. Be sure to use the **.DAT** extension for this file name.

The screen display will prompt you as to the steps that you must follow. Make sure that the computer interface is switched on, that the Spectronic 21 is properly connected to the mV1 position of the interface, that the spectrophotometer is switched on, and that the wavelength is set to 440 nm. Place a cuvette filled with

```
 1  CLEAR WHOLE SCREEN
 2  PRINT "In this experiment the rate constant" ON LINE 2
 3  PRINT "of a reaction will be determined by"
 4  PRINT "measuring the disappearance of color as"
 5  PRINT "a function of time."
 6  PRINT "Before starting experiment be sure" ON LINE 10
 7  PRINT "that the interface power is ON and"
 8  PRINT "the SPEC 21 is set to 440 nm."
 9  PAUSE
10  CLEAR TEXT SCREEN
11  PRINT "Place cuvette filled with DI H2O in the" ON LINE 2
12  PRINT "SPEC 21 and adjust the absorbance"
13  PRINT "to 0."
14  PRINT "Next, fill the cuvette with the chromate"
15  PRINT "solution and press ENTER to take the"
16  PRINT "initial mV reading."
17  PRINT "This value will be placed in column B"
18  PRINT "of the spreadsheet."
19  PAUSE
20  SEND INPUT FROM mV1 TO Mem
21  PAUSE
22  CLEAR TEXT SCREEN
23  PRINT "Remove cuvette and inject 10 microL" ON LINE 2
24  PRINT "of the alcohol being analyzed."
25  PRINT "Cover and shake the cuvette"
26  PRINT "vigorously for 20 seconds."
27  PRINT "Quickly place the cuvette in the"
28  PRINT "SPEC 21 and press ENTER to begin"
29  PRINT "collecting data."
30  PAUSE
31  CLEAR TEXT SCREEN
32  PRINT "Time and mV readings will now be" ON LINE 2
33  PRINT "taken and stored in column A"
34  PRINT "and column C."
35  PRINT "The experiment will take 70 minutes" ON LINE 15
36  PRINT "to complete."
37  START Timer1 FOR sec
38  SEND INPUT FROM Timer1 TO COL-A
39  SEND INPUT FROM Timer1 TO GraphX
40  SEND INPUT FROM mV1 TO COL-C
41  SEND INPUT FROM mV1 TO GraphY
42  SEND INPUT FROM Mem TO COL-B
43  DELAY 10 sec
44  IF INPUT FROM Timer1 < 600.000000 GOTO 38
45  WAIT UNTIL INPUT FROM Timer1 > 4200.000000
46  SEND INPUT FROM mV1 TO COL-D
47  SAVE DATA TO FILE
48  CLEAR TEXT SCREEN
49  PRINT "The experiment is now complete, press ENTER" ON LINE 10
50  PRINT "to return to the main menu."
51  STOP
```

Computer program for kinetics of the oxidation of an alcohol.

distilled water in the spectrophotometer and adjust the absorbance reading to zero, using the adjustment control located at the bottom of the front face of the instrument. When the adjustment has been made, fill the cuvette with the dichromate solution, place it in the spectrophotometer, and press the **ENTER** key to record the initial absorbance reading. Remove the cuvette and inject 10 microliters of the alcohol that you are analyzing. Cover the cuvette and shake it vigorously for about 20 seconds. Quickly replace the cuvette in the instrument, and press **ENTER** to begin collecting data. At this point the reaction will proceed, and the computer will begin to collect and display the data.

When 70 minutes have elapsed, the computer will terminate the experiment, record the final absorbance reading, and save the data to the spreadsheet. Your instructor may require you to perform a duplicate run. If so, repeat the experiment under precisely the same conditions used for the first run. Note that you will have to enter a *different* file name for the data collected during the second run.

From the Labworks main menu, select **Spreadsheet.** Locate the initial millivolt reading from the spectrophotometer (this will be stored in column B of the spreadsheet). Also locate the final millivolt reading (this will be found as the first entry of column D of the spreadsheet). Using a hand calculator, you will have to convert these millivolt readings into absorbance values. Use the following equation to perform the necessary conversion.

$$\text{Absorbance} = \log[1/(\text{millivolts}*0.001)]$$

From the Labworks editor, enter the following relationship into column E of the spreadsheet:

$$\text{LOG}(1/C*0.001)$$

The spreadsheet will automatically take the millivolt readings in column C and perform the conversion to absorbance values. Select **Analysis** and then select **Recompute Column** to fill column E with the calculated values.

In column F, enter

$$\frac{A_0 - A_\infty}{E - A_\infty}$$

where A_0 is the numerical value of the initial absorbance that you calculated and A_∞ is the numerical value of the final absorbance. Select **Analysis** and then select **Recompute Column** to fill column F with the calculated values.

In column G, enter

$$\text{LN(F)}$$

As before, it is necessary to allow the spreadsheet to compute the values in column F, by selecting **Analysis** and **Recompute.**

Select **Graph** from the spreadsheet menu. For the value of the X-axis, use the times that are recorded in column A of the spreadsheet. For the value of the Y-axis, use the values that have been computed for column G. Enter the title that you want displayed at the top of the graph. Finally, select **Plot** and then select **Linear Regression.** The display on the screen will show a plot of LN(F) on the Y-axis, and the time on the X-axis. A calculated linear regression line will be displayed as drawn through the experimental points. At the top of the graph, the value of the slope (which is equal to the rate constant, k) will be displayed, along with the value of the intercept of the linear regression line. In addition, a correlation value will also be shown; this value provides a measure of how well the calculated line fits the experimental points (a correlation value of 1.000 would mean that there was a perfect fit between the line and the data points).

You may print the graph by selecting **Print** from the spreadsheet menu. When you are finished with the calculations and other utilities, select **Quit** from the spreadsheet menu. Be sure that your data and experimental method have been saved. Select **Exit** from the main menu.

Analysis of Data. Prepare your report to the instructor. Be sure to include the rate constant (k) and the value of the half-life (τ) for each kinetic run that you performed and the average of the rate constants, if you did duplicate determinations. Your report to the instructor should include the plots of your data. If the instructor so desires, the results from the entire class may be compared.

REFERENCES

Lanes, R. M., and Lee, D. G. "Chromic Acid Oxidation of Alcohols," *Journal of Chemical Education, 45* (1968): 269.

Pavia, D. L., Lampman, G. M., and Kriz, G. S. *Introduction to Spectroscopy: A Guide for Students of Organic Chemistry,* 2nd ed. Philadelphia: Saunders, 1996. Chap. 6.

Westheimer, F. H., and Nicolaides, N. "Kinetics of the Oxidation of 2-Deuterio-2-propanol by Chromic Acid." *Journal of the American Chemical Society, 71* (1949): 25.

Westheimer, F. H. "The Mechanism of Chromic Acid Oxidations." *Chemical Reviews, 45* (1949): 419.

QUESTIONS

1. Plot the data given in the table. Determine the rate constant and the half-life for this example.

Oxidation of Ethanol

Time (min)	Absorbance (440 nm)	$A_t - A_\infty$	$\dfrac{A_0 - A_\infty}{A_t - A_\infty}$	$\log\left(\dfrac{A_0 - A_\infty}{A_t - A_\infty}\right)$
0.0	0.630	0.578	1.000	0.000
1.0	0.535	0.483	1.197	0.078
2.0	0.440	0.388	1.490	0.173
3.0	0.365	0.313	1.847	0.266
4.0	0.298	0.246	2.350	0.371
5.0	0.247	0.195	2.964	0.472
6.0	0.202	0.150	3.853	0.586
66.0 (∞)	0.052	0.000	. . .	. . .

2. Using data collected by the class, compare the relative rates of ethanol, 1-propanol, and 2-methoxyethanol. Explain the observed order of reactivities in terms of the mechanism of the oxidation reaction.

3. Using data collected by the class, compare the relative rates of 1-propanol and 2-propanol. Account for any differences that might be observed.

4. Balance the following oxidation–reduction reactions:

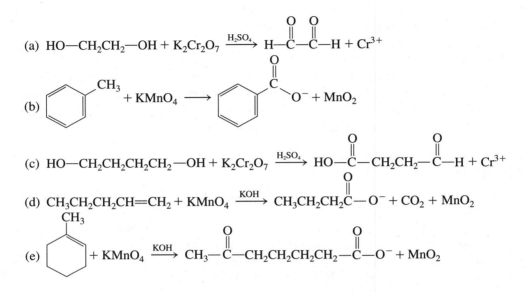

(a) $HO-CH_2CH_2-OH + K_2Cr_2O_7 \xrightarrow{H_2SO_4}$ H—C—C—H + Cr^{3+}

(b)

(c) $HO-CH_2CH_2CH_2CH_2-OH + K_2Cr_2O_7 \xrightarrow{H_2SO_4}$ HO—C—CH_2CH_2—C—H + Cr^{3+}

(d) $CH_3CH_2CH_2CH=CH_2 + KMnO_4 \xrightarrow{KOH} CH_3CH_2CH_2$C—$O^- + CO_2 + MnO_2$

(e)

E X P E R I M E N T   2 7

4-Methylcyclohexene

Preparation of an alkene
Dehydration of an alcohol
Distillation
Bromine and permanganate tests for unsaturation

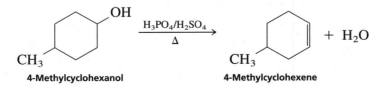

4-Methylcyclohexanol   4-Methylcyclohexene

Alcohol dehydration is an acid-catalyzed reaction performed by strong, concentrated mineral acids such as sulfuric and phosphoric acids. The acid protonates the alcoholic hydroxyl group, permitting it to dissociate as water. Loss of a proton from the intermediate (elimination) brings about an alkene. Since sulfuric acid often causes extensive charring in this reaction, phosphoric acid, which is comparatively free of this problem, is a better choice. In order to make the reaction proceed faster, however, a minimal amount of sulfuric acid will also be used.

The equilibrium that attends this reaction will be shifted in favor of the product, by distilling it from the reaction mixture as it is formed. The 4-methylcyclohexene (bp 101–102°C) will codistill with the water that is also formed. By continuously removing the products, one can obtain a high yield of 4-methylcyclohexene. Since the starting material, 4-methylcyclohexanol, also has a somewhat low boiling point (bp 171–173°C), the distillation must be done carefully so that the alcohol does not also distill.

Unavoidably, a small amount of phosphoric acid codistills with the product. It is removed by washing the distillate mixture with a saturated sodium chloride solution. This step also partially removes the water from the 4-methylcyclohexene layer; the drying process will be completed by allowing the product to stand over anhydrous sodium sulfate.

Compounds containing double bonds react with a bromine solution (red) to decolorize it. Similarly, they react with a solution of potassium permanganate (purple) to discharge its color and produce a brown precipitate (MnO_2). These reactions are often used as qualitative tests to determine the presence of a double bond in an organic molecule (see Experiment 57C). Both tests will be performed on the 4-methylcyclohexene formed in this experiment.

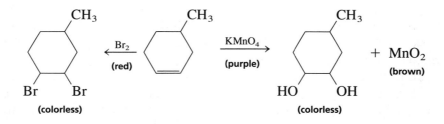

Required Reading

Review:	Techniques 1, 2	
	Technique 7	Sections 7.5, 7.7, and 7.8
New:	Technique 8	Simple Distillation

If performing the optional infrared spectroscopy, also read:

	Technique 19	Preparation of Samples for Spectroscopy, Part A
	Appendix 3	Infrared Spectroscopy

Special Instructions

Phosphoric and sulfuric acids are very corrosive. Do not allow either acid to touch your skin.

Waste Disposal

Dispose of aqueous wastes by pouring them into the container designated for aqueous wastes. Residues that remain after the first distillation may also be placed in the aqueous waste container. Discard the solutions that remain after the bromine test for unsaturation in an organic waste container designated for the disposal of *halogenated* wastes. The solutions that remain after the potassium permanganate test should be discarded into a waste container specifically marked for the disposal of potassium permanganate waste.

Procedure

Apparatus Assembly. Place 7.5 mL of 4-methylcyclohexanol ($MW = 114.2$) in a tared 50-mL round-bottom flask and reweigh the flask to determine an accurate weight for the alcohol. Add 2.0 mL of 85% phosphoric acid and 30 drops (0.40 mL) of concentrated sulfuric acid to the flask. Mix the liquids thoroughly using a glass stirring rod and add a boiling stone. Assemble a distillation apparatus as shown in Technique 8, Figure 8.4, p. 711, using a 25-mL flask as a receiver. Immerse the receiving flask in an ice-water bath to minimize the possibility that 4-methylcyclohexene vapors will escape into the laboratory.

Dehydration. Start circulating the cooling water in the condenser and heat the mixture with a heating mantle until the product begins to distill and collect in the receiver. The heating should be regulated so that the distillation requires about 30 minutes. Too rapid distillation leads to incomplete reaction and isolation of the starting material, 4-methylcyclohexanol. Continue the distillation until no more liquid is collected. The distillate contains 4-methylcyclohexene as well as water.

Isolation and Drying of the Product. Transfer the distillate to a centrifuge tube with the aid of 1 or 2 mL of saturated sodium chloride solution. Allow the layers to separate and remove the bottom aqueous layer with a Pasteur pipet (discard it). Using a dry Pasteur pipet, transfer the organic layer remaining in the centrifuge tube to an Erlenmeyer flask containing a small amount of granular anhydrous sodium sulfate. Place a stopper in the flask and set it aside for 10–15 minutes to remove the last traces of water. During this time, wash and dry the distillation apparatus, using small amounts of acetone and an air stream to aid the drying process.

Distillation. Transfer as much of the dried liquid as possible to the clean, dry 50-mL round-bottom flask, being careful to leave as much of the solid drying agent behind as possible. Add a boiling stone to the flask and assemble the distillation ap-

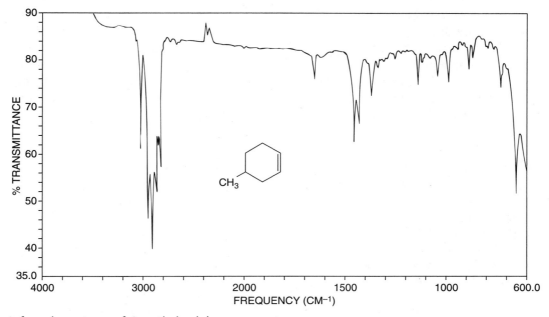

Infrared spectrum of 4-methylcyclohexene, neat.

paratus as before, using a *preweighed* 25-mL receiving flask. Because 4-methylcyclo-hexene is so volatile, you will recover more product if you cool the receiver in an ice-water bath. Using a heating mantle, distill the 4-methylcyclohexene, collecting the material that boils over the range 100–105°C. Record your observed boiling point range in your notebook. There will be little or no forerun, and very little liquid will remain in the distilling flask at the end of the distillation. Reweigh the receiving flask to determine how much 4-methylcyclohexene you prepared. Calculate the percentage yield of 4-methylcyclohexene (*MW* = 96.2).

Spectroscopy. At the instructor's option, obtain the infrared spectrum of 4-methylcyclohexene (Technique 19, Section 19.2, p. 843 or 19.3, p. 845). Because 4-methylcyclohexene is so volatile, you must work quickly to obtain a good spectrum using sodium chloride plates. Compare the spectrum with the one shown in this experiment. After performing the following tests, submit your sample, along with the report, to the instructor.

UNSATURATION TESTS

Place four to five drops of 4-methylcyclohexanol in each of two small test tubes. In each of another pair of small test tubes, place four to five drops of the 4-methylcyclohexene you prepared. Do not confuse the test tubes. Take one test tube from each group and add a solution of bromine in carbon tetrachloride or methylene chloride, drop by drop, to the contents of the test tube, until the red color is no longer discharged. Record the result in each case, including the number of drops required. Test the remaining two test tubes in a similar fashion with a solution of potassium permanganate. Because aqueous potassium permanganate is not miscible with organic compounds, you will have to add about 0.3 mL of 1,2-dimethoxyethane to each test tube before making the test. Record your results and explain them.

QUESTIONS

1. Outline a mechanism for the dehydration of 4-methylcyclohexanol catalyzed by phosphoric acid.
2. What major alkene product is produced by the dehydration of the following alcohols?
  (a) Cyclohexanol
  (b) 1-Methylcyclohexanol
  (c) 2-Methylcyclohexanol
  (d) 2,2-Dimethylcyclohexanol
  (e) 1,2-Cyclohexanediol (*Hint:* Consider keto-enol tautomerism.)
3. Compare and interpret the infrared spectra of 4-methylcyclohexene and 4-methylcyclohexanol.
4. Identify the C—H out-of-plane bending vibrations in the infrared spectrum of 4-methylcyclohexene. What structural information can be obtained from these bands?
5. In this experiment 1–2 mL of saturated sodium chloride is used to transfer the crude product after the initial distillation. Why is saturated sodium chloride, rather than pure water, used for this procedure?

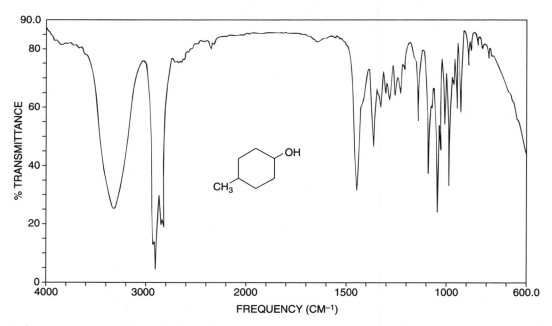

Infrared spectrum of 4-methylcyclohexanol, neat.

EXPERIMENT 28

Phase-Transfer Catalysis: Addition of Dichlorocarbene to Cyclohexene

Carbene formation
Phase-transfer catalysis

It has long been known that a haloform CHX_3 will react with a strong base to give a highly reactive carbene species CX_2 by Reactions 1 and 2. In the presence of an alkene, this carbene adds to the double bond to produce a cyclopropane ring (Reaction 3).

$$X-\overset{\overset{\displaystyle X}{|}}{\underset{\underset{\displaystyle X}{|}}{C}}-H + {}^-\!:\!\ddot{O}-H \;\rightleftharpoons\; X-\overset{\overset{\displaystyle X}{|}}{\underset{\underset{\displaystyle X}{|}}{C}}\!:^- + H\ddot{O}H \tag{1}$$

$$X-\overset{\overset{\displaystyle X}{|}}{\underset{\underset{\displaystyle X}{|}}{C}}\!:^- \;\overset{\text{slow}}{\rightleftharpoons}\; X-\overset{}{\underset{\underset{\displaystyle X}{|}}{C}}\!: + :\ddot{X}\!:^- \tag{2}$$

$$X-\overset{}{\underset{\underset{\displaystyle X}{|}}{C}}\!: + \;\;{}^{\diagdown}\!C\!=\!C^{\diagup}\; \longrightarrow \; -\overset{|}{C}\!\!\!\diagdown_{\underset{X}{}}\!\!\overset{C}{\underset{X}{}}\!\!\diagup^{|}\!C- \tag{3}$$

Traditionally, the reaction has been carried out in *one homogeneous phase* in anhydrous *t*-butyl alcohol solvent, using *t*-butoxide ion as the base [t-Bu = $C(CH_3)_3$].

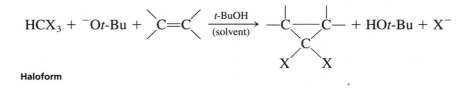

$$HCX_3 + {}^-Ot\text{-Bu} + \;{}^{\diagdown}\!C\!=\!C^{\diagup}\; \xrightarrow[\text{(solvent)}]{t\text{-BuOH}} \; -\overset{|}{C}\!\!\!\diagdown_{\underset{X}{}}\!\!\overset{C}{\underset{X}{}}\!\!\diagup^{|}\!C- + HOt\text{-Bu} + X^-$$

Haloform

Unfortunately, this technique requires time and effort to give good results. In addition, water must be avoided to prevent conversion of the haloform and carbene to formate ion and carbon monoxide by the undesirable base-catalyzed Reactions 4 and 5.

$$H-\underset{\underset{X}{|}}{\overset{\overset{X}{|}}{C}}-H + 2\ H_2O \longrightarrow H-\underset{\underset{O}{\|}}{C}-OH + 3\ HX \qquad (4)$$

$$H-\underset{\underset{O}{\|}}{C}-OH + {}^-OH \longrightarrow H-\underset{\underset{O}{\|}}{C}-O^- + H_2O$$

$$:CX_2 + H_2O \xrightarrow{\ {}^-OH\ } :C{\equiv}O + 2\ HX \qquad (5)$$

QUATERNARY AMMONIUM SALT CATALYSIS

As an alternative to a homogeneous reaction, a *two-phase* reaction can be considered when the organic phase contains the alkene and a haloform CHX_3, and the aqueous phase contains the base OH^-. Unfortunately, under these conditions the reaction will be very slow, because the two primary reactants, CHX_3 and OH^-, are in different phases. The reaction rate can be substantially increased, however, by adding a quaternary ammonium salt such as benzyltriethylammonium chloride as a **phase-transfer catalyst.**

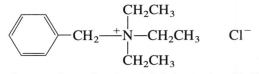

A phase-transfer catalyst: Benzyltriethylammonium chloride

Other common catalysts are tetrabutylammonium bisulfate, trioctylmethylammonium chloride, and cetyltrimethylammonium chloride. All these catalysts, including benzyltriethylammonium chloride, have at least 13 carbon atoms. The numerous carbon atoms give the catalyst organic character (hydrophobic) and allow it to be soluble in the organic phase. At the same time, the catalyst also has ionic character (hydrophilic) and can therefore be soluble in the aqueous phase.

Because of this *dual* nature, the large cation can cross the phase boundary efficiently and transport a hydroxide ion from the aqueous phase to the organic phase (see figure below). Once in the organic phase, the hydroxide ion will react with the haloform to give dihalocarbene by Reactions 1 and 2. Water, a product of the reaction, will move from the organic phase to the aqueous phase, thus keeping the water concentration in the organic phase at a very low level. Because the water content in the organic phase is low, it will not interfere with the desirable reaction of the carbene with an alkene by Reaction 3. Thus, the undesirable side Reactions 4 and 5 are minimized. Finally, the halide ion, which is also produced in Reactions 1 and 2, is transported to the aqueous phase by the tetraalkylammonium cation. In this way, electrical neutrality is maintained and the phase-transfer catalyst, R_4N^+, is returned to the aqueous phase, to repeat the whole procedure. The fig-

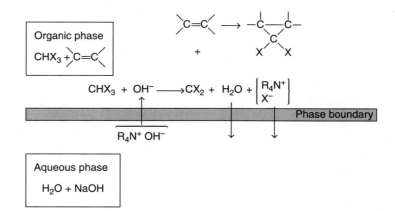

Organic phase

$$CHX_3 + \text{\Large{C=C}}$$

$$CHX_3 + OH^- \longrightarrow CX_2 + H_2O + \begin{bmatrix} R_4N^+ \\ X^- \end{bmatrix}$$

Phase boundary

$$\overline{R_4N^+ \, OH^-}$$

Aqueous phase

$$H_2O + NaOH$$

A two-phase reaction. The organic phase contains the alkene and the haloform CHX_3, while the aqueous phase contains the base OH^-.

ure above summarizes the overall process. This process probably goes on at the interface rather than in the bulk, organic phase.

There are numerous examples of other reactions that might be effectively accelerated by a quaternary ammonium salt or other phase-transfer catalyst (see references). These reactions often involve simple experimental techniques, give shorter reaction times than noncatalyzed reactions, and avoid relatively expensive aprotic solvents that have been widely used to give one phase. Examples of reactions are shown.

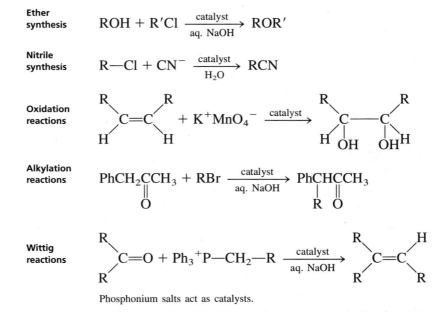

Ether synthesis

$$ROH + R'Cl \xrightarrow[\text{aq. NaOH}]{\text{catalyst}} ROR'$$

Nitrile synthesis

$$R-Cl + CN^- \xrightarrow[\text{H}_2\text{O}]{\text{catalyst}} RCN$$

Oxidation reactions

Alkylation reactions

$$PhCH_2CCH_3 + RBr \xrightarrow[\text{aq. NaOH}]{\text{catalyst}} PhCHCCH_3$$

Wittig reactions

Phosphonium salts act as catalysts.

Increased nucleophillicity   Anions are heavily solvated in an aqueous solvent and are therefore poor nucleophiles in some S_N2 reactions. When they are transported into the organic phase with the catalyst R_4N^+, X^- is no longer solvated with water and may have increased reactivity.

CROWN ETHER CATALYSIS

Another important class of phase-transfer catalysts includes the crown ethers (not used in this experiment). Crown ethers are used to dissolve organic and inorganic alkali metal salts in organic solvents. The crown ether complexes the cation and provides it with an organic exterior (hydrophobic) so that it is soluble in organic solvents. The anion is carried along into solution as the counterion. One example of a crown ether is dicyclohexyl-18-crown-6. Potassium permanganate $KMnO_4$ complexed to the crown ether is soluble in benzene and is known as purple benzene. It is useful in various oxidation reactions.

The crown ethers catalyze many of the same types of reactions listed in the preceding section on quaternary ammonium salt catalysis. Crown ethers are very expensive relative to ammonium salts and are not used as widely for large-scale reactions. In some cases, however, these ethers may be necessary to obtain an efficient and high-yield reaction.

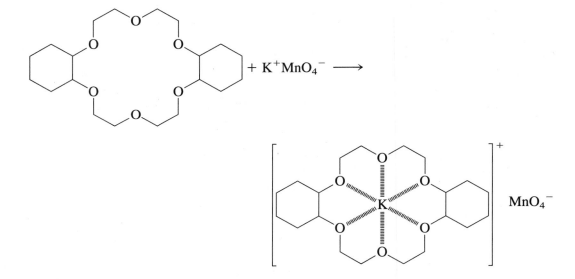

THE EXPERIMENT

In this experiment, you will prepare 7,7-dichlorobicyclo[4.1.0]heptane, also known as 7,7-dichloronorcarane, by the reaction

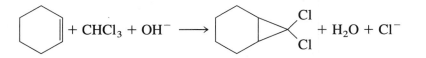

Chloroform $CHCl_3$ and base are used in excess in this reaction. Although most of the chloroform reacts to give the 7,7-dichloronorcarane via the carbene intermediate, a significant portion is hydrolyzed by the base to formate ion and carbon monoxide (Equations

4 and 5, p. 254). Bromoform $CHBr_3$ can be used to prepare the corresponding 7,7-dibromonorcarane via the dibromocarbene.

Required Reading

Review: Technique 7
          Technique 19   Sections 19.2, 19.9, and 19.10

New: Appendix 4   Nuclear Magnetic Resonance Spectroscopy
          Appendix 5   Carbon-13 Nuclear Magnetic Resonance Spectroscopy

Special Instructions

Caution: Chloroform is a suspected carcinogen; therefore, do not let it touch your skin, and avoid breathing the vapor.

This entire experiment must be performed in a hood. Avoid contact with the caustic 50% aqueous sodium hydroxide.

Waste Disposal

Dispose of all aqueous solutions produced in this experiment in the container for aqueous waste.

Procedure

Caution: Chloroform and cyclohexene should be kept in a hood. Avoid contact with these substances and do not breathe the vapors. Do your measuring and transferring operations in the hood.

Reaction Mixture.   Preweigh a 25-mL round-bottom flask with glass stopper and transfer 2.0 mL of cyclohexene ($MW = 82.2$) to the flask in a hood. Stopper the flask and reweigh it to determine the weight of cyclohexene. Add 5.0 mL of 50% aqueous sodium hydroxide[1] to the flask, being careful to avoid getting any solution on the glass

---

[1]This reagent should be prepared by the instructor. Dissolve 50 g of sodium hydroxide in 50 mL of water. Cool the solution to room temperature and store it in a plastic bottle.

joint. In a hood add 5.0 mL of chloroform ($MW = 119.4$, $d = 1.49$ g/mL) to the round-bottom flask. Add a magnetic stir bar to the flask. Weigh 0.20 g of the phase-transfer catalyst, benzyltriethylammonium chloride, on a smooth piece of paper and *reclose the bottle immediately*. (It is hygroscopic!)[2] Add the catalyst to the flask and stopper it.

Reaction Period. In a hood prepare a hot water bath at 40°C using a 250-mL beaker and hotplate. Assemble a reflux apparatus as shown in Technique 3, Figure 3.8, p. 619 Clamp the condenser so that the flask is immersed in the water bath. Stir the mixture as *rapidly* as possible for 1 hour. An emulsion forms during this time.

Extraction of Product. Following this reaction time, remove the flask from the water bath and allow the reaction mixture to cool to room temperature. Pour the reaction mixture into a small beaker and remove the magnetic stir bar. Transfer the mixture to a 125-mL separatory funnel. Add 8 mL of water and 5 mL of methylene chloride to the mixture. Shake the mixture with venting for about 30 seconds and allow the layers to separate. Swirl the funnel gently to help break up the emulsion. Drain the lower methylene chloride layer into a small Erlenmeyer flask. The small amount of emulsion that forms at the interface should be left behind with the aqueous layer. Add another 5 mL portion of methylene chloride, shake the mixture for 30 seconds, and allow the layers to separate. It may be necessary to swirl the funnel gently or tap the funnel gently with your finger to help break up the emulsion. Combine the lower organic layer with the first extract. Discard the remaining aqueous layer into the container for aqueous waste, avoiding contact with the basic solution. Rinse the funnel with water and pour the combined organic layers back into the separatory funnel. Extract the mixture with 10 mL of saturated sodium chloride solution. Drain the lower methylene chloride layer, avoiding any emulsion that might be present at the interface, into a *dry* Erlenmeyer flask containing 0.5 g of anhydrous sodium sulfate. Stopper the flask and swirl it occasionally for at least 10 minutes to dry the organic layer.

Evaporation of Solvent. Using a dry Pasteur pipet, transfer about half of the dried organic layer to a dry preweighed centrifuge tube. Evaporate the methylene chloride, together with any remaining cyclohexene and chloroform, in a hood in a warm water bath. Use a stream of dry air or nitrogen to aid the evaporation process. Evaporate the solvent until the volume is less than about 3 mL. Then, add the remainder of the liquid and continue evaporation. *Be very careful or you may also evaporate the product!* The product, 7,7-dichloronorcarane, is a liquid. Continue the evaporation until the level of the liquid is no longer changing (about 1–1.5 mL).

Analysis of Product. Following removal of methylene chloride, you are left with 7,7-dichloronorcarane of sufficient purity for spectroscopy. Weigh the centrifuge tube, determine the weight of product, and calculate the percentage yield. Obtain the infrared spectrum (Technique 19, Section 19.2, p. 843). At the option of the instructor, obtain the proton NMR spectrum. It may also be of interest to obtain the decoupled and coupled carbon-13 spectra of your product. Submit any remaining product in a labeled vial with your laboratory report.

---

[2]Note to the Instructor: The activity of benzyltriethylammonium chloride varies depending upon the source of the catalyst. Try the reaction in advance of the laboratory to make sure it works properly. We use Aldrich Chemical Co., #14,655-2.

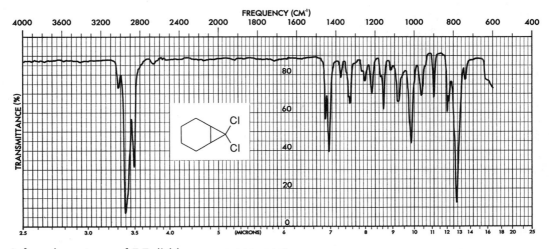

Infrared spectrum of 7,7-dichloronorcarane, neat.

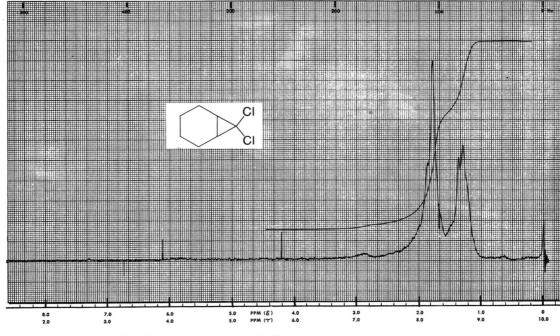

NMR spectrum of 7,7-dichloronorcarane, CDCl₃.

REFERENCES

Dehmlow, E. V. "Phase-Transfer Catalyzed Two-Phase Reactions in Preparative Organic Chemistry." *Angewandte Chemie, International Edition in English, 13* (1974): 170.

Dehmlow, E. V. "Advances in Phase-Transfer Catalysis." Ibid., *16* (1977): 493.

Dehmlow, E. V., and Dehmlow, S. S. *Phase-Transfer Catalysis,* 3rd ed. Weinheim: VCH Publishers, 1992.

Gokel, G. W., and Weber, W. P. "Phase Transfer Catalysis." *Journal of Chemical Education, 55* (1978): 350, 429.

Lee, A. W. M., and Yip, W. C. "Fabric Softeners as Phase-Transfer Catalysts in Organic Synthesis." *Journal of Chemical Education, 68* (1991): 69.

Makosza, M., and Wawrzyniewicz, M. "Catalytic Method for Preparation of Dichlorocyclopropane Derivatives in Aqueous Medium." *Tetrahedron Letters* (1969): 4659.

Starks, C. M. "Phase-Transfer Catalysis." *Journal of the American Chemical Society, 93* (1971): 195.

Starks, C. M., Liotta, C. L., and Halpern, M. *Phase-Transfer Catalysis.* New York: Chapman and Hall, 1993.

Weber, W. P., and Gokel, G. W. *Phase Transfer Catalysis in Organic Synthesis.* Berlin: Springer-Verlag, 1977.

QUESTIONS

1. Why did you need to stir the mixture vigorously during reaction?

2. Why did you wash the organic phase with saturated sodium chloride solution?

3. What short chemical test could you make on the product to indicate whether cyclohexene is present or absent?

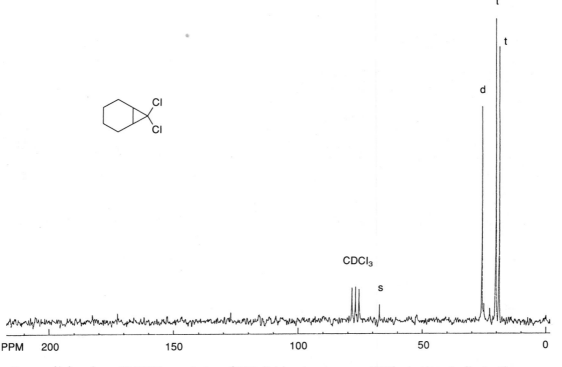

Decoupled carbon-13 NMR spectrum of 7,7-dichloronorcarane, CDCl₃. Letters indicate the appearance of spectrum when carbons are coupled to protons (s = singlet, d = doublet, t = triplet).

4. Would you expect 7,7-dichloronorcarane to give a positive sodium iodide in acetone test?

5. Assign the C—H stretch for the cyclopropane ring hydrogens in the infrared spectrum.

6. Suggest why it may be necessary to use a large excess of chloroform in this reaction.

7. A student obtained an NMR spectrum of the product isolated in this experiment. The spectrum shows peaks at about 7.3 and 5.6 ppm. What do you think these peaks indicate? Are they part of the 7,7-dichloronorcarane spectrum?

8. Draw the structures of the products that you would expect from the reactions of *cis-* and *trans-* 2-butene with dichlorocarbene.

9. Draw the structure of the expected dichlorocarbene adduct of methyl methacrylate (methyl 2-methylpropenoate). With compounds of this type, another product could have been obtained. It is the chloroform adduct to the double bond (Michael-type reaction). What would this structure look like?

10. Provide mechanisms for the following abnormal dichlorocarbene addition reactions. In both cases, the usual adduct is first obtained, and then a subsequent reaction occurs.

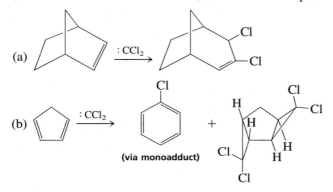

E X P E R I M E N T   2 9

Markovnikov and Anti-Markovnikov Hydration of 1-Methylcyclohexene

Addition to double bonds
Hydroboration–oxidation
Oxymercuration
Gas chromatography
Nuclear magnetic resonance
Regiospecificity–regioselectivity

The most characteristic reaction of alkenes is the **addition reaction.** This reaction can be described by the general equation

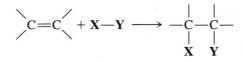

In this reaction, the π-bond of the alkene is broken and two σ-bonds are formed.

Because the double bond is an electron-rich center, it behaves like a Lewis base or a nucleophile, donating a pair of electrons to form a bond to a sufficiently electrophilic reagent. Some part of the attacking reagent **XY** must be electrophilic. As an example, in the reaction of an alkene with an acid **HA,** the first step of the mechanism involves the reaction of the double bond with the electrophilic species **H$^+$.** This step produces a cation:

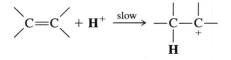

This intermediate cation can react rapidly with the anion **A$^-$** to yield the final product:

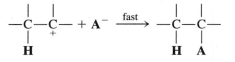

A complication arises when an unsymmetrical reagent **HA** adds to an unsymmetrically substituted double bond. In this case, there are two possible products:

$$R—CH{=}CH_2 + HA \longrightarrow \begin{array}{c} R—CH—CH_2 \\ \,| \quad\;\; | \\ H \quad\; A \end{array} \quad or \quad \begin{array}{c} R—CH—CH_2 \\ \,| \quad\;\; | \\ A \quad\; H \end{array}$$

An empirical rule, known as **Markovnikov's Rule,** was developed in 1868 by the Russian chemist V. V. Markovnikov to apply to such cases. Simply stated, the rule says

> In the ionic addition of an acid to the carbon–carbon double bond of an alkene, the hydrogen of the acid attaches itself to the carbon atom that already holds the *greater* number of hydrogens.

Chemists have determined that the reason Markovnikov's Rule holds is that addition according to the rule always leads to the more highly substituted, and hence the more stable, of the two possible cationic intermediates.

The addition of water to alkenes under acidic conditions (hydration) follows Markovnikov's Rule. The hydrogen of water attaches itself to the carbon atom that already carries the greater number of hydrogens:

$$R—CH{=}CH_2 + H_2O \longrightarrow \begin{array}{c} R—CH—CH_2 \\ \,| \quad\;\; | \\ OH \quad\; H \end{array} \quad \textbf{Markovnikov}\;\textbf{addition}$$

If the addition of water to the alkene had proceeded contrary to Markovnikov's Rule, the reaction would be said to have gone in an **anti-Markovnikov** fashion, and the product would be designated the anti-Markovnikov product:

$$R-CH{=}CH_2 + H_2O \longrightarrow R-\underset{\underset{\textstyle H}{|}}{C}H-\underset{\underset{\textstyle OH}{|}}{C}H_2 \quad \textbf{Anti-Markovnikov addition}$$

Our current experiment deals with two practical methods for carrying out the hydration of an alkene in both the Markovnikov and the anti-Markovnikov fashion. These methods are the **hydroboration–oxidation** and the **oxymercuration** reactions.

HYDROBORATION–OXIDATION

Boron hydrides can react with alkenes by addition of the boron–hydrogen bond across the carbon–carbon double bond. This reaction is called **hydroboration.** Although there are several types of boron hydrides that can be used in the hydroboration reaction, the most common approach is to use the simplest boron hydride, **borane,** which is sold commercially in the form of a complex with **tetrahydrofuran (THF).**

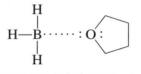

Borane–tetrahydrofuran complex

The boron atom of borane is electron-deficient because it has an incomplete octet of electrons. For this reason, the boron atom is a Lewis acid, or an electron-pair acceptor. An alkene will donate electrons to boron as shown in the first step of the following reaction. This creates a boron atom that is electron-rich and a carbon atom that is electron-deficient. This situation is remedied in the next step by the transfer of a hydride ion (H:⁻) to the carbon atom:

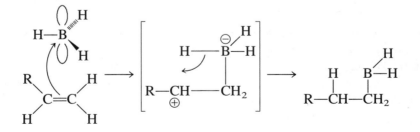

Experimental evidence shows that this entire process is **concerted** without the presence of intermediates and that the addition is stereospecific. The addition of borane to an alkene proceeds exclusively with **syn** stereochemistry. Both the boron and the hydrogen atoms add to the same side of the double bond:

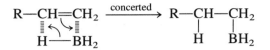

Notice that this addition is an **anti-Markovnikov** addition. In contrast to the situation that applies for the addition of an acid **HA**, the hydrogen from borane does not become attached to the carbon atom with the greater number of hydrogens. However, the electrophilic species that was added to the double bond in this case was not H^+ but the electron-deficient *boron* atom. The lowest-energy cationic intermediate would still predict the course of the reaction (even though it does not exist in hydroboration), and the hydrogen is transferred as $H:^-$ rather than as H^+. This addition occurs three times because there are three B—H bonds that can react; a trialkylborane is formed as a final product.

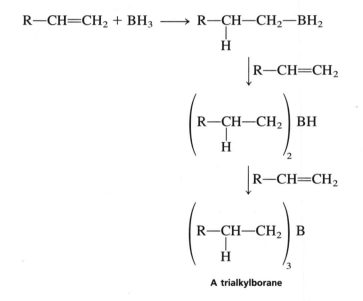

A trialkylborane

If the trialkylborane compound is treated with hydrogen peroxide, it is cleaved with oxidation to form 3 moles of an alcohol. The replacement of a boron atom by an oxygen atom is an intramolecular process that takes place via the migration of the alkyl group from boron to oxygen. The resulting trialkoxyborane is hydrolyzed under the conditions of the reaction to yield the product alcohol.

$$(R-CH_2-CH_2)_3B \xrightarrow[\text{NaOH}]{H_2O_2} (R-CH_2-CH_2-O)_3B$$

$$(R-CH_2-CH_2-O)_3B \xrightarrow[\text{H}_2\text{O}]{\text{NaOH}} 3\ R-CH_2-CH_2-OH + B(OH)_3$$

Hydroboration–oxidation of an alkene therefore leads to an alcohol that corresponds to the anti-Markovnikov addition of water across the double bond of an alkene. This seems to violate Markovnikov's Rule. However, analysis of the mechanism of the hydroboration–oxidation reactions shows that this is not the case:

$$3 \ R\!-\!CH\!=\!CH_2 \xrightarrow{\ BH_3\ } (R\!-\!CH_2CH_2)_3B \xrightarrow[\ OH^-\ ]{\ H_2O_2\ } 3 \ R\!-\!CH_2\!-\!CH_2\!-\!OH$$

Finally, the anti-Markovnikov addition of water across the double bond proceeds with *syn* stereochemistry; this can best be illustrated by using a cyclic alkene as an example. The stereospecificity results from the intramolecular nature of the hydroboration and the oxidation steps, as already described.

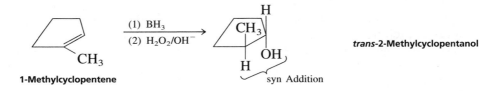

1-Methylcyclopentene   syn Addition   *trans*-**2-Methylcyclopentanol**

OXYMERCURATION

The second means of hydrating a double bond we consider is the **oxymercuration** reaction. In this reaction, mercuric acetate is added to an alkene to form an organomercury derivative:

$$RCH\!=\!CH_2 + Hg(OCOCH_3)_2 \xrightarrow{\ H_2O\ } \underset{\underset{OH}{|}}{RCH}\!-\!CH_2\!-\!HgOCOCH_3$$

The mechanism of this reaction involves the initial ionization of mercuric acetate, followed by addition of the mercury(II) ion across the double bond of the alkene to form a bridged intermediate.

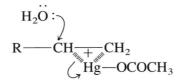

This intermediate is then opened by water, which is a nucleophile, to form the organomercury intermediate. This addition reaction is also stereospecific, proceeding with **anti** stereochemistry.

The organomercury intermediate is reduced with sodium borohydride, with the result that the mercury atom is replaced by hydrogen. The reduction of the organomercury intermediate is not stereospecific.

$$RCH-CH_2-HgOCOCH_3 \xrightarrow[\text{NaOH}]{\text{NaBH}_4} RCH-CH_3$$
$$\quad\;\; | \qquad\qquad\qquad\qquad\qquad\qquad | $$
$$\quad\;\; OH \qquad\qquad\qquad\qquad\qquad\quad OH$$

The product obtained from the oxymercuration of an alkene is the same as what would be expected for the hydration of an alkene according to Markovnikov's Rule. In this particular reaction, the mercury atom is the original electrophile, so it adds to the end of the double bond that bears the greater number of hydrogens. When the mercury atom is replaced by hydrogen in the reduction step of the reaction, hydrogen is then attached to the carbon atom in the manner the rule predicts.

In these two reactions, we see contrasting behavior as to the direction of addition of water. Although each reaction has been described as though it produced exclusively the product shown, in truth it must be mentioned that in some cases there may be a minor product with the opposite orientation. That is to say, while hydration of an alkene may produce the product predicted by Markovnikov's Rule as the principal product, the anti-Markovnikov alcohol may also be produced as a minor product. Reactions that produce a product with only one of several possible orientations are called **regiospecific.** Reactions that produce one substance as the predominant product and small amounts of isomers with other orientations are called **regioselective.** One of the objects of this experiment is to contrast the hydroboration–oxidation and the oxymercuration of an alkene to determine whether the reactions are regiospecific or regioselective.

E X P E R I M E N T   2 9 A

Hydroboration–Oxidation of 1-Methylcyclohexene

The balanced equation for the hydroboration of 1-methylcyclohexene, using a borane–tetrahydrofuran solution, is

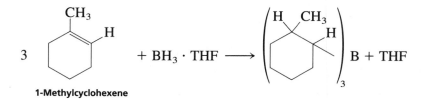

1-Methylcyclohexene

Once the organoborane intermediate is formed, it is oxidized and then hydrolyzed with a basic solution of hydrogen peroxide:

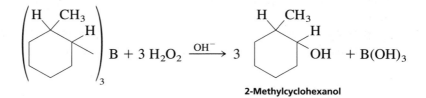

2-Methylcyclohexanol

As the addition of borane and the subsequent oxidation of the trialkylborane intermediate are both stereospecific processes, the product formed in this reaction should be exclusively the *trans* isomer. We can show the correct stereochemistry of the 2-methylcyclohexanol product as follows:

In the method used in this experiment, a solution of borane in tetrahydrofuran is added to the alkene 1-methylcyclohexene. The borane is consumed as it is added. The borane solution is a commercially available product. The addition method avoids the need for handling the potentially hazardous (toxic and flammable) diborane.

Required Reading

Review: Technique 3   Section 3.11
          Technique 7   Sections 7.4, 7.5, 7.7, 7.8, and 7.10
          Technique 15   Gas Chromatography
          Technique 19   Preparation of Samples for Spectroscopy, Part B
          Appendix 4   Nuclear Magnetic Resonance Spectroscopy

New: You should consult your organic chemistry lecture textbook for detailed information about the hydroboration–oxidation reaction.

Special Instructions

Because this reaction involves a somewhat lengthy period of preparation of apparatus and addition of reagents, it is important that you start the experiment at the beginning of the laboratory period. The best place to stop is when the ether extracts are stored over sodium sulfate, although the reaction can be stopped at any time after water has been added to the reaction, if necessary.

> **Caution:** This reaction involves the use of tetrahydrofuran, which is a potentially toxic and flammable solvent. Do not conduct this reaction in the presence of open flames. Avoid contact with 30% hydrogen peroxide, as it is a strong oxidant.

This experiment requires the use of syringes to measure and transfer reagents. It is advisable to use glass-bodied syringes with Teflon or glass plungers, as 1-methylcyclohexene may attack some plastic syringes. Glass plungers, in particular, slide very easily. They can expel reagents from the syringes faster than may be intended. In addition, they may slide out of the syringe barrels, fall, and break. Exercise care when using the syringes.

Waste Disposal

If you perform this experiment carefully, you should not have any leftover borane–tetrahydrofuran solution or 30% hydrogen peroxide. The entire amounts of these reagents should have been added to the reaction mixture. If you have any remaining diethyl ether or tetrahydrofuran, you should dispose of it in a container designated for the disposal of nonhalogenated organic wastes. Any remaining 2-methylcyclohexanol should be discarded in the same waste container. Discard aqueous solutions produced in this experiment into the container designated for aqueous waste.

Procedure

Hydroboration. Rinse a 20-mL round-bottom flask with acetone and dry it thoroughly by placing it in an oven at 110°C for at least 15 minutes. Also dry the glass parts of a hypodermic syringe. When the flask has cooled, weigh it and add 0.92 mL of 1-methylcyclohexene ($MW = 96.172$; $d = 0.813$ g/mL) to the flask. Reweigh the flask to determine an accurate weight of the alkene. Place a stir bar into the flask and seal it with a rubber septum cap, as shown in Figure 1.

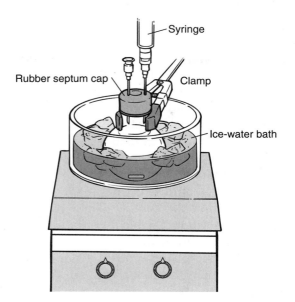

Figure 1. Apparatus for the addition of borane-THF to a reaction.

Insert a hypodermic syringe needle through the rubber septum, being careful to allow enough room for a second syringe needle, which will be inserted later. Place the flask in an ice-water bath on top of a magnetic stirrer. Using the dried syringe, draw 3.0 mL of a commercial borane–tetrahydrofuran solution from the stock bottle. Carefully insert the second syringe through the rubber septum. (It may be necessary to hold the syringe upright with a clamp attached to a ring stand).) Begin stirring the solution. Carefully add the borane–tetrahydrofuran solution to the reaction flask at a rate of about 1–2 drops per second. The addition will require approximately 15–20 minutes. An effective method of addition is to twist the plunger carefully down into the body of the syringe; this technique affords the fine control necessary to establish the correct rate of addition. After adding the borane–tetrahydrofuran solution, allow the reaction mixture to stand in the ice bath, and stir for 1 hour.

Oxidation. At the end of this period, remove the septum cap from the reaction flask. While maintaining the flask in the ice bath with stirring, slowly add 0.84 mL of 30% hydrogen peroxide, using a Pasteur pipet to add this solution. During this addition, maintain the pH of the solution near 8, by adding 3*M* sodium hydroxide, dropwise, as needed. Dip a spatula or glass stirring rod into the solution and touch it to pH paper in order to determine the pH. Add the sodium hydroxide carefully, as the reaction mixture may bubble very vigorously during this addition. It is very important to maintain the pH as near to 8 as possible to prevent premature precipitation of solid materials.

Isolation of Product. Slowly add 4 mL of water to the reaction mixture, followed by 4 mL of diethyl ether. Stir the mixture in the flask for about 10 minutes in order to mix the liquid and solid materials adequately. Remove the lower aqueous layer and transfer it to a 15-mL centrifuge tube. Extract the aqueous layer with three successive 4-mL portions of ether. After each extraction, transfer the ether layer to the round-bottom flask that contains the original ether layer. Transfer the combined ether layers to a separatory funnel and wash them with two successive 4-mL portions of saturated sodium bicarbonate solution. After each washing, remove the aqueous (lower) phase and discard it. Transfer the ether layer to a dry Erlenmeyer flask. Dry the combined ether layers with anhydrous magnesium sulfate. Allow the solution to stand for 10–15 minutes, while stirring occasionally. Be sure to keep the flask stoppered during this drying procedure.

Remove the ether solution from the drying agent and evaporate the ether under a stream of air while heating the container in a warm water bath. Be careful that you do not evaporate the product, which is a liquid. When the ether has been evaporated, remove any residual tetrahydrofuran that may remain in the product by evaporation under reduced pressure (Technique 3, Figure 3.18C, p. 631). **Do not heat the mixture until after the vacuum has been applied.** If you are using an aspirator as a source of vacuum, be sure to place a trap assembly between your apparatus and the aspirator (as shown in Technique 4, Figure 4.5, p. 641). Apply a *gentle* vacuum to the apparatus, while swirling the flask containing the organic material. Once the vacuum has been applied, *gently* heat the mixture so that the tetrahydrofuran slowly evaporates under reduced pressure.

When the bubbling in the flask ceases, remove the vacuum and weigh the residual alcohols to determine the percentage yield obtained from the reaction.[1] Analyze the mixture of alcohols according to the methods outlined in Experiment 29C.

EXPERIMENT 29B

Oxymercuration of 1-Methylcyclohexene

The balanced equation for the oxymercuration of 1-methylcyclohexene is

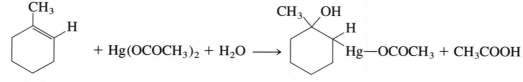

1-Methylcyclohexene

---

[1]Some unreacted alkene may be present in the mixture of alcohols. If an appreciable amount of alkene is present, use the percentage of alkene determined from the gas chromatogram to subtract the contribution of the alkene from the total weight of products to get the actual weight of the isomeric alcohols.

Reduction of the organomercury derivative with sodium borohydride yields the alcohol 1-methylcyclohexanol:

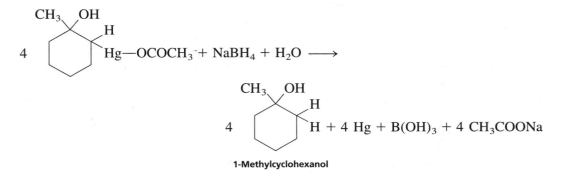

1-Methylcyclohexanol

Required Reading

Review: Technique 3   Section 3.11
          Technique 7   Sections 7.7 and 7.8
          Technique 15   Gas Chromatography
          Technique 19   Preparation of Samples for Spectroscopy, Part B
          Appendix 4   Nuclear Magnetic Resonance Spectroscopy

New: You should consult your organic chemistry lecture textbook for detailed information about the oxymercuration reaction.

Special Instructions

This experiment requires that the reaction mixture be allowed to stand overnight. As two lengthy periods of stirring are necessary, this experiment must be started at the beginning of the laboratory period.

> **Caution:** This experiment involves the use of mercuric acetate. Like all mercury compounds, mercuric acetate is very toxic. Wash your hands thoroughly after handling this substance. Do not get it on your hands or face.

Waste Disposal

All compounds of mercury are toxic and must be disposed of in a waste container designated for the disposal of heavy metals. The mercury produced in the final reduction step of this procedure must be disposed of in the same waste container.

> **Caution:** Do not allow mercury-containing materials to go into the sink.

If you have any remaining diethyl ether, dispose of it in a container designated for the disposal of nonhalogenated organic wastes. Any remaining 1-methylcyclohexanol should be discarded in the same waste container. Aqueous solutions produced in this experiment should be discarded in a container designated for the disposal of aqueous waste.

Procedure

> **Caution:** Mercuric acetate is a highly poisonous substance.

Oxymercuration. Set up an apparatus consisting of a 20-mL round-bottom flask equipped with a magnetic stirring bar. Place 1.20 g of mercuric acetate and 4 mL of water in the flask and attach a water-cooled condenser. Place the flask in a water bath (room temperature) and assemble all components on a magnetic stirrer. Stir the mixture vigorously. After the mercuric acetate has dissolved, add 2 mL of ether. Accurately measure 0.46 mL of 1-methylcyclohexene ($MW = 96.172$; $d = 0.813$ g/mL) to a centrifuge tube that contains 2 mL of ether. Calculate the weight of the alkene using the density. Add this solution dropwise with a Pasteur pipet down the reflux condenser and continue stirring vigorously at room temperature for 1 hour. After 1 hour, add 2 mL of 6M sodium hydroxide solution followed by a solution of 0.80 g of sodium borohydride in 4 mL of 3M sodium hydroxide.

Stir this mixture for 30 minutes, after which time most of the mercury should have settled to the bottom of the flask. Allow this mixture to settle until the next laboratory period.

Isolation of Product. After the mixture has settled, carefully withdraw the supernatant liquid from the mercury that has been deposited on the bottom of the flask using a Pasteur pipet. Transfer the supernatant liquid to a 15-mL centrifuge tube. Discard the mercury into the designated waste container. **Never discard the mercury down the sink drain.** Using a Pasteur pipet, transfer the aqueous layer (lower) to another 15-mL centrifuge tube and save the ether layer in the original centrifuge tube. Extract the aqueous layer with three successive 2-mL portions of ether. After each extraction, combine the ether layer with the ether solution in the first centrifuge tube. Using a dry Pasteur pipet, transfer the combined ether solutions to a dry test tube. Dry the combined ether layers with anhydrous magnesium sulfate. Allow the solution to stand for 10–15 minutes, while stirring occasionally. Be sure to keep the test tube stoppered during this drying procedure.

Remove the ether from the drying agent using a dry Pasteur pipet. It is important to draw the liquid into the pipet very carefully in order to avoid transferring particles of drying agent with the product. Transfer the ether to a clean, preweighed 15-mL centrifuge tube, and evaporate the ether under a stream of air in the hood, while heating the centrifuge tube in a water bath at about 40°C. Be careful that you do not evaporate the product, which is a liquid. This evaporation may have to be conducted on portions of the ether solution, as there may be more solution than will fit in a centrifuge tube. If the resulting product has a suspended fine precipitate, centrifuge the liquid before analyzing it by gas chromatography (Experiment 29C). Otherwise, the fine particles will likely plug the syringe used in the gas chromatographic analysis. Weigh the centrifuge tube to determine the yield of product. Calculate the percentage yield of alcohol isomers obtained.[2] Analyze the mixture of alcohols according to the methods outlined in Experiment 29C.

E X P E R I M E N T   2 9 C

Analysis Procedures

Procedure

Following your instructor's recommendations, analyze the product of your reaction using gas chromatography or proton NMR spectroscopy or both.

Gas Chromatography.   Analyze your mixture of alcohols using gas chromatography to determine the relative percentages of Markovnikov and anti-Markovnikov products (Technique 15, Section 15.11, p. 820). The correct conditions for the gas chromatographic analysis are[3]:

Column temperature: 120°C

Injection port temperature: 120°C

Detector temperature: 130°C

Column packing: 15-meter capillary column coated with a DB-WAX stationary liquid phase

Relative order of retention times: 1-methylcyclohexene, 1-methylcyclohexanol, 2-methylcyclohexanol

Proton NMR Spectroscopy.   You may use NMR spectroscopy to analyze the composition of the product mixture. The NMR spectra of 1-methylcyclohexanol and

---

[2]Some unreacted alkene may be obtained in the products. If desired, you can correct for the presence of alkene (see Footnote 1).

[3]Note to the Instructor: If other chromatography columns or conditions are used, run standard samples ahead of time so that the appropriate retention times are available for use by the class.

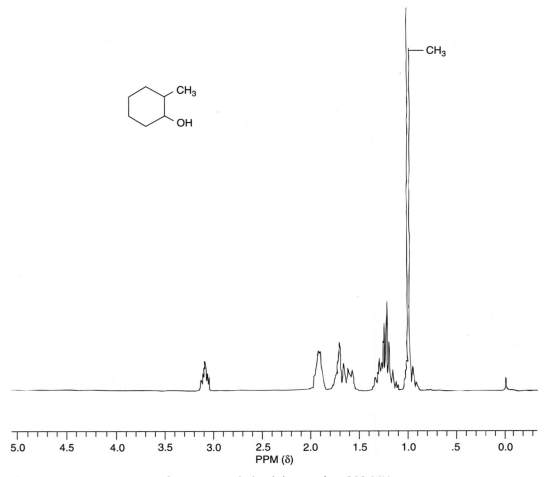

Figure 2.   NMR spectrum of *trans*-2-methylcyclohexanol at 300 MHz.

2-methylcyclohexanol are included for reference in Figures 2 and 3. Compare the integrals of the three hydrogens of the methyl groups in each of the two products to determine the relative percentages of the two isomers. Occasionally, the sample may be contaminated with residual tetrahydrofuran, which is difficult to remove under vacuum. If this is the case, peaks at about 1.5 and 3.5 ppm will also appear in the NMR spectrum of the product.

   Report.   Compare your results with the results obtained by students who completed the other procedure from the one you used. In your laboratory report, submit the gas chromatograms and NMR spectra you obtained. Calculate the percentage yield of alcohol isomers and the ratios of the two isomers from your chromatographic and spectral data. Comment on whether each reaction proceeded according to Markovnikov's Rule and whether the reactions were regiospecific or regioselective.

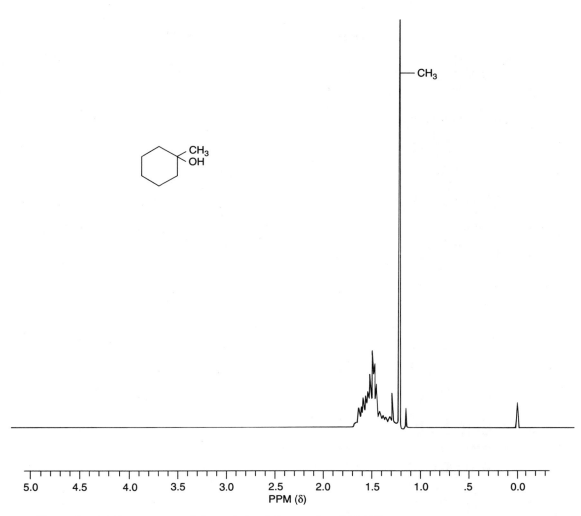

Figure 3.   NMR spectrum of 1-methylcyclohexanol at 300 MHz.

REFERENCES

Brown, H. C. "Hydride Reductions: A 40-Year Revolution in Organic Chemistry." *Chemical and Engineering News, 57* (March 5, 1979): 24.

Brown, H. C. *Hydroboration.* Reading, MA: Benjamin/Cummings, 1980.

Brown, H. C., and Zweifel, G. "Hydroboration. VII. Directive Effects in the Hydroboration of Olefins." *Journal of the American Chemical Society, 82* (1960): 4708.

Jerkunica, J. M., and Traylor, T. G. "Oxymercuration–Reduction: Alcohols from Olefins: 1-Methylcyclohexanol." *Organic Syntheses, 53* (1973): 94.

Zweifel, G., and Brown, H. C. "Hydroboration of Olefins: (+)-Isopinocampheol." *Organic Syntheses, 52* (1972): 59.

QUESTIONS

1. Compare the relative percentages of 1- and 2-methylcyclohexanols formed from 1-methyl-cyclohexene in each of the two reactions. Classify each reaction as regiospecific or regioselective.
2. Predict the products of the hydroboration–oxidation and the oxymercuration reactions for each of the following compounds. Include the correct stereochemistry, when appropriate. Assume that the reactions are regiospecific.

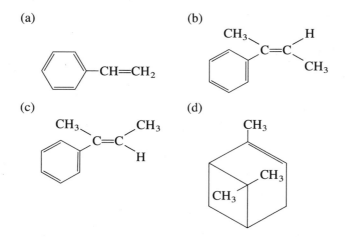

(a)

(b)

(c)

(d)

EXPERIMENT 30

Triphenylmethanol and Benzoic Acid

Grignard reaction
Extraction
Crystallization

In this experiment, you will prepare a Grignard reagent, or organomagnesium reagent. The reagent is phenylmagnesium bromide.

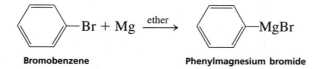

Bromobenzene   **Phenylmagnesium bromide**

This reagent will be converted to a tertiary alcohol or a carboxylic acid, depending on the experiment selected.

EXPERIMENT 30A

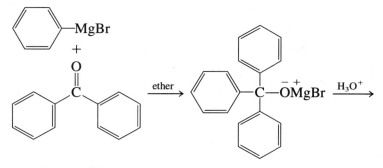

Benzophenone

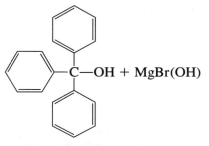

Triphenylmethanol

EXPERIMENT 30B

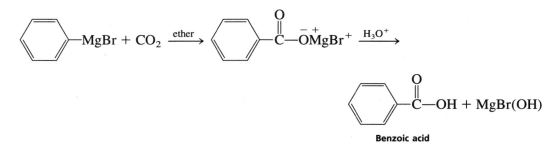

Benzoic acid

The alkyl portion of the Grignard reagent behaves as if it had the characteristics of a **carbanion.** We may write the structure of the reagent as a partially ionic compound:

$$\overset{\delta-}{R} \cdots \overset{\delta+}{MgX}$$

This partially bonded carbanion is a Lewis base. It reacts with strong acids, as you would expect, to give an alkane:

$$\overset{\delta-}{R} \cdots \overset{\delta+}{MgX} + HX \longrightarrow R{-}H + MgX_2$$

Any compound with a suitably acidic hydrogen will donate a proton to destroy the reagent. Water, alcohols, terminal acetylenes, phenols, and carboxylic acids are all acidic enough to bring about this reaction.

The Grignard reagent also functions as a good nucleophile in nucleophilic addition reactions of the carbonyl group. The carbonyl group has electrophilic character at its carbon atom (due to resonance), and a good nucleophile seeks out this center for addition.

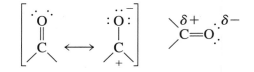

The magnesium salts produced form a complex with the addition product, an alkoxide salt. In a second step of the reaction, these must be hydrolyzed (protonated) by addition of dilute aqueous acid:

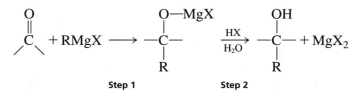

The Grignard reaction is used synthetically to prepare secondary alcohols from aldehydes and tertiary alcohols from ketones. The Grignard reagent will react with esters twice to give tertiary alcohols. Synthetically, it also can be allowed to react with carbon dioxide to give carboxylic acids and with oxygen to give hydroperoxides:

$$RMgX + O{=}C{=}O \longrightarrow R-\overset{\overset{\textstyle O}{\|}}{C}-OMgX \xrightarrow[H_2O]{HX} R-\overset{\overset{\textstyle O}{\|}}{C}-OH$$

Carboxylic acid

$$RMgX + O_2 \longrightarrow ROOMgX \xrightarrow[H_2O]{HX} ROOH$$

Hydroperoxide

Because the Grignard reagent reacts with water, carbon dioxide, and oxygen, it must be protected from air and moisture when it is used. The apparatus in which the reaction is to be conducted must be scrupulously dry (recall that 18 mL of H_2O is 1 mole), and the solvent must be free of water, or anhydrous. During the reaction, the flask must be protected by a calcium chloride drying tube. Oxygen should also be excluded. In practice this can be done by allowing the solvent ether to reflux. This blanket of solvent vapor keeps air from the surface of the reaction mixture.

In the experiment described here, the principal impurity is **biphenyl,** which is formed by a heat- or light-catalyzed coupling reaction of the Grignard reagent and unreacted bromobenzene. A high reaction temperature favors the formation of this product. Biphenyl is highly soluble in petroleum ether, and it is easily separated from triphenylmethanol. Biphenyl can be separated from benzoic acid by extraction.

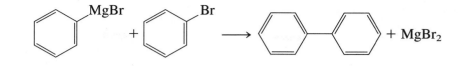

Required Reading

Review: Technique 4   Section 4.3
           Technique 5   Section 5.3
           Technique 7   Sections 7.4, 7.5, 7.8, 7.10
           Technique 19   Section 19.4

Special Instructions

This experiment must be conducted in one laboratory period either to the point after which benzophenone is added (Experiment 30A) or to the point after which the Grignard reagent is poured over dry ice (Experiment 30B). The Grignard reagent cannot be stored; you must react it before stopping. This experiment uses diethyl ether, which is extremely flammable. Be certain that no open flames are in your vicinity when you are using ether.

During this experiment, you will need to use *anhydrous* diethyl ether, which is usually contained in metal cans with a screwcap. You are instructed in the experiment to transfer a small portion of this solvent to a stoppered Erlenmeyer flask. Be certain to minimize exposure to atmospheric water during this transfer. Always recap the ether container after use. Do not use solvent-grade ether because it may contain some water.

All students will prepare the same Grignard reagent, phenylmagnesium bromide. At the option of the instructor, you should then proceed to either Experiment 30A (triphenylmethanol) or Experiment 30B (benzoic acid) when your reagent is ready.

Waste Disposal

All aqueous solutions should be placed in a container designated for aqueous waste. Be sure to decant these solutions away from any magnesium chips before placing them in the waste container. The unreacted magnesium chips that you separate should be placed in a solid waste container designated for that purpose. Place all ether solutions in the container for nonhalogenated liquid wastes. Likewise, the mother liquor from the crystallization using isopropyl alcohol (Experiment 30A) should also be placed in the container for nonhalogenated liquid wastes.

Notes to the Instructor

Whenever possible, you should require that your class wash and dry the necessary glassware *the period before this experiment is scheduled.* It is not a good idea to use glassware that has been washed earlier in the same period, even if it has been dried in the oven. When drying, be certain that no Teflon stopcocks, plastic stoppers, or plastic clips are placed in the oven.

Procedure

$$\text{C}_6\text{H}_5\text{—Br} + \text{Mg} \xrightarrow{\text{ether}} \text{C}_6\text{H}_5\text{—MgBr}$$

PREPARATION OF THE GRIGNARD REAGENT: PHENYLMAGNESIUM BROMIDE

Glassware. The following glassware is used:

100 mL round-bottom flask   Claisen head

125 mL separatory funnel   water-jacketed condenser

2 CaCl$_2$ drying tubes   2 50-mL Erlenmeyer flasks

10 mL graduated cylinder

Preparation of Glassware. If necessary, dry all the pieces of **glassware** (no plastic parts) listed above in an oven at 110°C for at least 30 minutes. This step can be omitted if your glassware is clean and has been unused in your drawer for at least two to three days. Otherwise, all glassware used in your Grignard reaction must be scrupulously dried. Surprisingly large amounts of water adhere to the walls of glassware, even when it is apparently dry. Glassware washed and dried the same day it is to be used can still give problems in starting a Grignard reaction.

> **Caution:** Do not place any plasticware, plastic connectors, or Teflon stoppers in the oven as they may melt, burn, or soften. Check with your instructor if in doubt.

Apparatus. Add a clean, dry stirring bar to the 100-mL round-bottom flask and assemble the apparatus as shown in the figure. Place drying tubes (filled with fresh calcium chloride) on both the separatory funnel and on the top of the condenser. A stirring hotplate will be used to stir and heat the reaction.[1,2] Make sure that the apparatus can be moved up and down easily on the ring stand. Movement up and down relative to the hotplate will be used to control the amount of heat applied to the reaction.

---

[1] A steam bath or steam cone may be used, but you will probably have to forgo any stirring, and use a boiling stone instead of a spin bar.

[2] A heating mantle could be used to heat the reaction. With a heating mantle, it is probably best to clamp the apparatus securely, and to support the heating mantle under the reaction flask with wooden blocks that can be added or removed. When the blocks are removed, the heating mantle can be lowered away from the flask.

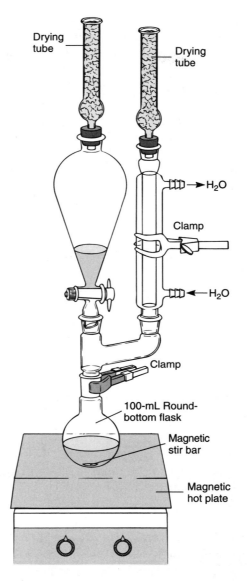

Apparatus for Grignard reactions.

Formation of the Grignard Reagent.   Using smooth paper or a small beaker, weigh about 0.5 g of magnesium turnings ($AW = 24.3$) and place them in the 100-mL round-bottom flask. Using a preweighed 10-mL graduated cylinder, measure approximately 2.1 mL of bromobenzene ($MW = 157.0$) and reweigh the cylinder to determine the exact mass of the bromobenzene. Transfer the bromobenzene to a stoppered 50-mL Erlenmeyer flask. Without cleaning the graduated cylinder, measure a 10-mL portion of anhydrous ether and transfer it to the same 50-mL Erlenmeyer flask containing the bromobenzene. Mix the solution (swirl) and then, using a dry, disposable Pasteur pipet, transfer about half of it into the round-bottom flask containing the magnesium turnings. Add the remainder of the solution to the 125-mL separatory funnel. Then add an additional 7.0 mL of anhydrous ether to the bromobenzene solution in the separatory funnel. At this point, make sure all joints are sealed and that the drying tubes are in place.

Position the apparatus just above the hotplate and stir the mixture *gently* so as to avoid throwing the magnesium out of the solution and onto the side of the flask. You should begin to notice the evolution of bubbles from the surface of the metal, which signals that the reaction is starting. It will probably be necessary to heat the mixture using your hotplate to start the reaction. The hotplate should be adjusted to its lowest setting. Since ether has a low boiling point (35°C), it should be sufficient to heat the reaction by placing the round-bottom flask just above the hotplate. Once the ether is boiling, check to see if the bubbling action continues after the apparatus is lifted above the hotplate. If the reaction continues to bubble without heating, the magnesium is reacting. You may have to repeat the heating several times to successfully start the reaction. After several tries at heating, the reaction should start, but if you are still experiencing difficulty, proceed to the next paragraph.

Optional Steps.   You may need to employ one or more of the following procedures if heating fails to start the reaction. If you are experiencing difficulty, remove the separatory funnel. Place a long, *dry* glass stirring rod into the flask and gently twist the stirring rod so as to crush the magnesium against the glass surface. *Be careful not to poke a hole in the bottom of the flask; do this gently!* Reattach the separatory funnel and heat the mixture again. Repeat the crushing procedure several times, if necessary, to start the reaction. If the crushing procedure fails to start the reaction, then add one small crystal of iodine to the flask. Again, heat the mixture *gently.* The most drastic action, other than starting the experiment over again, is to prepare a small sample of the Grignard reagent *externally* in a test tube. When this external reaction starts, add it to the main reaction mixture. This "booster shot" will react with any water that is present in the mixture and allow the reaction to get started.

Completing the Grignard Preparation.   When the reaction has started, you should observe the formation of a brownish-gray cloudy solution. Add the remaining solution of bromobenzene slowly over a period of 5 minutes at a rate that keeps the solution boiling gently. If the boiling stops, add more bromobenzene. It may be necessary to heat the mixture occasionally with the hotplate during the addition. If the reaction becomes too vigorous, slow the addition of the bromobenzene solution and raise the apparatus higher above the hotplate. Ideally, the mixture will boil without the application of external heat. *It is important that you heat the mixture if the reflux slows or stops.* As the reaction proceeds, you should observe the gradual disintegration of the magnesium metal. When all the bromobenzene has been added, place an additional 1.0 mL of *anhydrous* ether in the separatory funnel to rinse it, and add it to the reaction mixture. Remove the separatory funnel after making this addition and replace it with a stopper. Heat the solution under gentle reflux until most of the remaining magnesium dissolves (don't worry about a few small pieces). This should require about 15 minutes. Note the level of the solution in the flask. You should add additional anhydrous ether to replace any that is lost during the reflux period. During this reflux period, you can prepare any solutions needed for Part A or Part B. When the reflux is complete, allow the mixture to cool to room temperature. As your instructor designates, go on to either Experiment 30A or Experiment 30B.

EXPERIMENT 30A

Triphenylmethanol

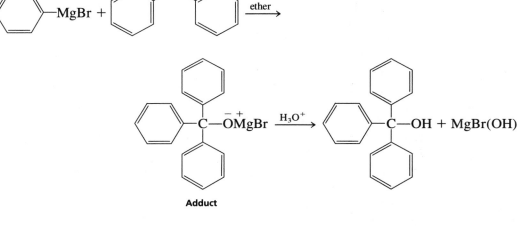

Adduct

Procedure

Addition of Benzophenone.   While the phenylmagnesium bromide solution is being heated and stirred under reflux, make a solution of 2.4 g benzophenone in 9.0 mL of *anhydrous* ether in a 50-mL Erlenmeyer flask. Stopper the flask until the reflux period is over. Once the Grignard reagent is cooled to room temperature, reattach the separatory funnel and transfer the benzophenone solution into it. Add this solution as rapidly as possible to the stirred Grignard reagent, but at such a rate that the solution does not reflux too vigorously. Rinse the Erlenmeyer flask that contained the benzophenone solution with about 5.0 mL of anhydrous ether and add it to the mixture. Once the addition has been completed, allow the mixture to cool to room temperature. The solution turns a rose color and then gradually solidifies as the adduct is formed. When magnetic stirring is no longer effective, stir the mixture with a spatula. Remove the reaction flask from the apparatus and stopper it. Occasionally stir the contents of the flask. The adduct should be fully formed after about 15 minutes. You may stop here.

Hydrolysis.   Add enough 6M hydrochloric acid (*dropwise at first*) to neutralize the reaction mixture (approximately 7.0 mL). Enough acid has been added when the lower aqueous layer turns blue litmus paper red. The acid converts the adduct to triphenylmethanol and inorganic compounds (MgX_2). Eventually, you should obtain two distinct phases: the upper ether layer will contain triphenylmethanol; the lower aqueous hydrochloric acid layer will contain the inorganic compounds. Use a spatula to break up the solid during the addition of hydrochloric acid. Swirl the flask occasionally to assure thorough mixing. Since the neutralization procedure evolves heat, some ether will be lost due to evaporation. You should add enough ether to maintain a 5- to 10-mL volume in the upper organic phase. Make sure that you have two distinct liquid phases before proceeding to separate the layers. More ether or hydrochloric acid may be added, if necessary, to dissolve any remaining solid.[3]

If some material stubbornly remains undissolved or if there are three layers, transfer all the liquids to a 250-mL Erlenmeyer flask. Add more ether and more hydrochloric acid to the flask, and swirl it to mix the contents. Continue adding small portions of ether and hydrochloric acid to the flask and swirl it until everything dissolves. At this point, you should have two clear layers.

Separation and Drying.   Transfer your mixture to a 125-mL separatory funnel, but avoid transferring the spin bar (or boiling stone). Shake and vent the mixture, then allow the layers to separate. If any unreacted magnesium metal is present, you will observe bubbles of hydrogen being formed. You may remove the aqueous layer even though the magnesium is still producing hydrogen. Drain off the lower aqueous phase and place it in a beaker for storage. Next, *save the upper ether layer* in an Erlenmeyer flask; it contains the triphenylmethanol product. Reextract the saved aqueous phase with 5.0 mL of ether. Remove the lower aqueous phase and discard it. Combine the remaining ether phase with the first ether extract. Transfer the combined ether layers to a dry Erlenmeyer flask and add about 1.0 g of granular anhydrous sodium sulfate to dry the solution.

Evaporation.   Decant the dried ether solution from the drying agent into a small Erlenmeyer flask, and rinse the drying agent with more diethyl ether. Evaporate the ether solvent in a hood by heating the flask in a warm water bath. Evaporation will occur more quickly if a stream of nitrogen or air is directed into the flask. You should be left with a mixture that varies from a brown oil to a colored solid mixed with an oil. This crude mixture contains the desired triphenylmethanol and the by-product biphenyl. Most of the biphenyl can be removed by adding about 10 mL of petroleum ether (bp 30–60°C). Petroleum ether is a mixture of hydrocarbons that easily dissolves the hydrocarbon biphenyl and leaves behind the alcohol triphenylmethanol. Do not confuse this solvent with diethyl ether ("ether"). Heat the mixture slightly, stir it, and then cool the mixture to room temperature. Collect the triphenylmethanol by vacuum filtration on a small Büchner funnel and rinse it with small portions of petroleum ether (Technique 4, Section 4.3, p. 640, and Fig. 4.5, p. 641). Air-dry the solid, weigh it, and calculate the percentage yield of the crude triphenylmethanol ($MW = 260.3$).

---

[3]In some cases, it may be necessary to add additional water instead of more hydrochloric acid.

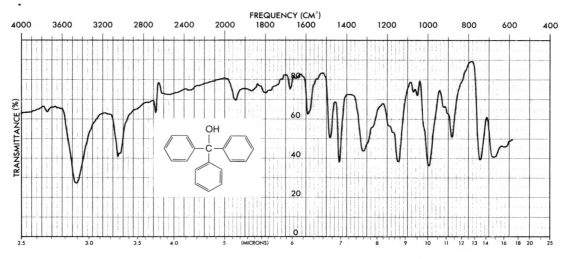

Infrared spectrum of triphenylmethanol, KBr.

Crystallization. Crystallize all your product from hot isopropyl alcohol and collect the crystals using a Büchner funnel (Technique 5, Section 5.3, p. 651, and Fig. 5.3, p. 652). Step 2 in Figure 5.3 (removal of insoluble impurities) should not be required in this crystallization. Set the crystals aside to air-dry. Report the melting point of the purified triphenylmethanol (literature value, 162°C) and the recovered yield in grams. Submit the sample to the instructor.

Spectroscopy. At the option of the instructor, determine the infrared spectrum of the purified material in a KBr pellet (Technique 19, Section 19.4, p. 846). Your instructor may assign certain tests on the product you prepared. These tests are described in the Instructor's Manual.

E X P E R I M E N T   3 0 B

Benzoic Acid

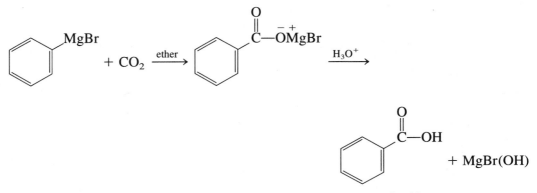

Procedure

Addition of Dry Ice.   When the phenylmagnesium bromide solution has cooled to room temperature, pour it as quickly as possible onto 10 g of crushed dry ice contained in a 250-mL beaker. The dry ice should be weighed as quickly as possible to avoid contact with atmospheric moisture. It need not be weighed precisely. Rinse the flask in which the phenylmagnesium bromide was prepared with 2 mL of anhydrous ether and add it to the beaker.

> **Caution:** Exercise caution in handling dry ice. Contact with the skin can cause severe frostbite. Always use gloves or tongs. The dry ice is best crushed by wrapping large pieces in a clean, dry towel and striking them with a hammer or a wooden block. It should be used as soon as possible after crushing it to avoid contact with atmospheric water.

Cover the reaction mixture with a watch glass and allow it to stand until the excess dry ice has completely sublimed. The Grignard addition compound will appear as a viscous glassy mass.

Hydrolysis.   Hydrolyze the Grignard adduct by slowly adding approximately 8 mL of 6M hydrochloric acid to the beaker and stirring the mixture with a glass rod or spatula. Any remaining magnesium chips will react with acid to evolve hydrogen. At this point, you should have two distinct liquid phases in the beaker. If you have solid present (other than magnesium), try adding a little more ether. If the solid is insoluble in ether, try adding a little more 6M hydrochloric acid solution or water. Benzoic acid is soluble in ether, while inorganic compounds (MgX_2) are soluble in the aqueous acid solution. Transfer the liquid phases to an Erlenmeyer flask, leaving behind any residual magnesium. Add more ether to the beaker to rinse it, and add this additional ether to the Erlenmeyer flask. You may stop here; stopper the flask with a cork, and continue with the experiment during the next laboratory period.

Isolation of the Product.   If you stored your product and the ether layer evaporated, add several milliliters of ether. If the solids do not dissolve on stirring, or if no water layer is apparent, try adding some water. Transfer your mixture to a 125-mL separatory funnel. If some material remains undissolved or if there are three layers, add more ether and hydrochloric acid to the separatory funnel, stopper it, shake it, and allow the layers to separate. Continue adding small portions of ether and hydrochloric acid to the separatory funnel and shake it until everything dissolves. After separation of the layers, remove the lower aqueous layer. The aqueous phase contains inorganic salts and may be discarded. The ether layer contains the product benzoic acid and the by-product biphenyl. Add 5.0 mL of 5% sodium hydroxide solution, restopper the funnel, and shake it. Allow the layers to separate, *remove the lower aqueous layer, and save this layer in a beaker.* This extraction removes benzoic

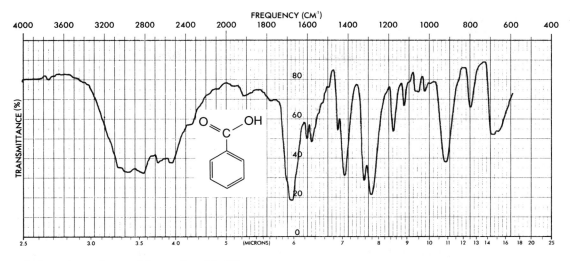

Infrared spectrum of benzoic acid, KBr.

acid from the ether layer by converting it to the water-soluble sodium benzoate. The by-product biphenyl stays in the ether layer along with some remaining benzoic acid. Again, shake the remaining ether phase in the separatory funnel with a second 5.0-mL portion of 5% sodium hydroxide and transfer the lower aqueous layer into the beaker with the first extract. Repeat the extraction process with a third portion (5.0 mL) of 5% sodium hydroxide and save the aqueous layer, as before. Discard the ether layer, which contains the biphenyl impurity, into the waste container designated for nonhalogenated organic wastes.

Heat the combined basic extracts while stirring on a hotplate (100 to 120°C) for about 5 minutes to remove any ether that may be dissolved in this aqueous phase. Ether is soluble in water to the extent of 7%. During this heating period you may observe slight bubbling, but the volume of liquid *will not decrease* substantially. Unless the ether is removed before the benzoic acid is precipitated, the product may appear as a waxy solid instead of crystals.

Cool the alkaline solution and precipitate the benzoic acid by adding 10.0 mL of 6*M* hydrochloric acid, while stirring. Cool the mixture in an ice bath. Collect the solid by vacuum filtration on a Büchner funnel (Technique 4, Section 4.3, p. 640, and Fig. 4.5, p. 641). The transfer may be aided and the solid washed with several small portions of cold water. Allow the crystals to dry thoroughly at room temperature at least overnight. Weigh the solid and calculate the percentage yield of benzoic acid (*MW* = 122.1).

Crystallization.   Crystallize your product from hot water using a Büchner funnel to collect the product by vacuum filtration (Technique 5, Section 5.3, p. 651, and Fig. 5.3, p. 652). Step 2 in Figure 5.3 (removal of insoluble impurities) should not be required in this crystallization. Set the crystals aside to air-dry at room temperature[4]

---

[4]If necessary, the crystals may be dried in a low temperature (ca. 50°C) oven for a *short* period of time. Be warned that benzoic acid sublimes, and heating it for a long time at elevated temperatures could result in loss of your product.

before determining the melting point of the purified benzoic acid (literature value, 122°C) and the recovered yield in grams. Submit your product to your instructor in a properly labeled vial.

Spectroscopy. At the option of the instructor, determine the infrared spectrum of the purified material in a KBr pellet (Technique 19, Section 19.4, p. 846). Your instructor may assign certain tests on the product you prepared. These tests are described in the Instructor's Manual.

QUESTIONS

1. Benzene is often produced as a side-product during Grignard reactions using phenylmagnesium bromide. How can its formation be explained? Give a balanced equation for its formation.

2. Write a balanced equation for the reaction of benzoic acid with hydroxide ion. Why is it necessary to extract the ether layer with sodium hydroxide?

3. Interpret the principal peaks in the infrared spectrum of either triphenylmethanol or benzoic acid, depending on the procedure used in this experiment.

4. Outline a separation scheme for isolating either triphenylmethanol or benzoic acid from the reaction mixture, depending on the procedure used in this experiment.

5. Provide methods for preparing the following compounds by the Grignard method:

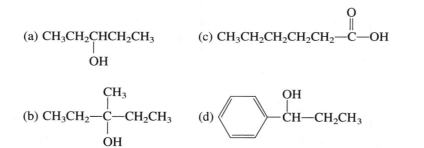

EXPERIMENT 31

Carbonation of an Unknown Aromatic Halide

Grignard reaction
Crystallization and melting point
Student designed project
Identification of an unknown
Carbon-13 and proton NMR (optional)

In this experiment you will be issued an unknown aromatic halide. Your project is to convert it to a carboxylic acid using the Grignard reaction.

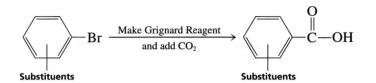

Convert the unknown halide to a carboxylic acid, purify the crude acid by crystal-lization, determine its melting point, and identify the starting compound based on the melting point of its acid derivative. At the option of your instructor, you might be asked to use NMR spectroscopy on either your product or your starting material.

You will be assigned one of the unknown starting materials shown in the following two lists. The compounds in List A can be identified by using the melting point of the carboxylic acid you obtain and no other data. If you have NMR available (especially carbon-13 NMR) your instructor may wish to expand the list of possible unknowns to include List B.

List A[1]		List B[1,2]	
Bromo compound	**mp of acid**	**Bromo compound**	**mp of acid**
2-bromoanisole	98–100	1-bromo-4-butylbenzene	100–113
3-bromoanisole	106–108	2-bromotoluene	103–105
1-bromo-2,4-dimethylbenzene	124–126	3-bromotoluene	108–110
1-bromo-2,5-dimethylbenzene	132–134	1-bromo-2,6-dimethylbenzene	114–116
1-bromo-4-propylbenzene	142–144	1-bromo-2,3-dimethylbenzene	145–147
1-bromo-2,4,6-trimethylbenzene	154–155	4-bromotoluene	180–182
1-bromo-4-t-butylbenzene	165–167		
1-bromo-3,5-dimethylbenzene	172–174		
4-bromoanisole	182–185		

[1]Except for the anisole- and toluene-based halides, the compounds in both lists are consistently named so that the bromo group is given the same priority as the carboxyl group that will replace it. In several cases, therefore, the name given is not necessarily the correct IUPAC name.

[2]These compounds can be used only if NMR is available. They cannot be distinguished from those in column A by melting point alone.

Required Reading

Review: Technique 5   Crystallization: Purification of Solids
          Technique 19   Preparation of Samples for Spectroscopy
                         Appendices 3, 4, 5

Special Instructions

If you are unable to identify your product based on its melting point, you may be required to use NMR spectroscopy.

Notes to the Instructor

This experiment will require a great deal of individual help with each student. As a result, it may be very difficult to use this experiment with a large class. It is a good idea to have the students prepare and present their procedures for approval before they are allowed to begin the experimental work. This experiment could require 3–4 periods.

Be sure that any halides you use have at least 90% purity. Avoid technical-grade chemicals or it will be difficult for students to achieve a good melting point. These compounds have a very strange type of naming in the Aldrich catalog (such as 3-bromo-*o*-xylene). Be sure you see the Instructor's Manual for correct ordering instructions.

Procedure

You are asked to devise the entire experimental procedure together with reagent quantities. You should also prepare a separation scheme. Your plan should be presented to your instructor for approval before you begin work. It may be helpful to consult Experiment 30B, which is closely related to this one.

You may assume that your unknown has a molecular weight of about 200 mass units, and you should use about 3 g of your starting halide. Be sure, however, that you check the stoichiometry and use a reasonable amount of magnesium. In order to determine an accurate melting point, the compound must be both pure and dry. It may be necessary to crystallize your product more than one time. Most carboxylic acids can be crystallized from water or an ethanol-water mixed solvent. It is recommended that you consult the *Merck Index, The Handbook of Chemistry and Physics,* or another handbook to determine the best solvent for your final crystallizations.

SPECTROSCOPY

Infrared Spectrum. An infrared spectrum should be determined to verify that the product is a carboxylic acid. The products are all solids, and the best method to determine the infrared spectrum would be through the use of a KBr pellet (Technique 19, Section 19.4, p. 846).

NMR Spectrum. If your instructor asks you to determine NMR spectra for your product, you should check its solubility in $CHCl_3$. If it dissolves in $CHCl_3$, it will probably dissolve in $CDCl_3$, the usual NMR solvent. However, many acids do not dissolve in $CDCl_3$. The commercial solvent called UNISOL (a mixture of $CDCl_3$ and DMSO-d_6) will dissolve most, but not all, carboxylic acids. If your product does not dissolve in UNISOL, a D_2O-NaOD solution (see p. 858) may be used. If possible, you should determine the carbon-13 NMR as well as the proton spectrum.

THE REPORT

The report should include a balanced equation for the preparation of your acid. You should calculate both the theoretical and percentage yields. Write out your complete procedure as you actually performed it. Include the actual results of your melting point determination(s) and compare it (them) to the expected result.

Include an infrared spectrum of your product and interpret the major absorption peaks. Attempt to use the overtone and out-of-plane absorption bands to explain the substitution pattern of your ring. If you determined NMR spectra, you should include them along with an interpretation of the peaks and splitting patterns. See if you can work out a complete tree diagram for the aromatic ring.

EXPERIMENT 32

The Preparation of Pinacol and the Pinacol Rearrangement

Synthesis of a 1,2-diol (pinacol)
Carbocation rearrangement
Infrared spectroscopy
Nuclear magnetic resonance spectroscopy

Ketones, such as acetone, can be converted to 1,2-diols by reduction with an active metal. The most common example of such a reaction is the reduction of acetone with magnesium and mercuric chloride to form **pinacol.**

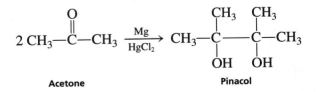

Acetone   **Pinacol**

This reaction proceeds by electron transfer from the metal to the ketone. The resulting species is a **radical-anion,** or a **ketyl.** Two ketyl species dimerize by pairing their lone electrons. The coupling step yields a dianion, which acquires protons during a subsequent hydrolysis step.

(1)
$$2 \ CH_3-\overset{\overset{O}{\|}}{C}-CH_3 + Mg \longrightarrow Mg^{2+} + 2 \ CH_3-\overset{\overset{\cdot\cdot}{:O:}^-}{\underset{\cdot}{C}}-CH_3$$

a ketyl

(2)
$$:\overset{\cdot\cdot}{O}-\overset{\overset{CH_3}{|}}{\underset{|}{C}}-CH_3 \cdot \ + \cdot \overset{\overset{CH_3}{|}}{\underset{|}{C}}-\overset{\cdot\cdot}{O}:^- \longrightarrow :\overset{\cdot\cdot}{O}-\overset{\overset{CH_3}{|}}{\underset{|}{C}}-\overset{\overset{CH_3}{|}}{\underset{|}{C}}-\overset{\cdot\cdot}{O}:^-$$

(3)
$$:^-\overset{\cdot\cdot}{O}-\overset{\overset{CH_3}{|}}{\underset{CH_3}{C}}-\overset{\overset{CH_3}{|}}{\underset{CH_3}{C}}-\overset{\cdot\cdot}{O}:^- + 2 \ H_2O \longrightarrow HO-\overset{\overset{CH_3}{|}}{\underset{CH_3}{C}}-\overset{\overset{CH_3}{|}}{\underset{CH_3}{C}}-OH + 2 \ OH^-$$

The most interesting chemical property of pinacol and similar compounds is their tendency to undergo a carbocation rearrangement when treated with an acid catalyst. This rearrangement is known as the **pinacol rearrangement.**

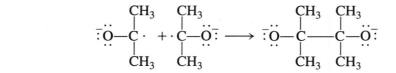

Pinacol

Pinacolone
(3,3-Dimethyl-2-butanone)

The mechanism of this rearrangement involves initial formation of a carbocation (Step 2), followed by 1,2-migration of a methyl group (Step 3). The oxonium ion that is formed as a result of the migration (Step 3) is sufficiently stable to drive the reaction to completion.

(1)
$$CH_3-\overset{\overset{CH_3}{|}}{\underset{\underset{OH}{|}}{C}}-\overset{\overset{CH_3}{|}}{\underset{\underset{OH}{|}}{C}}-CH_3 + H_2SO_4 \longrightarrow CH_3-\overset{\overset{CH_3}{|}}{\underset{\underset{\overset{OH_2}{+}}{|}}{C}}-\overset{\overset{CH_3}{|}}{\underset{\underset{OH}{|}}{C}}-CH_3 + HSO_4^-$$

(2)
$$CH_3-\overset{\overset{CH_3}{|}}{\underset{\underset{\overset{OH_2}{+}}{|}}{C}}-\overset{\overset{CH_3}{|}}{\underset{\underset{OH}{|}}{C}}-CH_3 \longrightarrow CH_3-\overset{\overset{CH_3}{|}}{\underset{+}{C}}-\overset{\overset{CH_3}{|}}{\underset{\underset{OH}{|}}{C}}-CH_3 + H_2O$$

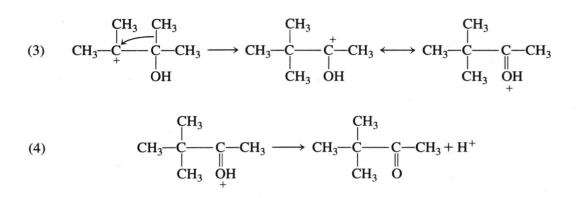

In this experiment, you will convert acetone to pinacol hexahydrate (Experiment 32A). Once the product is obtained you will complete the reaction sequence by performing the rearrangement step to isolate pinacolone (Experiment 32B).

EXPERIMENT 32A

Preparation of Pinacol Hexahydrate

Required Reading

Review: Technique 3 Sections 3.4, 3.7, 3.8
Technique 4 Section 4.3

Before beginning this experiment, you should read the material dealing with organometallic compounds and Grignard reagents in your lecture textbook.

Special Instructions

It is very important that all glassware be thoroughly dried before this experiment is started. We recommend that you rinse your glassware with acetone and place it in a drying oven one laboratory period before this experiment is scheduled to begin. Do not place any plastic, rubber, or Teflon pieces in the drying oven.

This experiment calls for the use of mercuric chloride, $HgCl_2$. As is the case with all mercury compounds, mercuric chloride is very toxic. Be careful not to get it on your hands or face. Wash your hands thoroughly after handling this substance.

Waste Disposal

All compounds of mercury are toxic and must be disposed of in a waste container designated for the disposal of heavy metals. Any magnesium residue remaining after the reaction will contain a significant proportion of mercuric chloride, and therefore these residues must also be placed in the same waste container.

All organic residues must be placed in the appropriate container for nonhalogenated organic waste. All aqueous solutions must be placed in the waste container specified for the disposal of aqueous waste.

Procedure

Place the following items of glassware in the drying oven: 100-mL round-bottom flask, condenser, Claisen head adapter, 5-mL glass syringe, a glass stirring rod, and one 10-mL graduate cylinder. Allow this equipment to dry overnight.

Remove the glassware from the oven and allow it to cool. Prepare a drying tube with calcium chloride. Place 1.65 g of dry magnesium turnings in the 100-mL round-bottom flask. Add 1.95 g mercuric chloride, one or two crystals of iodine, and 19 mL of dry toluene. Using the syringe, add 2 mL of dry acetone to the reaction mixture. Using a glass stirring rod, carefully crush the magnesium and stir the mixture until small bubbles begin to rise from the magnesium and it is clear that heat is beginning to evolve. If the reaction has not begun after a few minutes, simply continue with the procedure. (It will start after acetone is added and heating is applied in the next step.) Attach the round-bottom flask to the Claisen head. Attach the water condenser to one of the openings in the Claisen head, and clamp the apparatus to the ring stand. Use a rubber septum to close the other opening in the Claisen head. Place the drying tube on top of the condenser.

With a heating mantle, heat the reaction mixture to reflux and then add 9.2 mL of acetone through the rubber septum, using a syringe. Add the acetone as quickly as possible, while keeping the condensation ring within the condenser. Since the reaction is highly exothermic, very little heating will be required during the initial stages of the reaction. Allow the reaction to heat under reflux for one hour, while occasionally using the stirring rod to stir the reaction mixture and crush the magnesium turnings. Be careful not to leave the system open to the atmosphere any longer than is necessary.

After heating for one hour, add 4.5 mL of water and heat the mixture under reflux for an additional 30 minutes. At the end of this period, filter the hot mixture using vacuum filtration through a small Büchner funnel. Save the filtrate and transfer the solid portion back into the reaction flask, along with about 7.5 mL of toluene. Reflux the mixture for an additional 10 minutes and then vacuum filter the mixture. Combine the filtrates from the two steps. Discard the solid material into a container specified for heavy metal waste.

Pour the filtrate into a 125-mL Erlenmeyer flask and add 45 mL of low-boiling petroleum ether and 3 mL of water. Using an ice bath, swirl the solution while cooling it to less than 5°C. The product will crystallize as a hexahydrate. Filter the crystals by vacuum filtration using a small Büchner funnel. Allow the crystals to dry for 10–15 minutes and then transfer them to a stoppered vial. Do not allow the crystals to remain in an open container, as they will sublime.

Determine the melting point of the pinacol hexahydrate (literature value = 45.4°C) and calculate the percentage yield. At the option of the instructor, you should turn in your product in a labeled vial or save it for Experiment 32B.

E X P E R I M E N T   3 2 B

Preparation of Pinacolone

Required Reading

Review: Technique 7   Sections 7.5 and 7.7
          Appendices   3 and 4

Before beginning this experiment, you should read the material dealing with carbocation rearrangements in your lecture textbook.

Special Instructions

This experiment requires very little time and can be coscheduled with another, short experiment.

Waste Disposal

All organic residues must be placed in the appropriate container for nonhalogenated organic waste. All aqueous solutions must be placed in the waste container specified for the disposal of aqueous waste.

Procedure

Add 4.8 mL of concentrated sulfuric acid slowly and cautiously, with constant swirling, to 7.2 mL of water in a 25-mL round-bottom flask. Add 3.0 g of pinacol hexahydrate to the hot solution.

Attach a distillation head with a water-cooled condenser to the flask and distill the reaction mixture until the distillate appears clear (no oily droplets). Collect the distillate in a small glass centrifuge tube. There should be two phases of about equal volume in the centrifuge tube. The top layer contains the pinacolone.

Using a Pasteur pipet, remove the lower, aqueous layer and discard it. Using a clean, dry Pasteur pipet, transfer the organic layer to a small test tube. Add a small amount of sodium sulfate to the organic layer. Swirl the mixture and continue to add small portions of sodium sulfate until the drying agent no longer appears to clump together and flows freely.

Transfer the dried organic layer to a small tared vial. Weigh the vial with the pinacolone and determine the weight of product.

Calculate the percentage yield and determine the infrared and proton NMR spectra of the product. Submit your sample to your instructor in a labeled vial, along with your spectra. Interpret your spectra, showing how the results agree with the rearranged structure of the product.

QUESTIONS

1. Predict the product formed in the following pinacol rearrangement:

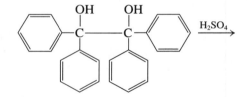

2. Outline a procedure for the preparation of the following spiro compound, starting from cyclopentanone:

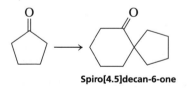

Spiro[4.5]decan-6-one

3. Will the product of the following reaction be optically active? Explain your answer.

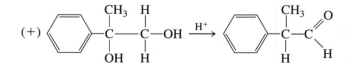

4. The reduction of 2-butanone with magnesium produces, after hydrolysis, two isomeric 1,2-diols. What are their structures?

EXPERIMENT 33

■

Nitration of Methyl Benzoate

Aromatic substitution
Crystallization
Vacuum filtration

The nitration of methyl benzoate to prepare methyl *m*-nitrobenzoate is an example of an electrophilic aromatic substitution reaction, in which a proton of the aromatic ring is replaced by a nitro group:

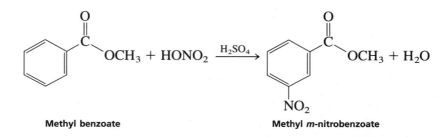

Methyl benzoate   Methyl *m*-nitrobenzoate

Many such aromatic substitution reactions are known to occur when an aromatic substrate is allowed to react with a suitable electrophilic reagent, and many other groups besides nitro may be introduced into the ring.

You may recall that alkenes (which are electron-rich due to an excess of electrons in the π system) can react with an electrophilic reagent. The intermediate formed is electron-deficient. It reacts with the nucleophile to complete the reaction. The overall sequence is called **electrophilic addition.** Addition of HX to cyclohexene is an example.

Nucleophile

Electrophile

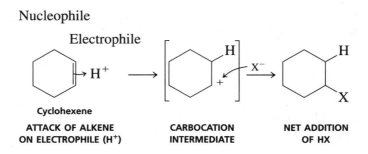

Cyclohexene

ATTACK OF ALKENE   **CARBOCATION**   **NET ADDITION**
ON ELECTROPHILE (H⁺)   **INTERMEDIATE**   **OF HX**

Aromatic compounds are not fundamentally different from cyclohexene. They can also react with electrophiles. However, due to resonance in the ring, the electrons of the π system are generally less available for addition reactions, since an addition would mean the loss of the stabilization that resonance provides. In practice, this means that aromatic compounds react only with *powerfully electrophilic reagents,* usually at somewhat elevated temperatures.

Benzene, for example, can be nitrated at 50°C with a mixture of concentrated nitric and sulfuric acids; the electrophile is NO_2^+ (nitronium ion), whose formation is promoted by action of the concentrated sulfuric acid on nitric acid:

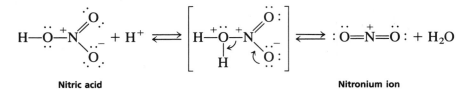

Nitric acid   **Nitronium ion**

The nitronium ion is sufficiently electrophilic to add to the benzene ring, *temporarily* interrupting ring resonance:

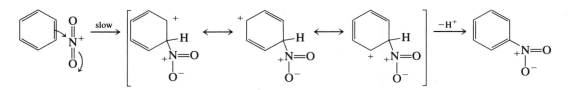

The intermediate first formed is somewhat stabilized by resonance and does not rapidly undergo reaction with a nucleophile; in this behavior, it is different from the unstabilized carbocation formed from cyclohexene plus an electrophile. In fact, aromaticity can be restored to the ring if *elimination* occurs instead. (Recall that elimination is often a reaction of carbocations). Removal of a proton, probably by HSO_4^-, from the sp^3-ring carbon *restores the aromatic system* and yields a net *substitution* wherein a hydrogen has been replaced by a nitro group. Many similar reactions are known, and they are called **electrophilic aromatic substitution reactions.**

The substitution of a nitro group for a ring hydrogen occurs with methyl benzoate in the same way as it does with benzene. In principle, one might expect that any hydrogen on the ring could be replaced by a nitro group. However, for reasons beyond our scope here (see your lecture textbook), the carbomethoxy group directs the aromatic substitution preferentially to those positions that are *meta* to it. As a result, methyl *m*-nitrobenzoate is the principal product formed. In addition, one might expect the nitration to occur more than once on the ring. However, both the carbomethoxy group and the nitro group that has just been attached to the ring *deactivate* the ring against further substitution. Consequently, the formation of a methyl dinitrobenzoate product is much less favorable than the formation of the mononitration product.

While the products described previously are the principal ones formed in the reaction, it is possible to obtain as impurities in the reaction small amounts of the ortho and para isomers of methyl *m*-nitrobenzoate and of the dinitration products. These side products are removed when the desired product is washed with methanol and purified by crystallization.

Water has a retarding effect on the nitration since it interferes with the nitric acid–sulfuric acid equilibria that form the nitronium ions. The smaller the amount of water present, the more active the nitrating mixture. Also, the reactivity of the nitrating mixture can be controlled by varying the amount of sulfuric acid used. This acid must protonate nitric acid, which is a *weak* base, and the larger the amount of acid available, the more numerous the protonated species (and hence NO_2^+) in the solution. Water interferes because it is a stronger base than H_2SO_4 or HNO_3. Temperature is also a factor in determining the extent of nitration. The higher the temperature, the greater will be the amounts of dinitration products formed in the reaction.

Required Reading

Review: Techniques 1–4
          Technique 19

New: Technique 5   Crystallization: Purification of Solids, Sections 5.3, 5.7, and 5.8
       Technique 6   Physical Constants: Part A, Melting Points

Special Instructions

It is important that the temperature of the reaction mixture be maintained at or below 15°C. Nitric acid and sulfuric acid, especially when mixed, are very corrosive substances. Be careful not to get these acids on your skin. If you do get some of these acids on your skin, flush the affected area liberally with water.

Waste Disposal

All aqueous solutions should be placed in a container specially designated for aqueous wastes. Place the methanol used to recrystallize the methyl nitrobenzoate in the container designated for nonhalogenated organic waste.

Procedure

In a 150-mL beaker cool 12 mL of concentrated sulfuric acid to about 0°C and add 6.1 g of methyl benzoate. Using an ice-salt bath (see Technique 2, Section 2.9, p. 611), cool the mixture to 0°C or below and add, VERY SLOWLY, using a Pasteur pipet, a cool mixture of 4 mL of concentrated sulfuric acid and 4 mL of concentrated nitric acid. During the addition of the acids, stir the mixture continuously and maintain the temperature of the reaction below 15°C. If the mixture rises above this temperature, the formation of by-product increases rapidly, bringing about a decrease in the yield of the desired product.

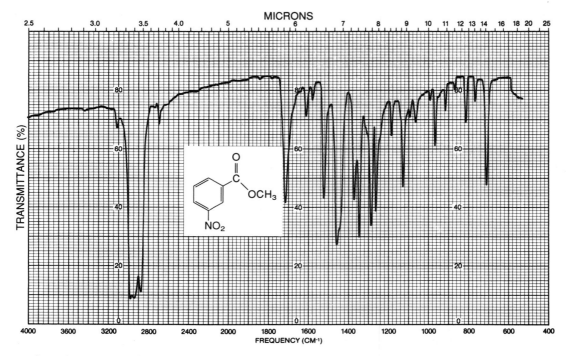

Infrared spectrum of methyl *m*-nitrobenzoate, Nujol mull (Nujol peaks: 2850–3000, 1460, and 1380 cm^{-1}).

After you have added all the acid, warm the mixture to room temperature. After 15 minutes, pour the acid mixture over 50 g of crushed ice in a 250-mL beaker. After the ice has melted, isolate the product by vacuum filtration through a Büchner funnel and wash it with two 25-mL portions of cold water and then with two 10-mL portions of ice cold methanol. Weigh the product and recrystallize it from an equal weight of methanol (Technique 5, Section 5.3, p. 651). The melting point of the recrystallized product should be 78°C. Determine the infrared spectrum of the product as a Nujol mull (Technique 19, Section 19.6, p. 853). Submit the product to your instructor in a labeled vial, along with your infrared spectrum. Compare the spectrum with the one shown here.

QUESTIONS

1. Why is methyl *m*-nitrobenzoate formed in this reaction instead of the *ortho* or *para* isomers?
2. Why does the amount of dinitration increase at high temperatures?
3. Why is it important to add the nitric acid-sulfuric acid mixture slowly?
4. Interpret the infrared spectrum of methyl *m*-nitrobenzoate.
5. Indicate the product formed on nitration of each of the following compounds: benzene, toluene, chlorobenzene, and benzoic acid.

EXPERIMENT 34

Relative Reactivities of Several Aromatic Compounds

Aromatic substitution
Relative activating ability of aromatic substituents
Crystallization

When substituted benzenes undergo electrophilic aromatic substitution reactions, both the reactivity and the orientation of the electrophilic attack are affected by the nature of the original group attached to the benzene ring. Substitutent groups that make the ring more reactive than benzene are called **activators.** Such groups are also said to be **ortho, para** directors because the products formed are those in which substitution occurs either ortho or para to the activating group. Various products may be formed depending on whether substitution occurs at the ortho or para position and the number of times substitution occurs on the same molecule. Some groups may activate the benzene ring so strongly that multiple substitution consistently occurs, while other groups may be moderate activators, and benzene rings containing such groups may undergo only a single substitution. The purpose of this experiment is to determine the relative activating effects of several substituent groups.

In this experiment, you will study the bromination of acetanilide, aniline, and anisole:

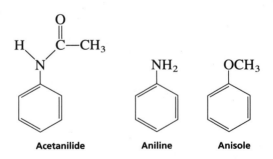

| Acetanilide | Aniline | Anisole |

The acetamido group, $-NHCOCH_3$, the amino group, $-NH_2$, and the methoxy group, $-OCH_3$, are all activators and ortho, para directors. Each student will carry out the bromination of one of these compounds and determine its melting point. By sharing your data, you will have information on the melting points of the brominated products for acetanilide, aniline, and anisole. Using the table on page 302, it will then be possible for you to rank the three substituents in order of activating strength.

The classical method of brominating an aromatic compound is to use Br_2 and a catalyst such as $FeBr_3$, which acts as a Lewis acid. The first step is the reaction between bromine and the Lewis acid:

$$Br_2 + FeBr_3 \longrightarrow [FeBr_4^- Br^+]$$

The positive bromine ion then reacts with the benzene ring in an aromatic electrophilic substitution reaction:

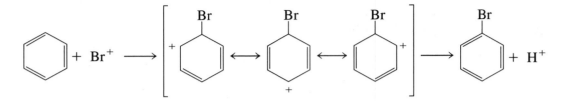

Aromatic compounds that contain activating groups can be brominated without the use of the Lewis acid catalyst since the π electrons in the benzene ring are more available and

Melting Points of Relevant Compounds

Compound	Melting Points (°C)
o-Bromoacetanilide	99
p-Bromoacetanilide	168
2,4-Dibromoacetanilide	145
2,6-Dibromoacetanilide	208
2,4,6-Tribromoacetanilide	232
o-Bromoaniline	32
p-Bromoaniline	66
2,4-Dibromoaniline	80
2,6-Dibromoaniline	87
2,4,6-Tribromoaniline	122
o-Bromoanisole	3
p-Bromoanisole	13
2,4-Dibromoanisole	60
2,6-Dibromoanisole	13
2,4,6-Tribromoanisole	87

polarize the bromine molecule sufficiently to produce the required electrophile Br^+. This is illustrated by the first step in the reaction between anisole and bromine:

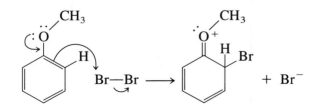

In this experiment, the brominating mixture consists of bromine, hydrobromic acid HBr, and acetic acid. The presence of bromide ion from the hydrobromic acid helps to solubilize the bromine and increase the concentration of the electrophile.

Required Reading

Review: Technique 5

You should review the chapters in your lecture textbook that deal with electrophilic aromatic substitution. Pay special attention to halogenation reactions and the effect of activating groups.

Special Instructions

Bromine is a skin irritant, and its vapors cause severe irritation to the respiratory tract. It will also oxidize many pieces of jewelry. Hydrobromic acid may cause skin or eye irritation. Aniline is highly toxic and a suspected teratogen. All bromoanilines are toxic. This experiment should be carried out in a fume hood or in a well-ventilated laboratory.

Each person will carry out the bromination of only one of the aromatic compounds according to your instructor's directions. The procedures are identical except for the initial compound used and the final recrystallization step.

Note to the Instructor: Prepare the brominating mixture in advance.

Waste Disposal

Dispose of the filtrate from the Hirsch funnel filtration of the crude product into a container specifically designated for this mixture. Place all other filtrates into the container for halogenated organic solvents.

Procedure

Running the Reaction.   To a tared 25-mL round-bottom flask, add the given amount of one of the following compounds: 0.45 g of acetanilide, 0.30 mL of aniline, or 0.35 mL of anisole. Reweigh the flask to determine the actual weight of the aromatic compound. Add 2.5 mL of glacial acetic acid and a magnetic stir bar to the round-bottom flask. Assemble the apparatus shown in the figure. Pack the drying

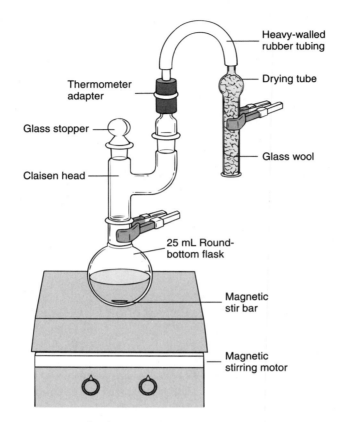

Apparatus for bromination.

tube loosely with glass wool. Add about 2.5 mL of 0.3 *M* sodium bisulfite dropwise to the glass wool until it is moistened but not soaked. This apparatus will capture any bromine given off during the following reaction. Stir the mixture until the aromatic compound is completely dissolved.

Caution: The procedure in the next paragraph must be carried out in a fume hood. Unclamp the apparatus shown in the figure and take it to the hood.

Under the hood, obtain 5.0 mL of the bromine/hydrobromic acid mixture[1] in a 10-mL graduated cylinder. Remove the glass stopper from the Claisen head. Pour the bromine/hydrobromic acid mixture through the Claisen head into the round-bottom flask.

---

[1]Note to the Instructor: The brominating mixture is prepared by adding 13.0 mL of bromine to 87.0 mL of 48% hydrobromic acid. This will provide enough solution for 20 students, assuming no waste of any type. This solution should be stored in the hood.

Caution: Be careful not to spill any of this mixture.

Place the stopper on the Claisen head before returning to your lab bench. Clamp the apparatus above the magnetic stirrer and stir the reaction mixture at room temperature for 20 minutes.

Crystallization and Isolation of Product.   When the reaction is complete, transfer the mixture to a 125-mL Erlenmeyer flask containing 25 mL of 0.3*M* sodium bisulfite solution. Stir this mixture with a glass stirring rod until the orange or yellow color of bromine disappears. If an oil has formed, it may be necessary to stir the mixture for several minutes to remove all of the color. Place the Erlenmeyer flask in an ice bath for 10 minutes. If the product does not solidify, scratch the bottom of the flask with a glass stirring rod to induce crystallization. It may take 10–15 minutes to induce crystallization of the brominated anisole product.[2] Filter the product on a Hirsch funnel with suction and rinse with several 5-mL portions of cold water. Air-dry the product on the funnel for about 10 minutes with the vacuum on.

Recrystallization and Melting Point of Product.   Recrystallize your product from the minimum amount of hot solvent (see Technique 5, Section 5.3, p. 651, and Figure 5.3, p. 652). Use 95% ethanol to recrystallize brominated aniline or brominated acetanilide; use hexane to recrystallize the brominated anisole product. Allow the crystals to air-dry and determine the weight and melting point.

Based on the melting point and the preceding table, you should be able to identify your product. Calculate the percentage yield and submit your product, along with your report, to your instructor.

REPORT

By collecting data from other students, you should be able to determine which product was obtained from the bromination of each of the three aromatic compounds. Using this information, arrange the three substituent groups (acetamido, amino, and methoxy) in order of decreasing ability to activate the benzene ring.

REFERENCE

Zaczek, N. M., and Tyszklewicz, R. B. "Relative Activating Ability of Various Ortho, Para-Directors." *Journal of Chemical Education, 63* (1986): 510.

QUESTIONS

1. Using resonance structures show why the amino group is activating. Consider an attack by the electrophile E^+ at the *para* position.

---

[2]If crystals fail to form after 15 minutes, it may be necessary to seed the mixture with a small crystal of product.

2. For the substituent in this experiment that was found to be least activating, explain why bromination took place at the position on the ring indicated by the experimental results.

3. What other experimental techniques (including spectroscopy) might be used to identify the products in this experiment?

E X P E R I M E N T   3 5

Friedel–Crafts Acylation

Aromatic substitution
Directive groups
Vacuum distillation
Infrared spectroscopy
NMR spectroscopy (proton/carbon-13)
Structure proof

In this experiment, a Friedel–Crafts acylation of an aromatic compound is undertaken, using acetyl chloride:

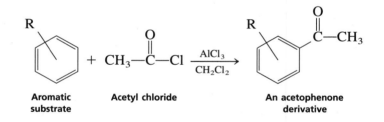

If benzene (R=H) were used as the substrate, the product would be a ketone, acetophenone. Instead of using benzene, however, you will perform the acylation on one of the following compounds:

Toluene   Ethylbenzene
o-Xylene ⎫   Mesitylene (1,3,5-trimethylbenzene)
m-Xylene ⎬ Dimethylbenzenes   Cumene (isopropylbenzene)
p-Xylene ⎭   Anisole (methoxybenzene)
p-Cymene (1-isopropyl-4-methylbenzene)

Each of these products will give a single product, a *substituted* acetophenone. You are to isolate this product by vacuum distillation and to determine its structure by infrared and NMR spectroscopy. That is, you are to determine at which position of the original compound the new acetyl group becomes attached.

This experiment is much the same kind that professional chemists perform every day. A standard procedure, Friedel–Crafts acylation, is applied to a new compound for which the results are not known (at least not to you). A chemist who knows reaction theory well should be able to predict the result in each case. However, once the reaction is completed, it must be proved that the expected product has actually been obtained. If it has not, and sometimes surprises do occur, then the structure of the unexpected product must be determined.

To determine the position of substitution, several features of the product's spectra should be examined closely. These include the following.

INFRARED SPECTRUM

- **The C—H out-of-plane bending modes found between 900 and 690 cm^{-1}.** The C—H out-of-plane absorptions (Appendix 3, Figure IR.7A, p. 902) often allow us to determine the type of ring substitution by their numbers, intensities, and positions.

- **The weak combination and overtone absorptions that occur between 2000 and 1667 cm^{-1}.** This set of combination bands (Figure IR.7B, p. 902), may not be as useful as the first set (mentioned above) because the spectral sample must be very concentrated for them to be visible. They are often weak. In addition, a broad carbonyl absorption may overlap and obscure this region, rendering it useless.

PROTON NMR SPECTRUM

- **The integral ratio of the downfield peaks in the aromatic ring resonances found between 6 and 8 ppm.** The acetyl group has a significant anisotropic effect, and those protons found *ortho* to this group on an aromatic ring usually have a greater chemical shift than the other ring protons (see Appendix 4, Section NMR.6, p. 919, and Section NMR.10, p. 925).

- **A splitting analysis of the patterns found in the 6–8 ppm region of the NMR spectrum.** The coupling constants for protons in an aromatic ring differ according to their positional relations:

ortho J = 6–10 Hz

meta J = 1–4 Hz

para J = 0–2 Hz

If complex second-order splitting interaction does not occur, a simple splitting diagram will often suffice to determine the positions of substitution for the protons on the ring. For several of these products, however, such an analysis will be difficult. In other cases, an easily interpretable pattern like those described in Section NMR.10 (p. 925) will be found.

CARBON-13 NMR SPECTRUM

- In *proton decoupled* carbon-13 spectra, the number of resonances for the aromatic ring carbons (at about 120–130 ppm) will give some help in deciding the substitution patterns of the ring. Ring carbons that are equivalent by symmetry will give rise to a single peak, thereby causing the number of aromatic carbon peaks to fall below the maximum of six. A *p*-disubstituted ring, for instance, will show only four resonances. Carbons that bear a hydrogen usually will have a larger intensity than "quaternary" carbons. (See Appendix 5, Carbon-13 Nuclear Magnetic Resonance Spectroscopy, p. 931.)
- In *proton coupled* carbon-13 spectra, the ring carbons that bear hydrogen atoms will be split into doublets, allowing them to be easily recognized.[1]

As a final note, you should not eschew using the library. Technique 20 (p. 861) outlines how to find several important types of information. Once you think you know the identity of your compound, you might well try to find whether it has been reported previously in the literature, and, if so, whether or not the reported data match your own findings. You may also wish to consult some spectroscopy books, such as Pavia, Lampman, and Kriz, *Introduction to Spectroscopy,* or one of the other textbooks listed at the end of either Appendix 3 (Infrared Spectroscopy) or Appendix 4 (Nuclear Magnetic Resonance Spectroscopy) for additional help in interpreting your spectra.

Required Reading

Review:	Techniques 1, 2, 7, 19	
	Technique 3	Sections 3.7 and 3.9
	Technique 6	Part B, Boiling Points
	Appendices 3, 4, 5	
New:	Technique 9	Vacuum Distillation, Manometers, Sections 9.1, 9.2, 9.8

Before you begin this experiment, you should review the chapters in your lecture text that deal with electrophilic aromatic substitution. Pay special attention to Friedel–Crafts acylation and to the explanations of directing groups. You should also review what you have learned about the infrared and NMR spectra of aromatic compounds.

Special Instructions

Both acetyl chloride and aluminum chloride are corrosive reagents. You should not allow them to come in contact with your skin, nor should you breathe them, because they generate HCl on hydrolysis. They may even react explosively on contact with water. When working with aluminum chloride, be especially careful to watch out for the powdered dust.

---

[1]Note to the Instructor: For those not equipped to perform carbon-13 NMR spectroscopy, carbon-13 NMR spectra can be found reproduced in the Instructor's Manual.

Weighing and dispensing operations should be carried out in a hood. The work-up procedure wherein excess aluminum chloride is decomposed with ice water should also be performed in the hood.

Your instructor will either assign you a compound or have you choose one yourself from the list given on page 306. While you will acetylate only one of these compounds, you should learn much more from this experiment by comparing results with other students.

Notice that the details of the vacuum distillation are left for you to figure out on your own. However, here are two hints. First, all the products boil between 100 and 150°C at 20 mm pressure. Second, if your chosen substrate is anisole, the product will be a *solid* with a low melting point and will solidify soon after the vacuum distillation is completed. The solid can be distilled, but you should not run any cooling water through the condenser. It will also be worthwhile to pre-weigh the receiving flask. It will be difficult to transfer all the solidified product to another container to determine a yield.

Waste Disposal

All aqueous solutions should be collected in a container specially marked for aqueous wastes. Place organic liquids in the container designated for nonhalogenated organic waste unless they contain methylene chloride. Waste materials that contain methylene chloride should be placed in the container designated for halogenated organic wastes.

Procedure

Apparatus.   Assemble the reaction apparatus shown in the figure. All the glassware must be *dry*, since both aluminum chloride and acetyl chloride react with water. Using a 500-mL round-bottom flask, attach a reflux condenser to the central neck, a separatory funnel to one side neck, and a stopper to the unused third neck. Place a filled calcium chloride drying tube on top of the addition (separatory) funnel. The entire apparatus, excluding the traps, should be assembled on a single ring stand so that it can be shaken from time to time. Connect a gas trap to the top of the reflux condenser by attaching a length of flexible rubber tubing from its exit to your aspirator trap. Connect the other side of the aspirator trap to a funnel that is inverted and placed about 2 mm *above* the surface of a quantity of water in a 250-mL beaker. The inverted funnel, which is a trap for acidic gases, may be supported by placing a one-hole rubber stopper on its stem so as to allow it to be clamped to a ring stand.

> **Caution:** Both aluminum chloride and acetyl chloride are corrosive and noxious. Avoid contact and conduct all weighings in a hood. On contact with water, either compound may react violently.

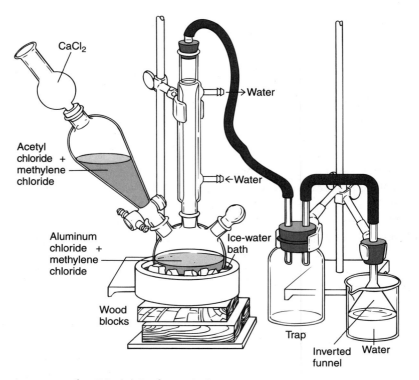

Apparatus for Friedel-Crafts acylation.

Starting the Reaction.   Measure 25 mL of dichloromethane (methylene chloride) in a graduated cylinder and have it at hand. Working quickly in a hood so as to avoid reaction with moisture in the air, weigh 14.0 g of anhydrous aluminum chloride into a 125-mL beaker. Using a powder funnel and a large spatula, transfer the aluminum chloride into the three-necked flask via the unused opening. Use the dichloromethane to transfer any last traces of powder into the flask and to rinse the neck of the flask. After adding all of the dichloromethane, replace the stopper and start the cooling water in the condenser. Place an ice-water bath under the three-necked flask and support it with wooden blocks. Mix and cool the suspension of aluminum chloride in the reaction flask by carefully rotating the entire ring stand so as to induce the contents of the flask to swirl gently.

Again working in a hood, use a pipet to transfer 8.0 g of acetyl chloride into a 125-mL Erlenmeyer flask. Using a graduated cylinder, add 15 mL of dichloromethane to this flask, and then transfer the mixture to the addition funnel attached to the reaction apparatus. Taking a total of about 15 minutes, slowly add the acetyl chloride solution to the suspension of aluminum chloride in the lower flask. (This should not be done hastily because the reaction is very exothermic; the mixture may boil quite vigorously.) During this addition period, frequently rotate the ring stand so as to mix and cool the contents of the flask, which should still be immersed in the ice-water bath.

After this addition is complete, dissolve 0.075 mole of your chosen aromatic compound in 10 mL of dichloromethane. Place this solution in the addition funnel and slowly add it to the cooled acylation mixture over about 30 minutes. Swirl the mixture occasionally, as you watch out for excessive bubbling from the liberation of hydrogen chloride gas.

After this second addition is complete, remove the ice-water bath and allow the mixture to stand at room temperature for an additional 30 minutes. Swirl the reaction mixture frequently during this period.

Isolation of Product.   Disconnect the gas trap, the condenser, and the separatory funnel, and take the three-necked flask to a hood. Pour the entire reaction mixture into a mixture of 50 g of ice and 25 mL of concentrated hydrochloric acid placed in a 400-mL beaker. Stir this mixture thoroughly for 10 to 15 minutes. Using a separatory funnel, separate the organic layer and save it. Extract the aqueous layer with 30 mL of dichloromethane and add the extract to the organic layer saved earlier. Wash the combined organic layers with 50 mL of saturated sodium bicarbonate solution. If a significant amount of acid is present, violent foaming will occur in this step due to evolution of CO_2. Continue mixing and venting until the carbon dioxide emission ceases. Extract with a second portion of sodium bicarbonate if necessary. Separate the organic layer and dry it over anhydrous sodium sulfate for 10–15 minutes. Decant or filter to remove the drying agent from the solution. You may stop here, and store your solution in a tightly stoppered flask.

Removal of Dichloromethane.   Assemble an apparatus for simple distillation (Technique 8, Figure 8.4, p. 711). Add a boiling stone and remove the dichloromethane by a distillation. Dichloromethane boils at a fairly low temperature (bp 40°C). Place the recovered dichloromethane in a container designated for this purpose. Allow the methylene chloride to be driven off completely, or it will cause foaming during the vacuum distillation. Monitor the evaporation by periodically checking the level in the flask and the amount of boiling action. When the volume is constant, the methylene chloride has been removed.

Note: Review Technique 9, Sections 9.1 and 9.2, before proceeding.

Vacuum Distillation.   Assemble an apparatus for vacuum distillation using an aspirator as shown in Figure 9.1, p. 718. A manometer should be attached as shown in Figure 9.12, p. 732. Using the manometer, check to see if the apparatus is vacuum tight and will achieve a good vacuum (less than 30–40 mm Hg) before you proceed. If you do not have sufficient vacuum, recheck all joints and tubing connections until you locate the difficulty. Once all problems are solved, you may distill your product mixture at reduced pressure to yield the aromatic ketone that is your final product.

Determination of Yield. Transfer the product to a pre-weighed storage container and determine its weight. Calculate the percentage yield. Determine the boiling point of your product using the standard-scale method (Technique 6, Section 6.10, p. 675).

Spectroscopy. Determine both the infrared and the NMR spectra (proton and carbon-13). The infrared spectra may be determined neat, using salt plates (Technique 19, Section 19.2, p. 843), except for the product from anisole, which is a solid. For this product, one of the solution spectrum techniques (Technique 19, Section 19.5, p. 850) should be used. The proton NMR spectra can be determined as described in Technique 19, Section 19.9, p. 855. The carbon-13 spectra can be run as described in Technique 19, Section 19.10, p. 859.

The Report. In the usual fashion, you should report the boiling point (or melting point) of your product, calculate the percentage yield, and construct a separation scheme diagram. You should also give the actual structure of your product. Include the infrared and NMR spectra and discuss carefully what you learned from each spectrum. If they did not help you determine the structure, explain why not. As many peaks as possible should be assigned on each spectrum and all important features explained, including the NMR splitting patterns, if possible. Consult a handbook for the boiling point (or melting point) of the possible products. Discuss any literature you consulted and compare the reported results with your own.

You should explain in terms of aromatic substitution theory why the substitution occurred at the position observed, and why a single substitution product was obtained. Could you have predicted the result in advance?

REFERENCE

Schatz, Paul F. "Friedel–Crafts Acylation." *Journal of Chemical Education, 56* (July 1979): 480.

QUESTIONS

1. The following are all relatively inexpensive aromatic compounds that could have been used as substrates in this reaction. Predict the product or products, if any, that would be obtained on acylation of each of them using acetyl chloride.

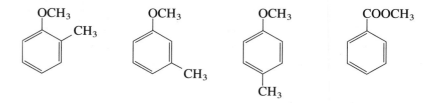

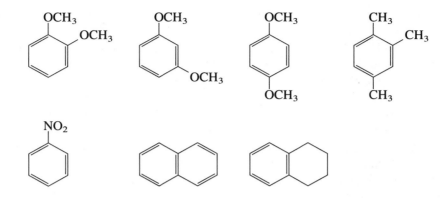

2. Why is it that only monosubstitution products are obtained in the acylation of the substrate compounds chosen for this experiment?

3. Draw a full mechanism for the acylation of the compound you chose for this experiment. Include attention to any relevent directive effects.

4. Why do none of the substrates given as choices for this experiment include any with meta-directing groups?

5. Acylation of *n*-propylbenzene gives an unexpected (?) side-product. Explain this occurrence and give a mechanism.

6. Write equations for what happens when aluminum chloride is hydrolyzed in water. Do the same for acetyl chloride.

7. Explain carefully, with a drawing, why the protons substituted ortho to an acetyl group normally have a greater chemical shift than the other protons on the ring.

8. The compounds shown are possible acylation products from 1,2,4-trimethylbenzene (pseudo-cumene). Explain the only way you could distinguish these two products by NMR spectroscopy.

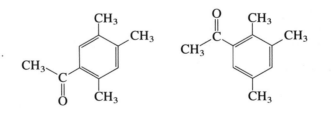

ESSAY

Synthetic Dyes

The practice of using dyes is an ancient art. There is substantial evidence that plant dyestuffs were known long before humans began to keep written history. Before the 20th century, practically all dyes were obtained from natural plant or animal sources; dyeing was a complicated and secret art passed from one generation to the next. Dyes were extracted from plants—mainly by macerating the roots, leaves, or berries in water. The extract was often boiled and then strained before use. In some cases, it was necessary to make the extraction mixture acidic or basic before the dye could be liberated from the plant tissues. Applying the dyes to cloth was also a complicated process. **Mordants** (to be discussed further in the essay, "Dyes and Fabrics") were used to fix the dyes to the cloth or even to modify their color.

Madder is one of the oldest known dyes. Alexander the Great was reputed to have used the dye to trick the Persians into overconfidence during a critical battle. Using madder, a root bearing a brilliant red dye, he simulated bloodstains on the tunics of his soldiers. The Persians, seeing the apparently incapacitated Greek army, became overconfident and, much to their surprise, were overwhelmingly defeated. Through modern chemical analysis, we now know the structure of the dye found in madder root. It is called **alizarin** (see structures) and is very similar in structure to another ancient red dye, **henna,** which has been responsible for a long line of synthetic redheads. Madder is obtained from the plant *Rubia tinctorum.* Henna is a dye prepared from the leaves of the Indian henna plant *(Lawsonia alba)* and an extract of *Acacia catechu.*

Indigo is another plant dyestuff with a long history. This dye, obtained from the plant *Indigofera tinctoria,* has been known in Asia for more than 4000 years. By the ancient process for producing indigo, the leaves of the indigo plant are cut and allowed to ferment in water. During the fermentation, **indican** (see page 315) is extracted into the solution, and the attached glucose molecule is split off to produce **indoxyl.** The fermented mixture is transferred to large open vats in which the liquid is beaten with bamboo sticks. During this process, the indoxyl is air-oxidized to indigo. Indigo, a strong blue dye, is insoluble in water, and it precipitates. Today, indigo is made synthetically, and its principal use is in dyeing denim to produce "blue jeans" material.

Many plants yield dyestuffs that will dye wool or silk, but there are few that dye cotton well. Also, most do not dye synthetic fibers like polyester or rayon. In addition, the natural dyes, with a few exceptions, do not cover a wide range of colors, nor do they yield "brilliant" colors. Even though some people prefer the softness of the "homespun" colors from natural dyes, the **synthetic dyes,** which give rise to deep, brilliant colors, are much in demand today. Also, synthetic dyes that will dye the popular synthetic fibers can now be manufactured. Thus, today we have available an almost infinite variety of colors as well as the ability to dye any type of fabric.

Before 1856, all dyes came from natural sources. However, an accidental discovery by W. H. Perkin, an English chemist, started the development of a huge synthetic dye industry, mostly in England and Germany. Perkin, then aged only 18, was trying to synthesize quinine. Structural organic chemistry was not very well developed at that time,

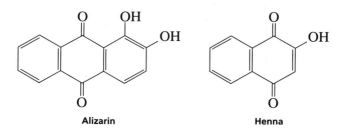

Alizarin   **Henna**

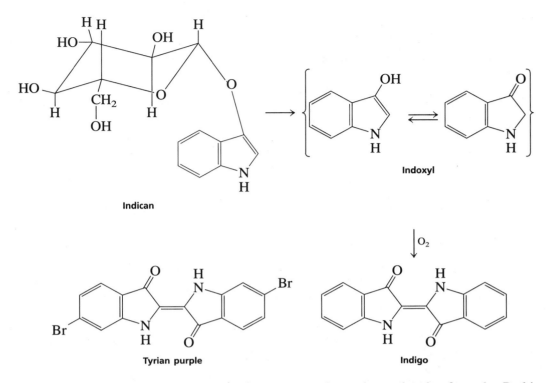

Indican

Indoxyl

Tyrian purple

Indigo

and the chief guide to the structure of a compound was its molecular formula. Perkin thought, judging from the formulas, that it might be possible to synthesize quinine by the oxidation of allyltoluidine:

$$2\ C_{10}H_{13}N + 3\ [O] \longrightarrow C_{20}H_{24}N_2O_2 + H_2O$$

Allyltoluidine   Quinine

He made allyltoluidine and oxidized it with potassium dichromate. The reaction was unsuccessful, because allyltoluidine bore no structural relation to quinine. He obtained no quinine, but he did recover a reddish-brown precipitate with properties that interested him. He decided to try the reaction with a simpler base, aniline. On treating aniline sulfate with potassium dichromate, he obtained a black precipitate, which could be extracted with ethanol to give a beautiful purple solution. This purple solution subsequently proved to be a good dye for fabrics. After receiving favorable comments from dyers, Perkin resigned his post at the Royal College and went on to found the British coal tar dye industry. He became a very successful industrialist and retired at age 36 (!) to devote full time to research. The dye he synthesized became known as **mauve.** The structure of mauve was not proved until much later. From the structure (see page 316) it is clear that the aniline Perkin used was not pure and that it also contained the *o-, m-,* and *p*-toluidines.

Mauve was the first synthetic dye, but soon (1859) the triphenylmethyl dyes pararosaniline, malachite green, and crystal violet (see structures) were discovered in France. These dyes were produced by treating mixtures of aniline or of the toluidines, or of both, with nitrobenzene, an oxidizing agent, and in a second step with concentrated hy-

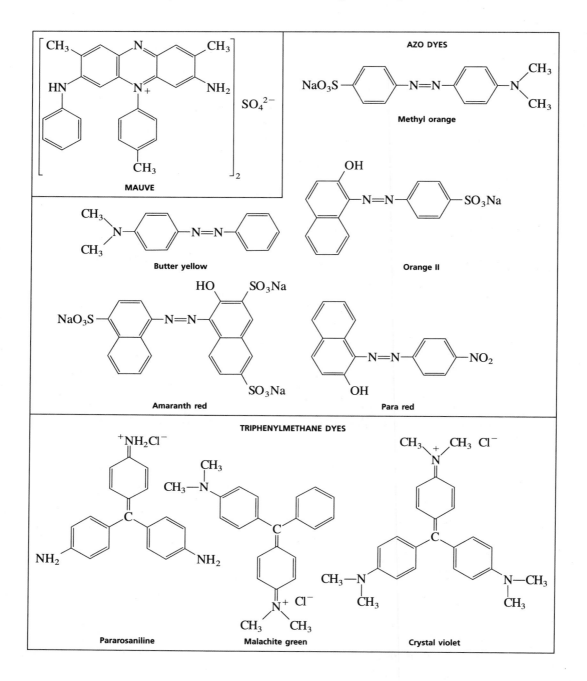

drochloric acid. The triphenylmethyl dyes were soon joined by *synthetic* alizarin (Lieberman, 1868), *synthetic* indigo (Baeyer, 1879), and the azo dyes (Griess, 1862). The azo dyes, also manufactured from aromatic amines, revolutionized the dye industry.

The azo dyes are one of the most common types of dye still in use today. They are used as dyes for clothing, as food dyes (see essay, "Food Colors," preceding Experiment

38), and as pigments in paints. In addition, they are used in printing inks and in certain color printing processes. Azo dyes have the basic structure

$$Ar-N=N-Ar'$$

Several of these dyes are illustrated on page 316. The unit containing the nitrogen–nitrogen bond is called an **azo** group, a strong chromophore that imparts a brilliant color to these compounds.

Producing an azo dye involves treating an aromatic amine with nitrous acid to give a **diazonium ion** intermediate. This process is called **diazotization:**

$$Ar-NH_2 + HNO_2 + HCl \longrightarrow Ar-\overset{+}{N}\equiv N: + Cl^- + 2H_2O$$

The diazonium ion is an electron-deficient (electrophilic) intermediate. A nucleophilic aromatic compound will react with the diazonium ion. The most common nucleophilic species are aromatic amines and phenols. Both these types of compounds are usually more nucleophilic at a ring carbon than at either nitrogen or oxygen. This is due to resonance of the following types:

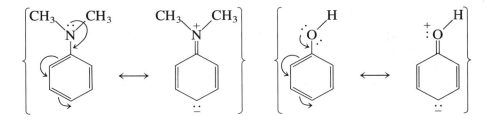

The addition of the amine or the phenol to the diazonium ion is called the **diazonium coupling** reaction, and it takes place as shown:

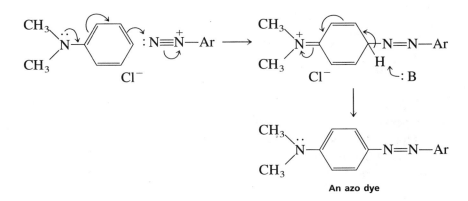

An azo dye

Azo dyes are both the largest and the most important group of synthetic dyes. In the formation of the azo linkage, many combinations of $ArNH_2$ and $Ar'NH_2$ (or $Ar'OH$) are possible. These combinations give rise to dyes with a broad range of colors, encompassing yellows, oranges, reds, browns, and blues. The preparation of an azo dye is given in Experiment 36.

The azo dyes, the triphenylmethyl dyes, and mauve are all synthesized from the anilines (aniline, *o-, m-,* and *p*-toluidine) and aromatic substances (benzene, naphthalene, anthracene). All these substances can be found in **coal tar,** a crude material that is obtained by distilling coal. Perkin's discovery led to a multimillion-dollar industry based on coal tar, a material that was once widely regarded as a foul-smelling nuisance. Today these same materials can be recovered from crude oil or from petroleum as by-products in the refining of gasoline. Although we no longer use coal tar, many of the dyes are still widely used.

REFERENCES

Abrahart, E. N. *Dyes and Their Intermediates.* London: Pergamon Press, 1968.

Allen, R. L. M. *Colour Chemistry.* New York: Appleton-Century-Crofts, 1971.

Decelles, C. "The Story of Dyes and Dyeing." *Journal of Chemical Education, 26* (1949): 583.

Jaffee, H. E., and Marshall, W. J. "The Origin of Color in Organic Compounds." *Chemistry, 37* (December 1964): 6.

Juster, N. J. "Color and Chemical Constitution." *Journal of Chemical Education, 39* (1962): 596.

Robinson, R. "Sir William Henry Perkin: Pioneer of Chemical Industry." *Journal of Chemical Education, 34* (1957): 54.

Sementsov, A. "The Medical Heritage from Dyes." *Chemistry, 39* (November 1966): 20.

Shaar, B. E. "Chance Favors the Prepared Mind. Part 1: Aniline Dyes." *Chemistry, 39* (January 1966): 12.

Travis, A. S. *The Rainbow Makers.* Cranbury, NJ: Associated University Presses, 1993.

EXPERIMENT 36

Methyl Orange

Diazotization
Diazonium coupling
Azo dyes

In this experiment the azo dye **methyl orange** is prepared by the diazo coupling reaction. It is prepared from sulfanilic acid and *N,N*-dimethylaniline. The first product obtained from the coupling is the bright red acid form of methyl orange, called **helianthin.** In base, helianthin is converted to the orange sodium salt, called methyl orange.

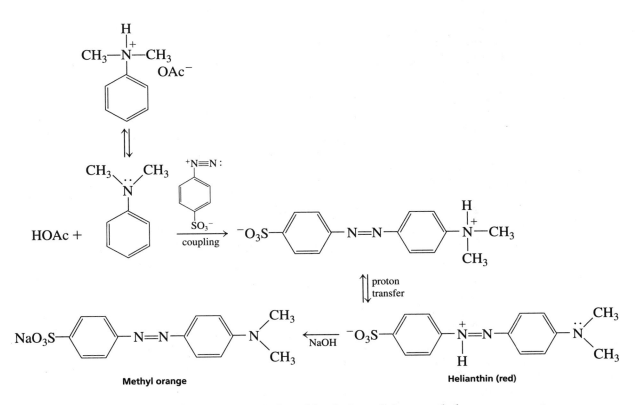

Methyl orange   **Helianthin (red)**

Although sulfanilic acid is insoluble in acid solutions, it is nevertheless necessary to carry out the diazotization reaction in an acid (HNO_2) solution. This problem can be avoided by precipitating sulfanilic acid from a solution in which it is initially soluble. The precipitate is a fine suspension and reacts instantly with nitrous acid. The first step is to dissolve sulfanilic acid in basic solution.

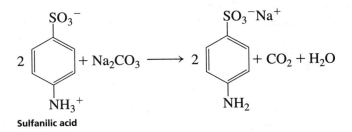

Sulfanilic acid

When the solution is acidified during the diazotization to form nitrous acid,

$$NaNO_2 + HCl \longrightarrow NaCl + HNO_2$$

the sulfanilic acid is precipitated out of solution as a finely divided solid, which is immediately diazotized. The finely divided diazonium salt is allowed to react immediately with dimethylaniline in the solution in which it was precipitated.

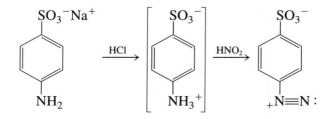

Methyl orange is often used as an acid-base indicator. In solutions that are more basic than pH 4.4, methyl orange exists almost entirely as the **yellow** negative ion. In solutions that are more acidic than pH 3.2, it is protonated to form a **red** dipolar ion.

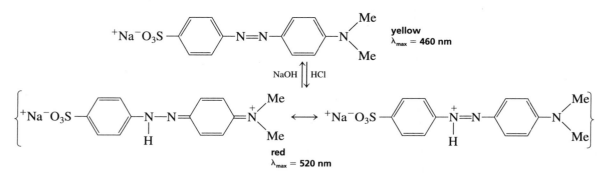

Thus, methyl orange can be used as an indicator for titrations that have their end points in the pH 3.2 to 4.4 region. The indicator is usually prepared as a 0.01% solution in water. In higher concentrations in basic solution, of course, methyl orange appears **orange.**

Azo compounds are easily reduced as the nitrogen–nitrogen double bond by reducing agents. Sodium hydrosulfite, $Na_2S_2O_4$, is often used to bleach azo compounds:

$$R'\!-\!N\!=\!N\!-\!R \xrightarrow{\ Na_2S_2O_4\ } R'\!-\!NH_2 + H_2N\!-\!R$$

Other good reducing agents, such as stannous chloride in concentrated hydrochloric acid, will also work.

Required Reading

Review: Techniques 1–6

New: Essay   Synthetic Dyes
        Essay   Dyes and Fabrics
        Essay   Food Colors

Special Instructions

Temperature is important in this reaction. You must keep the temperature of the reaction mixture below 10°C when the diazonium salt is formed. If the diazonium solution is stored for even a few minutes, it should be kept in an ice bath. If the temperature of the aqueous diazonium salt solution is allowed to rise, the diazonium salt will be hydrolyzed to a phenol, reducing the yield of the desired product.

Do not allow *N,N*-dimethylaniline to come in contact with your skin. When using this substance always wear rubber gloves and open the bottle in the hood. *N,N*-Dimethylaniline is a known carcinogen.

Waste Disposal

Dispose of any organic materials in the waste container specified for nonhalogenated organic waste. Dispose of any aqueous solutions in the waste container designated for the disposal of aqueous solutions.

Procedure

Diazotized Sulfanilic Acid. Dissolve 0.25 g of anhydrous sodium carbonate in 10 mL of water. Use a 50-mL Erlenmeyer flask. Add 0.8 g of sulfanilic acid monohydrate (0.7 g if anhydrous) to the solution and heat it on a steam bath or hotplate until it dissolves. A small amount of suspended material may render the solution cloudy. As a remedy, a small portion of Norit (approximately 0.02 g) can be added to the solution; the still-hot solution should then be gravity filtered into a 50-mL Erlenmeyer flask using fluted filter paper moistened with hot water. Rinse the filter paper with 1 to 2 mL of hot water. Cool the filtrate to *room temperature,* add 0.3 g of sodium nitrite, and stir until the solution is complete. Pour this mixture, with stirring, into a 100-mL beaker containing 5 mL of ice water to which 1 mL of concentrated hydrochloric acid has been added. The diazonium salt of sulfanilic acid should soon separate as a finely divided white precipitate. Keep this suspension cooled in an ice bath until it is to be used.

Methyl Orange. In a test tube, mix together 0.5 mL of *N,N*-dimethylaniline and

> **Caution:** Do not allow *N,N*-dimethylaniline to come in contact with your skin. When using this substance always wear rubber gloves and open the bottle in the hood.

0.40 mL of glacial acetic acid. Add this solution to the cooled suspension of diazotized sulfanilic acid in the 100-mL beaker. Stir the mixture vigorously with a stirring rod. In a few minutes, a red precipitate of helianthin should form. Keep the mixture cooled in an ice bath for about 15 minutes to ensure completion of the coupling reaction. Next, add 6 mL of a 10% aqueous sodium hydroxide solution. Do this slowly and with stirring as you continue to cool the beaker in an ice bath. Check with litmus or pH paper to make sure the solution is basic. If it is not, add extra base. Heat the mixture to boiling on a hotplate for 5 to 10 minutes to dissolve most of the newly formed methyl orange. (**Note:** The time required for this heating step will depend on the rate at which the methyl orange dissolves.) When all (or almost all) of the dye is dissolved, add 2 g of sodium chloride and cool the mixture in an ice bath. The methyl orange should crystallize. When the solution has cooled and the precipitation appears complete, collect the product by vacuum filtration, using a Büchner funnel. Rinse the beaker with two 2-mL portions of cold, saturated, aqueous sodium chloride solution and wash the filter cake with these rinse solutions.

Purification.   To purify the product, transfer the filter cake and paper to a beaker containing about 30 mL of boiling water. Maintain the solution at a gentle boil for a few minutes, stirring it constantly with a glass rod. Not all the dye will dissolve, but the salt with which it is contaminated will dissolve. Remove the filter paper and allow the solution to cool to room temperature. Cool the mixture in an ice bath, and when it is cold collect the product by vacuum filtration using a Büchner funnel (with filter paper). Rinse the beaker with 1- to 2-mL portions of ice water. Allow the product to dry, weigh it, and calculate the percentage yield. The product is a salt. Since salts do not generally have well-defined melting points, the melting point determination should not be attempted.

Recrystallization.   Your instructor may ask you to recrystallize a 1-g portion of the product isolated from the boiling water. To do this, prepare a saturated solution of the dye in boiling water by adding only enough water to permit the 1-g sample to go into solution. Do this in a 50-mL beaker. It may not be possible to get all the sample to dissolve. Remove any solid material by filtering the hot solution by gravity through a fluted filter placed in a funnel that has been preheated on a steam bath or hotplate. Set the filtered solution aside to cool and crystallize. Collect the product by vacuum filtration using a Hirsch funnel. Rinse the product with 1-mL aliquots of ice water to remove any solid that may remain in the beaker. Submit the purified sample, along with your laboratory report, to the instructor.

TESTS (OPTIONAL)

Obtain a square of Multifiber Fabric 10A from your instructor. This cloth contains alternate bands of acetate rayon, cotton, nylon, polyester, acrylic, and wool woven in sequence (see Experiment 37). Prepare a dye bath by dissolving 0.5 g of methyl orange (your crude material will suffice) in 300 mL of water to which 10 mL of a 15% aqueous sodium sulfate solution and 5 drops of concentrated sulfuric acid have been added. Heat the solution to just below its boiling point. Immerse the fabric in the bath for 5 to 10 minutes. Remove the fabric, rinse it well, and note the results.

Make the dye bath basic by adding sodium carbonate. Then add a solution of sodium dithionite (sodium hydrosulfite) until the color of the bath is discharged. Add a slight excess. Now place the bottom edge of the dyed fabric in the bath for a few minutes. Note the result.

INDICATOR ACTION (OPTIONAL)

Dissolve a few crystals of methyl orange in a small amount of water in a test tube. Alternately add a few drops of dilute hydrochloric acid and a few drops of dilute sodium hydroxide solution until the color changes are apparent in each case.

QUESTIONS

1. Why does the dimethylaniline couple with the diazonium salt at the para position of the ring?
2. Give a mechanism for producing a phenol from the diazonium salt that was prepared from sulfanilic acid.
3. What would be the result if cuprous chloride were added to the diazonium salt prepared in this reaction?
4. The diazonium coupling reaction is an electrophilic aromatic substitution reaction. Give a mechanism that clearly indicates this fact.
5. In the essay on food colors that precedes Experiment 38, the structures of several azo food colors are given. Indicate how you would synthesize each of these dyes by the diazo coupling reaction.
6. Immediately after the coupling reaction in this experiment, a proton transfer occurs. The proton is transferred from the dimethylamino group to the azo linkage. Why is the latter protonated form lower in energy than the product formed initially?

E S S A Y

Dyes and Fabrics

Not all dyes are suitable for every type of fabric. A dye must have the property of **fastness.** It must bond strongly to the fibers of the material and remain there even after repeated washings. It must not fade on exposure to light. Additionally, a dye must dye the fabric evenly if it is to be a commercially important dye. **Levelness** is the term used to refer to the uniformity of the dye on the fiber. A level dye, then, is one that colors the fabric evenly after its application. Just how a dye bonds to given fiber is a highly complex subject, since not all dyes and not all fibers are equivalent. As indicated in the essay, "Synthetic Dyes," a dye good for wool or silk may not dye cotton at all. Conversely, a dye that is fast and level on cotton may not be satisfactory for wool.

TABLE ONE. Fiber Types

NATURAL FIBERS

Wool (Polypeptide)
Silk

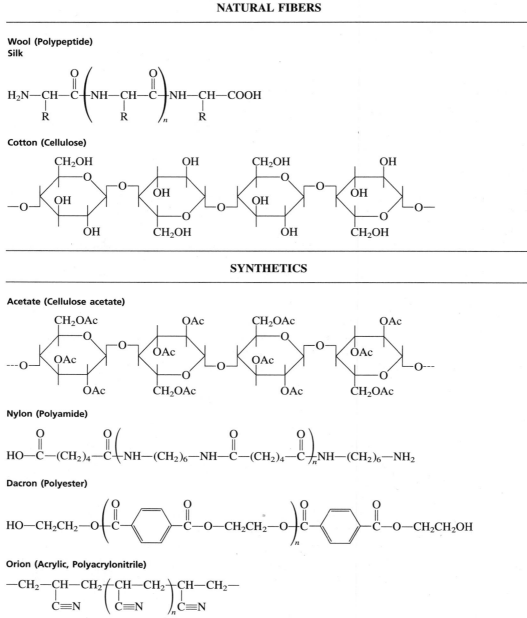

Cotton (Cellulose)

SYNTHETICS

Acetate (Cellulose acetate)

Nylon (Polyamide)

Dacron (Polyester)

Orion (Acrylic, Polyacrylonitrile)

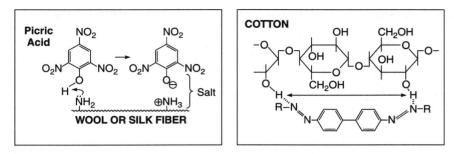

Figure 1. Direct dye.

Figure 2. Disazo dye.

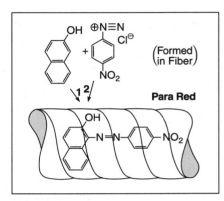

Figure 3. Ingrain dye.

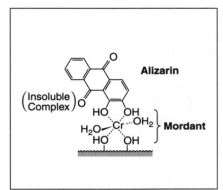

Figure 4. Mordant dye.

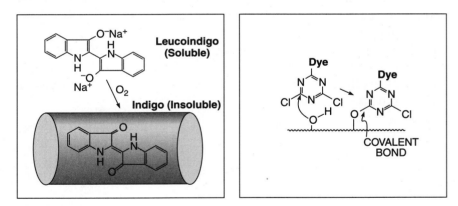

Figure 5. Vat dye.

Figure 6. Fiber-reactive dye.

To understand the mechanism by which a dye bonds to a fiber, you must know both the structure of the dye and the structure of the fiber. The main types of fabric fiber are illustrated in Table One.

The two natural fibers, wool and silk, are very similar in structure. They are both polypeptides, that is, polymers made of amino acid units. All amino acids have the same basic structure but differ in the nature of the substituent group, R. In naturally occurring amino acids, the various possible side-chain R groups consist of substituents that can be basic ($-NH_2$), neutral ($-CH_2OH$, $-CH_3$), or acidic ($-COOH$).

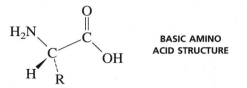

BASIC AMINO
ACID STRUCTURE

In wool and silk, there are numerous ways in which a dye can bind to the fiber. It may make chemical links either to the terminal $-NH_2$ or $-COOH$ groups of the polypeptide chain or to the functional groups present in the side chains of the component amino acids.

Dyes that attach themselves to a fiber by direct chemical interaction are called **direct dyes**. Picric acid is a direct dye for wool or silk. Since it is a strong acid, it interacts with the basic end groups or side chains in wool or silk to form a **salt linkage** between itself and the fiber. The picric acid gives up its proton to some basic group on the fiber and becomes an anion, which strongly bonds to a cationic group on the fiber by ionic interaction (salt formation). This interaction is illustrated in Figure 1.

At biological pH, most amino acids exist in their zwitterionic form:

$$\overset{+}{H_3N}-CH-COO^-$$
$$|$$
$$R$$

This means that in wool or silk there are $-NH_3^+$ and $-COO^-$ groups as well as $-NH_2$ and $-COOH$ groups. The $-NH_2$ and $-COO^-$ groups are basic, and the $-COOH$ and $-NH_3^+$ groups are acidic. Thus, in an alternative mechanism to the one given in Figure 1, picric acid might give up its proton to a $-COO^-$ group to form its anion. The picrate anion would then bond to some $-NH_3^+$ group already present on the chain.

Two other types of direct dyes for wool or silk are the **anionic** and **cationic** dyes. Anionic dyes are also called **acid dyes;** cationic dyes are also called **basic dyes.** Tartrazine, an azo dye, is an important yellow acid dye (anionic) for wool and silk.

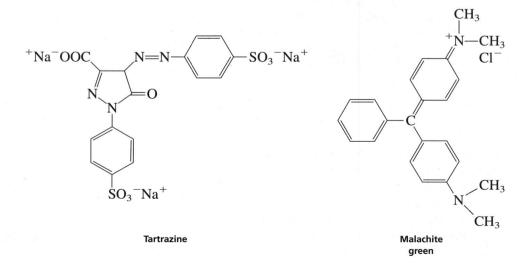

Tartrazine

Malachite
green

Malachite green, a triphenylmethane dye, is an important basic dye (cationic) for wool and silk. Anionic dyes interact with the acidic —NH_3^+ groups in the fiber.

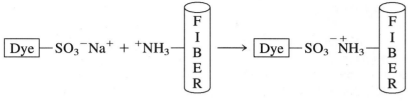

ANIONIC DYE

Cationic dyes interact with the —COO^- groups in the fiber.

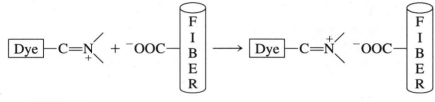

CATIONIC DYE

Of the synthetic fibers, nylon is most like wool and silk. It is constructed with "peptide" or amide linkages,

$$-NH-\underset{\underset{O}{\|}}{C}-$$

being formed from a diamine, NH_2—$(CH_2)_6$—NH_2, and a dicarboxylic acid, HOOC—$(CH_2)_4$—COOH. Nylon can be synthesized so that either —NH_3^+ or —COO^- groups predominate at the ends of the chains. It is more usual for nylon fabrics to contain excess —NH_3^+ groups. Therefore, nylon is usually dyed with anionic (acid) dyes.

Dacron (a polyester fiber), Orlon (an acrylic fiber), and cotton (cellulose) do not contain many anionic or cationic groups in their structures and are not easily dyed by picric acid or by dyes of the anionic and cationic types. Orlon is sometimes synthesized with —SO_3^- groups incorporated into the chain:

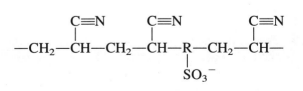

When modified in this way, Orlon can be dyed with cationic dyes.

Azo dyes are important for dyeing cotton fibers. Two such dyes are Congo Red and Para Red. Congo Red is a direct dye for cotton. Para Red is a **developed dye** for cotton. We shall explain both types in detail.

Cotton, which has only hydroxylic groups in its structure, does not dye well with picric acid or with anionic and cationic dyes. Simple azo dyes do not bond well to cotton, but **disazo** dyes are direct dyes for cotton. Congo Red is a disazo dye. A disazo dye has two azo linkages (R—N=N~~N=N—R). When the two azo groups are separated by 10.2 to 10.8 Å, a disazo dye is a fast, direct dye for cotton. The mode of attachment for these dyes is not well understood. Models show that the repeating length between structurally similar hydroxyl groups in cellulose is about the same distance (10.3 Å) as that required for a disazo dye to be a direct dye. In addition, acetate fibers do not dye with disazo dyes. Since the hydroxyl groups are absent in the acetate fiber, it is thought that hydrogen bonding between the hydroxyl groups and the azo linkages accounts for the attachment of a disazo dye to cotton. This is shown in Figure 2.

Although simple azo dyes are not direct dyes for cotton, they can be used to dye cotton as **developed dyes,** also called **ingrain dyes.** Ingrain dyes are synthesized right inside the fiber. Individually, the two components used to synthesize the dye will diffuse into the pores and spaces between the fibers in the fabric. The fully formed dye would be too large a molecule to do this and could bond only to the surface of the fibers. When the individual components react to form the dye, it is trapped **inside,** or "in the grain," of the fibers. Azo dyes are used as ingrain dyes for cotton. Para Red is a typical ingrain azo dye.

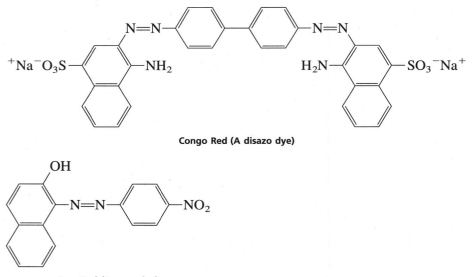

Congo Red (A disazo dye)

Para Red (An azo dye)

To form Para Red, the fabric is soaked first in a solution of the coupling component β-naphthol, and then soaked in a solution of diazotized *p*-nitroaniline. This process is depicted in Figure 3.

A third type of dye for cotton is the **mordant dye.** Alizarin, shown in Figure 4, is a typical mordant dye. Mordant dyes usually have incorporated in their structures groups capable of forming chelate complexes with heavy metals such as copper, chromium, tin, iron, and aluminum. Cotton, which has many hydroxyl groups, can also coordinate with these metals. For a fabric to be dyed with a mordant dye, the fabric is first treated with a

mordant. Mordants are heavy metal salts that can complex with the cotton fibers. Typical mordants are alum, copper sulfate, ferric chloride, stannous chloride, and potassium dichromate. After the fiber is treated with the mordant, it is dyed. The dye also complexes with the mordant, and the mordanting metal links the dye to the fabric. This process is depicted in Figure 4.

For a dye to be fast when a mordant is used, the complex formed among the dye, the fiber, and the mordanting metal must be very stable and **insoluble.** Different mordants (metals) often lead to different colors with the same dye.

When cationic dyes, such as malachite green, are used on cotton, the fabric must be treated with tannic acid as "mordant." In this case, the mordant (substance that binds the dye to the fiber) is not a metal. It is simply a substance that forms an insoluble precipitate when mixed with the dye. Tannic acid forms insoluble complexes with cationic dyes. In this process, the cotton is first impregnated with the mordant (tannic acid) and then treated with the dye.

A type of dye that can be used for all fibers, both natural and synthetic, is the **vat dye.** Vat dyes are normally water-soluble in their reduced form, but when oxidized, they become insoluble. Indigo is a vat dye. Indigo is an insoluble blue dye. It can be reduced by sodium dithionite (sodium hydrosulfite) to leucoindigo, a soluble form, which is colorless. In the vat process for indigo, the fabric is impregnated with the soluble leucoindigo in a hot dye bath or **vat.** Then the fabric is removed from the vat, and the leucoindigo is

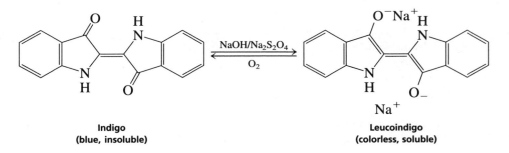

Indigo
(blue, insoluble)

Leucoindigo
(colorless, soluble)

allowed to air-oxidize to the insoluble indigo. The indigo precipitates both inside and on the surface of the fabric fibers. Since the indigo is insoluble in aqueous solutions, the dye is fast. The process is illustrated in Figure 5. Indigo is used for blue jeans.

On fibers like Dacron (polyester) and acetate, no groups are present that allow any direct dye-fiber interactions. These fibers are hard to dye. **Disperse dyes,** however, can be used. In the disperse process, a dye that is water-insoluble is used. It is allowed to penetrate the fiber with an organic solvent as "carrier." Often the dye is suspended in an aqueous solution, and a small amount of the carrier solvent or substance is added. The carrier dissolves the dye and carries it into the fiber where it is "dispersed." The dye becomes trapped within the fiber only because of water-insolubility. It is not bonded to the fibers but merely dispersed among them.

The newest types of dyes are the **fiber-reactive dyes.** The main class of these dyes, the Procion dyes, is based on cyanuric chloride. The chlorines in this compound can be replaced by nucleophilic substitution reactions. If a dye with a good nucleophilic group is allowed to react with cyanuric chloride in basic solution, the dye becomes bound to the cyanuric chloride ring by a **covalent** bond and a Procion dye is produced. Generally, the dye molecule must contain an —OH or an —NH$_2$ group to replace the chlorine in cyanuric chloride. If the dye replaces only one of the chlorines, two chlorines still remain that can be replaced. If the fiber to be dyed has —OH or —NH$_2$ groups, these can dis-

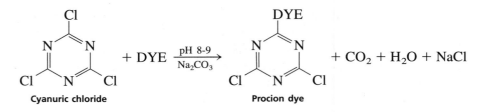

place the remaining chlorines, thus forming a **covalent** bond between the cyanuric chloride and the fiber. The net result is that the dye is attached to the fiber, through cyanuric chloride, by covalent bonds. This process is shown in Figure 6.

REFERENCES

Abrahart, E. N. *Dyes and Their Intermediates.* London: Pergamon Press, 1968.

Allen, R. L. M. *Colour Chemistry.* New York: Appleton-Century-Crofts, 1971.

Davidson, M. F. *The Dye Pot.* Published by the author. Route 1, Gatlinburg, Tenn. 37738.

Editorial Committee of the Brooklyn Botanic Garden. *Dye Plants and Dyeing—A Handbook.* Published by the Brooklyn Botanic Garden, 100 Washington Avenue, Brooklyn, New York 11225. A special printing of *Plants and Gardens,* Vol. 20, No. 3.

Giles, C. H. *A Laboratory Course in Dyeing.* Society of Dyers and Colorists Publication, 1957.

Harper, R. S., Jr., and Reinhardt, R. M. "Chemical Treatments of Textiles." *Journal of Chemical Education, 61* (April 1984): 368.

Maille, A. *Tie and Dye.* New York: Ballantine Books, 1971.

Nea, S. *Tie-Dye.* New York: Van Nostrand, 1971.

Torimoto, N. "An Indigo Plant as a Teaching Material." *Journal of Chemical Education, 64* (April 1987): 332.

E X P E R I M E N T   3 7

Dyes, Fabrics, and Dyeing

Dyes and fibers
Azo dyes, direct dyes, vat dyes, developed dyes, disperse dyes, substantive dyes
Mordants, wool, cotton, silk, nylon, polyester

In this experiment, we examine how dyes adhere to fibers in several fabrics—wool, cotton, polyester, and nylon (or silk). The various classes of dyes and their fastness are examined: direct (anionic and cationic), disazo, vat, ingrain (developed), disperse, and mordant. Finally, the action of mordants on several types of dyes is investigated. The theory of the experiment is discussed in the previous essay, "Dyes and Fabrics."

Special Instructions

This entire experiment can be completed in one lab period, but you must organize your time efficiently. It will be necessary to perform several experiments simultaneously. Since all the experiments involve waiting periods of as long as 15 to 20 minutes, you can use these waiting periods to begin the next experiment. To understand the experiment, it is necessary to read the preceding essay.

The experiment is written so as to include all the dyeing experiments. Your instructor may include only a few of these, however. In that case, you will be given specific instructions about which steps to perform and which materials to use.

Care should be taken not to immerse your hands in either the mordant or the dye baths. Many of these materials are toxic. In addition, all azo dyes should be suspected of possible carcinogenic activity. You will also find it socially uncomfortable to sport yellow, green, red, and blue digits.

Waste Disposal

Dispose of all aqueous solutions produced in this experiment in the container for aqueous waste.

Procedures

Obtain from the instructor or laboratory assistant a kit that contains the following materials:

DYES (in waxed envelopes)[1]

Picric acid	(0.5 g)	Alizarin	(0.1 g)
Indigo	(0.1 g)	Methyl orange	(0.1 g)
Congo Red	(0.1 g)	Malachite green	(0.1 g)
Eosin	(0.1 g)		

FABRIC[2]

Six 4-cm × 10-cm pieces of Multifiber Fabric 10A
(Alternatively, small squares [2 in. × 2 in.] of wool, cotton, polyester, and silk [or nylon] may be used.)

---

[1]The amounts of dye will be very approximate. The instructor should weigh the amount once or twice and then approximate all other envelopes by eye.

[2]Available from Testfabrics, Inc., P.O. Box 118, 200 Blackford Ave., Middlesex, NJ 08846.

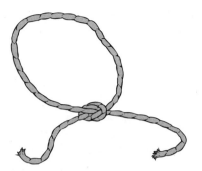

Figure 1.

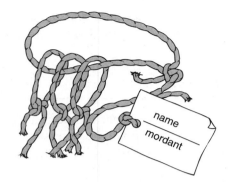

Figure 2.

YARNS

    Thirty-six 10-in. pieces of wool yarn[3]

OTHER

    Five 12-in. lengths of string
    An index card
    A 12-in. square of aluminum foil

Before beginning the experiments, tie the wool yarn pieces into 18 loops or hanks by the following method. Take two 10-in. strands of wool yarn, fold them in half, and tie an overhand knot in the center of the folded lengths. Prepare 18 such hanks. Tie the knot loosely, or the yarn will resist dyeing. See Figure 1, where this is illustrated for a *single* strand.

MORDANTING

    Since mordanting takes some time, it should be done before starting the initial dyeing experiments in Part A. The mordanted yarns will be used in Part B. Five large (2-L) mordant baths will be found in the laboratory. They should be kept warm on steam baths or hotplates. The five mordants to be used are alum (potassium aluminum sulfate), copper (cupric sulfate), chrome (potassium dichromate), tin (stannous chloride), and tannic acid. A fixing bath of tartar emetic (potassium antimony tartrate) will also be found. It is to be used after the tannic acid mordant. All the solutions are 0.1 M in the mordants. The samples should be mordanted 15 to 20 minutes in the mordant baths. The tin mordant is especially hard on wool, so keep the time brief for this mordant. On the other hand, if possible, the tannic acid mordant should be left for a longer time.

    It will be convenient to mordant several samples at once in the following way. Make five labels from an index card. Punch a hole in each label and tie a 12-in. length of string to it. Write your name and the name of one of each of the five mordants

---

[3]Wind the yarn around a 10-in. board and then cut at both ends with scissors. This will prepare many lengths quickly.

on these labels. The mordants are alum, copper, chrome, tin, and tannic acid. Then tie the following numbers of hanks of wool to the unfastened end of each label in the manner shown in Figure 2:

Alum	4
Copper	3
Chrome	3
Tin	3
Tannic acid	1

This will leave four unmordanted hanks of wool.

Place each set of samples in the appropriate bath. Use a long *wooden* stick (not your fingers) to make sure the samples are fully immersed and wetted. The label should not be immersed but should hang over the edge of the bath. Mordant the samples for 15 minutes in the hot baths. They should then be removed from the baths so as to allow excess mordant to drain back into the bath. In addition, squeeze out any excess mordant by pressing the sample against the side of the beaker with a spatula or stirring rod. Press each sample between a double thickness of paper toweling to remove any remaining mordant. After pressing, place all the samples, except that from the tannic acid bath, on a hot steam bath that has its top covered, first with a sheet of aluminum foil and then with a double thickness of paper towel. This will help to dry the samples somewhat. The last sample (from the tannic acid bath) should be fixed in a bath of tartar emetic for 5 minutes and then dried. Leave the samples to dry and start the next set of experiments. Experiments in Parts A and B may be done simultaneously since there are waiting periods in each set of procedures.

A. EXPERIMENTS WITH MULTIFIBER FABRIC (OR WITH MATERIAL PATCHES)

Table One outlines the experiments to be performed if patches of material are used. The entire set of experiments can be simplified if multifiber fabric is available. Multifiber Fabric 10A has six fibers—wool, acrylic, polyester, nylon, cotton, and acetate rayon—woven in sequence.[2] None of the experiments in this part involves mordants. All materials (fabrics) should have been laundered to remove sizing. If this has not been done, they should be washed in soapy water and thoroughly rinsed before dyeing. This is not necessary for the multifiber fabric.

1. Picric Acid (Direct to Wool, Silk, and Nylon)

Dissolve the picric acid (about 0.5 g) in about 60 mL of water to which 2 or 3 drops of concentrated sulfuric acid has been added. Use a small beaker and place it

TABLE ONE Experiments with Material Patches.

DYE TYPE	DYE	WOOL	COTTON	POLYESTER	SILK OR NYLON
Direct	Picric acid	✓	✓	✓	✓
Vat	Indigo	✓	✓		✓
Disazo (anionic)	Congo Red	✓	✓		✓
Azo	Para Red	✓	✓	✓	
Ingrain	Para Red	✓	✓		
Disperse	Para Red			✓	

on a steam bath for warming. Immerse the multifiber fabric or one piece each of wool, cotton, polyester, and silk (or nylon) fabric. Be sure that each piece is thoroughly wetted in the dye solution. Add a little more water if necessary. After about 2 minutes, remove the patches with a forceps and transfer them to a beaker containing warm water. Rinse the samples well and then discard the water. Repeat the rinsing until the dye no longer runs (colors the rinse water). Note the results.

2. Indigo (Vat Dye)

Placing the indigo powder (about 0.1 g) in a small Erlenmeyer flask containing a solution of 0.5 g of sodium dithionite (sodium hydrosulfite) and one sodium hydroxide pellet in 40 mL of warm water. Stopper the flask and shake it gently for several minutes until the indigo is dissolved and bleached. The leucoindigo solution will not be colorless but will appear somewhat green. Add more sodium dithionite if the indigo powder did not dissolve. Heat the solution on a steam bath. Unstopper the flask, and using a forceps or a glass rod, insert a patch of multifiber material into the leucoindigo solution. Stopper the flask and shake the solution to be sure that the material is thoroughly wetted. After about 30 seconds, remove the sample from the flask and squeeze it between a double thickness of paper toweling to remove the excess dye solution. Hang the sample to dry in the air and allow the leucoindigo to become oxidized. If a deeper blue is desired, repeat the dyeing. It may be necessary to add a bit more sodium dithionite to the indigo solution if it is still not bleached. If you are using separate material patches, repeat the dyeing, using wool, cotton, and silk (or nylon). Note and compare the results.

3. Congo Red (Disazo Dye for Cotton)

Dissolve the Congo Red (about 0.1 g) in a beaker containing 100 mL of hot water to which 1 mL of 10% aqueous sodium carbonate and 1 mL of 10% aqueous sodium sulfate have been added. Keep the solution warm on a steam bath. Introduce the multifiber fabric or pieces of wool, cotton, and silk (or nylon). After the samples have remained 10 minutes in the hot dye bath, remove them and wash them in warm water until no more dye is removed. Note the results.

4. Para Red (Developed or Ingrain Dye)

Solution 1—Diazotized *p*-Nitroaniline.   Place 1.4 g *p*-nitroaniline, 25 mL of water, and 5 mL of 10% hydrochloric acid in a small beaker. Heat the solution until most of the *p*-nitroaniline dissolves (add a bit more acid if necessary). Cool the solution to about 5°C in an ice bath. Add, all at once, a solution of 0.7 g of sodium nitrite dissolved in 10 mL of water. Stir the solution well, keeping it in the ice bath. If solid remains, the solution can be filtered by vacuum filtration using a Büchner funnel, but filtration is not necessary. Keep this solution in the ice bath, and prepare solution 2.

Solution 2—β-Naphthol.   Place 0.5 g of β-naphthol in a small beaker containing 100 mL of hot water. While stirring, add a solution of 10% sodium hydroxide **by drops** until most of the β-naphthol dissolves. Not all the β-naphthol will dissolve, and it is important not to add a large amount of base. The cotton will disintegrate in a strongly basic solution; the wool will fare even worse.

Place a sample of multifiber fabric in solution 2 and allow it to soak for only 2 or 3 minutes. Remove the sample from the bath and pat it dry between paper towels. Dilute solution 1 with about 100 mL of cold water and place into it the sample removed from solution 2. After several minutes, remove the sample and rinse it well in water. Note the result.

If you are using separate material patches, place a sample each of cotton and wool in solution 2 for 2 or 3 minutes. Remove them and dry them between paper towels. After this, immerse the sample in solution 1 for several minutes, and then remove it and rinse it well. Note the result.

After completing the ingrain dyeings, mix the two ingrain solutions (1 and 2) together and stir them well. Split the resulting dye mixture into two parts and proceed to parts 5 and 6.

5. Para Red (Azo Dye)

Use half of the Para Red solution prepared by mixing the two ingrain solutions (part 4). Add enough sulfuric acid to make the solution acidic to litmus or pH paper (about 3 drops of concentrated sulfuric acid). Immerse the multifiber fabric or pieces

TABLE TWO   Experiments with Yarn.

DYE TYPE	DYE	WOOL				
		UN	AL	CPR	CHR	TIN
Mordant	Eosin	✓	✓	✓	✓	✓
	Alizarin	✓	✓	✓	✓	✓
Azo (anionic)	Methyl orange	✓	✓	✓	✓	✓
		UN	AL	TAR		
Triphenyl-methane (cationic)	Malachite green	✓	✓	✓		

UN = unmordanted   AL = alum   CPR = copper   CHR = chrome
TIN = tin   TAR = tannic acid + tartar emetic

of wool, cotton, and polyester in the hot dye bath for about 5 minutes. Remove the samples and rinse them well. Note the results.

6. Para Red (Disperse)

In its insoluble, finely dispersed form, Para Red may be used as a disperse dye. Use half of the suspension of Para Red produced above by mixing the two ingrain solutions (part 4). Add to this solution 0.1 g of biphenyl (the carrier) and several drops of a surfactant (liquid detergent). Immerse a piece of the multifiber fabric or of the polyester cloth in the solution and heat it on a steam bath for 15 to 20 minutes. Stir the solution frequently. Remove the sample and note the results.

B.   EXPERIMENTS WITH YARN

Prepare four dye baths (use beakers) by dissolving each of the following dyes (0.1 g each) in 200 mL of hot water: eosin, alizarin, methyl orange, and malachite green. Add 0.1 g of sodium carbonate to the alizarin bath. Keep the dye baths hot on a steam bath. The experiments to be done are outlined in Table Two. Prepare a grid like Table Two on a piece of notebook paper. You will be able to keep your samples straight by placing them on this sheet. Perform the dyeing operations in the sequence given. This is necessary as some of the mordants have a tendency to precipitate the dyes.

1. Immerse one hank of **unmordanted** wool in each bath. After about 20 minutes, remove the samples and rinse them thoroughly in a beaker of warm water. Change the rinse water for each sample. Blot the samples dry between paper towels. Note the results.
2. Place one hank of the wool mordanted in alum in each dye bath. After about 15 minutes, remove the samples from the dye baths and rinse them well in warm water. Blot them dry between paper towels. Note the result.
3. Place the sample mordanted in tannic acid and fixed in tartar emetic in the malachite green bath. After 10 to 15 minutes, remove the sample and wash it well in warm water. Blot it dry between paper towels. Note the results. The malachite green bath will not be used for any further experiments and can be discarded.
4. Place one hank of the wool mordanted in copper in each of the three remaining dye baths: eosin, alizarin, and methyl orange. After 10 to 15 minutes, remove and dry these samples, rinsing as before.
5. In each of the three dye baths (eosin, alizarin, and methyl orange) place one hank of the wool mordanted in chrome. After 10 to 15 minutes, remove and dry these samples.
6. In each of the three dye baths (eosin, alizarin, and methyl orange) place one hank of the wool mordanted in tin. After 10 to 15 minutes, remove and dry these samples.

E S S A Y

Food Colors

Before 1850, most of the colors added to foods were derived from natural biological sources. Some of these natural colors are listed here.

Red	Alkanet root	Yellow	Annato seed (bixin)
	Beets (betanin)		Carrots (β-carotene)
	Cochineal insects (carminic acid)		Crocus stigmas (saffron) Turmeric (rhizome)
	Sandalwood	Green	Chlorophyll
Orange	Brazilwood	Blue	Purple grape skins (oenin)
Brown	Caramel (charred sugar)		

A wide variety of colors can be obtained from these natural sources, many of which are still used, but they have been largely supplanted by synthetic dyes.

After 1856, when Perkin succeeded in synthesizing mauve—the first coal tar dye (see the essay on synthetic dyes preceding Experiment 36)—and when chemists began to discover other new synthetic dyes, artificial colors began to find their way into foodstuffs with increasing regularity. Today, more than 90% of the coloring agents added to foods are synthetic.

The synthetic dyes have certain advantages over the natural coloring agents. Many natural dyes are sensitive to degradation by light and oxygen or by bacterial action; therefore, they are not stable or long-lasting. Synthetic colors that have a much longer shelf life can be devised. The synthetic dyes are also stronger and give more intense colors, and they can be used in smaller quantity to achieve a given color. Often the artificial coloring materials are cheaper than the natural colors. This fact of economics is especially true when the smaller amounts that are required are taken into account.

Why should artificial colors be added to foods at all? It is easier to answer this question from the point of view of the manufacturer rather than of the consumer. The manufacturer knows that, to a certain extent, the eye appeal of a product will affect its sales. For example, a consumer is more likely to buy an orange that has a bright orange skin than one with a mottled green and yellow skin. This is true, even though the flavor and nutritive value of the orange may not be affected at all by the color of the skin. Sometimes more than eye appeal is involved. The consumer is a creature of habit and is accustomed to having certain foodstuffs a particular color. How would you react to green margarine or blue steak? For obvious reasons, these products would not sell very well. Both butter and margarine are artificially colored yellow. Natural butter has a yellow color only in the summer; in the winter it is colorless, and manufacturers customarily add yellow coloring. Margarine must always be artificially colored yellow.

Thus, the colors that are added to foodstuffs are added for a different reason from what prompts the use of other types of food additives. Other additives may be added for either nutritional or technological reasons. Some of these additives can be justified by good arguments. For instance, during the processing of many foods, valuable vitamins and minerals are lost. Many manufacturers replace these lost nutrients by "enriching" their products. In another instance, preservatives are sometimes added to food to forestall spoilage from oxidation or the growth of bacteria, yeasts, and molds. With modern marketing practices, which involve the shipping and warehousing of products over long distances and periods, preservatives are often a virtual necessity. Other additives, such as thickeners and emulsifiers, are often added for technological reasons—for example, to improve the texture of the foodstuff.

There is no nutritional or technological necessity for the use of food colors, however. In fact, in some cases, dyes have been used to deceive customers. For instance, yellow dyes have been used in both cake mixes and egg noodles to suggest a higher egg content than what is actually present. On the grounds that synthetic food dyes are unnecessary and perhaps dangerous, many persons have advocated that their use be abandoned.

Of all the food additives, dyes have come under the heaviest attack. At the turn of the century, more than 90 dyes were used in foods. There were no governmental regulations, and the same dyes that were used for dyeing clothes could be used to color foodstuffs. The first legislation governing dyes was passed in 1906, when food colors known to be harmful were removed from the market. At that time, only seven colors were approved for use in food. In 1938, the law was extended, and any batch of dye destined for use in food had to be **certified** for chemical purity; previously, certification had been voluntary for the manufacturer. At that time, there were 15 food colors in general use and each was given a color and a Food, Drug, and Cosmetic (F,D&C) number designation rather than a chemical name. In 1950, when the number of dyes in use had expanded to 19, an unfortunate incident led to the discontinuation of three of the dyes: F,D&C Oranges Number 1 and Number 2 and F,D&C Red Number 32. These dyes were removed when several children became seriously ill after eating popcorn colored by them.

Since that time, research has revealed that many of these dyes are toxic, that they can cause birth defects or heart trouble, or that they are **carcinogenic** (cancer-inducing). Because of experimental evidence, mainly with chick embryos, rats, and dogs, Reds Numbers 1 and 4 and Yellows Numbers 1, 2, 3, and 4 were also removed from the approved list in 1960. Subsequently, Reds Numbers 4 and 32 were reinstated but restricted to particular uses. In 1965, the ban on Red Number 4 was partly lifted to allow it to be used to color maraschino cherries. This use was allowed because there was no substitute available that would dye cherries, and it was thought that since maraschino cherries are mainly decorative, they are not properly a foodstuff. This use of Red Number 4 was considered to be minor. Similarly, Red Number 32, which may not be used to color food to be eaten, is now called Citrus Red Number 2 and is allowed only for dyeing the skins of oranges.

The structures of the main food dyes are shown in the table on page 340. Note that many of them are azo dyes. Since many of the dyes with the azo linkage have been shown to be carcinogens, many persons suspect all such dyes. In 1960, the law was amended to require that any new dyes submitted for approval should undergo extensive scientific testing before they could be approved. They must be shown to be free from causing birth defects, organ dysfunction, and cancer. Old dyes may be subject to reconsideration if experimental evidence suggests that this is necessary.

Several recent studies have suggested that synthetic food dyes may be responsible, at least in part, for hyperkinetic activity in certain young children. It was shown that when these children were maintained on diets that excluded synthetic food dyes, many of them reverted to more normal behavior patterns. On the contrary, when they were administered a synthetic mixture of food dyes as a capsule along with this diet, the hyperkinetic syndrome would often manifest itself once again. Currently, several groups of researchers are involved in studying this apparent relationship.

Red Number 2 is the dye that has most recently been involved in a controversy concerning its safety. In many tests, some even performed by Food and Drug Administration (FDA) chemists, mounting evidence was found that this dye might be harmful, causing birth defects, spontaneous abortion of fetuses, and possibly cancer. However, the results of other researchers contradict these findings. Much controversy, involving the FDA, the opponents, the proponents, and the courts, ensued. Finally, in February 1976, this dye was banned for food use after the FDA and the courts decided that the bulk of the evidence argued for its discontinuation. More of this interesting story may be found in the references.

Nine Food Colors Approved by the FDA in 1975.*

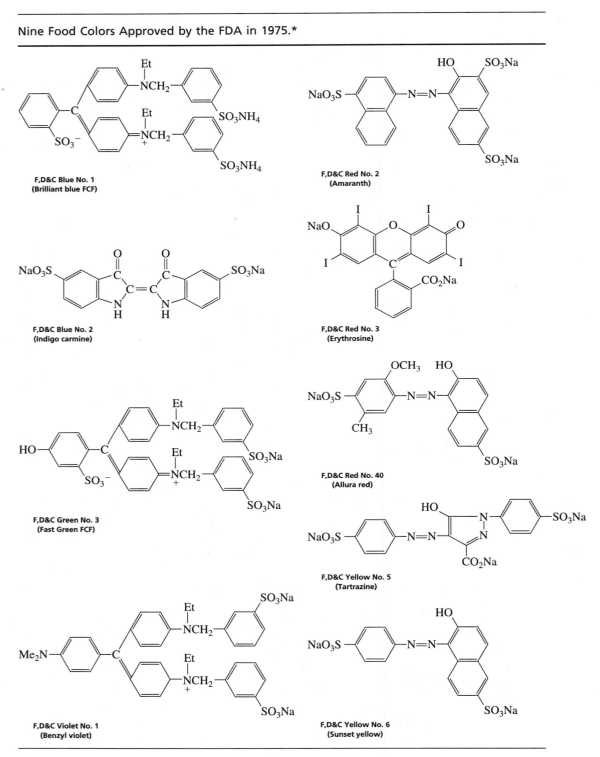

F,D&C Blue No. 1
(Brilliant blue FCF)

F,D&C Red No. 2
(Amaranth)

F,D&C Blue No. 2
(Indigo carmine)

F,D&C Red No. 3
(Erythrosine)

F,D&C Green No. 3
(Fast Green FCF)

F,D&C Red No. 40
(Allura red)

F,D&C Yellow No. 5
(Tartrazine)

F,D&C Violet No. 1
(Benzyl violet)

F,D&C Yellow No. 6
(Sunset yellow)

*All are still in use except for Red No. 2 (Amaranth), which was banned in 1976.

While use of Red Number 2 is proscribed in the United States, it is still approved for use in Canada and within the European Economic Community, and it may be found in products originating in those countries. Before the ban in the United States, Red Number 2 was the most widely used food dye in the industry, appearing in everything from ice cream to cherry soda. Fortunately, proscription of Red Number 2 has not been disastrous for the industry, because for most uses, either Red Number 3 or Red Number 40 is a ready substitute. This knowledge probably had much to do with the court decision finally to ban the dye.

Red Number 40, the most recently accepted food dye, was approved in 1971. Before gaining approval, the Allied Chemical Corporation, which holds exclusive patent rights to the dye, carried out the most thorough and expensive testing program ever performed on a food dye. These tests even included a study of possible birth defects. Red Number 40, called Allura red, seems destined to replace Red Number 2, since it has an extremely wide variety of applications, including the dyeing of maraschino cherries; in this it can replace the provisionally listed Red Number 4, which was also banned in 1976.

There are currently eight dyes allowed for food use. The structures of these eight approved food dyes are given in the accompanying table. The use of these eight dyes is unrestricted. In addition, two other dyes are approved for restricted uses. Citrus Red Number 2 (old Red Number 32) may be used to color the skins of oranges, and Orange B may be used to color the skins of sausages.

REFERENCES

Augustine, G. J., Jr., and Levitan, H. "Neurotransmitter Release from a Vertebrate Neuromuscular Synapse Affected by a Food Dye." *Science, 207* (March 28, 1980): 1489.

Boffey, P. M. "Color Additives: Botched Experiment Leads to Banning of Red Dye No. 2." *Science, 191* (February 6, 1976): 450.

Boffey, P. M. "Color Additives: Is Successor to Red Dye No. 2 Any Safer?" *Science, 191* (February 22, 1976): 832.

Feingold, B. F. *Why Your Child Is Hyperactive.* New York: Random House, 1975.

Mebane, R. C., and Rybolt, T. R. "Chemistry in the Dyeing of Eggs." *Journal of Chemical Education, 64* (April 1987): 291.

National Academy of Sciences, National Research Council. *Food Colors.* Printing and Publishing Office of the National Academy of Sciences, 2101 Constitution Avenue, NW, Washington, DC 20418 (1971).

Sanders, H. J. "Food Additives," *Chemical and Engineering News,* Part 1 (October 10, 1966): 100; Part 2 (October 17, 1966): 108.

Swanson, J. M., and Kinsbourne, M. "Food Dyes Impair Performance of Hyperactive Children on a Laboratory Learning Test." *Science, 207* (March 28, 1980): 1485.

Weiss, B., et al. "Behavioral Responses to Artificial Food Colors." *Science, 207* (March 28, 1980): 1487.

EXPERIMENT 38

Chromatography of Some Dye Mixtures

Thin-layer chromatography
Prepared plates
Paper chromatography

In this experiment, you will use two different types of chromatography, paper and thin-layer chromatography, to separate mixtures of dyes. Two types of dye mixtures are involved. The first type of mixture will be represented by the commercial food colors that can be bought in any grocery store (parts A and C). These are usually available in small packages containing bottles of red, yellow, blue, and green food dye mixtures. As the experiment will show, each color is rarely compounded of only a single dye. For instance, the blue dye usually has a small admixture of a red dye to make it more brilliant in color. A red dye is often added to the yellow food dye for a similar reason. The green dye is normally a mixture of blue and yellow dyes.

The second type of mixture consists of the dyes obtained from a commercial powdered drink mix, such as Kool-Aid (part B). In this case, you will be asked to try to identify the particular dyes used in its preparation.

For those students who are interested, the references listed at the end of this experiment give methods of extracting food dyes from various other types of foods.

Required Reading

New: Technique 12   Column Chromatography, Sections 12.1–12.4
Technique 14   Thin-Layer Chromatography
Essay   Food Colors

Special Instructions

The instructor may choose to have you perform part or all of this experiment. Several experiments may be performed at one time because much of the time is spent waiting for the solvent to ascend the chromatograms. To aid in your planning, an estimate of the amount of time required for development or separation is given at the beginning of each section. This time does not include preparation time for the solvents, development chambers, or spotting procedures.

Waste Disposal

At the option of your instructor, the developed chromatograms may be stapled or taped into your report, or they may be saved in your notebook. Be sure they are free of solvent before you attach them. All glass micropipets should be disposed of in a container designated for broken glass. Because the solvents contain ammonia, place used or excess solvents in a specially designated container rather than in the nonhalogenated organic waste container. Food dye or drink-mix solutions that are not used may be placed in a container specially designated for aqueous wastes.

Procedure

PART A.   PAPER CHROMATOGRAPHY OF FOOD COLORS
(DEVELOPMENT TIME: 40 MINUTES)

At least 12 capillary micropipets will be required for the experiment. Prepare them according to the method described and illustrated in Technique 14, Section 14.4, page 797.

Prepare about 90 mL of a development solvent consisting of

30-mL 2M NH_4OH (4-mL conc. NH_4OH + 26-mL H_2O)

30-mL 1-Pentanol (*n*-amyl or *n*-pentyl alcohol)

30-mL Absolute ethanol

The entire mixture may be prepared in a 100-mL graduated cylinder. Mix the solvent well and pour it into the development chamber for storage. A 32-oz wide-mouth screwcapped jar (or a Mason jar) is an appropriate development chamber. Cap the jar tightly to prevent loss of solvent from evaporation.

Next, obtain a 12-cm × 24-cm sheet of Whatman No. 1 paper. Using a pencil (not a pen), lightly draw a 24-cm-long line about 2 cm up from the long edge of the sheet. Using a centimeter ruler and the pencil, measure and mark off two dashed lines, each about 2 cm from each short end of the paper. Then make nine small marks at 2-cm intervals along the line on the long axis of the paper. These are the positions at which the samples will be spotted (see the illustration on page 344).

If they are available, starting from left to right, spot the reference dyes, F,D&C Red Number 3 (Erythrosine), F,D&C Red Number 40 (Allura red),[1] F,D&C Blue Number 1 (Erioglaucine), F,D&C Yellow Number 5 (Tartrazine), and F,D&C Yellow Number 6 (Sunset yellow). These dyes should be available in 2% aqueous solutions. It may be wise to practice the spotting technique on a small piece of Whatman No. 1 filter paper before trying to spot the actual chromatogram. The correct method of spotting is described in Technique 14, Section 14.4, page 797. It is important that the spots be

---

[1]In Canada or the United Kingdom, substitute F,D&C Red Number 2 (Amaranth) for Red Number 40.

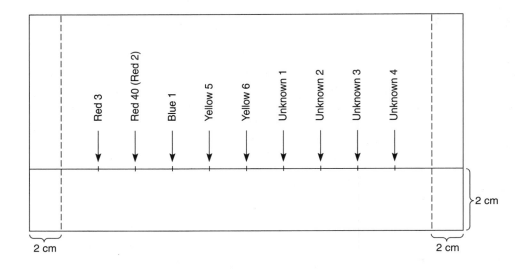

made as small as possible and that the paper not be overloaded. If either of these conditions is not met, the spots will tail and overlap after development. The applied spots should be 1–2 mm ($\frac{1}{16}$ in.) in diameter.

On the remaining four positions (nine if standards are not used) you may spot any dyes of your choice. Use of red, blue, green, and yellow dyes from a single manufacturer is suggested. If the dyes are supplied in screwcapped bottles, the pipets can be filled simply by dipping them into the bottles. If the dyes are supplied in squeeze bottles, however, it will be easier to place a drop of the dye on a microscope slide and to insert the pipet into the drop. One microscope slide should suffice for all the samples.

When the samples have been spotted, hold the paper upright with the spots at the bottom and coil it into a cylinder. Overlap the areas indicated by the dashed lines and fasten the cylinder together (spots inside) with a paper clip or a staple. When the spots have dried, place the cylinder, spotted edge down, in the development chamber. The solvent level should be below the spots, or the spots will dissolve in the solvent. Cap the jar and wait until the solvent ascends to the top of the paper. This will take about 40 minutes, and the remaining parts of the experiment (if required) can be done while you are waiting.

When the solvent has ascended to within 1 cm from the top of the paper, remove the cylinder, open it quickly, and mark the level of the solvent with a pencil. This uppermost level is the solvent front. Allow the chromatogram to dry. Then, using a ruler marked in millimeters, measure the distance that each spot has traveled relative to the solvent front; calculate its R_f value (see Technique 14, Section 14.9, p. 803). Using the list of approved food dyes in the essay "Food Colors" and the reference dyes (if used), try to determine which particular dyes were used to formulate the food colors you tested. Be sure to examine the dye package (or the bottles) to see whether the desired information is given. What conclusions can you draw? Include your chromatogram with your report.

PART B. PAPER CHROMATOGRAPHY OF THE DYES FROM A POWDERED DRINK MIX OR A GELATIN DESSERT (DEVELOPMENT TIME: 40 MINUTES)

Place a quantity of the powdered drink mix or gelatin dessert in a small test tube and add warm water dropwise until the sample just dissolves. Use this concentrated solution to spot the paper as described in the preceding section. Four drink mixes can be spotted on the same piece of Whatman No. 1 paper along with the five standards. Use a *pencil* to label each spot, and then develop the chromatogram in the solvent containing equal parts of 2M NH$_4$OH, pentanol, and ethanol as previously described. Try to identify which dyes are used in samples of several drink mixes (for example, black cherry, cherry, grape, lemon-lime, lime, orange, punch, raspberry, or strawberry). Calculate and compare the R_f values of the standards as well as those of the dyes from the drink mixes. Methods of treating other types of foods to extract and identify the dyes that have been added are described in the references listed at the end of this experiment.

PART C. SEPARATION OF FOOD COLORS USING PREPARED TLC PLATES (DEVELOPMENT TIME: 90 MINUTES)

Obtain from the instructor a 5-cm × 10-cm sheet of a prepared silica gel TLC plate (Eastman Chromatogram Sheet No. 13180 or No. 13181). These plates have a flexible backing, but they should not be bent excessively. They should be handled carefully, or the adsorbent may flake off them. In addition, they should be handled only by the edges. The surface should not be touched.

Using a lead pencil (not a pen), *lightly* draw a line across the short dimension of the plate about 1 cm from the bottom. Using a centimeter ruler, mark off four 1-cm intervals on the line (see figure). These are the points at which the samples will be spotted.

Prepare at least four capillary micropipets as described and illustrated in Technique 14, Section 14.4, page 797. Starting from left to right, spot first a red food

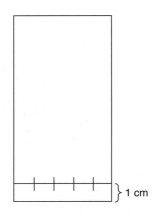
} 1 cm

dye, then a blue dye, a green dye, and a yellow dye. The correct method of spotting a TLC plate is described in Technique 14, Section 14.4. It is important that the spots be made as small as possible and that the plate not be overloaded. If either of these cautions is disregarded, the spots will tail and will overlap after development. The applied spots should be about 1–2 mm ($\frac{1}{16}$ in.) in diameter. If small scrap pieces of the TLC plates are available, it would be a good idea to practice spotting on them before using the actual sample plate.

Prepare a development chamber from an 8-oz wide-mouth screwcapped jar. It *should not* have a filter paper liner described in Technique 14, Section 14.5, page 799. These plates are very thin, and if they touch a liner at any point, solvent will begin to diffuse onto the plate from that point. The development solvent, which can be prepared in a 10-mL graduated cylinder, should be a 4:1 mixture of isopropyl alcohol (2-propanol) and concentrated ammonium hydroxide.[2] Mix the solvent well and pour enough into the development chamber to give a solvent depth of about 0.5 cm (or less). If the solvent level is too high, it will cover the spotted substances, and they will dissolve into the solvent reservoir.

Place the spotted TLC plate in the development chamber, cap the jar tightly, and wait for the solvent to rise almost to the top of the plate. When the solvent is close to the top edge, remove the plate, and using a pencil (not a pen), quickly mark the position of the solvent front. Allow the plate to dry. Using a ruler marked in millimeters, measure the distance that each spot has traveled relative to the solvent front and calculate its R_f value (see Technique 14, Section 14.9, p. 803).

At your instructor's option, and if the dyes are available, you may be asked to spot a second plate with a set of reference dyes. The reference dyes will include F,D&C Red Number 40 (Allura red),[3] F,D&C Blue Number 1 (Erioglaucine), F,D&C Yellow Number 5 (Tartrazine), and F,D&C Yellow Number 6 (Sunset yellow). If this second set of dyes is analyzed, it should be possible (using the list of approved dyes in the essay "Food Colors") to determine the identity of the dyes used to formulate the food colors tested on the first plate. Be sure to examine the package (or bottles) of the food dyes to determine if the desired information is given.

In your report, you should submit a sketch of your plates, showing the spots identified and labeled with their appropriate R_f values. In addition, at the option of your instructor, you may be asked to include your actual plates.

REFERENCES

McKone, H. T. "Identification of F,D&C Dyes by Visible Spectroscopy." *Journal of Chemical Education, 54* (June 1977): 376.

McKone, H. T., and Nelson, G. J. "Separation and Identification of Some F,D&C Dyes by TLC." *Journal of Chemical Education, 53* (November 1976): 722.

---

[2]An alternative solvent mixture suggested by McKone and Nelson (see references) is a 50:25:25:10 mixture of 1-butanol, ethanol, water, and concentrated ammonia.

[3]See footnote 1.

ESSAY

The Chemistry of Vision

An interesting and challenging topic for chemists to investigate is how the eye functions. What chemistry is involved in detection of light and transmission of that information to the brain? The first definitive studies on how the eye functions were begun in 1877 by Franz Boll, who demonstrated that the red color of the retina of a frog's eye could be bleached yellow by strong light. If the frog was then kept in the dark, the red color of the retina slowly returned. Boll recognized that a bleachable substance had to be connected somehow with the ability of the frog to perceive light.

Most of what is now known about the chemistry of vision is the result of the elegant work of George Wald, Harvard University; his studies, which began in 1933, ultimately resulted in his receiving the Nobel Prize in biology. Wald identified the sequence of chemical events during which light is converted into some form of electrical information that can be transmitted to the brain. Here is a brief outline of that process.

The retina of the eye is made up of two types of photoreceptor cells: **rods** and **cones.** The rods are responsible for vision in dim light, and the cones are responsible for color vision in bright light. The same principles apply to the chemical functioning of the rods and the cones; however, the details of functioning are less well understood for the cones than for the rods.

Each rod contains several million molecules of **rhodopsin.** Rhodopsin is a complex of a protein, **opsin,** and a molecule derived from Vitamin A, 11-*cis*-retinal (sometimes called **retinene**). Very little is known about the structure of opsin. The structure of 11-*cis*-retinal is shown here.

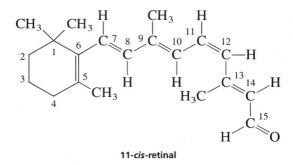

11-*cis*-retinal

The detection of light involves the initial conversion of 11-*cis*-retinal to its all-*trans* isomer. This is the only obvious role of light in this process. The high energy of a quantum of visible light promotes the fission of the π bond between carbons 11 and 12. When the π bond breaks, free rotation about the σ bond in the resulting radical is possible. When the π bond re-forms after such rotation, all-*trans*-retinal results. All-*trans*-retinal is more stable than 11-*cis*-retinal, which is why the isomerization proceeds spontaneously in the direction shown.

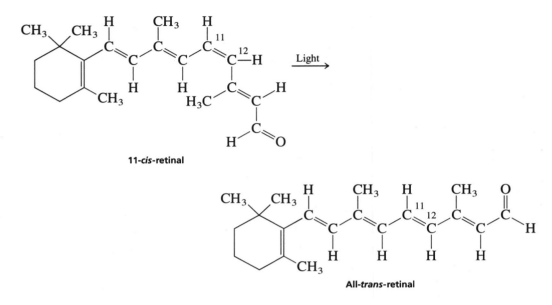

11-*cis*-retinal

All-*trans*-retinal

The two molecules have different shapes due to their different structures. The 11-*cis*-retinal has a fairly curved shape, and the parts of the molecule on either side of the *cis* double bond tend to lie in different planes. Because proteins have very complex and specific three-dimensional shapes (tertiary structures), 11-*cis*-retinal associates with the protein opsin in a particular manner. All-*trans*-retinal has an elongated shape, and the entire molecule tends to lie in a single plane. This different shape for the molecule, compared with the 11-*cis* isomer, means that all-*trans*-retinal will have a different association with the protein opsin.

In fact, all-*trans*-retinal associates very weakly with opsin because its shape does not fit the protein. Consequently, the next step after the isomerization of retinal is the dissociation of all-*trans*-retinal from opsin. The opsin protein undergoes a simultaneous change in conformation as the all-*trans*-retinal dissociates.

At some time after the 11-*cis*-retinal–opsin complex receives a photon, a message is received by the brain. It was originally thought that either the isomerization of 11-*cis*-retinal to all-*trans*-retinal or the conformational change of the opsin protein was an event that generated the electrical message sent to the brain. Current research, however, indicates that both these events occur too slowly relative to the speed with which the brain receives the message. Current hypotheses invoke involved quantum mechanical explanations, which hold it significant that the chromophores (light-absorbing groups) are arranged in a very precise geometrical pattern in the rods and cones, allowing the signal to be transmitted rapidly through space. The main physical and chemical events Wald discovered are illustrated in the figure for easy visualization. The question of how the electrical signal is transmitted still remains unsolved.

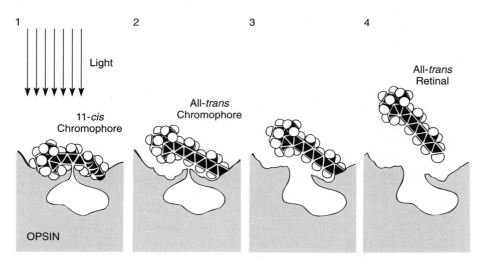

Wald was also able to explain the sequence of events by which the rhodopsin molecules are regenerated. After dissociation of all-*trans*-retinal from the protein, the following enzyme-mediated changes occur. All-*trans*-retinal is reduced to the alcohol all-*trans*-retinol, also called all-*trans*-Vitamin A.

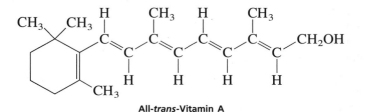

All-*trans*-Vitamin A

All-*trans*-Vitamin A is then isomerized to its 11-*cis*-Vitamin A isomer. Following the isomerization, the 11-*cis*-Vitamin A is oxidized back to 11-*cis*-retinal, which forthwith recombines with the opsin protein to form rhodopsin. The regenerated rhodopsin is then ready to begin the cycle anew, as illustrated in the accompanying diagram.

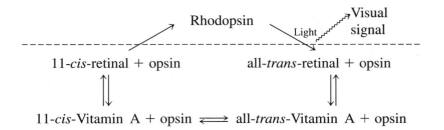

By this process, as little light as 10^{-14} of the number of photons emitted from a typical flashlight bulb can be detected. The conversion of light into isomerized retinal exhibits an extraordinarily high quantum efficiency. Virtually every quantum of light absorbed by a molecule of rhodopsin causes the isomerization of 11-*cis*-retinal to all-*trans*-retinal.

As you can see from the reaction scheme, the retinal derives from Vitamin A, which merely requires the oxidation of a —CH$_2$OH group to a —CHO group to be converted to retinal. The precursor in the diet that is transformed to Vitamin A is β-carotene. The β-carotene is the yellow pigment of carrots and is an example of a family of long-chain polyenes called **carotenoids.**

In 1907, Willstätter established the structure of carotene, but it was not known until 1931–1933 that there were actually three isomers of carotene. The α-carotene differs from β-carotene in that the α isomer has a double bond between C_4 and C_5 rather than between C_5 and C_6, as in the β isomer. The γ isomer has only one ring, identical to the ring in the β isomer, while the other ring is opened in the γ form between $C_{1'}$ and $C_{6'}$. The β isomer is by far the most common of the three.

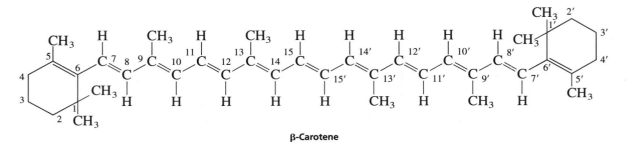

β-Carotene

The substance β-carotene is converted to Vitamin A in the liver. Theoretically, one molecule of β-carotene should give rise to two molecules of the vitamin by cleavage of the C_{15}–$C_{15'}$, double bond, but actually only one molecule of Vitamin A is produced from each molecule of the carotene. The Vitamin A thus produced is converted to 11-*cis*-retinal within the eye.

Along with the problem of how the electrical signal is transmitted, color perception is also currently under study. In the human eye there are three kinds of cone cells, which absorb light at 440, 535, and 575 nm, respectively. These cells discriminate among the primary colors. When combinations of them are stimulated, full color vision is the message received in the brain.

Since all these cone cells use 11-*cis*-retinal as a substrate-trigger, it has long been suspected that there must be three different opsin proteins. Recent work has begun to establish how the opsins vary the spectral sensitivity of the cone cells, even though all of them have the same kind of light-absorbing chromophore.

Retinal is an aldehyde, and it binds to the terminal amino group of a lysine residue in the opsin protein to form a Schiff base, or imine linkage ($>$C$=$N—). This imine linkage is believed to be protonated (with a plus charge) and to be stabilized by being located near a negatively charged amino acid residue of the protein chain. A second negatively charged group is thought to be located near the 11-*cis* double bond. Researchers have re-

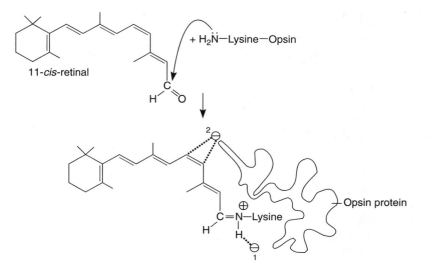

11-*cis*-retinal

+ H$_2$N—Lysine—Opsin

Opsin protein

C=N—Lysine

Rhodopsin.

cently shown, from synthetic models that use a simpler protein than opsin itself, that forcing these negatively charged groups to be located at different distances from the imine linkage causes the absorption maximum of the 11-*cis*-retinal chromophore to be varied over a wide enough range to explain color vision.

Whether there are actually three different opsin proteins, or whether there are just three different conformations of the same protein in the three types of cone cells, will not be known until further work is completed on the structure of the opsin or opsins.

REFERENCES

Borman, S. "New Light Shed on Mechanism of Human Color Vision." *Chemical and Engineering News* (April 6, 1992): 27.

Fox, J. L. "Chemical Model for Color Vision Resolved." *Chemical and Engineering News, 57* (46) (November 12, 1979): 25. A review of articles by Honig and Nakanishi in the *Journal of the American Chemical Society, 101* (1979): 7082, 7084, 7086.

Hubbard, R., and Kropf, A. "Molecular Isomers in Vision." *Scientific American, 216* (June 1967): 64.

Hubbard, R., and Wald, G. "Pauling and Carotenoid Stereochemistry." In A. Rich and N. Davidson, eds. *Structural Chemistry and Molecular Biology.* San Francisco: W. H. Freeman, 1968.

MacNichol, E. F., Jr., "Three Pigment Color Vision." *Scientific American, 211* (December 1964): 48.

"Model Mechanism May Detail Chemistry of Vision." *Chemical and Engineering News* (January 7, 1985): 40.

Rushton, W. A. H. "Visual Pigments in Man." *Scientific American, 207* (November 1962): 120.

Wald, G. "Life and Light." *Scientific American, 201* (October 1959): 92.

Zurer, P. S. "The Chemistry of Vision." *Chemical and Engineering News, 61* (November 28, 1983): 24.

EXPERIMENT 39

Isolation of Chlorophyll and Carotenoid Pigments from Spinach

Isolation of a natural product
Extraction
Column chromatography
Thin-layer chromatography

Photosynthesis in plants takes place in organelles called **chloroplasts.** Chloroplasts contain a number of colored compounds (pigments) which fall into two categories, **chlorophylls** and **carotenoids.**

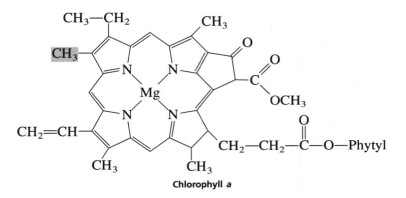

Chlorophyll a

$$Phytyl = -CH_2-CH=C\underset{\underset{CH_3}{|}}{}-CH_2-(CH_2-CH_2-CH\underset{\underset{CH_3}{|}}{}-CH_2)_2-CH_2-CH_2-CH\underset{\underset{CH_3}{|}}{}-CH_3$$

Carotenoids are yellow pigments that are also involved in the photosynthetic process. The structures of **α-** and **β-carotene** are given in the essay preceding this experiment. In addition, chloroplasts also contain several oxygen-containing derivatives of carotenes, called **xanthophylls.**

In this experiment, you will extract the chlorophyll and carotenoid pigments from spinach leaves using acetone as the solvent. The pigments will be separated by column chromatography using alumina as the adsorbent. Increasingly more polar solvents will be used to elute the various components from the column. The colored fractions collected will then be analyzed using thin-layer chromatography. It should be possible for you to identify most of the pigments already discussed on your thin-layer plate after development.

Chlorophylls are the green pigments that act as the principal photoreceptor molecules of plants. They are capable of absorbing certain wavelengths of visible light that are then converted by plants into chemical energy. Two different forms of these pigments found in plants are **chlorophyll *a*** and **chlorophyll *b*.** The two forms are identical, except that the methyl group that is shaded in the structural formula of chlorophyll *a* is replaced by a —CHO group in chlorophyll *b*. **Pheophytin *a*** and **pheophytin *b*** are identical to chlorophyll *a* and chlorophyll *β*, respectively, except that in each case the magnesium ion Mg^{2+} has been replaced by two hydrogen ions $2H^+$.

Required Reading

Review:	Techniques 1, 2	
	Technique 3	Reaction Methods, Section 3.11
	Technique 7	Extractions, Separations, and Drying Agents, Sections 7.7 and 7.8
	Technique 14	Thin-Layer Chromatography
New:	Technique 12	Column Chromatography
	Essay	The Chemistry of Vision

Special Instructions

Hexane and acetone are both highly flammable. Avoid the use of flames while working with these solvents. Perform the thin-layer chromatography in the hood. The procedure calls for a centrifuge tube with a tight-fitting cap. If this is not available, you can use a vortex mixer for mixing the liquids. Another alternative is to use a cork to stopper the tube; however, the cork will absorb some liquid.

Fresh spinach is preferable to frozen spinach. Because of handling, frozen spinach contains additional pigments that are difficult to identify. Since the pigments are light-sensitive and can undergo air oxidation, you should work quickly. Samples should be stored in closed containers and kept in the dark when possible. The column chromatography procedure takes less than 15 minutes to perform and cannot be stopped until it is completed. It is very important, therefore, that you have all the materials needed for this part of the experiment prepared in advance and that you are thoroughly familiar with the procedure before running the column. If you need to prepare the 70% hexane–30% acetone solvent mixture, be sure to mix it thoroughly before using.

Waste Disposal

Dispose of all organic solvents in the container for nonhalogenated organic solvents. Place the alumina in the container designated for wet alumina.

Notes to the Instructor

The column chromatography should be performed with activated alumina from EM Science (No. AX0612-1). The particle sizes are 80–200 mesh and the material is Type F-20. Dry the alumina overnight in an oven at 110°C and store it in a tightly sealed bottle. Alumina more than several years old may need to be dried for a longer time at a higher temperature. Depending on how dry the alumina is, solvents of different polarity will be required to elute the components from the column.

For thin-layer chromatography, use flexible silica gel plates from Whatman with a fluorescent indicator (No. 4410 222). If the TLC plates have not been purchased recently, place them in an oven at 100°C for 30 minutes and store them in a desiccator until used.

If you use different alumina or different thin-layer plates, try out the experiment before using it in class. Materials other than those specified here may give different results than indicated in this experiment.

Procedure

PART A. EXTRACTION OF THE PIGMENTS

Weigh about 0.5 g of fresh (or 0.25 g of frozen) spinach leaves (avoid using stems or thick veins). Fresh spinach is preferable, if available. If you must use frozen spinach, dry the thawed leaves by pressing them between several layers of paper towels. Cut or tear the spinach leaves into small pieces and place them in a mortar along with 1.0 mL of cold acetone. Grind with a pestle until the spinach leaves have been broken into particles too small to be seen clearly. If too much acetone has evaporated, you may need to add an additional portion of acetone (0.5–1.0 mL) to perform the following step. Using a Pasteur pipet, transfer the mixture to a centrifuge tube. Rinse the mortar and pestle with 1.0 mL of cold acetone and transfer the remaining mixture to the centrifuge tube. Centrifuge the mixture (be sure to balance the centrifuge). Using a Pasteur pipet, transfer the liquid to a centrifuge tube with a tight-fitting cap (see "Special Instructions," if one is not available).

Add 2.0 mL of hexane to the tube, cap the tube, and shake the mixture thoroughly. Then, add 2.0 mL of water and shake thoroughly with occasional venting. Centrifuge the mixture to break the emulsion, which usually appears as a cloudy, green layer in the middle of the mixture. Remove the bottom aqueous layer with a Pasteur pipet. Using a Pasteur pipet, prepare a column containing anhydrous sodium sulfate to dry the remaining hexane layer, which contains the dissolved pigments. Place a plug of cotton into a Pasteur pipet ($5\frac{3}{4}$-inch) and tamp it into position using a glass rod. The correct position of the cotton is shown in the figure. Add about 0.5 g of powdered or granular anhydrous sodium sulfate and tap the column with your finger to pack the material.

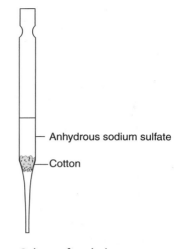

Anhydrous sodium sulfate

Cotton

Column for drying extract.

Clamp the column in a vertical position and place a dry test tube (13 × 100-mm) under the bottom of the column. Label this test tube with an **E** for extract so that you don't confuse it with the test tubes you will be working with later in this experiment. With a Pasteur pipet, transfer the hexane layer to the column. When all the solution has drained, add 0.5 mL of hexane to the column to extract all the pigments from the drying agent. Evaporate the solvent by placing the test tube in a warm water bath (40–60°C) and directing a stream of nitrogen gas (or dry air) into the tube. Dissolve the residue in 0.5 mL of hexane. Stopper the test tube and place it in your drawer until you are ready to run the alumina chromatography column.

PART B.   COLUMN CHROMATOGRAPHY

Introduction.   The pigments are separated on a column packed with alumina. Although there are many different components in your sample, they usually separate into two main bands on the column. The first band to pass through the column is yellow and consists of the carotenes. This band may be less than 1 mm wide and it may pass through the column very rapidly. It is easy to miss seeing the band as it passes through the alumina. The second band consists of all the other pigments discussed in the introduction to this experiment. Although it consists of both green and yellow pigments, it appears as a green band on the column. The green band spreads out on the column more than the yellow band and it moves more slowly. Occasionally, the yellow and green components in this band will separate as the band moves down the column. If this begins to occur, you should change to a solvent of higher polarity so that they come out as one band. As the samples elute from the column, collect the yellow band (carotenes) in one test tube and the green band in another test tube.

Because the moisture content of the alumina is difficult to control, different samples of alumina may have different activities. The activity of the alumina is an important factor in determining the polarity of the solvent required to elute each band of pigments. Several solvents with a range of polarities are used in this experiment. The solvents and their relative polarities follow:

Hexane

70% hexane–30% acetone

Acetone

80% acetone–20% methanol

increasing

polarity

A solvent of lower polarity elutes the yellow band; a solvent of higher polarity is required to elute the green band. In this procedure, you first try to elute the yellow band with hexane. If the yellow band does not move with hexane, you then add the next more polar solvent. Continue this process until you find a solvent that moves the yellow band. When you find the appropriate solvent, continue using it until the yellow band is eluted from the column. When the yellow band is eluted, change to the next more polar solvent. When you find a solvent that moves the green band, continue using it until the green band is eluted. Remember that occasionally a second yellow band will begin to move down the column before the green band moves. This yellow band will be much wider than the first one. If this occurs, change to a more polar solvent. This should bring all the components in the green band down at the same time.

Advance Preparation.   Before running the column, assemble the following glassware and liquids. Obtain five dry test tubes (16 × 100-mm) and number them 1 through 5. Prepare two dry Pasteur pipets with bulbs attached. Calibrate one of them to deliver a volume of about 0.25 mL (Technique 1, Section 1.4, p. 594). Place 10.0 mL of hexane, 6.0 mL of 70% hexane–30% acetone solution (by volume), 6.0 mL of acetone, and 6.0 mL of 80% acetone–20% methanol (by volume) into four separate containers. Clearly label each container.

Prepare a chromatography column packed with alumina. Place a *loose* plug of cotton in a Pasteur pipet ($5\frac{3}{4}$-inch) and push it *gently* into position using a glass rod (see figure on p. 355 for the correct position of the cotton). Add 1.25 g of alumina (EM Science, No. AX0612-1) to the pipet while tapping the column gently with your finger. When all the alumina has been added, tap the column with your finger for several seconds to ensure that the alumina is tightly packed. Clamp the column in a vertical position so that the bottom of the column is just above the height of the test tubes you will be using to collect the fractions. Place test tube #1 under the column.

> **Note:** Read the following procedure on running the column. The chromatography procedure takes less than 15 minutes, and you cannot stop until all the material is eluted from the column. You must have a good understanding of the whole procedure before running the column.

Running the Column.   Using a Pasteur pipet, slowly add about 3.0 mL of hexane to the column. The column must be completely moistened by the solvent. Drain the excess hexane until the level of hexane reaches the top of the alumina. Once you have added hexane to the alumina, the top of the column must not be allowed to run dry. If necessary, add more hexane.

> **Note:** It is essential that the liquid level not be allowed to drain below the surface of the alumina at any point during the procedure.

When the level of the hexane reaches the top of the alumina, add about half (0.25 mL) of the dissolved pigments to the column. Leave the remainder in the test tube for the thin-layer chromatography procedure. (Put a stopper on the tube and place it back in your drawer). Continue collecting the eluent in test tube #1. Just as the pigment solution penetrates the column, add 1 mL of hexane and drain until the surface of the liquid has reached the alumina.

Add about 4 mL of hexane. If the yellow band begins to separate from the green band, continue to add hexane until the yellow band passes through the column. If the yellow band does not separate from the green band, change to the next more polar solvent (70% hexane–30% acetone). When changing solvents, do not add the new solvent until the last solvent has nearly penetrated the alumina. When the appropriate solvent is found, add this solvent until the yellow band passes through the column. Just before the yellow band reaches the bottom of the column, place test tube #2 under the column. When the eluent becomes colorless again (the total volume of the yellow material should be less than 2 mL), place test tube #3 under the column.

Add several mL of the next more polar solvent when the level of the last solvent is almost at the top of the alumina. If the green band moves down the column, continue to add this solvent until the green band is eluted from the column. If the green band does not move or if a diffuse yellow band begins to move, change to the next more polar solvent. Change solvents again if necessary. Collect the green band in test tube #4. When there is little or no green color in the eluent, place test tube #5 under the column and stop the procedure.

Using a warm water bath (40–60°C) and a stream of nitrogen gas, evaporate the solvent from the tube containing the yellow band (tube #2), the tube containing the green band (tube #4), and the tube containing the original pigment solution (tube E). As soon as all the solvent has evaporated from each of the tubes, remove them from the water bath. Do not allow any of the tubes to remain in the water bath after the solvent has evaporated. Stopper the tubes and place them in your drawer.

PART C.   THIN-LAYER CHROMATOGRAPHY

Preparing the TLC Plate.   Technique 14 describes the procedures for thin-layer chromatography. Use a 10-cm × 3.3-cm TLC plate (Whatman Silica Gel Plates No. 4410

222). These plates have a flexible backing but should not be bent excessively. Handle them carefully, or the adsorbent may flake off them. Also, you should handle them only by the edges; the surface should not be touched. Using a lead pencil (not a pen) *lightly* draw a line across the plate (short dimension) about 1 cm from the bottom (see figure). Using a centimeter ruler, move its index about 0.6 cm in from the edge of the plate and lightly mark off three 1-cm intervals on the line. These are the points at which the samples will be spotted.

Prepare three micropipets to spot the plate. The preparation of these pipets is described and illustrated in Technique 14, Section 14.4, page 797. Prepare a TLC development chamber with 70% hexane–30% acetone (see Technique 14, Section 14.5, p. 799). A beaker covered with aluminum foil or a wide-mouth screwcap bottle is a suitable container to use (see Figure 14.5, p. 800). The backing on the TLC plates is very thin, so if they touch the filter paper liner of the development chamber *at any point,* solvent will begin to diffuse onto the absorbent surface at that point. To avoid this, be sure that the filter paper liner does not go completely around the inside of the container. A space about 2 inches wide must be provided.

Using a Pasteur pipet, add two drops of 70% hexane–30% acetone to each of the three test tubes containing dried pigments. Swirl the tubes so that the drops of solvent dissolve as much of the pigments as possible. The TLC plate should be spotted with three samples: the extract, the yellow band from the column, and the green band. For each of the three samples, use a different micropipet to spot the sample on the plate. The correct method of spotting a TLC plate is described in Technique 14, Section 14.4, page 797. Take up part of the sample in the pipet (don't use a bulb; capillary action will draw up the liquid). For the extract (tube labeled E) and the

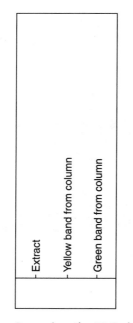

Preparing the TLC plate.

green band (tube #4), touch the plate once *lightly* and let the solvent evaporate. The spot should be no longer than 2 mm in diameter and should be a fairly dark green. For the yellow band (tube #2), repeat the spotting technique 5–10 times, until the spot is a definite yellow color. Allow the solvent to evaporate completely between successive applications, and spot the plate in exactly the same position each time. Save the samples in case you need to repeat the TLC.

Developing the TLC Plate. Place the TLC plate in the development chamber, making sure that the plate does not come in contact with the filter paper liner. Remove the plate when the solvent front is 1–2 cm from the top of the plate. Using a lead pencil, mark the position of the solvent front. As soon as the plates have dried, outline the spots with a pencil and indicate the colors. This is important to do soon after the plates have dried, because some of the pigments will change color when exposed to the air.

Analysis of the Results. In the crude extract you should be able to see the following components (in order of decreasing R_f values):

Carotenes (1 spot) (yellow-orange)

Pheophytin *a* (gray, may be nearly as intense as chlorophyll *b*)

Pheophytin *b* (gray, may not be visible)

Chlorophyll *a* (blue-green, more intense than chlorophyll *b*)

Chlorophyll *b* (green)

Xanthophylls (possibly 3 spots: yellow)

Depending on the spinach sample, the conditions of the experiment, and how much sample was spotted on the TLC plate, you may observe other pigments. These additional components can result from air oxidation, hydrolysis, or other chemical reactions involving the pigments discussed in this experiment. It is very common to observe other pigments in samples of frozen spinach. It is also common to observe components in the green band that were not present in the extract.

Identify as many of the spots in your samples as possible. Determine which pigments were present in the yellow band and in the green band. Draw a picture of the TLC plate in your notebook. Label each spot with its color and its identity, where possible. Calculate the R_f values for each spot produced by chromatography of the extract (see Technique 14, Section 14.9, p. 803. At the instructor's option, submit the TLC plate with your report.

QUESTIONS

1. Why are the chlorophylls less mobile on column chromatography and why do they have lower R_f values than the carotenes?

2. Propose structural formulas for pheophytin *a* and pheophytin *b*.

3. What would happen to the R_f values of the pigments if you were to increase the relative concentration of acetone in the developing solvent?

4. Using your results as a guide, comment on the purity of the material in the green and yellow bands.

ESSAY

Thiamine as a Coenzyme

Vitamin B_1, thiamine, as its pyrophosphate derivative, thiamine pyrophosphate, is a coenzyme universally present in all living systems. It was originally discovered as a required nutritional factor (vitamin) in humans by its link with the disease beriberi. **Beriberi** is a disease of the peripheral nervous system caused by a deficiency of Vitamin B_1 in the diet. Symptoms include pain and paralysis of the extremities, emaciation, or swelling of the body. The disease is most common in the Far East.

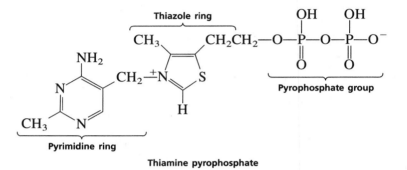

Thiamine pyrophosphate

Thiamine serves as a coenzyme (defined later) for three important types of enzymatic reactions:

1. Nonoxidative decarboxylations of α-keto acids

$$R-\overset{\overset{\displaystyle O}{\|}}{C}-COOH \xrightarrow{B_1} R-\overset{\overset{\displaystyle O}{\|}}{C}-H + CO_2$$

2. Oxidative decarboxylations of α-keto acids

$$R-\overset{\overset{\displaystyle O}{\|}}{C}-COOH \xrightarrow{B_1, O_2} R-\overset{\overset{\displaystyle O}{\|}}{C}-OH + CO_2$$

3. Formation of acyloins (α-hydroxy ketones)

$$R-\overset{\overset{\displaystyle O}{\|}}{C}-COOH + R-\overset{\overset{\displaystyle O}{\|}}{C}-H \xrightarrow{B_1} R-\overset{\overset{\displaystyle O}{\|}}{C}-\overset{\overset{\displaystyle OH}{|}}{C}H-R + CO_2$$

or

$$2\,R-\overset{\overset{\displaystyle O}{\|}}{C}-COOH \xrightarrow{B_1} R-\overset{\overset{\displaystyle O}{\|}}{C}-\overset{\overset{\displaystyle OH}{|}}{C}H-R + 2\,CO_2$$

or

$$2 \; R-\overset{\overset{\displaystyle O}{\|}}{C}-H \;\xrightarrow{\;B_1\;}\; R-\overset{\overset{\displaystyle O}{\|}}{C}-\overset{\overset{\displaystyle OH}{|}}{C}H-R$$

Most biochemical processes are no more than organic chemical reactions carried out under special conditions. It is easy to lose sight of this fact. Most of the steps of the ubiquitous metabolic pathways can, if they have been studied well enough, be explained mechanistically. Some simple organic reaction is a model for almost every biological process. Such reactions, however, are modified ingeniously through the intervention of a protein molecule ("enzyme") to make them more efficient (have greater yield), more selective in choice of substrate (molecule being acted on), more stereospecific in their result, and to enable them to occur under milder conditions (pH) than would normally be possible.

Experiment 40 is designed to illustrate the last circumstance. As a biological reagent, the coenzyme thiamine is used to carry out an organic reaction *without* resorting to an enzyme. The reaction is an acyloin condensation of benzaldehyde:

$$2 \; C_6H_5-CHO \;\longrightarrow\; C_6H_5-\overset{\overset{\displaystyle O}{\|}}{C}-\overset{\overset{\displaystyle OH}{|}}{C}H-C_6H_5$$

In the chemical view, the most important part of the entire thiamine molecule is the central ring—the thiazole ring—which contains nitrogen and sulfur. This ring constitutes the *reagent* portion of the coenzyme. The other portions of the molecule, although important in a biological sense, are not necessary to the chemistry that thiamine initiates. Undoubtedly, the pyrimidine ring and the pyrophosphate group have important ancillary functions, such as enabling the coenzyme to make the correct attachment to its associated protein molecule (enzyme) or enabling it to achieve the correct degree of polarity and the correct solubility properties necessary to allow free passage of the coenzyme across the cell membrane boundary (that is, to allow it to get to its site of action). These properties of thiamine are no less important to its biological functioning than to its chemical reagent abilities; only the latter is our concern here, however.

Experiments with the model compound 3,4-dimethylthiazolium bromide have explained how thiamine-catalyzed reactions work. It was found that this model thiazolium compound rapidly exchanged the C-2 proton for deuterium in D_2O solution. At a pD of 7 (no pH here), this proton was completely exchanged in seconds!

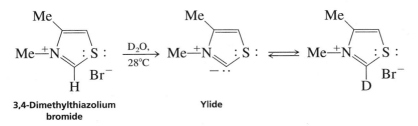

3,4-Dimethylthiazolium   Ylide
bromide

This indicates that the C-2 proton is more acidic than one would have expected. It is apparently easily removed because the conjugate base is a highly stabilized **ylide.** An ylide is a compound or intermediate with positive and negative formal charges on adjacent atoms.

The sulfur atom plays an important role in stabilizing this ylide. This was shown by comparing the rate of exchange of 1,3-dimethylimidazolium ion with the rate for the thiazolium ion shown above. The dinitrogen compound exchanged its C-2 proton more slowly than the sulfur-containing ion. Sulfur, being in the third row of the periodic chart, has d orbitals available for bonding to adjacent atoms. Thus, it has fewer geometrical restrictions than carbon and nitrogen atoms do and can form carbon–sulfur multiple bonds in situations in which carbon and nitrogen normally would not.

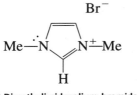

1,3-Dimethylimidazolium bromide

DECARBOXYLATION OF α-KETO ACIDS

From the knowledge just described, it is now thought that the active form of thiamine is its ylide. The system is interestingly constructed, as is seen in the decarboxylation of pyruvic acid by thiamine shown below. Notice especially how the positively charged nitrogen provides a site to accommodate the electron pair that is released on decarboxylation. Thiamine is regenerated by use of this same pair of electrons that become protonated in vinylogous fashion on carbon. The other product is the protonated form of acetaldehyde, the decarboxylation product of pyruvic acid.

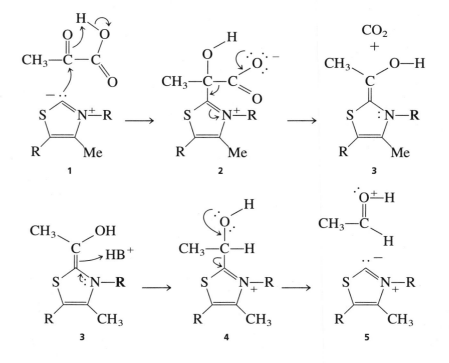

OXIDATIVE DECARBOXYLATION OF α-KETO ACIDS

In oxidative decarboxylations, two additional coenzymes—lipoic acid and coenzyme A—are involved. An example of this type of process, which characterizes all living organisms, is found in the metabolic process **glycolysis.** It is found in the steps that convert pyruvic acid to acetyl coenzyme A, which then enters the citric acid cycle (Krebs cycle, tricarboxylic acid cycle) to provide an energy source for the organism. In this process, the enamine intermediate **3** (see below) is first oxidized by lipoic acid and then transesterified by coenzyme A.

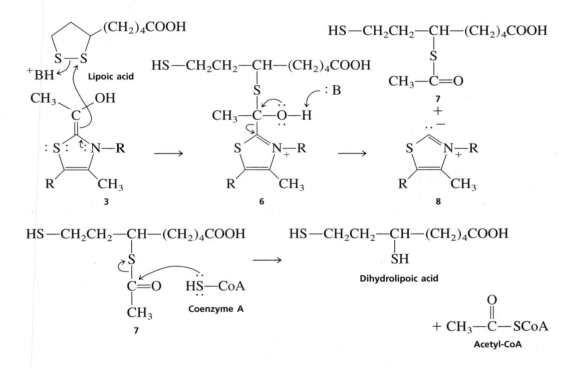

Following this sequence of events, the dihydrolipoic acid is oxidized (through a chain of events involving molecular oxygen) back to lipoic acid, and the acetyl coenzyme A is condensed with oxaloacetic acid to form citric acid. The formation of citric acid begins the citric acid cycle. Notice that acetyl coenzyme A is a thioester of acetic acid and could be hydrolyzed to give acetic acid, not an aldehyde. Thus, an oxidation has taken place in this sequence of events.

ACYLOIN CONDENSATIONS

The enamine intermediate **3** can also function much like the enolate partner in an acid-catalyzed aldol condensation. It can condense with a suitable carbonyl-containing ac-

ceptor to form a new carbon–carbon bond. Decomposition of the adduct **9** to regenerate
the thiamine ylide yields the protonated acyloin **10.**

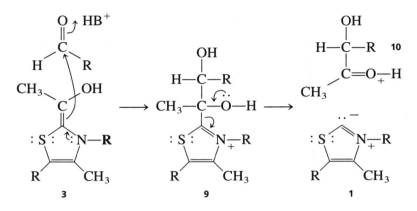

FUNCTION OF A COENZYME

In biological terminology, thiamine is a **coenzyme.** It must bind to an enzyme before the enzyme is activated. The enzyme also binds the substrate. The coenzyme reacts
with the substrate while they are both bound to the enzyme (a large protein). Without the
coenzyme thiamine, no chemical reaction would occur. The coenzyme is the **chemical
reagent.** The protein molecule (the enzyme) helps and mediates the reaction by control-

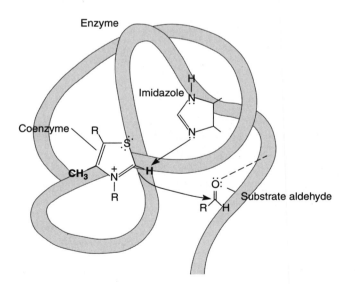

Thiamine (the coenzyme) and the substrate aldehyde are bound to the protein molecule,
here called an enzyme. A possible catalytic group (imidazole) is also shown.

ling stereochemical, energetic, and entropic factors, but, in this case, it is nonessential to the overall result (see Experiment 40). A special name is given to coenzymes that are essential to the nutrition of an organism. They are called **vitamins.** Many biological reactions are of this type, in which a chemical reagent (coenzyme) and a substrate are bound to an enzyme for reaction and, after the reaction, are again released into the medium.

REFERENCES

Bernhard, S. *The Structure and Function of Enzymes.* New York: W. A. Benjamin, 1968. Chap. 7, "Coenzymes and Cofactors."

Bruice, T. C., and Benkovic, S. *Bioorganic Mechanisms.* Vol. 2. New York: W. A. Benjamin, 1966. Chap. 8, "Thiamine Pyrophosphate and Pyridoxal-5'-Phosphate."

Lowe, J. N., and Ingraham, L. L. *An Introduction to Biochemical Reaction Mechanisms.* Englewood Cliffs, N.J.: Prentice-Hall. 1974, Chap. 5, "Coenzyme Function and Design."

EXPERIMENT 40

Coenzyme Synthesis of Benzoin

Coenzyme chemistry
Benzoin condensation

In this experiment, a benzoin condensation of benzaldehyde will be carried out with a biological coenzyme, thiamine hydrochloride, as the catalyst:

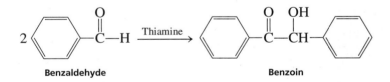

Benzaldehyde   Benzoin

The mechanistic information needed for understanding how thiamine accomplishes this reaction is given in the essay that precedes this experiment. The conversion of benzaldehyde to benzoin can also be accomplished with cyanide ion, an inorganic reagent, as the catalyst (see Question 7).

Required Reading

Review: Techniques 4, 5

New: Essay   Thiamine as a Coenzyme

Special Instructions

This experiment may be conducted concurrently with another experiment. It involves a few minutes at the beginning of a laboratory period for mixing reagents. The remaining portion of the period may be used for another experiment.

Waste Disposal

Pour all of the aqueous solutions produced in this experiment into a waste container designated for aqueous waste. The ethanolic mixtures obtained from the crystallization of crude benzoin should be poured into a waste container designated for nonhalogenated waste.

Notes to the Instructor

It is essential that the benzaldehyde used in this experiment is *pure*. Use a newly opened bottle that has been purchased recently. Benzaldehyde is easily oxidized in air, and crystals of benzoic acid are often visible in the reagent bottle. Even when benzaldehyde *appears* to be free of benzoic acid, you should check the purity of your benzaldehyde and thiamine by following the instructions given in the first paragraph of the procedure. If no *solid* benzoin appears after 2 days or after scratching the inside of the flask, then it is likely that there is a problem with the purity of the benzaldehyde.

Purify the benzaldehyde as follows: Shake the benzaldehyde in a separatory funnel with an equal volume of 5% aqueous sodium carbonate solution. Shake gently and occasionally open the stopcock of the funnel to vent carbon dioxide gas. An emulsion forms that may take 2–3 hours to separate. It is helpful to stir the mixture occasionally during this period to help break the emulsion. Remove the lower sodium carbonate layer, including any remaining emulsion. Add about $\frac{1}{4}$ volume of water to the benzaldehyde and shake the mixture gently so as to avoid an emulsion. Remove the cloudy *lower* organic layer and dry the benzaldehyde with calcium chloride until the next day. Any remaining cloudiness is removed by gravity filtration through fluted filter paper. The resulting *clear* purified benzaldehyde should be suitable for this experiment without vacuum distillation. You *must check* the purified benzaldehyde to see if it is suitable for the experiment by following the instructions in the first paragraph of the procedure.

It is advisable to use a fresh bottle of thiamine hydrochloride, which should be stored in the refrigerator. Fresh thiamine does not seem to be as important as fresh benzaldehyde for success in this experiment.

Procedure

Reaction Mixture.   Add 1.5 g thiamine hydrochloride to a 50-mL Erlenmeyer flask. Dissolve the solid in 2 mL of water by swirling the flask. Add 15 mL of 95% ethanol and swirl the solution until it is homogeneous. To this solution add 4.5 mL of an aqueous sodium hydroxide solution[1] and swirl the flask until the bright yellow color fades to a pale yellow color. Carefully measure 4.5 mL of benzaldehyde (density = 1.04 g/mL) and add it to the flask. Swirl the contents of the flask until it is homogeneous. Stopper the flask and allow it to stand in a dark place for at least two days.

Isolation of Crude Benzoin.   If crystals have not formed after two days, initiate crystallization by scratching the inside of the flask with a glass stirring rod. Allow about 5 minutes for the crystals of benzoin to form fully. Place the flask, with crystals, into an ice bath for 5–10 minutes.

If for some reason the product separates as an oil, it may be helpful to scratch the flask with a glass rod or seed the mixture by allowing a small amount of solution to dry on the end of a glass rod and then placing this into the mixture. Cool the mixture in an ice bath before filtering.

Break up the crystalline mass with a spatula, swirl the flask rapidly, and quickly transfer the benzoin to a Büchner funnel under vacuum (Technique Section 4.3, and Figure 4.5, p. 641). Wash the crystals with two 5-mL portions of ice cold water. Allow the benzoin to dry in the Büchner funnel by drawing air through the crystals for about 5 minutes. Transfer the benzoin to a watch glass and allow it to dry in air until the next laboratory period. The product may also be dried in a few minutes in an oven set at about 100°C.

Yield Calculation and Melting Point Determination.   Weigh the benzoin and calculate the percentage yield based on the amount of benzaldehyde used initially. Determine the melting point (pure benzoin melts between 134 and 135°C). Since your crude benzoin will normally melt between 129 and 132°C, the benzoin should be crystallized before the conversion to benzil (Experiment 41) or benzilic acid (Experiment 42).

Crystallization of Benzoin.   Purify the crude benzoin by crystallization from hot 95% ethanol (use 8 mL of alcohol/g of crude benzoin) using an Erlenmeyer flask for the crystallization (Technique 5, Section 5.3, p. 651; omit Step 2 shown in Figure 5.3). Upon cooling in an ice bath, collect the crystals on a Büchner funnel. The product may be dried in a few minutes in an oven set at about 100°C. Determine the melting point of the purified benzoin. If you are not scheduled to perform Experiments 41 or 42, submit the sample of benzoin, along with your report, to the instructor.

Spectroscopy.   Determine the infrared spectrum of the benzoin as a KBr pellet (Technique 19, Section 19.4, p. 846). A spectrum is shown here for comparison.

---

[1]Dissolve 40 g of NaOH in 500 mL water.

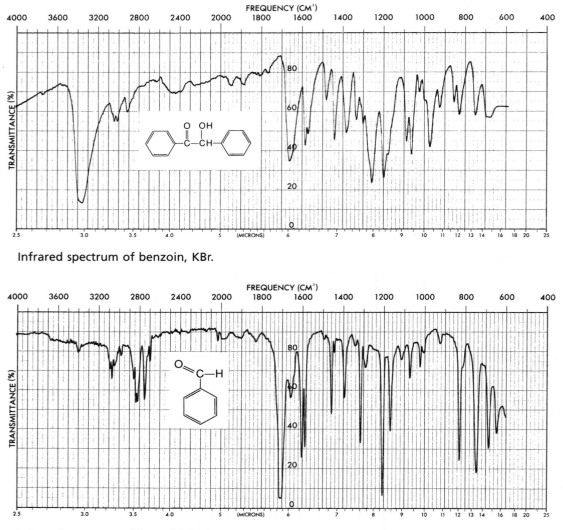

Infrared spectrum of benzoin, KBr.

Infrared spectrum of benzaldehyde, neat.

QUESTIONS

1. The infrared spectrum of benzoin and benzaldehyde are given in this experiment. Interpret the principal peaks in the spectra.

2. Why is sodium hydroxide added to the solution of thiamine hydrochloride?

3. Using the information given in the essay that precedes this experiment, formulate a complete mechanism for the thiamine-catalyzed conversion of benzaldehyde to benzoin.

4. How do you think the appropriate enzyme would have affected the reaction (degree of completion, yield, stereochemistry)?

5. What modifications of conditions would be appropriate if the enzyme were to be used?

6. Refer to the essay that precedes this experiment. It gives a structure for thiamine pyrophosphate. Using this structure as a guide, draw a structure for thiamine hydrochloride. The pyrophosphate group is absent in this compound.

7. Draw a mechanism for the cyanide-catalyzed conversion of benzaldehyde to benzoin. The intermediate, shown in brackets, is thought to be involved in the mechanism.

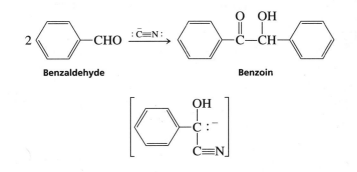

EXPERIMENT 41

Benzil

Oxidation
Crystallization

In this experiment an α-diketone, benzil, is prepared by the oxidation of an α-hydroxyketone, benzoin (Experiment 40).

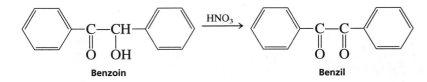

This oxidation can easily be done with mild oxidizing agents such as Fehling's solution (alkaline cupric tartrate complex) or copper sulfate in pyridine. In this experiment, the oxidation is performed with nitric acid.

Required Reading

Review: Techniques 4 and 5

Special Instructions

Nitric acid should be dispensed in a good fume hood to avoid the choking odor of this substance. The vapors will irritate your eyes. Avoid contact with your skin. During the reaction, considerable amounts of noxious nitrogen oxide gases are evolved. Be sure to run the reaction in a good fume hood.

Waste Disposal

Pour all of the aqueous solutions produced in this experiment into a waste container designated for aqueous waste. The ethanolic filtrates obtained from the crystallization of crude benzil should be poured into a waste container designated for nonhalogenated waste.

Procedure

Reaction Mixture.   Place 2.5 g benzoin (Experiment 40) in a 125-mL Erlenmeyer flask with 12 mL of concentrated nitric acid. Heat the mixture in a *good* fume hood using a steam bath or a hotplate. If a hotplate is used, carefully control the temperature near 100°C. Shake the mixture occasionally until nitrogen oxide gases (red) are no longer evolved (about 1 hour).

Isolation of Crude Benzil.   Pour the reaction mixture into 40 mL of cool water and stir the mixture vigorously until the oil crystallizes completely as a yellow solid. Scratching or seeding will be necessary to induce crystallization. Vacuum filter the crude benzil on a Büchner funnel and wash it well with cold water to remove the nitric acid. Allow the solid to dry thoroughly by drawing air through the filter. Weigh the crude benzil and calculate the percentage yield of the crude benzil.

Crystallization of Benzil.   Crystallize the crude benzil from hot 95% ethanol (6.5 mL/g) in an Erlenmeyer flask. Heat the mixture on a hotplate while swirling until the solid dissolves (try to avoid melting the solid). Remove the flask from the hotplate and allow the solution to cool slowly. As the solution cools, dip a spatula in the solution, remove it and allow the solvent to evaporate to create seed crystals on the spatula. Now, reinsert the spatula into the solution to seed it with solid benzil. Seeding allows the crystallization to occur more slowly. The solution will become supersaturated unless the solution is seeded. Yellow, needle-like crystals are formed. Cool the mixture in an ice bath to complete the crystallization. Vacuum filter the benzil on a Büchner funnel. Rinse the flask with three 2-mL portions of ice cold 95% ethanol, pouring each of these rinsings over the crystals in the funnel. Continue drawing air through the crystals on the Büchner funnel by suction for about 5 minutes. Allow the crystals to thoroughly air-dry.

Yield Calculation and Melting-Point Determination.   Weigh the dry benzil and calculate the percentage yield. Determine the melting-point range. The melting point of pure benzil is 95°C. The value obtained is often lower than this, ranging from a low value of about 87°C to a high value of about 91°C. This material is of sufficient purity for conversion to benzilic acid (Experiment 42) or tetraphenylcyclopentadienone (Experiment 46). Submit the benzil to the instructor unless it is to be used to prepare benzilic acid or tetraphenylcyclopentadienone. At the instructor's option,

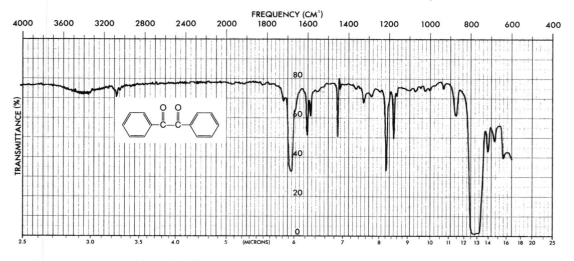

Infrared spectrum of benzil, CCl_4.

obtain the infrared spectrum of benzil in carbon tetrachloride or chloroform. Compare it with the infrared spectrum of benzil shown in this experiment. Also compare it with the infrared spectrum of benzoin shown in Experiment 40. What differences do you notice?

EXPERIMENT 42

Benzilic Acid

Anionic rearrangement

In this experiment, benzilic acid will be prepared by causing the rearrangement of the α-diketone benzil. Preparation of benzil is described in Experiment 41. The rearrangement of benzil proceeds in the following way:

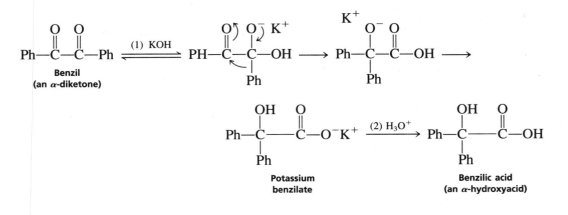

The driving force for the reaction is provided by the formation of a stable carboxylate salt (potassium benzilate). Once this salt is produced, acidification yields benzilic acid. The reaction can generally be used to convert aromatic α-diketones to aromatic α-hydroxy-acids. Other compounds, however, also will undergo benzilic acid type of rearrangement (see questions).

Required Reading

Review: Techniques 3, 4, and 5

Special Instructions

This experiment works best with pure benzil. The benzil prepared in Experiment 41 is usually of sufficient purity after it has been crystallized.

Waste Disposal

Pour all of the aqueous filtrates into the waste bottle designated for aqueous waste. Ethanolic filtrates should be put in the nonhalogenated organic waste bottle.

Procedure

Running the Reaction.   Add 2.00 g benzil and 6 mL 95% ethanol to a 25-mL round-bottom flask. Place a boiling stone in the flask and attach a reflux condenser. Be sure to use a thin film of stopcock grease when attaching the reflux condenser to the flask. Heat the mixture with a heating mantle or hotplate until the benzil has dissolved (see Technique 3, Figure 3.8, p. 619). Using a Pasteur pipet, add dropwise 5 mL of an aqueous potassium hydroxide solution[1] downward through the reflux condenser into the flask. Gently boil the mixture with occasional swirling of the contents of the flask. Heat the mixture at reflux for 15 minutes. The mixture will be blue-black in color. As the reaction proceeds, the color will turn to brown, and the solid should dissolve completely. Solid potassium benzilate may form during the reaction period. At the end of the heating period, remove the assembly from the heating device and allow it to cool for a few minutes.

Crystallization of Potassium Benzilate.   Detach the reflux condenser when the apparatus is cool enough to handle. Transfer the reaction mixture, which may contain some solid, with a Pasteur pipet into a small beaker. Allow the mixture to cool to room temperature and then cool in an ice-water bath for about 15 minutes until

[1]The aqueous potassium hydroxide solution should be prepared for the class by dissolving 55.0 g of potassium hydroxide in 120 mL of water. This will provide enough solution for 20 students, assuming little solution is wasted.

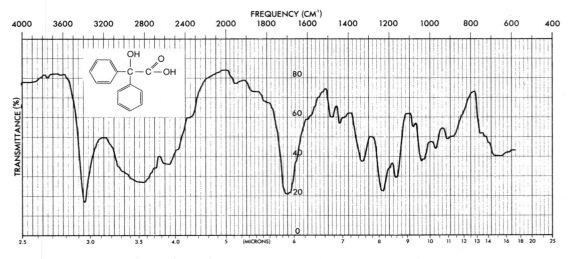

Infrared spectrum of benzilic acid, KBr.

crystallization is complete. It may be necessary to scratch the inside of the beaker with a glass stirring rod to induce crystallization. Crystallization is complete when virtually the entire mixture has solidified. Collect the crystals on a Büchner funnel by vacuum filtration (Technique 4, Section 4.3, p. 640) and wash the crystals thoroughly with three 4-mL portions of ice cold 95% ethanol. The solvent should remove most of the color from the crystals.

Transfer the solid, which is mainly potassium benzilate, to a 100-mL Erlenmeyer flask containing 60 mL of hot (70°C) water. Stir the mixture until all solid has dissolved or until it appears that the remaining solid will not dissolve. Any remaining solid will likely form a fine suspension. **If solid does remain in the flask,** gravity-filter the hot solution through fast fluted filter paper until the filtrate is clear (Technique 4, Section 4.1, p. 634). **If no solid remains in the flask,** the gravity filtration step may be omitted. In either case, proceed to the next step.

Formation of Benzilic Acid.   With swirling of the flask, slowly add dropwise 1.3 mL of concentrated hydrochloric acid to the warm solution of potassium benzilate. As the solution becomes acidic, solid benzilic acid will begin to precipitate. Keep adding the hydrochloric acid until the solid stays permanently, and then start monitoring the pH. The ideal pH should be about 2; if it is higher than this, add more acid and check the pH again. Allow the mixture to cool to room temperature and then complete the cooling in an ice bath. Collect the benzilic acid by vacuum filtration using a Büchner funnel. Wash the crystals with two 30-mL portions of ice cold water to remove potassium chloride salt that sometimes coprecipitates with benzilic acid during the neutralization with hydrochloric acid. Remove the wash water by drawing air through the filter. Dry the product thoroughly by allowing it to stand until the next laboratory period.

Melting Point and Crystallization of Benzilic Acid.   Weigh the dry benzilic acid and determine the percentage yield. Determine the melting point of the dry product. Pure benzilic acid melts at 150°C. If necessary, crystallize the product using the minimum amount of hot water needed to dissolve the solid (Technique 5, Section 5.3 and Figure 5.3, p. 652). If some impurities remain undissolved, gravity-filter the hot mixture through fast fluted filter paper (Technique 4, Section 4.1, p. 634). It will be necessary to keep the mixture hot during this gravity filtration step. Cool the solution and induce crystallization (Technique 5, Section 5.7, p. 661), if necessary, when the mixture has reached room temperature. Allow the mixture to stand at room temperature until crystallization is complete (about 15 minutes). Cool the mixture in an ice bath and collect the crystals by vacuum filtration on a Büchner funnel. Determine the melting point of the crystallized product after it is thoroughly dry.

At the instructor's option, determine the infrared spectrum of the benzilic acid in potassium bromide (Technique 19, Section 19.4, p. 846). Calculate the percentage yield. Submit the sample to your laboratory instructor in a labeled vial.

QUESTIONS

1. Show how to prepare the following compounds, starting from the appropriate aldehyde (see Experiments 40 and 41.

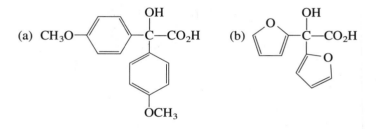

2. Give the mechanisms for the following transformations:

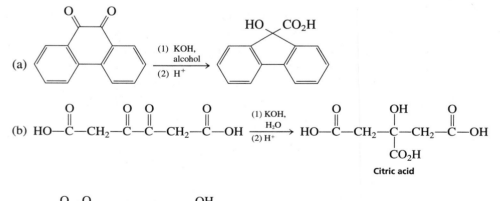

3. Interpret the infrared spectrum of benzilic acid.

ESSAY

Sulfa Drugs

The history of chemotherapy extends as far back as 1909 when Paul Ehrlich first used the term. Although Ehrlich's original definition of chemotherapy was limited, he is recognized as one of the giants of medicinal chemistry. **Chemotherapy** might be defined as "the treatment of disease by chemical reagents." It is preferable that these chemical reagents exhibit a toxicity toward only the pathogenic organism, and not toward both the organism and the host. A chemotherapeutic agent would not be useful if it poisoned the patient at the same time that it cured the patient's disease!

In 1932, the German dye manufacturing firm I. G. Farbenindustrie patented a new drug, Prontosil. Prontosil is a red azo dye, and it was first prepared for its dye properties. Remarkably, it was discovered that Prontosil showed antibacterial action when it was used to dye wool. This discovery led to studies of Prontosil as a drug capable of inhibiting the growth of bacteria. The following year, Prontosil was successfully used against staphylococcal septicemia, a blood infection. In 1935, Gerhard Domagk published the results of his research, which indicated that Prontosil was capable of curing streptococcal infections of mice and rabbits. Prontosil was shown to be active against a wide variety of bacteria in later work. This important discovery, which paved the way for a tremendous amount of research on the chemotherapy of bacterial infections, earned for Domagk the 1939 Nobel Prize in Medicine, but an order from Hitler prevented Domagk from accepting the honor.

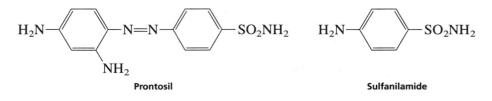

Prontosil   Sulfanilamide

Prontosil is an effective antibacterial substance **in vivo,** that is, when injected into a living animal. Prontosil is not medicinally active when the drug is tested **in vitro,** that is, on a bacterial culture grown in the laboratory. In 1935, the research group at the Pasteur Institute in Paris headed by J. Tréfouël learned that Prontosil is metabolized in animals to **sulfanilamide.** Sulfanilamide had been known since 1908. Experiments with sulfanilamide showed that it had the same action as Prontosil in vivo and that it was also active in vitro, where Prontosil was known to be inactive. It was concluded that the active portion of the Prontosil molecule was the sulfanilamide moiety. This discovery led to an explosion of interest in sulfonamide derivatives. Well over a thousand sulfonamide substances were prepared within a few years of these discoveries.

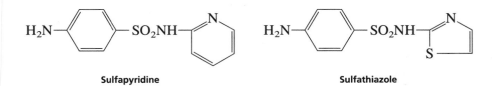

Sulfapyridine   Sulfathiazole

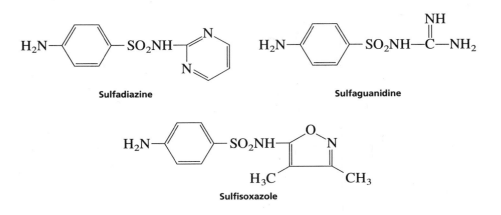

Sulfadiazine

Sulfaguanidine

Sulfisoxazole

Although many sulfonamide compounds were prepared, only a relative few showed useful antibacterial properties. As the first useful antibacterial drugs, these few medicinally active sulfonamides, or **sulfa drugs,** became the wonder drugs of their day. An antibacterial drug may be either **bacteriostatic** or **bactericidal.** A bacteriostatic drug suppresses the growth of bacteria; a bactericidal drug kills bacteria. Strictly speaking, the sulfa drugs are bacteriostatic. The structures of some of the most common sulfa drugs are shown here. These more complex sulfa drugs have various important applications. Although they do not have the simple structure characteristic of sulfanilamide, they tend to be less toxic than the simpler compound.

Sulfa drugs began to lose their importance as generalized antibacterial agents when production of antibiotics in large quantity began. In 1929, Sir Alexander Fleming made his famous discovery of **penicillin.** In 1941, penicillin was first used successfully on humans. Since that time, the study of antibiotics has spread to molecules that bear little or no structural similarity to the sulfonamides. Besides penicillin derivatives, antibiotics that are derivatives of **tetracycline,** including Aureomycin and Terramycin, were also discovered. These newer antibiotics have high activity against bacteria, and they do not usually have the severe unpleasant side effects of many of the sulfa drugs. Nevertheless, the sulfa drugs are still widely used in treating malaria, tuberculosis, leprosy, meningitis, pneumonia, scarlet fever, plague, respiratory infections, and infections of the intestinal and urinary tracts.

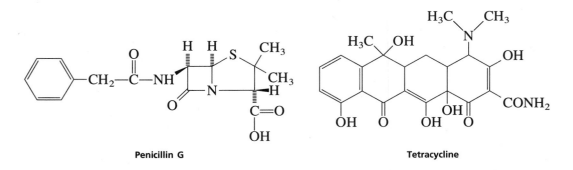

Penicillin G

Tetracycline

Even though the importance of sulfa drugs has declined, studies of how these materials act provide very interesting insights into how chemotherapeutic substances might behave. In 1940, Woods and Fildes discovered that *p*-aminobenzoic acid (PABA) inhibits the action of sulfanilamide. They concluded that sulfanilamide and PABA, because of their structural similarity, must compete with each other within the organism even though they cannot carry out the same chemical function. Further studies indicated that sulfanilamide does not kill bacteria but inhibits their growth. In order to grow, bacteria require an enzyme-catalyzed reaction that uses **folic acid** as a cofactor. Bacteria synthesize folic acid, using PABA as one of the components. When sulfanilamide is introduced into the bacterial cell, it competes with PABA for the active site of the enzyme that carries out the incorporation of PABA into the molecule of folic acid. Because sulfanilamide and PABA compete for an active site due to their structural similarity and because sulfanilamide cannot carry out the chemical transformations characteristic of PABA once it has formed a complex with the enzyme, sulfanilamide is called a **competitive inhibitor** of the enzyme. The enzyme, once it has formed a complex with sulfanilamide, is incapable of catalyzing the reaction required for the synthesis of folic acid. Without folic acid, the bacteria cannot synthesize the nucleic acids required for growth. As a result, bacterial growth is arrested until the body's immune system can respond and kill the bacteria.

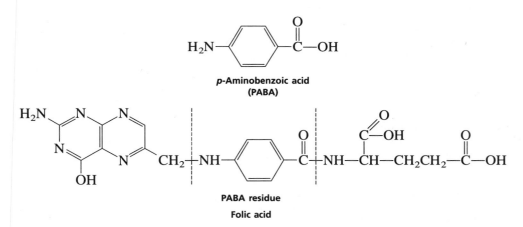

One might well ask the question, "Why, when someone takes sulfanilamide as a drug, doesn't it inhibit the growth of *all* cells, bacterial and human alike?" The answer is simple. Animal cells cannot synthesize folic acid. Folic acid must be a part of the diet of animals and is therefore an essential vitamin. Since animal cells receive their fully synthesized folic acid molecules through the diet, only the bacterial cells are affected by the sulfanilamide, and only their growth is inhibited.

For most drugs, a detailed picture of their mechanism of action is unavailable. The sulfa drugs, however, provide a rare example from which we can theorize how other therapeutic agents carry out their medicinal activity.

REFERENCES

Amundsen, L. H. "Sulfanilamide and Related Chemotherapeutic Agents." *Journal of Chemical Education, 19* (1942): 167.

Evans, R. M. *The Chemistry of Antibiotics Used in Medicine.* London: Pergamon Press, 1965.

Fieser, L. F., and Fieser, M. *Topics in Organic Chemistry.* New York: Reinhold, 1963. Chap. 7, "Chemotherapy."

Garrod, L. P., and O'Grady, F. *Antibiotic and Chemotherapy*. Edinburgh: E. and S. Livingstone, Ltd., 1968.

Goodman, L. S., and Gilman, A. *The Pharmacological Basis of Therapeutics*. 8th ed. New York: Pergamon Press, 1990. Chap. 45, "The Sulfonamides," by G. L. Mandell and M. A. Sande.

Sementsov, A. "The Medical Heritage from Dyes." *Chemistry, 39* (November 1966): 20.

Zahner, H., and Maas, W. K. *Biology of Antibiotics*. Berlin: Springer-Verlag, 1972.

E X P E R I M E N T   4 3

Sulfa Drugs: Preparation of Sulfanilamide

Crystallization
Protecting groups
Testing the action of drugs on bacteria
Preparation of a sulfonamide
Aromatic substitution

In this experiment, you will prepare the sulfa drug sulfanilamide by the following synthetic scheme. The synthesis involves converting acetanilide to the intermediate *p*-acetamidobenzenesulfonyl chloride in Step 1. This intermediate is converted to sulfanilamide by way of *p*-acetamidobenzenesulfonamide in Step 2.

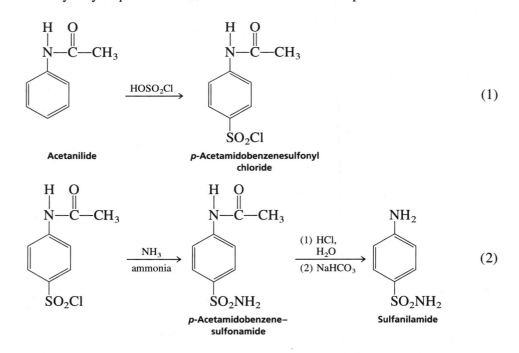

(1)

Acetanilide   *p*-Acetamidobenzenesulfonyl chloride

(2)

p-Acetamidobenzene–sulfonamide   Sulfanilamide

Acetanilide, which can easily be prepared from aniline, is allowed to react with chlorosulfonic acid to yield *p*-acetamidobenzenesulfonyl chloride. The acetamido group directs substitution almost totally to the *para* position. The reaction is an example of an electrophilic aromatic substitution reaction. Two problems would result if aniline itself were used in the reaction. First, the amino group in aniline would be protonated in strong acid to become a *meta* director; and, second, the chlorosulfonic acid would react with the amino group rather than with the ring, to give C_6H_5—$NHSO_3H$. For these reasons, the amino group has been "protected" by acetylation. The acetyl group will be removed in the final step, after it is no longer needed, to regenerate the free amino group present in sulfanilamide.

  p-Acetamidobenzenesulfonyl chloride is isolated by adding the reaction mixture to ice water, which decomposes the excess chlorosulfonic acid. This intermediate is fairly stable in water; nevertheless, it is converted slowly to the corresponding sulfonic acid (Ar—SO_3H). Thus, it should be isolated as soon as possible from the aqueous medium by filtration.

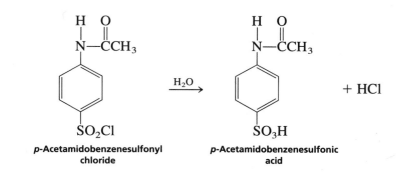

The intermediate sulfonyl chloride is converted to *p*-acetamidobenzenesulfonamide by a reaction with aqueous ammonia (Step 2). Excess ammonia neutralizes the hydrogen chloride produced. The only side reaction is the hydrolysis of the sulfonyl chloride to *p*-acetamidobenzenesulfonic acid.

  The protecting acetyl group is removed by acid-catalyzed hydrolysis to generate the hydrochloride salt of the product, sulfanilamide. Notice that of the two amide linkages present, only the carboxylic acid amide (acetamido group) was cleaved, not the sulfonic acid amide (sulfonamide). The salt of the sulfa drug is converted to sulfanilamide when the base, sodium bicarbonate, is added.

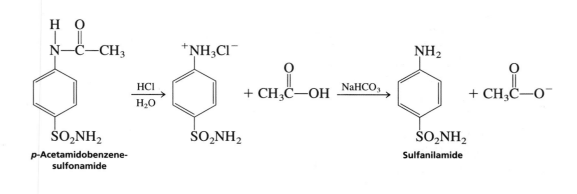

Required Reading

Review:	Technique 3	Sections 3.4 and 3.9A
	Technique 4	Section 4.3
	Technique 5	Section 5.3
	Technique 19	Section 19.4
New:	Essay	Sulfa Drugs

Special Instructions

If possible, all of this experiment should be completed in a fume hood. Otherwise, a hood must be used where indicated in the procedure.

Chlorosulfonic acid must be handled with care because it is a corrosive liquid and reacts violently with water. Be very careful when washing any glassware that has come in contact with chlorosulfonic acid. Even a small amount of the acid will react vigorously with water.

The *p*-acetamidobenzenesulfonyl chloride should be used during the same laboratory period in which it is prepared. It is unstable and will not survive long storage. The sulfa drug may be tested on several kinds of bacteria (Instructor's Manual).

Waste Disposal

Dispose of all aqueous filtrates in the container for aqueous waste. Place organic wastes in the nonhalogenated organic waste container. Place the glass wool that has been moistened with 0.1 M sodium hydroxide into the container designated for this material.

Procedure

PART A.   *p*-ACETAMIDOBENZENESULFONYL CHLORIDE

The Reaction Apparatus.   Assemble the apparatus as shown in the figure. Prepare the side-arm test tube for use as a gas trap by packing the tube loosely with dry glass wool wrapped around the glass tube. Add about 2.5 mL of 0.1 *M* sodium hydroxide dropwise to the glass wool until it is moistened but not soaked. This apparatus will capture any hydrogen chloride that is evolved in the reaction. Attach the Erlenmeyer flask after the acetanilide and chlorosulfonic acid have been added, as directed in the following paragraph.

Reaction of Acetanilide with Chlorosulfonic Acid.   Place 1.80 g of acetanilide in the *dry* 50-mL Erlenmeyer flask. Melt the acetanilide (mp 113°C) by heating gently with a flame. Remove the flask from the heat and swirl the heavy oil so that it is deposited uniformly on the lower wall and bottom of the flask. Allow the flask to cool to room temperature and then cool it further in an ice-water bath. Keep the flask in the ice bath until you are instructed to remove it.

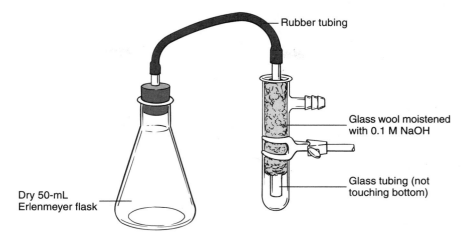

Apparatus for making *p*-acetamidobenzenesulfonyl chloride.

Caution: Chlorosulfonic acid is an extremely noxious and corrosive chemical and should be handled with care. Use only dry glassware with this reagent. Should the chlorosulfonic acid be spilled on your skin, wash it off immediately with water. Be very careful when washing any glassware that has come in contact with chlorosulfonic acid. Even a small amount of the acid will react vigorously with water and may splatter. Wear safety glasses.

In a hood, transfer 5.0 mL of chlorosulfonic acid $ClSO_2OH$ ($MW = 116.5$, $d = 1.77$ g/mL) to the acetanilide in the flask. Attach the trap to the flask at your laboratory bench, remove the flask from the ice bath, and swirl it. Hydrogen chloride gas is evolved vigorously, so be certain that the rubber stopper is securely placed in the neck of the flask. The reaction mixture usually will not have to be cooled. If the reaction becomes too vigorous, however, slight cooling may be necessary. After 10 minutes, the reaction should have subsided and only a small amount of acetanilide should remain. Heat the flask for an additional 10 minutes on the steam bath or in a hot water bath at 70°C to complete the reaction (continue to use the trap). After this time, remove the trap assembly and cool the flask in an ice bath.

Isolation of *p*-Acetamidobenzenesulfonyl Chloride.   The operations described in this paragraph should be conducted as rapidly as possible since the *p*-acetamidobenzenesulfonyl chloride reacts with water. Add 30 g of crushed ice to a 250-mL beaker. In a hood, transfer the cooled reaction mixture slowly (it may splatter somewhat) with a Pasteur pipet onto the ice while stirring the mixture with a glass stirring rod. (The remaining operations in this paragraph may be completed at your laboratory bench.) Rinse the flask with 5 mL of cold water and transfer the contents to the beaker containing the ice. Stir the precipitate to break up the lumps and then filter the *p*-acetamidobenzenesulfonyl chloride on a Büchner funnel (Technique 4, Section 4.3, p. 640, and Figure 4.5, p. 641). Rinse the flask and beaker with two

5-mL portions of ice water. Use the rinse water to wash the crude product on the funnel. Do not stop here. Convert the solid into *p*-acetamidobenzenesulfonamide in the same laboratory period.

PART B.   SULFANILAMIDE

Preparation of *p*-Acetamidobenzenesulfonamide.   In a hood, prepare a hot water bath at 70°C using a 250-mL beaker. Place the crude *p*-acetamidobenzenesulfonyl chloride into a 50-mL Erlenmeyer flask and add 11 mL of dilute ammonium hydroxide solution.[1] Stir the mixture well with a stirring rod to break up the lumps. Heat the mixture in the hot water bath for 10 minutes, stirring occasionally. Allow the flask to cool to the touch and place it in an ice water bath for several minutes. The remainder of this experiment may be completed at your laboratory bench. Collect the *p*-acetamidobenzenesulfonamide on a Büchner funnel and rinse the flask and product with about 10 mL of ice water. You may stop here.

Hydrolysis of *p*-Acetamidobenzenesulfonamide.   Transfer the solid into a 25-mL round-bottom flask and add 5.3 mL of dilute hydrochloric acid solution[2] and a boiling stone. Attach a reflux condenser to the flask. Using a heating mantle, heat the mixture under reflux until the solid has dissolved (about 10 minutes) and then reflux for an additional 5 minutes. Allow the mixture to cool to room temperature. If a solid (unreacted starting material) appears, bring the mixture to a boil again for several minutes. When the flask has cooled to room temperature, no further solids should appear.

Isolation of Sulfanilamide.   With a Pasteur pipet transfer the solution to a 100-mL beaker. While stirring with a glass rod, cautiously add dropwise a slurry of 5 g of sodium bicarbonate in about 10 mL of water to the mixture in the beaker. Foaming will occur after each addition of the bicarbonate mixture because of carbon dioxide evolution. Allow gas evolution to cease before making the next addition. Eventually, sulfanilamide will begin to precipitate. At this point, begin to check the pH of the solution. Add the aqueous sodium bicarbonate until the pH of the solution is between 4 and 6. Cool the mixture thoroughly in an ice water bath. Collect the sulfanilamide on a Büchner funnel and rinse the beaker and solid with about 5 mL of cold water. Allow the solid to air-dry on the Büchner funnel for several minutes using suction.

Crystallization of Sulfanilamide.   Weigh the crude product and crystallize it from hot water, using about 10 to 12 mL water per gram of crude product. Allow the purified product to dry until the next laboratory period.

Yield Calculation, Melting Point, and Infrared Spectrum.   Weigh the dry sulfanilamide and calculate the percentage yield (*MW* = 172.2). Determine the melting point (pure sulfanilamide melts at 163–164°C) and obtain the infrared spectrum in potassium bromide (Technique 19, Section 19.4, p. 846). Submit the sulfanilamide to the instructor in a labeled vial or save it for the tests with bacteria (Instructor's Manual). Your infrared spectrum can be compared to the one reproduced here.

---

[1]Prepared by mixing 110 mL of concentrated ammonium hydroxide with 110 mL of water.
[2]Prepared by mixing 70 mL of water with 36 mL of concentrated hydrochloric acid.

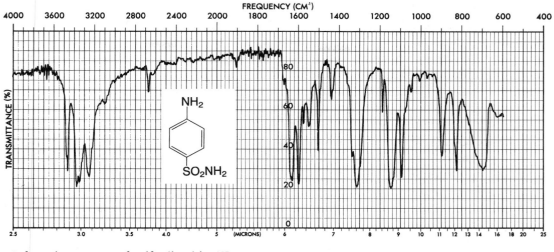

Infrared spectrum of sulfanilamide, KBr.

QUESTIONS

1. Write an equation showing how excess chlorosulfonic acid is decomposed in water.

2. In the preparation of sulfanilamide, why was aqueous sodium bicarbonate, rather than aqueous sodium hydroxide, used to neutralize the solution in the final step?

3. At first glance, it might seem possible to prepare sulfanilamide from sulfanilic acid by the set of reactions shown here.

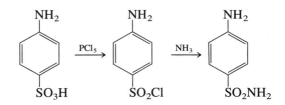

When the reaction is conducted in this way, however, a polymeric product is produced after the first step. What is the structure of the polymer? Why does *p*-acetamidobenzenesulfonyl chloride not produce a polymer?

EXPERIMENT 44

p-Aminobenzoic Acid

Synthesis of a vitamin
Amide formation
Permanganate oxidation
Amide hydrolysis

The object of this set of chemical reactions is to prepare the vitamin (for bacteria) **_p_-aminobenzoic acid,** or PABA. For a description of the importance of PABA in biological processes, read the essay on the sulfa drugs immediately preceding Experiment 43. Because PABA can absorb the ultraviolet component of solar radiation, it also finds an important application in sunscreen preparations.

The synthesis of *p*-aminobenzoic acid involves three reactions, which are outlined in the following paragraphs. The first of these reactions is the conversion of the commercially available *p*-**toluidine** into *N*-**acetyl**-*p*-**toluidine** (or *p*-acetotoluidide), the corresponding amide. The reaction is carried out by treating the amine, *p*-toluidine, with acetic anhydride. Such a procedure is a standard method for preparing amides. The reason for acetylating the amine in the initial step is to protect it during the second step, a permanganate oxidation. If the oxidation of the methyl group to the corresponding carboxyl group were carried out directly with *p*-toluidine, the highly reactive amino group would also be oxidized in the reaction. To prevent this undesired oxidation, a **protective group** is used. A protective group is a functional group that is added during a reaction sequence to shield a particular position on a molecule from undesired reactions. A good protective group is one that is easily added to the substrate molecule,

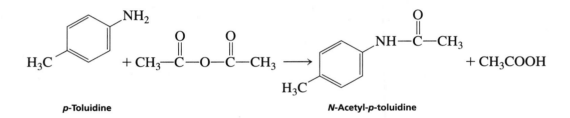

| *p*-Toluidine | | *N*-Acetyl-*p*-toluidine |

does not permit the protected group to undergo reactions, and is easily removed after the protective role has been played. This last point is important, since if the protective group cannot be removed, the original functional group cannot be used. In the example used in this experiment, the protective group is the **acetyl group.** The amide formed in this initial reaction is stable toward the oxidation conditions used in the second step. As a result, the methyl group can be oxidized without destroying the amino function in the process.

The second step of this reaction sequence is the oxidation of the methyl group to the corresponding carboxyl group, with potassium permanganate serving as the oxidizing agent. Because of the great stability of substituted benzoic acids, alkyl groups

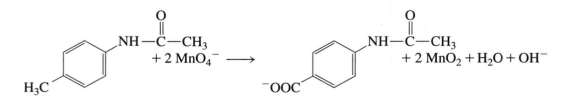

attached to aromatic rings can be oxidized easily to the corresponding benzoic acids. During the oxidation, the violet solution of permanganate ion is converted to a brown precipitate, manganese dioxide, as Mn(VII) is reduced to Mn(IV). The product of the reaction is not the carboxylic acid but rather its salt that is produced directly from the reaction. Acidification of the reaction mixture yields the carboxylic acid, which precipitates from solution.

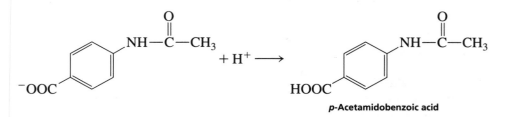

p-Acetamidobenzoic acid

The final step of this procedure is the hydrolysis of the amide functional group to remove the protecting acetyl group; *p*-aminobenzoic acid is thus produced. The reaction proceeds easily in dilute aqueous acid, and it is a characteristic reaction of amides in general. The product is crystallized from dilute aqueous acetic acid.

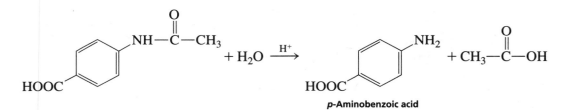

p-Aminobenzoic acid

Required Reading

Review: Techniques 1–6

        Essay   Sulfa Drugs

Special Instructions

This experiment should be conducted in two parts. It is essential that the procedure be carried out through placing the crystals of *p*-acetamidobenzoic acid in the drying oven in the first laboratory period. Between 1 and 1.5 hours of reaction time are needed to reach this point, along with considerable time spent on other operations.

p-Toluidine is toxic and is suspected to be a carcinogen. Do not allow it to come in contact with your skin. Wear protective gloves and open the bottle in the hood.

Waste Disposal

Dispose of any organic materials in the waste container specified for nonhalogenated organic waste. Dispose of any aqueous solutions in the waste container designated for the disposal of aqueous waste. Dispose of solid inorganic materials in the container identified for the disposal of inorganic solids.

Procedure

N-Acetyl-p-toluidine.   Place 0.80 g of powdered *p*-toluidine in a 50-mL Erlenmeyer flask. Add 20 mL of water and 0.8 mL of concentrated hydrochloric acid. Warm the mixture in a water bath on a hotplate, with stirring, to facilitate solution. If the solution is dark, add a small amount of decolorizing charcoal (Norit), stir it for several minutes, and filter by gravity through fluted filter paper.

Prepare a solution of 1.2 g of sodium acetate trihydrate in 2 mL of water. If necessary, use a steam bath or a hotplate to warm the solution until all the solid has dissolved.

Warm the decolorized solution of *p*-toluidine hydrochloride to 50°C. Add 0.85 mL of acetic anhydride (density = 1.08 g/mL), stir rapidly, and immediately add the previously prepared sodium acetate solution. Mix the solution thoroughly, and cool the mixture in an ice bath. A white solid should appear at this point. Filter the mixture by vacuum, using a Büchner funnel, wash the crystals three times with cold water, and allow the crystals to stand in the filter to air-dry while the vacuum is maintained. These crystals will not be isolated and dried but will be used directly in the next step.

p-Acetamidobenzoic Acid.   Place the previously prepared, wet *N*-acetyl-*p*-toluidine in a 100-mL round-bottom flask, along with 1.4 g of potassium permanganate and 30 mL of water. Add a stir bar to the flask, and attach a reflux condenser. Heat the solution to boiling with vigorous stirring. Use a heating mantle with a magnetic stirrer or a large sand bath (Technique 2, Section 2.9, page 611) on a hotplate/stirrer to heat the solution. Throughout the reaction, pay careful attention to ensure that the reaction mixture does not boil too vigorously. If frothing occurs, it may force the mixture into the condenser.

After 20 minutes of heating, the pink color should have disappeared and the reaction mixture should appear brown. Remove the apparatus from the heat and allow it to cool for a few minutes. Add an additional 1.4 g of potassium permanganate to the flask, and heat the mixture again under reflux, with stirring, for another 20 minutes, or until no pink color remains.

Vacuum filter the *hot* solution through a bed of Celite (Technique 4, Section 4.4, page 642). Wash the precipitated manganese dioxide with a small amount of hot water. If the filtrate shows the presence of excess permanganate by its pink color, add *not more than* 0.10 mL of ethanol, heat the solution under reflux for another 10 minutes, and filter the *hot* solution through fluted filter paper. Cool the pale yellow filtrate and acidify with excess 20% sulfuric acid solution. A white solid should form at this point. Filter the solid by vacuum filtration and dry it in an oven. The yield based on *p*-toluidine and the melting point should now be determined. The melting point of the pure material is 250 to 252°C. In some cases an in-

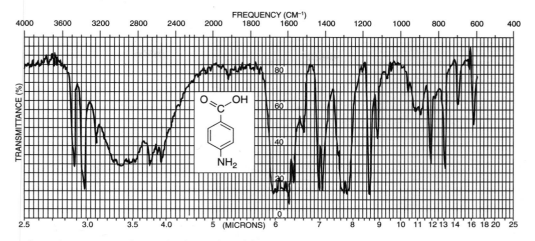

Infrared spectrum of *p*-aminobenzoic acid, KBr.

organic salt may be produced. If a salt is produced, simply go to the next step of the procedure.

p-Aminobenzoic Acid. Prepare a dilute solution of hydrochloric acid by mixing 2.4 mL of concentrated hydrochloric acid and 2.4 mL of water. Place the previously prepared *p*-acetamidobenzoic acid in a 25-mL round-bottom flask that has a reflux condenser attached to it. Add the hydrochloric acid solution. Heat the mixture gently under reflux for 30 minutes. Allow the reaction mixture to cool, transfer it to a 50-mL Erlenmeyer flask, add 1.5 mL of water, and make the solution just alkaline (pH 8–9; use pH paper) with dilute ammonia solution. For each 3 mL of final solution, add 0.5 mL of glacial acetic acid, chill the solution in an ice bath, and initiate crystallization by scratching the inside of the flask with a glass rod. Filter the crystals by vacuum on a Hirsch funnel and allow them to dry.

Determine the yield for this step, basing the overall yield on *p*-toluidine. Determine the melting point of the product. The melting point of the pure *p*-aminobenzoic acid is 186 to 187°C. Frequently, the melting point of the product is somewhat lower. Attempts to recrystallize the product are not sufficiently promising to be recommended. Recrystallization is not necessary before the *p*-aminobenzoic acid is used in Experiment 45. At the option of the instructor, an infrared spectrum of the *p*-aminobenzoic acid may be determined (Technique 19, Section 19.4, page 846 and Appendix 3).

The *p*-aminobenzoic acid may be saved for use in Experiment 45, the preparation of benzocaine. If it is not to be used for the later experiment, submit the sample of *p*-aminobenzoic acid, in a labeled vial, to the instructor.

REFERENCE

Kremer, C. B. "The Laboratory Preparation of a Simple Vitamin: *p*-Aminobenzoic Acid." *Journal of Chemical Education, 33* (1956): 71.

QUESTIONS

1. Write a mechanism for the reaction of *p*-toluidine with acetic anhydride. Why is sodium acetate added to this reaction?

2. In the oxidation step, if excess permanganate remains after the reaction period, a small amount of ethanol is added to discharge the purple color. Write a chemical equation that describes the reaction of permanganate with ethanol.

3. Write a mechanism for the acid-catalyzed hydrolysis of *p*-acetamidobenzoic acid to form *p*-aminobenzoic acid.

4. Interpret the principal absorption bands in the infrared spectrum of *p*-aminobenzoic acid.

E S S A Y

Local Anesthetics

Local anesthetics, or "painkillers," are a well-studied class of compounds with which chemists have shown their ability to study the essential features of a naturally occurring drug and to improve on them by substituting totally new, synthetic surrogates. Often such substitutes are superior in desired medical effects and in lack of unwanted side effects or hazards.

The coca shrub *(Erythroxylon coca)* grows wild in Peru, specifically in the Andes Mountains, at elevations of 1,500 to 6,000 ft above sea level. The natives of South America have long chewed these leaves for their stimulant effects. Leaves of the coca shrub have even been found in pre-Inca Peruvian burial urns. The leaves bring about a definite sense of mental and physical well-being and have the power to increase endurance. For

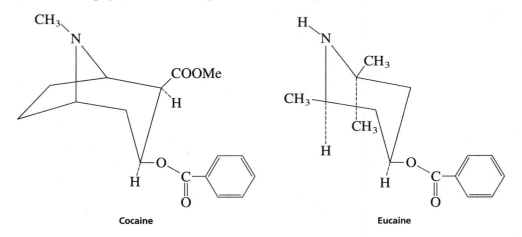

Cocaine   Eucaine

chewing, the Indians smear the coca leaves with lime and roll them. The lime $Ca(OH)_2$ apparently releases the free alkaloid components; it is remarkable that the Indians learned this subtlety long ago by some empirical means. The pure alkaloid responsible for the properties of the coca leaves is **cocaine.**

The amounts of cocaine consumed in this way by the Indians are extremely small. Without such a crutch of central-nervous-system stimulation, the natives of the Andes would probably find it more difficult to perform the nearly Herculean tasks of their daily lives, such as carrying heavy loads over the rugged mountainous terrain. Unfortunately, overindulgence in cocaine can lead to mental and physical deterioration and eventually an unpleasant death.

The pure alkaloid in large quantities is a common drug of addiction. Sigmund Freud first made a detailed study of cocaine in 1884. He was particularly impressed by the ability of the drug to stimulate the central nervous system, and he used it as a replacement drug to wean one of his addicted colleagues from morphine. This attempt was successful, but unhappily, the colleague became the world's first known cocaine addict.

An extract from coca leaves was one of the original ingredients in Coca-Cola. However, early in the present century, government officials, with much legal difficulty, forced the manufacturer to omit coca from its beverage. The company has managed to this day to maintain the *coca* in its trademarked title even though "Coke" contains none.

Our interest in cocaine lies in its anesthetic properties. The pure alkaloid was isolated in 1862 by Niemann, who noted that it had a bitter taste and produced a queer numbing sensation on the tongue, rendering it almost devoid of sensation. (Oh, those brave, but foolish chemists of yore who used to taste everything!) In 1880, Von Anrep found that the skin was made numb and insensitive to the prick of a pin when cocaine was injected subcutaneously. Freud and his assistant Karl Koller, having failed at attempts to rehabilitate morphine addicts, turned to a study of the anesthetizing properties of cocaine. Eye surgery is made difficult by involuntary reflex movements of the eye in response to even the slightest touch. Koller found that a few drops of a solution of cocaine would overcome this problem. Not only can cocaine serve as a local anesthetic, but it can also be used to produce **mydriasis** (dilation of the pupil). The ability of cocaine to block signal conduction in nerves (particularly of pain) led to its rapid medical use in spite of its dangers. It soon found use as a "local" in both dentistry (1884) and in surgery (1885). In this type of application, it was injected directly into the particular nerves it was intended to deaden.

Soon after the structure of cocaine was established, chemists began to search for a substitute. Cocaine has several drawbacks for wide medical use as an anesthetic. In eye surgery it also produces mydriasis; it can become a drug of addiction; and finally, it has a dangerous effect on the central nervous system.

The first totally synthetic substitute was eucaine. It was synthesized by Harries in 1918 and retains many of the essential skeletal features of the cocaine molecule. The development of this new anesthetic partly confirmed the portion of the cocaine structure essential for local anesthetic action. The advantage of eucaine over cocaine is that it does not produce mydriasis and is not habit-forming. Unfortunately, it is highly toxic.

A further attempt at simplification led to piperocaine. The molecular portion common to cocaine and eucaine is outlined by dotted lines in the structure shown. Piperocaine is only a third as toxic as cocaine itself.

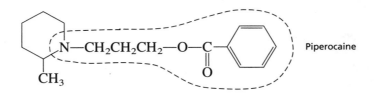

Piperocaine

Local anesthetics.

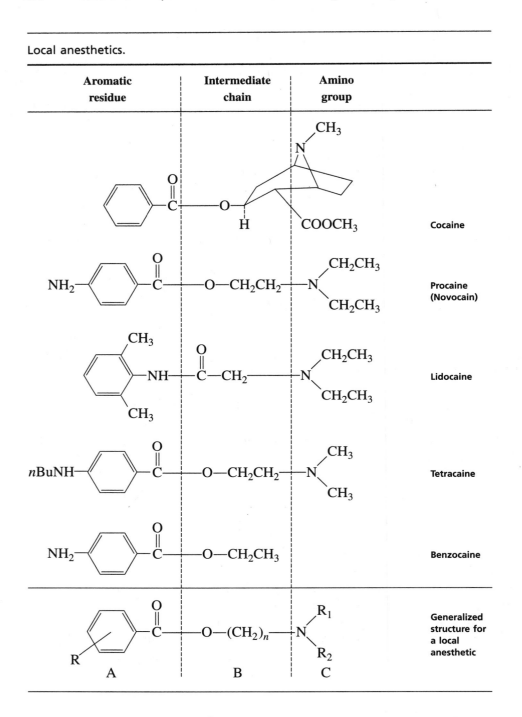

Aromatic residue	Intermediate chain	Amino group	
			Cocaine
			Procaine (Novocain)
			Lidocaine
			Tetracaine
			Benzocaine
A	B	C	Generalized structure for a local anesthetic

The most successful synthetic for many years was the drug procaine, known more commonly by its trade name Novocain (see table). Novocain is only a fourth as toxic as cocaine, giving a better margin of safety in its use. The toxic dose is almost ten times the effective amount, and it is not a habit-forming drug.

Over the years, hundreds of new local anesthetics have been synthesized and tested. For one reason or another, most have not come into general use. The search for the perfect local anesthetic is still under way. All the drugs found to be active have certain structural features in common. At one end of the molecule is an aromatic ring. At the other is a secondary or tertiary amine. These two essential features are separated by a central chain of atoms usually one to four units long. The aromatic part is usually an ester of an aromatic acid. The ester group is important to the bodily detoxification of these compounds. The first step in deactivating them is a hydrolysis of this ester linkage, a process that occurs in the bloodstream. Compounds that do not have the ester link are both longer lasting in their effects and generally more toxic. An exception is lidocaine, which is an amide. The tertiary amino group is apparently necessary to enhance the solubility of the compounds in the injection solvent. Most of these compounds are used in their hydrochloride salt forms, which can be dissolved in water for injection. Benzocaine, in contrast, is active as a local anesthetic but is not used for injection. It does not suffuse well into tissue and is not water-soluble. It is used primarily in skin preparations, in which it can be included in an ointment or salve for direct application. It is an ingredient of many sunburn preparations.

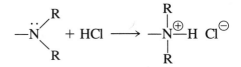

How these drugs act to stop pain conduction is not well understood. Their main site of action is at the nerve membrane. They seem to compete with calcium at some receptor site, altering the permeability of the membrane and keeping the nerve slightly depolarized electrically.

REFERENCES

Foye, W. O. *Principles of Medicinal Chemistry.* Philadelphia: Lea & Febiger, 1974. Chap. 14, "Local Anesthetics."

Goodman, L. S., and Gilman, A. *The Pharmacological Basis of Therapeutics.* 8th ed. New York: Pergamon Press, 1990. Chap. 15, "Cocaine, Procaine and Other Synthetic Local Anesthetics," by J. M. Ritchie, et al.

Ray, O. S. and Ksir, C. *Drugs, Society, and Human Behavior.* 7th ed. St. Louis: C. V. Mosby, 1996, Chap. 7, "Stimulants."

Snyder, S. H. "The Brain's Own Opiates." *Chemical and Engineering News* (November 28, 1977): 26–35.

Taylor, N. *Narcotics: Nature's Dangerous Gifts.* New York: Dell, 1970. Paperbound revision of *Flight from Reality.* Chap. 3, "The Divine Plant of the Incas."

Taylor, N. *Plant Drugs That Changed the World.* New York: Dodd, Mead, 1965. Pp. 14–18.

Wilson, C. O., Gisvold, O., and Doerge, R. F. *Textbook of Organic Medicinal and Pharmaceutical Chemistry.* 6th ed. Philadelphia: J. B. Lippincott, 1971. Chap. 22, "Local Anesthetic Agents," R. F. Doerge.

EXPERIMENT 45

Benzocaine

Esterification

In this experiment, a procedure is given for the preparation of a local anesthetic, ben-zocaine, by the direct esterification of *p*-aminobenzoic acid with ethanol. At the instruc-tor's option, you may test the prepared anesthetic on a frog's leg muscle.

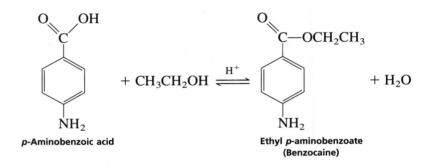

Required Reading

Review: Filtration   Section 4.3
          Crystallization   Sections 5.3 and 5.9

New: Essay   Local Anesthetics

Special Instructions[1]

Sulfuric acid is very corrosive. Do not allow it to come in contact with your skin. Use a calibrated Pasteur pipet to transfer the liquid. If you are using *p*-aminobenzoic acid pre-pared in Experiment 44, you will need to combine your product with that of several other students in order to have enough material to run this experiment at the scale indicated.

Waste Disposal

Dispose of all aqueous filtrates into the container designated for these wastes. The methanol filtrate from the crystallization should be poured into the nonhalogenated or-ganic waste container.

---

[1]Note to the Instructor: Benzocaine may be tested for its effect on a frog's leg muscle. See Instructor's Manual for in-structions.

Procedure

Running the Reaction.   Place 1.2 g of *p*-aminobenzoic acid and 12 mL of absolute ethanol in a 100-mL round-bottom flask. Swirl the mixture until the solid dissolves completely. While gently swirling, add dropwise 1.0 mL of concentrated sulfuric acid from a calibrated Pasteur pipet. A large amount of precipitate forms when you add the sulfuric acid, but this solid will slowly dissolve during the reflux that follows. Add boiling stones to the flask, attach a reflux condenser, and heat the mixture at a gentle reflux for 60–75 minutes using a heating mantle. Occasionally swirl the reaction mixture during this period to help avoid bumping.

Precipitation of Benzocaine.   At the end of the reaction time, remove the apparatus from the heating mantle and allow the reaction mixture to cool for several minutes. Using a Pasteur pipet, transfer the contents of the flask to a beaker containing 30 mL of water. When the liquid has cooled to room temperature, add a 10% sodium carbonate solution (about 10 mL needed) dropwise to neutralize the mixture. Stir the contents of the beaker with a stirring rod or spatula. After each addition of the sodium carbonate solution, extensive gas evolution (frothing) will be perceptible until the mixture is nearly neutralized. As the pH increases, a white precipitate of benzocaine is produced. When gas no longer evolves as you add a drop of sodium carbonate, check the pH of the solution and add further portions of sodium carbonate until the pH is about 8.

Collect the benzocaine by vacuum filtration using a Büchner funnel. Use three 10-mL portions of water to aid in the transfer and to wash the product in the funnel. Be sure that the solid is rinsed thoroughly with water so that any sodium sulfate formed during the neutralization will be washed out of your product. After the product has dried overnight, weigh it, calculate the percentage yield, and determine its melting point. The melting point of pure benzocaine is 92°C.

Recrystallization and Characterization of Benzocaine.   Although the product should be fairly pure, it may be recrystallized by the mixed solvent method using methanol and water (Technique 5, Section 5.9, p. 663). Place the product in a small Erlenmeyer flask and add hot methanol until the solid dissolves; swirl the mixture to help dissolve the solid. After the solid has dissolved, add hot water dropwise until the mixture turns cloudy or a white precipitate forms. Add a few more drops of hot methanol until the solid or oil redissolves completely. Allow the solution to cool slowly to room temperature. Scratch the inside of the flask as the contents cool to help crystallize the benzocaine; otherwise an oil may form. Complete the crystallization by cooling the mixture in an ice bath and collect the crystals by vacuum filtration. Use a minimum amount of ice cold methanol to aid the transfer of the solid from the flask to the filter. When the benzocaine is thoroughly dry, weigh the purified benzocaine. Again, calculate the percentage yield of benzocaine and determine its melting point.

At the option of the instructor, obtain the infrared spectrum in chloroform (Technique 19, Section 19.5, p. 850) and the NMR spectrum in $CDCl_3$ (Section 19.9, p. 855). Submit the sample in a labeled vial to the instructor.

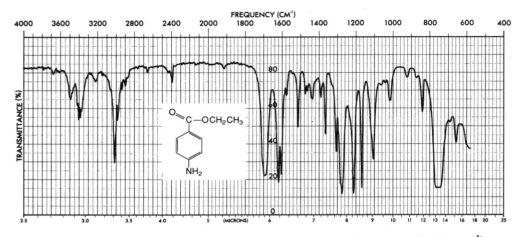

Infrared spectrum of benzocaine, $CHCl_3$. ($CHCl_3$ solvent: 3030, 1220, and 750 cm^{-1}).

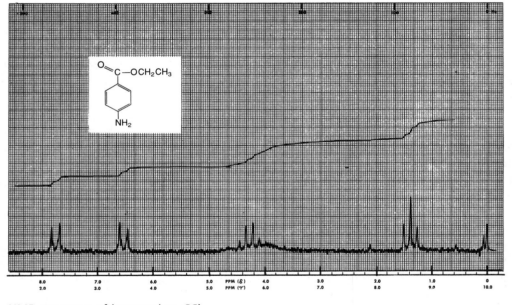

NMR spectrum of benzocaine, CCl_4.

QUESTIONS

1. Interpret the infrared and NMR spectra of benzocaine.

2. What is the structure of the precipitate that forms after the sulfuric acid has been added?

3. When 10% sodium carbonate solution is added, a gas evolves. What is the gas? Give a balanced equation for this reaction.

4. Explain why benzocaine precipitates during the neutralization.

5. Refer to the structure of procaine in the table in the essay "Local Anesthetics." Using *p*-aminobenzoic acid, give equations showing how procaine and procaine monohydrochloride could be prepared. Which of the two possible amino functional groups in procaine will be protonated first? Defend your choice. (*Hint:* Consider resonance.)

EXPERIMENT 46

Tetraphenylcyclopentadienone

Aldol condensation

In this experiment, tetraphenylcyclopentadienone will be prepared by the reaction of dibenzyl ketone (1,3-diphenyl-2-propanone) with benzil (Experiment 41) in the presence of base.

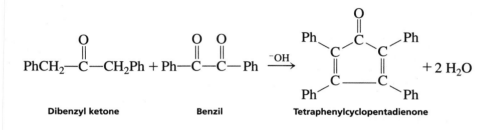

Dibenzyl ketone   **Benzil**   **Tetraphenylcyclopentadienone**

This reaction proceeds via an aldol condensation reaction, with dehydration giving the purple unsaturated cyclic ketone. A stepwise mechanism for the reaction may proceed as follows:

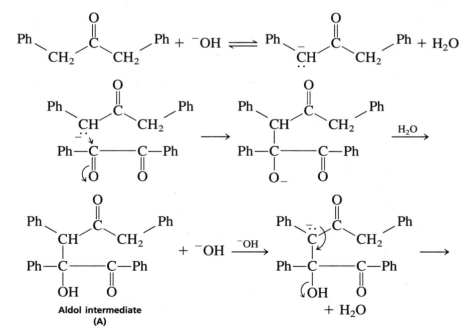

Aldol intermediate
(A)

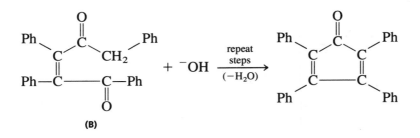

(B)

The aldol intermediate **A** readily loses water to give the highly conjugated system **B,** which reacts further to form a ring by an intramolecular aldol condensation. After a dehydration step (loss of water), the dienone product is formed.

Required Reading

Review: Technique 4   Section 4.3
           Technique 5   Section 5.3

Special Instructions

This reaction can be completed in 1 hour. The product can then be converted to 1,2,3,4-tetraphenylnaphthalene (Experiment 52). The ethanolic potassium hydroxide should be prepared in advance by the instructor.

Waste Disposal

Place all organic wastes in the nonhalogenated organic waste container.

Procedure

Running the Reaction. Add 1.5 g of benzil (Experiment 41), 1.5 g of dibenzyl ketone (1,3-diphenyl-2-propanone, 1,3-diphenylacetone), and 12 mL of absolute ethanol to a 50-mL round-bottom flask. Place a magnetic stir bar in the flask. Apply a small amount of stopcock grease to the inner joint of a condenser and attach the condenser to the round-bottom flask. Prepare a hot water bath at 70°C. Heat the mixture in the water bath with stirring until the solids dissolve.

Raise the temperature of the hot water bath to about 80°C. Continue to stir the mixture. Using a 9-inch Pasteur pipet, add dropwise 2.25 mL of ethanolic potassium hydroxide solution[1] downward through the condenser into the flask.

[1]Note to the Instructor: This solution is prepared by dissolving 6 g of potassium hydroxide in 60 mL of absolute ethanol. It will take about 30 minutes, with vigorous stirring, for the solid to dissolve. As the solid dissolves, crush the pieces with a spatula to aid in the solution process. This will provide enough solution for 20 students, assuming little material is wasted.

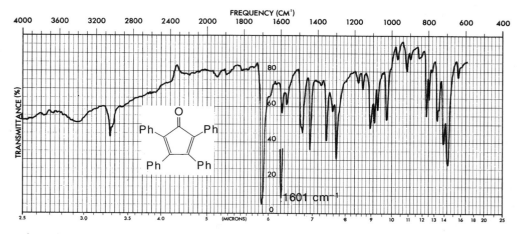

Infrared spectrum of tetraphenylcyclopentadienone, KBr.

Caution: Foaming may occur.

The mixture will immediately turn deep purple. Once you add the potassium hydroxide, raise the temperature of the hot water bath to about 85°C. Heat the mixture with stirring for 15 minutes.

Isolation of Product. At the end of the heating period, remove the flask from the hot water bath. Allow the mixture to cool to room temperature. Then place the flask in an ice-water bath for 5 minutes to complete crystallization of the product. Collect the deep purple crystals on a Büchner funnel. Wash the crystals with three 4-mL portions of cold 95% ethanol. The rinse solvent can also be used to aid in transferring crystals from the round-bottom flask to the Büchner funnel. Dry the tetraphenylcyclopentadienone in an oven for 30 minutes or in air overnight.

Yield Calculation and Melting Point Determination. The crude product is pure enough (mp 218–220°C) for the preparation of 1,2,3,4-tetraphenylnaphthalene (Experiment 52). Weigh the product and calculate the percentage yield. Determine the melting point. A small portion may be crystallized, if desired, from a 1:1 mixture of 95% ethanol and toluene (12 mL/0.5 g; mp 219–220°C). At the instructor's option, determine the infrared spectrum of tetraphenylcyclopentadienone in potassium bromide (Technique 19, Section 19.4, p. 846). Submit the product to the instructor in a labeled vial or save it for Experiment 52.

QUESTIONS

1. Interpret the infrared spectrum of tetraphenylcyclopentadienone.

2. Draw the structure of the product you would expect from the reaction of benzaldehyde and acetophenone with base.

3. Suggest several possible by-products of this reaction.

EXPERIMENT 47

Enamine Reactions: 2-Acetylcyclohexanone

Enamine reaction
Column chromatography
Keto-enol tautomerism
Infrared and NMR spectroscopy

Hydrogens on the α-carbon of ketones, aldehydes, and other carbonyl compounds are weakly acidic and are removed in a basic solution (Equation 1). Although resonance stabilizes the conjugate base **A** in such a reaction, the equilibrium is still unfavorable because of the high pK_a (about 20) of a carbonyl compound.

$$R-CH_2\overset{O}{\overset{\|}{C}}CH_2R + {}^-OH \rightleftharpoons \left[RCH_2-\overset{O}{\overset{\|}{C}}-\overset{..}{C}HR \longleftrightarrow RCH_2-\overset{O^-}{\overset{|}{C}}=CHR\right] + H_2O \qquad (1)$$

A

Typically, carbonyl compounds are alkylated (Equation 2) or acylated (Equation 4) only with difficulty in the presence of aqueous sodium hydroxide because of more important secondary side reactions (Equations 3, 5, and 6). In effect, the concentration of the nucleophilic conjugate base species (**A** in Equation 1) is low because of the unfavorable equilibrium (Equation 1), while the concentration of the competing nucleophile (OH$^-$) is very high. A significant side reaction occurs when hydroxide ion reacts with an alkyl halide by Equation 3 or acyl halide by Equation 5. In addition, the conjugate base can react with unreacted carbonyl compound by an aldol condensation reaction (Equation 6). Enamine reactions, described in the next section, avoid many of the problems described here.

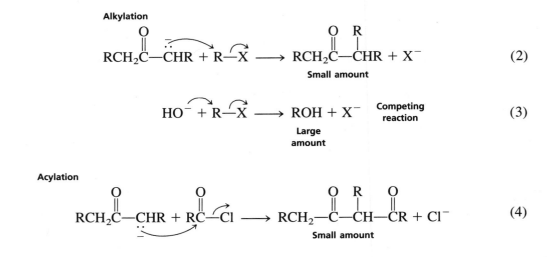

Alkylation

$$RCH_2\overset{O}{\overset{\|}{C}}-\overset{..}{C}HR + R-X \longrightarrow RCH_2\overset{O}{\overset{\|}{C}}-\overset{R}{\overset{|}{C}}HR + X^- \qquad (2)$$

Small amount

$$HO^- + R-X \longrightarrow ROH + X^- \qquad \text{**Competing reaction**} \qquad (3)$$

Large amount

Acylation

$$RCH_2\overset{O}{\overset{\|}{C}}-\overset{..}{C}HR + R\overset{O}{\overset{\|}{C}}-Cl \longrightarrow RCH_2-\overset{O}{\overset{\|}{C}}-\overset{R}{\overset{|}{C}}H-\overset{O}{\overset{\|}{C}}R + Cl^- \qquad (4)$$

Small amount

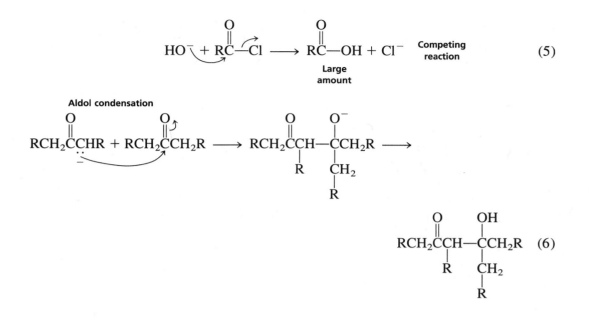

$$\text{HO}^- + \text{RC}-\text{Cl} \longrightarrow \text{RC}-\text{OH} + \text{Cl}^- \quad \text{Competing reaction} \quad (5)$$

Aldol condensation

$$\text{RCH}_2\text{CCHR} + \text{RCH}_2\text{CCH}_2\text{R} \longrightarrow \text{RCH}_2\text{CCH}-\text{CCH}_2\text{R} \longrightarrow$$

$$\text{RCH}_2\text{CCH}-\text{CCH}_2\text{R} \quad (6)$$

FORMATION AND REACTIVITY OF ENAMINES

Enamines are prepared easily from carbonyl compounds (for example, cyclohexanone) and a secondary amine (for example, pyrrolidine) by an acid-catalyzed addition-elimination reaction. An excess of the amine can be used to shift the equilibrium to the right:

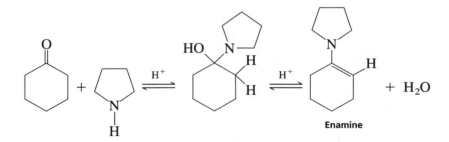

An enamine has the desirable property of being nucleophilic (carbon alkylation is more important than nitrogen alkylation) and is easily alkylated:

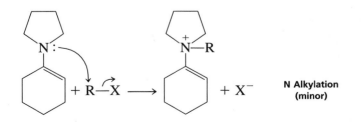

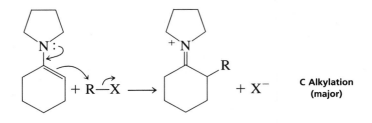

The *key point* is that the resonance hybrid **B** is like the resonance hybrid **A** shown in Equation 1. However, **B** has been produced under nearly neutral conditions so that it is the *only* nucleophile present.

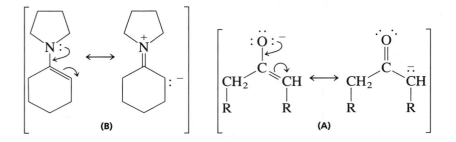

Contrast this situation to the one in Equation 1 where hydroxide ion, present in a large amount, produces undesirable side reactions (Equations 3 and 5).

The alkylation step is followed by removal of the secondary amine by an acid-catalyzed hydrolysis:

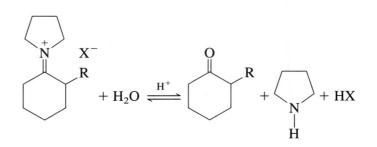

EXAMPLES OF ENAMINE REACTIONS

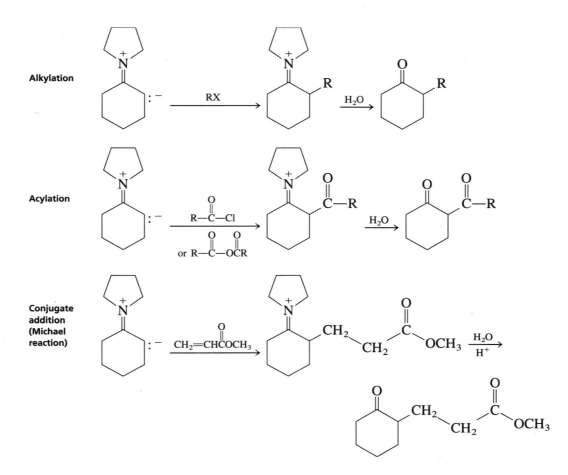

ROBINSON ANNELATION (RING-FORMATION) REACTION

Reactions that combine the Michael addition reaction and aldol condensation to form a six-membered ring fused on another ring are well known in the steroid field. These reactions are known as **Robinson annelation reactions.** An example is the formation of $\Delta^{1,9}$-2-octalone.

Michael addition
(conjugate addition)

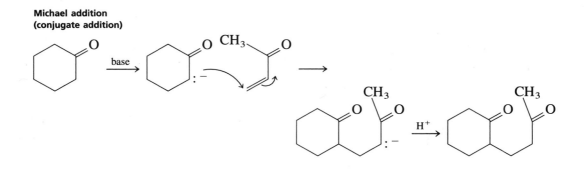

Aldol condensation

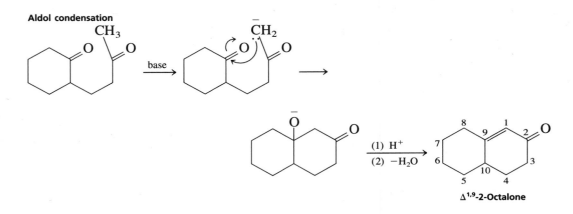

$\Delta^{1,9}$-2-Octalone

Robinson annelation reactions can also be conducted by enamine chemistry. One advantage of enamines is that the unsaturated ketones are not easily polymerized under the mild conditions of this reaction. Base-catalyzed reactions often give large amounts of polymer.

THE EXPERIMENT

In this experiment, pyrrolidine reacts with cyclohexanone to give the enamine. This enamine is used to prepare 2-acetylcyclohexanone.

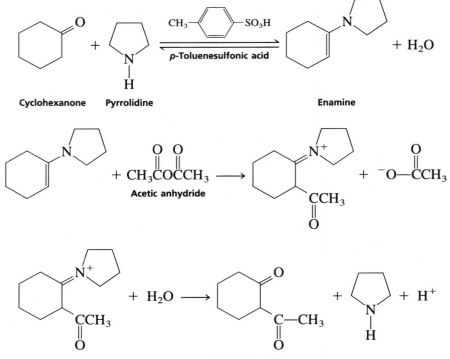

Required Reading

Review: Technique 3   Section 3.4
          Technique 7   Sections 7.4 and 7.10
          Technique 8   Section 8.3
          Technique 12   Section 12.6–12.9

Special Instructions

Pyrrolidine and acetic anhydride are toxic and noxious. You must measure and transfer these substances in a hood. If you are not careful, the entire room will be filled with vapors of pyrrolidine, and it will not be pleasant to work in the laboratory.

The enamine should be made during the first part of the laboratory period and used as soon as possible. Once the acetic anhydride has been added, the reaction mixture must be allowed to stand in your drawer for at least 48 hours to complete the reaction. The second period is used for the work-up and column chromatography. The yields in these reactions are low (less than 50%), partly due to reduced reaction periods necessary to fit the experiment into convenient 3-hour laboratory periods.

Waste Disposal

Dispose of the distillates into the waste container designated for nonhalogenated organic solvents. All aqueous solutions produced in this experiment should be disposed of by pouring them into the container designated for aqueous waste.

Procedure

PART A.   PREPARATION OF ENAMINE

Running the Reaction.   Place 3.2 mL of cyclohexanone ($MW = 98.1$) into a preweighed 50-mL round-bottom flask and determine the weight of the material transferred. Add 15 mL of toluene to the flask. Place about 0.1 g of p-toluenesulfonic acid monohydrate in the mixture. In a hood, transfer 4.0 mL of pyrrolidine ($MW = 71.1$, $d = 0.85$ g/mL) to this flask from a bottle that has been cooled in ice to reduce its volatility. Add a boiling stone and assemble the apparatus shown in the figure. The purpose of the trap is to control the release of pyrrolidine into the laboratory room. Using a heating mantle, heat under reflux for 30 minutes.

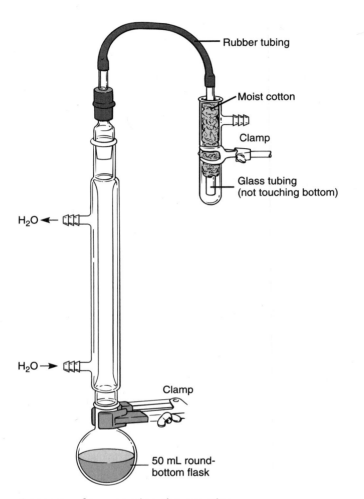

- Rubber tubing
- Moist cotton
- Clamp
- Glass tubing (not touching bottom)

H_2O ←

H_2O →

Clamp

50 mL round-bottom flask

Apparatus for preparing the enamine.

Distillation. Allow the reaction to cool somewhat and reassemble the apparatus for simple distillation (Technique 8, Figure 8.4, p. 711). Cool the receiving flask in an ice bath to prevent the noxious vapors of pyrrolidine from being released into the room. Distill the mixture until the temperature reaches 108 to 110°C (boiling point of toluene) and then stop the distillation. At this point most of the remaining pyrrolidine and water have been removed. The enamine and toluene solvent remain in the distilling flask. *Be sure you save this liquid for the next step.* Allow the reaction mixture in the distilling flask to cool to room temperature. Remove the flask and prepare 2-acetylcyclohexanone as described in the next section. Proceed to the next step during this laboratory period. Pour the distillate, containing pyrrolidine, toluene, and water, into a suitable waste container.

PART B.   PREPARATION OF 2-ACETYLCYCLOHEXANONE

Running the Reaction.   In a hood dissolve 3.2 mL of acetic anhydride (*MW* = 102.1, *d* = 1.08 g/mL) in 5.0 mL of toluene in a small beaker. Add this solution to the enamine solution contained in the round-bottom flask. Place a glass stopper in the flask, swirl it for a few minutes at room temperature, and allow the mixture to stand for at least 48 hours.

Following this period, add 5.0 mL of water. Attach a condenser and heat the mixture under reflux for 30 minutes. Cool the flask to room temperature. Transfer the liquid to a separatory funnel. Add another 5.0 mL of water, stopper the funnel, shake it, and allow the layers to separate. The 2-acetylcyclohexane is contained in the upper toluene layer. Remove the lower aqueous layer and discard it.

Extraction.   Add 10 mL of 6*M* hydrochloric acid to the toluene layer remaining in the funnel and shake the mixture to extract any nitrogen-containing contaminants from the organic phase. After allowing the layers to separate, remove the lower aqueous layer and discard it. Finally, shake the organic phase with 5.0 mL of water, remove the lower aqueous layer, and discard it. Pour the organic layer into a small Erlenmeyer flask and add 1 g of magnesium sulfate to dry the organic phase. When the liquid has become clear, transfer the dried organic phase from the drying agent and place it in a 25-mL round-bottom flask. Set up a simple distillation apparatus and remove most of the toluene (bp 110°C) by distillation. Stop the distillation when the temperature goes above 110°C. Pour the distillate containing toluene into a suitable waste container. Transfer the remaining liquid in the distilling flask into a centrifuge tube.

Evaporate the toluene in a water bath at about 70°C, using a stream of dry air or nitrogen. *Watch the liquid carefully during this procedure or your product may evaporate.* When the toluene has all been removed, the volume of liquid will remain constant (2 to 2.5 mL). Save the yellow liquid residue for purification by column chromatography.

Column Chromatography.   Prepare a column for column chromatography using a 20-cm length of 10-mm glass tubing with a constriction at one end (Technique 12, Section 12.8, p. 777). Place a piece of cotton in the column and gently push it into the constriction. Pour 5.0 g of alumina[1] slowly into the column while tapping the column constantly with a pencil or finger. Obtain about 20 mL of methylene chloride. Use the methylene chloride to prepare the column, dissolve the crude product, and elute the purified product as described in the next paragraph.

Dissolve the crude product in 2.5 mL of methylene chloride. Clamp the column above a 50-mL Erlenmeyer flask. Using a Pasteur pipet, add about 5 mL of the methylene chloride to the column and allow it to percolate through the alumina. Allow the solvent to drain until the solvent surface just begins to enter the alumina. Add

---

[1]EM Science (No. AX 0612-1). The particle sizes are 80-200 mesh and the material is Type F-20.

the crude product to the top of the column and allow the mixture to pass onto the column. Use about 5 mL of methylene chloride to rinse the centrifuge tube that contained the crude product. When the first batch of crude product has drained so that the surface of the liquid just begins to enter the alumina, add the methylene chloride rinse to the column.

When the solvent level has again reached the top of the alumina, add more methylene chloride with a Pasteur pipet to elute the product into the flask. Continue adding methylene chloride to the column until all the colored material has eluted off the column. Collect all the liquid that passes through the column as one fraction.

Evaporation of Solvent. Pre-weigh a 15-mL centrifuge tube and transfer about half the liquid in the Erlenmeyer flask to the tube. Place the centrifuge tube in a warm water bath (about 50°C) and evaporate the methylene chloride with a light stream of air or nitrogen in a hood until the volume is about 3 mL. Transfer the remaining liquid in the Erlenmeyer flask to the centrifuge tube and continue evaporating the methylene chloride to give the 2-acetylcyclohexanone as a yellow liquid. When the solvent has been removed, reweigh the tube to determine the weight of product. Calculate the percentage yield (*MW* = 140.2).

At the option of the instructor, determine the infrared spectrum and/or the NMR spectrum. The NMR spectrum may be used to determine the percentage enol content for 2-acetylcyclohexanone. This compound is highly enolic, giving calculated values between 25 and 70%. The enol content depends upon the time delay between when the compound was synthesized and when the enol content is measured. Solvent effects also influence the enol content. Submit the remaining sample to the instructor in a labeled vial with your laboratory report.

> **Note:** The percentage enol content can be calculated using the 60-MHz NMR spectrum reproduced in this experiment. The offset peak is assigned to the enolic hydrogen (integral height, 10 mm). The remaining absorptions at 1.5 to 2.85 ppm (integral height, 155 mm) are assigned to the 11 protons remaining in the enol structure and the 12 protons in the keto structure. Thus, 110 mm (10 × 11) of the 155 mm integral height is assigned to the enol structure. Enol % = 110/155 = 71; keto % = 45/155 = 29. At 300 MHz, one can integrate the enol hydrogen at 16 ppm and compare it to the methyl hydrogens at 2.15 ppm. At 60 MHz the methyl groups are not clearly resolved.

REFERENCES

Augustine, R. L., and Caputa, J. A. "$\Delta^{1,9}$-2-Octalone." *Organic Syntheses,* Coll. Vol. 5 (1973): 869.
Cook, A. G., ed. *Enamines: Synthesis, Structure, and Reactions.* New York: Marcel Dekker, 1969.
Dyke, S. F. *The Chemistry of Enamines.* London: Cambridge Univ. Press, 1973.

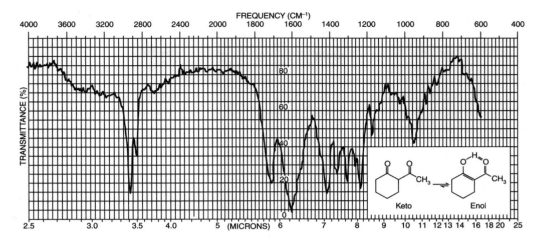

Infrared spectrum of 2-acetylcyclohexanone, neat.

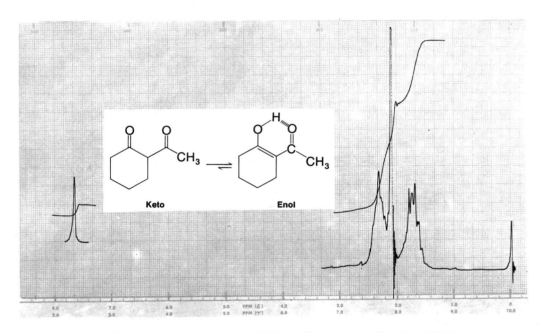

NMR spectrum of 2-acetylcyclohexanone, CDCl₃, offset peak shifted by 500 Hz.

Mundy, B. P. "The Synthesis of Fused Cycloalkenones via Annelation Methods." *Journal of Chemical Education, 50* (1973): 110.

Stork, G., Brizzolara, A., Landesman, H., Szmuszkovicz, J., and Terrell, R. "The Enamine Alkylation and Acylation of Carbonyl Compounds." *Journal of the American Chemical Society, 85* (1963): 207.

QUESTIONS

1. Draw a mechanism for the enamine synthesis of $\Delta^{1,9}$-2-octalone. Why is this octalone rather than the $\Delta^{9,10}$-2-octalone the main product in the reaction? On the other hand, why is there a substantial amount of the $\Delta^{9,10}$-2-octalone produced in the reaction?

2. (a) The enamine formed from pyrrolidine and 2-methylcyclohexanone has the **A** structure. What reason can you give for the less substituted enamine being formed instead of the more substituted enamine **B**? (*Hint:* Consider steric effects.)

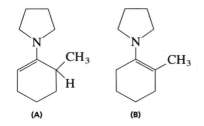

<table>
<tr><td>(A)</td><td>(B)</td></tr>
</table>

(b) Draw the structure of the product that would result from the reaction of enamine **A** with methyl vinyl ketone. Compare its structure with the product obtained in Question 3.

3. (a) The enolate formed from 2-methylcyclohexanone has the following structure. What is the structure of the other possible enolate, and why is it not as stable as the one shown here?

(b) Draw the structure of the product that would result from the reaction with methyl vinyl ketone. Compare its structure with the product obtained in Question 2.

4. Draw the structures of the Robinson annelation products that would result from the following reactions.

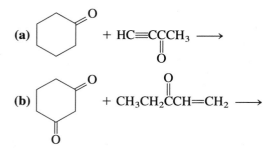

5. Draw the structures of the products that would result from the following enamine reactions. Use pyrrolidine as the amine and write equations for the reaction sequence.

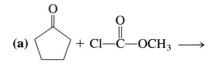

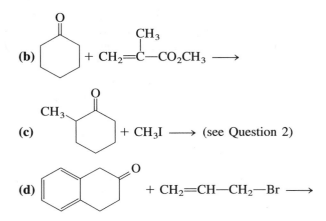

(b) [cyclohexanone] + CH_2=$\overset{\underset{|}{CH_3}}{C}$—$CO_2CH_3$ ⟶

(c) [2-methylcyclohexanone] + CH_3I ⟶ (see Question 2)

(d) [3,4-dihydronaphthalen-2(1H)-one] + CH_2=CH—CH_2—Br ⟶

6. Interpret the infrared spectrum of 2-acetylcyclohexanone, especially in the O—H and C=O stretch regions of the spectrum.

7. Write equations showing how one could carry out the following multistep transformation, starting from the indicated materials. One need not use an enamine synthesis.

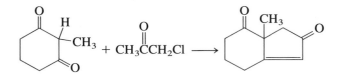

[2-methyl-1,3-cyclohexanedione] + $CH_3\overset{\overset{O}{\|}}{C}CH_2Cl$ ⟶ [bicyclic diketone product]

EXPERIMENT 48

The Aldol Condensation Reaction: Preparation of Benzalacetophenones (Chalcones)

Aldol condensation
Crystallization

Benzaldehyde reacts with a ketone in the presence of base to give α,β-unsaturated ketones. This reaction is an example of a crossed aldol condensation where the intermediate dehydrates to produce the resonance-stabilized unsaturated ketone.

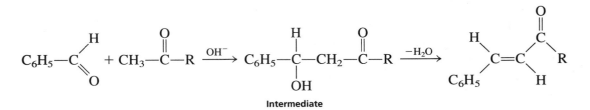

Intermediate

Crossed aldol condensations of this type proceed in high yield since benzaldehyde cannot react with itself by an aldol condensation reaction because it has no α-hydrogen. Likewise, ketones do not react easily with themselves in aqueous base. Therefore, the only possibility is for a ketone to react with benzaldehyde.

In this experiment, procedures are given for the preparation of benzalacetophenones (chalcones). You should choose one of the substituted benzaldehydes and react it with the ketone, acetophenone. All the products are solids that can be recrystallized easily.

Benzalacetophenones (chalcones) are prepared by the reaction of a substituted benzaldehyde with acetophenone in aqueous base. Piperonaldehyde, *p*-anisaldehyde, and 3-nitrobenzaldehyde are used.

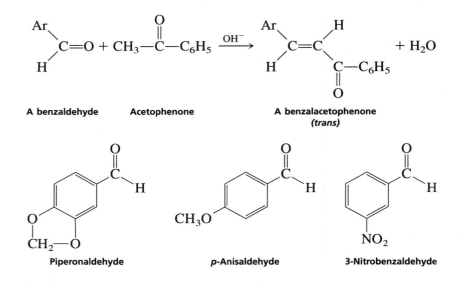

A benzaldehyde   Acetophenone   A benzalacetophenone
                                                    (trans)

Piperonaldehyde   *p*-Anisaldehyde   3-Nitrobenzaldehyde

Required Reading

Review: Technique 4   Section 4.3
          Technique 5   Section 5.3

Special Instructions

Before beginning this experiment, you should select one of the substituted benzaldehydes. Alternatively, your instructor may assign a particular compound to you.

Waste Disposal

All filtrates should be poured into a waste container designated for nonhalogenated organic waste.

Procedure

Running the Reaction.   Choose one of three aldehydes for this experiment: piperonaldehyde (solid), 3-nitrobenzaldehyde (solid), or *p*-anisaldehyde (liquid). Place 0.75 g of piperonaldehyde (3,4-methylenedioxybenzaldehyde, *MW* = 150.1) or 0.75 g of 3-nitrobenzaldehyde (*MW* = 151.1) into a 50-mL Erlenmeyer flask. Alternatively, transfer 0.65 mL of *p*-anisaldehyde (4-methoxybenzaldehyde, *MW* = 136.2) to a *tared* 50-mL Erlenmeyer flask and reweigh the flask to determine the weight of material transferred.

Add 0.60 mL of acetophenone (*MW* = 120.2, *d* = 1.03 g/mL) and 4.0 mL of 95% ethanol to the flask containing your choice of aldehyde. Swirl the flask to mix the reagents and dissolve any solids present. It may be necessary to warm the mixture on a steam bath or hotplate to dissolve the solids. If this is necessary, then the solution should be cooled to room temperature before proceeding with the next step.

Add 0.5 mL of sodium hydroxide solution[1] to the aldehyde/acetophenone mixture. Swirl the mixture until it solidifies or until the entire mixture becomes very cloudy (approximately 3 minutes).

Isolation of the Crude Product.   Add 10 mL of ice water to the flask. If a solid is present at this point, stir the mixture with a spatula to break up the solid mass. If an oil is present, stir the mixture until the oil solidifies. Transfer the mixture to a beaker with 15 mL of ice water. Stir the precipitate to break it up and then collect the solid on a Büchner funnel. Wash the product with cold water. Allow the solid to air-dry for about 30 minutes. Weigh the solid and determine the percentage yield.

Crystallization of the Benzalacetophenone (Chalcone).   Crystallize all or part of the chalcone as follows:

3,4-methylenedioxychalcone (from piperonaldehyde). Crystallize all of the sample from hot 95% ethanol. Use about 12.5 mL of ethanol per gram of solid. The literature melting point is 122°C.

4-methoxychalcone (from *p*-anisaldehyde). Crystallize all of the sample from hot 95% ethanol. Use about 4 mL of ethanol per gram of solid. Scratch the flask to induce crystallization while cooling. The literature melting point is 74°C.

3-nitrochalcone (from 3-nitrobenzaldehyde). Crystallize a 0.50 g sample from about 20 mL of hot methanol. Scratch the flask gently to induce crystallization while cooling. The literature melting point is 146°C.

Laboratory Report.   Determine the melting point of your purified product. At the option of the instructor, obtain the proton and/or carbon-13 NMR spectrum. Include a balanced equation for the reaction in your report. Submit the crude and purified samples to the instructor in labeled vials.

---

[1]This reagent should be prepared in advance by the instructor in the ratio of 6.0 g of sodium hydroxide to 10 mL of water.

QUESTIONS

1. Give a mechanism for the preparation of the appropriate benzalacetophenone using the alde-hyde that you selected in this experiment.

2. Draw the structure of the *cis* and *trans* isomers of the compound that you prepared. Why did you obtain the *trans* isomer?

3. Using proton NMR, how could you experimentally determine that you have the *trans* isomer rather than the *cis* one? (*Hint:* Consider the use of coupling constants for the vinyl hydrogens.)

4. Provide the starting materials needed to prepare the following compounds:

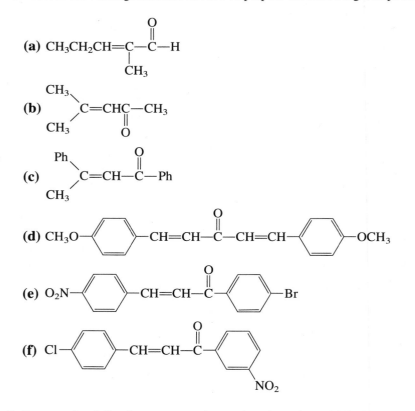

5. Prepare the following compounds starting from benzaldehyde and the appropriate ketone. Provide reactions for preparing the ketones starting from aromatic hydrocarbon compounds (see Experiment 35).

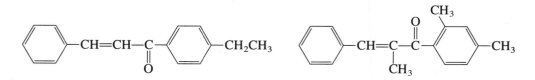

EXPERIMENT 49

Preparation of an α,β-Unsaturated Ketone *via* Michael and Aldol Condensation Reactions

Crystallization
Michael reaction (conjugate addition)
Aldol condensation reaction

This experiment illustrates how two important synthetic reactions can be combined to prepare an α,β-unsaturated ketone, 6-ethoxycarbonyl-3,5-diphenyl-2-cyclohexenone. The first step in this synthesis is a sodium hydroxide–catalyzed conjugate addition of ethyl acetoacetate to *trans*-chalcone (a Michael addition reaction). Sodium hydroxide serves as a source of hydroxide ion to catalyze the reaction.[1] In the reactions that follow, Ph and Et are abbreviations for the phenyl and ethyl groups, respectively.

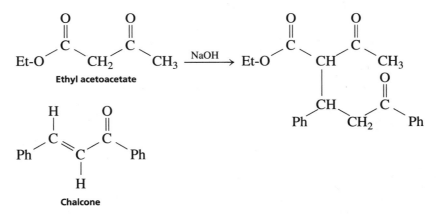

The second step of the synthesis is a base-catalyzed aldol condensation reaction. The methyl group loses a proton in the presence of base and the resulting methylene carbanion nucleophilically attacks the carbonyl group. A stable 6-membered ring is formed. Ethanol supplies a proton to yield the aldol intermediate.

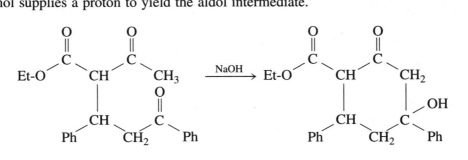

---

[1]Barium hydroxide has also been used as a catalyst (see references).

Finally, the aldol intermediate is dehydrated to form the final product, 6-ethoxycarbonyl-3,5-diphenyl-2-cyclohexenone. The α,β-unsaturated ketone that is formed is very stable because of the conjugation of the double bond with both the carbonyl group and a phenyl group.

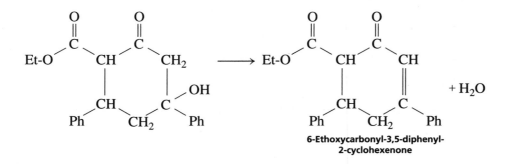

6-Ethoxycarbonyl-3,5-diphenyl-
2-cyclohexenone

Required Reading

Review: Techniques 3, 4, 5, and 7

Special Instructions

The sodium hydroxide catalyst used in this experiment must be kept dry. Be sure to keep the top on the bottle when not in use.

Waste Disposal

Dispose of all aqueous wastes containing ethanol in the bottle designated for aqueous wastes. Ethanolic filtrates from the crystallization of the product should be poured into the nonhalogenated organic waste container.

Notes to the Instructor

The *trans*-chalcone (Aldrich Chemical Co., #13,612-3) should be finely ground for use by the class. The 95% ethanol used in this experiment contains 5% water.

Procedure

Assembling the Apparatus.   To a 50-mL round-bottom flask add 1.2 g of finely ground *trans*-chalcone, 0.75 g of ethyl acetoacetate, and 25 mL 95% ethanol. Swirl the flask until the solid dissolves and place a boiling stone in the flask. Add 1 pellet (between 0.090 and 0.120 g) of sodium hydroxide. Weigh the pellet quickly before it begins to absorb water. Attach a reflux condenser to the round-bottom flask and heat the mixture to reflux using a hotplate or heating mantle. Once the mixture has been brought to a gentle boil, continue to reflux the mixture for at least one hour. During this reflux, the mixture will become very cloudy.

Isolation of the Crude Product.   After the end of the reflux period, allow the mixture to cool to room temperature. Add 10 mL of water and scratch the inside of the flask with a glass stirring rod to induce crystallization (an oil may form; scratch vigorously). Place the flask in an ice bath for a minimum of 30 minutes. It is essential to cool the mixture thoroughly in order to completely crystallize the product. Since the product may precipitate slowly, you should also scratch the inside of the flask occasionally over the 30-minute period as well as cooling it in an ice bath.

Vacuum filter the crystals on a Büchner funnel, using 4 mL of ice cold water to aid in the transfer. Then rinse the round-bottom flask with 3 mL of ice cold 95% ethanol to complete the transfer of the remaining solid from the flask to the Büchner funnel. Allow the crystals to air-dry overnight. Alternatively, the crystals may be dried for 30 minutes in an oven set at 75–80°C. Weigh the dry product. The solid contains some sodium hydroxide and sodium carbonate which are removed in the next step.

Removal of Catalyst.   Place the solid product in a 100-mL beaker. Add 7 mL of reagent-grade acetone and stir the mixture with a spatula. Most of the solid dissolves in acetone, but do not expect all of it to dissolve. Using a Pasteur pipet, remove the liquid and transfer it into a glass centrifuge tube, leaving as much solid as possible behind in the beaker. It is impossible to avoid drawing some solid up into the pipet, so the transferred liquid will contain suspended solids and the solution will be very cloudy. You should not be concerned about the suspended solids in the cloudy acetone extract since the centrifugation step will clear the liquid completely. Centrifuge the acetone extract for approximately 2–3 minutes, or until the liquid clears. Using a clean and dry Pasteur pipet, transfer the *clear* acetone extract from the centrifuge tube to a *dry preweighed* 50-mL Erlenmeyer flask. If the transfer operation is done carefully, you should be able to leave the solid behind in the centrifuge tube. The solids left behind in the beaker and centrifuge tube are inorganic materials related to the sodium hydroxide originally used as the catalyst.

Evaporate the acetone solvent by carefully heating the flask in a hot water bath while directing a light stream of dry air or nitrogen into the flask. Use a *slow* stream of gas to avoid blowing your product out of the flask. When the acetone has evaporated, you may be left with an oily solid in the bottom of the flask. Scratch the oily product with a spatula to induce crystallization. You may need to redirect air or nitrogen into the flask to remove all traces of acetone. Reweigh the flask to determine the yield of this partially purified product.

Crystallization of Product.   Crystallize the product using a minimum amount (approximately 9 mL) of boiling 95% ethanol. After all the solid has dissolved, allow the flask to cool slightly. Scratch the inside of the flask with a glass stirring rod until crystals appear. Allow the flask to sit undisturbed at room temperature for a few minutes. Then place the flask in an ice-water bath for at least 15 minutes.

Collect the crystals by vacuum filtration on a Büchner funnel. Use three 1-mL portions of ice cold 95% ethanol to aid in the transfer. Allow the crystals to dry until the next laboratory period or dry them for 30 minutes in a 75–80°C oven. Weigh the dry 6-ethoxycarbonyl-3,5-diphenyl-2-cyclohexenone and calculate the percentage yield. Determine the melting point of the product (literature value, 111–112°C). Submit the sample to the instructor in a labeled vial.

Spectroscopy.   At the option of the instructor, obtain the infrared spectrum in Nujol (Technique 19, Section 19.6, p. 853). You should observe absorbances at 1734 and 1660 cm^{-1} for the ester carbonyl and enone groups, respectively. Compare your spectrum to that shown in this experiment. Your instructor may also want you to determine the proton and carbon NMR spectra. These may be run in CDCl$_3$ solvent.[2]

REFERENCES

García-Raso, A., García-Raso, J., Campaner, B., Mestres, R., and Sinisterra, J.V. "An Improved Procedure for the Michael Reaction of Chalcones." *Synthesis,* 1982: 1037.

García-Raso, A., García-Raso, J., Sinisterra, J.V., and Mestres, R. "Michael Addition and Aldol Condensation: A Simple Teaching Model for Organic Laboratory." *Journal of Chemical Education, 63* (May 1986): 443.

QUESTIONS

1. Why was it possible to separate the product from sodium hydroxide using acetone?

2. The white solid that remains in the centrifuge tube after acetone extraction fizzes when hydrochloric acid is added, suggesting that sodium carbonate is present. How did this substance form? Give a balanced equation for its formation. Also give an equation for the reaction of sodium carbonate with hydrochloric acid.

3. Draw a mechanism for each of the three steps in the preparation of the 6-ethoxycarbonyl-3,5-diphenyl-2-cyclohexenone. You may assume that sodium hydroxide functions as a base and ethanol serves as a proton source.

4. Indicate how you could synthesis *trans*-chalcone. (*Hint:* See Experiment 48).

---

[2]Proton NMR determined at 300 MHz: 1.05 ppm (triplet, 3H, J = 7.1 Hz), 2.95–3.05 ppm (multiplet, 1 H), 3.05–3.15 ppm (multiplet, 1 H), 3.80 ppm (multiplet, 2 H), 4.05 ppm (quartet, 2 H, J = 7.1 Hz), 6.57 ppm (doublet, 1 H, J = 2.0 Hz), and 7.30–7.45 (multiplets, 10 H). Carbon NMR determined at 75 MHz: 17 peaks; 14.1, 36.3, 44.3, 59.8, 61.1, 124.3, 126.4, 127.5, 127.7, 129.0, 129.1, 130.7, 137.9, 141.2, 158.8, 169.5, and 194.3 ppm.

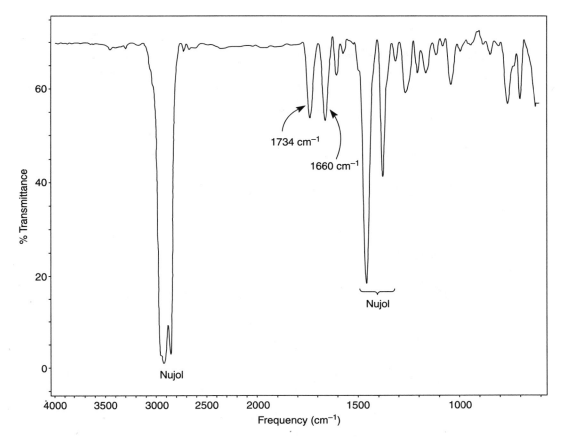

Infrared spectrum of 6-ethoxycarbonyl-3,5-diphenyl-2-cyclohexenone.

ESSAY

Molecular Modeling and Molecular Mechanics

Since the beginnings of organic chemistry, somewhere in the middle of the 19th century, chemists have sought to visualize the three-dimensional characteristics of the all-but-invisible molecules that participate in chemical reactions. Concrete models that could be held in the hand were developed. Many kinds of model sets, such as framework, ball-and-stick, and space-filling models, were devised to allow one to see the spatial and directional relationships within molecules. These hand-held models were interactive, and they could be readily manipulated in space.

Today we can also use the computer to help us visualize these molecules. The computer images are also completely interactive, allowing us to rotate, scale, and change the type of model viewed at the press of a button or the click of a mouse. In addition, the computer can perform rapid calculation of many of the properties of the molecules that we view. This combination of visualization and calculation is often called **computational chemistry** or, more colloquially, **molecular modeling.**

There are two distinct methods of molecular modeling commonly used by organic chemists today. The first of these is **quantum mechanics,** and it involves the calculation of orbitals and their energies using solutions of the Schrödinger equation. The second method is not based on orbitals at all, but is founded on our knowledge of the way in which the bonds and angles in a molecule behave. Classical equations that describe the stretching of bonds and the bending of angles are used. This second approach is called **molecular mechanics.** The two types of calculation are used for different purposes and do not calculate the same types of molecular properties. In this essay, molecular mechanics will be discussed.

MOLECULAR MECHANICS

Molecular mechanics (MM) was first developed in the early 1970s by two groups of chemical researchers: the Engler, Andose and Schleyer group on one hand, and the Allinger group on the other. In molecular mechanics, a mechanical **force field** is defined which is used to calculate an energy for the molecule under study. The energy calculated is often called the **strain energy** of the molecule. The force field is composed of several components, such as bond stretching energy, angle bending energy, and bond torsion energy. A typical force field expression[1] might be represented by the following composite expression:

$$E_{strain} = E_{stretch} + E_{angle} + E_{torsion} + E_{oop} + E_{vdW} + E_{dipole}$$

To calculate the final strain energy for a molecule, the computer systematically changes every bond length, bond angle, and torsional angle in the molecule, recalculating the strain energy each time, keeping each change that minimizes the total energy, and rejecting those that increase the energy. In other words, all of the bond lengths and angles are changed until the energy of the molecule is *minimized.*

Each of the terms contained in the composite expression (E_{strain}) is defined in Table One. All of these terms come from classical physics, not quantum mechanics. We will not discuss every term, but take $E_{stretch}$ as an illustrative example. Classical mechanics says that a bond behaves like a spring. Each type of bond in a molecule can be assigned a normal bond length, x_o. If the bond is stretched or compressed, its potential energy will increase, and there will be a restoring force that attempts to restore the bond to its normal length. According to Hooke's Law, the restoring force is proportional to the size of the displacement

$$F = -k_i(x_f - x_o) \text{ or } F = -k_i(\Delta x)$$

---

[1]Other force fields may be found that include more terms than this one, and which contain more sophisticated calculational methods than those shown here.

TABLE ONE   Some of the Factors Contributing to a Molecular Force Field*

Type of Contribution	Illustration	Typical Equation
$E_{stretch}$ (bond stretching)		$$E_{stretch} = \sum_{i=1}^{n\_bonds} (k_i/2)(x_i - x_o)^2$$
E_{angle} (angle bending)		$$E_{angle} = \sum_{j=1}^{n\_angles} (k_j/2)(\theta_j - \theta_o)^2$$
$E_{torsion}$ (bond torsion)		$$E_{torsion} = \sum_{k=1}^{n\_torsions} (k_k/2)[1 + sp_k(\cos p_k\theta)]$$
E_{oop} (out of plane bending)		$$E_{oop} = \sum_{m=1}^{n\_oops} (k_m/2)d_m^2$$
E_{vdW} (van der Waals repulsion)		$$E_{vdW} = \sum_{i=1}^{n\_atoms} \sum_{j=1}^{n\_atoms} (E_iE_j)^{1/2} \frac{1}{a_{ij}^{12}} - \frac{2}{a_{ij}^{6}}$$ $$a_{ij} = r_{ij}/(R_i + R_j)$$
E_{dipole} (electric dipole repulsion)		$$E_{vdW} = K \sum_{i=1}^{n\_atoms} \sum_{j=i+1}^{n\_atoms} Q_iQ_j / r_{ij}^2$$

*The factors selected here are similar to those in the "Tripos force field" used in the Alchemy III molecular modeling program.

where k_i is the **force constant** of the bond being studied (that is, the "stiffness" of the spring) and Δx is the change in bond length from the bond's normal length (x_o). The actual energy term that is minimized is given in Table One. This equation indicates that all of the bonds in the molecule contribute to the strain; it is a sum (Σ) starting with the first bond's contribution (n = 1) and proceeding through all of the other bonds' (n_bonds) contributions.

These calculations are based on empirical data. In order to perform these calculations, the system must be **parameterized** with experimental data. To parameterize, a table of the normal bond lengths (x_o) and force constants (k_i) for every type of bond in the molecule has to be created. The program uses these experimental parameters to perform its calculations. The quality of the results from any molecular mechanics approach directly depends on how well the parameterization has been performed for each type of atom and bond that has to be considered. The MM procedure requires each of the factors in Table One to have its own parameter table.

Each of the first four terms in Table One is treated as a spring in the same manner as discussed for bond stretching. For instance, an angle also has a force constant k that resists a change in the size of the angle θ. In effect, in the first four terms, the molecule is treated as a collection of interacting springs, and the energy of this collection of springs must be minimized. In contrast, the last two terms are based on electrostatic or "Coulomb" repulsions. Without going into detail for these terms, it should be understood that these terms must also be minimized.

MINIMIZATION AND CONFORMATION

The object of the minimization of the strain energy is to find the lowest energy *conformation* of a molecule. It turns out that molecular mechanics does a very good job of finding conformations, since it varies bond distances, bond angles, torsional angles, and the positions of atoms in space. However, most minimizers have some limitations that

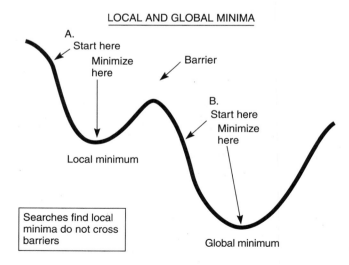

Global and local energy minima.

users must be aware of. Many of the programs use a minimization procedure that will locate a local minimum in the energy, but will not necessarily find a global minimum. The figure illustrates the problem.

In the figure, the molecule under consideration has two conformations that represent energy minima for the molecule. Many minimizers will not automatically find the lowest energy conformation, the **global minimum.** The global minimum will be found only when the structure of your starting molecule is already close to the global minimum's conformation. For instance, if the starting structure corresponds to the point labeled B on the curve in Figure One, then the global minimum will be found. However, if your starting molecule is not close to the global minimum in structure, a **local minimum** (one nearby) may be found. In the figure, if your starting structure corresponds to the point labeled A, then a local minimum will be found, instead of the global minimum. Some of the more expensive programs always find the global minimum, since they use more sophisticated minimization procedures that depend on random (Monte Carlo) changes instead of sequential ones. However, unless the program has specifically dealt with this problem, the user must be careful to avoid finding a false local minimum when the global minimum is expected. It may be necessary to use several different starting structures to discover the global minimum for a given molecule.

LIMITATIONS OF MOLECULAR MECHANICS

From our discussion thus far it should be obvious that molecular mechanics was developed to find the lowest energy conformation of a given molecule, or to compare the energies of several conformations of the same molecule. Molecular mechanics calculates a "strain energy," not a thermodynamic energy such as a heat of formation. Procedures based on quantum mechanics and statistical mechanics are required to calculate thermodynamic energies. Therefore, it is very dangerous to compare the strain energies of two *different* molecules. For instance, molecular mechanics can make a good evaluation of the relative energies of *anti-* and *gauche*-butane conformations, but it cannot fruitfully compare butane and cyclobutane. Isomers can be compared only if they are very closely related. The *cis-* and *trans*-isomers of 1,2-dimethylcyclohexane, or those of 2-butene, can be compared. However, the isomers 1-butene and 2-butene cannot be compared; one is a monosubstituted alkene while the other is disubstituted.

Molecular mechanics will perform the following tasks quite well:

1. It will give good estimates for the actual bond lengths and angles in a molecule.
2. It will find the best conformation for a molecule, but you must watch out for local minima!

Molecular mechanics will not calculate the following properties:

1. It will not calculate thermodynamic properties such as the heat of formation of a molecule.

2. It will not calculate electron distributions, charges, or dipole moments.
3. It will not calculate molecular orbitals or their energies.
4. It will not calculate infrared, NMR, or ultraviolet spectra.

CURRENT IMPLEMENTATIONS

With time, the most popular version of molecular mechanics has become that developed by Norman Allinger and his research group. The original program from this group was called MM1. The program has undergone constant revisions and improvements, and the current Allinger versions are now designated MM2 and MM3. However, many other versions of molecular mechanics are now available from both private and commercial sources. Some popular commercial programs that now incorporate their own force fields and parameters include: Alchemy III, Alchemy 2000, CAChe, Personal CAChe, HyperChem, Insight II, PC Model, MacroModel, Spartan, PC Spartan, MacSpartan, and Sybyl. You should also realize, however, that there are many modeling programs that do not have molecular mechanics or minimization. These programs will "clean up" a structure that you create by attempting to make every bond length and angle "perfect." With these programs every sp^3 carbon will have 109° angles, and every sp^2 carbon will have perfect 120° angles. Using one of these programs is equivalent to using a standard model set that has connectors and bond with perfect angles or lengths. If you intend to find a molecule's preferred conformation, be sure you use a program that has a force field and does a true minimization procedure. Also remember that you may have to control the starting structure's geometry in order to find the correct result.

REFERENCES

Casanova, J. "Computer-Based Molecular Modeling in the Curriculum," Computer Series 155. *Journal of Chemical Education, 70* (November 1993): 904.

Clark, T. *A Handbook of Computational Chemistry—A Practical Guide to Chemical Structure and Energy Calculations.* New York: John Wiley and Sons, 1985.

Lipkowitz, K. B. "Molecular Modeling in Organic Chemistry—Correlating Odors with Molecular Structure." *Journal of Chemical Education, 66* (April 1989): 275.

Tripos Associates. *Alchemy III—User's Guide.* St. Louis: Tripos Associates, 1992.

Ulrich, B., and Allinger, N. L. *Molecular Mechanics,* ACS Monograph 177. Washington, D. C.: American Chemical Society, 1982.

EXPERIMENT 50

An Introduction to Molecular Modeling

Molecular modeling
Molecular mechanics

Required Reading

Review: The sections of your lecture textbook dealing with:
1. conformations of cyclic and acyclic compounds,
2. the energies of alkenes with respect to degree of substitution, and
3. the relative energies of *cis*- and *trans*-alkenes.

New: Essay   Molecular Modeling and Molecular Mechanics

Special Instructions

In order to perform this experiment, you must use computer software that has the ability to perform molecular mechanics (MM2 or MM3) calculations with minimization of the strain energy. Either your instructor will provide instruction in the use of the software, or you will be provided with a handout giving instructions.

Notes to the Instructor

This molecular mechanics experiment was devised using the modeling program Alchemy III; however, it should be possible to use many other implementations of molecular mechanics. Some of the other capable programs available are: Alchemy 2000, Spartan, PC Spartan, MacSpartan, CAChe and Personal CAChe, PCModel, Insight II, Nemesis, and Sybyl. You will have to provide your students with an introduction to your specific implementation. The introduction should show students how to build a molecule, how to minimize its energy, and how to load and save files. Students will also need to be able to measure bond lengths and bond angles.

EXPERIMENT 50A

The Conformations of *n*-Butane: Local Minima

The acyclic butane molecule has several conformations derived by rotation about the C2-C3 bond. The relative energies of these conformations have been well established experimentally and are listed in the table below.

Conformation	Torsional Angle	Relative Energy (Kcal/mol)	Relative Energy (KJ/mol)	Types of Strain
Syn	0°	6.0	25.0	Steric/Torsional
Gauche	60°	1.0	4.2	Steric
Eclipsed-120	120°	3.4	14.2	Torsional
Anti	180°	0	0	No Strain

In this section we will show that, while molecular mechanics does not calculate the precise thermodynamic energies for the conformations of butane, it will give strain energies that predict the *order* of stability correctly. We will also investigate the difference between a local minimum and a global minimum.

When you construct a butane skeleton, you might expect the minimizer to always arrive at the *anti* conformation (lowest energy). In fact, for most molecular mechanics programs, this will happen only if you bias the minimizer by starting with a butane skeleton that closely resembles the *anti* conformation. If this is done, the minimizer will find the *anti* conformation (the *global* minimum). However, if a skeleton is constructed that does not closely resemble the *anti* conformation, the butane will usually minimize to the *gauche* conformation (the nearest *local* minimum), and not proceed to the global minimum. For the two staggered conformations, you will begin by constructing your starting butane molecules with torsional angles slightly removed from the two minima. The eclipsed conformations, however, will be set on the exact angles to see if they will minimize. Your data should be recorded in a table with the following headings: Starting Angle, Minimized Angle, Final Conformation, and Minimized Energy.

Your program should have a feature that allows you to set bond lengths, bond angles, and torsional angles.[1] If it does, you can merely select the torsion angle C1-C2-C3-C4 and specify 160° to set the first starting shape. Select the minimizer and allow it to run until it stops. Did it find the *anti* conformation (180°)? Record the energy.

---

[1]If your program does not have this feature, you can approximate the angles specified by constructing your starting molecules on the screen in a Z-shape for one, and in a U-shape for the other.

Repeat the process, starting with torsion angles of 0°, 45°, and 120° for the butane skeleton. Record the strain energies and report the final conformations that are formed in each case. What are your conclusions? Do your final results agree with those in the table?

If your minimizer rotated the two eclipsed conformations (0° and 120°) to their closest staggered minima, you may have to restrict the minimizer to a single iteration in order to calculate their energies. This restriction calculates a **single point energy,** and the energy of the structure is not minimized. If necessary, calculate the single point energies of the eclipsed conformations and record your results.

The lesson here is that you may have to try several starting points to find the correct structure for the lowest energy conformation of a molecule! Do not blindly accept your first result, but look at it with the skeptical eye of a practiced chemist, and test it further.

Optional.   Record the single point energies for every 30° rotation, starting at 0° and ending at 360°. When these energies are plotted against their angle, the plot should resemble the rotational energy curve shown for butane in most organic textbooks.

EXPERIMENT 50B

Cyclohexane Chair and Boat Conformations

In this exercise, we will investigate the chair and boat conformations of cyclohexane. Many programs will have these stored on disk as templates or fragments. If they are available as templates or fragments, you will only have to add hydrogens to the template. The chair is not difficult to build if you construct your cyclohexane on the screen in such a way that it suggests a chair (that is, just as you might draw it on paper). This crude construct will usually minimize to a chair. The boat is more difficult to construct. When you draw a crude boat on the screen, it will minimize to a *twist* boat, instead of the desired symmetrical boat.

Before you construct any cyclohexanes, construct a propane molecule. Minimize it and measure the CH and CC bond lengths, and the CCC bond angle. Record these values; we will use them for reference.

Now construct a cyclohexane chair and minimize it. Measure the CH and CC bond lengths, and the CCC angle in the ring. Compare these values to those of propane. What do you conclude? Rotate the molecule so that you view it end-on, looking down two of the bonds simultaneously (as in a Newman projection). Are all the hydrogens staggered? Rotate the chair and look at it from a different end-on angle. Are all the hydrogens still staggered? The van der Waals radius of a hydrogen atom is 1.20 Ångstroms. Hydrogen atoms that are closer than 2.40 Ångstroms apart will "touch" each other and create steric strain. Are any of the hydrogens in the cyclohexane chair close enough to cause steric stain? What are your conclusions?

Now construct a cyclohexane boat (from a template) and do not minimize it.[1] Measure the CH and CC bond lengths, and the CCC bond angles at both the peaks and the lower corner of the ring. Compare these values to those of propane. Rotate the molecule so that you view it end-on looking down the two parallel bonds on the sides of the boat. Are the hydrogens eclipsed or staggered? Now measure the distances between the various hydrogens on the ring, including the bowsprit-flagpole hydrogens and the axial and equatorial hydrogens on the side of the ring. Are any of the hydrogens generating steric strain?

Now minimize the boat to a twist boat and repeat all of the measurements. Write all of your conclusions about chairs, boats, and twist boats in your report.

EXPERIMENT 50C

Substituted Cyclohexane Rings

Using a cyclohexane template, construct *cis*(a,a)-1,3-dimethylcyclohexane, *cis*(e,e)-1,3-dimethylcyclohexane, and *trans*(a,e)-1,3-dimethylcyclohexane and measure their energies. In the diaxial isomer, measure the distance between the two diaxial methyl groups. What do you conclude?

Optional.   Similar comparisons can be made for the *cis*- and *trans*-1,2-dimethyl-cyclohexanes and the *cis*- and *trans*-1,4-dimethylcyclohexanes.

Now construct *cis*(e,a)-1,4-di-*tert*-butylcyclohexane in a chair conformation, minimize it, and record its energy. Now compare it to *cis*(e,e)-1,4-di-*tert*-butylcyclohexane in a boat conformation. Minimize this isomer to a twist boat and record its energy.

EXPERIMENT 50D

cis- and *trans*-2-Butene

Heats of hydrogenation for the three isomers of butene are given in the table below. Construct both *cis*- and *trans*-2-butene, minimize them, and report their energies. Which of these isomers has the lowest energy? Can you determine why?

Compound	ΔH (Kcal/mol)	ΔH (KJ/mol)
trans-2-butene	−27.6	−115
cis-2-butene	−28.6	−120
1-butene	−30.3	−126

---

[1]A single point energy (see page 425) may be obtained, if you desire.

Now construct and minimize 1-butene. Record its energy. Obviously 1-butene does not fit with the hydrogenation data. Molecular mechanics works quite well for *cis-* and *trans-* 2-butene because they are very similar isomers. Both are 1,2-disubstituted alkenes. However, 1-butene is a monosubstituted alkene, and direct comparison to the 2-butenes cannot be made. The differences in the stability of mono- and di-substituted alkenes require that factors other than those used in molecular mechanics be used. These factors are due to electronic and resonance differences. The molecular orbitals of the methyl groups interact with the pi bonds of the disubstituted alkenes (hyperconjugation) and help to stabilize them. Two such groups (as in 2-butene) are better than one (as in 1-butene). Therefore, while the bond lengths and angles come out pretty well for 1-butene, the energy derived for 1-butene does not directly compare to the energies of the 2-butenes. Molecular mechanics does not include terms that allow these factors to be included; it is necessary to use either semiempirical or *ab initio* quantum mechanical methods, which are based on molecular orbitals.

ESSAY

Diels–Alder Reaction and Insecticides

Since the 1930s, it has been known that the addition of an unsaturated molecule across a diene system forms a substituted cyclohexene. The original research dealing with this type of reaction was performed by Otto Diels and Kurt Alder in Germany, and the reaction is known as the **Diels–Alder reaction.** The Diels–Alder reaction is the reaction of a **diene** with a species capable of reacting with the diene, the **dienophile.**

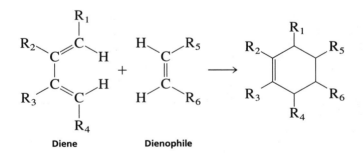

Diene   Dienophile

The product of the Diels–Alder reaction is usually a structure that contains a cyclohexene ring system. If the substituents as shown are simply alkyl groups or hydrogen atoms, the reaction proceeds only under extreme conditions of temperature and pressure. With more complex substituents, however, the Diels–Alder reaction may go at low temperatures and under mild conditions. The reaction of cyclopentadiene with maleic anhydride (Experiment 51) and the reaction of tetraphenylcyclopentadiene with benzyne (Experiment 52) are examples of Diels–Alder reactions carried out under reasonably mild conditions.

A commercially important use of the Diels–Alder reaction involves using hexachlorocyclopentadiene as the diene. Depending on the dienophile, a variety of chlorine-containing addition products may be synthesized. Nearly all these products are powerful **insecticides.** Three insecticides synthesized by the Diels–Alder reaction are shown here.

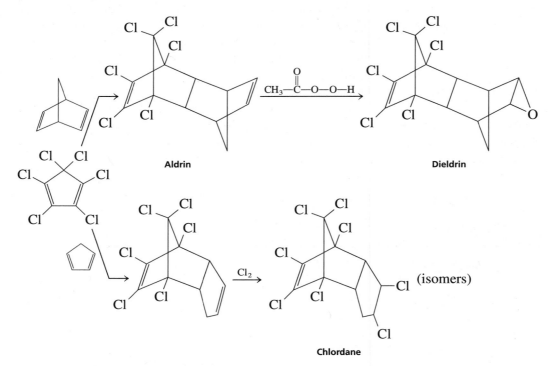

Dieldrin and Aldrin are named after Diels and Alder. These insecticides have been used against the insect pests of fruits, vegetables, and cotton; against soil insects, termites, and moths; and in treating seeds. Chlordane has been used in veterinary medicine against insect pests of animals, including fleas, ticks, and lice.

The best known insecticide, DDT, is not prepared by the Diels–Alder reaction but is nevertheless the best illustration of the difficulties experienced when insecticides are used indiscriminately. DDT was first synthesized in 1874, and its insecticidal properties were first demonstrated in 1939. It is easily synthesized commercially, with inexpensive reagents.

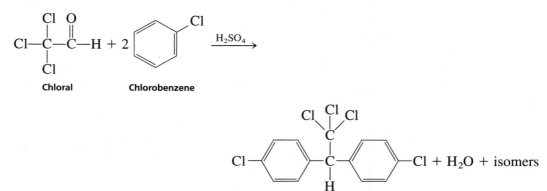

At the time DDT was introduced, it was an important boon to humankind. It was effective in controlling lice, fleas, and malaria-carrying mosquitoes and thus helped to control human and animal disease. The use of DDT rapidly spread to the control of hundreds of insects that damage fruit, vegetable, and grain crops.

Pesticides that persist in the environment for a long time after application are called "hard" pesticides. Beginning in the 1960s, some of the harmful effects of such "hard" pesticides as DDT and the other chlorocarbon materials became known. DDT is a fat-soluble material and is therefore likely to collect in the fat, nerve, and brain tissues of animals. The concentration of DDT in tissues increases in animals high in the food chain. Thus, birds that eat poisoned insects accumulate large quantities of DDT. Animals that feed on the birds accumulate even more DDT. In birds, at least two undesirable effects of DDT have been recognized. First, birds whose tissues contain large amounts of DDT have been observed to lay eggs having shells too thin to survive until young birds are hatched. Second, large quantities of DDT in the tissues seem to interfere with normal reproductive cycles. The massive destruction of bird populations that sometimes occurs after heavy spraying with DDT has become an issue of great concern. The brown pelican and the bald eagle are in danger of extinction. The use of chlorocarbon insecticides has been identified as the principal reason for the decline in the numbers of these birds.

Because DDT is chemically inert, it persists in the environment without decomposing to harmless materials. It can decompose very slowly, but the decomposition products are every bit as harmful as DDT itself. Consequently, each application of DDT means that still more DDT will pass from species to species, from food source to predator, until it concentrates in the higher animals, possibly endangering their existence. Even humans may be threatened. As a result of evidence of the harmful effects of DDT, the Environmental Protection Agency banned general use of DDT in the early 1970s; it may still be used for certain purposes, although permission of the Environmental Protection Agency is required. In 1974, the EPA granted permission to use DDT against the tussock moth in the forests of Washington and Oregon.

Because the life cycles of insects are short, they are able to evolve an immunity to insecticides within a short period. As early as 1948, several strains of DDT-resistant insects were identified. Today, the malaria-bearing mosquitoes are almost completely resistant to DDT, an ironic development. Other chlorocarbon insecticides have been used as alternatives to DDT against resistant insects. Examples of other chlorocarbon materials include Dieldrin, Aldrin, Chlordane, and the substances whose structures are shown here. Heptachlor and Mirex are also prepared using Diels–Alder reactions.

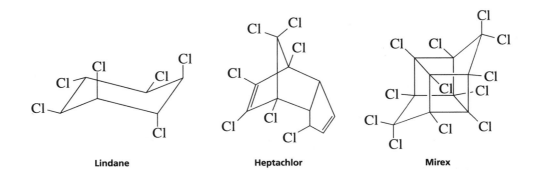

Lindane   Heptachlor   Mirex

In spite of structural similarity, Chlordane and Heptachlor show different behavior. Compared with Heptachlor, Chlordane is short-lived and less toxic to mammals. Nevertheless, all the chlorocarbon insecticides have been the objects of much suspicion. A ban on the use of Dieldrin and Aldrin has also been ordered by the Environmental Protection Agency. In addition, strains of insects resistant to Dieldrin, Aldrin, and other materials have been observed. Some insects become addicted to a chlorocarbon insecticide and thrive on it!

The problems associated with chlorocarbon materials have led to the development of "soft" insecticides. These usually are organophosphorus or carbamate derivatives, and they are characterized by a short lifetime before they are decomposed to harmless materials in the environment.

The organic structures of some organophosphorus insecticides are shown below.

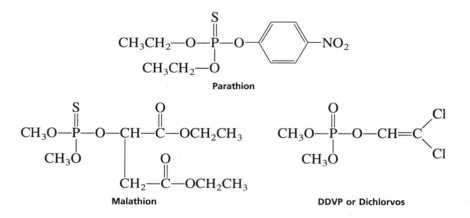

Parathion

Malathion

DDVP or Dichlorvos

Parathion and Malathion are used widely for agriculture. DDVP is used in "pest strips," which are used for combating household insect pests. The organophosphorus materials do not persist in the environment, so they are not passed between species up the food chain, as the chlorocarbon compounds are. However, the organophosphorus compounds are highly toxic to humans. Some migrant and other agricultural workers have lost their lives because of accidents involving these materials. Stringent safety precautions must be applied when organophosphorus insecticides are being used.

The carbamate derivatives, including Carbaryl, tend to be less toxic than the organophosphorus compounds. They are also readily degraded to harmless materials. Nevertheless, insects resistant to "soft" insecticides have also been observed. Furthermore, the organophosphorus and carbamate derivatives destroy many more nontarget pests than the chlorocarbon compounds do. The danger to earthworms, mammals, and birds is very high.

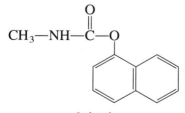

Carbaryl

ALTERNATIVES TO INSECTICIDES

Several alternatives to the massive application of insecticides have recently been explored. Insect attractants, including the pheromones (see the essay preceding Experiment 16), have been used in localized traps. Such methods have been effective against the gypsy moth. A "confusion technique," whereby a pheromone is sprayed into the air in such high concentrations that male insects are no longer able to locate females, has been studied. These methods are specific to the target pest and do not cause repercussions in the general environment.

Recent research has been focused on using an insect's own biochemical processes to control pests. Experiments with **juvenile hormone** have shown promise. Juvenile hormone is one of three internal secretions used by insects to regulate growth and metamorphosis from larva to pupa and thence to the adult. At certain stages in the metamorphosis from larva to pupa, juvenile hormone must be secreted; at other stages it must be absent, or the insect will either develop abnormally or fail to mature. Juvenile hormone is important in maintaining the juvenile, or larval, stage of the growing insect. The male cecropia moth, which is the mature form of the silkworm, has been used as a source of juvenile hormone. The structure of the cecropia juvenile hormone is shown below. This material has been found to prevent the maturation of yellow-fever mosquitoes and human body lice. Because insects are not expected to develop a resistance to their own hormones, it is hoped that insects will be unlikely to develop a resistance to juvenile hormone.

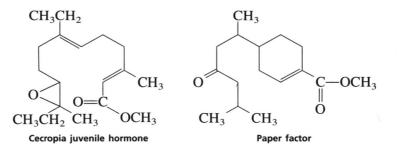

Cecropia juvenile hormone   Paper factor

While it is very difficult to get enough of the natural substance for use in agriculture, synthetic analogues have been prepared, and they have been shown to be similar in properties and effectiveness to the natural substance. Williams, Sláma, and Bowers have identified and characterized a substance found in the American balsam fir (*Abies balsamea*), known as **paper factor,** which is active against the linden bug, *Pyrrhocoris apterus,* a European cotton pest. This substance is merely one of thousands of terpenoid materials synthesized by the fir tree. Other terpenoid substances are being investigated as potential juvenile hormone analogues.

Certain plants are capable of synthesizing substances that protect them against insects. Included among these natural insecticides are the **pyrethrins** and derivatives of **nicotine.**

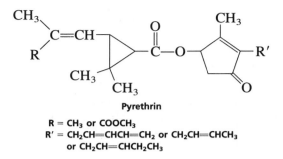

Pyrethrin

R = CH$_3$ or COOCH$_3$
R′ = CH$_2$CH═CHCH═CH$_2$ or CH$_2$CH═CHCH$_3$
or CH$_2$CH═CHCH$_2$CH$_3$

The search for environmentally suitable means of controlling agricultural pests continues with a great sense of urgency. Insects cause billions of dollars of damage to food crops each year. With food becoming increasingly scarce and with the world's population growing at an exponential rate, preventing such losses to food crops becomes absolutely essential.

REFERENCES

Berkoff, C. E. "Insect Hormones and Insect Control." *Journal of Chemical Education, 48* (1971): 577.

Bowers, W. S., and Nishida, R. "Juvocimenes: Potent Juvenile Hormone Mimics from Sweet Basil." *Science, 209* (1980): 1030.

Carson, R. *Silent Spring.* Boston: Houghton Mifflin, 1962.

Keller, E. "The DDT Story." *Chemistry, 43* (February 1970): 8.

O'Brien, R. D. *Insecticides: Action and Metabolism.* New York: Academic Press, 1967.

Peakall, D. B. "Pesticides and the Reproduction of Birds." *Scientific American, 222* (April 1970): 72.

Saunders, H. J. "New Weapons Against Insects." *Chemical and Engineering News, 53* (July 28, 1975): 18.

Williams, C. M. "Third-Generation Pesticides." *Scientific American, 217* (July 1967): 13.

Williams, W. G., Kennedy, G. G., Yamamoto, R. T., Thacker, J. D., and Bordner, J. "2-Tridecanone: A Naturally Occurring Insecticide from the Wild Tomato." *Science, 207* (1980): 888.

E X P E R I M E N T   5 1

The Diels–Alder Reaction of Cyclopentadiene with Maleic Anhydride

Diels–Alder reaction
Fractional distillation

Cyclopentadiene and maleic anhydride react readily in a Diels–Alder reaction to form the adduct, *cis*-norbornene-5,6-*endo*-dicarboxylic anhydride:

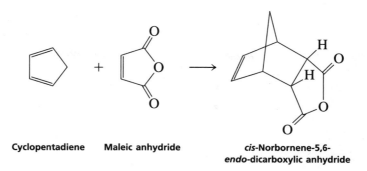

Cyclopentadiene   Maleic anhydride   *cis*-Norbornene-5,6-
                                               endo-dicarboxylic anhydride

Because two molecules of cyclopentadiene can also undergo a Diels–Alder reaction to form dicyclopentadiene, it is not possible to store cyclopentadiene in the monomeric form. Therefore, it is necessary to first "crack" dicyclopentadiene to produce cyclopentadiene for use in this experiment. This will be accomplished by heating the dicyclopentadiene to a boil and collecting the cyclopentadiene as it is formed by fractional distillation. The cyclopentadiene must be kept cold and used fairly soon in order to keep it from dimerizing.

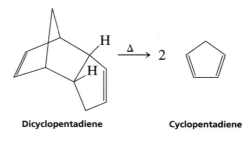

Dicyclopentadiene   Cyclopentadiene

Required Reading

Review: Technique 5   Section 5.3

New: Essay   Diels–Alder Reaction and Insecticides

Special Instructions

The cracking of dicyclopentadiene should be performed by the instructor or laboratory assistant. If a flame is used for this, be sure that there are no leaks in the system since both cyclopentadiene and the dimer are highly flammable.

Waste Disposal

Dispose of the mother liquor from the crystallization in the container designated for nonhalogenated organic solvents.

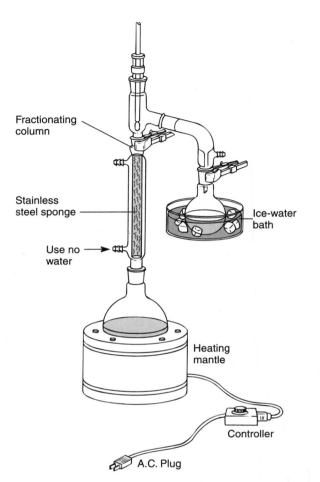

Fractionating column

Stainless steel sponge

Use no water

Ice-water bath

Heating mantle

Controller

A.C. Plug

Fractional distillation apparatus for cracking dicyclopentadiene.

Notes to the Instructor

Working in a hood, assemble a fractional distillation apparatus, as shown in the figure. Although the required temperature control can best be obtained with a micro burner, using a heating mantle lessens the possibility of a fire occurring. Place several boiling stones and dicyclopentadiene in the distilling flask. The amount of dicyclopentadiene and the size of the distilling flask should be determined by the amount of cyclopentadiene required by your class. The volume of cyclopentadiene recovered will be 50–75% of the initial volume of dicyclopentadiene, depending on the volume distilled and the size of the fractionating column. Control the heat source so that the cyclopentadiene distills at 40–43° C. If the cyclopentadiene is cloudy, dry the liquid over granular anhydrous sodium sulfate. Store the product in a sealed container and keep it cooled in an ice-water bath until all students have taken their portions. It must be used within a few hours to keep it from dimerizing.

Procedure

Preparation of the Adduct. Add 1.00 g of maleic anhydride and 4.0 mL of ethyl acetate to a 25-mL Erlenmeyer flask. Swirl the flask to dissolve the solid (slight heating on a hotplate may be necessary). Add 4.0 mL of ligroin (bp 60–90°C) and swirl the flask to mix the solvents and reactant thoroughly. Add 1.0 mL of cyclopentadiene and mix thoroughly until no visible layers of liquid are present. Since this reaction is exothermic, the temperature of the mixture will likely become high enough to keep the product in solution. However, if a solid does form at this point, it will be necessary to heat the mixture on a hotplate to dissolve any solids present.

Crystallization of Product. Allow the mixture to cool slowly to room temperature. Better crystal formation can be achieved by seeding the solution before it cools to room temperature. To seed the solution, dip a spatula or glass stirring rod into the solution after it has cooled for about 5 minutes. Allow the solvent to evaporate so that a small amount of solid forms on the surface of the spatula or glass rod. Place the spatula or stirring rod back into the solution for a few seconds to induce crystallization. When crystallization is complete at room temperature, cool the mixture in an ice bath for several minutes.

Isolate the crystals by filtration on a Hirsch funnel or a small Büchner funnel, and allow the crystals to air-dry. Determine the weight and the melting point (164°C).

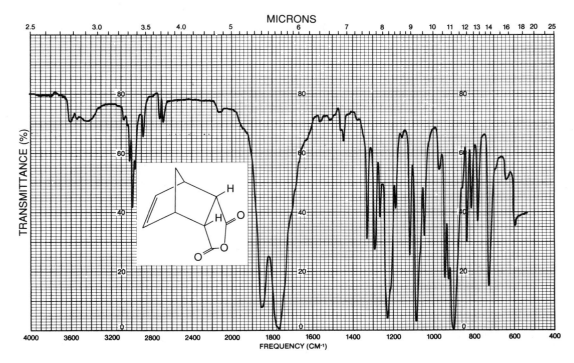

Infrared spectrum of *cis*-norbornene-5,6-*endo*-dicarboxylic anhydride, KBr.

At the instructor's option, determine the infrared spectrum of the adduct in potassium bromide. Calculate the percentage yield and submit the product to the instructor in a labeled vial. Compare your infrared spectrum to the one reproduced here.

QUESTIONS

1. Draw a structure for the exo product formed by cyclopentadiene and maleic anhydride.
2. Since the exo form is more stable than the endo form, why is the endo product formed almost exclusively in this reaction?
3. In addition to the main product, what are two side reactions that could occur in this experiment?
4. The infrared spectrum of the adduct is given in this experiment. Interpret the principal peaks.

E X P E R I M E N T   5 2

Benzyne Formation and the Diels–Alder Reaction: Preparation of 1,2,3,4-Tetraphenylnaphthalene

Benzyne formation
Diels–Alder reaction

In this experiment, you will prepare 1,2,3,4-tetraphenylnaphthalene. In Step 1, benzyne is produced via the unstable diazonium salt. Benzyne is also unstable and cannot be isolated. In Step 2, tetraphenylcyclopentadienone (Experiment 46) traps the reactive benzyne as it is formed by a Diels–Alder reaction, to give an unstable intermediate. This intermediate readily loses carbon monoxide and yields the fully aromatized naphthalene system. The reaction can be followed easily, because the reaction mixture changes from a purple to a yellow-orange solution when the tetraphenylcyclopentadienone is consumed, and 1,2,3,4-tetraphenylnaphthalene is produced.

Step 1 BENZYNE FORMATION

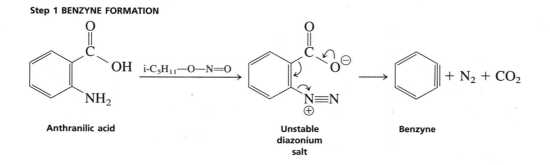

Step 2 DIELS–ALDER REACTION

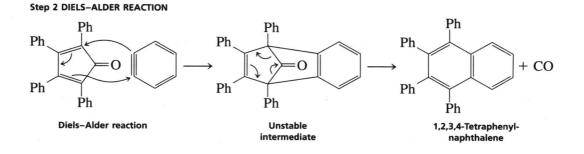

Diels–Alder reaction   Unstable
                              intermediate

1,2,3,4-Tetraphenyl-
naphthalene

+ CO

Required Reading

Review: Technique 3   Section 3.8
          Technique 4   Section 4.3
          Technique 19   Sections 19.5A and 19.9

New: Essay   Diels–Alder Reaction and Insecticides

Special Instructions

Special care should be taken to avoid breathing isopentyl nitrite (isoamyl nitrite) as it is a powerful heart stimulant. The isopentyl nitrite must be stored in a refrigerator when not in use. Restopper the bottle after the liquid has been removed to minimize contact with air.

A small amount of carbon monoxide is produced in this reaction. Although it may be advisable to run the reaction in a hood, the reaction can be conducted on a bench if the laboratory has reasonable ventilation.[1]

Occasional explosions have been observed with unstable diazonium salts. This has usually happened when a large concentration of the diazonium salt was present. In this experiment, however, the salt reacts quickly to form benzyne, which is trapped rapidly with tetraphenylcyclopentadienone.

The 1,2-dimethoxyethane (also known as ethylene glycol dimethyl ether and mono-glyme) used in small quantities as a solvent for the reaction may cause reproductive disorders, based on tests with laboratory animals. Dispense this solvent in a hood. Use polyethylene gloves, if available. Do not inhale this substance. Rinse your hands if it comes in contact with your skin.

---

[1]The amount of carbon monoxide formed in this reaction is very small (see Question 1). It may be trapped by placing a solution of cuprous chloride and ammonium chloride in aqueous ammonia in a test tube. Gases that are evolved are led from the condenser to the test tube (Fig. 3.14, p. 627, omit the moistened glass wool). Immerse the tip of the tubing just below the surface of the liquid so that if the pressure changes, the trapping solution will not be pulled back into the vial. You must remove the trap from the top of the condenser when the isopentyl nitrite is added. A large amount of trapping agent is prepared as follows: Dissolve 20 g of cuprous chloride and 25 g of ammonium chloride in 70 mL of water. Add to the solution a third of its volume of concentrated (28%) ammonium hydroxide.

Waste Disposal

The methanol and isopropyl alcohol filtrates should be placed in the waste containers designated for disposal of nonhalogenated wastes. Any residual 1,2-dimethoxyethane waste should also be placed in the same waste container.

Procedure

Reaction Mixture. Place 1.5 g of tetraphenylcyclopentadienone (*MW* = 384.5, Experiment 46) and 12 mL of 1,2-dimethoxyethane into a 100 mL round-bottom flask. Add a Claisen adapter, addition funnel, and reflux condenser to the round-bottom flask, and bring the solution to reflux, using a steam bath or a heating mantle to heat (see illustration). Since carbon monoxide is evolved in this reaction, the reaction should be conducted in a hood, if possible.[1] Dissolve 0.675 g of anthranilic acid (*MW* = 137.1) in 6 mL of 1,2-dimethoxyethane and place it in the addition funnel. Using the pipet provided with the reagent bottle, transfer 1 mL of isopentyl nitrite (isoamyl nitrite, *MW* = 117.2, *d* = 0.875 g/mL) to 7 mL of 1,2-dimethoxyethane in a small Erlenmeyer flask.

> **Caution:** Do not breathe the isopentyl nitrite vapor as it is a powerful heart stimulant.

Stopper the Erlenmeyer flask to prevent loss by evaporation. Replace the lid on the reagent bottle as soon as possible to minimize exposure to air.[2]

Running the Reaction. When the solution begins to boil, add the solution of isopentyl nitrite *dropwise* through the top of the condenser with an eye dropper or a Pasteur pipet. (Wear your safety glasses!) *Add simultaneously* the anthranilic acid solution from the addition funnel and adjust the rates of addition so that the materials are added at about the same rate. Add the two solutions over a 45- to 60-minute period (three to four drops per minute). When adding isopentyl nitrite, make sure that the pipet is inserted deep into the condenser so that the solution is added directly into the reaction flask. When all of the isopentyl nitrite has been added, use a few drops of 1,2-dimethoxyethane to rinse the empty Erlenmeyer flask and add this solution to the reaction mixture.

Continue to boil the mixture until the color changes from the deep purple color of the tetraphenylcyclopentadienone to a yellow-orange solution formed after the dienone is consumed (usually less than 10 minutes). If the color has not changed after about 15 minutes, add about 10 drops of pure isopentyl nitrite (no solvent, pipet ex-

---

[2]Isopentyl nitrite (isoamyl nitrite) must be stored in a refrigerator. It decomposes in the presence of light and air. This reaction gives the best results if the material has been bought recently. Material from Aldrich Chemical Co. (#15,049-5) works well.

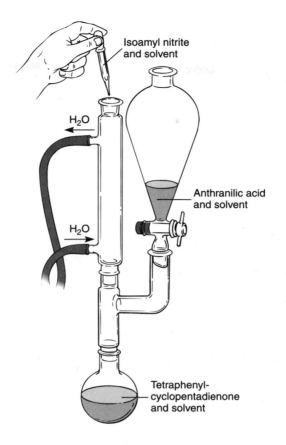

Isoamyl nitrite
and solvent

H_2O

H_2O

Anthranilic acid
and solvent

Tetraphenyl-
cyclopentadienone
and solvent

tended down into the condenser) and continue to boil the solution for an additional 10-minute period until the solution is yellow orange. Repeat the addition if necessary.[3]

Isolation of Crude Tetraphenylnapththalene. After the color changes, cool the mixture to room temperature and transfer the solution to a beaker containing 75 mL of water and 30 mL of methanol. Stir the mixture well to break up the precipitate. Collect the solid on a Büchner funnel under vacuum (Technique 4, Section 4.3, and Fig. 4.6, p. 643). In some cases, the filtration process may be slow because the solid plugs the filter paper. If this occurs, add a little ice cold methanol to the Büchner funnel while the mixture is being filtered. Use 20 mL of ice cold methanol to aid the transfer of the solid remaining in the beaker and to wash the solid collected in the Büchner funnel. Additional product precipitates in the filter flask. Collect this material and add it to the solid in the Büchner funnel. Rinse the combined solids on the Büchner funnel with 20 mL of ice cold methanol. Weigh the crude product.

---

[3]If the color has not changed after the additions of the extra isopentyl nitrite, add a small amount (about 0.10 g) of anthranilic acid. Reflux the mixture for another 15 minutes, or until the color changes.

Crystallization of the Tetraphenylnaphthalene. Purify about 1 g of your product by crystallization from hot isopropyl alcohol (2-propanol) in order to remove the remaining colored impurities. Place the crude tetraphenylnaphthalene in an Erlenmeyer flask and dissolve it in boiling isopropyl alcohol. It will require approximately 140 mL of isopropyl alcohol to dissolve one gram of your crude product (140 mL/1 g). This is a rough estimate, and you may use more or less than this amount. The solution process is enhanced by breaking up any lumps with a spatula. Once your crude product has dissolved in the boiling solvent, cool the mixture in an ice bath. When it has cooled somewhat, scratch the side of the flask with a stirring rod. The product slowly crystallizes. Allow the mixture to cool for at least 30 minutes in an ice bath. Collect the product on a Büchner funnel under vacuum, and wash it with a small amount of ice cold isopropyl alcohol. Allow the solid to dry completely, weigh the product, and calculate the percentage yield.

Determination of Melting Point. When the solid has dried completely, determine the melting point. Pure 1,2,3,4-tetraphenylnaphthalene melts at 196–199°C. When the material has melted, remove the capillary tube from the melting-point apparatus and cool the tube until the material solidifies. Redetermine the melting point (literature 203–205°C).[4] The tetraphenylnaphthalene exists in two crystalline forms, each with a different melting point. Submit the sample to the instructor in a labeled vial.

Spectroscopy. At the option of the instructor, obtain the infrared spectrum in CCl_4 (Technique 19, Section 19.5, Method A, p. 850) or the proton NMR spectrum in CCl_4 or $CDCl_3$ (Technique 19, Section 19.9, p. 855).

REFERENCES

Dougherty, C. M., Baumgarten, R. L., Sweeney, A., and Concepcion, E. "Phthalimide, Anthranilic Acid and Benzyne." *Journal of Chemical Education, 54* (1977): 643.

Fieser, L. F., and Haddadin, M. J. "Isobenzofurane, A Transient Intermediate." *Canadian Journal of Chemistry, 43* (1965): 1599.

QUESTIONS

1. Calculate the number of moles and milliliters of carbon monoxide gas theoretically produced in the reaction performed in this experiment at 25°C and 1 atmosphere pressure.

---

[4]This compound exhibits a double melting point. The initial melting point varies according to its particle size and is not a reliable index of purity. The remelt melting point is more reproducible and reliable.

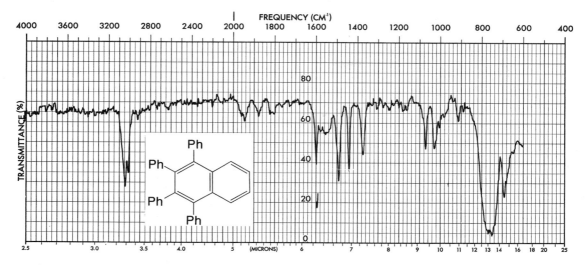

Infrared spectrum of 1,2,3,4-tetraphenylnaphthalene, CCl$_4$.

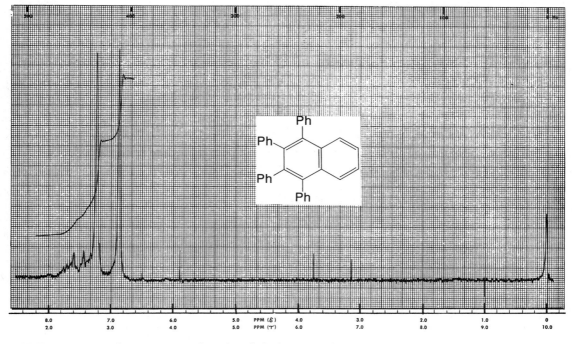

NMR spectrum of 1,2,3,4-tetraphenylnaphthalene, CDCl$_3$.

2. Draw the structures of the products that would result from the following reactions.

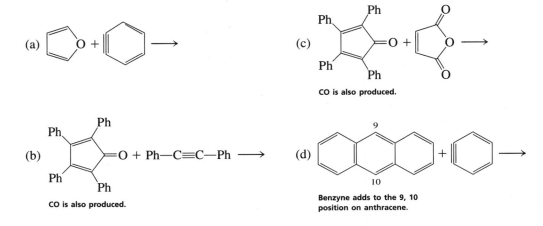

(a)

(b)

CO is also produced.

(c)

CO is also produced.

(d)

Benzyne adds to the 9, 10 position on anthracene.

3. Interpret the principal absorption bands in the infrared spectrum of 1,2,3,4-tetraphenylnaphthalene.

4. Interpret the NMR spectrum of 1,2,3,4-tetraphenylnaphthalene. In interpreting the NMR spectrum, notice that the molecule is symmetrical and that each of the singlets integrates for 10 hydrogens. The multiplet at 7.2 to 7.8 ppm represents four hydrogens.

5. Draw a mechanism for the formation of the diazonium salt from anthranilic acid and isopentyl nitrite.

6. What is the ultimate result of the reaction of the isopentyl group from the isopentyl nitrite? That is, what compound or compounds are formed?

EXPERIMENT 53

Photoreduction of Benzophenone and Rearrangement of Benzpinacol to Benzopinacolone

Photochemistry
Photoreduction
Energy transfer
Pinacol rearrangement

This experiment consists of two parts. In the first part (Experiment 53A) benzophenone will be subjected to **photoreduction,** a dimerization brought about by exposing a solution of benzophenone in isopropyl alcohol to natural sunlight. The product of this photoreaction is benzpinacol. In part two (Experiment 53B) benzpinacol will be induced to undergo an acid catalyzed rearrangement called the **pinacol rearrangement** (see also Experiment 32B, page 295). The product of the rearrangement is benzopinacolone.

EXPERIMENT 53A

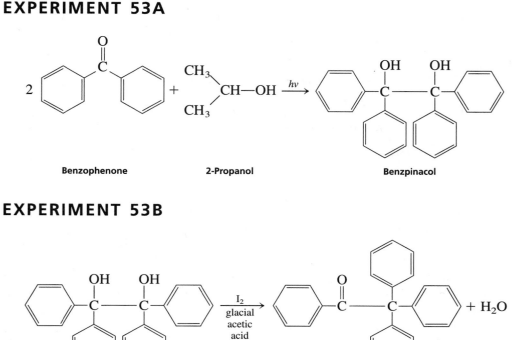

Benzophenone   2-Propanol   Benzpinacol

EXPERIMENT 53B

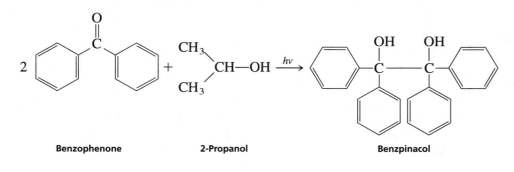

Benzpinacol   Benzopinacolone

E X P E R I M E N T   5 3 A

Photoreduction of Benzophenone

The photoreduction of benzophenone is one of the oldest and most thoroughly studied photochemical reactions. Early in the history of photochemistry, it was discovered that solutions of benzophenone are unstable in light when certain solvents are used. If benzophenone is dissolved in a "hydrogen-donor" solvent, such as 2-propanol, and exposed to ultraviolet light, an insoluble dimeric product, benzpinacol, will form.

Benzophenone   2-Propanol   Benzpinacol

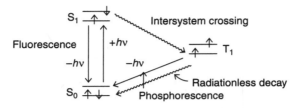

Electronic states of a typical molecule and the possible interconversions. In each state (S_0, S_1, T_1), the lower line represents the highest occupied orbital and the upper line represents the lowest unoccupied orbital of the unexcited molecule. Straight lines represent processes in which a photon is absorbed or emitted. Wavy lines represent radiationless processes—those that occur *without* emission or absorption of a photon.

To understand this reaction, let's review some simple photochemistry as it relates to aromatic ketones. In the typical organic molecule, all the electrons are paired in the occupied orbitals. When such a molecule absorbs ultraviolet light of the appropriate wavelength, an electron from one of the occupied orbitals, usually the one of highest energy, is excited to an unoccupied molecular orbital, usually to the one of lowest energy. During this transition, the electron must retain its spin value, because a change of spin is a quantum-mechanically forbidden process during an electronic transition. Therefore, just as the two electrons in the highest occupied orbital of the molecule originally had their spins paired (opposite), so they will retain paired spins in the first electronically excited state of the molecule. This is true even though the two electrons will be in *different* orbitals after the transition. This first excited state of a molecule is called a **singlet state** (S_1) because its spin multiplicity ($2S + 1$) is one. The original unexcited state of the molecule is also a singlet state because its electrons are paired, and it is called the **ground-state singlet** state (S_0) of the molecule.

The excited state singlet S_1 may return to the ground state S_0 by reemission of the absorbed photon of energy. This process is called **fluorescence.** Alternatively, the excited electron may undergo a change of spin to give a state of higher multiplicity, the excited **triplet state,** so called because its spin multiplicity ($2S + 1$) is three. The conversion from the first excited singlet state to the triplet state is called **intersystem crossing.** Because the triplet state has a higher multiplicity, it inevitably has a lower energy state than the excited singlet state (Hund's rule). Normally, this change of spin (intersystem crossing) is a quantum-mechanically forbidden process, just as a direct excitation of the ground state (S_0) to the triplet state (T_1) is forbidden. However, in those molecules in which the singlet and triplet states lie close to one another in energy, the two states inevitably have several overlapping vibrational states, that is, states in common, a situation that allows the "forbidden" transition. In many molecules in which S_1 and T_1 have similar energy ($\Delta E < 10$ kcal/mole), intersystem crossing occurs faster than fluorescence, and the molecule is rapidly converted from its excited singlet state to its triplet state. In benzophenone, S_1 undergoes intersystem crossing to T_1 with a rate of $k_{isc} = 10^{10}$ sec^{-1}, meaning that the lifetime of S_1 is only 10^{-10} second. The rate of fluorescence for benzophenone is $k_f = 10^6$ sec^{-1}, meaning that intersystem crossing occurs at a rate that is 10^4 times faster than fluorescence. Thus, the conversion of S_1 to T_1 in benzophenone is essentially a quantitative process. In molecules that have a wide energy gap between S_1 and T_1, this situation would be reversed. As you will see shortly, the naphthalene molecule presents a reversed situation.

Because the excited triplet state is lower in energy than the excited singlet state, the molecule cannot easily return to the excited singlet state. Nor can it easily return to the ground state by returning the excited electron to its original orbital. Once again, the transition $T_1 \rightarrow S_0$ would require a change of spin for the electron, and this is a forbidden process. Hence, the triplet excited state usually has a long lifetime (relative to other excited states) because it generally has nowhere to which it can easily go. Even though the process is forbidden, the triplet T_1 may eventually return to the ground state (S_0) by a process called a **radiationless transition.** In this process, the excess energy of the triplet is lost to the surrounding solution as heat, thereby "relaxing" the triplet back to the ground state (S_0). This process is the study of much current research and is not well understood. In the second process, in which a triplet state may revert to the ground state, **phosphorescence,** the excited triplet emits a photon to dissipate the excess energy and returns directly to the ground state. Although this process is "forbidden," it nevertheless occurs when there is no other open pathway by which the molecule can dissipate its excess energy. In benzophenone, radiationless decay is the faster process, with rate $k_d = 10^5 \text{ sec}^{-1}$, and phosphorescence, which is not observed, has a lower rate of $k_p = 10^2 \text{ sec}^{-1}$.

Benzophenone is a ketone. Ketones have *two* possible excited singlet states and, consequently, two excited triplet states as well. This occurs because two relatively low-energy transitions are possible in benzophenone. It is possible to excite one of the π electrons in the carbonyl π bond to the lowest-energy unoccupied orbital, a π^* orbital. It is also possible to excite one of the nonbonded or n electrons on oxygen to the same orbital. The first type of transition is called a $\pi-\pi^*$ transition, while the second is called an $n-\pi^*$ transition. These transitions and the states that result are illustrated pictorially.

Spectroscopic studies show that for benzophenone and most other ketones, the $n-\pi^*$ excited states S_1 and T_1 are of lower energy than the $\pi-\pi^*$ excited states. An energy diagram depicting the excited states of benzophenone (along with one that depicts those of naphthalene) is shown.

It is now known that the photoreduction of benzophenone is a reaction of the $n-\pi^*$ triplet state (T_1) of benzophenone. The $n-\pi^*$ excited states have radical character at the carbonyl oxygen atom because of the unpaired electron in the nonbonding orbital. Thus, the radical-like and energetic T_1 excited state species can abstract a hydrogen atom from

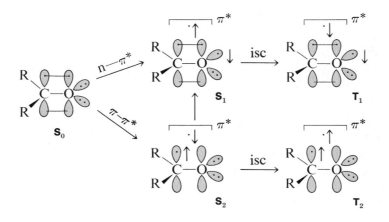

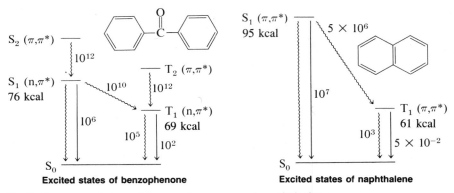

Excited energy states of benzophenone and naphthalene.

a suitable donor molecule to form the diphenylhydroxymethyl radical. Two of these radicals, once formed, may couple to form benzpinacol. The complete mechanism for photoreduction is outlined in the steps that follow.

$$Ph_2C{=}O \xrightarrow{h\nu} Ph_2\overset{\cdot}{C}{-}O{\cdot}(S_1)$$

$$Ph_2\overset{\cdot}{C}{-}O{\cdot}(S_1) \xrightarrow{\text{isc}} Ph_2\overset{\cdot}{C}{-}O{\cdot}(T_1)$$

$$Ph_2\overset{\cdot}{C}{-}O{\cdot} \curvearrowright H{-}\underset{\underset{CH_3}{|}}{\overset{\overset{CH_3}{|}}{C}}{-}OH \longrightarrow Ph_2\overset{\cdot}{C}{-}OH + {\cdot}\underset{\underset{CH_3}{|}}{\overset{\overset{CH_3}{|}}{C}}{-}OH$$
$$(T_1)$$

$$Ph_2\overset{\cdot}{C}{-}O{\cdot} \curvearrowright HO{-}\underset{\underset{CH_3}{|}}{\overset{\overset{CH_3}{|}}{C}}{\cdot} \longrightarrow Ph_2\overset{\cdot}{C}{-}OH + O{=}C\overset{\diagup CH_3}{\diagdown CH_3}$$
$$(T_1)$$

$$2\ Ph_2\overset{\cdot}{C}{-}OH \longrightarrow Ph{-}\underset{\underset{Ph}{|}}{\overset{\overset{OH}{|}}{C}}{-}\underset{\underset{Ph}{|}}{\overset{\overset{OH}{|}}{C}}{-}Ph$$

Many photochemical reactions must be carried out in a quartz apparatus because they require ultraviolet radiation of shorter wavelengths (higher energy) than the wavelengths that can pass through Pyrex. Benzophenone, however, requires radiation of approximately 350 nm to become excited to its $n{-}\pi^*$ singlet state S_1, a wavelength that readily passes through Pyrex. In the figure shown on page 447, the ultraviolet absorption spectra of benzophenone and naphthalene are given. Superimposed on their spectra are two curves, which show the wavelengths that can be transmitted by Pyrex and quartz, respectively. Pyrex will not allow any radiation of wavelength shorter than approximately 300 nm to pass, whereas quartz allows wavelengths as short as 200 nm to pass. Thus,

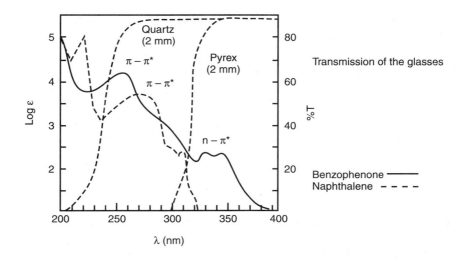

when benzophenone is placed in a Pyrex flask, the only electronic transition possible is the $n-\pi^*$ transition, which occurs at 350 nm.

However, even if it were possible to supply benzophenone with radiation of the appropriate wavelength to produce the second excited singlet state of the molecule, this singlet would rapidly convert to the lowest singlet state (S_1). The state S_2 has a lifetime of less than 10^{-12} second. The conversion process $S_2 \rightarrow S_1$ is called an **internal conversion.** Internal conversions are processes of conversion between excited states of the same multiplicity (singlet–singlet or triplet–triplet), and they usually are very rapid. Thus, when a S_2 or T_2 is formed, it readily converts to S_1 or T_1, respectively. As a consequence of their very short lifetimes, very little is known about the properties or the exact energies of S_2 and T_2 of benzophenone.

ENERGY TRANSFER

Using a simple **energy-transfer** experiment, one can show that the photoreduction of benzophenone proceeds via the T_1 excited state of benzophenone, rather than the S_1 excited state. If naphthalene is added to the reaction, the photoreduction is stopped because the excitation energy of the benzophenone triplet is transferred to naphthalene. The naphthalene is said to have **quenched** the reaction. This occurs in the following way.

When the excited states of molecules have long enough lifetimes, they often can transfer their excitation energy to another molecule. The mechanisms of these transfers are complex and cannot be explained here; however, the essential requirements can be outlined. First, for two molecules to exchange their respective states of excitation, the process must occur with an overall decrease in energy. Second, the spin multiplicity of the total system must not change. These two features can be illustrated by the two most common examples of energy transfer—singlet transfer and triplet transfer. In these two examples, the superscript 1 denotes an excited singlet state, the superscript 3 denotes a triplet state, and the subscript 0 denotes a ground-state molecule. The designations A and B represent different molecules.

$$A^1 + B_0 \longrightarrow B^1 + A_0 \qquad \text{Singlet energy transfer}$$

$$A^3 + B_0 \longrightarrow B^3 + A_0 \qquad \text{Triplet energy transfer}$$

In singlet energy transfer, excitation energy is transferred from the excited singlet state of A to a ground-state molecule of B, converting B to its excited singlet state and returning A to its ground state. In triplet energy transfer, there is a similar interconversion of excited state and ground state. Singlet energy is transferred through space by a dipole–dipole coupling mechanism, but triplet energy transfer requires the two molecules involved in the transfer to collide. In the usual organic medium, about 10^9 collisions occur per second. Thus, if a triplet state A^3 has a lifetime longer than 10^{-9} second, and if an acceptor molecule B_0, which has a lower triplet energy than that of A^3 is available, energy transfer can be expected. If the triplet A^3 undergoes a reaction (like photoreduction) at a rate lower than the rate of collisions in the solution, and if an acceptor molecule is added to the solution, the reaction can be *quenched*. The acceptor molecule, which is called a **quencher,** deactivates, or "quenches," the triplet before it has a chance to react. Naphthalene has the ability to quench benzophenone triplets in this way and to stop the photoreduction.

Naphthalene cannot quench the excited-state singlet S_1 of benzophenone because its own singlet has an energy (95 kcal/mol) that is higher than the energy of benzophenone (76 kcal/mol). In addition, the conversion $S_1 \rightarrow T_1$ is very rapid (10^{-10} second) in benzophenone. Thus, naphthalene can intercept only the triplet state of benzophenone. The triplet excitation energy of benzophenone (69 kcal/mol) is transferred to naphthalene ($T_1 = 61$ kcal/mol) in an exothermic collision. Finally, the naphthalene molecule does not absorb light of the wavelengths transmitted by Pyrex (see spectra on p. 447); therefore, benzophenone is not inhibited from absorbing energy when naphthalene is present in solution. Thus, since naphthalene quenches the photoreduction reaction of benzophenone, we can infer that this reaction proceeds via the triplet state T_1 of benzophenone. If naphthalene did not quench the reaction, the singlet state of benzophenone would be indicated as the reactive intermediate. In the following experiment, the photoreduction of benzophenone is attempted both in the presence and in the absence of added naphthalene.

Required Reading

Review: Technique 4   Section 4.3

Special Instructions

This experiment may be performed concurrently with some other experiment. It requires only 15 minutes during the first laboratory period and only about 15 minutes in a subsequent laboratory period about 1 week later (or at the end of the laboratory period if you use a sunlamp).

Using Direct Sunlight. It is important that the reaction mixture be left where it will receive direct sunlight. If it does not, the reaction will be slow and may need more than 1 week for completion. It is also important that the room temperature not be too low, or the benzophenone will precipitate. If you perform this experiment in the winter and the laboratory is not heated at night, you must shake the solutions every morning to redissolve the benzophenone. Benzpinacol should not redissolve easily.

Using a Sunlamp. If you wish, you may use a 275-W sunlamp instead of direct sunlight. Place the lamp in a hood that has had its window covered with aluminum foil (shiny side in). The lamp (or lamps) should be mounted in a ceramic socket attached to a ring stand with a three-pronged clamp.

> **Caution:** The purpose of the aluminum foil is to protect the eyes of persons in the laboratory. You should not view a sunlamp directly, or damage to the eyes may result. Take all possible viewing precautions.

Attach samples to a ring stand placed at least 18 inches from the sunlamp. Placing them at this distance will avoid their being heated by the lamp. Heating may cause loss of the solvent. It is a good idea to agitate the samples every 30 minutes. With a sunlamp, the reaction will be complete in 3–4 hours.

Waste Disposal

Dispose of the filtrate from the vacuum filtration procedure in the container designated for nonhalogenated organic wastes.

Procedure

Label two 13 × 100-mm test tubes near the top of the tubes. The labels should have your name and "No. 1" and "No. 2" written on them. Place 0.50 g of benzophenone in the first tube. Place 0.50 g of benzophenone and 0.05 g of naphthalene in the second tube. Add about 2 mL of 2-propanol (isopropyl alcohol) to each tube and warm them in a beaker of warm water to dissolve the solids. When the solids have dissolved, add one small drop (Pasteur pipet) of glacial acetic acid to each tube and then fill each tube nearly to the top with more 2-propanol. Stopper the tubes tightly with rubber stoppers, shake them well, and place them in a beaker on a windowsill where they will receive direct sunlight.

> **Note:** You may be directed by your instructor to use a sunlamp instead of direct sunlight (see Special Instructions).

The reaction requires about 1 week for completion (3 hours with a sunlamp). If the reaction has occurred during this period, the product will have crystallized from the solution. Observe the result in each test tube. Collect the product by vacuum filtration using a small Büchner or Hirsch funnel (Technique 4, Section 4.3, p. 640) and allow it to dry. Weigh the product and determine its melting point and percentage yield. At your instructor's option, determine the infrared spectrum of the benzpinacol as a KBr mull (Technique 19, Section 19.4, p. 846). Submit the product to your instructor with the report.

REFERENCE

Vogler, A., and Kunkely, H. "Photochemistry and Beer." *Journal of Chemical Education, 59* (January 1982): 25.

EXPERIMENT 53B

Synthesis of β-Benzopinacolone: The Acid-Catalyzed Rearrangement of Benzpinacol

The ability of carbocations to rearrange represents an important concept in organic chemistry. In Experiment 32B, pinacol was allowed to rearrange to pinacolone, under the influence of a strong acid. In this experiment, a similar rearrangement will be accomplished. The benzpinacol, prepared in Experiment 53A, will rearrange to **benzopinacolone (2,2,2-triphenylacetophenone)** under the influence of iodine in glacial acetic acid.

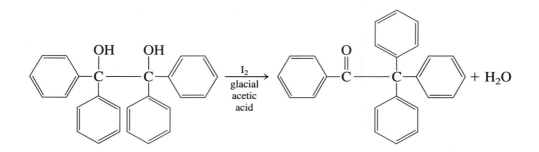

The product is isolated as a crystalline, white solid. Benzopinacolone is known to crystallize in two different crystalline forms, each with a different melting point. The **alpha** form has a melting point of 206–207°C, while the **beta** form melts at 182°C. The product formed in this experiment is the β-benzopinacolone.

Required Reading

Review: Technique 3   Section 3.4
          Technique 5   Section 5.3
          Experiment 32
          Appendices 3 and 4

Before beginning this experiment, you should read the material dealing with carbocation rearrangements in your lecture textbook.

Special Instructions

This experiment requires very little time and can be coscheduled with another short experiment.

Waste Disposal

All organic residues must be placed in the appropriate container for nonhalogenated organic waste.

Procedure

In a 25-mL round-bottom flask, add 5 mL of a 0.015 M solution of iodine dissolved in glacial acetic acid. Add 1 gram of benzpinacol and attach a water-cooled condenser. Using a small heating mantle, allow the solution to heat under reflux for 5 minutes. Crystals may begin to appear from the solution during this heating period.

Remove the heat source and allow the solution to cool slowly. The product will crystallize from the solution as it cools. When the solution has cooled to room temperature, collect the crystals by vacuum filtration using a small Büchner funnel. Rinse the crystals with three 2-mL portions of cold, glacial acetic acid. Allow the crystals to dry in the air overnight. Weigh the dried product and determine its melting point. Pure β-benzopinacolone melts at 182°C. Obtain an infrared and proton NMR spectrum of your product as a KBr pellet.

Calculate the percentage yield. Submit the product to your instructor in a labeled vial, along with your spectra. Interpret your spectra, showing how they are consistent with the rearranged structure of the product.

QUESTIONS

1. Can you think of a way to produce the benzophenone $n–\pi^*$ triplet T_1 *without* having benzophenone pass through its first singlet state? Explain.

2. A reaction similar to the one described here occurs when benzophenone is treated with the metal magnesium (pinacol reduction).

$$2\ Ph_2{=}CO \xrightarrow{Mg} \overset{\displaystyle \overset{OH}{|}\ \overset{OH}{|}}{Ph_2C{-}CPh_2}$$

Compare the mechanism of this reaction with the photoreduction mechanism. What are the differences?

3. Which of the following molecules do you expect would be useful in quenching benzophenone photoreduction? Explain.

Oxygen	$(S_1 = 22\ kcal/mol)$	Biphenyl $(T_1 = 66\ kcal/mol)$
9,10-Diphenylanthracene	$(T_1 = 42\ kcal/mol)$	Toluene $(T_1 = 83\ kcal/mol)$
trans-1,3-Pentadiene	$(T_1 = 59\ kcal/mol)$	Benzene $(T_1 = 84\ kcal/mol)$
Naphthalene	$(T_1 = 61\ kcal/mol)$	

E S S A Y

Fireflies and Photochemistry

The production of light as a result of a chemical reaction is called **chemiluminescence.** A chemiluminescent reaction generally produces one of the product molecules in an electronically excited state. The excited state emits a photon, and light is produced. If a reaction that produces light is biochemical, occurring in a living organism, the phenomenon is called **bioluminescence.**

The light produced by fireflies and other bioluminescent organisms has fascinated observers for many years. Many different organisms have developed the ability to emit light. They include bacteria, fungi, protozoans, hydras, marine worms, sponges, corals, jellyfish, crustaceans, clams, snails, squids, fish, and insects. Curiously, among the higher forms of life, only fish are included on the list. Amphibians, reptiles, birds, mammals, and the higher plants are excluded. Among the marine species, none is a freshwater organism. The excellent *Scientific American* article by McElroy and Seliger (see references) delineates the natural history, characteristics, and habits of many bioluminescent organisms.

The first significant studies of a bioluminescent organism were performed by the French physiologist Raphael Dubois in 1887. He studied the mollusk *Pholas dactylis,* a bioluminescent clam indigenous to the Mediterranean Sea. Dubois found that a cold-water extract of the clam was able to emit light for several minutes following the extraction. When the light emission ceased, it could be restored, he found, by a material extracted from the clam by hot water. A hot-water extract of the clam alone did not produce the luminescence. Reasoning carefully, Dubois concluded that there was an enzyme in the cold-water extract that was destroyed in hot water. The luminescent compound, however, could be extracted without destruction in either hot or cold water. He called the luminescent material **luciferin,** and the enzyme that induced it to emit light **luciferase;** both names were derived from *Lucifer,* a Latin name meaning "bearer of light." Today the luminescent materials from all organisms are called luciferins, and the associated enzymes are called luciferases.

The most extensively studied bioluminescent organism is the firefly. Fireflies are found in many parts of the world and probably represent the most familiar example of bioluminescence. In such areas, on a typical summer evening, fireflies, or "lightning bugs," can frequently be seen to emit flashes of light as they cavort over the lawn or in the garden. It is now universally accepted that the luminescence of fireflies is a mating device. The male firefly flies about 2 feet above the ground and emits flashes of light at regular intervals. The female, who remains stationary on the ground, waits a characteristic interval and then flashes a response. In return, the male reorients his direction of flight toward her and flashes a signal once again. The entire cycle is rarely repeated more than 5 to 10 times before the male reaches the female. Fireflies of different species can recognize one another by their flash patterns, which vary in number, rate, and duration among species.

Although the total structure of the luciferase enzyme of the American firefly *Photinus pyralis* is unknown, the structure of the luciferin has been established. In spite of a large amount of experimental work, however, the complete nature of the chemical reactions that produce the light is still subject to some controversy. It is possible, nevertheless, to outline the most salient details of the reaction.

$$\text{Luciferase} + \text{ATP} \longrightarrow \text{Luciferase}-\text{ATP}$$

Besides the luciferin and the luciferase, other substances—magnesium(II), ATP (adenosine triphosphate), and molecular oxygen—are needed to produce the luminescence. In the postulated first step of the reaction, the luciferase complexes with an ATP molecule. In the second step, the luciferin binds to the luciferase and reacts with the already bound ATP molecule to become "primed." In this reaction, pyrophosphate ion is expelled, and AMP (adenosine monophosphate) becomes attached to the carboxyl group of the luciferin. In the third step, the luciferin–AMP complex is oxidized by molecular oxygen to form a hydroperoxide; this cyclizes with the carboxyl group, expelling AMP and forming the cyclic endoperoxide. This reaction would be difficult if the carboxyl group of the luciferin had not been primed with ATP. The endoperoxide is unstable and readily decarboxylates, producing decarboxyketoluciferin in an *electronically excited state,* which is deactivated by the emission of a photon (fluorescence). Thus, it is the cleavage of the four-membered-ring endoperoxide that leads to the electronically excited molecule and hence the bioluminescence.

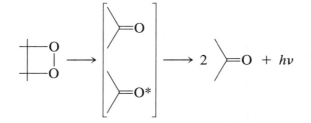

That one of the two carbonyl groups, either that of the decarboxyketoluciferin or that of the carbon dioxide, should be formed in an excited state can be readily predicted from the orbital symmetry conservation principles of Woodward and Hoffmann. This reaction is formally like the decomposition of a cyclobutane ring and yields two ethylene molecules. In analyzing the forward course of that reaction, that is, 2 ethylene → cyclobutane, one can easily show that the reaction, which involves four π electrons, is forbidden for two ground-state ethylenes but allowed for only one ethylene in the ground state and the other in an excited state. This suggests that, in the reverse process, one of the ethylene molecules should be formed in an excited state. Extending these arguments to the endoperoxide also suggests that one of the two carbonyl groups should be formed in its excited state.

The emitting molecule, decarboxyketoluciferin, has been isolated and synthesized. When it is excited photochemically by photon absorption in basic solution (pH > 7.5–8.0), it fluoresces, giving a fluorescence emission spectrum that is identical to the emission spectrum produced by the interaction of firefly luciferin and firefly luciferase. The emitting form of decarboxyketoluciferin has thus been identified as the **enol dianion.** In neutral or acidic solution, the emission spectrum of decarboxyketoluciferin does not match the emission spectrum of the bioluminescent system.

The exact function of the enzyme firefly luciferase is not yet known, but it is clear that all these reactions occur while luciferin is bound to the enzyme as a substrate. Also, because the enzyme undoubtedly has several basic groups (—COO⁻, —NH₂, and so on), the buffering action of those groups would easily explain why the enol dianion is also the emitting form of decarboxyketoluciferin in the biological system.

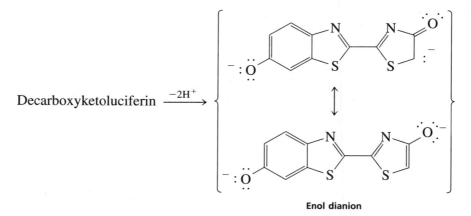

Enol dianion

Most chemiluminescent and bioluminescent reactions require oxygen. Likewise, most produce an electronically excited emitting species through the decomposition of a **peroxide** of one sort or another. In the experiment that follows, a **chemiluminescent** reaction that involves the decomposition of a peroxide intermediate is described.

REFERENCES

Clayton, R. K. *Light and Living Matter.* Vol. 2: *The Biological Part.* New York: McGraw-Hill, 1971. Chap. 6, "The Luminescence of Fireflies and Other Living Things."

Fox, J. L. "Theory May Explain Firefly Luminescence." *Chemical and Engineering News, 56* (March 6, 1978): 17.

Harvey, E. N. *Bioluminescence.* New York: Academic Press, 1952.

Hastings, J. W. "Bioluminescence." *Annual Review of Biochemistry, 37* (1968): 597.

McCapra, F. "Chemical Mechanisms in Bioluminescence." *Accounts of Chemical Research, 9* (1976): 201.

McElroy, W. D., and Seliger, H. H. "Biological Luminescence." *Scientific American, 207* (December 1962): 76.

McElroy, W. D., Seliger H. H., and White, E. H. "Mechanisms of Bioluminescence, Chemiluminescence and Enzyme Function in the Oxidation of Firefly Luciferin." *Photochemistry and Photobiology, 10* (1969): 153.

Seliger, H. H., and McElroy, W. D. *Light: Physical and Biological Action.* New York: Academic Press, 1965.

EXPERIMENT 54

Luminol

Chemiluminescence
Energy transfer
Reduction of a nitro group
Amide formation

In this experiment, the chemiluminescent compound **luminol,** or **5-aminophthal-hydrazide,** will be synthesized from 3-nitrophthalic acid.

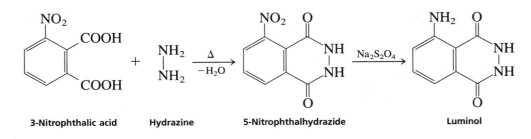

3-Nitrophthalic acid   Hydrazine   5-Nitrophthalhydrazide   Luminol

The first step of the synthesis is the simple formation of a cyclic diamide, 5-nitrophthalhydrazide, by reaction of 3-nitrophthalic acid with hydrazine. Reduction of the nitro group with sodium dithionite affords luminol.

In neutral solution, luminol exists largely as a dipolar anion (zwitterion). This dipolar ion exhibits a weak blue fluorescence after being exposed to light. However, in alkaline solution, luminol is converted to its dianion, which may be oxidized by molecular oxygen to give an intermediate that is chemiluminescent. The reaction is thought to have the following sequence:

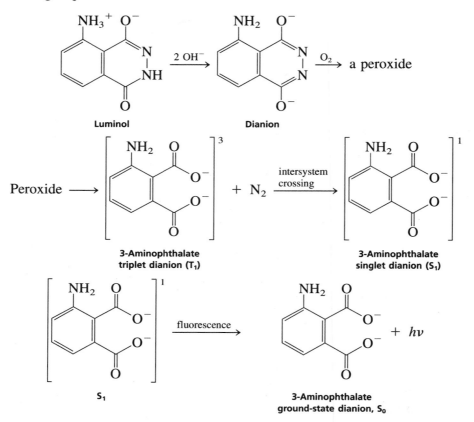

The dianion of luminol undergoes a reaction with molecular oxygen to form a peroxide of unknown structure. This peroxide is unstable and decomposes with the evolution of nitrogen gas, producing the 3-aminophthalate dianion in an electronically excited state. The excited dianion emits a photon that is visible as light. One very attractive hypothesis for the structure of the peroxide postulates a cyclic endoperoxide that decomposes by the following mechanism:

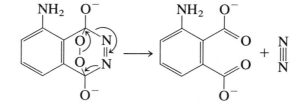

Certain experimental facts argue against this intermediate, however. For instance, certain acyclic hydrazides that cannot form a similar intermediate have also been found to be chemiluminescent.

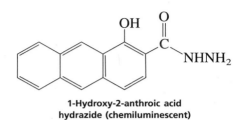

1-Hydroxy-2-anthroic acid
hydrazide (chemiluminescent)

Although the nature of the peroxide is still debatable, the remainder of the reaction is well understood. The chemical products of the reaction have been shown to be the 3-aminophthalate dianion and molecular nitrogen. The intermediate that emits light has been identified definitely as the *excited state singlet* of the 3-aminophthalate dianion.[1] Thus, the fluorescence emission spectrum of the 3-aminophthalate dianion (produced by photon absorption) is identical to the spectrum of the light emitted from the chemiluminescent reaction. However, for numerous complicated reasons, it is believed that the 3-aminophthalate dianion is formed first as a vibrationally excited triplet state molecule, which makes the intersystem crossing to the singlet state before emission of a photon.

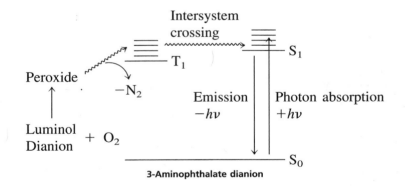

3-Aminophthalate dianion

The excited state of the 3-aminophthalate dianion may be quenched by suitable acceptor molecules, or the energy (about 50–80 kcal/mol) may be transferred to give emission from the acceptor molecules. Several such experiments are described in the following procedure.

The system chosen for the chemiluminescence studies of luminol in this experiment uses dimethylsulfoxide $(CH_3)_2SO$ as the solvent, potassium hydroxide as the base required for the formation of the dianion of luminol, and molecular oxygen. Several alternative systems have been used, substituting hydrogen peroxide and an oxidizing agent for molecular oxygen. An aqueous system using potassium ferricyanide and hydrogen peroxide is an alternative system used frequently.

---

[1]The terms *singlet, triplet, intersystem crossing, energy transfer,* and *quenching* are explained in Experiment 53.

REFERENCES

Rahaut, M. M. "Chemiluminescence from Concerted Peroxide Decomposition Reactions." *Accounts of Chemical Research,* 2 (1969): 80.

White, E. H., and Roswell, D. F. "The Chemiluminescence of Organic Hydrazides." *Accounts of Chemical Research,* 3 (1970): 54.

Required Reading

Review: Technique 3   Section 3.11

New: Essay   Fireflies and Photochemistry

Special Instructions

This entire experiment can be completed in about 1 hour. When you are working with hydrazine, you should remember that it is toxic and should not be spilled on the skin. It is also a suspected carcinogen. Dimethylsulfoxide may also be toxic; avoid breathing the vapors or spilling it on your skin.

A darkened room is required to observe adequately the chemiluminescence of luminol. A darkened hood that has had its window covered with butcher paper or aluminum foil also works well. Other fluorescent dyes besides those mentioned (for instance, 9,10-diphenylanthracene) can also be used for the energy-transfer experiments. The dyes selected may depend on what is immediately available. The instructor may have each student use one dye for the energy-transfer experiments, with one student making a comparison experiment without a dye.

Waste Disposal

Dispose of the filtrate from the vacuum filtration of 5-nitrophthalhydrazide in the waste container designated for nonhalogenated organic solvents. The filtrate from the vacuum filtration of 5-aminophthalhydrazide may be diluted with water and poured into the waste container designated for aqueous waste. The mixture containing potassium hydroxide, dimethylsulfoxide, and luminol should be placed in the special container designated for this material.

Procedure

PART A.   5-NITROPHTHALHYDRAZIDE

Place 0.60 g of 3-nitrophthalic acid and 0.8 mL of a 10% aqueous solution of hydrazine (use gloves) in a small (15 × 125-mm) sidearm test tube.[2] At the same time, heat

---

[2]A 10% aqueous solution of hydrazine can be prepared by diluting 15.6 g of a commercial 64% hydrazine solution to a volume of 100 mL using water.

8 mL of water in a beaker on a hotplate to about 80°C. Heat the test tube over a micro burner until the solid dissolves. Add 1.6 mL of triethylene glycol and clamp the test tube in an upright position on a ring stand. Place a thermometer (do not seal the system) and a boiling stone in the test tube and attach a piece of pressure tubing to the sidearm. Connect this tubing to an aspirator (use a trap). The thermometer bulb should be in the liquid as much as possible. Heat the solution with a micro burner until the liquid boils vigorously and the refluxing water vapor is drawn away by the aspirator vacuum (the temperature will rise to about 120°C). Continue heating and allow the temperature to increase rapidly until it rises just above 200°C. This heating requires 2–3 minutes, and you must watch the temperature closely to avoid heating the mixture well above 200°C. Remove the burner briefly when this temperature has been achieved and then resume gentle heating to maintain a fairly constant temperature of 220–230°C for about 3 minutes. Allow the test tube to cool to about 100°C, add the 8 mL of hot water that was prepared previously, and cool the test tube to room temperature by allowing tap water to flow over the outside of the test tube. Collect the brown crystals of 5-nitrophthalhydrazide by vacuum filtration, using a small Hirsch funnel. It is not necessary to dry the product before you go on with the next reaction step.

PART B.   LUMINOL (5-AMINOPHTHALHYDRAZIDE)

Transfer the moist 5-nitrophthalhydrazide to a 20 × 150-mm test tube. Add 2.6 mL of a 10% sodium hydroxide solution and agitate the mixture until the hydrazide dissolves. Add 1.6 g of sodium dithionite dihydrate (sodium hydrosulfite dihydrate, $Na_2S_2O_4 \cdot 2 H_2O$). Using a Pasteur pipet, add 2–4 mL of water to wash the solid from the walls of the test tube. Add a boiling stone to the test tube. Heat the test tube until the solution boils. Agitate the solution and maintain the boiling, continuing the agitation for at least 5 minutes. Add 1.0 mL of glacial acetic acid and cool the test tube to room temperature by allowing tap water to flow over the outside of it. Agitate the mixture during the cooling step. Collect the light yellow or gold crystals of luminol by vacuum filtration, using a small Hirsch funnel. Save a small sample of this product, allow it to dry overnight, and determine its melting point (mp 319–320°C). The remainder of the luminol may be used without drying for the chemiluminescence experiments. When drying the luminol, it is best to use a vacuum desiccator charged with calcium sulfate drying agent.

PART C.   CHEMILUMINESCENCE EXPERIMENTS

Caution: Be careful not to allow any of the mixture to touch your skin while shaking the flask. Hold the stopper securely.

Cover the bottom of a 10-mL Erlenmeyer flask with a layer of potassium hydroxide pellets. Add enough dimethylsulfoxide to cover the pellets. Add about

0.025 g of the moist luminol to the flask, stopper it, and shake it vigorously to mix air into the solution.[3] In a dark room a faint glow of bluish white light will be visible. The intensity of the glow will increase with continued shaking of the flask and occasional removal of the stopper to admit more air.

To observe energy transfer to a fluorescent dye, dissolve one or two crystals of the indicator dye in about 0.25 mL of water. Add the dye solution to the dimethylsulfoxide solution of luminol, stopper the flask, and shake the mixture vigorously. Observe the intensity and the color of the light produced.

A table of some dyes and the colors produced when they are mixed with luminol follows. Other dyes not included on this list may also be tested in this experiment.

Fluorescent Dye	Color
No dye	Faint bluish white
2,6-Dichloroindophenol	Blue
9-Aminoacridine	Blue-green
Eosin	Salmon pink
Fluorescein	Yellow-green
Dichlorofluorescein	Yellow-orange
Rhodamine B	Green
Phenolphthalein	Purple

EXPERIMENT 55

1,4-Diphenyl-1,3-butadiene

Wittig reaction
Working with sodium ethoxide
Thin-layer chromatography
UV/NMR spectroscopy

The Wittig reaction is often used to form alkenes from carbonyl compounds. In this experiment, the isomeric dienes *cis,trans,* and *trans,trans*-1,4-diphenyl-1,3-butadiene will

---

[3]An alternative method for demonstrating chemiluminescence, using potassium ferricyanide and hydrogen peroxide as oxidizing agents, is described in E. H. Huntress, L. N. Stanley, and A. S. Parker, *Journal of Chemical Education, 11* (1934): 142.

be formed from cinnamaldehyde and benzyltriphenylphosphonium chloride Wittig reagent. Only the *trans,trans* isomer will be isolated.

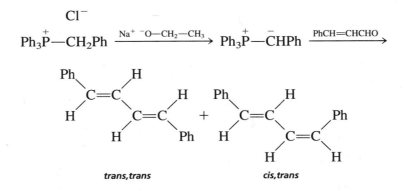

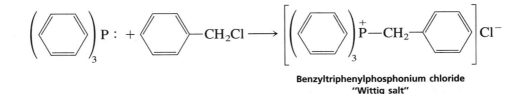

trans,trans   *cis,trans*

The reaction is carried out in two steps. First, the phosphonium salt is formed by the reaction of triphenylphosphine with benzyl chloride. The reaction is a simple nucleophilic displacement of chloride ion by triphenylphosphine. The salt that is formed is called the "Wittig reagent" or "Wittig salt."

Benzyltriphenylphosphonium chloride
"Wittig salt"

When treated with base, the Wittig salt forms an **ylide.** An ylide is a species having adjacent atoms oppositely charged. The ylide is stabilized due to the ability of phosphorus to accept more than eight electrons in its valence shell. Phosphorus uses its 3d orbitals to form the overlap with the 2p orbital of carbon that is necessary for resonance stabilization. Resonance stabilizes the carbanion.

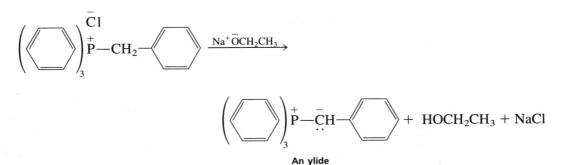

An ylide

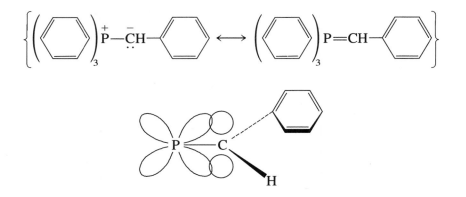

The ylide is a carbanion that acts as a nucleophile, and it adds to the carbonyl group in the first step of the mechanism. Following the initial nucleophilic addition, a remarkable sequence of events occurs, as outlined in the following mechanism:

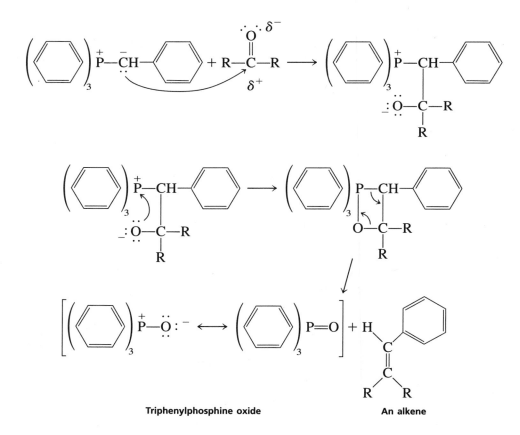

Triphenylphosphine oxide   An alkene

The addition intermediate, formed from the ylide and the carbonyl compound, cyclizes to form a four-membered-ring intermediate. This new intermediate is unstable and fragments into an alkene and triphenylphosphine oxide. Notice that the ring breaks open in a different way than it was formed. The driving force for this ring opening process is the formation of a very stable substance, triphenylphosphine oxide. A large decrease in potential energy is achieved upon the formation of this thermodynamically stable compound.

In this experiment, cinnamaldehyde is used as the carbonyl compound and yields mainly the *trans,trans*-1,4-diphenyl-1,3-butadiene which is obtained as a solid. The *cis,trans* isomer is formed in smaller amounts, but it is an oil that is not isolated in this experiment. The *trans,trans* isomer is the more stable isomer and is formed preferentially.

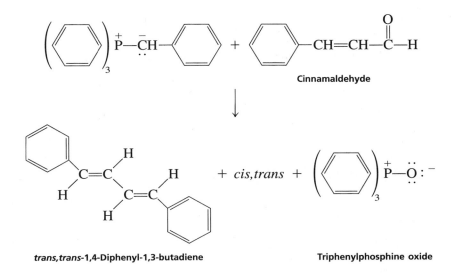

Cinnamaldehyde

trans,trans-1,4-Diphenyl-1,3-butadiene   Triphenylphosphine oxide

Required Reading

Review: Technique 4   Section 4.3
                Technique 14

Special Instructions

Your instructor may ask you to prepare 1,4-diphenyl-1,3-butadiene starting with commercially available benzyltriphenylphosphonium chloride. If so, start with Part B of this experiment. The prepared sodium ethoxide solution must be kept tightly stoppered when not in use, as it reacts readily with atmospheric water. Fresh cinnamaldehyde should be used in this experiment. Old cinnamaldehyde should be checked by infrared spectroscopy to be certain that it does not contain any cinnamic acid.

If your instructor asks you to prepare benzyltriphenylphosphonium chloride in the first part of this experiment, you can conduct another experiment concurrently during the 1.5-hour reflux period. Triphenylphosphine is rather toxic. Be careful not to inhale the dust. Benzyl chloride is a skin irritant and a lachrymator. It should be handled in the hood with care.

Waste Disposal

Place the alcohol, petroleum ether, and xylene wastes into the waste container for nonhalogenated organic solvents. Aqueous mixtures should be poured into the waste bottle designated for aqueous wastes.

Procedure

PART A.   BENZYLTRIPHENYLPHOSPHONIUM CHLORIDE (WITTIG SALT)

Place 2.2 g of triphenylphosphine (*MW* = 262.3) into a 100-mL round-bottom flask. In a hood transfer 1.44 mL of benzyl choride (*MW* = 126.6, *d* = 1.10 g/mL) to the flask and add 8 mL of xylenes (mixture of *o*-, *m*-, and *p*-isomers).

> **Caution:** Benzyl choride is a lachrymator, a tear-producing substance.

Add a magnetic stir bar to the flask and attach a water-cooled condenser. Using a heating mantle placed on top of a magnetic stirrer, boil the mixture for at least 1.5 hours. An increased yield may be expected when the mixture is heated for longer periods. The solution will be homogeneous at first, and then the Wittig salt will begin to precipitate. Maintain the stirring during the entire heating period, or bumping may occur. Following the reflux, remove the apparatus from the heating mantle and allow it to cool for a few minutes. Remove the flask and cool it thoroughly in an ice bath for about 5 minutes.

Collect the Wittig salt by vacuum filtration using a Büchner funnel. Use three 4-mL portions of cold petroleum ether (bp 60–90°C) to aid the transfer and to wash the crystals free of the xylene solvent. Dry the crystals, weigh them, and calculate the percentage yield of the Wittig salt. At the option of the instructor, obtain the proton NMR spectrum of the salt in $CDCl_3$. The methylene group appears as a doublet (J = 14 Hz) at 5.5 ppm because of ^{1}H-^{31}P coupling.

PART B.   1,4-DIPHENYL-1,3-BUTADIENE

In the following operations, stopper the round-bottom flask whenever possible to avoid contact with moisture from the atmosphere. If you prepared your own benzyltriphenylphosphonium chloride in Part A, you may need to supplement your yield in this part of the experiment.

Preparation of the Ylide. Place 1.92 g benzyltriphenylphosphonium chloride (*MW* = 388.9) in a *dry* 50-mL round-bottom flask. Add a magnetic stir bar. Transfer 8.0 mL absolute (anhydrous) ethanol to the flask and stir the mixture to dissolve the phosphonium salt (Wittig salt). Add 3.0 mL of sodium ethoxide solution[1] to the flask using a dry pipet, while stirring continuously. Stopper the flask and stir the mixture for 15 minutes. During this period, the cloudy solution acquires the characteristic yellow color of the ylide.

Reaction of the Ylide with Cinnamaldehyde.   Measure 0.60 mL of *pure* cinnamaldehyde (*MW* = 132.2, *d* = 1.11 g/mL) and place it in a small test tube. To the cinnamaldehyde, add 2.0 mL of absolute ethanol. Stopper the test tube until it is needed. After the 15-minute period, use a Pasteur pipet to mix the cinnamaldehyde with the ethanol and add this solution to the ylide in the round-bottom flask. A color change should be observed as the ylide reacts with the aldehyde and the product precipitates. Stir the mixture for 10 minutes.

Separation of the Isomers of 1,4-Diphenyl-1,3-butadiene.   Cool the flask thoroughly in an ice-water bath (10 min), stir the mixture with a spatula, and transfer the material from the flask to a small Büchner funnel under vacuum. Use two 4-mL portions of ice cold absolute ethanol to aid the transfer and to rinse the product. Dry the crystalline *trans,trans*-1,4-diphenyl-1,3-butadiene by drawing air through the solid. The product contains a small amount of sodium chloride that is removed as described in the next paragraph. The cloudy material in the filter flask contains triphenylphosphine oxide, the *cis,trans* isomer, and some *trans,trans* product. Pour the filtrate into a breaker and save it for the thin-layer chromatography experiment described in the next section.

Remove the *trans,trans*-1,4-diphenyl-1,3-butadiene from the filter paper, place the solid in a beaker, and add 12 mL of water. Stir the mixture and filter it on a Büchner funnel, under vacuum, to collect the nearly colorless crystalline *trans,trans* product. Use a minimum of water to aid the transfer. Allow the solid to dry thoroughly.

Analysis of the Filtrate.   Use thin-layer chromatography to analyze the filtrate that you saved in the previous section. This mixture must be analyzed as soon as possible so that the *cis,trans* isomer will not be photochemically converted to the *trans,trans* compound. Use a 2 × 8-cm silica gel TLC plate that has a fluorescent indicator (Eastman Chromatogram Sheet, No. 13181). At one position on the TLC plate, spot the filtrate, as is, without dilution. Dissolve a few crystals of the *trans,trans*-1,4-diphenyl-1,3-butadiene in a few drops of acetone and spot it at another position on the plate. Use petroleum ether (bp 60–90°C) as a solvent to develop (run) the plate.

---

[1]This reagent is prepared in advance by the instructor and will serve about 12 students. Carefully dry a 250-mL Erlenmeyer flask and insert a drying tube filled with calcium chloride into a one-hole rubber stopper. Obtain a large piece of sodium, clean it by cutting off the oxidized surface, weigh out a 2.30-g piece, cut it into 20 smaller pieces, and store it under xylene. Using tweezers remove each piece, wipe off the xylene, and add the sodium slowly over a period of about 30 minutes to 40 mL of absolute (anhydrous) ethanol in the 250-mL Erlenmeyer flask. After the addition of each piece, replace the stopper. The ethanol will warm as the sodium reacts, but do not cool the flask. After the sodium has been added, warm the solution and shake it *gently* until all the sodium reacts. Cool the sodium ethoxide solution to room temperature. This reagent may be prepared in advance of the laboratory period, but it must be stored in a refrigerator between laboratory periods. When it is stored in a refrigerator, it may be kept for about 3 days. Before using this reagent, bring it to room temperature and swirl it gently in order to redissolve any precipitated sodium ethoxide. Keep the flask stoppered between each use.

Visualize the spots with a UV lamp using both the long and short wavelength settings. The order of increasing R_f values is as follows: triphenylphosphine oxide, *trans,trans*-diene, *cis,trans*-diene. It is easy to identify the spot for the *trans,trans* isomer because it fluoresces brilliantly. What conclusion can you make about the contents of the filtrate and the purity of the *trans,trans* product? Report the results that you obtain, including R_f values and the appearance of the spots under illumination. Discard the filtrate in the container designated for nonhalogenated waste.

Yield Calculation and Melting Point Determination.   When the *trans,trans*-1,4-diphenyl-1,3-butadiene is dry, determine the melting point (literature, 152°C). Weigh the solid and determine the percentage yield. If the melting point is below 145°C, recrystallize a portion of the compound from hot 95% ethanol. Redetermine the melting point.

Spectroscopy (Optional).   Obtain the proton NMR spectrum in $CDCl_3$ or the UV spectrum in hexane. For the UV spectrum of the product, dissolve a 10-mg sample in 100 mL of hexane in a volumetric flask. Remove 10 mL of this solution and dilute it to 100 mL in another volumetric flask. This concentration should be adequate for analysis. The *trans,trans* isomer absorbs at 328 nm and possesses fine structure, while the *cis,trans* isomer absorbs at 313 nm and has a smooth curve.[2] See if your spectrum is consistent with these observations. Submit the spectral data with your laboratory report.

QUESTIONS

1. There is an additional isomer of 1,4-diphenyl-1,3-butadiene (mp 70°C), which has not been shown in this experiment. Draw the structure and name it. Why is it not produced in this experiment?

2. Why should the *trans,trans* isomer be the thermodynamically most stable one?

3. A lower yield of phosphonium salt is obtained in refluxing benzene than in xylene. Look up the boiling points for these solvents and explain why the difference in boiling points might influence the yield.

4. Outline a synthesis for *cis* and *trans* stilbene (the 1,2-diphenylethenes) using the Wittig reaction.

5. The sex attractant of the female housefly (*Musca domestica*) is called **muscalure,** and its structure follows. Outline a synthesis of muscalure, using the Wittig reaction. Will your synthesis lead to the required *cis* isomer?

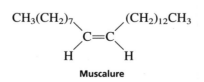

Muscalure

---

[2]The comparative study of the stereoisomeric 1,4-diphenyl-1,3-butadienes has been published: J. H. Pinkard, B. Wille, and L. Zechmeister, *Journal of the American Chemical Society, 70* (1948): 1938.

ESSAY

Polymers and Plastics

Chemically, plastics are composed of chain-like molecules of high molecular weight called **polymers.** Polymers have been built up from simpler chemicals called **monomers.** The word *poly* is defined as "many," *mono* means "one," and *mer* indicates "units." Thus, many monomers are combined to give a polymer. A different monomer or combination of monomers is used to manufacture each type or family of polymers. There are two broad classes of polymers: addition and condensation. Both types are described here.

Many polymers (plastics) produced in the past were of such low quality that they gained a bad reputation. The plastics industry now produces high-quality materials that are increasingly replacing metals in many applications. They find use in many products such as clothes, toys, furniture, machine components, paints, boats, automobile parts, and even artificial organs. In the automobile industry, metals have been replaced with plastics to help reduce the overall weight of the car and to help reduce corrosion. This reduction in weight helps to improve gas mileage. Epoxy resins can even replace metal in engine parts.

CHEMICAL STRUCTURES OF POLYMERS

Basically, a polymer is made up of many repeating molecular units formed by sequential addition of monomer molecules to one another. Many monomer molecules of A, say 1,000 to 1 million, can be linked to form a gigantic polymeric molecule:

$$\text{Many A} \longrightarrow \text{etc.—A-A-A-A-A—etc.} \qquad \text{or} \qquad \text{(A)}_n$$

Monomer   Polymer
molecules   molecule

Monomers that are different can also be linked to form a polymer with an alternating structure. This type of polymer is called a **copolymer.**

$$\text{Many A + many B} \longrightarrow \text{etc.—A-B-A-B-A-B—etc.} \qquad \text{or} \qquad \text{(A-B)}_n$$

Monomer   Polymer
molecules   molecule

TYPES OF POLYMERS

For convenience, chemists classify polymers in several main groups, depending on the method of synthesis.

1. **Addition polymers** are formed by a reaction in which monomer units simply add to one another to form a long-chain (generally linear or branched) polymer. The monomers usu-

ally contain carbon–carbon double bonds. Examples of synthetic addition polymers include polystyrene (Styrofoam), polytetrafluoroethylene (Teflon), polyethylene, polypropylene, polyacrylonitrile (Orlon, Acrilan, Creslan), poly(vinyl chloride) (PVC), and poly(methyl methacrylate) (Lucite, Plexiglas). The process can be represented as follows:

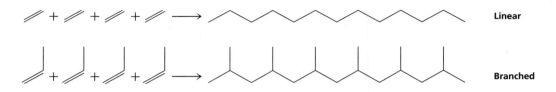

2. **Condensation polymers** are formed by the reaction of bifunctional or polyfunctional molecules, with the elimination of some small molecule (such as water, ammonia, or hydrogen chloride) as a by-product. Familiar examples of synthetic condensation polymers include polyesters (Dacron, Mylar), polyamides (nylon), polyurethanes, and epoxy resin. Natural condensation polymers include polyamino acids (protein), cellulose, and starch. The process can be represented as follows:

3. **Cross-linked polymers** are formed when long chains are linked in one gigantic, three-dimensional structure with tremendous rigidity. Addition and condensation polymers can exist with a cross-linked network, depending on the monomers used in the synthesis. Familiar examples of cross-linked polymers are Bakelite, rubber, and casting (boat) resin. The process can be represented as follows:

~~~~~~~~~~~~~~~   ⟶   ‡‡‡

Linear   **Cross-linked**

</div>

THERMAL CLASSIFICATION OF POLYMERS

Industrialists and technologists often classify polymers as either thermoplastics or thermoset plastics rather than as addition or condensation polymers. This classification takes into account their thermal properties.

1. **Thermal properties of thermoplastics.** Most addition polymers and many condensation polymers can be softened (melted) by heat and reformed (molded) into other shapes. Industrialists and technologists often refer to these types of polymers as **thermoplastics.** Weaker, noncovalent bonds (dipole–dipole and London dispersion) are broken during the heating. Technically, thermoplastics are the materials we call plastics. Thermoplastics may be repeatedly melted and recast into new shapes. They may be recycled as long as degradation does not occur during reprocessing.

Some addition polymers, such as poly(vinyl chloride), are difficult to melt and process. Liquids with high boiling points, such as dibutyl phthalate, are added to the polymer to separate the chains from each other. These compounds are called **plasticizers.** In effect, they act as lubricants that neutralize the attractions that exist between chains. As a result, the polymer can be melted at a lower temperature to aid in processing. In addition, the polymer becomes more flexible at room temperature. By varying the amount of plasticizer, poly(vinyl chloride) can range from a very flexible, rubber-like material to a very hard substance.

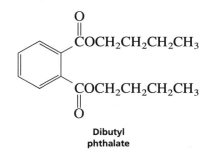

Dibutyl phthalate

Phthalate plasticizers are volatile compounds of low molecular weight. Part of the new car smell comes from the odor of these materials as they evaporate from the vinyl upholstery. The vapor often condenses on the windshield as an oily film. After some time, the vinyl material may lose enough plasticizer to cause it to crack.

2. **Thermal properties of thermoset plastics.** Industrialists use the term **thermoset** plastics to describe materials that melt initially but on further heating become permanently hardened. Once formed, thermoset materials cannot be softened and remolded without destruction of the polymer, because covalent bonds are broken. Thermoset plastics cannot be recycled. Chemically, thermoset plastics are cross-linked polymers. They are formed when long chains are linked in one gigantic three-dimensional structure with tremendous rigidity.

Polymers can also be classified in other ways; for example, many varieties of rubber are often referred to as elastomers. Dacron is a fiber, and poly(vinyl acetate) is an adhesive. The addition and condensation classifications are used in this essay.

ADDITION POLYMERS

By volume, most of the polymers prepared in industry are of the addition type. The monomers generally contain a carbon–carbon double bond. The most important example of an addition polymer is the well-known polyethylene, for which the monomer is ethylene. Countless numbers (n) of ethylene molecules are linked in long-chain polymeric molecules by breaking the pi bond and creating two new single bonds between the monomer units. The number of recurring units may be large or small, depending on the polymerization conditions.

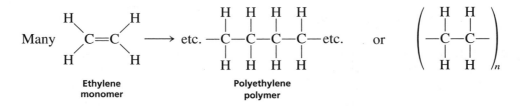

Ethylene monomer   **Polyethylene polymer**

This reaction can be promoted by heat, pressure, and a chemical catalyst. The molecules produced in a typical reaction vary in the number of carbon atoms in their chains. In other words, a mixture of polymers of varying length, rather than a pure compound, is produced.

Polyethylenes with linear structures can pack together easily and are referred to as high-density polyethylenes. They are fairly rigid materials. Low-density polyethylenes consist of branched-chain molecules, with some cross-linking in the chains. They are more flexible than the high-density polyethylenes. The reaction conditions and the catalysts that produce polyethylenes of low and high density are quite different. The monomer, however, is the same in each case.

Another example of an addition polymer is polypropylene. In this case, the monomer is propylene. The polymer that results has a branched methyl on alternate carbon atoms of the chain.

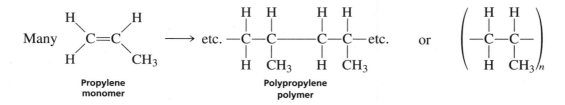

Propylene monomer   **Polypropylene polymer**

A number of common addition polymers are shown in Table One. Some of their principal uses are also listed. The last three entries in the table all have a carbon–carbon double bond remaining after the polymer is formed. These bonds activate or participate in a further reaction to form cross-linked polymers called elastomers; this term is almost synonymous with *rubber*, because elastomers designate materials with common characteristics.

CONDENSATION POLYMERS

Condensation polymers, for which the monomers contain more than one type of functional group, are more complex than addition polymers. Moreover, most condensation polymers are copolymers made from more than one type of monomer. Recall that addition polymers, in contrast, are all prepared from substituted ethylene molecules. The single functional group in each case is one or more double bonds, and a single type of monomer is generally used.

TABLE ONE Addition Polymers

Example	Monomer(s)	Polymer	Use
Polyethylene	$CH_2{=}CH_2$	$-CH_2-CH_2-$	Most common and important polymer; bags, insulation for wires, squeeze bottles

Polypropylene	$CH_2{=}CH$
	       $\mid$
	      CH_3
		$-CH_2-CH-$
		         $\mid$
		        CH_3
			Fibers, indoor–outdoor carpets, bottles

Polystyrene	$CH_2{=}CH$	$-CH_2-CH-$	Styrofoam, inexpensive household goods, inexpensive molded objects

Poly(vinyl chloride) (PVC)	$CH_2{=}CH$
	       $\mid$
	      Cl
		$-CH_2-CH-$
		         $\mid$
		        Cl
			Synthetic leather, clear bottles, floor covering, water pipe

Polytetrafluoroethylene (Teflon)	$CF_2{=}CF_2$	$-CF_2-CF_2-$	Nonstick surfaces, chemically resistant films

Poly(methyl methacrylate) (Lucite, Plexiglas)	CO_2CH_3
	       $\mid$
	$CH_2{=}C$
	      $\mid$
	     CH_3
		CO_2CH_3
		         $\mid$
		$-CH_2-C-$
		         $\mid$
		        CH_3
			Unbreakable "glass," latex paints

Polyacrylonitrile (Orlon, Acrilan, Creslan)	$CH_2{=}CH$
	       $\mid$
	      CN
		$-CH_2-CH-$
		         $\mid$
		        CN
			Fiber used in sweaters, blankets, carpets

Poly(vinyl acetate) (PVA)	$CH_2{=}CH$
	       $\mid$
	      $OCCH_3$
	        $\parallel$
	       O
		$-CH_2-CH-$
		         $\mid$
		        $OCCH_3$
		          $\parallel$
		        O
			Adhesives, latex paints, chewing gum, textile coatings

Natural rubber	CH_3
	       $\mid$
	$CH_2{=}CCH{=}CH_2$
		CH_3
		         $\mid$
		$-CH_2-C{=}CH-CH_2-$
			The polymer is cross-linked with sulfur (vulcanization)

Polychloroprene (neoprene rubber)	Cl
	       $\mid$
	$CH_2{=}CCH{=}CH_2$
		Cl
		         $\mid$
		$-CH_2-C{=}CH-CH_2-$
			Cross-linked with ZnO; resistant to oil and gasoline

Styrene butadiene rubber (SBR)	$CH_2{=}CH$ $CH_2{=}CHCH{=}CH_2$	$-CH_2CH-CH_2CH{=}CHCH_2-$	Cross-linked with peroxides; most common rubber; used for tires; 25% styrene, 75% butadiene

Dacron, a polyester, can be prepared by causing a dicarboxylic acid to react with a bifunctional alcohol (a diol):

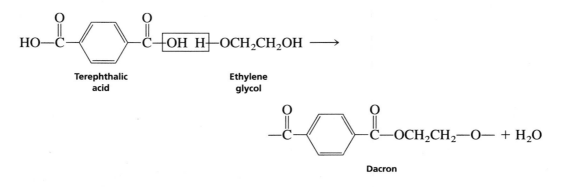

Terephthalic
acid

Ethylene
glycol

Dacron

Nylon 6-6, a polyamide, can be prepared by causing a dicarboxylic acid to react with a bifunctional amine.

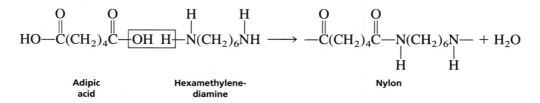

Adipic
acid

Hexamethylene-
diamine

Nylon

Notice, in each case, that a small molecule, water, is eliminated as a product of the reaction. Several other condensation polymers are listed in Table Two. Linear (or branched) chain polymers as well as cross-linked polymers are produced in condensation reactions.

The nylon structure contains the amide linkage at regular intervals,

This type of linkage is extremely important in nature because of its presence in proteins and polypeptides. Proteins are gigantic polymeric substances made up of monomer units of amino acids. They are linked by the peptide (amide) bond.

Other important natural condensation polymers are starch and cellulose. They are polymeric materials made up of the sugar monomer glucose. Another important natural condensation polymer is the DNA molecule. A DNA molecule is made up of the sugar deoxyribose linked with phosphates to form the backbone of the molecule. A portion of a DNA molecule is shown in the essay that precedes Experiment 5.

TABLE TWO Condensation Polymers

Example	Monomers	Polymer	Uses
Polyamides (nylon)	$HOC(CH_2)_nCOH$ (with two C=O) $H_2N(CH_2)_nNH_2$	$-C(CH_2)_nC-NH(CH_2)_nNH-$ (with two C=O)	Fibers, molded objects
Polyesters (Dacron, Mylar, Fortrel)	$HOC-\langle benzene\rangle-COH$ (two C=O) $HO(CH_2)_nOH$	$-C-\langle benzene\rangle-C-O(CH_2)_nO-$ (two C=O)	Linear polyesters, fibers, recording tape
Polyesters (Glyptal resin)	phthalic anhydride $HOCH_2CHCH_2OH$ with OH	$-COCH_2CHCH_2O-$ (with C=O groups)	Cross-linked polyester, paints
Polyesters (casting resin)	$HOCCH=CHCOH$ (two C=O) $HO(CH_2)_nOH$	$-CCH=CHC-O(CH_2)_nO-$ (two C=O)	Cross-linked with styrene and peroxide, fiberglass boat resin
Phenol-formaldehyde resin (Bakelite)	phenol (OH on benzene) $CH_2=O$	cross-linked $-CH_2-$ phenol network with OH, CH_2 bridges	Mixed with fillers, molded electrical goods, adhesives, laminates, varnishes
Cellulose acetate*	glucose ring with CH_2OH, OH, OH CH_3COOH	glucose ring with CH_2OAc, OAc, OAc	Photographic film
Silicones	$Cl-Si-Cl$ (with CH_3 above and below) H_2O	$-O-Si-O-$ (with CH_3 above and below)	Water-repellent coatings, temperature-resistant fluids and rubbers (CH_3SiCl_3 cross-links in water)
Polyurethanes	toluene diisocyanate (CH_3, $N=C=O$, $N=C=O$) $HO(CH_2)_nOH$	$NHC-O(CH_2)_nO-$ (with CH_3 and two urethane linkages, C=O)	Rigid and flexible foams, fibers

*Cellulose, a polymer of glucose, is used as the monomer.

DISPOSABILITY PROBLEMS

What do we do with all our wastes? Currently, the most popular method is to bury our garbage in sanitary landfills. However, as we run out of good places to bury our garbage, incineration has become a more attractive method for solving the solid waste

TABLE THREE Code System for Plastic Materials

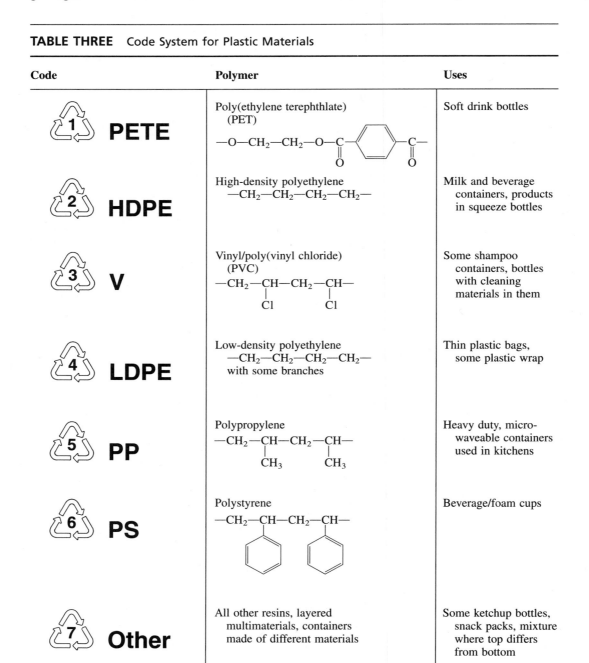

Code	Polymer	Uses
1 **PETE**	Poly(ethylene terephthlate) (PET)	Soft drink bottles
2 **HDPE**	High-density polyethylene	Milk and beverage containers, products in squeeze bottles
3 **V**	Vinyl/poly(vinyl chloride) (PVC)	Some shampoo containers, bottles with cleaning materials in them
4 **LDPE**	Low-density polyethylene with some branches	Thin plastic bags, some plastic wrap
5 **PP**	Polypropylene	Heavy duty, micro-waveable containers used in kitchens
6 **PS**	Polystyrene	Beverage/foam cups
7 **Other**	All other resins, layered multimaterials, containers made of different materials	Some ketchup bottles, snack packs, mixture where top differs from bottom

problem. Plastics, which compose about 2% of our garbage, burn readily. The new high-temperature incinerators are extremely efficient and can be operated with very little air pollution. It should also be possible to burn our garbage and generate electrical power from it.

Ideally, we should either recycle all our wastes or not produce the waste in the first place. Plastic waste consists of about 55% polyethylene and polypropylene, 20% polystyrene, and 11% PVC. All these polymers are thermoplastics and can be recycled. They can be resoftened and remolded into new goods. Unfortunately, thermosetting plastics (cross-linked polymers) cannot be remelted. They decompose on high-temperature heating. Thus, thermosetting plastics should not be used for "disposable" purposes. To recycle plastics effectively, we must sort the materials according to the various types. The plastics industry has introduced a code system consisting of seven categories for the common plastic used in packaging. The code is conveniently stamped on the bottom of the container. Using these codes, consumers can separate the plastics into groups for recycling purposes. These codes are listed in Table Three, together with the most common uses around the home. Notice that the seventh category is a miscellaneous one, called Other.

It is quite amazing that so few plastics are in packaging. The most common ones are polyethylene (low and high density), polypropylene, polystyrene, and poly(ethylene terephthlate). All these materials can easily be recycled because they are thermoplastics. Incidently, vinyls (polyvinyl chloride) are becoming less common in packaging. The "other" category, Code 7, is virtually nonexistent and usually consists of packaging where the top is made of a different material than the bottom. This dilemma should be easy to solve by placing the appropriate code on each part of the container.

Polymers, if they are well made, will not corrode or rust, and they last almost indefinitely. Unfortunately, these desirable properties also lead to a problem when plastics are buried in a landfill or thrown on the landscape—they do not decompose. Research is being undertaken to discover plastics that are biodegradable or photodegradable, so that either microorganisms or light from the sun can decompose our litter and garbage. While there are some advantages to this approach, it is probably better to eliminate packaging at the source or to engage in an effective recycling program. We must learn to use plastics wisely.

REFERENCES

Ainsworth, S. J. "Plastics Additives." *Chemical Engineering News, 70* (August 31, 1992): 34–55.

Burfield, D. R. "Polymer Glass Transition Temperatures." *Journal of Chemical Education, 64* (1987): 875.

Carraher, C. E., Jr., Hess, G., and Sperling, L. H. "Polymer Nomenclature—Or What's in a Name?" *Journal of Chemical Education, 64* (1987): 36.

Carraher, C. E., Jr., and Seymour, R. B. "Physical Aspects of Polymer Structure: A Dictionary of Terms." *Journal of Chemical Education, 63* (1986): 418.

Carraher, C. E., Jr., and Seymour, R. B. "Polymer Properties and Testing—Definitions." *Journal of Chemical Education, 64* (1987): 866.

Carraher, C. E., Jr., and Seymour, R. B. "Polymer Structure—Organic Aspects (Definitions)" *Journal of Chemical Education, 65* (1988): 314.

Fried, J. R. "Elastomers and Thermosets." *Plastics Engineering* (March 1983): 67–73.

Fried, J. R. "Molecular Weight and Its Relation to Properties." *Plastics Engineering* (August 1982): 27–33.

Fried, J. R. "The Polymers of Commercial Plastics." *Plastic Engineering* (June 1982): 49–55.

Fried, J. R. "Polymer Properties in the Solid State." *Plastics Engineering* (July 1982): 27–37.

Fried, J. R., and Yeh, E. B. "Polymers and Computer Alchemy." *Chemtech, 23* (March 1993): 35–40.

Goodall, B. L. "The History and Current State of the Art of Propylene Polymerization Catalysts." *Journal of Chemical Education, 63* (1986): 191.

Harris, F. W., et al. "State of the Art: Polymer Chemistry." *Journal of Chemical Education, 58* (November 1981). This issue contains 17 papers on polymer chemistry. The series covers structures, properties, mechanisms of formation, methods of preparation, stereochemistry, molecular weight distribution, rheological behavior of polymer melts, mechanical properties, rubber elasticity, block and graft copolymers, organometallic polymers, fibers, ionic polymers, and polymer compatibility.

Jordan, R. F. "Cationic Metal—Alkyl Olefin Polymerization Catalysts." *Journal of Chemical Education, 65* (1988): 285.

Kauffman, G. B. "Rayon: The First Semi-Synthetic Fiber Product." *Journal of Chemical Education, 70* (1993): 887.

Kauffman, G. B. "Wallace Hume Carothers and Nylon, The First Completely Synthetic Fiber." *Journal of Chemical Education, 65* (1988): 803.

Kauffman, G. B., and Seymour, R. B. "Elastomers I. Natural Rubber." *Journal of Chemical Education, 67* (1990): 422.

Kauffman, G. B., and Seymour, R. B. "Elastomers II. Synthetic Rubbers." *Journal of Chemical Education, 68* (1991): 217.

Marvel, C. S. "The Development of Polymer Chemistry in America—The Early Days." *Journal of Chemical Education, 58* (1981): 535.

Quirk, R. P. "Stereochemistry and Macromolecules: Principles and Applications." *Journal of Chemical Education, 58* (1981): 540.

Seymour, R. B. "Alkenes and Their Derivatives: The Alchemists' Dream Come True." *Journal of Chemical Education, 66* (1989): 670.

Seymour, R. B. "Polymers Are Everywhere." *Journal of Chemical Education, 65* (1988): 327.

Seymour, R. B., and Kauffman, G. B. "Polymer Blends: Superior Products from Inferior Materials." *Journal of Chemical Education, 69* (1992): 646.

Seymour, R. B., and Kauffman, G. B. "Polyurethanes: A Class of Modern Versatile Materials." *Journal of Chemical Education, 69* (1992): 909.

Seymour, R. B., and Kauffman, G. B. "The Rise and Fall of Celluloid." *Journal of Chemical Education 69* (1992): 311.

Seymour, R. B., and Kauffman, G. B. "Thermoplastic Elastomers." *Journal of Chemical Education, 69* (1992): 967.

Stevens, M. P. "Polymer Additives: Mechanical Property Modifiers." *Journal of Chemical Education, 70* (1993): 444.

Stevens, M. P. "Polymer Additives: Chemical and Aesthetic Property Modifiers." *Journal of Chemical Education, 70* (1993): 535.

Stevens, M. P. "Polymer Additives: Surface Property and Processing Modifiers." *Journal of Chemical Education, 70* (1993): 713.

Waller, F. J. "Fluoropolymers." *Journal of Chemical Education, 66* (1989): 487.

Webster, O. W. "Living Polymerization Methods." *Science, 251* (1991): 887.

EXPERIMENT 56

Preparation and Properties of Polymers: Polyester, Nylon, and Polystyrene

Condensation polymers
Addition polymers
Cross-linked polymers
Infrared spectroscopy

In this experiment, the syntheses of two polyesters (Experiment 56A), nylon (Experiment 56B), and polystyrene (Experiment 56C) will be described. These polymers represent important commercial plastics. They also represent the main classes of polymers: condensation (linear polyester, nylon), addition (polystyrene), and cross-linked (Glyptal polyester). Infrared spectroscopy is used in Experiment 56D to determine the structure of polymers.

Required Reading

Review: Appendix 3

New: Essay   Polymers and Plastics

Special Instructions

Experiments 56A, 56B, and 56C all involve toxic vapors. Each experiment should be conducted in a good hood. The styrene used in Experiment 56C irritates the skin and eyes. Avoid breathing its vapors. Styrene must be dispensed and stored in a hood. Benzoyl peroxide is flammable and may detonate on impact or on heating.

Waste Disposal

The test tubes containing the polyester polymers from Experiment 56A should be placed in a box designated for disposal of these samples. The nylon from Experiment 56B should be washed thoroughly with water and placed in a waste basket. The liquid wastes from Experiment 56B (nylon) should be poured into a container designated for disposal of these wastes. The polystyrene prepared in Experiment 56C should be placed in the container designated for solid wastes.

EXPERIMENT 56A

Polyesters

Linear and cross-linked polyesters will be prepared in this experiment. Both are examples of condensation polymers. The linear polyester is prepared as follows:

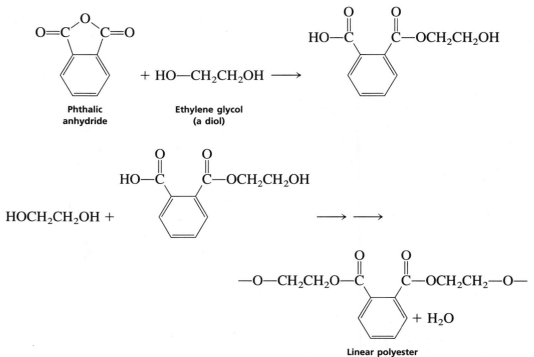

Phthalic anhydride   **Ethylene glycol (a diol)**

Linear polyester

This linear polyester is isomeric with Dacron, which is prepared from terephthalic acid and ethylene glycol (see the preceding essay). Dacron and the linear polyester made in this experiment are both thermoplastics.

If more than two functional groups are present in one of the monomers, the polymer chains can be linked to one another (cross-linked) to form a three-dimensional network. Such structures are usually more rigid than linear structures and are useful in making paints and coatings. They may be classified as thermosetting plastics. The polyester Glyptal is prepared as follows:

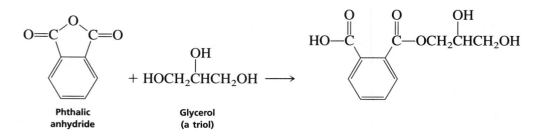

Phthalic anhydride   **Glycerol (a triol)**

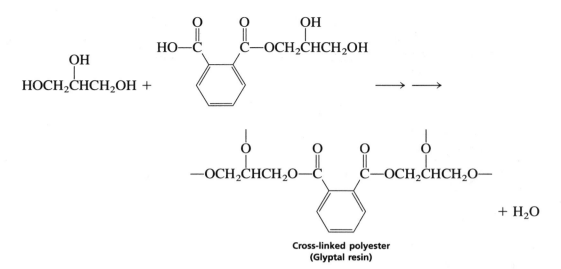

**Cross-linked polyester
(Glyptal resin)**

The reaction of phthalic anhydride with a diol (ethylene glycol) is described in the procedure. This linear polyester is compared with the cross-linked polyester (Glyptal) prepared from phthalic anhydride and a triol (glycerol).

Procedure

Place 1 g of phthalic anhydride and 0.05 g of sodium acetate in each of two test tubes. To one tube add 0.4 mL of ethylene glycol and to the other add 0.4 mL of glycerol. Clamp both tubes so that they can be heated simultaneously with a flame. Heat the tubes gently until the solutions appear to boil (water is eliminated during the esterification); then continue heating for 5 minutes.

If you are performing the optional infrared analysis of the polymer, immediately save a sample of the polymer formed from ethylene glycol only. After removing a sample for infrared spectroscopy, allow the two test tubes to cool and compare the viscosity and brittleness of the two polymers. The test tubes cannot be cleaned.

Infrared Spectroscopy (Optional). Lightly coat a watch glass with stopcock grease. Pour some of the *hot* polymer from the tube containing ethylene glycol; use a wooden applicator stick to spread the polymer on the surface so as to create a thin film of the polymer. Remove the polymer from the watch glass and save it for Experiment 56D.

E X P E R I M E N T   5 6 B

Polyamide (Nylon)

Reaction of a dicarboxylic acid, or one of its derivatives, with a diamine leads to a linear polyamide through a condensation reaction. Commercially, nylon 6-6 (so called because each monomer has six carbons) is made from adipic acid and hexamethylenediamine. In this experiment, you will use the acid chloride instead of adipic acid. The acid chloride is dissolved in cyclohexane and this is added *carefully* to hexamethylenediamine dissolved in water. These liquids do not mix, so two layers will form. The polymer can then be drawn out continuously to form a long strand of nylon. Imagine how many molecules have been linked in this long strand! It is a fantastic number.

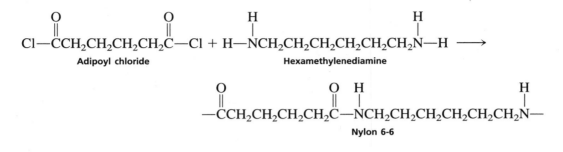

Procedure

Pour 10 mL of a 5% aqueous solution of hexamethylenediamine (1,6-hexanediamine) into a 50-mL beaker. Add 10 drops of 20% sodium hydroxide solution. Carefully add 10 mL of a 5% solution of adipoyl chloride in cyclohexane to the solution by pour-

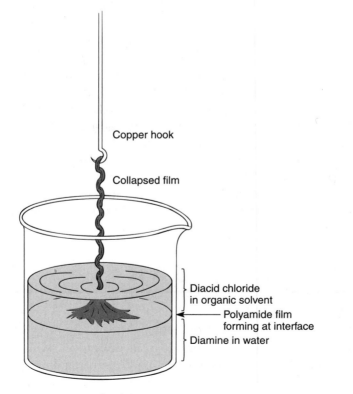

Preparation of nylon.

ing it down the wall of the slightly tilted beaker. Two layers will form (see figure on page 480), and there will be an immediate formation of a polymer film at the liquid–liquid interface. Using a copper-wire hook (a 6-inch piece of wire bent at one end), gently free the walls of the beaker from polymer strings. Then hook the mass at the center and slowly raise the wire so that polyamide forms continuously, producing a rope that can be drawn out for many feet. The strand can be broken by pulling it faster. Rinse the rope several times with water and lay it on a paper towel to dry. With the piece of wire, vigorously stir the remainder of the two-phase system to form additional polymer. Decant the liquid and wash the polymer thoroughly with water. Allow the polymer to dry. Do not discard the nylon in the sink. Use a waste container.

EXPERIMENT 56C

Polystyrene

An addition polymer, polystyrene, will be prepared in this experiment. Reaction can be brought about by free-radical, cationic, or anionic catalysts, the first of these being the most common. In this experiment, polystyrene is prepared by free-radical-catalyzed polymerization.

The reaction is initiated by a free-radical source. The initiator will be benzoyl peroxide, a relatively unstable molecule, which at 80–90°C decomposes with homolytic cleavage of the oxygen–oxygen bond:

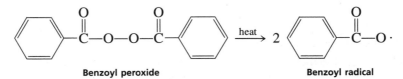

Benzoyl peroxide   **Benzoyl radical**

If an unsaturated monomer is present, the catalyst radical adds to it, initiating a chain reaction by producing a new free radical. If we let R stand for the catalyst radical, the reaction with styrene can be represented as

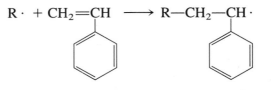

The chain continues to grow:

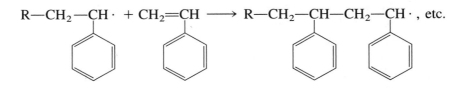

The chain can be terminated by causing two radicals to combine (either both polymer radicals or one polymer radical and one initiator radical) or by causing a hydrogen atom to become abstracted from another molecule.

Procedure

Because it is difficult to clean the glassware, this experiment is best performed by the laboratory instructor. One large batch should be made for the entire class (at least 10 times the amounts given). After the polystyrene is prepared, a small amount will be dispensed to each student. The students will provide their own watch glass for this purpose. Perform the experiment in a hood. Place several thicknesses of newspaper in the hood.

> **Caution:** Styrene vapor is very irritating to the eyes, mucous membranes, and upper respiratory tract. Do not breathe the vapor and do not get it on your skin. Exposure can cause nausea and headaches. All operations with styrene must be conducted in a hood.
>
> Benzoyl peroxide is flammable and may detonate on impact or on heating (or grinding). It should be weighed on a glassine (glazed, not ordinary) paper. Clean all spills with water. Wash the glassine paper with water before discarding it.

Place 12–15 mL of styrene monomer in a 100-mL beaker and add 0.35 g of benzoyl peroxide. Heat the mixture on a hotplate until the mixture turns yellow. When the color disappears and bubbles begin to appear, immediately take the beaker of styrene off the hotplate because the reaction is exothermic (use tongs or an insulated glove). After the reaction subsides, put the beaker of styrene back on the hotplate and continue heating it until the liquid becomes very syrupy. With a stirring rod, draw out a long filament of material from the beaker. If this filament can be cleanly snapped after a few seconds of cooling, the polystyrene is ready to be poured. If the filament does not break, continue heating the mixture and repeat this process until the filament breaks easily.

If you are performing the optional infrared analysis of the polymer, immediately save a sample of the polymer. After removing a sample for infrared spectroscopy, pour the remainder of the syrupy liquid on a watch glass that has been lightly coated with stopcock grease. After being cooled, the polystyrene can be lifted from the glass surface by gently prying with a spatula.

Infrared Spectroscopy (Optional). Pour a small amount of the *hot* polymer from the beaker onto a warm watch glass (no grease) and spread the polymer with a wooden applicator stick so as to create a thin film of the polymer. Peel the polymer from the watch glass and save it for Experiment 56D.

EXPERIMENT 56D

Infrared Spectra of Polymer Samples

Infrared spectroscopy is an excellent technique for determining the structure of a polymer. For example, polyethylene and polypropylene have relatively simple spectra because they are saturated hydrocarbons. Polyesters have stretching frequencies associated with the C=O and C—O groups in the polymer chain. Polyamides (nylon) show absorptions that are characteristic for the C=O stretch and N—H stretch. Polystyrene has characteristic features of a monosubstituted aromatic compound (see Fig. 19.11 on p. 855). You may determine the infrared spectra of the linear polyester from Procedure 56A and polystyrene from Experiment 56C in this part of the experiment. Your instructor may ask you to analyze a sample that you bring to the laboratory or one supplied to you.

Procedure

Mounting the Samples. Prepare cardboard mounts for your polymer samples. Cut 3 × 5-inch index cards so that they fit into the sample cell holder of your infrared spectrometer. Then cut a 0.5 wide × 1-inch high rectangular hole in the center of the card stock. Attach a polymer sample on the cardboard mount with tape.

Choices of Polymer Samples. If you have completed Experiments 56A and 56C, you can obtain the spectra of your polyester or polystyrene. Alternatively, your instructor may provide you with known or unknown polymer samples for you to analyze.

Your instructor may ask you to bring a polymer sample of your own choice. If possible, these samples should be clear and as thin as possible (similar to the thickness of plastic sandwich wrap). Good choices of plastic materials include windows from envelopes, plastic sandwich wrap, sandwich bags, soft drink bottles, milk containers, shampoo bottles, candy wrappers, and shrink-wrap. If necessary, the samples can be heated in an oven and stretched to obtain thinner samples. If you are bringing a sample cut from a plastic container, obtain the recycling code from the bottom of the container, if one is given.

Running the Infrared Spectrum. Insert the cardboard mount into the cell holder in the spectrometer so that your polymer sample is centered in the infrared beam of the instrument. Find the thinnest place in your polymer sample. Determine the infrared spectrum of your sample. Because of the thickness of your polymer sample, many absorptions are so strong that you will not be able to see individual bands. To obtain a better spectrum, try moving the sample to a new position in the beam and rerun the spectrum.

Analyzing the Infrared Spectrum. You can use the essay "Polymers and Plastics" and Appendix 3 with your spectrum to help determine the structure of the polymer. Most likely, the polymers will consist of plastic materials listed in Table Three of the essay (p. 474). This table lists the recycling codes for a number of household plastics used in packaging. Submit the infrared spectrum along with the structure of the polymer to your instructor. Do your spectrum and structure agree with the recycling code?

Label the spectrum with the important absorption bands consistent with the structure of the polymer.

Using a Polymer Library.   If your particular instrument has a polymer library, you can search the library for a match. Do this after you have made a preliminary "educated guess" as to the structure of the polymer. The library search should help confirm the structure that you determined.

QUESTIONS

1. Ethylene dichloride $ClCH_2CH_2Cl$ and sodium polysulfide Na_2S_4 react to form a chemically resistant rubber, Thiokol A. Draw the structure of the rubber.

2. Draw the structure for the polymer produced from the monomer, vinylidene chloride $(CH_2=CCl_2)$.

3. Draw the structure of the copolymer produced from vinyl acetate and vinyl chloride. This copolymer is employed in some paints, adhesives, and paper coatings.

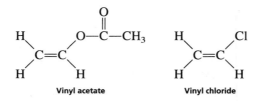

Vinyl acetate   Vinyl chloride

4. Isobutylene $CH_2=C(CH_3)_2$ is used to prepare cold-flow rubber. Draw a structure for the addition polymer formed from this alkene.

5. Kel-F is an addition polymer with the following partial structure. What is the monomer used to prepare it?

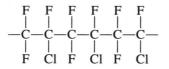

6. Maleic anhydride reacts with ethylene glycol to produce an alkyd resin. Draw the structure of the condensation polymer produced.

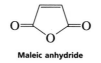

Maleic anhydride

7. Kodel is a condensation polymer made from terephthalic acid and 1,4-cyclohexanedimethanol. Write the structure of the resulting polymer.

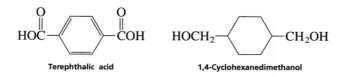

Terephthalic acid   1,4-Cyclohexanedimethanol

EXPERIMENT 57

Identification of Unknowns

Qualitative organic analysis, the identification and characterization of unknown compounds, is an important part of organic chemistry. Every chemist must learn the appropriate methods for establishing the identity of a compound. In this experiment, you will be issued an unknown compound and asked to identify it through chemical and spectroscopic methods. Your instructor may give you a general unknown or a specific unknown. With a **general unknown,** you must first determine the class of compound to which the unknown belongs, that is, identify its main functional group; then you must determine the specific compound in that class that corresponds to the unknown. With a **specific unknown,** you will know the class of compound (ketone, alcohol, amine, and so on) in advance, and it will be necessary to determine only whatever specific member of that class was issued to you as an unknown. This experiment is designed so that the instructor can issue several general unknowns or as many as six successive specific unknowns, each having a different main functional group.

Although there are several million organic compounds that an organic chemist might be called upon to identify, the scope of this experiment is necessarily limited. In this textbook, just over 300 compounds are included in the tables of possible unknowns given for the experiment (see Appendix 1). Your instructor may wish to expand the list of possible unknowns, however. In such a case, you will have to consult more extensive tables, such as those found in the work compiled by Rappoport (see references). In addition, the experiment is restricted to include only seven important functional groups:

Aldehydes	Carboxylic acids	Amines	Esters
Ketones	Phenols	Alcohols	

Even though this list of functional groups omits some of the important types of compounds (alkyl halides, alkenes, alkynes, aromatics, ethers, amides, mercaptans, nitriles, acid chlorides, acid anhydrides, nitro compounds, and so on), the methods introduced here can be applied equally well to other classes of compounds. The list is sufficiently broad to illustrate all the principles involved in identifying an unknown compound.

In addition, although many of the functional groups listed as being excluded will not appear as the major functional group in a compound, several of them will frequently appear as secondary, or subsidiary, functional groups. Three examples of this are presented here.

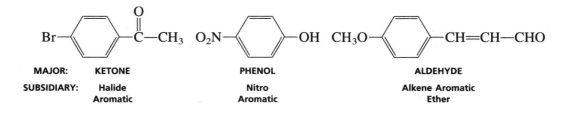

MAJOR:	KETONE	PHENOL	ALDEHYDE
SUBSIDIARY:	Halide	Nitro	Alkene Aromatic
	Aromatic	Aromatic	Ether

The groups included that have subsidiary status are

—Cl	Chloro	—NO$_2$	Nitro	C=C	Double Bond
—Br	Bromo	—C≡N	Cyano	C≡C	Triple Bond
—I	Iodo	—OR	Alkoxy	⬡	Aromatic

The experiment presents all the chief chemical and spectroscopic methods of determining the main functional groups, and it includes methods for verifying the presence of the subsidiary functional groups as well. It will usually not be necessary to determine the presence of the subsidiary functional groups to identify the unknown compound correctly. *Every* piece of information helps the identification, however, and if these groups can be detected easily, you should not hesitate to determine them. Finally, complex bifunctional compounds are generally avoided in this experiment; only a few are included.

How to Proceed

Fortunately, we can detail a fairly straightforward procedure for determining all the necessary pieces of information. This procedure consists of the following steps:

Part One: Chemical Classifications

1. Preliminary classification by physical state, color, and odor.
2. Melting-point or boiling-point determination; other physical data.
3. Purification, if necessary.
4. Determination of solubility behavior in water and in acids and bases.
5. Simple preliminary test: Beilstein, ignition (combustion).
6. Application of relevant chemical classification tests.

Part Two: Spectroscopy

7. Determination of infrared and NMR spectra.

Part Three: Optional Procedures

8. Elemental analysis, if necessary.
9. Preparation of derivatives.
10. Confirmation of identity.

Each of these steps is discussed briefly in the sections below.

PRELIMINARY CLASSIFICATION

Note the physical characteristics of the unknown, including its color, its odor, and its physical state (liquid, solid, crystalline form). Many compounds have characteristic colors or odors, or they crystallize with a specific crystal structure. This information can often be found in a handbook and can be checked later. Compounds with a high degree of conjugation are frequently yellow to red. Amines often have a fishlike odor. Esters have a pleasant fruity or floral odor. Acids have a sharp and pungent odor. A part of the training of every good chemist includes cultivating the ability to recognize familiar or typical odors. As a note of caution, many compounds have distinctly unpleasant or nauseating odors. Some have corrosive vapors. Sniff any unknown substance with the greatest caution. As a first step, open the container, hold it away from you, and using your hand, carefully waft the vapors toward your nose. If you get past this stage, a closer inspection will be possible.

MELTING-POINT OR BOILING-POINT DETERMINATION

The single most useful piece of information to have for an unknown compound is its melting point or boiling point. Either piece of data will drastically limit the compounds that are possible. The electric melting-point apparatus gives a rapid and accurate measurement (see Technique 6, Sections 6.5 and 6.7). To save time, you can often determine two separate melting points. The first determination can be made rapidly to get an approximate value. Then, you can determine the second melting point more carefully.

The boiling point is obtained by a simple distillation of the unknown (Technique 8, Section 8.3), by a micro boiling-point determination (Technique 6, Section 6.11), or by refluxing the unknown in a test tube (Technique 6, Section 6.10). The simple distillation has the advantage that it also purifies the compound. The smallest distilling flask available should be used if a simple distillation is performed, and you should be sure that the thermometer bulb is fully immersed in the vapor of the distilling liquid. For an accurate boiling-point value, the liquid should be distilled rapidly. The micro boiling-point method requires the least amount of unknown, but the refluxing method is more reliable and requires much less liquid than performing a distillation.

If the solid is high-melting ($>200°C$), or the liquid high-boiling ($>200°C$), a thermometer correction may be needed (Technique 6, Sections 6.12 and 6.13). In any event, allowance should be made for errors of as large as $\pm5°C$ in these values.

PURIFICATION

If the melting point of a solid has a wide range ($>4–5°C$), it should be recrystallized and the melting point redetermined.

If a liquid was highly colored before distillation, if it yielded a wide boiling-point range, or if the temperature did not hold constant during the distillation, it should be redistilled to determine a new temperature range. A reduced-pressure distillation is in order for high-boiling liquids or for those that show any sign of decomposition on heating.

Occasionally, column chromatography may be necessary to purify solids that have large amounts of impurities and do not yield satisfactory results on crystallization.

Acidic or basic impurities that contaminate a neutral compound may often be removed by dissolving the compound in a low-boiling solvent, such as CH_2Cl_2 or ether, and extracting with 5% $NaHCO_3$ or 5% HCl, respectively. Conversely, acidic or basic compounds can be purified by dissolving them in 5% $NaHCO_3$ or 5% HCl, respectively, and extracting them with a low-boiling organic solvent to remove impurities. After neutralization of the aqueous solution, the desired compound can be recovered by extraction.

SOLUBILITY BEHAVIOR

Tests on solubility are described fully in Experiment 57A. They are extremely important. Determine the solubility of small amounts of the unknown in water, 5% HCl, 5% $NaHCO_3$, 5% NaOH, concentrated H_2SO_4, and organic solvents. This information reveals whether a compound is an acid, a base, or a neutral substance. The sulfuric acid test reveals whether a neutral compound has a functional group that contains an oxygen, a nitrogen, or a sulfur atom that can be protonated. This information allows you to eliminate or to choose various functional-group possibilities. The solubility tests must be made on *all* unknowns.

PRELIMINARY TESTS

The two combustion tests, the Beilstein test (Experiment 57B) and the ignition test (Experiment 57C), can be performed easily and quickly, and they often give valuable information. It is recommended that they be performed on all unknowns.

CHEMICAL CLASSIFICATION TESTS

The solubility tests usually suggest or eliminate several possible functional groups. The chemical classification tests listed in Experiments 57D to 57I allow you to distinguish among the possible choices. Choose only those tests that the solubility tests suggest might be meaningful. Time will be wasted performing unnecessary tests. There is no substitute for a firsthand, thorough knowledge of these tests. Study each of the sections carefully until you understand the significance of each test. Also, it will be helpful to actually try the tests on **known** substances. In this way, it will be easier to recognize a positive test. Appropriate test compounds are listed for many of the tests. When you are performing a test that is new to you, it is always good practice to run the test separately on both a known substance and the unknown *at the same time*. This practice lets you compare results directly.

Once the melting or boiling point, the solubilities, and the main chemical tests have been made, it will be possible to identify the class of compound. At this stage, with the melting point or boiling point as a guide, it will be possible to compile a list of possible compounds. Inspection of this list will suggest additional tests that must be performed to distinguish among the possibilities. For instance, one compound may be a methyl ketone and the other may not. The iodoform test is called for to distinguish the two possibilities. The tests for the subsidiary functional groups may also be required. These tests are described in Experiments 57B and 57C. These tests should also be studied carefully; there is no substitute for firsthand knowledge about these either.

Do not perform the chemical tests either haphazardly or in a methodical, comprehensive sequence. Instead, use the tests selectively. Solubility tests automatically eliminate the need for some of the chemical tests. Each successive test will either eliminate the need for another test or dictate its use. You should also examine the tables of unknowns carefully. The boiling point or the melting point of the unknown may eliminate the need for many of the tests. For instance, the possible compounds may simply not include one with a double bond. Efficiency is the key word here. You should not waste time performing nonsensical or unnecessary tests. Many possibilities can be eliminated on the basis of logic alone.

How you proceed with the following steps may be limited by your instructor's wishes. Many instructors may restrict your access to infrared and NMR spectra until you have narrowed your choices to a few compounds *all within the same class.* Others may have you determine these data routinely. Some instructors may want students to perform elemental analysis on all unknowns; others may restrict it to only the most essential situations. Most unknowns can be identified without either spectroscopy or elemental analysis. Again, some instructors may require derivatives as a final confirmation of the compound's identity; others may not wish to use them at all.

SPECTROSCOPY

Spectroscopy is probably the most powerful and modern tool available to the chemist for determining the structure of an unknown compound. It is often possible to determine structure through spectroscopy alone. On the other hand, there are also situations for which spectroscopy is not of much help and the traditional methods must be relied on. For this reason, you should use spectroscopy not to the exclusion of the more traditional tests but rather as a confirmation of those results. Nevertheless, the main functional groups and their immediate environmental features can be determined quickly and accurately with spectroscopy.

ELEMENTAL ANALYSIS

Elemental analysis, which allows you to determine the presence of nitrogen, sulfur, or a specific halogen atom (Cl, Br, I) in a compound is often useful; however, other information often renders these tests unnecessary. A compound identified as an amine by solubility tests obviously contains nitrogen. Many nitrogen-containing groups (for instance, nitro groups) can be identified by infrared spectroscopy. Finally, it is not usually necessary to identify a specific halogen. The simple information that the compound contains a halogen (any halogen) may be enough information to distinguish between two compounds. A simple Beilstein test provides this information.

DERIVATIVES

One of the principal tests for the correct identification of an unknown compound comes in trying to convert the compound by a chemical reaction to another known compound. This second compound is called a **derivative.** The best derivatives are solid compounds, since the melting point of a solid provides an accurate and reliable identification of most compounds. Solids are also easily purified through crystallization. The derivative provides a way of distinguishing two otherwise very similar compounds. Usually, they will have derivatives (both prepared by the same reaction) that have different melting points. Tables of unknowns and derivatives are listed in Appendix 1. Procedures for preparing derivatives are given in Appendix 2.

CONFIRMATION OF IDENTITY

A rigid and final test for identifying an unknown can be made if an "authentic" sample of the compound is available for comparison. One can compare infrared and NMR spectra of the unknown compound with the spectra of the known compound. If the spectra match, peak for peak, then the identity is probably certain. Other physical and chemical properties can also be compared. If the compound is a solid, a convenient test is the mixed melting point (Technique 6, Section 6.4). Thin-layer or gas chromatographic comparisons may also be useful. For thin-layer analysis, however, it may be necessary to experiment with several different development solvents to reach a satisfactory conclusion about the identity of the substance in question.

While we cannot be complete in this experiment in terms of the functional groups covered, or the tests described, the experiment should provide a good introduction to the methods and the techniques chemists use to identify unknown compounds. Textbooks that cover the subject more thoroughly are listed in the references. You are encouraged to consult these for more information, including specific methods and classification tests.

REFERENCES

Comprehensive Textbooks
Cheronis, N. D., and Entrikin, J. B. *Identification of Organic Compounds.* New York: Wiley-Interscience, 1963.
Pasto, D. J., and Johnson, C. R. *Laboratory Text for Organic Chemistry.* Englewood Cliffs, NJ: Prentice-Hall, 1979.
Shriner, R. L., Fuson, R. C., Curtin, D. Y., and Morrill, T. C. *The Systematic Identification of Organic Compounds.* 6th ed. New York: Wiley, 1980.

Spectroscopy

Bellamy, L. J. *The Infra-red Spectra of Complex Molecules.* 3rd ed. New York: Methuen, 1975.

Colthup, N. B., Daly, L. H., and Wiberly, S. E. *Introduction to Infrared and Raman Spectroscopy.* 3rd ed. San Diego, CA: Academic Press, 1990.

Dyer, J. R. *Applications of Absorption Spectroscopy of Organic Compounds.* Englewood Cliffs, NJ: Prentice-Hall, 1965.

Lin-Vien, D., Colthup, N. B., Fateley, W. B., and Grasselli, J. G. *The Handbook of Infrared and Raman Characteristic Frequencies of Organic Molecules.* San Diego, CA: Academic Press, 1991.

Nakanishi, K. *Infrared Absorption Spectroscopy.* San Francisco: Holden-Day, 1962.

Pavia, D. L., Lampman, G. M., and Kriz, G. S. *Introduction to Spectroscopy: A Guide for Students of Organic Chemistry.* 2nd ed. Philadelphia: W. B. Saunders, 1996.

Silverstein, R. M., Bassler, G. C., and Morrill, T. C. *Spectrometric Identification of Organic Compounds.* 5th ed. New York: Wiley, 1991.

Extensive Tables of Compounds and Derivatives

Rappoport, Z., ed. *Handbook of Tables for Organic Compound Identification.* Cleveland: Chemical Rubber Co., 1967.

EXPERIMENT 57A

Solubility Tests

Solubility tests should be performed on *every unknown.* They are extremely important in determining the nature of the main functional group of the unknown compound. The tests are very simple and require only small amounts of the unknown. In addition, solubility tests reveal whether the compound is a strong base (amine), a weak acid (phenol), a strong acid (carboxylic acid), or a neutral substance (aldehyde, ketone, alcohol, ester). The common solvents used to determine solubility types are

5% HCl	Concentrated H_2SO_4
5% $NaHCO_3$	Water
5% NaOH	Organic solvents

The solubility chart on page 492 indicates solvents in which compounds containing the various functional groups are likely to dissolve. The summary charts in Experiments 57D through 57I repeat this information for each functional group included in this experiment. In this section, the correct procedure for determining whether a compound is soluble in a test solvent is given. Also given is a series of explanations detailing the reasons that compounds having specific functional groups are soluble in only specific solvents. This is accomplished by indicating the type of chemistry or the type of chemical interaction that is possible in each solvent.

Waste Disposal

Dispose of all aqueous solutions in the container designated for aqueous waste. Any remaining organic compounds must be disposed of in the appropriate organic waste container.

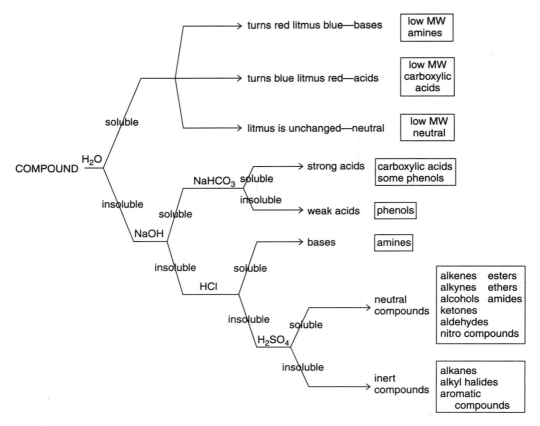

Solubility chart.

SOLUBILITY TESTS

Procedure. Place about 1 mL of the solvent in a small test tube. Add one drop of an unknown liquid from a Pasteur pipet or a few crystals of an unknown solid from the end of a spatula, directly into the solvent. Gently tap the test tube with your finger to ensure mixing and then observe whether any mixing lines appear in the solution. The disappearance of the liquid or solid or the appearance of the mixing lines indicates that solution is taking place. Add several more drops of the liquid or a few more crystals of the solid to determine the extent of the compound's solubility. A common mistake in determining the solubility of a compound is testing with a quantity of the unknown too large to dissolve in the chosen solvent. Use small amounts. It may take several minutes to dissolve solids. Compounds in the form of large crystals need more time to dissolve than powders or very small crystals. In some cases, it is helpful to pulverize a compound with large crystals using a mortar and pestle. Sometimes gentle heating helps, but strong heating is discouraged, as it often leads to reaction. When colored compounds dissolve, the solution often assumes the color.

Using the preceding procedure, determine the solubility of the unknown in each of the following solvents: water, 5% HCl, 5% NaHCO$_3$, 5% NaOH, and concentrated H$_2$SO$_4$. With sulfuric acid, a color change may be observed rather than solution. A color change should be regarded as a positive solubility test. Solid unknowns that do not dissolve in any of the test solvents may be inorganic substances. To eliminate this possibility, determine the solubility of the unknown in several organic solvents, like ether. If the compound is organic, a solvent that will dissolve it can usually be found.

If a compound is found to dissolve in water, the pH of the aqueous solution should be estimated with pH paper or litmus. Compounds soluble in water are usually soluble in *all* the aqueous solvents. If a compound is only slightly soluble in water, it may be *more* soluble in another aqueous solvent. For instance, a carboxylic acid may be only slightly soluble in water but very soluble in dilute base. It often will not be necessary to determine the solubility of the unknown in every solvent.

Test Compounds.   Five solubility unknowns can be found on the supply shelf. The five unknowns include a base, a weak acid, a strong acid, a neutral substance with an oxygen-containing functional group, and a neutral substance that is inert. Using solubility tests, distinguish these unknowns by type. Verify your answer with the instructor.

Solubility in Water.   Compounds that contain four or fewer carbons and also contain oxygen, nitrogen, or sulfur are often soluble in water. Almost any functional group containing these elements will lead to water solubility for low-molecular-weight (C$_4$) compounds. Compounds having five or six carbons and any of those elements are often insoluble in water or have borderline solubility. Branching of the alkyl chain in a compound lowers the intermolecular forces between its molecules. This is usually reflected in a lowered boiling point or melting point and a greater solubility in water for the branched compound than for the corresponding straight-chain compound. This occurs simply because the molecules of the branched compound are more easily separated from one another. Thus, *t*-butyl alcohol would be expected to be more soluble in water than *n*-butyl alcohol.

When the ratio of the oxygen, nitrogen, or sulfur atoms in a compound to the carbon atoms is increased, the solubility of that compound in water often increases. This is due to the increased number of polar functional groups. Thus, 1,5-pentanediol would be expected to be more soluble in water than 1-pentanol.

As the size of the alkyl chain of a compound is increased beyond about four carbons, the influence of a polar functional group is diminished, and the water solubility begins to decrease. A few examples of these generalizations are given here.

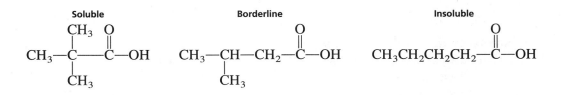

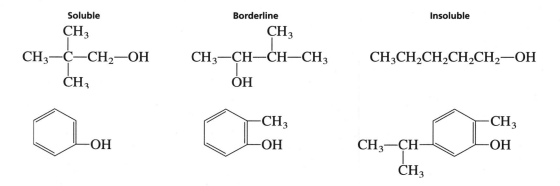

Soluble **Borderline** **Insoluble**

Solubility in 5% HCl. The possibility of an amine should be considered immediately if a compound is soluble in dilute acid (5% HCl). Aliphatic amines (RNH_2, R_2NH, R_3N) are basic compounds that readily dissolve in acid because they form hydrochloride salts that are soluble in the aqueous medium:

$$R—NH_2 + HCl \longrightarrow R—NH_3^+ + Cl^-$$

The substitution of an aromatic ring Ar for an alkyl group R reduces the basicity of an amine somewhat, but the amine will still protonate, and it will still generally be soluble in dilute acid. The reduction in basicity in an aromatic amine is due to the resonance delocalization of the unshared electrons on the amino nitrogen of the free base. The delocalization is lost on protonation, a problem that does not exist for aliphatic amines. The substitution of two or three aromatic rings on an amine nitrogen reduces the basicity of the amine even further. Diaryl and triaryl amines do not dissolve in dilute HCl because they do not protonate easily. Thus, Ar_2NH and Ar_3N are insoluble in dilute acid. Some amines of very high molecular weight, like tribromoaniline ($MW = 330$), may also be insoluble in dilute acid.

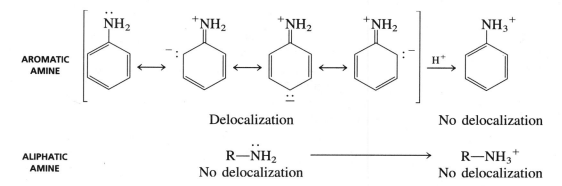

AROMATIC AMINE

Delocalization   No delocalization

ALIPHATIC AMINE $R—NH_2$ $R—NH_3^+$

No delocalization   No delocalization

Solubility in 5% NaCHO$_3$ and 5% NaOH. Compounds that dissolve in sodium bicarbonate, a weak base, are strong acids. Compounds that dissolve in sodium hydroxide, a strong base, may be either strong or weak acids. Thus, one can distinguish weak and strong acids by determining their solubility in both strong (NaOH) and weak (NaHCO$_3$) base. The classification of some functional groups as either weak or strong acids is given in the table on page 495.

Strong Acids (Soluble in both NaOH and NaHCO$_3$)		Weak Acids (Soluble in NaOH but not NaHCO$_3$)	

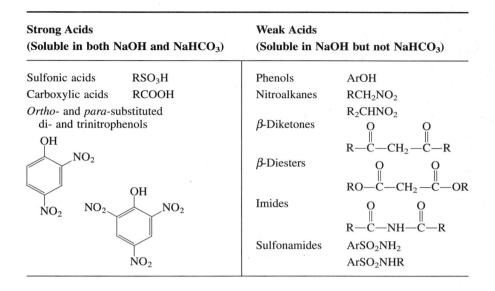

In this experiment, carboxylic acids (pK_a ~ 5) are generally indicated when a compound is soluble in both bases, while phenols (pK_a ~ 10) are indicated when it is soluble in NaOH only.

Compounds dissolve in base because they form sodium salts that are soluble in the aqueous medium. The salts of some high-molecular-weight compounds are not soluble, however, and precipitate. The salts of the long-chain carboxylic acids, such as myristic acid C_{14}, palmitic acid C_{16}, and stearic acid C_{18}, which form soaps, are in this category. Some phenols also produce insoluble sodium salts, and often these are colored due to resonance in the anion.

Both phenols and carboxylic acids produce resonance-stabilized conjugate bases. Thus, bases of the appropriate strength may easily remove their acidic protons to form the sodium salts.

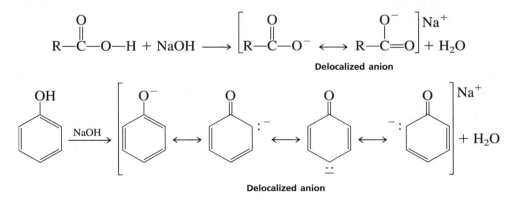

In phenols, substitution of nitro groups in the *ortho* and *para* positions of the ring increases the acidity. Nitro groups in these positions provide additional delocalization in the conjugate anion. Phenols that have two or three nitro groups in the *ortho* and *para* positions often dissolve in *both* sodium hydroxide and sodium bicarbonate solutions.

Solubility in Concentrated Sulfuric Acid. Many compounds are soluble in cold concentrated sulfuric acid. Of the compounds included in this experiment, alcohols, ketones, aldehydes, and esters are in this category. Other compounds that also dissolve include alkenes, alkynes, ethers, nitroaromatics, and amides. Since several different kinds of compounds are soluble in sulfuric acid, further chemical tests and spectroscopy will be needed to differentiate among them.

Compounds that are soluble in concentrated sulfuric acid but not in dilute acid are extremely weak bases. Almost any compound containing a nitrogen, an oxygen, or a sulfur atom can be protonated in concentrated sulfuric acid. The ions produced are soluble in the medium.

$$R-O-H + H_2SO_4 \longrightarrow R-\overset{+}{\underset{H}{O}}-H + HSO_4^- \longrightarrow R^+ + H_2O + HSO_4^-$$

$$R-\overset{O}{\overset{\|}{C}}-R + H_2SO_4 \longrightarrow R-\overset{\overset{+}{O}-H}{\overset{\|}{C}}-R + HSO_4^-$$

$$R-\overset{O}{\overset{\|}{C}}-OR + H_2SO_4 \longrightarrow R-\overset{\overset{+}{O}-H}{\overset{\|}{C}}-OR + HSO_4^-$$

$$\underset{R}{\overset{R}{>}}C=C\underset{R}{\overset{R}{<}} + H_2SO_4 \longrightarrow R-\overset{R}{\underset{H}{\overset{|}{C}}}-\overset{R}{\underset{+}{\overset{|}{C}}}-R + HSO_4^-$$

Inert Compounds. Compounds not soluble in concentrated sulfuric acid or any of the other solvents are said to be inert. Compounds not soluble in concentrated sulfuric acid include the alkanes, most simple aromatics, and the alkyl halides. Some examples of inert compounds are hexane, benzene, chlorobenzene, chlorohexane, and toluene.

E X P E R I M E N T   5 7 B

Tests for the Elements (N, S, X)

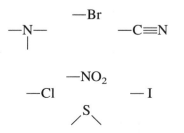

Except for amines (Experiment 57 G), which are easily detected by their solubility behavior, all compounds issued in this experiment will contain heteroelements (N, S, Cl, Br, or I) only as *secondary* functional group. These will be subsidiary to some other important functional group. Thus, no alkyl or aryl halides, nitro compounds, thiols, or thioethers will be issued. However, some of the unknowns may contain a halogen or a nitro group. Less frequently, they may contain a sulfur atom or a cyano group.

Consider as an example *p*-bromobenzaldehyde, an **aldehyde** that contains bromine as a ring substituent. The identification of this compound would hinge on whether the investigator could identify it as an aldehyde. It could probably be identified *without* proving the existence of bromine in the molecule. That information, however, could make the identification easier. In this experiment, methods are given for identifying the presence of a halogen or a nitro group in an unknown compound. Also given is a general method (sodium fusion) for detecting the principal heteroelements that may exist in organic molecules.

CLASSIFICATION TESTS		
Halides	**Nitro Groups**	**N, S, X(Cl, Br, I)**
Beilstein test Silver nitrate Sodium iodide/acetone	Ferrous hydroxide	Sodium fusion

Waste Disposal

Dispose of all solutions containing silver into a waste container designated for this purpose. Any other aqueous solutions should be disposed of in the container designated for aqueous waste. Any remaining organic compounds must be disposed of in the appropriate organic waste container under the hood. This is particularly true of any solution containing benzyl bromide, which is a lachrymator.

Tests for a Halide

BEILSTEIN TEST

Procedure. Bend a small loop in the end of a short length of copper wire. Heat the loop end of the wire in a Bunsen burner flame. After cooling, dip the wire directly into a small sample of the unknown. Now, heat the wire in the Bunsen burner flame again. The compound will first burn. After the burning, a green flame will be produced if a halogen is present.

Test Compounds. Try this test on bromobenzene and benzoic acid.

Halogens can be detected easily and reliably by the Beilstein test. It is the simplest method for determining the presence of a halogen, but it does not differentiate among chlorine, bromine, and iodine, any one of which will give a positive test. However, when the identity of the unknown has been narrowed to two choices, of which one has a halogen and one does not, the Beilstein test will often be enough to distinguish between the two.

A positive Beilstein test results from the production of a volatile copper halide when an organic halide is heated with copper oxide. The copper halide imparts a blue-green color to the flame.

This test can be very sensitive to small amounts of halide impurities in some compounds. Therefore, use caution in interpreting the results of the test if you obtain only a weak color.

SILVER NITRATE TEST

Procedure. Add one drop of a liquid or five drops of a concentrated ethanolic solution of a solid unknown to 2 mL of a 2% ethanolic silver nitrate solution. If no reaction is observed after 5 minutes at room temperature, heat the solution in a hot water bath at about 100°C and note whether a precipitate forms. If a precipitate forms, add two drops of 5% nitric acid and note whether the precipitate dissolves. Carboxylic acids may give a false test by precipitating in silver nitrate, but they dissolve when nitric acid is added. Silver halides, on the other hand, do not dissolve in nitric acid.

Test Compounds. Apply this test to benzyl bromide (α-bromotoluene) and bromobenzene. Discard all waste reagents in a suitable waste container in the hood, since benzyl bromide is a lachrymator.

This test depends on the formation of a white or off-white precipitate of silver halide when silver nitrate is allowed to react with a sufficiently reactive halide.

$$RX + Ag^+NO_3^- \longrightarrow \underset{\textbf{Precipitate}}{\textbf{AgX}} + R^+NO_3^- \xrightarrow{CH_3CH_2OH} R-O-CH_2CH_3$$

The test does not distinguish among chlorides, bromides, and iodides but does distinguish **labile** (reactive) halides from halides that are unreactive. Halides substituted on an aromatic ring will not usually give a positive silver nitrate test; however, alkyl halides of many types will give a positive test.

The most reactive compounds are those able to form stable carbocations in solution and those equipped with good leaving groups (X = I, Br, Cl). Benzyl, allyl, and tertiary halides react immediately with silver nitrate. Secondary and primary halides do not react at room temperature but react readily when heated. Aryl and vinyl halides do not react at all, even at elevated temperatures. This pattern of reactivity fits the stability order for various carbocations quite well. Compounds that produce stable carbocations react at higher rates than those that do not.

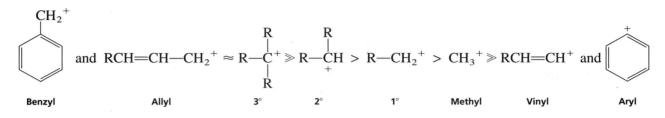

Benzyl   **Allyl**   **3°**   **2°**   **1°**   **Methyl**   **Vinyl**   **Aryl**

The fast reaction of benzylic and allylic halides is a result of the resonance stabilization that is available to the intermediate carbocations formed. Tertiary halides are more reactive than secondary halides, which are in turn more reactive than primary or methyl halides because alkyl substituents are able to stabilize the intermediate carbocations by an electron-releasing effect. The methyl carbocations have no alkyl groups and are the least stable of all the carbocations mentioned thus far. Vinyl and aryl carbocations are extremely unstable because the charge is localized on an sp^2-hybridized carbon (double-bond carbon) rather than on one that is sp^3-hybridized.

SODIUM IODIDE IN ACETONE

Procedure.   This test is described in Experiment 22.

Detection of Nitro Groups

Although nitro compounds will not be issued as distinct unknowns, many of the unknowns may have a nitro group as a secondary functional group. The presence of a nitro group, and hence nitrogen, in an unknown compound is best determined by infrared spectroscopy. However, many nitro compounds give a positive result in the following test, although other functional groups may give false results.

FERROUS HYDROXIDE TEST

Procedure.   Place 1.5 mL of freshly prepared 5% aqueous ferrous ammonium sulfate in a small test tube and add about 10 mg of the unknown compound. Mix the solution well and then add first one drop of 2*M* sulfuric acid, and then 1 mL of 2*M* potassium hydroxide in methanol. Stopper the test tube and shake it vigorously. A positive test is indicated by the formation of a red-brown precipitate, usually within 1 minute.

Test Compound.   Try this test with 2-nitrotoluene

Most nitro compounds oxidize ferrous hydroxide to ferric hydroxide, which is a red-brown solid. A precipitate indicates a positive test.

$$R\!-\!NO_2 + 4\ H_2O + 6\ Fe(OH)_2 \longrightarrow R\!-\!NH_2 + 6\ Fe(OH)_3$$

INFRARED SPECTROSCOPY

The nitro group gives two strong bands near 1560 and 1350 cm^{-1}.

Detection of a Cyano Group

Although nitriles will not be given as unknowns in this experiment, the cyano group may be a subsidiary functional group whose presence or absence is important to the final identification of an unknown compound. The cyano group can be hydrolyzed in a strong base, by heating vigorously, to give a carboxylic acid and ammonia gas:

$$R\!-\!C\!\equiv\!N + 2\ H_2O \xrightarrow[\Delta]{\text{NaOH}} R\!-\!COOH + NH_3$$

The ammonia can be detected by its odor or by moist pH paper. However, this method is somewhat difficult, and the presence of a nitrile group is confirmed most easily by infrared spectroscopy. No other functional groups (except some C≡C) absorb in the same region of the spectrum as C≡N.

INFRARED SPECTROSCOPY

C≡N stretch is a very sharp band of medium intensity near 2250 cm^{-1}.

Sodium Fusion Tests (Detection of N, S, and X) (Optional)

When an organic compound containing nitrogen, sulfur, or halide atoms is fused with sodium metal, there is a reductive decomposition of the compound, which converts these atoms to the sodium salts of the inorganic ions CN^-, S^{2-}, and X^-.

$$[N,\ S,\ X] \xrightarrow[\Delta]{\text{Na}} NaCN,\ Na_2S,\ NaX$$

When the fusion mixture is dissolved in distilled water, the cyanide, sulfide, and halide ions can be detected by standard qualitative inorganic tests.

> **Caution:** Always remember to manipulate the sodium metal with a knife or a forceps. Do not touch it with your fingers. Keep sodium away from water. Destroy all waste sodium with 1-butanol or ethanol. Wear safety glasses.

PREPARATION OF STOCK SOLUTION

General Method

Procedure.   Using a forceps and a knife, take some sodium from the storage container, cut a small piece about the size of a small pea (3 mm on a side), and dry it on a paper towel. Place this small piece of sodium in a clean and dry small test tube (10 × 75 mm). Clamp the test tube to a ring stand and heat the bottom of the tube with a microburner until the sodium melts and its metallic vapor can be seen to rise about a third of the way up the tube. The bottom of the tube will probably have a dull red glow. Remove the burner and *immediately* drop the sample directly into the tube. Use about 10 mg of a solid placed on the end of a spatula or two to three drops of a liquid. Be sure to drop the sample directly down the center of the tube so that it touches the hot sodium metal and does not adhere to the side of the test tube. There will usually be a flash or a small explosion if the fusion is successful. If the reaction is not successful, heat the tube to red heat for a few seconds to ensure complete reaction.

Allow the test tube to cool to room temperature and then carefully add 10 drops of methanol, a drop at a time, to the fusion mixture. Using a spatula or a long glass rod, reach into the test tube and stir the mixture to ensure complete reaction of any excess sodium metal. The fusion will have destroyed the test tube for other uses. Thus, the easiest way to recover the fusion mixture is to crush the test tube into a small beaker containing 5–10 mL of *distilled* water. The tube is easily crushed if it is placed in the angle of a clamp holder. Tighten the clamp until the tube is securely held near its bottom, and then—standing back from the beaker and holding the clamp at its opposite end—continue tightening the clamp until the test tube breaks and the pieces fall into the beaker. Stir the solution well, heat it to boiling, and then filter it by gravity through a fluted filter (Fig. 4.3, p. 637). Portions of this solution will be used for the tests to detect nitrogen, sulfur, and the halogens.

Alternative Method

Procedure. With some volatile liquids, the previous method will not work. The compounds volatilize before they reach the sodium vapors. For such compounds, place four or five drops of the pure liquid in the clean and dry test tube, clamp it, and cautiously add the small piece of sodium metal. If there is any reaction, wait until it subsides. Then heat the test tube to red heat and continue according to the instructions in the second paragraph of the preceding procedure.

NITROGEN TEST

Procedure. Using pH paper and a 10% sodium hydroxide solution, adjust the pH of about 1 mL of the stock solution to pH 13. Add two drops of saturated ferrous ammonium sulfate solution and two drops of 30% potassium fluoride solution. Boil the solution for about 30 seconds. Then acidify the hot solution by adding 30% sulfuric acid dropwise until the iron hydroxides dissolve. Avoid using excess acid. If nitrogen is present, a dark blue (not green) precipitate of Prussian blue $NaFe_2(CN)_6$ will form or the solution will assume a dark blue color.

Reagents. Dissolve 5 g of ferrous ammonium sulfate in 100 mL of water and 30 g of potassium fluoride in 100 mL of water.

SULFUR TEST

Procedure. Acidify about 1 mL of the test solution with acetic acid and add a few drops of a 1% lead acetate solution. The presence of sulfur is indicated by a black precipitate of lead sulfide PbS.

> **Caution:** Many compounds of lead(II) are suspected carcinogens (see p. 21) and should be handled with care. Avoid contact.

HALIDE TESTS

Procedure. Cyanide and sulfide ions interfere with the test for halides. If such ions are present, they must be removed. To accomplish this, acidify the solution with dilute nitric acid and boil it for about 2 minutes. This will drive off any HCN or H_2S that is formed. When the solution cools, add a few drops of a 5% silver nitrate solution. A *voluminous* precipitate indicates a halide. A faint turbidity *does not* mean a positive test. Silver chloride is white. Silver bromide is off-white. Silver iodide is yellow. Silver chloride will readily dissolve in concentrated ammonium hydroxide, whereas silver bromide is only slightly soluble.

DIFFERENTIATION OF CHLORIDE, BROMIDE, AND IODIDE

Procedure. Acidify 2 mL of the test solution with 10% sulfuric acid and boil it for about 2 minutes. Cool the solution and add about 0.5 mL of methylene chloride. Add a few drops of chlorine water or 2–4 mg of calcium hypochlorite.[1] Check to be sure that the solution is still acidic. Then stopper the tube, shake it vigorously, and set it aside to allow the layers to separate. An orange to brown color in the methylene chloride layer indicates bromine. Violet indicates iodine. No color or a *light* yellow indicates chlorine.

E X P E R I M E N T   5 7 C

Tests for Unsaturation

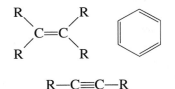

The unknowns to be issued for this experiment have neither a double bond nor a triple bond as their *only* functional group. Hence, simple alkenes and alkynes can be ruled out as possible compounds. Some of the unknowns may have a double or a triple bond, however, *in addition to* another more important functional group. The tests described allow you to determine the presence of a double bond or a triple bond (unsaturation) in such compounds.

CLASSIFICATION TESTS	
Unsaturation	**Aromaticity**
Bromine–carbon tetrachloride Potassium permanganate	Ignition test

---

[1]Clorox, the commercial bleach, is a permissible substitution for chlorine water, as is any other brand of bleach, provided that it is based on sodium hypochlorite.

Waste Disposal

Test reagents that contain bromine should be discarded into a special waste container designated for this purpose. Methylene chloride and carbon tetrachloride must be placed in the organic waste container designated for the disposal of halogenated organic wastes. Dispose of all other aqueous solutions in the container designated for aqueous waste. Any remaining organic compounds must be disposed of in the appropriate organic waste container.

Tests for Simple Multiple Bonds

BROMINE IN CARBON TETRACHLORIDE OR METHYLENE CHLORIDE

Procedure. Dissolve 50 mg of a solid unknown or two drops of a liquid unknown in 1 mL of carbon tetrachloride (or 1,2-dimethoxyethane). Add a 2% (by volume) solution of bromine in carbon tetrachloride, dropwise, and shake it until the bromine color persists. The test is positive if more than five drops of the bromine solution are needed so that the color remains for 1 minute. Usually, many drops of the bromine solution will be needed if unsaturation is present. Hydrogen bromide should not be evolved. If hydrogen bromide gas is evolved, you will note a "fog" while you blow across the mouth of the test tube. The HBr can also be detected by a moistened piece of litmus or pH paper. If hydrogen bromide is evolved, the reaction is a **substitution reaction** and not an **addition reaction,** and a double or triple bond is probably not present.

Methylene Chloride. Even though carbon tetrachloride is used in very small quantities in this test, it poses certain health hazards (see p. 19), and another solvent may be preferable. Methylene chloride (dichloromethane) can be substituted for carbon tetrachloride. Certain problems arise, however, because methylene chloride slowly reacts with bromine, presumably by a light-induced free radical process, to produce HBr. After about 1 week, the color of a 2% solution of bromine in methylene chloride fades noticeably, and the odor of HBr can be detected in the reagent. Although the decolorization tests still work satisfactorily, the presence of HBr makes it difficult to distinguish between addition and substitution reactions. A freshly prepared solution of bromine in methylene chloride must be used to make this distinction. Deterioration of the reagent can be forestalled by storing it in a brown glass bottle. Most other substitute solvents also present problems. Ethers, for instance, react slowly in the same way as methylene chloride, and hydrocarbons, like hexane, are not general enough solvents to be able to dissolve all the possible test compounds.

Test Compounds. Try this test with cyclohexene, cyclohexane, toluene, and acetone.

A successful test depends on the addition of bromine, a red liquid, to a double or a triple bond to give a colorless dibromide:

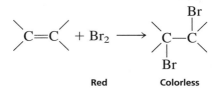

Not all double bonds react with bromine–carbon tetrachloride solution. Only those that are electron-rich are sufficiently reactive nucleophiles to initiate the reaction. A double bond that is substituted by electron-withdrawing groups often fails to react or reacts slowly. Fumaric acid is an example of a compound that fails to give the reaction.

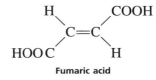

Aromatic compounds either do not react with bromine–carbon tetrachloride reagent or they react by **substitution.** Only the aromatic rings that have activating groups as substituents (—OH, —OR, or —NR$_2$) give the substitution reaction.

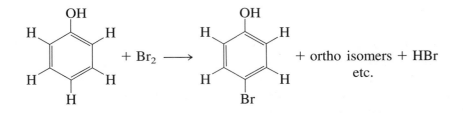

Some ketones and aldehydes react with bromine to give a **substitution product,** but this reaction is slow except for acetone and some aldehydes and ketones that have a high enol content. When substitution occurs, not only is the bromine color discharged, but hydrogen bromide gas is also evolved.

POTASSIUM PERMANGANATE (BAEYER TEST)

Procedure. Dissolve 25 mg of a solid unknown or two drops of a liquid unknown in 2 mL of water or 95% ethanol (1,2-dimethoxyethane may also be used). Slowly add a 1% aqueous solution (weight/volume) of potassium permanganate, drop by drop while shaking, to the unknown. In a positive test, the purple color of the reagent is discharged, and a brown precipitate of manganese dioxide forms, usually within 1 minute. If alcohol was the solvent, the solution should not be allowed to

stand for more than 5 minutes because oxidation of the alcohol will begin slowly. Because permanganate solutions undergo some decomposition to manganese dioxide on standing, any small amount of precipitate should be interpreted with caution.
Test Compounds. Try this test on cyclohexene and toluene.

This test is positive for double and triple bonds but not for aromatic rings. It depends on the conversion of the purple ion MnO_4^- to a brown precipitate of MnO_2 following the oxidation of an unsaturated compound.

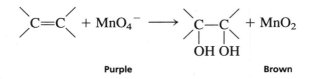

Purple   **Brown**

Other easily oxidized compounds also give a positive test with potassium permanganate solution. These substances include aldehydes, some alcohols, phenols, and aromatic amines. If you suspect that any of these functional groups are present, you should interpret the test with caution.

SPECTROSCOPY

Infrared

DOUBLE BONDS (C=C)

C=C stretch usually occurs near 1680–1620 cm^{-1}. Symmetrical alkenes may have no absorption.

C—H stretch of vinyl hydrogens occurs >3000 cm^{-1}, but usually not higher than 3150 cm^{-1}.

C—H out-of-plane bending occurs near 1000–700 cm^{-1} (see Appendix 3).

TRIPLE BONDS (C≡C)

C≡C stretch usually occurs near 2250–2100 cm^{-1}. The peak is usually sharp. Symmetrical alkynes show no absorption.

C—H stretch of terminal acetylenes occurs near 3310–3200 cm^{-1}.

Nuclear Magnetic Resonance

Vinyl hydrogens have resonance near 5–7 ppm and have coupling values as follows: J_{trans} = 11–18 Hz, J_{cis} = 6–15 Hz, $J_{geminal}$ = 0–5 Hz. Allylic hydrogens have resonance near 2 ppm. Acetylenic hydrogens have resonance near 2.8–3.0 ppm.

Tests for Aromaticity

None of the unknowns to be issued for this experiment will be simple aromatic hydrocarbons. All aromatic compounds will have a principal functional group as a part of their structure. Nevertheless, in many cases it will be useful to be able to recognize the presence of an aromatic ring. Although spectroscopy provides the easiest method of determining aromatic systems, often they can be detected by a simple ignition test.

IGNITION TEST

Procedure.   In a hood, place a small amount of the compound on a spatula and place it in the flame of a Bunsen burner. Observe whether a sooty flame is the result. Compounds giving a sooty yellow flame have a high degree of unsaturation and may be aromatic.

Test Compound.   Try this test with naphthalene.

The presence of an aromatic ring or other centers of unsaturation will lead to the production of a sooty yellow flame in this test. Compounds that contain little oxygen and have a high carbon-to-hydrogen ratio, burn at a low temperature with a yellow flame. Much carbon is produced when they are burned. Compounds that contain oxygen generally burn at a higher temperature with a clean blue flame.

SPECTROSCOPY

Infrared

$C=C$ aromatic ring double bonds appear in the 1600–1450 cm^{-1} region. There are often four sharp absorptions that occur in pairs near 1600 cm^{-1} and 1450 cm^{-1} which are characteristic of an aromatic ring.

Special ring absorptions: There are often weak ring absorptions around 2000–1600 cm^{-1}. These are often obscured, but when they can be observed, the relative shapes and numbers of these peaks can often be used to ascertain the type of ring substitution (see Appendix 3).

$=C-H$ stretch, aromatic ring: The aromatic C—H stretch always occurs at a higher frequency than 3000 cm^{-1}.

$=C-H$ out-of-plane bending peaks appear in the region 900–690 cm^{-1}. The number and position of these peaks can be used to determine the substitution pattern of the ring (see Appendix 3).

Nuclear Magnetic Resonance

Hydrogens attached to an aromatic ring usually have resonance near 7 ppm. Monosubstituted rings not substituted by anisotropic or electronegative groups usually give a single resonance for all the ring hydrogens. Monosubstituted rings with anisotropic or electronegative groups usually have the aromatic resonances split into two groups integrating either 3:2 or 2:3. A nonsymmetric, *para*-disubstituted ring has a characteristic four-peak splitting pattern (see Appendix 4).

EXPERIMENT 57D

Aldehydes and Ketones

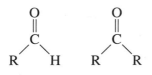

Compounds containing the carbonyl function group $\ce{C=O}$, where it has only hydrogen atoms or alkyl groups as substituents, are called aldehydes RCHO or ketones RCOR'. The chemistry of these compounds is primarily due to the chemistry of the carbonyl functional groups. These compounds are identified by the distinctive reactions of the carbonyl function.

SOLUBILITY CHARACTERISTICS	CLASSIFICATION TESTS
HCl NaHCO$_3$ NaOH H$_2$SO$_4$ Ether (−) (−) (−) (+) (+) Water: <C$_5$ and some C$_6$ (+) 　　　　　>C$_5$ (−)	**Aldehydes and ketones** 　2,4-Dinitrophenylhydrazine **Aldehydes only**　　　　**Methyl ketones** 　Chromic acid　　　　　　Iodoform test 　Tollens reagent **Compounds with high enol content** 　Ferric chloride test

Waste Disposal

Solutions containing 2,4-dinitrophenylhydrazine or derivatives formed from it should be placed in a waste container designated for these compounds. Any solution containing chromium must be disposed of in a waste container specifically identified for the disposal of chromium wastes. Dispose of all solutions containing silver by acidifying them with 5% hydrochloric acid and then placing them into a waste container designated for this purpose. Dispose of all other aqueous solutions in the container designated for aqueous waste. Any remaining organic compounds must be disposed of in the appropriate organic waste container.

Classification Tests

Most aldehydes and ketones give a solid, yellow-to-red precipitate when mixed with 2,4-dinitrophenylhydrazine. However, only aldehydes will reduce chromium(VI) or silver(I). By this difference in behavior, you can differentiate between aldehydes and ketones.

2,4-DINITROPHENYLHYDRAZINE

Procedure. Place one drop of the liquid unknown in a small test tube and add 1 mL of the 2,4-dinitrophenylhydrazine reagent. If the unknown is a solid, dissolve about 10 mg (estimate) in a minimum amount of 95% ethanol or bis(2-ethoxyethyl) ether before adding the reagent. Shake the mixture vigorously. Most aldehydes and ketones will give a yellow-to-red precipitate immediately. However, some compounds will require up to 15 minutes, or even *gentle* heating, to give a precipitate. A precipitate indicates a positive test.

Test Compounds. Try this test on cyclohexanone, benzaldehyde, and benzophenone.

> **Caution:** Many derivatives of phenylhydrazine are suspected carcinogens (see p. 21) and should be handled with care. Avoid contact.

Reagent. Dissolve 3.0 g of 2,4-dinitrophenylhydrazine in 15 mL of concentrated sulfuric acid. In a beaker mix 20 mL of water and 70 mL of 95% ethanol. While stirring vigorously, add the 2,4-dinitrophenylhydrazine solution to the aqueous ethanol mixture slowly. After thorough mixing, filter the solution by gravity through a fluted filter.

Aldehyde or ketone   2,4-Dinitrophenylhydrazine

2,4-Dinitrophenylhydrazone

Most aldehydes and ketones give a precipitate, but esters generally do not give this result. Thus, an ester usually can be eliminated by this test. The color of the 2,4-dinitrophenylhydrazone (precipitate) formed is often a guide to the amount of conjugation in the original aldehyde or ketone. Unconjugated ketones, such as cyclohexanone, give yellow precipitates, whereas conjugated ketones, such as benzophenone, give orange-to-red precipitates. Compounds that are highly conjugated give red precipitates. However, the 2,4-dinitrophenylhydrazine reagent is itself orange-red, and the color of any precipitate must be judged cautiously. Occasionally, compounds that are either strongly basic or strongly acidic precipitate the unreacted reagent.

Some allylic and benzylic alcohols give this test result because the reagent can oxidize them to aldehydes and ketones, which subsequently react. Some alcohols may be contaminated with carbonyl impurities, either because of their method of synthesis (reduction) or because they have become air-oxidized. A precipitate formed from small amounts of impurity in the solution will be formed in small amount. With some caution, a test that gives only a slight amount of precipitate can usually be ignored. The infrared spectrum of the compound should establish its identity and identify any impurities present.

CHROMIC ACID TEST

Procedure. Dissolve one drop of a liquid or 10 mg (approximate) of a solid aldehyde in 1 mL of *reagent-grade* acetone. Add several drops of the chromic acid reagent, a drop at a time while shaking the mixture. A positive test is indicated by a green precipitate and a loss of the orange color in the reagent. With aliphatic aldehydes RCHO, the solution turns cloudy within 5 seconds and a precipitate appears within 30 seconds. With aromatic aldehydes ArCHO, it generally takes 30–120 seconds for a precipitate to form, but with some it may take even longer.

In a negative test, there is usually no precipitate. In some cases, however, a precipitate forms, but the solution remains orange.

In performing this test, make quite sure that the acetone used for the solvent does not give a positive test with the reagent. Add several drops of the chromic acid reagent to a few drops of the reagent acetone contained in a small test tube. Allow this mixture to stand for 3–5 minutes. If no reaction has occurred by this time, the acetone is pure enough to use as a solvent for the test. If a positive test resulted, try another bottle of acetone, or distill some acetone from potassium permanganate to purify it.

Test Compounds. Try this test on benzaldehyde, butanal (butyraldehyde), and cyclohexanone.

Caution: Many compounds of chromium(VI) are suspected carcinogens (see p. 21) and should be handled with care. Avoid contact.

Reagent.   Dissolve 1.0 g of chromic oxide CrO_3 in 1 mL of concentrated sulfuric acid. Then dilute this mixture carefully with 3 mL of water.

This test has as its basis the fact that aldehydes are easily oxidized to the corresponding carboxylic acid by chromic acid. The green precipitate is due to chromium(III) sulfate.

$$2\ CrO_3 + 2\ H_2O \underset{}{\overset{H^+}{\rightleftharpoons}} 2\ H_2CrO_4 \underset{}{\overset{H^+}{\rightleftharpoons}} H_2Cr_2O_7 + H_2O$$

$$\underset{\text{Orange}}{3\ RCHO + H_2Cr_2O_7 + 3\ H_2SO_4} \longrightarrow \underset{\text{Green}}{3\ RCOOH + Cr_2(SO_4)_3 + 4\ H_2O}$$

Primary and secondary alcohols are also oxidized by this reagent (see Experiment 57H). Therefore, this test is not useful in identifying aldehydes *unless* a positive identification of the carbonyl group has already been made. Aldehydes give a 2,4-dinitrophenylhydrazine test result, whereas alcohols do not.

There are numerous other tests used to detect the aldehyde functional group. Most are based on an easily detectable oxidation of the aldehyde to a carboxylic acid. The most common tests are the Tollens, Fehling, and Benedict tests. The Benedict test is described in Experiment 58. Only the Tollens test is described here.

TOLLENS TEST

Procedure.   The reagent must be prepared immediately before use. To prepare the reagent, mix 1 mL of Tollens solution A with 1 mL of Tollens solution B. A precipitate of silver oxide will form. Add enough dilute (10%) ammonia solution (dropwise) to the mixture to dissolve the silver oxide *just barely.* The reagent so prepared can be used immediately for the following test.

Dissolve one drop of a liquid aldehyde or 10 mg (approximate) of a solid aldehyde in the minimum amount of bis(2-ethoxyethyl) ether. Add this solution, a little at a time, to the 2–3 mL of reagent contained in a small test tube. Shake the solution well. If a mirror of silver is deposited on the inner walls of the test tube, the test is positive. In some cases, it may be necessary to warm the test tube in a bath of warm water.

> **Caution:** The reagent should be prepared immediately before use and all residues disposed of immediately after use. Dispose of any residues by acidifying them with 5% hydrochloric acid and then placing them in a waste container designated for this purpose. On standing, the reagent tends to form silver fulminate, a *very explosive* substance. Solutions containing the mixed Tollens reagent should never be stored.

Test Compounds. Try the test on acetone and benzaldehyde.

Reagents. Solution A: Dissolve 3.0 g of silver nitrate in 30 mL of water. Solution B: Prepare a 10% sodium hydroxide solution.

Most aldehydes reduce ammoniacal silver nitrate solution to give a precipitate of silver metal. The aldehyde is oxidized to a carboxylic acid:

$$RCHO + 2\,Ag(NH_3)_2OH \longrightarrow 2\,Ag + RCOO^-NH_4^+ + H_2O + NH_3$$

Ordinary ketones do not give a positive result in this test. The test should be used only if it has already been shown that the unknown compound is either an aldehyde or a ketone.

IODOFORM TEST

Procedure. Prepare a 60 to 70°C water bath in a beaker. Using a Pasteur pipet, add six drops of a liquid unknown to a 15 × 100-mm or 15 × 125-mm test tube. Alternatively, 0.06 g of a solid unknown may be used. Dissolve the liquid or solid unknown compound in 2 mL of 1,2-dimethoxyethane. Add 2 mL of 10% aqueous sodium hydroxide solution and place the test tube in the hot water bath. Next, add 4 mL of iodine–potassium iodide solution in 1-mL portions to the test tube. *Cork* the test tube and shake it after adding each portion of iodine reagent. Heat the mixture in the hot water bath for about 5 minutes, shaking the test tube occasionally. It is likely that some or all the dark color of the iodine reagent will be discharged.

If the dark color of the iodine reagent is still apparent following heating, add 10% sodium hydroxide solution until the dark color of the iodine reagent has been discharged. Shake the mixture in the test tube (corked) during the addition of the sodium hydroxide. Care need not be taken to avoid adding excess sodium hydroxide.

After the dark iodine color of the solution has been discharged, fill the test tube with water to within 2 cm of the top. Cork the test tube and shake it vigorously. Allow the tube to stand for at least 15 minutes at room temperature. The appearance of a pale yellow precipitate of iodoform CHI_3 constitutes a positive test, indicating that the unknown is a methyl ketone or a compound that is easily oxidized to a methyl ketone, such as a 2-alkanol. Other ketones will also decolorize the iodine solution, but they will not give a precipitate of iodoform *unless* there is an impurity of a methyl ketone present in the unknown.

The yellow precipitate usually settles out slowly onto the bottom of the test tube. Sometimes the yellow color of iodoform is masked by a dark substance. If this is the case, cork the test tube and shake it vigorously. If the dark color persists, add more sodium hydroxide solution and shake the test tube again. Then allow the tube to stand for at least 15 minutes. If there is some doubt as to whether the solid is iodoform, collect the precipitate on a Hirsch funnel and dry it. Iodoform melts at 119–121°C.

Test Compounds. Try the test on 2-heptanone, 4-heptanone (dipropyl ketone), and 2-pentanol.

Reagents. The iodine reagent is prepared by dissolving 20 g of potassium iodide and 10 g of iodine in 100 mL of water. The aqueous sodium hydroxide solution is prepared by dissolving 10 g of sodium hydroxide in 100 mL of water.

The basis of this test is the ability of certain compounds to form a precipitate of iodoform when treated with a basic solution of iodine. Methyl ketones are the most common types of compounds that give a positive result in this test. However, acetaldehyde CH_3CHO and alcohols with the hydroxyl group at the 2-position of the chain also give a precipitate of iodoform. 2-Alkanols of the type described are easily oxidized to methyl ketone under the conditions of the reaction. The other product of the reaction, besides iodoform, is the sodium or potassium salt of a carboxylic acid.

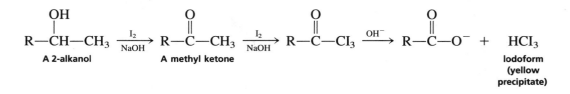

FERRIC CHLORIDE TEST

Procedure. Some aldehydes and ketones, those that have a high **enol content,** give a positive ferric chloride test, as described for phenols in Experiment 57F.

SPECTROSCOPY

Infrared

The carbonyl group is usually one of the strongest-absorbing groups in the infrared spectrum, with a very broad range: 1800–1650 cm^{-1}. The aldehyde functional group has *very characteristic* C—H stretch absorptions: two sharp peaks that lie *far outside* the usual region for —C—H, =C—H, or ≡C—H.

ALDEHYDES

C=O stretch at approximately 1725 cm^{-1}
   is normal. 1725–1685 cm^{-1}.[2]
C—H stretch (aldehyde–CHO) has two
   weak bands at about 2750 cm^{-1} and
   2850 cm^{-1}.

KETONES

C=O stretch at approximately 1715 cm^{-1}
   is normal. 1780–1665 cm^{-1}.[3]

---

[2]**Conjugation** moves the absorption to lower frequencies. **Ring strain** (cyclic ketones) moves the absorption to higher frequencies.
[3]See Footnote 2.

Nuclear Magnetic Resonance

Hydrogens alpha to a carbonyl group have resonance in the region between 2 and 3 ppm. The hydrogen of an aldehyde group has a characteristic resonance between 9 and 10 ppm. In aldehydes, there is coupling between the aldehyde hydrogen and any alpha hydrogens ($J = 1–3$ Hz).

DERIVATIVES

The most common derivatives of aldehydes and ketones are the 2,4-dinitrophenyl-hydrazones, oximes, and semicarbazones. Procedures for preparing these derivatives are given in Appendix 2.

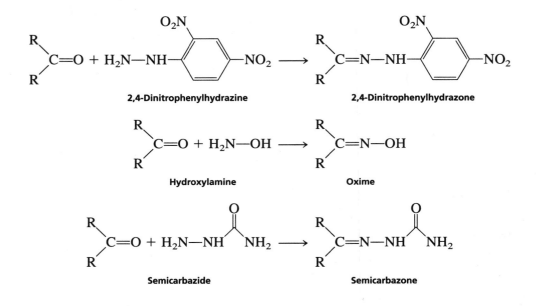

EXPERIMENT 57E

Carboxylic Acids

$$\underset{R}{\overset{\displaystyle O}{\underset{\displaystyle }{\overset{\displaystyle \|}{C}}}}\!\!-\!OH$$

Carboxylic acids are detectable mainly by their solubility characteristics. They are soluble in *both* dilute sodium hydroxide and sodium bicarbonate solutions.

SOLUBILITY CHARACTERISTICS					CLASSIFICATION TESTS
HCl	NaHCO$_3$	NaOH	H$_2$SO$_4$	Ether	pH of an aqueous solution
(−)	(+)	(+)	(+)	(+)	Sodium bicarbonate
Water: <C$_6$ (+)					Silver nitrate
>C$_6$ (−)					Neutralization equivalent

Waste Disposal

Dispose of all aqueous solutions in the container designated for aqueous waste. Any remaining organic compounds must be disposed of in the appropriate organic waste container.

Classification Tests

pH OF AN AQUEOUS SOLUTION

Procedure. If the compound is soluble in water, simply prepare an aqueous solution and check the pH with pH paper. If the compound is an acid, the solution will have a low pH.

Compounds that are insoluble in water can be dissolved in ethanol (or methanol) and water. First dissolve the compound in the alcohol and then add water until the solution *just* becomes cloudy. Clarify the solution by adding a few drops of the alcohol, and then determine its pH using pH paper.

SODIUM BICARBONATE

Procedure. Dissolve a small amount of the compound in a 5% aqueous sodium bicarbonate solution. Observe the solution carefully. If the compound is an acid, you will see bubbles of carbon dioxide form.

$$RCOOH + NaHCO_3 \longrightarrow RCOO^-Na^+ + H_2CO_3 \text{ (unstable)}$$

$$H_2CO_3 \longrightarrow CO_2 + H_2O$$

SILVER NITRATE

Procedure. Acids may give a false silver nitrate test, as described in Experiment 57B.

NEUTRALIZATION EQUIVALENT

Procedure.   Accurately weigh (three significant figures) approximately 0.2 g of the acid and place in a 125-mL Erlenmeyer flask. Dissolve the acid in about 50 mL of water or aqueous ethanol (the acid need not dissolve completely, because it will dissolve as it is titrated). Titrate the acid, using a solution of sodium hydroxide of known molarity (about 0.1M) and a phenolphthalein indicator.

Calculate the neutralization equivalent (NE) from the equation

$$NE = \frac{mg\ acid}{molarity\ of\ NaOH \times mL\ of\ NaOH\ added}$$

The NE is identical to the equivalent weight of the acid. If the acid has only one carboxyl group, the neutralization equivalent and the molecular weight of the acid are identical. If the acid has more than one carboxyl group, the neutralization equivalent equals the molecular weight of the acid divided by the number of carboxyl groups, that is, the equivalent weight. The NE can be used much like a derivative to identify a specific acid.

Many phenols are acidic enough to behave much like carboxylic acids. This is especially true of those substituted with electron-withdrawing groups at the *ortho* and *para* ring positions. These phenols, however, can usually be eliminated either by the ferric chloride test (Experiment 57F) or by spectroscopy (phenols have no carbonyl group).

SPECTROSCOPY

Infrared

C=O stretch is very strong and often broad in the region between 1725 cm^{-1} and 1690 cm^{-1}.

O—H stretch is a very broad absorption in the region between 3300 cm^{-1} and 2500 cm^{-1}; it usually overlaps the CH stretch region.

Nuclear Magnetic Resonance

The acid proton of a —COOH group usually has resonance near 12.0 ppm.

DERIVATIVES

Derivatives of acids are usually amides. They are prepared via the corresponding acid chloride:

$$R-\overset{\overset{\displaystyle O}{\|}}{C}-OH + SOCl_2 \longrightarrow R-\overset{\overset{\displaystyle O}{\|}}{C}-Cl + SO_2 + HCl$$

The most common derivatives are the amides, the anilides, and the *p*-toluidides.

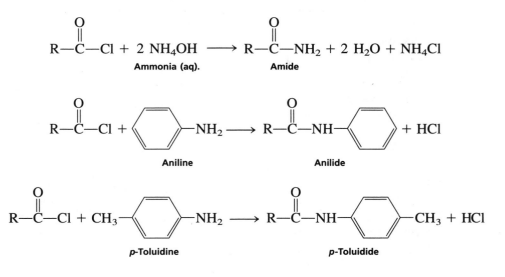

Ammonia (aq).   Amide

Aniline   Anilide

p-Toluidine   *p*-Toluidide

Procedures for the preparation of these derivatives are given in Appendix 2.

EXPERIMENT 57F

Phenols

Like carboxylic acids, phenols are acidic compounds. However, except for the nitrosubstituted phenols (discussed in the section covering solubilities), they are not as acidic as the carboxylic acids. The pK_a of a typical phenol is 10, whereas the pK_a of a carboxylic acid is usually near 5. Hence, phenols are generally not soluble in the weakly basic sodium bicarbonate solution, but they dissolve in sodium hydroxide solution, which is more strongly basic.

SOLUBILITY CHARACTERISTICS					CLASSIFICATION TESTS
HCl	NaHCO$_3$	NaOH	H$_2$SO$_4$	Ether	Colored phenolate anion
(−)	(−)	(+)	(+)	(+)	Ferric chloride
Water: Most are insoluble, although phenol itself and the nitrophenols are soluble					Bromine/water

Waste Disposal

Dispose of all aqueous solutions in the container designated for aqueous waste. Any remaining organic compounds must be disposed of in the appropriate organic waste container.

Classification Tests

SODIUM HYDROXIDE SOLUTION

With phenols that have a high degree of conjugation possible in their conjugate base (phenolate ion), the anion is often colored. To observe the color, dissolve a small amount of the phenol in 10% aqueous sodium hydroxide solution. Some phenols do not give a color. Others have an insoluble anion and give a precipitate. The more acidic phenols, like the nitrophenols, tend more toward colored anions.

FERRIC CHLORIDE

Procedure 1 (Water-Soluble Phenols). Add several drops of a 2.5% aqueous solution of ferric chloride to 1 mL of a dilute aqueous solution (about 1–3% by weight of the phenol). Most phenols produce an intense red, blue, purple, or green color. Some colors are transient, and it may be necessary to observe the solution carefully just as the solutions are mixed. The formation of a color is usually immediate, but the color may not last over any great period. Some phenols do not give a positive result in this test, so a negative test must not be taken as significant without other adequate evidence.

Test Compound. Try this test on phenol.

Procedure 2 (Water-Insoluble Phenols). Many phenols do not give a positive result when Procedure 1 is used. However, Procedure 2 often gives a positive result. Dissolve or suspend 20 mg of a solid phenol or one drop of a liquid phenol in 1 mL of chloroform. Add one drop of pyridine and three to five drops of a 1% (weight/volume) solution of ferric chloride in chloroform.

The colors observed in this test result from the formation of a complex of the phenols with Fe(III) ion. Carbonyl compounds that have a high enol content also give a positive result in this test.

BROMINE/WATER

Procedure. Prepare a 1% aqueous solution of the unknown and then add a saturated solution of bromine in water to it, drop by drop while shaking, until the bromine color is no longer discharged. A positive test is indicated by the precipitation of a substitution product at the same time that the bromine color of the reagent is discharged.

Test Compound. Try this test on phenol.

Aromatic compounds with ring-activating substituents give a positive test with bromine in water. The reaction is an aromatic substitution reaction which introduces bromine atoms into the aromatic ring at the positions *ortho* and *para* to the hydroxyl group. All available positions are usually substituted. The precipitate is the brominated phenol, which is generally insoluble because of its large molecular weight.

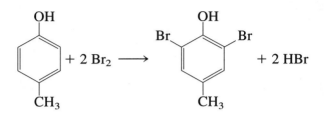

Other compounds that give a positive result with this test include aromatic compounds that have activating substituents other than hydroxyl. These compounds include anilines and alkoxyaromatics.

SPECTROSCOPY

Infrared

O—H stretch is observed near 3600 cm^{-1}.
C—O stretch is observed near 1200 cm^{-1}.
The typical aromatic ring absorptions between 1600 cm^{-1} and 1450 cm^{-1} are also found.
Aromatic C—H is observed near 3100 cm^{-1}.

Nuclear Magnetic Resonance

Aromatic protons are observed near 7 ppm. The hydroxyl proton has a resonance position that is concentration-dependent.

DERIVATIVES

Phenols form the same derivatives as alcohols do (Experiment 57H). They form ure-thanes by reaction with isocyanates. Phenylurethanes are used for alcohols, and the α-

naphthylurethanes are more useful for phenols. Like alcohols, phenols yield 3,5-dinitrobenzoates.

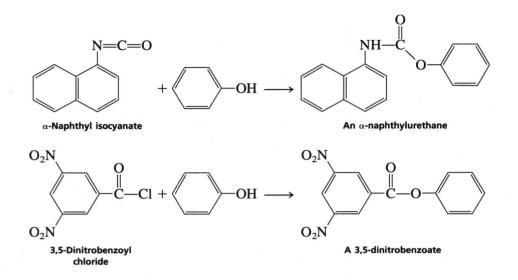

α-Naphthyl isocyanate   An α-naphthylurethane

3,5-Dinitrobenzoyl chloride   A 3,5-dinitrobenzoate

The bromine–water reagent yields solid bromo derivatives of phenols in several cases. These solid derivatives can be used to characterize an unknown phenol. Procedures for preparing these derivatives are given in Appendix 2.

EXPERIMENT 57G

Amines

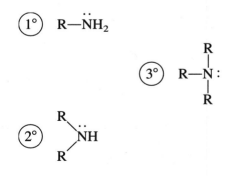

Amines are detected best by their solubility behavior and their basicity. They are the only basic compounds that will be issued for this experiment. Hence, once the compound has been identified as an amine, the main problem that remains is to decide whether it is primary (1°), secondary (2°), or tertiary (3°). This can usually be decided either by the nitrous acid tests or by infrared spectroscopy.

SOLUBILITY CHARACTERISTICS	CLASSIFICATION TESTS
HCl NaHCO$_3$ NaOH H$_2$SO$_4$ Ether (+) (−) (−) (+) (+) Water: <C$_6$ (+) >C$_6$ (−)	pH of an aqueous solution Hinsberg test Nitrous acid test Acetyl chloride

Waste Disposal

Residues from the nitrous acid test should be poured into a waste container containing 6N hydrochloric acid. Dispose of all other aqueous solutions in the container designated for aqueous waste. Any remaining organic compounds must be disposed of in the appropriate organic waste container.

Classification Tests

NITROUS ACID TEST

Procedure. In a large test tube, add 2 mL of water and 8 drops of concentrated sulfuric acid. Dissolve 0.1 g of an amine in this solution (a precipitate of the amine salt may form at this point, but it will not interfere with the test). Cool the solution to 5°C or less in an ice bath. Also cool 2 mL of 10% aqueous sodium nitrite in another test tube. In a third test tube, prepare a solution of 0.1-g β-naphthol in 2 mL of aqueous 10% sodium hydroxide, and place it in an ice bath to cool. Add the cold sodium nitrite solution, drop by drop while shaking, to the cooled solution of the amine. Look for bubbles of nitrogen gas. Be careful not to confuse the evolution of the *colorless* nitrogen gas with an evolution of *brown* nitrogen oxide gas. Substantial evolution of nitrogen gas at 5°C or below indicates a primary aliphatic amine RNH$_2$. The formation of a yellow oil or a yellow solid usually indicates a secondary amine R$_2$NH. Either tertiary amines do not react, or they behave like secondary amines.

If little or no gas evolves at 5°C, take *half* the solution and warm it gently to about room temperature. Nitrogen gas bubbles at this elevated temperature indicate that the original compound was a **primary aromatic amine** ArNH$_2$. Take the remaining solution and drop by drop add the solution of β-naphthol in base. If a red dye precipitates, the unknown has been conclusively shown to be a primary aromatic amine ArNH$_2$.

Test Compounds.   Try this test with aniline, *N*-methylaniline, and butylamine.

> **Caution:** The products of this reaction may include nitrosamines. Nitrosamines are suspected carcinogens. Avoid contact and dispose of all residues by pouring them into a waste container that contains 6*N* hydrochloric acid.

Before you make this test, it should definitely be proved by some other method that the unknown is an amine. Many other compounds react with nitrous acid (phenols, ketones, thiols, amides), and a positive result with one of these could lead to an incorrect interpretation.

The test is best used to distinguish *primary* aromatic and *primary* aliphatic amines from secondary and tertiary amines. It also differentiates aromatic and aliphatic primary amines. It cannot distinguish between secondary and tertiary amines. Primary aliphatic amines lose nitrogen gas at low temperatures under the conditions of this test. Aromatic amines yield a more stable diazonium salt and do not lose nitrogen until the temperature is elevated. In addition, aromatic diazonium salts produce a red azo dye when β-naphthol is added. Secondary and tertiary amines produce yellow nitroso compounds, which may be soluble or may be oils or solids. Many nitroso compounds have been shown to be carcinogenic. Avoid contact and immediately dispose of all such solutions in an appropriate waste container.

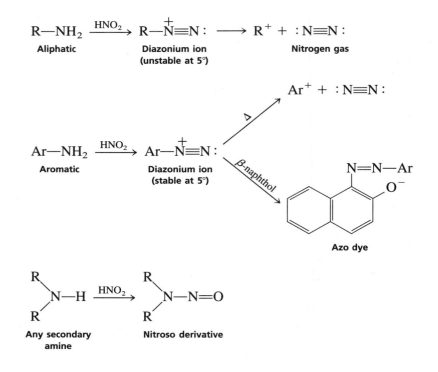

HINSBERG TEST

A traditional method for classifying amines is the **Hinsberg test.** A discussion of this test can be found in the comprehensive textbooks listed on page 490. We have found that infrared spectroscopy is a more reliable method of distinguishing between primary, secondary, and tertiary amines.

pH OF AN AQUEOUS SOLUTION

Procedure. If the compound is soluble in water, simply prepare an aqueous solution and check the pH with pH paper. If the compound is an amine, it will be basic, and the solution will have a high pH. Compounds that are insoluble in water can be dissolved in ethanol–water or 1,2-dimethoxyethane–water.

ACETYL CHLORIDE

Procedure. Primary and secondary amines give a positive acetyl chloride test result (liberation of heat). This test is described for alcohols in Experiment 57H. When the test mixture is diluted with water, primary and secondary amines often give a solid acetamide derivative; tertiary amines do not.

SPECTROSCOPY

Infrared

N—H stretch. Both aliphatic and aromatic primary amines show two absorptions (doublet due to symmetric and asymmetric stretches) in the region 3500–3300 cm^{-1}. Secondary amines show a single absorption in this region. Tertiary amines have no N—H bonds.

N—H bend. Primary amines have a strong absorption at 1640–1560 cm^{-1}. Secondary amines have an absorption at 1580–1490 cm^{-1}.

Aromatic amines show bands typical for the aromatic ring in the region 1600–1450 cm^{-1}. Aromatic C—H is observed near 3100 cm^{-1}.

Nuclear Magnetic Resonance

The resonance position of amino hydrogens is extremely variable. The resonance may also be very broad (quadrupole broadening). Aromatic amines give resonances near 7 ppm due to the aromatic ring hydrogens.

DERIVATIVES

The derivatives of amines that are most easily prepared are the acetamides and the benzamides. These derivatives work well for both primary and secondary amines but not for tertiary amines.

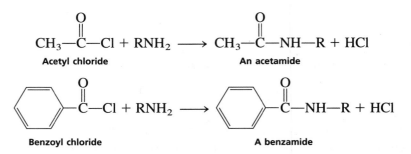

The most general derivative that can be prepared is the picric acid salt, or picrate, of an amine. This derivative can be used for primary, secondary, and tertiary amines.

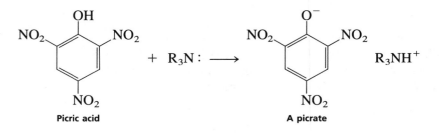

For tertiary amines, the methiodide salt is often useful.

$$CH_3I + R_3N: \longrightarrow CH_3-NR_3{}^+I^-$$
A methiodide

Procedures for preparing derivatives from amines can be found in Appendix 2.

EXPERIMENT 57H

Alcohols

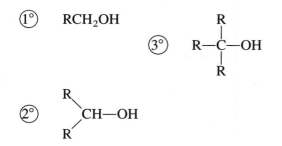

Alcohols are neutral compounds. The only other classes of neutral compounds used in this experiment are the aldehydes and ketones and the esters. Alcohols and esters usually do not give a positive 2,4-dinitrophenylhydrazine test; aldehydes and ketones do. Esters do not react with acetyl chloride or with Lucas reagent, as alcohols do, and they are easily distinguished from alcohols on this basis. Primary and secondary alcohols are easily oxidized; esters and tertiary alcohols are not. A combination of the Lucas test and the chromic acid test will differentiate among primary, secondary, and tertiary alcohols.

SOLUBILITY CHARACTERISTICS					CLASSIFICATION TESTS
HCl	NaHCO$_3$	NaOH	H$_2$SO$_4$	Ether	Acetyl chloride
(−)	(−)	(−)	(+)	(+)	Lucas test
Water: <C$_6$ (+)					Chromic acid test
>C$_6$ (−)					Iodoform test

Waste Disposal

Any solution containing chromium must be disposed of by placing it in a waste container specifically identified for the disposal of chromium wastes. Dispose of all other aqueous solutions in the container designated for aqueous waste. Any remaining organic compounds must be disposed of in the appropriate organic waste container.

Classification Tests

ACETYL CHLORIDE

Procedure. Cautiously add about 10–15 drops of acetyl chloride, drop by drop, to about 0.5 mL of the liquid alcohol contained in a small test tube. Evolution of heat and hydrogen chloride gas indicates a positive reaction. Addition of water will sometimes precipitate the acetate.

Acid chlorides react with alcohols to form esters. Acetyl chloride forms acetate esters.

$$CH_3-\overset{\overset{\displaystyle O}{\|}}{C}-Cl + ROH \longrightarrow CH_3-\overset{\overset{\displaystyle O}{\|}}{C}-O-R + HCl$$

Usually the reaction is exothermic, and the heat evolved is easily detected. Phenols react with acid chlorides somewhat as alcohols do. Hence, phenols should be eliminated as possibilities before this test is attempted. Amines also react with acetyl chloride to evolve heat (see Experiment 57G). This test does not work well with solid alcohols.

LUCAS TEST

Procedure.   Place 2 mL of Lucas reagent in a small test tube and add three to four drops of the alcohol. Stopper the test tube and shake it vigorously. Tertiary (3°), benzylic, and allylic alcohols give an immediate cloudiness in the solution as the insoluble alkyl halide separates from the aqueous solution. After a short time, the immiscible alkyl halide will form a separate layer. Secondary (2°) alcohols produce a cloudiness after 2–5 minutes. Primary (1°) alcohols dissolve in the reagent to give a clear solution. Some secondary alcohols may have to be heated slightly to encourage reaction with the reagent.

> **NOTE:** This test works only for alcohols that are soluble in the reagent. This often means that alcohols with more than six carbon atoms cannot be tested.

Test Compounds.   Try this test with 1-butanol (*n*-butyl alcohol), 2-butanol (*sec*-butyl alcohol), and 2-methyl-2-propanol (*t*-butyl alcohol).

Reagent.   Cool 10 mL of concentrated hydrochloric acid in a beaker, using an ice bath. While still cooling and while stirring, dissolve 16 g of anhydrous zinc chloride in the acid.

This test depends on the appearance of an alkyl chloride as an insoluble second layer when an alcohol is treated with a mixture of hydrochloric acid and zinc chloride (Lucas reagent):

$$R\text{—}OH + HCl \xrightarrow{ZnCl_2} R\text{—}Cl + H_2O$$

Primary alcohols do not react at room temperature; therefore, the alcohol is seen simply to dissolve. Secondary alcohols react slowly, whereas tertiary, benzylic, and allylic alcohols react instantly. These relative reactivities are explained on the same basis as the silver nitrate reaction, which is discussed in Experiment 57B. Primary carbocations are unstable and do not form under the conditions of this test. Hence, no results are observed for primary alcohols.

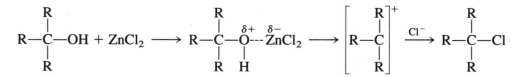

The Lucas test does not work well with solid alcohols.

CHROMIC ACID TEST

Procedure. Dissolve one drop of a liquid or about 10 mg of a solid alcohol in 1 mL of *reagent-grade* acetone. Add one drop of the chromic acid reagent and note the result that occurs within 2 seconds. A positive test for a primary or a secondary alcohol is the appearance of a blue-green color. Tertiary alcohols do not produce the test result within 2 seconds, and the solution remains orange. To make sure that the acetone solvent is pure and does not give a positive test result, add one drop of chromic acid to 1 mL of acetone that does not have an unknown dissolved in it. The orange color of the reagent should persist for *at least* 3 seconds. If it does not, a new bottle of acetone should be used.

Test Compounds. Try this test with 1-butanol (*n*-butyl alcohol), 2-butanol (*sec*-butyl alcohol), and 2-methyl-2-propanol (*t*-butyl alcohol).

> **Caution:** Many compounds of chromium(VI) are suspected carcinogens (see p. 21) and should be handled with care. Avoid contact.

Reagent. Dissolve 1 g of chromic oxide CrO_3 in 1 mL of concentrated sulfuric acid. Carefully add the mixture to 3 mL of water.

This test is based on the reduction of chromium(VI), which is orange, to chromium(III), which is green, when an alcohol is oxidized by the reagent. A change in color of the reagent from orange to green represents a positive test. Primary alcohols are oxidized by the reagent to carboxylic acids; secondary alcohols are oxidized to ketones.

$$2\ CrO_3 + 2\ H_2O \xrightarrow{H^+} 2\ H_2CrO_4 \xrightarrow{H^+} H_2Cr_2O_7 + H_2O$$

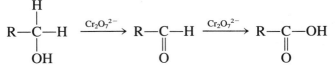

Primary alcohols

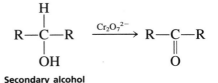

Secondary alcohol

Although primary alcohols are first oxidized to aldehydes, the aldehydes are further oxidized to carboxylic acids. The ability of chromic acid to oxidize aldehydes but not ketones is taken advantage of in a test that uses chromic acid to distinguish between aldehydes and ketones (Experiment 57D). Secondary alcohols are oxidized to ketones, but no further. Tertiary alcohols are not oxidized at all by the reagent. Hence, this test can be used to distinguish primary and secondary alcohols from tertiary alcohols. Unlike the Lucas test, this test can be used with all alcohols regardless of molecular weight and solubility.

IODOFORM TEST

Alcohols with the hydroxyl group at the 2-position of the chain give a positive iodoform test. See the discussion in Experiment 57D, Aldehydes and Ketones.

SPECTROSCOPY

Infrared

O—H stretch. A medium to strong, and usually broad, absorption comes in the region 3600–3200 cm^{-1}. In dilute solutions or with little hydrogen bonding, there is a sharp absorption near 3600 cm^{-1}. In more concentrated solutions, or with considerable hydrogen bonding, there is a broad absorption near 3400 cm^{-1}. Sometimes both bands appear.

C—O stretch. There is a strong absorption in the region 1200–1050 cm^{-1}. Primary alcohols absorb nearer 1050 cm^{-1}; tertiary alcohols and phenols absorb nearer 1200 cm^{-1}. Secondary alcohols absorb in the middle of this range.

NUCLEAR MAGNETIC RESONANCE

The hydroxyl resonance is extremely concentration-dependent, but it is usually found between 1 and 5 ppm. Under normal conditions, the hydroxyl proton does not couple with protons on adjacent carbon atoms.

DERIVATIVES

The most common derivatives for alcohols are the 3,5-dinitrobenzoate esters and the phenylurethanes. Occasionally, the α-naphthylurethanes (Experiment 57F) are also prepared, but these latter derivatives are more often used for phenols.

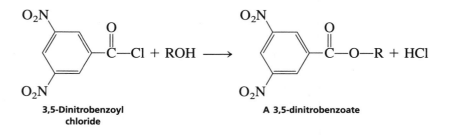

3,5-Dinitrobenzoyl
chloride   A 3,5-dinitrobenzoate

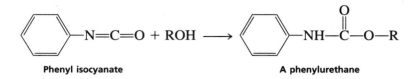

Phenyl isocyanate   **A phenylurethane**

Procedures for preparing these derivatives are given in Appendix 2.

EXPERIMENT 57I

Esters

Esters are formally considered "derivatives" of the corresponding carboxylic acid. They are frequently synthesized from the carboxylic acid and the appropriate alcohol:

$$R-COOH + R'-OH \underset{}{\overset{H^+}{\rightleftharpoons}} R-COOR' + H_2O$$

Thus, esters are sometimes referred to as though they were composed of an acid part and an alcohol part.

Although esters, like aldehydes and ketones, are neutral compounds that have a carbonyl group, they do not usually give a positive 2,4-dinitrophenylhydrazine test result. The two most common tests for identifying esters are the basic hydrolysis and ferric hydroxamate tests.

SOLUBILITY CHARACTERISTICS					CLASSIFICATION TESTS
HCl	NaHCO$_3$	NaOH	H$_2$SO$_4$	Ether	Ferric hydroxamate test
(−)	(−)	(−)	(+)	(+)	Basic hydrolysis
Water: <C$_4$ (+)					
>C$_5$ (−)					

Waste Disposal

Solutions containing hydroxylamine or derivatives formed from it should be poured into a waste container containing 6N hydrochloric acid. Dispose of any other aqueous solutions in the container designated for aqueous waste. Any remaining organic compounds must be disposed of in the appropriate organic waste container.

Classification Tests

FERRIC HYDROXAMATE TEST

Procedure. Before starting, you must determine whether the compound to be tested already has enough enolic character in acid solution to give a positive ferric chloride test. Dissolve one or two drops of the unknown liquid or a few crystals of the unknown solid in 1 mL of 95% ethanol and add 1 mL of $1M$ hydrochloric acid. Add a drop or two of 5% ferric chloride solution. If a definite color, except yellow, appears, the ferric hydroxamate test cannot be used.

If the compound did not show enolic character, continue as follows. Dissolve five or six drops of a liquid ester, or about 40 mg of a solid ester, in a mixture of 1 mL of $0.5M$ hydroxylamine hydrochloride (dissolved in 95% ethanol) and 0.4 mL of $6M$ sodium hydroxide. Heat the mixture to boiling for a few minutes. Cool the solution and then add 2 mL of $1M$ hydrochloric acid. If the solution becomes cloudy, add 2 mL of 95% ethanol to clarify it. Add a drop of 5% ferric chloride solution and note whether a color is produced. If the color fades, continue to add ferric chloride until the color persists. A positive test should give a deep burgundy or magenta color.

Test Compound. Try this test on ethyl butanoate (ethyl butyrate).

On being heated with hydroxylamine, esters are converted to the corresponding hydroxamic acids:

The hydroxamic acids form strong, colored complexes with ferric ion.

BASIC HYDROLYSIS

Procedure. Place 1.0 g of ester in a 25-mL round-bottom flask with 10 mL of 25% aqueous sodium hydroxide. Add a boiling stone and attach a water condenser.

Use a small amount of stopcock grease to lubricate the ground-glass joint. Boil the mixture for about 30 minutes. Stop the heating and observe the solution to determine whether the oily ester layer has disappeared or whether the odor of the ester (usually pleasant) has disappeared. Low-boiling esters (below 110°C) usually dissolve within 30 minutes if the alcohol part has a low molecular weight. If the ester has not dissolved, reheat the mixture to reflux for 1–2 hours. After that time, the oily ester layer should have disappeared along with the characteristic odor. Esters with boiling points up to 200°C should hydrolyze during this time. Compounds remaining after this extended period of heating are either unreactive esters or are *not* esters at all.

For esters derived from solid acids, the acid part can, if desired, be recovered after hydrolysis. Extract the basic solution with ether to remove any unreacted ester (even if it appears to be gone), acidify the basic solution with hydrochloric acid, and extract the acidic phase with ether to remove the acid. Dry the ether layer over anhydrous sodium sulfate, decant, and evaporate the solvent to obtain the parent acid from the original ester. The melting point of the parent acid can provide valuable information in the identification process.

This procedure converts the ester to its separate acid and alcohol parts. The ester dissolves because the alcohol part (if small) is usually soluble in the aqueous medium, as is the sodium salt of the acid. Acidification produces the parent acid:

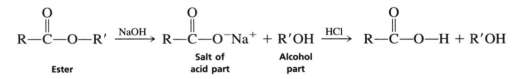

All derivatives of carboxylic acids are converted to the parent acid on basic hydrolysis. Thus, amides, which are not covered in this experiment, would also dissolve in this test, liberating the free amine and the sodium salt of the carboxylic acid.

SPECTROSCOPY

Infrared

The ester–carbonyl group (C=O) peak is usually a strong absorption, as is the absorption of the carbonyl–oxygen link (C—O) to the alcohol part. C=O stretch at approximately 1735 cm^{-1} is normal.[1] C—O stretch usually gives two or more absorptions, one stronger than the others, in the region 1280–1050 cm^{-1}.

---

[1]Conjugation with the carbonyl group moves the carbonyl absorption to lower frequencies. Conjugation with the alcohol oxygen raises the carbonyl absorption to higher frequencies. Ring strain (lactones) moves the carbonyl absorption to higher frequencies.

Nuclear Magnetic Resonance

Hydrogens that are alpha to an ester carbonyl group have resonance in the region 2–3 ppm. Hydrogens alpha to the alcohol oxygen of an ester have resonance in the region 3–5 ppm.

DERIVATIVES

Esters present a double problem when trying to prepare derivatives. To characterize an ester completely, you need to prepare derivatives of *both* the acid part and the alcohol part.

Acid Part.   The most common derivative of the acid is the *N*-benzylamide derivative.

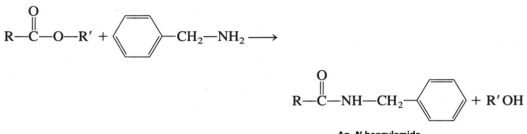

An *N*-benzylamide

The reaction does not proceed well unless R′ is methyl or ethyl. For alcohol portions that are larger, the ester must be transesterified to a methyl or an ethyl ester before preparing the derivative.

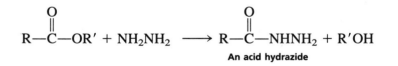

Hydrazine also reacts well with methyl and ethyl esters to give acid hydrazides.

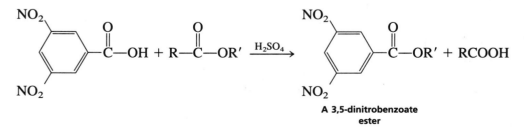

An acid hydrazide

Alcohol Part.   The best derivative of the alcohol part of an ester is the 3,5-dinitrobenzoate ester, which is prepared by an acyl interchange reaction:

A 3,5-dinitrobenzoate ester

Most esters are composed of very simple acid and alkyl portions. For this reason, spectroscopy is usually a better method of identification than is the preparation of derivatives. Not only is it necessary to prepare two derivatives with an ester, but all esters with the same acid portion, or all those with the same alcohol portion, give identical derivatives of those portions.

E S S A Y

The Chemistry of Sweeteners

Americans, like no other nationality in the world, possess a particularly demanding sweet tooth. Our craving for sugar, either added to food or included in candies and desserts, is astounding. Even when we are choosing a food that is not considered to be sweet, we ingest large quantities of sugar. A casual examination of the *Nutrition Facts* label and the list of ingredients of virtually any processed food will reveal that sugar is generally one of the principal ingredients.

Americans, paradoxically, are also obsessed by the need to diet. As a result, the search for noncaloric substitutes for natural sugar represents a multi-million dollar business in this country. There is a ready market for foods that taste sweet but don't contain sugar.

For a molecule to taste sweet, it must fit into a taste bud site where a nerve impulse can carry the message of sweetness from the tongue to the brain. Not all natural sugars, however, trigger an equivalent neural response. Some sugars, such as glucose, have a relatively bland taste, while others, such as fructose, taste very sweet. Fructose, in fact, has a sweeter taste than common table sugar, or sucrose. Furthermore, individuals vary in their ability to perceive sweet substances. The relationship between perceived sweetness and molecular structure is very complicated and, to date, it is rather poorly understood.

The most common sweetener is, of course, common table sugar, or **sucrose.** Sucrose is a disaccharide, consisting of a unit of glucose and a unit of fructose, connected by a 1,2-glycosidic linkage. Sucrose is purified and crystallized from the syrups that are extracted from such plants as sugar cane and sugar beets.

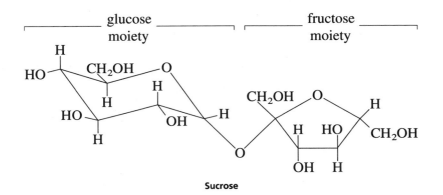

Sucrose

When sucrose is hydrolyzed, it yields one molecule of D-fructose and one of D-glucose. This hydrolysis is catalyzed by an enzyme, **invertase,** and produces a mixture known as **invert sugar.** Invert sugar derives its name from the fact that the mixture is levorotatory while sucrose is dextrorotatory. Thus the sign of the rotation has been "inverted" in the course of the hydrolysis. Invert sugar is somewhat sweeter than sucrose, owing to the presence of free fructose. **Honey** is composed mostly of invert sugar, which is why it has such a sweet taste.

Persons who suffer from diabetes are urged to avoid sugar in their diets. Nevertheless, those individuals also have a craving for sweet foods. A substitute sweetener that is used for food items recommended for diabetics is **sorbitol,** which is an alcohol formed by the catalytic hydrogenation of glucose. Sorbitol has about 60% the sweetness of sucrose. It is a common ingredient in products such as sugarless chewing gum. Even though sorbitol is a different substance from sucrose, it still possesses about the same number of calories per gram. Therefore, sorbitol is not a suitable sweetener for diet foods or beverages.

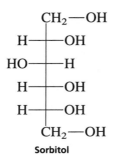

Sorbitol

As sucrose and honey are implicated in problems of tooth decay, as well as being culprits in the continuing battle against obesity, an active field of study is the search for new, noncaloric, noncarbohydrate sweeteners. Even if such a non-nutritive sweetener possessed some calories, if it were very sweet, it would not be necessary to use as much of the sweetener; therefore, the impact on dental hygiene and on diet would be less.

The first artificial sweetener to be used extensively was **saccharin,** which is used commonly as its more soluble sodium salt. Saccharin is about 300 times sweeter than sucrose. The discovery of saccharin was hailed as a great benefit for diabetics, since it could be used as an alternative to sugar. As a pure substance, the sodium salt of saccharin has a very intense, sweet taste, with a somewhat bitter aftertaste. Since it has such an intense taste, it can be used in very small amounts to achieve the desired effect. In some preparations, sorbitol is added to ameliorate the bitter aftertaste. Prolonged studies on laboratory animals have shown that saccharin is a possible carcinogen. In spite of this health risk, the government has permitted saccharin to be used in foods that are intended to be used by diabetics.

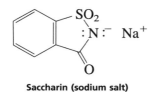

Saccharin (sodium salt)

Another artificial sweetener, which gained wide use in the 1960s and 1970s is **sodium cyclamate.** Sodium cyclamate, which is 33 times as sweet as sucrose, belongs to the class of compounds known as the **sulfamates.** The sweet taste of many of the sulfamates has been known since 1937, when Sweda accidentally discovered that sodium cyclamate had a powerfully sweet taste. The availability of sodium cyclamate spurred the popularity of diet soft drinks. Unfortunately, in the 1970s, research showed that a metabolite of sodium cyclamate, cyclohexylamine, posed some potentially serious health risks, including a risk of cancer. This sweetener has been withdrawn from the market.

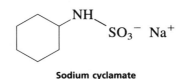

Sodium cyclamate

The most widely used artificial sweetener available today is a dipeptide, consisting of a unit of aspartic acid linked to a unit of phenylalanine. The carboxyl group of the phenylalanine moiety has been converted to the methyl ester. This substance is known commercially as **aspartame,** but it is also sold under the trade names, **NutraSweet** and **Equal.** Aspartame is about 200 times more sweet than sucrose. It is found in diet soft drinks, puddings, juices, and many other foods. Unfortunately, aspartame is not stable when heated, so it is not suitable as an ingredient in cooking. Other dipeptides that have structures similar to that of aspartame are many thousands of times sweeter than sucrose.

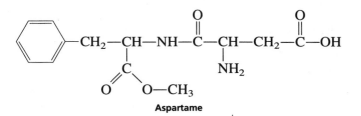

Aspartame

When aspartame was being developed as a commercial product, concern was raised over potential health hazards associated with its use. The potential cancer-causing effects of aspartame, along with other potential adverse side effects, were considered. Extensive testing of this product demonstrated that it met health risk criteria established by the Food and Drug Administration, which granted approval for the sale of aspartame as a food additive in 1974.

The search for new substances that can serve as sweeteners continues. There is a great deal of interest in substances that are naturally occurring and which can be isolated from various plants. In addition, research, including molecular modeling studies and spectroscopic investigations, is attempting to clarify exactly what structural features are required for a sweet taste. Armed with that information, chemists will then be able to synthesize molecules that will be designed specifically for their sweet taste.

REFERENCES

Bragg, R. W., Chow, Y., Dennis, L., Ferguson, L. N., Howell, S., Morga, G., Ogino, C., Pugh, H., and Winters, M. "Sweet Organic Chemistry." *Journal of Chemical Education, 55* (1978): 281.

Sharon, N. "Carbohydrates." *Scientific American, 243* (November 1980): 90.

Crammer, B., and Ikan, R. "Sweet Glycosides from the *Stevia* Plant." *Chemistry in Britain, 22* (1986): 915.

Barker, S. A., Garegg, P. J., Bucke, C., Rastall, R. A., Sharon, N., Lis, H., and Hounsell, E. F. "Contemporary Carbohydrate Chemistry." *Chemistry in Britain, 26* (1990): 663. (This is a series of five articles, each written by one or two of the cited authors, compiled as part of a series of articles on carbohydrate chemistry.)

EXPERIMENT 58

Carbohydrates

In this experiment, you will perform tests that distinguish among various carbohydrates. The carbohydrates included and the classes they represent are as follows:

Aldopentoses: xylose and arabinose

Aldohexoses: glucose and galactose

Ketohexoses: fructose

Disaccharides: lactose and sucrose

Polysaccharides: starch and glycogen

The structures of these carbohydrates can be found in your lecture textbook. The tests are classified in the following groups:

(A) Tests based on the production of furfural or a furfural derivative: Molisch's test, Bial's test, and Seliwanoff's test

(B) Tests based on the reducing property of a carbohydrate (sugar): Benedict's test and Barfoed's test

(C) Osazone formation

(D) Iodine test for starch

(E) Hydrolysis of sucrose

(F) Mucic acid test for galactose and lactose

(G) Tests on unknowns

Required Reading

New: Read the sections in your lecture textbook that give the structures and describe the chemistry of aldopentoses, aldohexoses, ketohexoses, disaccharides, and polysaccharides.

Special Instructions

All the procedures in this experiment involve simple test-tube reactions. Most of the tests are short; however, Seliwanoff's test, osazone formation, and the mucic acid test take relatively longer to complete. You will need a minimum of ten test tubes (15 × 125 mm) numbered in order. Clean them carefully each time they are used. The laboratory instructor will prepare the 1% solutions of carbohydrates and the reagents needed for the tests in advance. Be sure to shake the starch solution before using it.

Phenylhydrazine, which is used for the osazone formation procedure, is considered to be a potential carcinogen. It is important to wear protective gloves when using this reagent. Wash your hands thoroughly in case this substance accidentally comes in contact with your skin.

Waste Disposal

The reagents used in this experiment are relatively harmless aqueous solutions. They can be discarded safely by diluting them and pouring them into the sink. Residues that contain copper should be placed in a designated waste container. Phenylhydrazine, which is used for the osazone formation procedure, must be dissolved in $6N$ hydrochloric acid. The resulting solution may then be diluted with water and poured into a waste container marked specifically for the disposal of phenylhydrazine.

PART A. TESTS BASED ON PRODUCTION OF FURFURAL OR A FURFURAL DERIVATIVE

Under acidic conditions, aldopentoses and ketopentoses *rapidly* undergo dehydration to give furfural (Eq. 1). Ketohexoses *rapidly* yield 5-hydroxymethylfurfural (Eq. 2). Disaccharides and polysaccharides can first be hydrolyzed in an acid medium to produce monosaccharides, which then react to give furfural or 5-hydroxymethylfurfural.

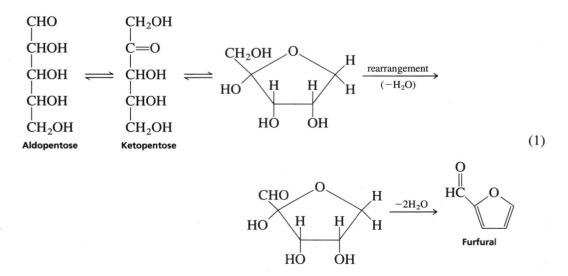

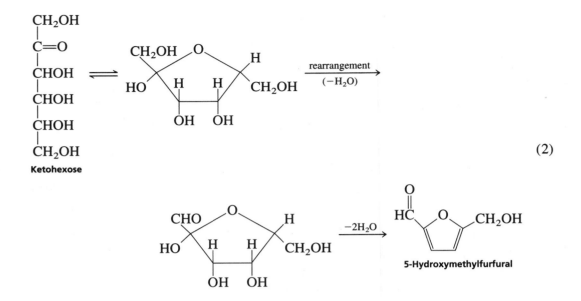

(2)

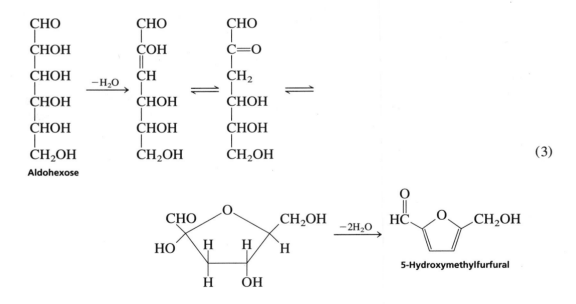

Aldohexoses are *slowly* dehydrated to 5-hydroxymethylfurfural. One possible mechanism is shown in Equation 3. The mechanism is different from that given in Equations 1 and 2 in that dehydration occurs at an early step and the rearrangement step is absent.

(3)

Once furfural or 5-hydroxymethylfurfural is produced by Equations 1, 2, or 3, either will then react with a phenol to produce a colored condensation product. The substance α-naphthol is used in Molisch's test, orcinol in Bial's test, and resorcinol in Seliwanoff's test.

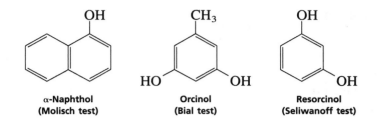

α-Naphthol
(Molisch test)

Orcinol
(Bial test)

Resorcinol
(Seliwanoff test)

The colors and the rates of formation of these colors are used to differentiate between the carbohydrates. The various color tests are discussed in Sections 1, 2, and 3. A typical colored product formed from furfural and α-naphthol (Molisch's test) is the following (Eq. 4):

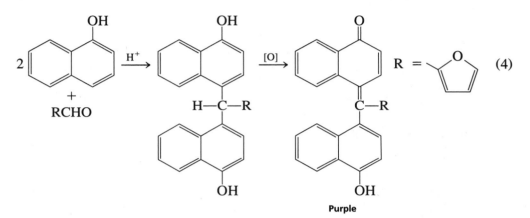

Purple

1. Molisch Test for Carbohydrates

The Molisch test is a *general* test for carbohydrates. Most carbohydrates are dehydrated with concentrated sulfuric acid to form furfural or 5-hydroxymethylfurfural. These furfurals react with the α-naphthol in the test reagent to give a purple product. Compounds other than carbohydrates may react with the reagent to give a positive test. A negative test usually indicates that there is no carbohydrate.

Procedure for the Molisch Test.   Place 1 mL of each of the following 1% carbohydrate solutions in nine separate test tubes: xylose, arabinose, glucose, galactose, fructose, lactose, sucrose, starch (shake it), and glycogen. Also add 1 mL of distilled water to another tube to serve as a control.

Add two drops of the Molisch reagent[1] to each test tube and thoroughly mix the contents of the tube. Tilt each test tube slightly, and cautiously add 1 mL of concentrated sulfuric acid down the sides of the tubes. An acid layer forms at the bottom of the tubes. Note and record the color at the interface between the two layers in each tube. A purple color constitutes a positive test.

---

[1]Dissolve 2.5 g of α-naphthol in 50 mL of 95% ethanol.

2. Bial Test for Pentoses

The Bial test is used to differentiate pentose sugars from hexose sugars. Pentose sugars yield furfural on dehydration in acidic solution. Furfural reacts with orcinol and ferric chloride to give a blue-green condensation product. Hexose sugars give 5-hydroxymethylfurfural, which reacts with the reagent to yield colors such as green, brown, and reddish brown.

Procedure for Bial's Test.   Place 1 mL of each of the following 1% carbohydrate solutions in separate test tubes: xylose, arabinose, glucose, galactose, fructose, lactose, sucrose, starch (shake it), and glycogen. Also add 1 mL of distilled water to another tube to serve as a control.

Add 1 mL of Bial's reagent[2] to each test tube. Carefully heat each tube over a Bunsen burner flame until the mixture just begins to boil. Note and record the color produced in each test tube. If the color is not distinct, add 2.5 mL of water and 0.5 mL of 1-pentanol to the test tube. After shaking the test tubes, again observe and record the color. The colored condensation product will be concentrated in the 1-pentanol layer.

3. Seliwanoff Test for Ketohexoses

The Seliwanoff test depends on the relative rates of dehydration of carbohydrates. A ketohexose reacts rapidly by Equation 2 to give 5-hydroxymethylfurfural, whereas an aldohexose reacts more slowly by Equation 3 to give the same product. Once 5-hydroxymethylfurfural is produced, it reacts with resorcinol to give a dark red condensation product. If the reaction is followed for some time, you will observe that sucrose hydrolyzes to give fructose, which eventually reacts to produce a dark red color.

Procedure for Seliwanoff's Test.   Prepare a boiling water bath for this experiment. Place 0.5 mL of each of the following 1% carbohydrate solutions in separate test tubes: xylose, arabinose, glucose, galactose, fructose, lactose, sucrose, starch (shake it), and glycogen. Add 0.5 mL of distilled water to another tube to act as a control.

Add 2 mL of Seliwanoff's reagent[3] to each test tube. Place all ten tubes in a beaker of boiling water for *60 seconds.* Remove them and note the results in the notebook.

For the remainder of Seliwanoff's test, it is convenient to place a group of three or four tubes in the boiling water bath and to complete the observations before going on to the next group of tubes. Place three or four tubes in the boiling water bath. Observe the color in each of the tubes at 1-minute intervals for 5 minutes beyond the original minute. Record the results at each 1-minute interval. Leave the

---

[2]Dissolve 3 g of orcinol in 1 L of concentrated hydrochloric acid and add 3 mL of 10% aqueous ferric chloride.
[3]Dissolve 0.5 g of resorcinol in 1 L of dilute hydrochloric acid (one volume of concentrated hydrochloric acid and two volumes of distilled water).

tubes in the boiling water bath during the entire 5-minute period. After the first group has been observed, remove that set of test tubes, and place the next group of three or four tubes in the bath. Follow the color changes as before. Finally, place the last group of tubes in the bath and follow the color changes over the 5-minute period.

PART B. TESTS BASED ON THE REDUCING PROPERTY OF A CARBOHYDRATE (SUGAR)

Monosaccharides and those disaccharides that have a potential aldehyde group will reduce reagents such as Benedict's solution to produce a red precipitate of copper(I) oxide:

$$RCHO + 2\ Cu^{2+} + 4\ OH^- \longrightarrow RCOOH + Cu_2O + 2\ H_2O$$

**Red
precipitate**

Glucose, for example, is a typical aldohexose, showing reducing properties. The two diastereomic α- and β-D-glucoses are in equilibrium with each other in aqueous solution. The α-D-glucose opens at the anomeric carbon atom (hemiacetal) to produce the free aldehyde. This aldehyde rapidly closes to give β-D-glucose, and a new hemiacetal is produced. It is the presence of this free aldehyde that makes glucose a reducing carbohydrate (sugar). It reacts with Benedict's reagent to produce a red precipitate, the basis of the test. Carbohydrates that have the hemiacetal functional group show reducing properties.

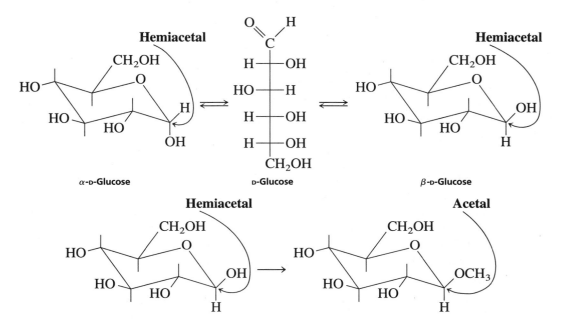

If the hemiacetal is converted to an acetal by methylation, the carbohydrate (sugar) will no longer reduce Benedict's reagent.

With disaccharides, two situations may arise. If the anomeric carbon atoms are bonded (head to head) to give an acetal, then the sugar will not reduce Benedict's reagent. If, however, the sugar molecules are joined head to tail, then one end will still be able to equilibrate through the free aldehyde form (hemiacetal). Examples of a reducing and a nonreducing disaccharide follow.

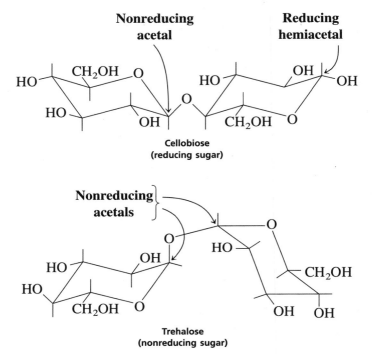

1. Benedict Test for Reducing Sugars

Benedict's test is performed under mildly basic conditions. The reagent reacts with all reducing sugars to produce the red precipitate copper(I) oxide, as shown on page 541. It also reacts with water-soluble aldehydes that are not sugars. Ketoses, such as fructose, also react with Benedict's reagent. Benedict's test is considered one of the classical tests for determining the presence of an aldehyde functional group.

Procedure for Benedict's Test. Prepare a boiling water bath for this experiment. Place 0.5 mL of each of the following 1% carbohydrate solutions in separate test tubes: xylose, arabinose, glucose, galactose, fructose, lactose, sucrose, starch (shake it), and glycogen. Add 0.5 mL of distilled water to another tube to serve as a control.

Add 2 mL of Benedict's reagent[4] to each test tube. Place the test tubes in a boiling water bath for 2–3 minutes. Remove the tubes and note the results in a note-

[4]Dissolve 173 g of hydrated sodium citrate and 100 g of anhydrous sodium carbonate in 800 mL of distilled water, while heating. Filter the solution. Add to it a solution of 17.3 g of copper(II) sulfate ($CuSO_4 \cdot 5H_2O$) dissolved in 100 mL of distilled water. Dilute the combined solutions to 1 L.

book. A red, brown, or yellow precipitate indicates a positive test for a reducing sugar. Ignore a change in the color of the solution. A precipitate must form for the test to be positive.

2. Barfoed Test for Reducing Monosaccharides

Barfoed's test distinguishes reducing monosaccharides and reducing disaccharides by a difference in the rate of reaction. The reagent consists of copper(II) ions, like Benedict's reagent. In this test, however, Barfoed's reagent reacts with reducing mono-saccharides to produce copper(II) oxide faster than with reducing disaccharides.

$$RCHO + 2\ Cu^{2+} + 2\ H_2O \longrightarrow RCOOH + Cu_2O + 4H^+$$

Red
precipitate

Procedure for Barfoed's Test. Place 0.5 mL of each of the following 1% carbo-hydrate solutions in separate test tubes: xylose, arabinose, glucose, galactose, fruc-tose, lactose, sucrose, starch (shake it), and glycogen. Add 0.5 mL of distilled water to another tube to function as a control.

Add 2 mL of Barfoed's reagent[5] to each test tube. Place the tubes in a boiling water bath for 10 minutes. Remove the tubes and note the results in a notebook.

PART C. OSAZONE FORMATION

Carbohydrates react with phenylhydrazine to form crystalline derivatives called **os-azones.**

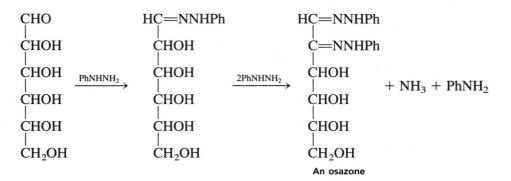

An osazone

An osazone can be isolated as a derivative and its melting point determined. However, some of the monosaccharides give *identical* osazones (glucose, fructose, and mannose). Also, the melting points of different osazones are often in the same range. This limits the usefulness of an isolation of the osazone derivative.

---

[5]Dissolve 66.6 g of copper(II) acetate in 1 L of distilled water. Filter the solution, if necessary, and add 9 mL of glacial acetic acid.

A good experimental use for the osazone is to observe its rate of formation. The rates of reaction vary greatly even though the *same* osazone may be produced from different sugars. For example, fructose forms a precipitate in about 2 minutes, whereas glucose forms a precipitate about 5 minutes later. The osazone is the same in each case. The crystal structure of the osazone is often distinctive. Arabinose, for example, produces a fine precipitate; glucose produces a coarse precipitate.

> **Caution:** Phenylhydrazine is a suspected carcinogen. Handle with gloves.

Procedure for Osazone Formation.   A boiling water bath is needed for this experiment. Place 0.5 mL of each of the following *10%* carbohydrate solutions in separate test tubes: xylose, arabinose, glucose, galactose, fructose, lactose, sucrose, starch (shake it), and glycogen. Add 2 mL of phenylhydrazine reagent[6] to each tube. Place the tubes in a boiling water bath simultaneously. Watch for a precipitate or, in some cases, cloudiness. Note the time at which the precipitate begins to form. After 30 minutes, cool the tubes and record the crystalline form of the precipitates. Reducing disaccharides will not precipitate until the tubes are cooled. Nonreducing disaccharides will hydrolyze first, and then the osazones will precipitate.

PART D. IODINE TEST FOR STARCH

Starch forms a typical blue color with iodine. This color is due to the absorption of iodine into the open spaces of the amylose molecules (helices) present in starch. Amylopectins, which are the other types of molecules present in starch, form a red to purple color with iodine.

Procedure for the Iodine Test.   Place 1 mL of each of the following 1% carbohydrate solutions in three separate test tubes: glucose, starch (shake it), and glycogen. Add 1 mL of distilled water to another tube to act as a control.

Add one drop of iodine solution to each test tube and observe the results.[7] Add a few drops of sodium thiosulfate to the solutions and note the results.[8]

---

[6]Dissolve 50 g of phenylhydrazine hydrochloride and 75 g of sodium acetate trihydrate in 500 mL of distilled water. The reagent deteriorates over time and should be prepared fresh.

[7]The iodine solution is prepared as follows. Dissolve 1 g of potassium iodide in 25 mL of distilled water. Add 0.5 g of iodine and shake the solution until the iodine dissolves. Dilute the solution to 50 mL.

[8]The sodium thiosulfate solution is prepared as follows. Dissolve 1.25 g of sodium thiosulfate in 50 mL of water.

PART E. HYDROLYSIS OF SUCROSE

Sucrose can be hydrolyzed in acid solution to its component parts, fructose and glucose. The component parts can then be tested with Benedict's reagent.

Procedure for the Hydrolysis of Sucrose.   Place 1 mL of a 1% solution of sucrose in a test tube. Add two drops of concentrated hydrochloric acid and heat the tube in a boiling water bath for 10 minutes. Cool the tube and neutralize the contents with 10% sodium hydroxide solution until the mixture is just basic to litmus (about 12 drops are needed). Test the mixture with Benedict's reagent (Part B). Note the results and compare them with the results obtained on sucrose that has not been hydrolyzed.

PART F. MUCIC ACID TEST FOR GALACTOSE AND LACTOSE

Procedures are given in Experiment 60 for the oxidation of galactose and lactose to mucic acid. This test confirms the presence of galactose or a galactose unit in a carbohydrate (sugar).

PART G. TESTS ON UNKNOWNS

Procedure.   Obtain an unknown solid carbohydrate from the laboratory instructor or assistant. The unknown will be one of the following carbohydrates: xylose, arabinose, glucose, galactose, fructose, lactose, sucrose, starch, or glycogen. Carefully dissolve part of the unknown in distilled water to prepare a 1% solution (0.060 g carbohydrate in 6 mL water). Also prepare a 10% solution by dissolving 0.1 g of carbohydrate in 1 mL of water. Save the remainder of the solid for the mucic acid test. Apply whatever tests are necessary to identify the unknown.

At the instructor's option, the optical rotation can be determined as part of the experiment. Experimental details are given in Experiment 61 and Technique 17. Optical rotation data and decomposition points for carbohydrates and osazones are given in the standard reference works on the identification of organic compounds (Experiment 57).

QUESTIONS

1. Find the structures for the following carbohydrates (sugars) in a reference work or a textbook and decide whether they are reducing or nonreducing carbohydrates (sugars): sorbose, mannose, ribose, maltose, raffinose, and cellulose.

2. Mannose gives the same osazone as glucose. Explain.

3. Predict the results of the following tests with the carbohydrates listed in Question 1: Molisch, Bial, Seliwanoff (after 1 minute and 6 minutes), Barfoed, and mucic acid tests.

4. Give a mechanism for the hydrolysis of the acetal linkage in sucrose.

5. The rearrangement in Equations 1 and 2 can be considered a type of pinacol rearrangement. Give a mechanism for that step.

6. Give a mechanism for the acid-catalyzed condensation of furfural with two moles of α-naphthol, shown in Equation 4.

EXPERIMENT 59

Analysis of a Diet Soft Drink by HPLC

High-performance liquid chromatography

In this experiment, high-performance liquid chromatography (HPLC) will be used to identify the artificial additives present in a sample of commercial diet soft drink. The experiment uses HPLC as an analytical tool for the separation and identification of the additive substances. The method uses a reversed-phase column and eluent system, with isochratic elution. Detection is accomplished by measuring the absorbance of ultraviolet radiation at 254 nm by the solution as it is eluted from the column. The mobile phase that will be used is a mixture of 80% $1M$ acetic acid and 20% acetonitrile, buffered to pH 4.2.

Diet soft drinks contain many chemical additives, including several substances that can be used as artificial sweeteners. Among these additives are the four substances that we will be detecting in this experiment: caffeine, saccharine, benzoic acid, and aspartame. The structure of these compounds are shown here.

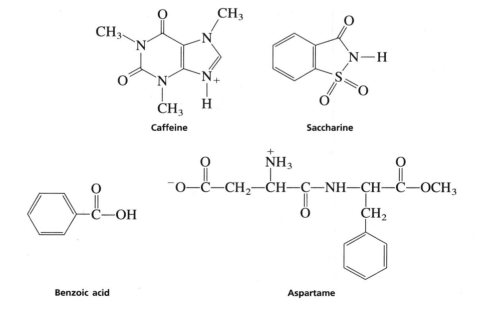

Caffeine   Saccharine

Benzoic acid   Aspartame

You will identify each compound in a sample of diet soft drink by its retention time on the HPLC column. You will be provided with data for a reference mixture of each substance in a test mixture, in order to compare retention times in your test sample with a set of standards.

Required Reading

New: Technique 13 High-Performance Liquid Chromatography (HPLC)

Special Instructions

The instructor will provide specific instruction in the operation of the particular HPLC instrument being used in your laboratory. The instructions that follow indicate the general procedure.

Waste Disposal

Discard the excess acetic acid–methanol solvent in the organic waste container designated for the disposal of nonhalogenated organic wastes. The acetonitrile–acetic acid solvent mixture should be collected in a specially designated container so that it may be either safely discarded or reused.

Procedure

Following your instructor's directions, form a small group of students to perform this experiment. Each small group will analyze a different diet soft drink, and the results obtained by each group will be shared among all students in the class.

The instructor will prepare a mixed standard of the four components, consisting of 200 mg of aspartame, 40 mg of benzoic acid, 40 mg of saccharine, and 20 mg of caffeine in 100 mL of solvent. The solvent for these standards is a mixture of 80% acetic acid and 20% methanol, buffered to pH 4.2 with 50% sodium hydroxide. The lab instructor will also run an HPLC of this standard mixture beforehand, and you should obtain a copy of the results. Some of the steps described in the next two paragraphs may be completed in advance by your instructor.

You may select from a variety of diet soft drinks with different chemical compositions.[1] Select a soft drink from the supply shelf and dispense approximately 50 mL into a small flask.

---

[1]Note to the Instructor: The experiment will be more interesting if the diet soft drink TAB is included among the choices. TAB is one of the few readily available diet soft drinks that contains substantial amounts of saccharine.

Completely remove the carbon dioxide gas, which causes the bubbles in the soft drink, before examining the sample by HPLC. The bubbles will affect the retention times of the compounds and possibly cause damage to the expensive HPLC columns. Most of the gas can be eliminated by allowing the containers of soft drinks to remain open overnight. To remove the final traces of dissolved gases, set up a filtering flask with a Büchner funnel and connect it to a vacuum line. Place a 4-μm filter in the Büchner funnel. (*Note:* Be sure to use a piece of filter paper, not one of the colored spacers that are placed between the pieces of filter paper. The spacers are normally blue.) Filter the soda sample by vacuum filtration through the 4-μm filter and place the filtered sample in a *clean* 4-dram snap-cap vial.

Before using the HPLC instrument, be certain that you have obtained specific instruction in the operation of the instrument in your laboratory. Alternatively, your instructor may have someone operate the instrument for you. Before your sample is analyzed on the HPLC instrument, it should be filtered one more time, this time through a 0.2-μm filter. The recommended sample size for analysis is 10 μL. The solvent system used for this analysis is a mixture of 80% 1M acetic acid and 20% acetonitrile, buffered to pH 4.2. The instrument will be operated in an isochratic mode.

When you examine the chart obtained from the analysis, you may find that the peak corresponding to aspartame appears to be rather small. The peak is small because aspartame absorbs ultraviolet radiation most efficiently at 220 nm, while the detector is set to measure the absorption of light at 254 nm. Nevertheless, the observed retention time of aspartame will not depend upon the setting of the detector, and the interpretation of the results should not be affected. The expected order of elution is: saccharine (first), caffeine, aspartame, benzoic acid. Another interesting point is that, while the caffeine peak appears to be quite large in this analysis, it is nevertheless quite small when compared with the peak that would be obtained if you injected coffee into the HPLC. For a caffeine peak from coffee to fit onto your graph, you would have to dilute the coffee *at least* tenfold. Even decaffeinated coffee usually has more caffeine in it than most sodas (decaffeinated coffee is required to be only 95–96% decaffeinated).

When you have completed your experiment, report your results by preparing a table showing the retention times of each of the four standard substances. In your report, be sure to identify the diet soft drink that you used and to identify the substances that you found in that sample. Also report the substances that were found in each of the other soft drink samples that were tested by other groups in your class.

REFERENCES

Bidlingmeyer, B. A., and Schmitz, S. "The Analysis of Artificial Sweeteners and Additives in Beverages by HPLC." *Journal of Chemical Education, 68* (August 1991): A195.

ESSAY

Chemistry of Milk

Milk is a food of exceptional interest. Not only is milk an excellent food for the very young, but humans have also adopted milk, specifically cow's milk, as a food substance for persons of all ages. Many specialized milk products like cheese, yogurt, butter, and ice cream are staples of our diet.

Milk is probably the most nutritionally complete food that can be found in nature. This property is important for milk, since it is the only food young mammals consume in the nutritionally significant weeks following birth. Whole milk contains vitamins (principally thiamine, riboflavin, pantothenic acid, and vitamins A, D, and K), minerals (calcium, potassium, sodium, phosphorus, and trace metals), proteins (which include all the essential amino acids), carbohydrates (chiefly lactose), and lipids (fats). The only important elements in which milk is seriously deficient are iron and Vitamin C. Infants are usually born with a storage supply of iron large enough to meet their needs for several weeks. Vitamin C is easily secured through an orange juice supplement. The average composition of the milk of each of several mammals follows.

Average Percentage Composition of Milk from Various Mammals

	Cow	Human	Goat	Sheep	Horse
Water	87.1	87.4	87.0	82.6	90.6
Protein	3.4	1.4	3.3	5.5	2.0
Fats	3.9	4.0	4.2	6.5	1.1
Carbohydrates	4.9	7.0	4.8	4.5	5.9
Minerals	0.7	0.2	0.7	0.9	0.4

FATS

Whole milk is an oil–water type of emulsion, containing about 4% fat dispersed as very small (5–10 microns in diameter) globules. The globules are so small that a drop of milk contains about a million of them. Because the fat in milk is so finely dispersed, it is digested more easily than fat from any other source. The fat emulsion is stabilized to some extent by complex phospholipids and proteins that are absorbed on the surfaces of the globules. The fat globules, which are lighter than water, coalesce on standing and eventually rise to the surface of the milk, forming a layer of **cream.** Because vitamins A and D are fat-soluble vitamins, they are carried to the surface with the cream. Commercially, the cream is often removed by centrifugation and skimming and is either diluted to form coffee cream ("half and half"), sold as **whipping cream,** converted to **butter,** or converted to **ice cream.** The milk that remains is called **skimmed milk.** Skimmed milk, except for lacking the fats and vitamins A and D, has approximately the same composition as whole milk. If milk is **homogenized,** its fatty content will not separate. Milk is homogenized by forcing it through a small hole. This breaks up the fat globules and reduces their size to about 1–2 μ in diameter.

The structure of fats and oils is discussed in the essay that precedes Experiment 12. The fats in milk are primarily triglycerides. For the saturated fatty acids, the following percentages have been reported:

C_2 (3%) C_8 (2.7%) C_{14} (25.3%) $>C_{18}$ (~5%)
C_4 (1.4%) C_{10} (3.7%) C_{16} (9.2%)
C_6 (1.5%) C_{12} (12.1%) C_{18} (1.3%)

Thus, about two-thirds of all the fatty acids in milk are saturated, and about one-third are unsaturated. Milk is unusual in that about 12% of the fatty acids are *short*-chain fatty acids (C_2–C_{10}) like butyric, caproic, and caprylic acids.

Additional lipids (fats and oils) in milk include small amounts of cholesterol, phospholipids, and lecithins (phospholipids conjugated with choline). The structures of phospholipids and lecithins are shown. The phospholipids help to stabilize the whole milk emulsion; the phosphate groups help to achieve partial water solubility for the fat globules. All the fat can be removed from milk by extraction with petroleum ether or a similar organic solvent.

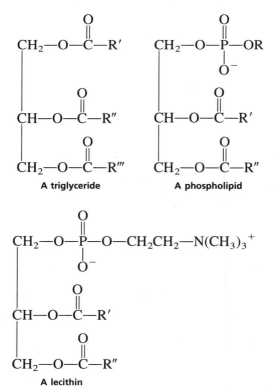

A triglyceride   A phospholipid

A lecithin

PROTEINS

Proteins may be classified broadly in two general categories: fibrous and globular. Globular proteins are those that tend to fold back on themselves into compact units that approach nearly spheroidal shapes. These types of proteins do not form intermolecular interactions between protein units (hydrogen bonds and so on) as fibrous proteins do, and they are more easily solubilized as colloidal suspensions. There are three kinds of proteins in milk: **caseins, lactalbumins,** and **lactoglobulins.** All are globular.

Casein is a phosphoprotein, meaning that phosphate groups are attached to some of the amino acid side chains. These are attached mainly to the hydroxyl groups of the serine and threonine moieties. Actually, casein is a mixture of at least three similar proteins, principally α, β, and κ caseins. These three proteins differ primarily in molecular weight and amount of phosphorus they contain (number of phosphate groups).

Casein	MW	Phosphate Groups/Molecule
α	27,300	~9
β	24,100	~4–5
κ	~8,000	~1.5

Casein exists in milk as the calcium salt, **calcium caseinate.** This salt has a complex structure. It is composed of α, β, and κ caseins, which form a **micelle,** or a solubilized unit. Neither the α nor the β casein is soluble in milk, and neither is soluble either singly or in combination. If κ casein is added to either one, or to a combination of the two, however, the result is a casein complex that is soluble owing to the formation of the micelle.

A structure proposed for the casein micelle is shown in the figure on page 552. The κ casein is thought to stabilize the micelle. Since both α and β casein are phosphoproteins, they are precipitated by calcium ions. Recall that $Ca_3(PO_4)_2$ is fairly insoluble.

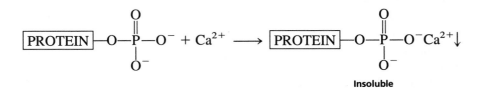

The κ casein protein, however, has fewer phosphate groups and a high content of carbohydrate bound to it. It is also thought to have all its serine and threonine residues (which have hydroxyl groups), as well as its bound carbohydrates, on only one side of its outer surfaces. This portion of its outer surface is easily solubilized in water because these po-

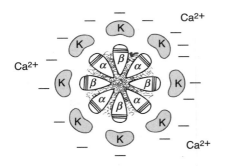

A casein micelle (average diameter, 1200 Å).

lar groups are present. The other portion of its surface binds well to the water-insoluble α and β caseins and solubilizes them by forming a protective colloid or micelle around them. Because the entire outer surface of the micelle can be solubilized in water, the unit is solubilized *as a whole,* thus bringing the α and β caseins, as well as κ casein, into solution.

Calcium caseinate has its isoelectric (neutrality) point at pH 4.6. Therefore, it is insoluble in solutions of pH less than 4.6. The pH of milk is about 6.6; therefore, casein has a negative charge at this pH and is solubilized as a salt. If acid is added to milk, the negative charges on the outer surface of the micelle are neutralized (the phosphate groups are protonated) and the neutral protein precipitates:

$$Ca^{2+} \text{ Caseinate } + 2 \text{ HCl } \longrightarrow \text{ Casein } \downarrow + CaCl_2$$

The calcium ions remain in solution. When milk sours, lactic acid is produced by bacterial action (see equations on p. 554), and the consequent lowering of the pH causes the same *clotting* reaction. The isolation of casein from milk is described in Experiment 60.

The casein in milk can also be clotted by the action of an enzyme called **rennin.** Rennin is found in the fourth stomach of young calves. However, both the nature of the clot and the mechanism of clotting differ when rennin is used. The clot formed using rennin, **calcium paracaseinate,** contains calcium.

$$Ca^{2+} \text{ Caseinate } \xrightarrow{\text{rennin}} Ca^{2+} \text{ Paracaseinate } + \text{ a small peptide}$$

Rennin is a hydrolytic enzyme (peptidase) and acts specifically to cleave peptide bonds between phenylalanine and methionine residues. It attacks the κ casein, breaking the peptide chain so as to release a small segment of it. This destroys the water-solubilizing surface of the κ casein, which protects the inner α and κ caseins, and causes the entire micelle to precipitate as calcium paracaseinate. Milk can be decalcified by treatment with oxalate ion, which forms an insoluble calcium salt. If the calcium ions are removed from milk, a clot will not be formed when the milk is treated with rennin.

The clot, or **curd,** formed by the action of rennin is sold commercially as **cottage cheese.** The liquid remaining is called the **whey.** The curd can also be used in producing various types of **cheese.** It is washed, pressed to remove any excess whey, and chopped. After this treatment, it is melted, hardened, and ground. The ground curd is then salted, pressed into molds, and set aside to age.

Albumins are globular proteins that are soluble in water and in dilute salt solutions. They are, however, denatured and coagulated by heat. The second most abundant protein types in milk are the **lactalbumins.** Once the caseins have been removed, and the solution has been made acidic, the lactalbumins can be isolated by heating the mixture to precipitate them. The typical albumin has a molecular weight of about 41,000.

A third type of protein in milk is the **lactoglobulins.** They are present in smaller amounts than the albumins and generally denature and precipitate under the same conditions as the albumins. The lactoglobulins carry the immunological properties of milk. They protect the young mammal until its own immune systems have developed.

CARBOHYDRATES

When the fats and the proteins have been removed from milk, the carbohydrates remain, as they are soluble in aqueous solution. The main carbohydrate in milk is lactose.

Lactose, a disaccharide, is the *only* carbohydrate that mammals synthesize. Hydrolyzed, it yields one molecule of D-glucose and one of D-galactose. It is synthesized in the mammary glands. In this process, one molecule of glucose is converted to galactose and joined to another of glucose. The galactose is apparently needed by the developing infant to build developing brain and nervous tissue. Brain cells contain **glycolipids** as a part of their structure. A glycolipid is a triglyceride in which one of the fatty acid groups has been replaced by a sugar, in this case galactose. Galactose is more stable (to metabolic oxidation) than glucose and affords a better material for forming structural units in cells.

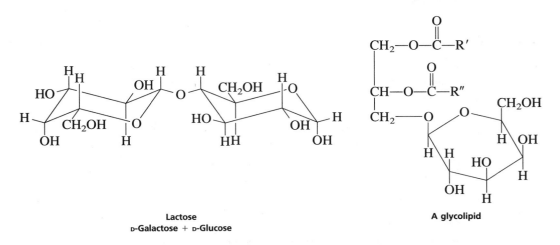

Lactose
D-Galactose + D-Glucose

A glycolipid

Although almost all human infants can digest lactose, some adults lose this ability on reaching maturity, because milk is no longer an important part of their diet. An enzyme called **lactase** is necessary to digest lactose. Lactase is secreted by the cells of the small intestine, and it cleaves lactose into its two component sugars, which are easily digested. Persons lacking the enzyme lactase do not digest lactose properly. Because it is poorly absorbed by the small intestine, it remains in the digestive tract, where its osmotic potential causes an influx of water. This results in cramps and diarrhea for the affected individual.

Persons with a lactase deficiency cannot tolerate more than one glass of milk a day. The deficiency is most common among blacks, but it is also quite common among older Caucasians.

Lactose can be removed from whey by adding ethanol. Lactose is insoluble in ethanol, and when the ethanol is mixed with the aqueous solution, the lactose is forced to crystallize. The isolation of lactose from milk is described in Experiment 60.

When milk is allowed to stand at room temperature, it sours. Many bacteria are present in milk, particularly **lactobacilli.** These bacteria act on the lactose in milk to produce the sour **lactic acid.** These microorganisms actually **hydrolyze** lactose and produce lactic acid only from the galactose portion of the lactose. Since the production of the lactic acid also lowers the pH of the milk, the milk clots when it sours:

$$C_{12}H_{22}O_{11} + H_2O \longrightarrow C_6H_{12}O_6 + C_6H_{12}O_6$$

Lactose   Galactose   Glucose

$$C_6H_{12}O_6 \xrightarrow{\text{lactobacilli}} CH_3-CH-COOH$$
$$|$$
$$OH$$

Galactose   Lactic acid

Many "cultured" milk products are manufactured by allowing milk to sour before it is processed. For instance, milk or cream is usually allowed to sour somewhat by lactic acid bacteria before it is churned to make butter. The fluid left after the milk is churned is sour and is called **buttermilk.** Other cultured milk products include sour cream, yogurt, and certain types of cheese.

REFERENCES

Boyer, R. F. "Purification of Milk Whey α-Lactalbumin by Immobilized Metal-Ion Affinity Chromatography." *Journal of Chemical Education, 68* (May 1991): 430.

Fox, B. A., and Cameron, A. G. *Food Science—A Chemical Approach.* New York: Crane, Russak, 1973. Chap. 6, "Oils, Fats, and Colloids."

Kleiner, I. S., and Orten, J. M. *Biochemistry.* 7th ed. St. Louis: C. V. Mosby, 1966. Chap. 7, "Milk."

McKenzie, H. A., ed. *Milk Proteins.* 2 volumes. New York: Academic Press, 1970.

Oberg, C. J., "Curdling Chemistry—Coagulated Milk Products." *Journal of Chemical Education, 63* (September 1986): 770.

E X P E R I M E N T   6 0

Isolation of Casein and Lactose from Milk

Isolation of a protein
Isolation of a sugar

In this experiment, you will isolate several of the chemical substances found in milk. First, you will isolate a phosphorus-containing protein, casein (Experiment 60A). The remaining milk mixture will then be used as a source of a sugar, α-lactose (Experiment

60B). After you isolate the milk sugar, you will make several chemical tests on this material. Fats, which are present in whole milk, are not isolated in this experiment because powdered nonfat milk is used.

Here is the procedure you will follow. First, the casein is precipitated by warming the powdered milk and adding dilute acetic acid. It is important that the heating not be excessive or the acid too strong, because these conditions also hydrolyze lactose into its components, glucose and galactose. After the casein has been removed, the excess acetic acid is neutralized with calcium carbonate, and the solution is heated to its boiling point to precipitate the initially soluble protein, albumin. The liquid containing the lactose is poured away from the albumin. Alcohol is added to the solution, and any remaining protein is removed by centrifugation. α-Lactose crystallizes on cooling.

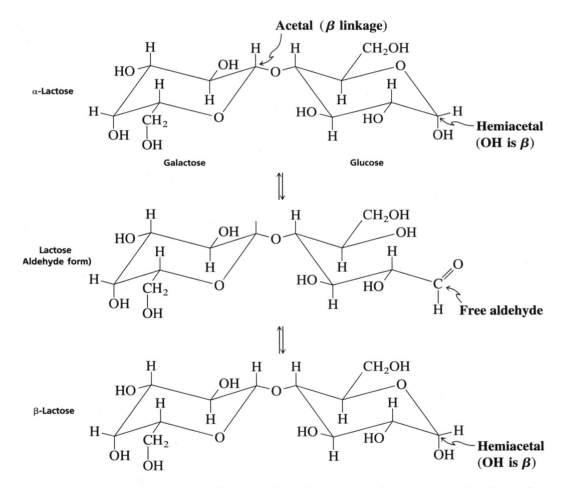

Lactose is an example of a disaccharide. It is made up of two sugar molecules: galactose and glucose. In the preceding structures, the galactose portion is on the left and glucose is on the right. Galactose is bonded through an acetal linkage to glucose.

Notice that the glucose portion can exist in one of two isomeric hemiacetal structures: α-lactose and β-lactose. Glucose can also exist in a free aldehyde form. This aldehyde form (open form) is an intermediate in the equilibration (interconversion) of α- and β-lactose. Very little of this free aldehyde form exists in the equilibrium mixture. The isomeric α- and β-lactose are diastereomers because they differ in the configuration at one carbon atom, called the anomeric carbon atom.

The sugar α-lactose is easily obtainable by crystallization from a water–ethanol mixture at room temperature. On the other hand, β-lactose must be obtained by a more difficult process, which involves crystallization from a concentrated solution of lactose at temperatures about 93.5°C. In the present experiment, α-lactose is isolated because the experimental procedure is easier.

α-Lactose undergoes numerous interesting reactions. First, α-lactose interconverts, via the free aldehyde form, to a large extent, to the β-isomer in aqueous solution. This causes a change in the rotation of polarized light from +92.6° to +52.3° with increasing time. The process that causes the change in optical rotation with time is called **mutarotation.** Mutarotation of lactose can be studied in Experiment 61.

A second reaction of lactose is the oxidation of the free aldehyde form by Benedict's reagent. Lactose is referred to as a reducing sugar because it reduces Benedict's reagent (copper(II) ion to copper(I) ion) and produces a red precipitate (Cu_2O). In the process, the aldehyde group is oxidized to a carboxyl group. The reaction that takes place in Benedict's test is

$$R-CHO + 2\ Cu^{2+} + 4\ OH^- \longrightarrow RCOOH + Cu_2O + 2\ H_2O$$

A third reaction of lactose is the oxidation of the galactose part by the mucic acid test. In this test, the acetal linkage between galactose and glucose units is cleaved by the acidic medium to give free galactose and glucose. Galactose is oxidized with nitric acid to the dicarboxylic acid, galactaric acid (mucic acid). Mucic acid is an insoluble, high-melting solid that precipitates from the reaction mixture. On the other hand, glucose is oxidized to a diacid (glucaric acid), which is more soluble in the oxidizing medium and does not precipitate.

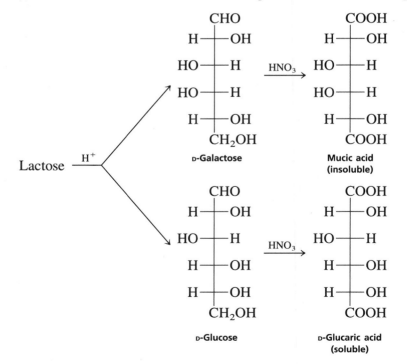

Required Reading

Review: Technique 4

New: Essay   Chemistry of Milk

Special Instructions

Experiments 60A and 60B should both be performed during one laboratory period. The lactose solution must be allowed to stand until the following laboratory period. Casein must dry for 2 or 3 days.

Waste Disposal

Place residues left from the Benedict's test in the container designated for the disposal of copper wastes. Dispose of solid materials obtained from milk in a trash container. Aqueous wastes, including those that contain ethanol, should be poured into the aqueous waste container.

EXPERIMENT 60A

Isolation of Casein from Milk

Procedure

Precipitation of Casein. Place 4.0 g of powdered milk and 10 mL of water into a 100-mL beaker. Heat the mixture in a water bath to 40°C. Check the temperature of the milk solution with a thermometer. Place 1.0 mL of dilute acetic acid solution[1] in a small flask for temporary storage. When the mixture has reached 40°C, add the dilute acetic acid dropwise to the warm milk. After every 5 drops, stir the mixture gently using a glass stirring rod with an attached rubber policeman. Using the rubber policeman, push the casein up onto the side of the beaker so that most of the liquid drains from the solid. Then transfer the congealed casein to another small beaker in portions. If any liquid separates from the casein in the small beaker, use a Pasteur pipet to transfer the liquid back into the reaction mixture. Continue to add dropwise the remainder of the 1.0 mL of dilute acetic acid to the milk mixture in the

---

[1]The laboratory instructor should prepare a large batch for the class in the ratio of 2 mL glacial acetic acid to 20 mL of water.

100-mL beaker to precipitate the casein fully. Remove as much of the casein as possible and transfer it to the small beaker. Avoid adding an excess of acetic acid to the milk solution, as this will cause the lactose in the milk to hydrolyze into glucose and galactose.

When you have removed most of the casein from the milk solution, add 0.2 g of calcium carbonate to the milk in the 100-mL beaker. Stir the mixture for a few minutes and save it for use in Experiment 60B. Use this mixture as soon as possible during the laboratory period. This beaker contains lactose and albumins.

Isolation of Casein.   Transfer the casein from the beaker to a Büchner funnel (Technique 4, Section 4.3, p. 640, and Fig. 4.5, p. 641). Draw a vacuum on the casein to remove as much liquid as possible (about 5 minutes). Press the casein with a spatula during this time. Transfer the casein to a piece of filter paper (about 7 cm). Using a spatula, move the solid around on the paper so that the remaining liquid is absorbed into the filter paper. When most of the liquid has been removed, transfer the solid to a watch glass to complete the drying operation. Allow the casein to air-dry completely for 2 to 3 days before weighing the product. You must remove the casein from the filter paper or it will become "glued" to the paper. (You have nearly prepared white glue.) Submit the casein in a labeled vial to the instructor. Calculate the weight percent of the casein isolated from the powdered milk.

E X P E R I M E N T   6 0 B

Isolation of Lactose from Milk

Procedure

Precipitation of Albumins.   Heat the mixture directly on a hot plate in the 100-mL beaker that you saved from Experiment 60A to about 75°C for about 5 minutes. This heating operation results in a nearly complete separation of the albumins from the solution. Decant the liquid in the beaker away from the solid into a clean centrifuge tube. You may need to hold the solid with a spatula while transferring the liquid. Press the albumins with a spatula to remove as much liquid as possible and pour the liquid into the centrifuge tube (save the albumins in the original beaker for the procedure in the next paragraph). You should have about 7 mL of liquid. When the liquid has cooled to about room temperature, centrifuge the contents of the tube for 2–3 minutes. Be sure to place another tube in the centrifuge to balance the unit. Following centrifugation, decant the liquid away from the solid into a beaker and save it for use in the next section (Precipitation of Lactose).

Allow the albumins to dry for 2–3 days in the original beaker. Break up the solid and weigh it. Calculate the weight percent of albumins isolated from the powdered milk.

Precipitation of Lactose.   Add 15 mL of 95% ethanol to the beaker containing the centrifuged and decanted liquid. Solids will precipitate. Heat this mixture to about 60°C, placing it directly on the hotplate, to dissolve some of the solid. Pour the *hot* liquid into a 40-mL centrifuge tube (or two 15-mL tubes) and centrifuge the hot solution as soon as possible before the solution cools appreciably. Centrifuge the mixture for 2–3 minutes. Be sure to place another tube in the centrifuge to balance the unit. It is important to centrifuge this mixture while it is warm to prevent premature crystallization of the lactose. A considerable quantity of solid forms in the bottom of the centrifuge tube. This solid is not lactose.

Remove the warm supernatant liquid from the tube using a Pasteur pipet and transfer the liquid to a small Erlenmeyer flask. Discard the solid remaining in the centrifuge tube. Stopper the flask and allow the lactose to crystallize for at least 2 days. Granular crystals will form during this time.

Isolation of Lactose.   Collect the lactose by vacuum filtration on a Büchner funnel. Use about 3 mL of 95% ethanol to aid the transfer and to wash the product. α-Lactose crystallizes with one water of hydration, $C_{12}H_{22}O_{11} \cdot H_2O$. Weigh the product after it is thoroughly dry. Submit the α-lactose in a labeled vial to the instructor unless it is to be used in the following optional tests or in Experiment 61. Calculate the weight percent of the lactose isolated from the powdered milk.

Benedict's Test (Optional).   Prepare a hot water bath (above 90°C) for this experiment. Dissolve about 0.01 g of your lactose in 1 mL of water in a test tube. Heat the mixture to dissolve most of the lactose (some cloudiness remains). Place about 1 mL each of 1% solutions of glucose and galactose in separate test tubes. Add to each of the three test tubes 2 mL of Benedict's reagent.[1] Place the test tubes in the hot water bath for 2 minutes. Remove the tubes and note the results. The formation of an orange to brownish-red precipitate indicates a positive test for a reducing sugar. This test is described on pages 541–543 in Experiment 58.

Mucic Acid Test (Optional).   Prepare a hot water bath (above 90°C) for this experiment or use the one prepared for the Benedict's test. Place 0.1 g of the isolated lactose, 0.05 g of glucose (dextrose), and 0.05 g of galactose in three separate test tubes. Add 1 mL of water to each tube and dissolve the solids, with heating if necessary. The lactose solution may be somewhat cloudy but will clear when the nitric acid is added. Add 1 mL of concentrated nitric acid to each tube. Heat the tubes in the hot water bath for 1 hour in a hood (nitrogen oxide gases are evolved). Remove the tubes and allow them to cool slowly after the heating period. Scratch the test tubes with clean stirring rods to induce crystallization. After the test tubes are cooled to room temperature, place them in an ice bath. A fine precipitate of mucic acid should begin to form in the galactose and lactose tubes about 30 minutes after the tubes are removed from the water bath. Allow the test tubes to stand until the next laboratory period to complete the crystallization. Confirm the insolubility of the solid formed by adding about 1 mL of water and then shaking the resulting mixture. If the solid remains, it is mucic acid.

---

[1]Dissolve 34.6 g of hydrated sodium citrate and 20.0 g of anhydrous sodium carbonate in 160 mL of distilled water by heating. Filter the solution, if necessary. Add to it a solution of 3.46 g of copper(II) sulfate $CuSO_4 \cdot 5H_2O$ dissolved in 20 mL of distilled water. Dilute the combined solutions to 200 mL.

QUESTIONS

1. A student decided to determine the optical rotation of mucic acid. What should be expected as a value? Why?

2. Draw a mechanism for the acid-catalyzed hydrolysis of the acetal bond in lactose.

3. β-Lactose is present to a larger extent in an aqueous solution when the solution is at equilibrium. Why is this to be expected?

4. Very little of the free aldehyde form is present in an equilibrium mixture of lactose. However, a positive test is obtained with Benedict's reagent. Explain.

5. Outline a separation scheme for isolating casein, albumin, and lactose from milk. Use a flowchart like that shown in the Advance Preparation and Laboratory Records section at the beginning of the book.

EXPERIMENT 61 ▪

Mutarotation of Lactose

Polarimetry

In this experiment, you will study the mutarotation of lactose by polarimetry. The disaccharide α-lactose, made up of galactose and glucose, can be isolated from milk (Experiment 60). As you can see in the structures drawn in Experiment 60, the glucose unit of lactose can exist in one of two isomeric hemiacetal structures, yielding α- and β-lactose. These isomers are diastereomers because they differ in configuration at one carbon atom. The glucose unit can also exist in a free aldehyde form. This aldehyde form (open form in the equation below) is an intermediate in the equilibration of α- and β-lactose. Very little of this free aldehyde form exists in the equilibrium mixture.

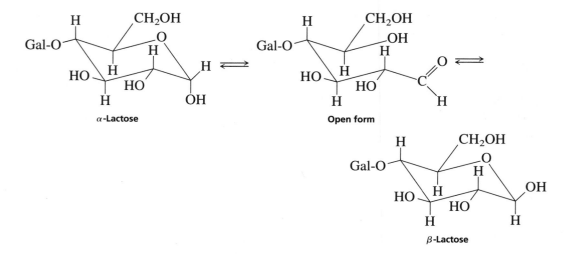

α-Lactose has a specific rotation at 20°C of +92.6°. However, when it is placed in water, the optical rotation **decreases** until it reaches an equilibrium value of +52.3°. β-Lactose has a specific rotation of +34°. The optical rotation of β-lactose *increases* in water until it reaches the same equilibrium value obtained for α-lactose. At the equilibrium point, both the α and β isomers are present. However, since the equilibrium rotation is closer in value to the initial rotation of β-lactose, the mixture must contain more of this isomer. The process, which results in a change in optical rotation over time to approach an equilibrium value, is called **mutarotation.**

Required Reading

Review: Essay   Chemistry of Milk

New: Technique 17   Polarimetry

Special Instructions

The procedure for preparing the cells and for operating the instrument are those appropriate for the Zeiss polarimeter. Your instructor will provide instructions for the use of another type of polarimeter if a Zeiss instrument is not available. A 2-dm cell is used for this experiment. If a cell of a different path length is used, adjust the concentrations appropriately. About 1 hour is required to complete the mutarotation study.

If you are using the lactose you isolated in Experiment 60, you will need to combine your product with one other student's so that you will have an adequate amount of lactose with which to perform this experiment. One person can measure the rotations while the other student records the data. Student-prepared lactose may yield a cloudy solution when the sample is dissolved in water. This should not be of concern as the cloudiness is caused by a trace of protein that remains in the lactose.

Waste Disposal

Dispose of all aqueous solutions in the container designated for aqueous waste.

Notes to the Instructor

If you intend to use commercially obtained α-lactose for this experiment, you should check out the experiment in advance. We have found that, in many cases, commercially obtained α-lactose has already undergone substantial mutarotation (despite catalogue specifications). You can obtain pure α-lactose from the commercial material by reprecipitating it according to the method described on page 559.

Procedure

Preparation. Turn on the polarimeter to warm the sodium lamp. After about 10 minutes, adjust the instrument so that the scale reads about +9°C. This scale reading provides an adjustment of the instrument to the approximate range of rotation that will be observed at the initial reading. Set the timer to zero. Clean and dry a 2-dm cell (Technique 17, Fig. 17.6, p. 833). Weigh 1.25 g of α-lactose and transfer all of it to a *dry* 25-mL volumetric flask.

The operations described in the next section should be studied carefully before starting this part of the experiment. It is essential to complete carefully the described operations in *2 minutes* or less. The reason for speedy operation is that the α-lactose immediately begins to mutarotate when it comes in contact with water. The initial rotations obtained are necessary to get a precise value of the rotation at zero time. You should practice with the necessary equipment in a place near the polarimeter before performing the actual operations. Study the scale on the polarimeter so that you can read it rapidly.

Optical Rotation Procedure. Add about half the volume of distilled water to the volumetric flask containing the α-lactose and swirl it to dissolve the solid. When about half the solid is dissolved (a rough estimate), start the timer. As soon as the solid is dissolved (about 20–25 seconds), carefully fill the flask to the mark with distilled water. Use a Pasteur pipet to finish adding the water. Stopper the flask and invert it six to eight times to mix the contents. Using a funnel, fill the polarimeter cell with the lactose solution. Screw the end piece on the cell and tilt it to transfer any remaining bubbles to the enlarged ring. Place the cell in the polarimeter, close the cover, and adjust the analyzer until the split field is of uniform density (Technique 17, Fig. 17.7, p. 834). Record the time and rotation in your notebook.

Obtain the optical rotation at 1-minute intervals for 8 additional minutes (10 minutes total from the time of initial mixing) and record these values, along with the times at which they were determined. After the 10-minute period, obtain readings at 2-minute intervals for the next 20 minutes. Record the optical rotations and times.

Remove the cell from the polarimeter and add two drops of concentrated ammonium hydroxide to the lactose solution. The ammonia rapidly catalyzes the mutarotation of lactose to its equilibrium value. If the ammonia is not added, the equilibrium value will not be obtained until after about 22 hours. Shake the tube and replace it in the polarimeter. Follow the decrease in rotation until there is no longer a change with time. This final value, which is the equilibrium optical rotation, is a value that will remain constant for at least 5 minutes. Place a thermometer in the polarimeter and determine the temperature in the cell compartment.

Graphing the Data. Plot the data on a piece of graph paper ruled in millimeters with the optical rotation plotted on the vertical axis and time plotted (up to 30 minutes) on the horizontal axis. Alternatively, you may use a spreadsheet computer program to plot the data. Draw the best possible curved line through the points and extrapolate the line to $t = 0$. Remember that there may be some scattering of points about the line, especially at the values for the longer times. The extrapolated values at $t = 0$ correspond to the optical rotation of α-lactose at the time of initial mixing.

Calculations.   Using the equation in Technique 17, Section 17.2, page 832, calculate the specific rotation, $[\alpha]_D$ of α-lactose at $t = 0$. Likewise, calculate the specific rotation of the equilibrium mixture of α- and β-lactose.

Calculate the percentage of each of the diastereomers at equilibrium, using the experimentally determined specific rotation values for α-lactose and the equilibrium mixture and the literature value for the specific rotation of β-lactose (+34°). Assume a linear relation between the specific rotations and the concentrations of the species.

QUESTIONS

1. Explain why β-lactose predominates in the equilibrium mixture of α- and β-lactose.
2. The following rotation data have been obtained for D-glucose at 20°C.

α-D-glucose	+112.2°
β-D-glucose	+18.7°
Equilibrium mixture	+52.7°

Using these values, calculate the percentage composition of the α and β isomers at equilibrium. Inspect the structures of α- and β-D-glucose in Experiment 58 and rationalize the values obtained in the calculations.

EXPERIMENT 62

Resolution of (±)-α-Phenylethylamine and Determination of Optical Purity

Resolution of enantiomers
Use of a separatory funnel
Polarimetry
NMR spectroscopy
Chiral resolving agent
Diastereomeric methyl groups

Although racemic (±)-α-phenylethylamine is readily available from commercial sources, the pure enantiomers are more difficult to obtain. In this experiment, you will isolate one of the enantiomers, the levorotatory one, in a high state of optical purity. A **resolution,** or separation, of enantiomers will be performed, using (+)-tartaric acid as the resolving agent. After obtaining the product, you will test its purity, either by the classical method, using a polarimeter, or by the more modern approach, using NMR spectroscopy and a chiral resolving agent.

RESOLUTION OF ENANTIOMERS

The resolving agent to be used is (+)-tartaric acid, which forms diastereomeric salts with racemic α-phenylethylamine. The important reactions for this experiment are indicated below.

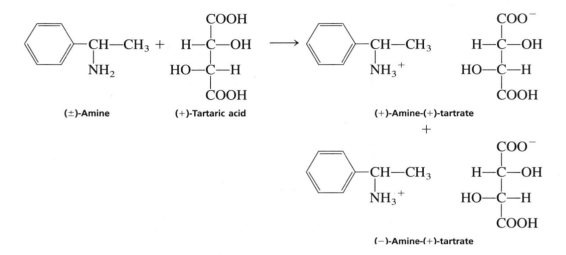

(±)-Amine   (+)-Tartaric acid   (+)-Amine-(+)-tartrate

+

(−)-Amine-(+)-tartrate

Optically pure (+)-tartaric acid is quite abundant in nature. It is frequently obtained as a by-product of wine-making. The separation depends on the fact that diastereomers usually have different physical and chemical properties. The (−)-amine-(+)-tartrate salt has a lower solubility than its diastereomeric counterpart, the (+)-amine-(+)-tartrate salt. With some care, the (−)-amine-(+)-tartrate salt can be induced to crystallize, leaving (+)-amine-(+)-tartrate in solution. The crystals are removed by filtration and purified. The (−)-amine can be obtained from the crystals by treating them with base. This breaks apart the salt by removing the proton, and it regenerates the free, unprotonated (−)-amine.

NMR DETERMINATION OF OPTICAL PURITY

A polarimeter can be used to measure the observed rotation α of the resolved amine sample. From this value, you can calculate the specific rotation $[\alpha]_D$ and the optical purity of the amine. An alternate, and perhaps more accurate, means of determining the optical purity of the sample makes use of NMR spectroscopy. A group attached to a chiral carbon normally has the same chemical shift whether the chiral carbon has either R or S configuration. However, it can be made diastereotopic in the NMR spectrum (have different chemical shifts) when the racemic parent compound is treated with an optically pure chiral resolving agent to produce diastereomers. In this case, the group is no longer found in two enantiomers, but rather in two different diastereomers, and its chemical shift will be different in each environment.

In this experiment, the partly resolved amine (containing both R and S enantiomers) is mixed with optically pure S-(+)-O-acetylmandelic acid in an NMR tube containing $CDCl_3$. Two diastereomers are formed:

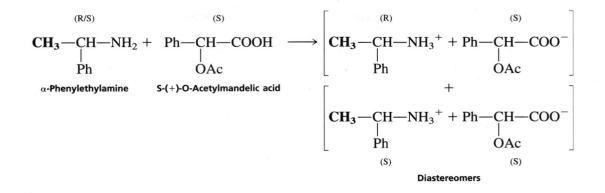

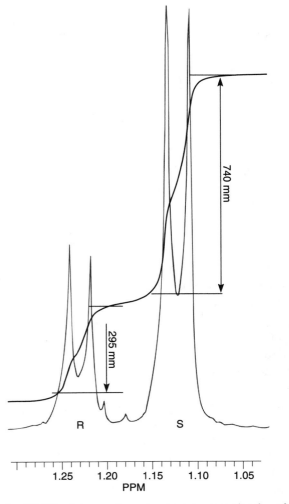

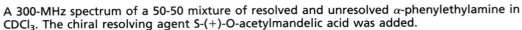

A 300-MHz spectrum of a 50-50 mixture of resolved and unresolved α-phenylethylamine in CDCl₃. The chiral resolving agent S-(+)-O-acetylmandelic acid was added.

The methyl groups in the amine portions of the two diastereomeric salts are attached to a chiral center, S in one case and R in the other. As a result, the methyl groups themselves become diastereomeric, and they have different chemical shifts. In this case, the R-isomer is downfield, and the S-isomer is upfield. These methyl groups appear at approximately (varies) 1.1 and 1.2 ppm, respectively, in the proton NMR spectrum of the mixture. Since the methyl groups are adjacent to a methine (CH) group, they appear as doublets. These doublets may be integrated in order to determine the percentage of the R and S amines in the resolved α-phenylethylamine. In the example, the NMR spectrum was determined with a mixture made by dissolving equal quantities (50/50 mixture) of the original unresolved ($\pm$)-α-phenylethylamine and a student's resolved product, which contained predominantly S-($-$)-α-phenylethylamine.

E X P E R I M E N T   6 2 A

Resolution of ($\pm$)-α-Phenylethylamine

In this procedure, you will resolve racemic ($\pm$)-α-phenylethylamine, using ($+$)-tartaric acid as the resolving agent.

Required Reading

Review: Technique 4   Section 4.3
          Technique 7   Sections 7.4 and 7.8
          Technique 17

Special Instructions

α-Phenylethylamine readily reacts with carbon dioxide in the air to form a white solid, the N-carboxyl amine derivative. Every effort should be taken to avoid prolonged exposure of the amine to air. Be sure to close the bottle tightly after you have measured the rotation of your amine, and be sure to place your sample quickly into the flask where you will perform the resolution. This flask should also be stoppered. Use a cork stopper, because a rubber stopper will dissolve somewhat and discolor your solution. The crystalline salt will not react with carbon dioxide until you decompose it to recover the resolved amine. Then you must be careful once again.

The observed rotation for a sample isolated by a single student may be only a few degrees, which limits the precision of the optical purity determination. Better results can be obtained if four students combine their resolved amine products for the polarimetric analysis. If you have allowed your amine to have excessive exposure to air, the polarimetry solution may be cloudy. This will make it difficult to obtain an accurate determination of the optical rotation.

Waste Disposal

Place the mother liquor solution from the crystallization, which contains (+)-α-phenylethylamine, (+)-tartaric acid, and methanol, in the special container provided for this purpose. Dispose of all other aqueous solutions in the container designated for aqueous waste. When you are finished with polarimetry, depending on the wishes of your instructor, you should either place your resolved S-(−)-α-phenylethylamine in a special container marked for this purpose, or you should submit it to your instructor in a suitably labeled container that includes the names of those people who have combined their samples.

Procedure

Note to the Instructor: This experiment is designed for students to work individually but to combine their products with three other students for polarimetry.

Preparations.   Place 7.8 g of L-(+)-tartaric acid and 125 mL of methanol in a 250-mL Erlenmeyer flask. Heat this mixture on a hotplate until the solution is nearly boiling. Remove the flask from the hotplate and *slowly* add 6.25 g of racemic α-phenylethylamine (α-methylbenzylamine) to this hot solution.

Caution: At this step, the mixture is very likely to froth and boil over.

Crystallization.   Stopper the flask and allow it to stand overnight. The crystals that form should be prismatic. If needles form, they are not optically pure enough to give a complete resolution of the enantiomers—*prisms must form.* Needles should be dissolved (by careful heating) and cooled slowly to crystallize once again. When you recrystallize, you can "seed" the mixture with a prismatic crystal, if one is available. If it appears that you have prisms, but that they are overgrown (covered) with needles, the mixture may be heated until *most* of the solid has dissolved. The needle crystals dissolve easily, and usually a small amount of the prismatic crystals remains to seed the solution. After dissolving the needles, allow the solution to cool slowly and form prismatic crystals from the seeds.

Workup.   Filter the crystals, using a Büchner funnel (Technique 4, Section 4.3, and Fig. 4.5, p. 641), and rinse them with a few portions of cold methanol. Partially dissolve the crystalline amine-tartrate salt in 25 mL of water, add 4 mL of 50% sodium hydroxide, and extract this mixture with three 10-mL portions of methylene chloride using a separatory funnel (Technique 7, Section 7.4, p. 690). Combine the organic layers from each extraction in a stoppered flask and dry them over about 1 gram of an-hydrous sodium sulfate for about 10 minutes. During this waiting period, pre-weigh a clean, dry 25-mL Erlenmeyer flask, stoppered with a cork. You will need this pre-weighed, stoppered flask in the next section. Decant the dried solution into a 50-mL beaker and evaporate the methylene chloride on a hotplate (about 60°C) in the hood. A stream of nitrogen or air can be directed into the beaker to increase the rate of evaporation. The product is a **liquid** (2–3 mL will be obtained). Some solid amine car-bonate will form on the sides of the beaker during this operation. If you prolong this step, a large amount of the white solid will form.

Yield Calculation and Storage.   Carefully transfer the liquid amine to the pre-weighed 25-mL Erlenmeyer flask. If possible, avoid transferring any of the white solid. Stopper the flask and weigh it to determine the yield. Also calculate the percentage yield of the S-(−)-amine based on the amount you started with.

Polarimetry.   Combine your product with the products obtained by three other students. If anyone's product is cloudy and opaque, do not use it. If there is floating solid in anyone's sample, try to avoid transferring it. Be sure to keep the pure amine tightly stoppered. Mix the combined liquids, and use them to fill a pre-weighed 10-mL volumetric flask. Weigh the flask to determine the weight of amine and calculate the density (concentration) in g/mL. You should obtain a value of about 0.94 g/mL. This should give you a sufficient amount of material to proceed with the polarimetry measurements that follow without diluting your sample. If, however, your combined products do not amount to more than 10 mL of the amine, you may have to dilute your sample with methanol (check with your instructor).

If you have less than 10 mL of product, weigh the flask to determine the amount of the amine present. Then fill the volumetric flask to the mark with absolute methanol and mix the solution thoroughly. The concentration of your solution in grams per milliliter is easily calculated.

Transfer the solution to a 0.5-dm polarimeter tube and determine its observed rotation. Your instructor will show you how to use the polarimeter. Report the val-ues of the observed rotation, specific rotation, and optical purity to the instructor. The published value for the specific rotation is $[\alpha]_D^{22} = -40.3°$. Calculate the per-centage of *each* of the enantiomers in the sample (Technique 17, Section 17.5, p. 835) and include the figures in your report.

EXPERIMENT 62B

Determination of Optical Purity Using NMR and a Chiral Resolving Agent

In this procedure, you will use NMR spectroscopy with the chiral resolving agent S-(+)-O-acetylmandelic acid to determine the optical purity of the S-(−)-α-phenylethyl-amine you isolated in Experiment 62A.

Required Reading

New: Technique 19   Preparation of Samples for Spectroscopy
   Appendix 4   Nuclear Magnetic Resonance Spectroscopy

Special Instructions

Be sure to use a clean Pasteur pipet or syringe whenever you remove $CDCl_3$ from its supply bottle. Avoid contamination of the stock of NMR solvent! Also be sure to fill and empty the pipet or syringe several times before attempting to remove the solvent from the bottle. If you bypass this equilibration technique, the volatile solvent may squirt out of the tube before you can transfer it successfully to another container.

Waste Disposal

When you dispose of your NMR sample, which contains $CDCl_3$, place it in the container designated for halogenated wastes.

Procedure

Using a small test tube, weigh approximately 0.05 mmole (0.006 g, $MW = 121$) of your resolved amine by adding it drop by drop from a Pasteur pipet. Cork the test tube to protect it from atmospheric carbon dioxide. Carbon dioxide reacts with the amine to form an amine carbonate (white solid). Using a weighing paper, weigh approximately 0.06 mmole (0.012 g, $MW = 194$) of S-(+)-O-acetylmandelic acid and add it to the amine in the test tube. Using a clean Pasteur pipet, add about 0.25 mL of $CDCl_3$ to dissolve everything. If the solid does not dissolve completely, you can mix the solution by drawing it several times into your Pasteur pipet and redelivering it back into the test tube. When everything is dissolved, transfer the mixture to an NMR tube using a Pasteur pipet. Using a clean Pasteur pipet, add enough $CDCl_3$ to bring the total height of the solution in the NMR tube to 35 mm.

Determine the proton NMR spectrum, preferably at 300 MHz, using a method that expands and integrates the peaks of interest. Using the integrals, calculate the percentages of the R and S isomers in the sample, and its optical purity.[1] Compare your results from this NMR determination to those you obtained by polarimetry (Experiment 62A).

---

[1]Note to the Instructor: In some cases, the resolution is so successful that it is very difficult to detect the doublet arising from the R-(+)-α-phenylethylamine + S-(+)-O-acetylmandelic acid diastereomer. If this occurs, it is useful to have the students add a single drop of *racemic* α-phenylethylamine to the NMR tube and redetermine the spectrum. In this way, both diastereomers can be clearly seen.

REFERENCES

Ault, A. "Resolution of D,L-α-Phenylethylamine." *Journal of Chemical Education, 42* (1965): 269.

Jacobus, J., and Raban, M. "An NMR Determination of Optical Purity." *Journal of Chemical Education, 46* (1969): 351.

Parker, D., and Taylor, R. J. "Direct ¹H NMR Assay of the Enantiomeric Composition of Amines and β-Amino Alcohols Using O-Acetyl Mandelic Acid as a Chiral Solvating Agent." *Tetrahedron, 43,* No. 22 (1987): 5451.

QUESTIONS

1. Using a reference textbook, find examples of reagents used in performing chemical resolutions of acidic, basic, and neutral racemic compounds.

2. Propose methods of resolving each of the following racemic compounds:

(a)

$$CH_3{-}CH{-}\underset{\underset{Br}{|}}{\overset{\overset{O}{\|}}{C}}{-}OH$$

(b)

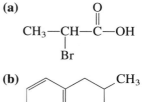

3. Explain how you would proceed to isolate R-(+)-α-phenylethylamine from the *mother liquor* that remained after you crystallized S-(−)-α-phenylethylamine.

4. What is the white solid that forms when α-phenylethylamine comes in contact with carbon dioxide? Write an equation for its formation.

5. Which method, polarimetry or NMR spectroscopy, gives the more accurate results in this experiment? Explain.

6. Draw the three-dimensional structure of S-(−)-α-phenylethylamine.

7. Draw the three-dimensional structure of the diastereomer formed when S-(−)-α-phenylethylamine is reacted with S-(+)-O-acetylmandelic acid.

EXPERIMENT 63

Chiral Reduction of Ethyl Acetoacetate; Optical Purity Determination Using a Chiral Shift Reagent

Fermentation
Stereochemistry
Reduction with yeast
Polarimetry

Use of a separatory funnel
Nuclear magnetic resonance
Chemical shift reagents
Optical purity determination

The experiment described in Experiment 63A uses common baker's yeast as a chiral reducing agent to transform an achiral starting material, ethyl acetoacetate, into a chiral product, (S)-(+)-ethyl 3-hydroxybutanoate. In Experiment 63B, you will use nuclear magnetic resonance spectroscopy to determine the optical purity of the product. This experiment requires the use of a chiral shift reagent.

E X P E R I M E N T   6 3 A

Chiral Reduction of Ethyl Acetoacetate

In this experiment, you will use baker's yeast as a chiral reducing agent to convert ethyl acetoacetate into a chiral product (S)-(+)-ethyl 3-hydroxybutanoate. The chiral product is used as an important building block in the laboratory synthesis of natural products.

The product ethyl 3-hydroxybutanoate is formed principally as the enantiomer with the (S) configuration. The reaction does produce a small amount (generally less than 10%) of the opposite enantiomer (R)-(−)-ethyl 3-hydroxybutanoate.

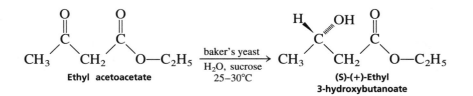

Required Reading

Review:	Technique 3	Section 3.10
	Technique 4	Section 4.4
	Technique 7	Section 7.4
New:	Technique 12	Column Chromatography
	Technique 17	Polarimetry

Special Instructions

The fermentation requires at least 3 days; the experiment should be begun in advance of the time set aside for product isolation and polarimetry. If the fermentation is allowed to continue for longer than 3 days, the optical purity of the product will increase. Under these conditions, however, it may become difficult to identify the minor enantiomer in Experiment 63B. The observed rotation for a sample isolated by a single student may be only a few degrees, which limits the precision of the optical purity determination. Better results can be obtained if several students combine their products for the polarimetric analysis.

Waste Disposal

Most of the solutions used in this experiment are dilute aqueous solutions; they should be discarded by pouring them into a waste container marked for the disposal of aqueous waste. Slurries that contain yeast may also be safely discarded by diluting them and pouring them into the same waste container. Discard any remaining ethyl acetoacetate or ethyl 3-hydroxybutanoate by pouring it into the waste container designated for nonhalogenated organic waste. The filtration residue that contains Filter Aid and yeast can be discarded by placing it into a trash can. To discard the polarimetry solution, which contains methylene chloride, pour it into a waste container designated for halogenated organic waste.

Procedure

Fermentation.   Equip a 500-mL Erlenmeyer flask with a magnetic stirring bar and a one-hole rubber stopper with a glass tube leading to a beaker or a test tube containing a solution of barium hydroxide. Protect the barium hydroxide from air by adding some mineral oil or xylene to form a layer above the barium hydroxide. The figure on page 573 depicts the apparatus for this experiment. A precipitate of barium carbonate will form, indicating that carbon dioxide is being evolved during the course of the reaction. Oxygen from the atmosphere is excluded through the use of the trap.

Add 100 mL of tap water, 30 g of sucrose, and about 3.5 g (one package) of dry baker's yeast to the flask. Add these materials, while stirring, in the order indicated. Attach the trap to the fermentation flask. Stir this mixture for about 1 hour, preferably in a warm location. Add 4.0 grams of ethyl acetoacetate and allow the fermenting mixture to stand at room temperature until the next laboratory period, stirring vigorously. If your laboratory is equipped with a shaker, place your flask in the shaker until the next laboratory period.

After this time, prepare a second warm (about 40°C) solution of 30 g sucrose in 100-mL tap water. Add this solution, along with 3.5 g (one package) of dry baker's yeast to the fermenting mixture and allow it to stir for 48 hours (with the trap attached) at room temperature.

Isolation of Product.   Place about 8 g of Filter Aid (Johns-Manville Celite) in a beaker with about 20 mL of water. Stir the mixture vigorously and then pour the contents into a small Büchner funnel (with filter paper) while applying a *gentle* vacuum, as in a vacuum filtration. Be careful not to let the Filter Aid dry completely. This procedure will cause a thin layer of Filter Aid to be deposited on the filter paper. Discard the water that passes through this filter. Decant as much of the clear su-

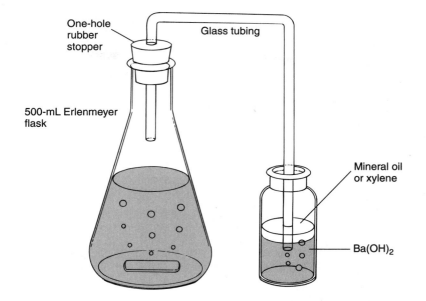

Apparatus for the fementation of ethyl acetoacetate.

pernatant fluid as possible and pass it through this filter, using *very gentle* suction. Also add the solid residue to the Büchner funnel. The extremely tiny yeast particles are trapped in the pores of the Filter Aid (Technique 4, Section 4.4, p. 642). Wash the residue with 20 mL of water, allowing the water to pass into the flask containing the filtered reaction mixture. Add 30 g of sodium chloride to the filtrate and stir the mixture vigorously for 5 minutes. Extract the aqueous solution with three separate 30-mL portions of diethyl ether using a 250-mL separatory funnel (Technique 7, Section 7.4, p. 690). Be careful not to shake the separatory funnel too vigorously to prevent the formation of emulsions. If an emulsion should develop, drain the aqueous solution from the separatory funnel up to the level of the emulsion. Add 2–3 mL of water to the separatory funnel and swirl the mixture to break up the emulsion. Drain the remaining water from the separatory funnel.

Collect the ether extracts in a 125-mL Erlenmeyer flask, add 1 gram of anhydrous magnesium sulfate, stopper the flask, and allow the solution to dry for at least 5 minutes. Decant the liquid into a beaker, add a boiling stone, and evaporate the ether using a warm water bath in the hood and a stream of air or nitrogen to recover the liquid ester. You should recover about 2–3 mL of liquid.

Column Chromatography. Prepare a small chromatography column in the following manner. Place a small plug of cotton in a $5\frac{3}{4}$-inch Pasteur pipet. Tamp the cotton to form a loose plug. Add alumina on top of the cotton plug to form a column 1 cm high. Tap the pipet with your finger to pack the alumina. Using a second Pasteur pipet, add the crude hydroxyester to the column. Rinse the remaining crude product onto the column using 1–2 mL of methylene chloride. Collect the eluted product in a 10-mL Erlenmeyer flask. Use a dropper bulb to force the liquid material through the chromatography column. Dry the organic layer over anhydrous magnesium sulfate for about 10 minutes. Transfer the dried solution with a Pasteur pipet into a pre-weighed test tube. Evaporate the solvent in a warm water bath (at about 60°C) using a gentle stream of air or nitrogen. Weigh the test tube again in order to determine the weight of the pure hydroxyester obtained.

When you have isolated your product, it is important that you check it for purity before proceeding to complete the experiment. Obtain an infrared spectrum of your product. Make sure that you observe an O—H stretching peak at around 3200–3500 cm^{-1} and that the C=O stretching peak of the ketone functional group (at about 1715 cm^{-1}) has disappeared.[1] If the C=O stretching peak remains, or if the O—H stretching peak is absent, your reduction did not take place, and you will have to repeat the fermentation procedure.

If you are going to perform Experiment 63B, remove a 0.030-g portion of your product, place it in an NMR tube, and set it aside. Combine the remainder of your product with the products of three other students in order to proceed with the polarimetry part of this experiment. Using a Pasteur pipet, transfer the hydroxyester to a pre-weighed 10-mL volumetric flask. Transfer each student's product carefully to the volumetric flask. Weigh the volumetric flask again in order to determine the concentration of the sample. Fill the volumetric flask to the mark with methylene chloride. Stopper the volumetric flask and invert it ten to fifteen times to mix the solution thoroughly. The concentration in grams per milliliter of this solution can be determined. Transfer the solution to a 0.5-dm polarimeter tube and determine its observed rotation. The published value of the *specific* rotation of (+)-ethyl 3-hydroxybutanoate is $[\alpha]_D^{25} = +43.5°$. Report the value of the specific rotation and the optical purity to the instructor. Calculate the percentage of *each* of the enantiomers in the sample (Technique 17, Section 17.5, p. 835).

EXPERIMENT 63B ▪

NMR Determination of the Optical Purity of (S)-(+)-Ethyl 3-Hydroxybutanoate

In Experiment 63A, a method for the chiral reduction of ethyl acetoacetate was given. This reduction produces a product that is predominantly the (S)-(+) enantiomer of ethyl 3-hydroxybutanoate. In this procedure, we will use NMR to determine the actual optical purity of the product. An NMR spectrum of racemic ethyl 3-hydroxybutanoate is shown on page 575. In this spectrum there is no discernible difference between the two enantiomers. The methyl hydrogens on carbons 2′ and 4 appear together at about 1.25 ppm, the methylene hydrogens on carbon 2 appear at 2.4 ppm, the hydroxyl proton appears at

---

[1]There will still be a C=O stretching peak from the *ester* functional group at about 1735 cm^{-1}.

3.6 ppm, and the methylene hydrogens on carbon 1′ and the methine hydrogen on carbon 3 appear together at about 4.2 ppm.

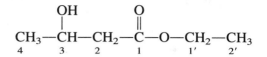

Although the normal spectrum shows no visible difference for the two enantiomers, there is a method that will allow the spectra of the two enantiomers to be distinguished. This method uses a chiral shift reagent. A general discussion of the chemical shift reagents is found in Appendix 4, Section NMR.13. These reagents "spread out" the resonances of the compound with which they are used, increasing the chemical shifts of the protons that are nearest the center of the metal complex by the largest amount. Since the spectra of both (+)-and (−)-ethyl 3-hydroxybutanoate are identical, the usual chemical shift reagent would not help our analysis. However, if one uses a chemical shift reagent that is itself chiral, one can begin to distinguish the two enantiomers by their NMR spectra. The two

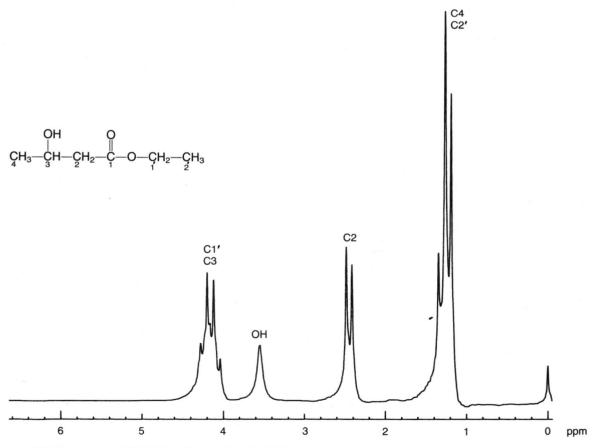

NMR spectrum (90 MHz) of racemic ethyl 3-hydroxybutanoate.

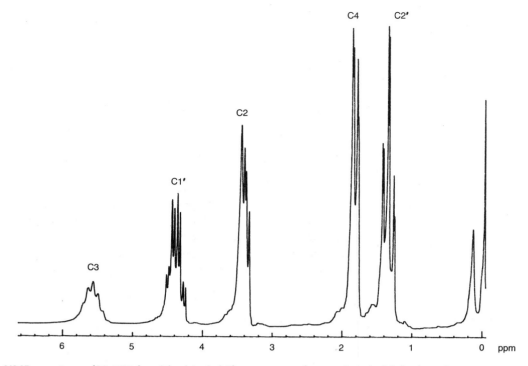

NMR spectrum (90 MHz), with chiral shift reagent, of racemic ethyl 3-hydroxybutanoate. (Note: The OH resonance is off-scale.)

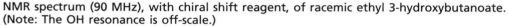

$R_{quartet} = RQ = \dfrac{7+6}{2} = 6.5 \text{ mm}$ $\qquad R_{doublet} = RD = \dfrac{19+19}{2} = 19 \text{ mm}$

$S_{quartet} = SQ = \dfrac{45+46}{2} = 45.5 \text{ mm}$ $\qquad S_{doublet} = SD = \dfrac{92+86}{2} = 89 \text{ mm}$

$\text{Ratio} = \dfrac{RQ}{SQ} = \dfrac{6.5}{45.5} = 0.14$ $\qquad \text{Ratio} = \dfrac{RD}{SD} = \dfrac{19}{89} = 0.21$

$\%(S) - \text{enantiomer} = \left(\dfrac{SQ}{RQ + SQ}\right) \times 100$ $\qquad \%(S) - \text{enantiomer} = \left(\dfrac{SD}{RD + SD}\right) \times 100$

$\qquad\qquad = \left(\dfrac{45.5}{6.5 + 45.5}\right) \times 100 = \underline{88\%}$ $\qquad = \left(\dfrac{89}{89 + 19}\right) \times 100 = \underline{82\%}$

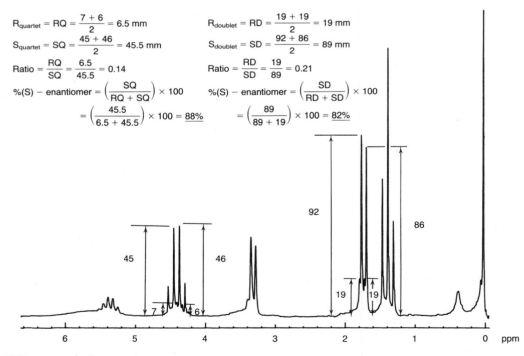

NMR spectrum (90 MHz), with chiral shift reagent, of chiral reduction product of ethyl acetoacetate.

enantiomers, which are chiral, will interact differently with the chiral shift reagent. The complexes formed from the (R) and (S) isomers and with (+)-camphor-containing shift reagent will be diastereomers. Diastereomers usually have different physical properties, and the NMR spectra are no exception. The two complexes will be formed with slightly differing geometries. Although the effect is small, it is large enough to begin to see differences in the NMR spectra of the two enantiomers. In particular, the originally superimposed methylene and methine multiplets will begin to be resolved (see upper spectrum on p. 576).

The chiral shift reagent used in this experiment is *tris*-[3-(heptafluoropropylhydroxymethylene)-(+)-camphorato]europium(III), or Eu(hfc)$_3$. In this complex, the europium is in a chiral environment because it is complexed to camphor, which is a chiral molecule.

Eu(hfc)$_3$ has the following structure:

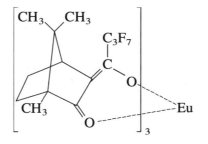

Required Reading

Review: Appendix 4   Nuclear Magnetic Resonance Spectroscopy

Special Instructions

This experiment requires use of an NMR spectrometer. It is a short experiment, designed as an option to Experiment 63A.

Waste Disposal

Discard the remaining solution from your NMR tube into the container designated for the disposal of halogenated organic waste.

Procedure

Place approximately 0.030 g of ethyl 3-hydroxybutanoate (prepared in Experiment 63A) in an NMR tube. Use a Pasteur pipet and an analytical balance to perform this operation. It is not important to weigh an exact quantity of the ester; any amount from 0.025 to 0.05 g will suffice, but you must know the exact weight.

Divide the quantity of ester that you weighed by 1.35 to determine the amount of shift reagent you will need. Using smooth weighing paper, use the analytical balance to weigh out this quantity of shift reagent. Again, it is not necessary to be perfectly exact, but you must record the amount. Carefully add this shift reagent to the NMR sample. Add a small quantity of $CDCl_3$ solvent, which contains tetramethylsilane (TMS), but do not add more than double the initial volume of the sample of the ester. Allow the sample to stand for 20 minutes.

Determine the NMR spectrum of the sample.[2] The peaks of interest are the methyl hydrogens on carbon 4 and the methylene hydrogens on carbon 1′. You should be able to see two sets of overlapping multiplets. If you do not see this pattern, you may not have added enough shift reagent, the amount of one of the enantiomers may be too small, or it may just appear as shoulders on the base of the peaks from the larger multiplet. If you wish, add a second portion of shift reagent, similar to the portion added originally.

Determine the percentage of each isomer in both of your samples in the following manner. Compare the heights of the two inner peaks of the quartets that correspond to the methylene hydrogens on carbon 1′. Determine the ratios of the heights of these peaks. Repeat the comparison, using the two peaks of the doublets that correspond to the methyl hydrogens on carbon 4. Refer to the sample determination illustrated on the NMR spectrum shown on page 576. Average all the ratios that you have determined and calculate the percentages of each enantiomer from this average ratio. On a typical NMR spectrometer, this comparison-of-peak-heights method gives results that are accurate to within 2–3% when measuring an accurately prepared reference sample.

REFERENCES

Seebach, D., Sutter, M. A., Weber, R. H., and Züger, M. F. "Yeast Reduction of Ethyl Acetoacetate: (S)-(+)-Ethyl 3-Hydroxybutanoate." *Organic Syntheses, 63* (1984): 1.

---

[2]Note to the Instructor: It is a good idea to test the ability of your instrumentation by preparing a reference sample containing equal quantities of racemic ethyl 3-hydroxybutanoate and a sample obtained form Experiment 63A. This mixed sample should contain about 75% (S)-(+)-isomer and 25% (R)-(−)-isomer. Use about 0.030 g of this mixed sample. The results from this mixed sample can be used to tell how well the method is working and also to help students assign the upfield and downfield peaks to the correct enantiomers.

EXPERIMENT 64

Esterification Reactions of Vanillin: The Use of NMR to Solve a Structure Proof Problem[1]

Esterification
Crystallization
Use of a Craig tube
Nuclear magnetic resonance

The reaction of vanillin with acetic anhydride, in the presence of base, is an example of the esterification of a phenol. The product, which is a white solid, can be characterized easily by its infrared and NMR spectra.

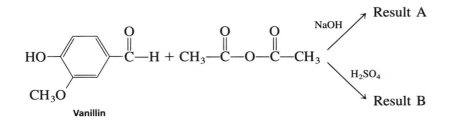

Vanillin

When vanillin is esterified with acetic anhydride under acidic conditions, however, the product that is isolated has a different melting point and different spectra. The object of this experiment is to identify the products formed in each of these reactions and to propose mechanisms that will explain why the reaction proceeds differently under basic and acidic conditions.

Required Reading

Review: Techniques 3, 4, 5, and 19
           Appendices 3 and 4

You should also read the sections in your lecture textbook that deal with the formation of esters and nucleophilic addition reactions of aldehydes.

---

[1]This experiment is based on a paper presented at the 12th Biennial Conference on Chemical Education, Davis, California, August 2–7, 1992, by Professor Rosemary Fowler, Cottey College, Nevada, Missouri. The authors are very grateful to Professor Fowler for her generosity in sharing her ideas.

Special Instructions

Sulfuric acid is very corrosive. Do not allow it to touch your skin.

Waste Disposal

Dispose of all organic residues in the waste container specified for the disposal of nonhalogenated organic waste. All aqueous solutions produced in this experiment should be disposed of in the container for aqueous waste. Dispose of solutions used for NMR spectroscopy in the waste container designated for the disposal of halogenated materials.

Procedure

Preparation of 4-Acetoxy-3-Methoxybenzaldehyde (Vanillyl Acetate). Dissolve 1.50 g of vanillin in 25 mL of 10% sodium hydroxide in a 250-mL Erlenmeyer flask. Add 30 g of crushed ice and 4.0 mL of acetic anhydride. Stopper the flask with a cork and shake it several times over a 20-minute period. A cloudy, milky-white precipitate will form immediately upon adding the acetic anhydride. Filter the precipitate, using a Hirsch funnel or small Büchner funnel, and wash the solid with three 5-mL portions of ice cold water.

Recrystallize the solid from a solvent composed of 50% ethanol and 50% water. Allow the crystals to dry overnight. Weigh the dried crystals, calculate the percentage yield, and determine the melting point (literature value = 77–79°C). Determine the infrared spectrum of the product as a KBr pellet. Determine the proton NMR spectrum of the product in CDCl$_3$ solution. Using the spectral data, confirm that the structure of the product is consistent with the predicted result.

Esterification of Vanillin in the Presence of Acid. Dissolve 1.50 g of vanillin in 10 mL of acetic anhydride in a 125-mL Erlenmeyer flask. Place a magnetic spin bar in the flask and add one drop of 4.5M sulfuric acid to the reaction mixture. Stopper the flask and stir at room temperature for 1 hour. During this period, you may notice that the solution turns light yellow or pale orange in color. However, if the reaction mixture turns dark red-orange, this is an indication that you added too much sulfuric acid. If this happens, discard the mixture into the waste container designated for nonhalogenated organic waste and start this part of the experiment again.

At the end of the reaction period, cool the flask in an ice-water bath for 3–4 minutes. Add a mixture of 20 mL of cold water and 15 g of ice to the flask. Stopper the flask and shake it vigorously. Continue to cool and shake the flask to induce crystallization. Filter the product on a Hirsch or small Büchner funnel and wash the solid with three 5-mL portions of ice cold water.

Recrystallize the solid from 95% ethanol. Allow the crystals to dry overnight. Weigh the dried crystals, calculate the percentage yield, and determine the melting point (literature value = 90–91°C). Determine the infrared spectrum of the product as a KBr pellet. Determine the proton NMR spectrum of the product in $CDCl_3$ solution.

REPORT

Compare the two sets of spectra obtained for the base- and acid-promoted reactions. Using the spectra, identify the structures of the compounds formed in each reaction. Outline mechanistic pathways to account for the formation of both products isolated in this experiment.

EXPERIMENT 65

The Fatty Acids: An Exercise in Molecular Modeling

Molecular modeling

In the essay, "Fats and Oils," we saw that fatty acids are essentially chains of carbon atoms with a carboxyl group at one end that can react with some type of alcohol to produce an ester. In fats or oils, fatty acids are combined with glycerol in the proportion of three fatty acid units to one glycerol molecule. These esters are known as **triglycerols,** or **triglycerides.**

Fatty acids are found with different chain lengths, ranging from four carbons to 24 carbons. These fatty acids are either **saturated** (with no carbon–carbon double bonds) or **unsaturated** (containing one or more carbon–carbon double bonds). The double bonds in naturally occurring fatty acids are found in the *cis* configuration, imparting an overall curved shape to the long chains of carbon atoms. Owing to their curved structure, the fatty acids cannot easily pack into a crystal lattice at room temperature; as a result, fats containing unsaturated fatty acids tend to be liquids. Saturated fatty acids, on the other hand, tend to be more linear in shape. The linear molecules pack more easily into a crystal lattice at room temperature. Fats containing saturated fatty acids have a greater tendency to be solids at room temperature. When naturally occurring *cis* fatty acids are altered by partial hydrogenation, the long chains of carbon atoms change from a curved shape to a linear shape; hydrogenation of an unsaturated fat converts that material from a liquid to a solid.

Unfortunately, the partial hydrogenation of unsaturated fats results in the **isomerization** of a percentage of the fats from the natural *cis* configuration to the unnatural *trans* configuration. The change of configuration from a *cis* double bond to a *trans* double bond results in a dramatic change in the overall shape of the molecule. Since biochemical processes are often sensitive to molecular shape, it is reasonable to expect that *trans* fatty acids will behave differently from *cis* fatty acids when they become an important component of the diet. Many semi-solid forms of fats and oils, including solid shortening and margarine, contain significant amounts of *trans* fatty acids. When *trans* fatty acids are deposited in cell membranes, they may disrupt cellular function. Among the adverse effects on human health of *trans* fatty acids that have been reported, we may include a lowering of HDL levels in blood serum (HDL is considered to be the "good" form of cholesterol), an elevation of the LDL levels in blood serum (LDL is considered to be the "bad" form of cholesterol), and a series of adverse effects on cell membrane functions.

The purpose of this experiment is to demonstrate that the shapes of the *cis*-unsaturated, *trans*-unsaturated, and saturated fatty acids differ from one another. We will use molecular modeling to arrive at the correct three-dimensional shapes of each of these types of fatty acids.

Required Reading

Review: Essay   Fats and Oils
          Essay   Molecular Modeling and Molecular Mechanics
          Experiment 50

Special Instructions

In order to perform this experiment, you must use computer software that has the ability to perform molecular mechanics (MM2 or MM3) calculations with minimization of the strain energy. Your instructor will either provide instruction in the use of the software, or you will be provided with a handout containing the necessary instructions.

This experiment has been conceived as a type of open-ended project for students interested in more advanced practice in molecular modeling. The student should expect that "incorrect" structures may be produced that result from convergence to local minima. It will be necessary to apply corrections manually to calculated structures and to recalculate the minimized structures. With each correction, it is important to note the value of the energy and to strive for the structure with the *lowest* energy.

Notes to the Instructor

This molecular modeling experiment was devised using the modeling program Alchemy III. It should, however, be possible to use many other implementations of molecular mechanics. Some of the other capable programs available are: PCModel; Chem 3D Pro; Spartan (SGI Version), MacSpartan, or PC-Spartan; CaChe and Personal CaChe; Insight II (SGI); Nemesis; and Sybyl. You will have to provide your students with an introduction to your specific implementation. Specific instructions for this experiment using Alchemy III may be found in the Instructor's Manual. The introduction should show students how to build a molecule, how to minimize its energy, and how to load and save files. Students will also need to be able to measure bond lengths and bond angles.

Procedure

PART A. OLEIC ACID [(Z)-9-OCTADECENOIC ACID]

Building the Carbon Chain. Select an sp³-hybridized carbon from the menu of atoms. Using this carbon atom, build a chain of seven carbon atoms connected to each other. In order to optimize the geometry of this fragment, select the minimizer and allow it to calculate the conformation of lowest energy for this fragment. In order to avoid arriving at a structure that represents a local minimum, rather than the true structure with the lowest energy, it is necessary that this seven-carbon fragment be arranged on the screen as a planar zig-zag. Save this carbon chain as a *fragment*. You may have to specify that this fragment contains unfilled valences before saving it. Notice that we have not considered any hydrogen atoms that may be attached to the carbons at this point.[1]

Addition of the Double Bond. Select an sp²-hybridized carbon from the menu of atoms. Attach an sp²-hybridized carbon to the end of the seven-carbon chain that you just created, and then attach a second sp²-hybridized carbon to the first one by matching their double-bond valences. Recall the seven-carbon chain fragment that you saved previously, and attach it to the second sp²-hybridized carbon of the double bond. In order to have the correct configuration, you should attach this chain to the double bond in order to have a *cis* configuration.

Completion of the Structure. From the menu of atoms, select an sp³-hybridized carbon and attach it to one end of the molecule that you have been building. Select an sp²-hybridized carbon and attach it to the opposite end of the molecule. To complete the carboxyl function group, select an sp²-hybridized *oxygen* atom and attach it to the double-bond portion of the sp²-hybridized carbon. Select an sp³-hybridized oxygen atom and attach it to the single-bond valence of the sp²-hybridized carbon atom. To place hydrogens at all of the available valence sites of the molecule, select the **Add Hydrogens** menu item. Finally, select the minimizer and allow it to calculate the lowest energy structure of this completed molecule. When the minimization has been completed, save your molecule. You now have a completed structure of oleic acid. How would you describe its overall shape? You should redisplay the structure of oleic acid as a space-filling model and make a hard copy of the structure.

PART B. *trans*-OLEIC ACID [(E)-9-OCTADECENOIC ACID]

To build the structure of the *trans* isomer of oleic acid, you will use your minimized structure of oleic acid as a starting point. Select the **Alter Torsion Angles** menu

---

[1]Some programs may "crash" if you try to minimize an "open" structure (i.e., one without hydrogens and having *unfilled valences*). You may have to add hydrogens to your seven-carbon chain and then delete them later in order to obtain a minimized structure for this seven-carbon chain.

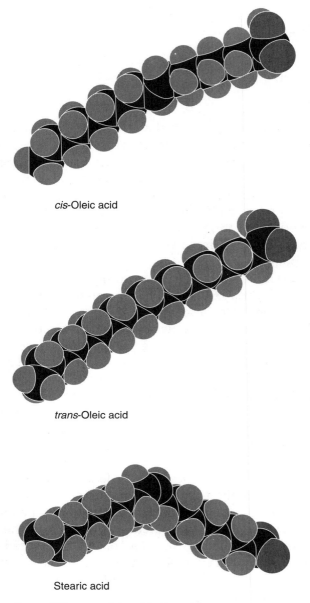

cis-Oleic acid

trans-Oleic acid

Stearic acid

Space-filling models of fatty acids.

item. Identify carbons 8, 9, 10, and 11 by clicking the mouse on each of them in order. In the dialog box, indicate that the torsional angle should be changed to 180°. When you select **OK,** the double bond will be converted from the *cis* configuration to the *trans* configuration. Select the minimizer and allow it to calculate the lowest energy structure for *trans*-oleic acid. When the minimization has been completed, save your molecule (be sure to use a different file name from that used for oleic acid, in Part A!). How would you describe the overall shape of this molecule? Redisplay the structure of *trans*-oleic acid as a space-filling model and make a hard copy of it.

PART C.   STEARIC ACID [OCTADECANOIC ACID]

To build the structure of the saturated fatty acid, stearic acid, you will use your minimized structure of *trans*-oleic acid as a template. Select the **Delete Atoms** menu item. Identify carbons 9 and 10 by clicking the mouse on each of them. You will probably have to delete each atom separately. Select the **Delete Hydrogens** menu item. This will remove all of the hydrogens from the entire molecule. Then select the **Add** menu item and select an sp³-hybridized carbon atom from the menu. Attach a carbon atom to each half of the carbon chain, to replace carbons 9 and 10, which you just deleted. Connect these two carbon atoms together by selecting the **Connect** menu item and then clicking on the two valences that are to be connected. Restore the hydrogen atoms to this molecule by selecting **Add Hydrogens**[2]. Select the minimizer and allow it to calculate the structure of lowest energy from stearic acid. When the minimization has been completed, save your molecule, using a new file name. How would you describe the overall shape of stearic acid? How does it compare with the *cis*- and *trans*-isomers of oleic acid? Are there any similarities? Any differences? Redisplay the structure of stearic acid as a space-filling model and make a hard copy of it.

Finally, run a final minimization on all three fatty acids. Record the energy of each minimized structure. In your laboratory report, you should include the three print-outs of the space-filling models of the three fatty acids, along with a table showing the relative strain energies of the three compounds. The figure shows how the correct structures should appear.

QUESTIONS

1. Which is more stable (i.e., which has the lower strain energy), the *trans* isomer or the *cis* isomer of oleic acid? How can you account for this difference in terms of molecular structure?

2. Write a chemical equation for the conversion of oleic acid to stearic acid.

3. Propose a mechanism to account for the conversion of the *cis* isomer of oleic acid to its *trans* isomer during the course of the hydrogenation of the double bond.

4. Based on the shapes of the three molecules that you calculated, propose an explanation for the harmful effects of *trans*-oleic acid in human health.

5. Would you predict that *trans*-oleic acid should be a solid or a liquid at room temperature?

---

[2]You may have to readjust the torsional angles in order to restore the planar zig-zag structure of the carbon chain before proceeding to the minimization step.

PART TWO

The Techniques

TECHNIQUE 1 ◼

Measurement of Volume and Weight

Performing successful organic chemistry experiments requires the ability to measure solids and liquids accurately. This ability involves both selecting the proper measuring device and using this device correctly.

Liquids to be used for an experiment will usually be found in small containers in a hood. For most of the experiments in this textbook, a graduated cylinder, a dispensing pump, or a graduated pipet will be used for measuring the volume of a liquid. You will usually transfer the required volume of liquid to a round-bottom flask or an Erlenmeyer flask.

In cases where the liquid is a limiting reagent, you should pre-weigh **(tare)** the container before dispensing the liquid into the container. When the container is reweighed, you obtain the actual weight for the volume of liquid you have dispensed. The laboratory procedure usually specifies when you should weigh the liquid. When transferring the liquid to a round-bottom flask, place the flask in a beaker and tare both the flask and the beaker. The beaker keeps the round-bottom flask in an upright position and prevents spills from occurring. It is also advisable to tare a glass stopper, along with the beaker and flask, so that you can stopper the flask immediately after dispensing the liquid and prevent vapors from escaping into the air.

In cases where the liquid is not the limiting reagent, you may calculate the weight of the liquid from the volume you have delivered and the density of the liquid. Usually, densities are provided in the experimental procedures. You may calculate the weight from the following relationship:

$$\text{Weight (g)} = \text{Density (g/mL)} \times \text{Volume (mL)}$$

When using a graduated cylinder to measure small volumes (< 3 mL) of a limiting reagent, it is important to transfer the liquid **quantitatively** from the cylinder to the reaction vessel. This means that all of the liquid should be transferred. When the reagent is poured from the graduated cylinder, a small amount of liquid will remain in the cylinder. This remaining liquid can be transferred by pouring a small amount of the solvent being used in the reaction into the cylinder and then pouring this solution into the reaction vessel. If no solvent is used in the reaction, most of the remaining liquid in the cylinder can be transferred by using a Pasteur pipet.

Using a small amount of solvent to transfer a liquid quantitatively can also be applied in other situations. For example, if your product is dissolved in a solvent and the procedure instructs you to transfer the reaction mixture from a round-bottom flask to a separatory funnel, after pouring most of the liquid into the funnel, a small amount of solvent could be used to transfer the rest of the product quantitatively.

Solids to be used for an experiment will usually be found near the balance. When an accurate measurement is required, solids must be weighed on a balance that reads at least to the nearest decigram (0.01 g). To weigh a solid, place your round-bottom flask or

Erlenmeyer flask in a small beaker and take these with you to the balance. Place a piece of paper that has been folded once on the balance pan. The folded paper will enable you to pour the solid into the flask without spilling. Use your spatula to aid the transfer of the solid to the paper. Never weigh directly into the flask and never pour, dump, or shake a material from a bottle. While still at the balance, carefully transfer the solid from the paper to your flask. The flask should be placed in a beaker while transferring the solid. The beaker acts as a trap for any material that fails to make it into the container. It also supports the round-bottom flask so that it does not fall over. It is not necessary to obtain the exact amount specified in the experimental procedure, and trying to be exact requires too much time at the balance. For example, if you obtained 1.52 g of a solid, rather than the 1.50 g specified in a procedure, the actual amount weighed should be recorded in your notebook. Use the amount you weighed to calculate the theoretical yield, if this solid is the limiting agent.

Careless dispensing of liquids and solids is a hazard in any laboratory. When reagents are spilled, you may be subjected to an unnecessary health or fire hazard. In addition, you may waste expensive chemicals, destroy balance pans and clothing, and damage the environment. Always clean up any spills immediately.

1.1 GRADUATED CYLINDERS

Graduated cylinders are most often used to measure liquids for the experiments in this book (see Fig. 1.1). The most common sizes are 10 mL, 25 mL, 50 mL, and 100 mL, but it is possible that not all of these will be available in your laboratory. Volumes from

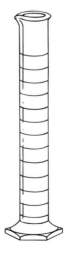

Figure 1.1   Graduated cylinder.

about 2 to 100 mL can be measured with reasonably good accuracy provided that the correct cylinder is used. You should use the *smallest* cylinder available that can hold all of the liquid that is being measured. For example, if a procedure calls for 4.5 mL of a reagent, use a 10 mL graduated cylinder. Using a large cylinder in this case will result in a less accurate measurement. Furthermore, using any cylinder to measure less than 10% of the total capacity of that cylinder will likely result in an inaccurate measurement. Always remember that whenever using a graduated cylinder to measure the volume of a limiting reagent, you must weigh the liquid to determine the amount used accurately. You should use a graduated pipet, a dispensing pump, or an automatic pipet for accurate transfer of liquids with a volume of less than 2 mL.

If the storage container is reasonably small (< 1.0 L) and has a narrow neck, you may pour most of the liquid into the graduated cylinder and use a Pasteur pipet to adjust to the final line. If the storage container is large (> 1.0 L) or has a wide mouth, two strategies are possible. First, you may use a pipet to transfer the liquid to the graduated cylinder. Alternatively, you may pour some of the liquid into a beaker first and then pour this liquid into a graduated cylinder. Use a Pasteur pipet to adjust to the final line. Remember that you should not take more than you need. Excess material should never be returned to the storage bottle. Unless you can convince someone else to take it, it must be poured into the appropriate waste container. You should be frugal in your estimation of amounts needed.

1.2 DISPENSING PUMPS

Dispensing pumps are simple to operate, chemically inert, and quite accurate. Since the plunger assembly is made of Teflon, the dispensing pump may be used with most corrosive liquids and organic solvents. Dispensing pumps come in a variety of sizes, ranging from 1 mL to 300 mL. When used correctly, dispensing pumps can be used to deliver accurate volumes ranging from 0.1 mL to the maximum capacity of the pump. The pump is attached to a bottle containing the liquid being dispensed. The liquid is drawn up from this reservoir into the pump assembly through a piece of inert plastic tubing.

Dispensing pumps are somewhat difficult to adjust to the proper volume. Normally, the instructor or assistant will carefully adjust the unit to deliver the proper amount of liquid. As shown in Figure 1.2, the plunger is pulled up as far as it will travel to draw in the liquid from the glass reservoir. To expel the liquid from the spout into a container, you slowly guide the plunger down. With low-viscosity liquids, the weight of the plunger will expel the liquid. With more viscous liquids, however, you may need to push the plunger gently to deliver the liquid into a container. Remove the last drop of liquid on the end of the spout by touching the tip on the interior wall of the container. When the liquid being transferred is a limiting reagent or when you need to know the weight precisely, you should weigh the liquid to determine the amount accurately.

As you pull up the plunger, look to see if the liquid is being drawn up into the pump unit. Some volatile liquids may not be drawn up in the expected manner, and you will observe an air bubble. Air bubbles are commonly observed when the pump has not been

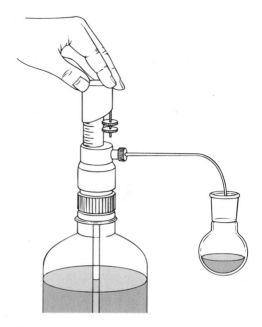

Figure 1.2   Use of a dispensing pump.

used for a while. The air bubble can be removed from the pump by dispensing, and discarding, several volumes of liquid to "reprime" the dispensing pump. Also check to see if the spout is filled completely with liquid. An accurate volume will not be dispensed unless the spout is filled with liquid before you lift up the plunger.

1.3 GRADUATED PIPETS

A widely used measuring device is the graduated serological pipet. These *glass* pipets are available commercially in a number of sizes. "Disposable" graduated pipets may be used many times and discarded only when the graduations become too faint to be seen. A good assortment of these pipets consists of the following:

1.00-mL pipets calibrated in 0.01-mL divisions (1 in 1/100 mL)

2.00-mL pipets calibrated in 0.01-mL divisions (2 in 1/100 mL)

5.0-mL pipets calibrated in 0.1-mL divisions (5 in 1/10 mL)

Never draw liquids into the pipets using mouth suction. A pipet pump or a pipet bulb, not a rubber dropper bulb, must be used to fill pipets. Two types of pipet pumps and

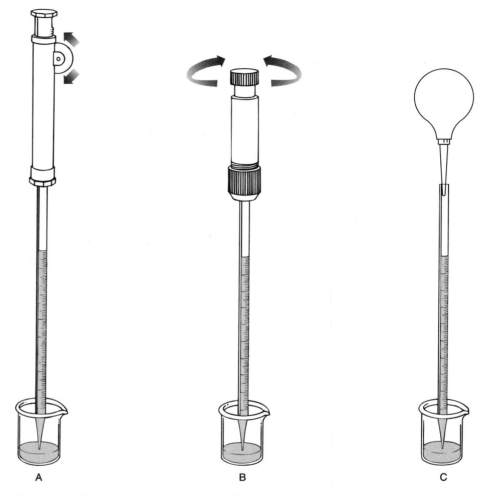

Figure 1.3   Pipet pumps and pipet bulb.

a pipet bulb are shown in Figure 1.3. A pipet fits snugly into the pipet pump, and the pump can be controlled to deliver precise volumes of liquids. Control of the pipet pump is accomplished by rotating a knob on the pump. Suction created when the knob is turned draws the liquid into the pipet. Liquid is expelled from the pipet by turning the knob in the opposite direction. The pump works satisfactorily with organic, as well as aqueous, solutions.

The style of pipet pump shown in Figure 1.3A is available in four sizes. The top of the pipet must be inserted securely into the pump and held there with one hand to obtain an adequate seal. The other hand is used to load and release the liquid. The pipet pump shown in Figure 1.3B may also be used with graduated pipets. With this style of pipet, the top of the pipet is held securely by a rubber O-ring, and it is easily handled with one hand. You should be certain that the pipet is held securely by the O-ring before using it. Disposable pipets may not fit tightly in the O-ring because they often have smaller diameters than nondisposable pipets.

An alternative, and less expensive, approach is to use a rubber pipet bulb, shown in Figure 1.3C. Use of the pipet bulb is made more convenient by inserting a plastic automatic pipet tip into a rubber pipet bulb.[1] The tapered end of the pipet tip fits snugly into the end of a pipet. Drawing the liquid into the pipet is made easy, and it is also convenient to remove the pipet bulb and place a finger over the pipet opening to control the flow of liquid.

The calibrations printed on graduated pipets are reasonably accurate, but you should practice using the pipets in order to achieve this accuracy. When accurate quantities of liquids are required, the best technique is to weigh the reagent that has been delivered from the pipet.

The following description, along with Figure 1.4, illustrates how to use a graduated pipet. Insert the end of the pipet firmly into the pipet pump. Rotate the knob of the pipet pump in the correct direction (counterclockwise or up) to fill the pipet. Fill the pipet to a point just above the uppermost mark and then reverse the direction of rotation of the knob to allow the liquid to drain from the pipet until the meniscus is adjusted to the 0.00-mL mark. Move the pipet to the receiving vessel. Rotate the knob of the pipet pump (clockwise or down) to force the liquid from the pipet. Allow the liquid to drain from the pipet until the meniscus arrives at the mark corresponding to the volume that you wish to dispense. Be sure to touch the tip of the pipet to the inside of the container before withdrawing the pipet. Remove the pipet and drain the remaining liquid into a waste receiver. Avoid transferring the entire contents of the pipet when measuring volumes with a pipet. Remember that in order to achieve the greatest possible accuracy with this method, you should deliver volumes as a *difference* between two marked calibrations.

Pipets may be obtained in a number of styles, but only three types will be described here (Figure 1.5). One type of graduated pipet is calibrated "to deliver" (TD) its total capacity when the last drop is blown out. This style of pipet, shown in Figure 1.5A, is probably the most common type of graduated pipet in use in the laboratory; it is designated by two rings at the top. Of course, one does not need to transfer the entire volume to a container. In order to deliver a more accurate volume, you should transfer an amount less than the total capacity of the pipet using the graduations on the pipet as a guide.

Another type of graduated pipet is shown in Figure 1.5B. This pipet is calibrated to deliver its total capacity when the meniscus is located on the last graduation mark near the bottom of the pipet. For example, the pipet shown in the Figure 1.5B delivers 10.0 mL of liquid when it has been drained to the point where the meniscus is located on the 10.0-mL mark. With this type of pipet, you must not drain the entire pipet or blow it out. In contrast, notice that the pipet discussed in Figure 1.5A has its last graduation at 0.90 mL. The last 0.10-mL volume is blown out to give the 1.00-mL volume.

A nongraduated volumetric pipet is shown in Figure 1.5C. It is easily identified by the large bulb in the center of the pipet. This pipet is calibrated so that it will retain its last drop after the tip is touched on the side of the container. It must not be blown out.

---

[1]This technique was described in Deckey, G. "A Versatile and Inexpensive Pipet Bulb." *Journal of Chemical Education, 57* (July 1980): 526.

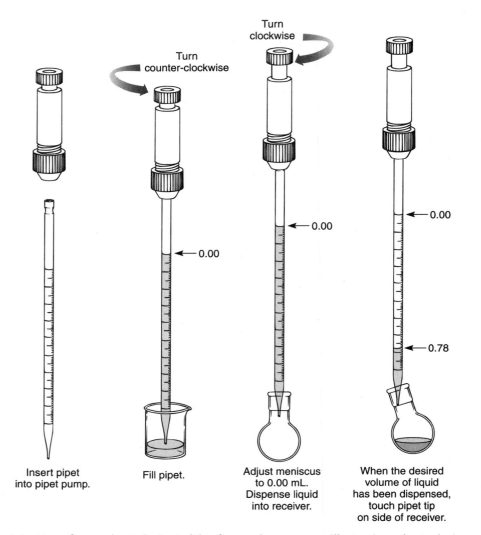

Figure 1.4 Use of a graduated pipet. (The figure shows, as an illustration, the technique required to deliver a volume of 0.78 mL from a 1.00-mL pipet.)

These pipets often have a single colored band at the top that identifies it as a "touch-off" pipet. The color of the band is keyed to its total volume. This type of pipet is commonly used in analytical chemistry.

1.4 PASTEUR PIPETS

The Pasteur pipet is shown in Figure 1.6A with a 2-mL rubber bulb attached. There are two sizes of Pasteur pipets: a short one ($5\frac{3}{4}$ inch), which is shown in the figure, and a long one (9 inch). It is important that the pipet bulb fit securely. You should not use a medicine dropper bulb because of its small capacity. A Pasteur pipet is an indispensable

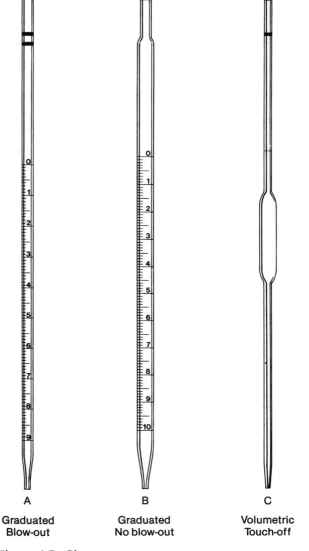

A	B	C
Graduated	Graduated	Volumetric
Blow-out	No blow-out	Touch-off

Figure 1.5   Pipets.

piece of equipment for the routine transfer of liquids. It is also used for separations (Technique 7). Pasteur pipets may be packed with cotton for use in gravity filtration (Technique 4) or packed with an adsorbent for small-scale column chromatography (Technique 12). Although they are considered disposable, you should be able to clean them for reuse as long as the tip remains unchipped.

A Pasteur pipet may be supplied by your instructor for dropwise addition of a particular reagent to a reaction mixture. For example, concentrated sulfuric acid is often dispensed in this way. When sulfuric acid is transferred, you should take care to avoid getting the acid into the rubber or latex dropper bulb.

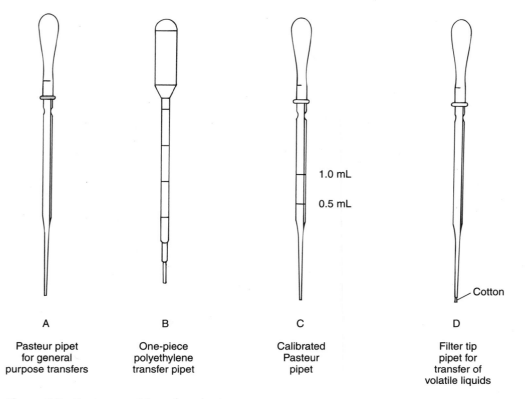

A	B	C	D
Pasteur pipet for general purpose transfers	One-piece polyethylene transfer pipet	Calibrated Pasteur pipet	Filter tip pipet for transfer of volatile liquids

Figure 1.6   Pasteur and transfer pipets.

One may avoid the rubber dropper bulb entirely by using one-piece transfer pipets made entirely of polyethylene (Figure 1.6B). These plastic pipets are available in 1- or 2-mL sizes. They come from the manufacturers with approximate calibration marks stamped on them. These pipets can be used with all aqueous solutions and most organic liquids. They cannot be used with a few organic solvents.

Pasteur pipets may be calibrated for use in operations where the volume does not need to be known precisely. Examples include measurement of solvents needed for extraction and for washing a solid obtained following crystallization. A calibrated Pasteur pipet is shown in Figure 1.6C. It is suggested that you calibrate several $5\frac{3}{4}$-inch pipets using the following procedure. On a balance, weigh 0.5 grams (0.5 mL) of water into a small test tube. Select a short Pasteur pipet and attach a rubber bulb. Squeeze the rubber bulb before inserting the tip of the pipet into the water. Try to control how much you depress the bulb so that, when the pipet is placed into the water and the bulb is completely released, only the desired amount of liquid is drawn into the pipet. When the water has been drawn up, place a mark with an indelible marking pen at the position of the meniscus. A more durable mark can be made by scoring the pipet with a file. Repeat this procedure with 1.0 gram of water, and make a 1-mL mark on the same pipet.

Your instructor may provide you with a calibrated Pasteur pipet and bulb for transferring liquids where an accurate volume is not required. The pipet may be used to transfer a volume of 1.5 mL or less. You may find that the instructor has taped a test tube to the side of the storage bottle. The pipet is stored in the test tube with that particular reagent.

Note: You should not assume that a certain number of drops equals a 1-mL volume. The common rule that 20 drops equal 1 mL, often used for a buret, does not hold true for a Pasteur pipet!

A Pasteur pipet may be packed with cotton to create a filter-tip pipet as shown in Figure 1.6D. This pipet is prepared by the instructions given in Technique 4, Section 4.6, page 645. Pipets of this type are very useful in transferring volatile solvents during extractions and in filtering small amounts of solid impurities from solutions. A filter-tip pipet is very useful for removing small particles from a solution of a sample prepared for NMR analysis.

1.5 SYRINGES

Syringes may be used to add a pure liquid or a solution to a reaction mixture. They are especially useful when anhydrous conditions must be maintained. The needle is inserted through a septum, and the liquid is added to the reaction mixture. Caution should be used with disposable syringes as they often use solvent-soluble rubber gaskets on the plungers. A syringe should be cleaned carefully after each use by drawing acetone or another volatile solvent into it and expelling the solvent with the plunger. Repeat this procedure several times to clean the syringe thoroughly. Draw air through the barrel with an aspirator to dry the syringe.

Syringes are usually supplied with volume graduations inscribed on the barrel. Large-volume syringes are not accurate enough to be used for measuring liquids in small scale experiments. A small microliter syringe, such as that used in gas chromatography, delivers a very precise volume.

1.6 AUTOMATIC PIPETS

Automatic pipets are commonly used in microscale organic laboratories and in biochemistry laboratories. One type of adjustable automatic pipet is shown in Figure 1.7. The automatic pipet is very accurate with aqueous solutions, but it is not as accurate with organic liquids. They are available in different sizes and can deliver accurate volumes ranging from 0.10 mL to 1.0 mL. These pipets are very expensive and must be shared by the entire laboratory. If they are used in your laboratory, your instructor will give directions for the correct use of the automatic pipet.

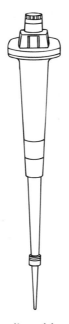

Figure 1.7   The adjustable automatic pipet.

1.7 MEASURING VOLUMES WITH BEAKERS AND ERLENMEYER FLASKS

Beakers and Erlenmeyer flasks both have graduations inscribed on them. Beakers and flasks can be used to give only a *crude* approximation of the volume. They are much less precise than graduated cylinders for measuring volume. You should use a graduated cylinder, dispensing pump, or graduated pipet for accurate measurement of liquids.

1.8 BALANCES

Solids and some liquids will need to be weighed on a balance that reads to at least the nearest decigram (0.01 g). A top-loading balance (see Fig. 1.8) works well if the balance pan is covered with a plastic draft shield. The shield has a flap that opens to allow access to the balance pan. An analytical balance (see Fig. 1.9) may also be used. This type of balance will weigh to the nearest tenth of a milligram (0.0001 g) when provided with a glass draft shield.

Modern electronic balances have a tare device that automatically subtracts the weight of a container or a piece of paper from the combined weight to give the weight of the sample. With solids, it is easy to place a piece of paper on the balance pan, press the tare device so that the paper appears to have zero weight, and then add your solid until the balance gives the weight you desire. You can then transfer the weighed solid to a container. You should al-

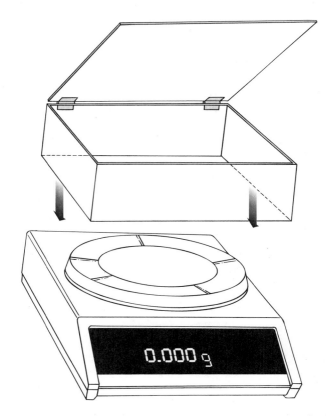

Figure 1.8 A top-loading balance with plastic draft shield.

ways use a spatula to transfer a solid and never pour material from a bottle. In addition, solids must be weighed on paper and not directly on the balance pan. Remember to clean any spills.

With liquids, you should weigh the flask to determine the tare weight, transfer the liquid with a graduated cylinder, dispensing pump, or graduated pipet into the flask, and then reweigh it. With liquids, it is usually necessary to weigh only the limiting reagent. The other liquids may be transferred using a graduated cylinder, dispensing pump, or graduated pipet. Their weights can be calculated by knowing the volumes and densities of the liquids.

PROBLEMS

1. What measuring device would you use to measure the volume under each of the conditions described below? In some cases, there may be more than one answer to the question.

   (a) 25 mL of a solvent needed for a crystallization

   (b) 2.4 mL of a liquid needed for a reaction

   (c) 1 mL of a solvent needed for an extraction

2. Assume that the liquid used in Question 1b is a limiting reagent for a reaction. What should you do after measuring the volume?

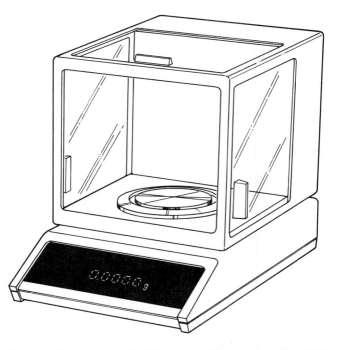

Figure 1.9   An analytical balance with glass draft shield.

3. Calculate the weight of a 2.5-mL sample of each of the following liquids:
   (a) Diethyl ether (ether)
   (b) Methylene chloride (dichloromethane)
   (c) Acetone

4. A laboratory procedure calls for 1.46 g of acetic anhydride. Calculate the volume of this reagent needed in the reaction.

T E C H N I Q U E   2

Heating and Cooling Methods

Most organic reaction mixtures need to be heated in order to complete the reaction. In general chemistry, you used a Bunsen burner for heating because nonflammable aqueous solutions were used. In an organic laboratory, however, the student must heat non-aqueous solutions that may contain *highly flammable* solvents. You *should not heat organic mixtures with a Bunsen burner* unless you are directed by your laboratory instructor. Open flames present a potential fire hazard. Whenever possible you should use one of the alternative heating methods, as described in the following sections.

2.1 HEATING MANTLES

A useful source of heat for most experiments in this textbook is the heating mantle, illustrated in Figure 2.1. The heating mantle shown here consists of a ceramic heating shell with electric heating coils embedded within the shell. The temperature of a heating mantle is regulated with the heat controller. Although it is difficult to monitor the actual temperature of the heating mantle, the controller is calibrated so that it is fairly easy to duplicate approximate heating levels after one has gained some experience with this apparatus. Reactions or distillations requiring relatively high temperatures can be easily performed with a heating mantle. For temperatures in the range of 50–80°C, you should use a water bath (Section 2.3) or a steam bath (Section 2.8).

In the center of the heating mantle shown in Figure 2.1 is a well which can accomodate round-bottom flasks of several different sizes. Some heating mantles, however, are designed to fit only specific sizes of round-bottom flasks. Some heating mantles are also made to be used with a magnetic stirrer so that the reaction mixture can be heated and stirred at the same time. Figure 2.2 shows a reaction mixture being heated with a heating mantle.

Heating mantles are very easy to use and safe to operate. The metal housing is grounded to prevent electrical shock if liquid is spilled into the well; however, flammable liquids may ignite if spilled into the well of a hot heating mantle.

> **Caution:** You should be very careful to avoid spilling liquids into the well of the heating mantle. The surface of the ceramic shell may be very hot and could cause the liquid to ignite.

Raising and lowering the apparatus is a much more rapid method of changing the temperature within the flask than changing the temperature with the controller. For this

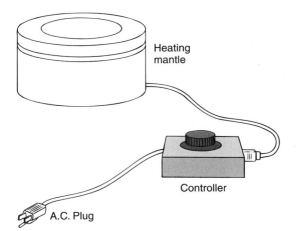

Figure 2.1 Heating mantle.

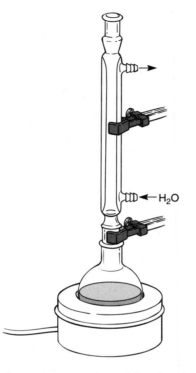

Figure 2.2   Heating with a heating mantle.

reason, the entire apparatus should be clamped above the heating mantle so that it can be raised quickly if overheating occurs. Some laboratories may provide a lab jack or blocks of wood which can be placed under the heating mantle. In this case, the heating mantle itself is lowered and the apparatus remains clamped in the same position.

There are two situations in which it is relatively easy to overheat the reaction mixture. The first situation occurs when a larger heating mantle is used to heat a relatively small flask. You should be very careful when doing this. Many laboratories provide heating mantles of different sizes to prevent this situation from happening. The second situation occurs when one is first bringing the reaction mixture to a boil. In order to bring the mixture to a boil as rapidly as possible, the heat controller is often turned up higher than it will need to be set in order to keep the mixture boiling. When the mixture begins boiling very rapidly, turn the controller to a lower setting and raise the apparatus until the mixture boils less rapidly. As the temperature of the heating mantle cools down, lower the apparatus until the flask is resting on the bottom of the well.

2.2 HOTPLATES

Hotplates are a very convenient source of heat; however, it is difficult to monitor the actual temperature, and changes in temperature occur somewhat slowly. Care must be

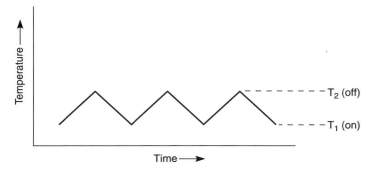

Figure 2.3   Temperature response for a hotplate with a thermostat.

taken with flammable solvents to ensure against fires caused by "flashing," when solvent vapors come into contact with the hotplate surface. One should never evaporate large quantities of a solvent by this method; the fire hazard is too large.

Some hotplates *heat constantly* at a given setting. They have no thermostat and you will have to control the temperature manually, either by removing the container being heated, or by adjusting the temperature up or down until a balance point is found. Some hotplates have a thermostat to control the temperature. A good thermostat will maintain a very even temperature. With many hotplates, however, the temperature may vary greatly (>10°–20°C), depending upon whether the heater is in its "on" cycle or its "off" cycle. These hotplates will have a cycling (or oscillating) temperature, as shown in Figure 2.3. They too will have to be adjusted continually to maintain even heat.

Some hotplates also have built-in magnetic stirring motors that enable the reaction mixture to be stirred and heated at the same time. Their use is described in Section 3.5, p. 621.

2.3 WATER BATH WITH HOTPLATE/STIRRER

A hot water bath is a very effective heat source when a temperature below 80°C is required. A beaker (250 mL or 400 mL) is partially filled with water and heated on a hot plate. A thermometer is clamped into position in the water bath. You may need to cover the water bath with aluminum foil to prevent evaporation, especially at higher temperatures. The water bath is illustrated in Figure 2.4. A mixture can be stirred with a magnetic stir bar (Technique 3, Section 3.5, p. 621). A hot water bath has some advantage over a heating mantle in that the temperature in the bath is uniform. In addition, it is sometimes easier to establish a lower temperature with a water bath than with other heating devices. Finally, the temperature of the reaction mixture will be closer to the temperature of the water, which allows for more precise control of the reaction conditions.

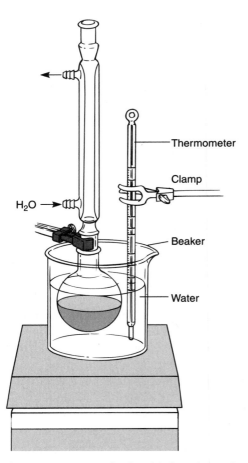

Figure 2.4 Water bath with hotplate/stirrer.

2.4 OIL BATH WITH HOTPLATE/STIRRER

In some laboratories oil baths may be available. An oil bath can be used when one is carrying out a distillation or is heating a reaction mixture that needs a temperature above 100°C. An oil bath can be heated most conveniently with a hotplate, and a *heavy-wall* beaker[1] provides a suitable container for the oil. A thermometer is clamped into position in the oil bath. In some laboratories the oil may be heated electrically by an immersion coil. Because oil baths have a high heat capacity and heat slowly, it is advisable to heat the oil bath partially before the actual time at which it is to be used.

An oil bath with ordinary mineral oil cannot be used above 200 to 220°C. Above this temperature the oil bath may "flash," or suddenly burst into flame. A hot oil fire is not extinguished easily. If the oil starts smoking, it may be near its flash temperature; dis-

---

[1]It is very dangerous to use a thin-wall beaker for an oil bath. Breakage due to heating can occur, spilling hot oil everywhere!

continue heating. Old oil, which is dark, is more likely to flash than new oil is. Also, hot oil causes bad burns. Water should be kept away from a hot oil bath, since water in the oil will cause it to splatter. Never use an oil bath when it is obvious that there is water in the oil. If there should be water present, replace the oil before using the heating bath. An oil bath has only a finite lifetime. New oil is clear and colorless but, after extended use, becomes dark brown and gummy from oxidation.

Besides ordinary mineral oil, a variety of other types of oils can be used in an oil bath. Silicone oil does not begin to decompose at as low a temperature as does mineral oil. When silicone oil is heated high enough to decompose, however, its vapors are far more hazardous than mineral oil vapors. The polyethylene glycols may be used in oil baths. They are water-soluble, which makes cleaning up after using an oil bath much easier than with mineral oil. One may select any one of a variety of polymer sizes of polyethylene glycol, depending on the temperature range required. The polymers of large molecular weight are often solid at room temperature. Wax may also be used for higher temperatures, but this material also becomes solid at room temperature. Some workers prefer to use a material that solidifies when not in use since it minimizes both storage and spillage problems. Vegetable shortening is occasionally used in heating baths.

2.5 ALUMINUM BLOCK WITH HOTPLATE/STIRRER

Although aluminum blocks are most commonly used in microscale organic laboratories, they can also be used with the smaller round-bottom flasks used in standard-scale experiments.[2] The aluminum block shown in Figure 2.5A can be used to hold 25-, 50-, or 100-mL round-bottom flasks, as well as a thermometer. Heating will occur more rapidly if the flask fits all the way into the hole; however, heating is also effective if the flask only partially fits into the hole. The aluminum block with smaller holes, as shown in Figure 2.5B, is designed for microscale glassware. It will hold a conical vial, a Craig tube or small test tubes, and a thermometer.

There are several advantages to heating with an aluminum block. The metal heats very quickly, high temperatures can be obtained, and you can cool the aluminum rapidly by removing it with crucible tongs and immersing it in cold water. Aluminum blocks are also inexpensive or can be fabricated readily in a machine shop.

Figure 2.6 shows a reaction mixture being heated with an aluminum block on a hotplate/stirrer unit. The thermometer in the figure is used to determine the temperature of the aluminum block. *Do not use a mercury thermometer:* use a thermometer containing a liquid other than mercury or use a metal dial thermometer that can be inserted into a smaller diameter hole drilled into the side of the block.[3] Make sure that the thermometer fits loosely in the hole, or it may break. Secure the thermometer with a clamp.

---

[2]The use of solid aluminum heating devices was developed by Siegfried Lodwig at Centralia College, Centralia, WA: Lodwig, S. N., *Journal of Chemical Education, 66* (1989): 77.
[3]Garner, C. M. "A Mercury-Free Alternative for Temperature Measurement in Aluminum Blocks." *Journal of Chemical Education, 68* (1991): A244.

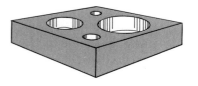

A. Large holes for 25-, 50- or
   100-mL round-bottom flasks.

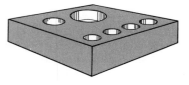

B. Small holes for Craig tube and 3-mL and
   5-mL conical vials, and small test tubes.

Figure 2.5 Aluminum heating blocks.

As already mentioned, aluminum blocks are often used in the microscale organic laboratory. The use of an aluminum block to heat a microscale reflux apparatus is shown in Figure 2.7. The reaction vessel in the figure is a conical vial, which is used in many microscale experiments. Also shown in Figure 2.7 is a split aluminum collar that may be used when very high temperatures are required. The collar is split to facilitate easy placement around a 5-mL conical vial. The collar helps to distribute heat further up the wall of the vial.

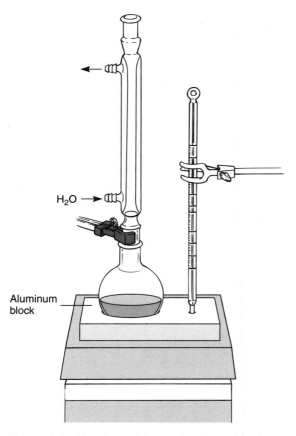

H_2O →

Aluminum
block

Figure 2.6 Heating with an aluminum block.

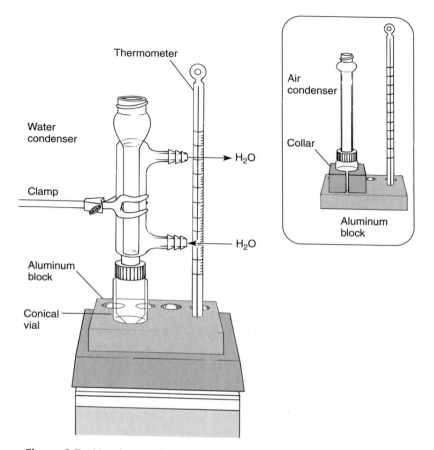

Figure 2.7   Heating with an aluminum block (microscale).

You should first calibrate the aluminum block so that you have an approximate idea where to set the control on the hotplate to achieve a desired temperature. Place the aluminum block on the hotplate and insert a thermometer into the small hole in the block. Select five equally spaced temperature settings, including the lowest and highest settings, on the heating control of the hotplate. Set the dial to the first of these settings and monitor the temperature recorded on the thermometer. When the thermometer reading arrives at a constant value,[4] record this final temperature, along with the dial setting. Repeat this procedure with the remaining four settings. Using these data, prepare a calibration curve for future reference.

It is a good idea to use the same hotplate each time, as it is very likely that two hotplates of the same type may give different temperatures with identical settings. Record the identification number printed on the unit that you are using in your notebook to ensure that you always use the same hotplate.

---

[4]See, however, Technique 2, Section 2.2, p. 602.

For many experiments, you can determine what the approximate setting on the hot-plate should be from the boiling point of the liquid being heated. Because the temperature inside the flask is lower than the aluminum block temperature, you should add at least 20°C to the boiling point of the liquid and set the aluminum block at this higher temperature. In fact, you may need to raise the temperature even higher than this value in order to bring the liquid to a boil.

Many organic mixtures need to be stirred as well as heated to achieve satisfactory results. To stir a mixture, place a magnetic stir bar (Technique 3, Fig. 3.10A, p. 621) in a round-bottom flask containing the reaction mixture as shown in Figure 2.8A. If the mixture is to be heated as well as stirred, attach a water condenser as shown in Figure 2.6. With the combination stirrer/hotplate unit, it is possible to stir and heat a mixture simultaneously. Many reactions in this textbook are stirred continuously during the course of the reaction. With conical vials, a magnetic spin vane must be used to stir mixtures (Fig. 3.10B, p. 621). This is shown in Figure 2.8B. More uniform stirring will be obtained if the flask or vial is placed in the aluminum block so that it is centered on the hotplate. Mixing may also be achieved by boiling the mixture. A boiling stone (Section 3.6, p. 621) should be added when a mixture is boiled without magnetic stirring.

2.6 SAND BATH WITH HOTPLATE/STIRRER

The sand bath is used in some microscale laboratories to heat organic mixtures. It can also be used as a heat source in some standard-scale experiments. Sand provides a clean way of distributing heat to a reaction mixture. To prepare a sand bath, place about a 1-cm depth of sand in a crystallizing dish and then set the dish on a hotplate/stirrer unit. The apparatus is shown in Figure 2.9. Clamp the thermometer into position in the sand bath. You should calibrate the sand bath in a manner similar to that used with the aluminum block (see previous section). Because sand heats more slowly than an aluminum block, you will need to begin heating the sand bath well before using it.

Do not heat the sand bath much above 200°C or you may break the dish. If you need to heat at very high temperatures, you should use a heating mantle or an aluminum block rather than a sand bath. With sand baths, it may be necessary to cover the dish with alu-

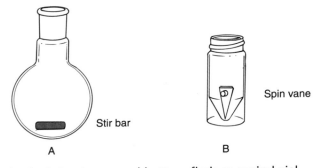

A   B

Figure 2.8   Methods of stirring in a round-bottom flask or conical vial.

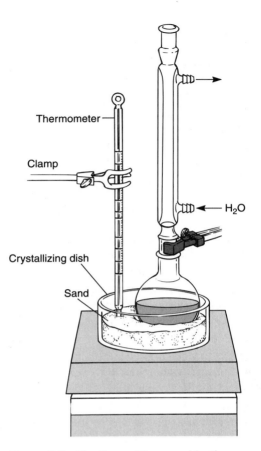

Figure 2.9   Heating with a sand bath.

minum foil to achieve a temperature near 200°C. Because of the relatively poor heat conductivity of sand, a temperature gradient is established within the sand bath. It is warmer near the bottom of the sand bath and cooler near the top for a given setting on the hotplate. To make use of this gradient, you may find it convenient to bury the flask or vial in the sand to heat a mixture more rapidly. Once the mixture is boiling, you can then slow the rate of heating by raising the flask or vial. These adjustments may be made easily and do not require a change in the setting on the hotplate.

2.7 FLAMES

The simplest technique for heating mixtures is to use a Bunsen burner. Because of the high danger of fires, however, the use of the Bunsen burner should be strictly limited to those cases for which the danger of fire is low or for which no reasonable alternative source of heat is available. A flame should generally be used only to heat aqueous solutions or solutions with very high boiling points. You should always check with your instructor about using a burner. If you use a burner at your bench, great care should be taken to ensure that others in the vicinity are not using flammable solvents.

In heating a flask with a Bunsen burner, you will find that using a wire gauze can produce more even heating over a broader area. The wire gauze, when placed under the object being heated, spreads the flame to keep the flask from being heated in one small area only.

Bunsen burners may be used to prepare capillary micropipets for thin-layer chromatography or to prepare other pieces of glassware requiring an open flame. For these purposes, burners should be used in designated areas in the laboratory and not at your laboratory bench.

2.8 STEAM BATHS

The steam cone or steam bath is a good source of heat when temperatures around 100°C are needed. Steam baths are used to heat reaction mixtures and solvents needed for crystallization. A steam cone and a portable steam bath are shown in Figure 2.10. These methods of heating have the disadvantage that water vapor may be introduced, through condensation of steam, into the mixture being heated. A slow flow of steam may minimize this difficulty.

Because water condenses in the steam line when it is not in use, it is necessary to purge the line of water before the steam will begin to flow. This purging should be accomplished before the flask is placed on the steam bath. The steam flow should be started with a high rate to purge the line; then, the flow should be reduced to the desired rate. When using a portable steam bath, be certain that condensate (water) is drained into a sink. Once the steam bath or cone is heated, a slow steam flow will maintain the temperature of the mixture being heated. There is no advantage to having a Vesuvius on your desk! An excessive steam flow may cause problems with condensation in the flask. This condensation problem can often be avoided by selecting the correct place at which to locate the flask on top of the steam bath.

The top of the steam bath consists of several flat concentric rings. The amount of heat delivered to the flask being heated can be controlled by selecting the correct sizes of

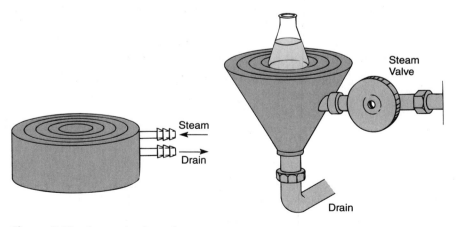

Figure 2.10   Steam bath and steam cone.

these rings. Heating is most efficient when the largest opening that will still support the flask is used. Heating large flasks on a steam bath while using the smallest opening leads to slow heating and wastes laboratory time.

2.9 COLD BATHS

At times, you may need to cool an Erlenmeyer flask or round-bottom flask below room temperature. A cold bath is used for this purpose. The most common cold bath is an **ice bath,** which is a highly convenient source of 0°C temperatures. An ice bath requires water along with ice to work well. If an ice bath is made up of only ice, it is not a very efficient cooler since the large pieces of ice do not make good contact with the flask. Enough water should be present with ice so that the flask is surrounded by water but not so much that the temperature is no longer maintained at 0°C. In addition, if too much water is present, the buoyancy of a flask resting in the ice bath may cause it to tip over. There should be enough ice in the bath to allow the flask to rest firmly.

For temperatures somewhat below 0°C, you may add some solid sodium chloride to the ice-water bath. The ionic salt lowers the freezing point of the ice, so that temperatures in the range of 0 to −10°C can be reached. The lowest temperatures are reached with ice-water mixtures that contain relatively little water.

A temperature of −78.5°C can be obtained with solid carbon dioxide or dry ice. However, large chunks of dry ice do not provide uniform contact with a flask being cooled. A liquid such as isopropyl alcohol is mixed with small pieces of dry ice to provide an efficient cooling mixture. Acetone and ethanol can be used in place of isopropyl alcohol. Be careful when handling dry ice because it can inflict severe frostbite. Extremely low temperatures can be obtained with liquid nitrogen (−195.8°C).

PROBLEMS

1. What would be the preferred heating device(s) in each of the following situations?
   (a) Reflux a solvent with a 56°C boiling point
   (b) Reflux a solvent with a 110°C boiling point
   (c) Distillation of a substance that boils at 220°C

2. Obtain the boiling points for the following compounds by using a handbook (Technique 20, Section 20.1, p. 861). In each case, suggest a heating device(s) that should be used for refluxing the substance.
   (a) Butyl benzoate
   (b) 1-Pentanol
   (c) 1-Chloropropane

3. What type of bath would you use to get a temperature of −10°C?

4. Obtain the melting point and boiling point for benzene and ammonia from a handbook (Technique 20, Section 20.1 p. 861) and answer the following questions.
   (a) A reaction was conducted in benzene as the solvent. Because the reaction was very exothermic, the mixture was cooled in a salt-ice bath. This was a bad choice. Why?
   (b) What bath should be used for a reaction that is conducted in liquid ammonia as the solvent?

TECHNIQUE 3

Reaction Methods

The successful completion of an organic reaction requires the chemist to be familiar with a variety of laboratory methods. These methods include operating safely, choosing and handling solvents correctly, heating reaction mixtures, adding liquid reagents, maintaining anhydrous conditions in the reaction, and collecting gaseous products. Several techniques that are used in bringing a reaction to a successful conclusion are treated in this chapter.

3.1 SOLVENTS

Organic solvents must be handled safely. Always remember that organic solvents are all at least mildly toxic and that many are flammable. You should become thoroughly familiar with the introductory chapter, "Laboratory Safety."

Read "Laboratory Safety," pages 5–22.

TABLE 3.1.   Common Organic Solvents

Solvent	bp (°C)	Solvent	bp (°C)
Hydrocarbons		Ethers	
Pentane	36	**Ether** (Diethyl)	35
Hexane	69	**Dioxane***	101
Benzene*	80	**1,2-Dimethoxyethane**	83
Toluene	111	Others	
Hydrocarbon Mixtures		Acetic acid	118
Petroleum ether	30–60	Acetic anhydride	140
Ligroin	60–90	**Pyridine**	115
Chlorocarbons		**Acetone**	56
Methylene chloride	40	**Ethyl acetate**	77
Chloroform*	61	Dimethylformamide	153
Carbon tetrachloride*	77	Dimethylsulfoxide	189
Alcohols			
Methanol	65		
Ethanol	78		
Isopropyl alcohol	82		

NOTE: Boldface type indicates flammability.
*Suspect carcinogen (see p. 21).

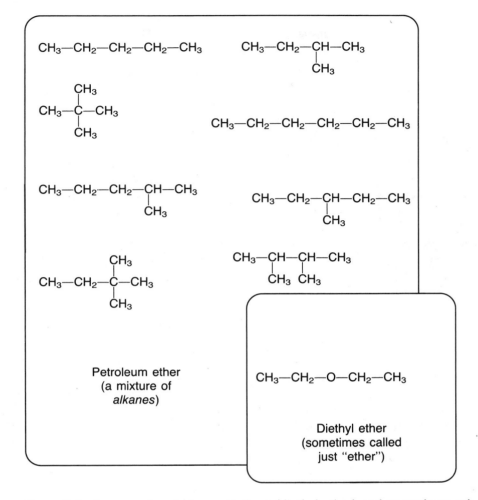

Figure 3.1   A comparison between "ether" (diethyl ether) and "petroleum ether."

The most common organic solvents are listed in Table 3.1 along with their boiling points. Solvents marked in boldface type will burn. Ether, pentane, and hexane are especially dangerous; if they are combined with the correct amount of air they will explode.

The terms **petroleum ether** and **ligroin** are often confusing. Petroleum ether is a mixture of hydrocarbons with isomers of formulas C_5H_{12} and C_6H_{14} predominating. Petroleum ether is not an ether at all because there are no oxygen-bearing compounds in the mixture. In organic chemistry, an ether is usually a compound containing an oxygen atom to which two alkyl groups are attached. Figure 3.1 shows some of the hydrocarbons that appear commonly in petroleum ether. Use special care when instructions call for either **ether** or **petroleum ether;** the two must not become accidentally confused. Confusion is particularly easy when one is selecting a container of solvent from the supply shelf.

Ligroin, or high-boiling petroleum ether, is like petroleum ether in composition except that compared with petroleum ether, ligroin generally includes higher-boiling alkane isomers. Depending on the supplier, ligroin may have different boiling ranges. While some brands of ligroin have boiling points ranging from about 60°C to about 90°C, other brands have boiling points ranging from about 60°C to about 75°C. The boiling point ranges of petroleum ether and ligroin are often included on the labels of the containers.

3.2 GROUND-GLASS JOINTS

The glassware in your organic kit has **standard-taper ground-glass joints.** For example, the Claisen head in Figure 3.2 consists of an inner (male) ground-glass joint at the bottom and two outer (female) joints at the top. Each end is ground to a precise size which is designated by the symbol ⊺ followed by two numbers. A common joint size in standard-scale glassware is ⊺ 19/22. The first number indicates the diameter (in millimeters) of the joint at its widest point, and the second number refers to its length (see Fig. 3.2). One advantage of standard-taper joints is that the pieces fit together snugly and form a good seal. In addition, standard-taper joints allow all glassware components with the same joint size to be connected, thus permitting the assembly of a wide variety of apparatus. One disadvantage of glassware with ground-glass joints, however, is that it is very expensive.

It is important to make sure no solid is on the joint surfaces. Such material will lessen the efficiency of the seal, and the joints may leak. Also, if the apparatus is to be heated, material caught between the joint surfaces will increase the tendency for the joints to stick. If the joint surfaces are coated with liquid or adhering solid, you should wipe them with a cloth or lint-free paper towel before assembling. It is sometimes advisable to lubricate joints lightly with stopcock grease before assembling them (see p. 34).

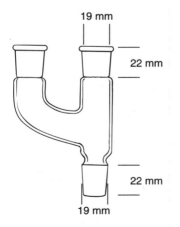

Figure 3.2   Illustration of ⊺ 19/22 inner and outer joints showing dimensions.

The most important thing you can do to prevent ground-glass joints from becoming "frozen" or stuck together is to disassemble the glassware as soon as possible after a procedure is completed. Even when this precaution is followed, ground-glass joints may become stuck tightly together. The same is true of glass stoppers in bottles or flasks. If the pieces do not separate easily, you must be careful when you try to pull them apart. The best way is to hold the two pieces, with both hands touching, as close as possible to the joint. With a firm grasp, try to loosen the joint with a slight twisting motion (do not twist very hard). If this does not work, try to pull your hands apart without pushing sideways on the glassware. If it is not possible to pull the pieces apart, see page 34 for other options.

3.3 ASSEMBLING THE APPARATUS

Care must be taken when assembling the glass components into the desired apparatus. You should always remember that Newtonian physics applies to chemical apparatus, and unsecured pieces of glassware are certain to respond to gravity.

Assembling an apparatus in the correct manner requires that the individual pieces of glassware are connected to each other securely and the entire apparatus is held in the correct position. This can be accomplished by using **adjustable metal clamps** or a combination of adjustable metal clamps and **plastic joint clips.**

Two types of adjustable metal clamps are shown in Figure 3.3. Although these two types of clamps can usually be interchanged, the extension clamp is more commonly used to hold round-bottom flasks in place and the three-finger clamp is frequently used to clamp condensers. Both types of clamps must be attached to a ring stand using a clamp holder, shown in Figure 3.3C.

Securing Standard-Scale Apparatus Assemblies

It is possible to assemble an apparatus using only adjustable metal clamps. An apparatus used to perform a distillation is shown in Figure 3.4. It is held together securely

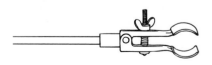

A. Extension clamp

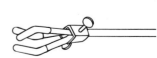

B. Three finger clamp

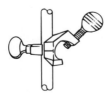

C. Clamp holder

Figure 3.3   Adjustable metal clamps.

with three metal clamps. Because of the size of the apparatus and its geometry, the various clamps would likely be attached to three different ring stands. This apparatus would be somewhat difficult to assemble, because one must ensure that the individual pieces stay together while securing and adjusting the clamps required to hold the entire apparatus in place. In addition, one must be very careful not to bump any part of the apparatus or the ring stands after the apparatus is assembled.

A more convenient alternative is to use a combination of metal clamps and plastic joint clips. A plastic joint clip is shown in Figure 3.5A. These clips are very easy to use (they just "clip" on), will withstand temperatures up to 140°C, and are quite durable. They hold together two pieces of glassware that are connected by ground glass joints, as shown in Figure 3.5B. These clips come in different sizes to fit ground glass joints of different sizes and they are color-coded for each size.

When used in combination with metal clamps, the plastic joint clips make it much easier to assemble most apparatus in a secure manner. There is less chance of dropping the glassware while assembling the apparatus, and once the apparatus is set up it is more secure. Figure 3.6 shows the same distillation apparatus held in place with both adjustable metal clamps and plastic joint clips.

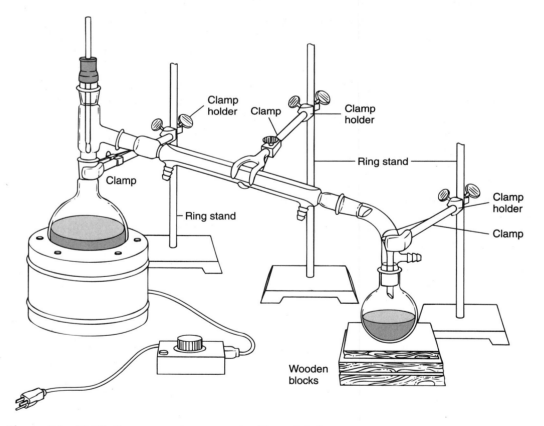

Figure 3.4 Distillation apparatus secured with metal clamps.

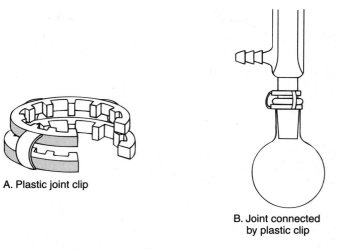

A. Plastic joint clip

B. Joint connected
by plastic clip

Figure 3.5   Plastic joint clip.

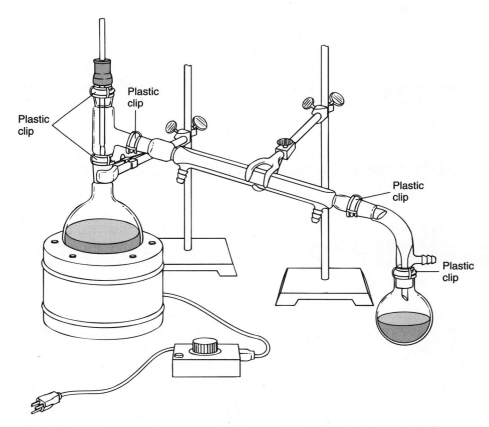

Figure 3.6   Distillation apparatus secured with metal clamps and plastic joint clips.

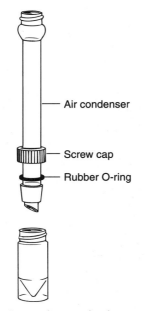

Figure 3.7   A microscale standard-taper joint assembly.

To assemble this apparatus, one would first connect all of the individual pieces together using the plastic clips. The entire apparatus is then connected to the ring stands using the adjustable metal clamps. Note that only two ring stands are required and the wooden blocks are not needed.

Securing Microscale Apparatus Assemblies

The glassware in most microscale kits is made with standard-taper ground joints. The most common joint size is ⊤ 14/10. Some microscale glassware with ground-glass joints also has threads cast into the outside surface of the outer joints (see top of air condenser in Figure 3.7). The threaded joint allows the use of a plastic screw cap with a hole in the top to fasten two pieces of glassware together securely. The plastic cap is slipped over the inner joint of the upper piece of glassware, followed by a rubber O-ring (see Figure 3.7). The O-ring should be pushed down so that it fits snugly on top of the ground-glass joint. The inner ground-glass joint is then fitted into the outer joint of the bottom piece of glassware. The screw cap is tightened, without excessive force, to attach the entire apparatus firmly together. The O-ring provides an additional seal that makes this joint air-tight. With this connecting system, it is unnecessary to use any type of grease to seal the joint. The O-ring *must be used* to obtain a good seal and to lessen the chances of breaking the glassware when you tighten the plastic cap.

Microscale glassware connected together in this fashion can be assembled very easily. The entire apparatus is held together securely, and usually only one metal clamp is required to hold the apparatus onto a ring stand.

3.4 HEATING UNDER REFLUX

Often we wish to heat a mixture for a long time and to leave it untended. A **reflux apparatus** (see Fig. 3.8) allows such heating. It also keeps the solvent from being lost by evaporation. A condenser is attached to the boiling flask.

Condenser.   The **water-jacketed condenser** shown in the figure consists of two concentric tubes with the outer cooling tube sealed onto the inner tube. The vapors rise within the inner tube and water circulates through the outer tube. The circulating water removes heat from the vapors and condenses them. The figure also shows a typical microscale apparatus for heating small quantities of material under reflux (Fig. 3.8B).

When using a water-jacketed condenser, the direction of the water flow should be such that the condenser will fill with cooling water. The water should enter the bottom of the condenser and leave from the top. The water should flow fast enough to withstand any changes in pressure in the water lines, but it should not flow any faster than absolutely necessary. An excessive flow rate greatly increases the chance of a flood, and high water pressure may force the hose from the condenser. Cooling water should be flowing before heating is begun! If the water is to remain flowing overnight, it is advisable to fasten the

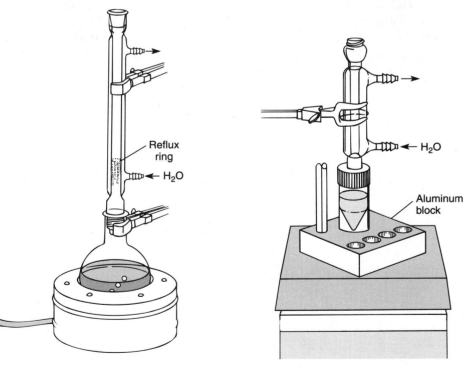

A. Reflux apparatus for standard-scale reactions, using a heating mantle and water-jacketed condenser.

B. Reflux apparatus for microscale reactions, using a hotplate, aluminum block, and water-jacketed condenser.

Figure 3.8   Heating under reflux.

rubber tubing securely with wire to the condenser. If a flame is used as a source of heat (see Cautions, p. 609), it is wise to use a wire gauze beneath the flask to provide an even distribution of heat from the flame. In most cases, a heating mantle, water bath, oil bath, aluminum block, sand bath, or steam bath is preferred over a flame.

Stirring.   When heating a solution, always use a magnetic stirrer or a boiling stone (see Sections 3.5 and 3.6) to keep the solution from "bumping" (see next section).

Rate of Heating.   If the heating rate has been correctly adjusted, the liquid being heated under reflux will travel only part way up the condenser tube before condensing. Below the condensation point, solvent will be seen running back into the flask; above it, the interior of the condenser will appear dry. The boundary between the two zones will be clearly demarcated, and a **reflux ring** or a ring of liquid will appear there. The reflux ring can be seen in Figure 3.8A. In heating under reflux, the rate of heating should be adjusted so that the reflux ring is no higher than a third to a half the distance to the top of the condenser. With microscale experiments, the quantities of vapor rising in the condenser frequently are so small that a clear reflux ring cannot be seen. In those cases, the heating rate must be adjusted so that the liquid boils smoothly but not so rapidly that solvent can escape the condenser. With such small volumes, the loss of even a small amount of solvent can affect the reaction. With standard-scale reactions, the reflux ring is much easier to see, and one can adjust the heating rate more easily.

Tended Reflux.   It is possible to heat small amounts of a solvent under reflux in an Erlenmeyer flask. By heating gently, the evaporated solvent will condense in the relatively cold neck of the flask and return to the solution. This technique (see Fig. 3.9) requires constant attention. The flask must be swirled frequently and removed from the heating source for a short period if the boiling becomes too vigorous. When heating is in progress, the reflux ring should not be allowed to rise into the neck of the flask.

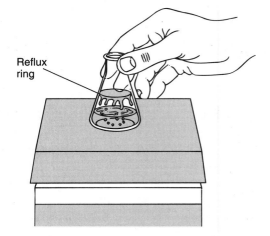

Reflux
ring

Figure 3.9   Tended reflux of small quantities on a hotplate.

3.5 STIRRING METHODS

When a solution is heated, there is a danger that it may become superheated. When this happens, very large bubbles sometimes erupt violently from the solution; this is called **bumping.** Bumping must be avoided because it brings with it the risk that material may be lost from the apparatus, that a fire might start, or that the apparatus may break.

Magnetic stirrers are used to prevent bumping because they produce turbulence in the solution. The turbulence breaks up the large bubbles that form in boiling solutions. An additional purpose for using a magnetic stirrer is to stir the reaction to ensure that all the reagents are thoroughly mixed. A magnetic stirring system consists of a magnet that is rotated by an electric motor. The rate at which this magnet rotates can be adjusted by a potentiometric control. A small magnet, which is coated with a nonreactive material such as Teflon or glass, is placed in the flask. The magnet within the flask rotates in response to the rotating magnetic field caused by the motor-driven magnet. The result is that the inner magnet stirs the solution as it rotates. A very common type of magnetic stirrer includes the stirring system within a hotplate. This type of hotplate-stirrer permits one to heat the reaction and stir it simultaneously.

Magnetic stirring bars are available in several sizes and shapes. For standard-scale apparatus, magnetic stirring bars of various sizes and shapes are available. For microscale apparatus, a **magnetic spin vane** is often used. It is designed to contain a tiny bar magnet and to have a shape that conforms to the conical bottom of a reaction vial. A small Teflon-coated magnetic stirring bar works well with very small round-bottom boiling flasks. Small stirring bars of this type (often sold as "disposable" stirring bars) can be obtained very cheaply. A variety of magnetic stirring bars is illustrated in Figure 3.10.

There is also a variety of simple techniques that may be used to stir a liquid mixture in a centrifuge tube or conical vial. A thorough mixing of the components of a liquid can be achieved by repeatedly drawing the liquid into a Pasteur pipet and then ejecting the liquid back into the container by pressing sharply on the dropper bulb. Liquids can also be stirred effectively by placing the flattened end of a spatula into the container and twirling it rapidly.

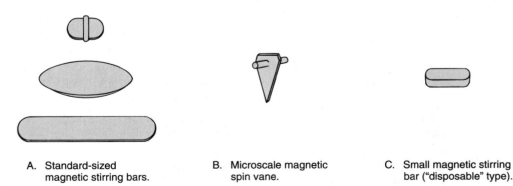

A. Standard-sized
    magnetic stirring bars.

B. Microscale magnetic
    spin vane.

C. Small magnetic stirring
    bar ("disposable" type).

Figure 3.10 Magnetic stirring bars.

3.6 BOILING STONES

A **boiling stone,** also known as a **boiling chip** or **Boileezer,** is a small lump of porous material that produces a steady stream of fine air bubbles when it is heated in a solvent. This stream of bubbles and the turbulence that accompanies it breaks up the large bubbles of gases in the liquid. In this way, it reduces the tendency of the liquid to become superheated, and it promotes the smooth boiling of the liquid. The boiling stone decreases the chances for bumping.

Two common types of boiling stones are carborundum and marble chip. Carborundum boiling stones are more inert and the pieces are usually quite small, suitable for most small-scale applications. If available, carborundum boiling stones are preferred for most purposes. Marble chips may dissolve in strong acid solutions and the pieces are larger. The advantage of marble chips is that they are cheaper.

Because boiling stones act to promote the smooth boiling of liquids, you should always make certain that a boiling stone has been placed in a liquid *before* heating is begun. If you wait until the liquid is hot, it may have become superheated. Adding a boiling stone to a superheated liquid will cause all the liquid to try to boil at once. The liquid, as a result, would erupt entirely out of the flask or froth violently.

As soon as boiling ceases in a liquid containing a boiling stone, the liquid is drawn into the pores of the boiling stone. When this happens, the boiling stone no longer can produce a fine stream of bubbles; it is spent. You may have to add a new boiling stone if you have allowed boiling to stop for a long period.

Wooden applicator sticks are used in some applications. They function in the same manner as boiling stones. Occasionally, glass beads are used. Their presence also causes sufficient turbulence in the liquid to prevent bumping.

3.7 ADDITION OF LIQUID REAGENTS

Liquid reagents and solutions are added to a reaction by several means, some of which are shown in Figure 3.11. The most common type of assembly for standard-scale experiments is shown in Figure 3.11A. In this apparatus, a separatory funnel is attached to the sidearm of a Claisen head adapter. The separatory funnel must be equipped with a standard-taper, ground-glass joint to be used in this manner. The liquid is stored in the separatory funnel (which is called an **addition funnel** in this application) and is added to the reaction. The rate of addition is controlled by adjusting the stopcock. When it is being used as an addition funnel, the upper opening must be kept open to the atmosphere. If the upper hole is stoppered, a vacuum will develop in the funnel and will prevent the liquid from passing into the reaction vessel. Because the funnel is open to the atmosphere, there is a danger that atmospheric moisture can contaminate the liquid reagent as it is being added. To prevent this outcome, a drying tube (see Section 3.8) may be attached to the upper opening of the addition funnel. The drying tube allows the funnel to maintain atmospheric pressure without allowing the passage of water vapor into the reaction. For

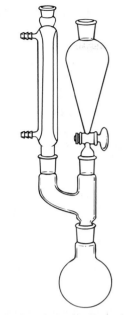

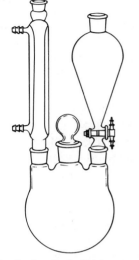

A. Standard-scale equipment, using a
   separatory funnel as an addition funnel.

B. Standard-scale, for larger amounts.

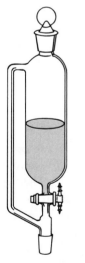

C. A pressure equalizing addition funnel.

D. Addition with a hypodermic
   syringe inserted through a
   rubber septum.

Figure 3.11   Methods for adding liquid reagents to a reaction.

reactions that are particularly sensitive to moisture, it is also advisable to attach a second drying tube to the top of the condenser.

Another standard-scale assembly, suitable for larger amounts of material, is shown in Figure 3.11B. Drying tubes may also be used with this apparatus to prevent contamination from atmospheric moisture.

Figure 3.11C shows an alternative type of addition funnel that is useful for reactions that must be maintained under an atmosphere of inert gas. This is the **pressure-equalizing addition funnel.** With this glassware, the upper opening is stoppered. The sidearm allows the pressure above the liquid in the funnel to be in equilibrium with the pressure in the rest of the apparatus, and it allows the inert gas to flow over the top of the liquid as it is being added.

With either type of standard-scale addition funnel, you can control the rate of addition of the liquid by carefully adjusting the stopcock. Even after careful adjustment, changes in pressure can occur, causing the flow rate to change. In some cases, the stopcock can become clogged. It is important, therefore, to monitor the addition rate carefully and to refine the adjustment of the stopcock as needed to maintain the desired rate of addition.

A fourth method, shown in Figure 3.11D, is suitable for use in microscale and some standard-scale experiments where the reaction should be kept isolated from the atmosphere. In this approach, the liquid is kept in a hypodermic syringe. The syringe needle is inserted through a rubber septum, and the liquid is added dropwise from the syringe. The septum seals the apparatus from the atmosphere, which makes this technique useful for reactions that are conducted under an atmosphere of inert gas or where anhydrous conditions must be maintained. The drying tube is used to protect the reaction mixture from atmospheric moisture.

3.8 DRYING TUBES

With certain reactions, atmospheric moisture must be prevented from entering the reaction vessel. A **drying tube** can be used to maintain anhydrous conditions within the apparatus. Two types of drying tubes are shown in Figure 3.12. The typical drying tube is prepared by placing a small, loose plug of glass wool or cotton into the constriction at the end of the tube nearest the ground-glass joint or hose connection. The plug is tamped gently with a glass rod or piece of wire to place it in the correct position. A drying agent, typically calcium sulfate ("Drierite") or calcium chloride (see Technique 7, Section 7.8, p. 696), is poured on top of the plug to the approximate depth shown in Figure 3.12. Another loose plug of glass wool or cotton is placed on top of the drying agent to prevent the solid material from falling out of the drying tube. The drying tube is then attached to the flask or condenser.

Air that enters the apparatus must pass through the drying tube. The drying agent absorbs any moisture from air passing through it, so that air entering the reaction vessel has had the water vapor removed from it.

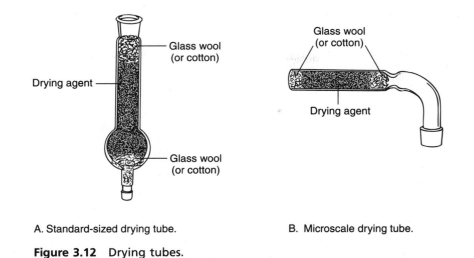

A. Standard-sized drying tube.   B. Microscale drying tube.

Figure 3.12   Drying tubes.

3.9 CAPTURING NOXIOUS GASES

Many organic reactions involve the production of a noxious gaseous product. The gas may be corrosive, such as hydrogen chloride, hydrogen bromide, or sulfur dioxide; or it may be toxic, such as carbon monoxide. The safest way to avoid exposure to these gases is to conduct the reaction in a ventilated hood where the gases can be safely drawn away by the ventilation system.

In many instances, however, it is quite safe and efficient to conduct the experiment on the laboratory bench, away from the hood. This is particularly true when the gases are soluble in water. Some techniques for capturing noxious gases are presented in this section.

A.   EXTERNAL GAS TRAPS

One approach to capturing gases is to prepare a trap that is separate from the reaction apparatus. The gases are carried from the reaction to the trap by means of tubing. There are several variations on this type of trap. With standard-scale reactions, a trap using an inverted funnel placed in a beaker of water is used. A piece of glass tubing, inserted through a thermometer adapter attached to the reaction apparatus, is connected to flexible tubing. The tubing is attached to a conical funnel. The funnel is clamped in place inverted over a beaker of water. The funnel is clamped so that its lip *almost touches* the water surface, but is not placed below the surface of the water. With this arrangement, water cannot be sucked back into the reaction if the pressure in the reaction vessel changes suddenly. This type of trap can also be used in microscale applications. An example of the inverted funnel type of gas trap is shown in Figure 3.13.

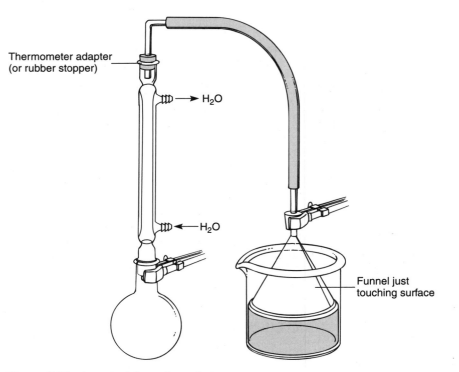

Thermometer adapter
(or rubber stopper)

H_2O

H_2O

Funnel just
touching surface

Figure 3.13 Inverted-funnel gas trap.

One method that works well for standard-scale and microscale experiments is to place a thermometer adapter into the opening in the reaction apparatus. A Pasteur pipet is inserted upside-down through the adapter, and a piece of flexible tubing is fitted over the narrow tip. It might be helpful to break the Pasteur pipet before using it for this purpose, so that only the narrow tip and a short section of the barrel is used. The other end of the flexible tubing is placed through a large plug of moistened glass wool in a test tube. The water in the glass wool absorbs the water-soluble gases. This method is shown in Figure 3.14.

B. DRYING-TUBE METHOD

Some standard-scale and most microscale experiments have the advantage that the amounts of gases produced are very small. Hence, it is easy to trap them and prevent them from escaping into the laboratory room. You can take advantage of the water solubility of corrosive gases such as hydrogen chloride, hydrogen bromide, and sulfur dioxide. A simple technique is to attach the drying tube (see Fig. 3.12) to the top of the reaction flask or condenser. The drying tube is filled with moistened glass wool. The moisture in the glass wool absorbs the gas, preventing its escape. To prepare this type of gas trap, fill the drying tube with glass wool and then add water dropwise to the glass wool until it has

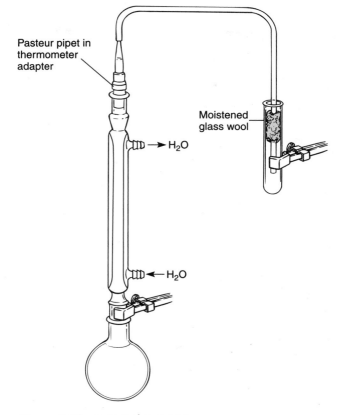

Figure 3.14   External gas trap.

been moistened to the desired degree. Moistened cotton can also be used, although cotton will absorb so much water that it is easy to plug the drying tube.

When using glass wool in a drying tube, moisture from the glass wool must not be allowed to drain from the drying tube into the reaction. It is best to use a drying tube that has a constriction between the part where the glass wool is placed and the neck, where the joint is attached (see Fig. 3.12B). The constriction acts as a partial barrier preventing the water from leaking into the neck of the drying tube. Make certain not to make the glass wool too moist. When it is necessary to use the drying tube shown in Figure 3.12A as a gas trap and it is essential that water not be allowed to enter the reaction flask, the modification shown in Figure 3.15 should be used. The rubber tubing between the thermometer adapter and the drying tube should be heavy enough to prevent crimping.

C.   REMOVAL OF NOXIOUS GASES USING AN ASPIRATOR

An aspirator can be used to remove noxious gases from the reaction. The simplest approach is to clamp a disposable Pasteur pipet so that its tip is placed well into the condenser atop the reaction flask. An inverted funnel clamped over the apparatus can also be used. The

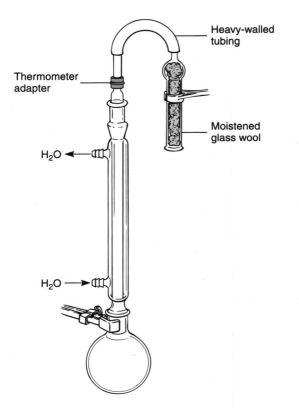

Figure 3.15 Drying tube used to capture evolved gases.

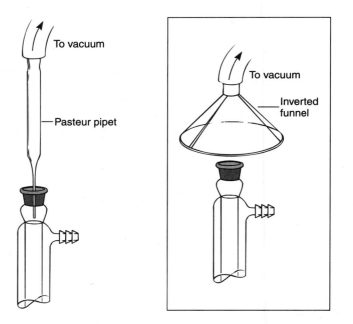

Figure 3.16 Removal of noxious gases under vacuum. (The inset shows an alternative assembly, using an inverted funnel in place of the Pasteur pipet.)

pipet or funnel is attached to an aspirator with flexible tubing. A trap should be placed between the pipet or funnel and the aspirator. As gases are liberated from the reaction, they rise into the condenser. The vacuum draws the gases away from the apparatus. Both types of systems are shown in Figure 3.16. In the special case where the noxious gases are soluble in water, connecting a water aspirator to the pipet or funnel removes the gases from the reaction and traps them in the flowing water without the need for a separate gas trap.

3.10 COLLECTING GASEOUS PRODUCTS

In Section 3.9, means of removing unwanted gaseous products from the reaction system were examined. Some experiments produce gaseous products that you must collect and analyze. Methods to collect gaseous products are all based on the same principle. The gas is carried through tubing from the reaction to the opening of a flask or a test tube, which has been filled with water and is inverted in a container of water. The gas is allowed to bubble into the inverted collection tube (or flask). As the collection tube fills with gas, the water is displaced into the water container. If the collection tube is graduated, as in a graduated cylinder or a centrifuge tube, you can monitor the quantity of gas produced in the reaction.

If the inverted gas collection tube is constructed from a piece of glass tubing, a rubber septum can be used to close the upper end of the container. This type of collection tube is shown in Figure 3.17. A sample of the gas can be removed using a gas-tight syringe equipped with a needle. The gas that is removed can be analyzed by gas chromatography (see Technique 14).

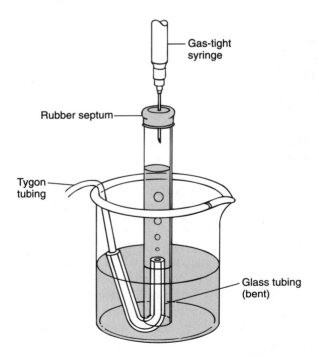

Figure 3.17   Gas collection tube, with rubber septum.

In Figure 3.17, a piece of glass tubing is attached to the free end of the flexible hose. This piece of glass tubing sometimes makes it easier to fix the open end in the proper position in the opening of the collection tube or flask. The other end of the flexible tubing is attached to a piece of glass tubing or a Pasteur pipet that has been inserted into a Thermometer adapter.

3.11 EVAPORATION OF SOLVENTS

In many experiments, it is necessary to remove excess solvent from a solution. An obvious approach is to allow the container to stand unstoppered in the hood for several hours until the solvent has evaporated. This method is generally not practical, however, and a quicker, more efficient means of evaporating solvents must be used.

Caution: You must always evaporate solvents in the hood.

This can be done by evaporating the solvent from an open Erlenmeyer flask (Fig. 3.18A and B). Such an evaporation must be conducted in a hood, since many solvent vapors are toxic or flammable. A boiling stone must be used. A gentle stream of air directed toward the surface of the liquid will remove vapors that are in equilibrium with the solution and accelerate the evaporation. A Pasteur pipet connected by a short piece of rubber tubing to the compressed air line will act as a convenient air nozzle (Fig. 3.18A). A tube or an inverted funnel connected to an aspirator may also be used (Fig. 3.18B). In this case, vapors are removed by suction. It is better to use an Erlenmeyer flask than a beaker for this procedure, since deposits of solid will usually build up on the sides of the beaker where the solvent evaporates. The refluxing action in an Erlenmeyer flask does not allow this buildup. If a hotplate is used as the heat source, care must be taken with flammable solvents to ensure against fires caused by "flashing," when solvent vapors come into contact with the hotplate surface.

It is also possible to remove low-boiling solvents under reduced pressure (Figure 3.18C). In this method, the solution is placed in a filter flask, along with a wooden applicator stick. The flask is stoppered, and the sidearm is connected to an aspirator (by a trap), as described in Technique 4, Section 4.3, p. 640. Under reduced pressure, the solvent begins to boil. The wooden stick serves the same function as a boiling stone. By this method, solvent can be evaporated from a solution without using much heat. This technique is often used when heating the solution might decompose thermally sensitive substances. The method has the disadvantage that, when low-boiling solvents are used, solvent evaporation cools the flask below the freezing point of water. When this happens, a layer of frost forms on the outside of the flask. Since frost is insulating, it must be removed to keep evaporation proceeding at a reasonable rate. Frost is best removed by one of two methods: either the flask is placed in a bath of warm water (with constant swirling) or it is heated on the steam bath (again with swirling). Either method promotes efficient heat transfer.

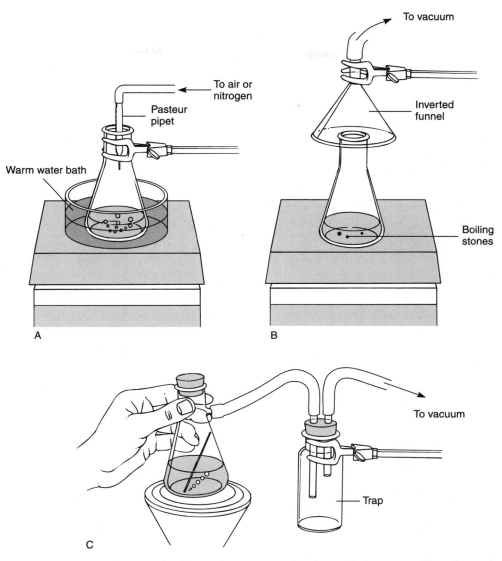

Figure 3.18 Evaporation of solvents (heat source can be varied among those shown.)

Large amounts of a solvent should be removed by distillation (see Technique 8). **One should never evaporate ether solutions to dryness,** except on a steam bath or by the reduced-pressure method. The tendency of ether to form explosive peroxides is a serious potential hazard. If peroxides should be present, the large and rapid temperature increase in the flask once the ether evaporates could bring about the detonation of any residual peroxides. The temperature of a steam bath is not high enough to cause such a detonation.

Small-Scale Methods

A simple means of evaporating a small amount of solvent is to place a centrifuge tube in a warm water bath. The heat from the water bath will warm the solvent to a temperature at which it can evaporate within a short time. The heat from the water can be adjusted to provide the best rate of evaporation, but the liquid should not be allowed to boil vigorously. The evaporation rate can be increased by allowing a stream of dry air or nitrogen to be directed into the centrifuge tube (Fig. 3.19A). The moving gas stream will sweep the vapors from the tube and accelerate the evaporation. As an alternative, a vacuum can be applied above the tube to draw away solvent vapors.

A convenient water bath suitable for microscale methods can be constructed by placing the aluminum collars, which are generally used with aluminum heating blocks, into a 150-mL beaker (Fig. 3.19B). In some cases, it may be necessary to round off the sharp edges of the collars with a file in order to allow them to fit properly into the beaker. Held by the aluminum collars, the conical vial will stand securely in the beaker. This assembly can be filled with water and placed on a hotplate for use in the evaporation of small amounts of solvent.

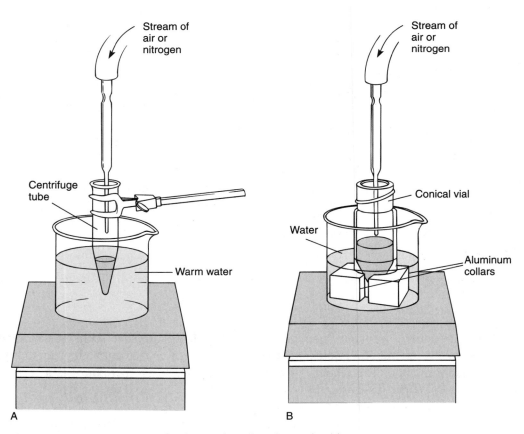

Figure 3.19 Evaporation of solvents (small-scale methods).

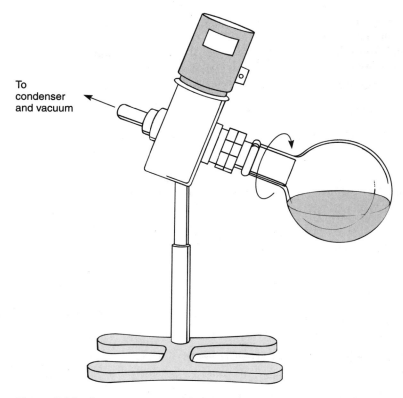

Figure 3.20   Rotary evaporator.

3.12 ROTARY EVAPORATOR

In the research laboratory, solvents are evaporated under reduced pressure using a **rotary evaporator.** This is a motor-driven device, which is designed for rapid evaporation of solvents, with heating, while minimizing the possibility of bumping. A vacuum is applied to the flask, and the motor spins the flask. The rotation of the flask spreads a thin film of the liquid over the surface of the glass. This accelerates evaporation. The rotation also agitates the solution sufficiently to reduce the problem of bumping. A water bath can be placed under the flask to warm the solution and increase the vapor pressure of the solvent. One can select the speed at which the flask is rotated and the temperature of the water bath to attain the desired evaporation rate. A rotary evaporator is shown in Figure 3.20.

PROBLEMS

1. What is the difference between
   (a) ether and petroleum ether?
   (b) ether and diethyl ether?
   (c) ligroin and petroleum ether?

2. What is the best type of stirring device to use for stirring a reaction that takes place in
   (a) a conical vial?
   (b) a 10-mL round-bottom flask?
   (c) a 250-mL round-bottom flask?
3. Should you use a drying tube for the following reaction? Explain.

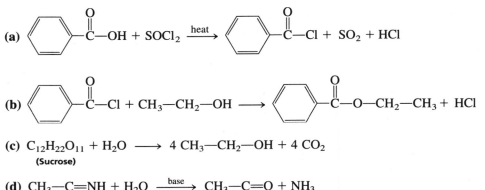

4. For which of the following reactions should you use a trap to collect noxious gases?

(a) (structure) C_6H_5—C(=O)—OH + SOCl$_2$ $\xrightarrow{\text{heat}}$ (structure) —C(=O)—Cl + SO$_2$ + HCl

(b) (structure) —C(=O)—Cl + CH$_3$—CH$_2$—OH $\longrightarrow$ (structure) —C(=O)—O—CH$_2$—CH$_3$ + HCl

(c) $C_{12}H_{22}O_{11}$ + H$_2$O $\longrightarrow$ 4 CH$_3$—CH$_2$—OH + 4 CO$_2$
(Sucrose)

(d) CH$_3$—C(H)=NH + H$_2$O $\xrightarrow[\text{heat}]{\text{base}}$ CH$_3$—C(H)=O + NH$_3$

TECHNIQUE 4

Filtration

Filtration is a technique used for two main purposes. The first is to remove solid impurities from a liquid. The second is to collect a desired solid from the solution from which it was precipitated or crystallized. Several different kinds of filtration are commonly used: two general methods include gravity filtration and vacuum (or suction) filtration. The various filtration techniques and their applications are summarized in Table 4.1. These techniques are discussed in more detail in the following sections.

4.1 GRAVITY FILTRATION

The most familiar filtration technique is probably filtration of a solution through a paper filter held in a funnel, allowing gravity to draw the liquid through the paper. This technique is useful only when the volume of mixture to be filtered is greater than 10 mL. For

TABLE 4.1. Filtration Methods

Method	Application	Section
GRAVITY FILTRATION		
Filter cones	The volume of liquid to be filtered is about 10 mL or greater, and the solid collected in the filter is saved.	4.1A
Fluted filters	The volume of liquid to be filtered is greater than about 10 mL, and solid impurities are removed from a solution; often used in crystallization procedures.	4.1B
Filtering pipets	Used with volumes less than about 10 mL to remove solid impurities from a liquid.	4.1C
VACUUM FILTRATION		
Büchner funnels	Primarily used to collect a desired solid from a liquid when the volume is greater than about 10 mL; used frequently to collect the crystals obtained from crystallization.	4.3
Hirsch funnels	Used in the same way as Büchner funnels, except the volume of liquid is usually smaller (1–10 mL)	4.3
FILTERING MEDIA	Used to remove finely divided impurities.	4.4
FILTER TIP PIPETS	May be used to remove a small amount of solid impurities from a small volume (1–2 mL) of liquid; also useful for pipetting volatile liquids, especially in extraction procedures.	4.6
CRAIG TUBES	Used to collect a small amount of crystals resulting from crystallizations in which the volume of the solution is less than 2 mL.	4.7
CENTRIFUGATION	Although not strictly a filtration technique, centrifugation may be used to remove suspended impurities from a liquid (1–25 mL).	4.8

some small-scale procedures a more suitable technique, which also makes use of gravity, is to use a Pasteur (or disposable) pipet with a cotton or glass wool plug (called a filtering pipet).

A. FILTER CONES

This filtration technique is most useful when the solid material being filtered from a mixture is to be collected and used later. The filter cone, because of its smooth sides, can easily be scraped free of collected solids. Because of the many folds, fluted filter paper, described in the next section, cannot be scraped easily. The filter cone is likely to be used only when a relatively large volume (greater than 10 mL) is being filtered and when a Büchner funnel (Section 4.3) is not appropriate.

The filter cone is prepared as indicated in Figure 4.1. It is then placed into a funnel of an appropriate size. With filtrations using a simple filter cone, solvent may form seals between the filter and the funnel and between the funnel and the lip of the receiving flask. When a seal forms, the filtration stops because the displaced air has no possibility of escaping. To avoid the solvent seal, you can insert a small piece of paper, a paper clip, or some other bent wire between the funnel and the lip of the flask to let the displaced air escape. As an alter-

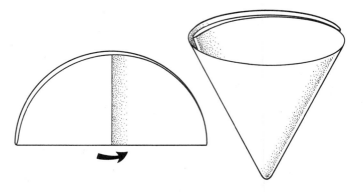

Figure 4.1   Folding a filter cone.

native, you can support the funnel by a clamp fixed *above* the flask, rather than by placing it on the neck of the flask. A gravity filtration using a filter cone is shown in Figure 4.2.

B.   FLUTED FILTERS

This filtration method is also most useful when filtering a relatively large amount of liquid. Because a fluted filter is used when the desired material is expected to remain in solution, this filter is used to remove undesired solid materials, such as dirt particles, decolorizing charcoal, and undissolved impure crystals. A fluted filter is often used to filter a hot solution saturated with a solute during a crystallization procedure.

The technique for folding a fluted filter paper is shown in Figure 4.3. An advantage of a fluted filter is that it increases the speed of filtration, which occurs for two reasons.

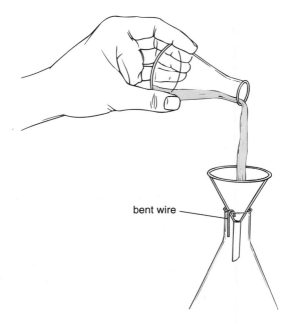

bent wire

Figure 4.2   Gravity filtration with a filter cone.

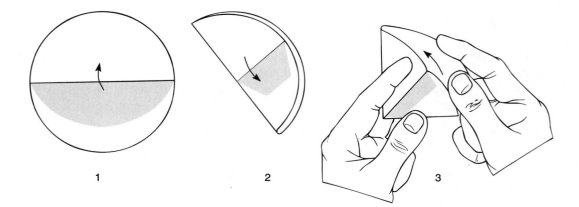

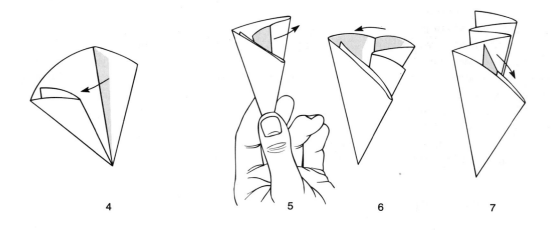

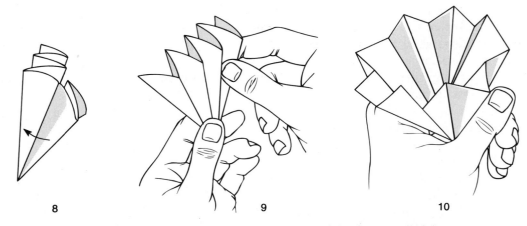

Figure 4.3 Folding a fluted filter paper, or origami at work in the organic lab.

First, it increases the surface area of the filter paper through which the solvent seeps; second, it allows air to enter the flask along its sides to permit rapid pressure equalization. If pressure builds up in the flask from hot vapors, filtering slows down. This problem is especially pronounced with filter cones. The fluted filter tends to reduce this problem considerably, but it may be a good idea to clamp the funnel above the receiving flask or to use a piece of paper, paper clip, or wire between the funnel and the lip of the flask as an added precaution against solvent seals.

Filtration with a fluted filter is relatively easy to perform when the mixture is at room temperature. However, when it is necessary to filter a hot solution saturated with a dissolved solute, a number of steps must be taken to ensure that the filter does not become clogged by solid material accumulated in the stem of the funnel or in the filter paper. When the hot saturated solution comes in contact with a relatively cold funnel (or a cold flask, for that matter), the solution is cooled and may become supersaturated. If crystallization then occurs in the filter, either the crystals will fail to pass through the filter paper or they will clog the stem of the funnel.

To keep the filter from clogging, use one of the following four methods. The first is to use a short-stemmed or a stemless funnel. With these funnels, it is less likely that the stem of the funnel will become clogged by solid material. The second method is to keep the liquid to be filtered at or near its boiling point at all times. The third way is to preheat the funnel by pouring hot solvent through it before the actual filtration. This keeps the cold glass from causing instantaneous crystallization. And fourth, it is helpful to keep the **filtrate** (filtered solution) in the receiver hot enough to continue boiling *slightly* (by setting it on a hotplate, for example). The refluxing solvent heats the receiving flask and the funnel stem and washes them clean of solids. This boiling of the filtrate also keeps the liquid in the funnel warm.

C. FILTERING PIPETS

A filtering pipet is a microscale technique most often used to remove solid impurities from a liquid with a volume less than 10 mL. It is important that the mixture being filtered be at or near room temperature because it is difficult to prevent premature crystallization in a hot solution saturated with a solute.

To prepare this filtration device, a small piece of cotton is inserted into the top of a Pasteur (disposable) pipet and pushed down to the beginning of the lower constriction in the pipet, as shown in Figure 4.4. It is important that enough cotton is used to collect all the solid being filtered; however, the amount of cotton used should not be so large that the flow rate through the pipet is significantly restricted. For the same reason, the cotton should not be packed too tightly. The cotton plug can be pushed down with a long thin object such as a glass stirring rod or a wooden applicator stick. It is advisable to wash the cotton plug by passing about 1 mL of solvent (usually the same solvent that is to be filtered) through the filter.

In some cases, such as when filtering a strongly acidic mixture or when performing a very rapid filtration to remove dirt or impurities of large particle size from a solution, it may be better to use glass wool in place of the cotton. The disadvantage in using glass

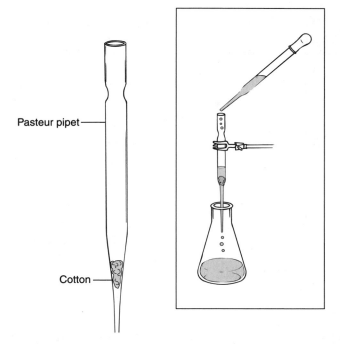

Figure 4.4   Filtering pipet.

wool is that the fibers do not pack together as tightly, and small particles will pass through the filter more easily.

To conduct a filtration (with either a cotton or glass wool plug), the filtering pipet is clamped so that the filtrate will drain into an appropriate container. The mixture to be filtered is usually transferred to the filtering pipet with another Pasteur pipet. If a small volume of liquid is being filtered (less than 1 or 2 mL), it is advisable to rinse the filter and plug with a small amount of solvent after the last of the filtrate has passed through the filter. The rinse solvent is then combined with the original filtrate. If desired, the rate of filtration can be increased by gently applying pressure to the top of the pipet using a pipet bulb.

Depending on the amount of solid being filtered and the size of the particles (small particles are more difficult to remove by filtration), it may be necessary to put the filtrate through a second filtering pipet. This should be done with a new filtering pipet rather than the one already used.

4.2 FILTER PAPER

Many kinds and grades of filter paper are available. The paper must be correct for a given application. In choosing filter paper, you should be aware of its various properties. **Porosity** is a measure of the size of the particles that can pass through the paper.

TABLE 4.2.   Some Common Qualitative Filter Paper Types and Approximate Relative Speeds and Retentivities

Fine   High   Slow

Porosity   Retentivity   Speed

Speed	Type (by number)		
	E&D	*S&S*	*Whatman*
Very slow	610	576	5
Slow	613	602	3
Medium	615	597	2
Fast	617	595	1
Very fast	—	604	4

Coarse   Low   Fast

Highly porous paper does not remove small particles from solution; paper with low porosity removes very small particles. **Retentivity** is a property that is the opposite of porosity. Paper with low retentivity does not remove small particles from the filtrate. The **speed** of filter paper is a measure of the time it takes a liquid to drain through the filter. Fast paper allows the liquid to drain quickly; with slow paper, it takes much longer to complete the filtration. Since all these properties are related, fast filter paper usually has a low retentivity and high porosity, and slow filter paper usually has high retentivity and low porosity.

Table 4.2 compares some commonly available qualitative filter paper types and ranks them according to porosity, retentivity, and speed. Eaton-Dikeman (E&D), Schleicher and Schuell (S&S), and Whatman are the most common brands of filter paper. The numbers in the table refer to the grades of paper used by each company.

4.3 VACUUM FILTRATION

Vacuum, or suction, filtration is more rapid than gravity filtration and is most often used to collect solid products resulting from precipitation or crystallization. This technique is used primarily when the volume of liquid being filtered is more than 1–2 mL. With smaller volumes, use of the Craig tube (Section 4.7) is the preferred technique. In a vacuum filtration, a receiver flask with a sidearm, a **filter flask,** is used. For small-scale laboratory work, the most useful sizes of filter flasks range from 50 mL to 500 mL, depending on the volume of liquid being filtered. The sidearm is connected by *heavy-walled* rubber tubing (see Technique 9, Fig. 9.2, p. 720) to a source of vacuum. Thin-walled tubing will collapse under vacuum, due to atmospheric pressure on its outside walls, and will seal the vacuum source from the flask. Because this apparatus is unstable and can tip over easily, it should be clamped, as shown in Figure 4.5.

> **Caution:** It is essential that the filter flask be clamped.

Two types of funnels are useful for vacuum filtration, the Büchner funnel and the Hirsch funnel. The **Büchner funnel** is used for filtering larger amounts of solid from solution. Büchner funnels are usually made from polypropylene or porcelain. A Büchner funnel (see Fig. 4.5) is sealed to the filter flask by a rubber stopper or a filter (Neoprene) adapter. The flat bottom of the Büchner funnel is covered with an unfolded piece of circular filter paper. To prevent the escape of solid materials from the funnel, you must be certain that the filter paper fits the funnel exactly. It must cover all the holes in the bottom of the funnel but not extend up the sides. Before beginning the filtration, it is advisable to moisten the paper with a small amount of solvent. The moistened filter paper adheres more strongly to the bottom of the funnel and prevents unfiltered mixture from passing around the edges of the filter paper.

The **Hirsch funnel,** which is shown in Figure 4.5B and C, operates on the same principle as the Büchner funnel, but it is usually smaller and its sides are sloped rather

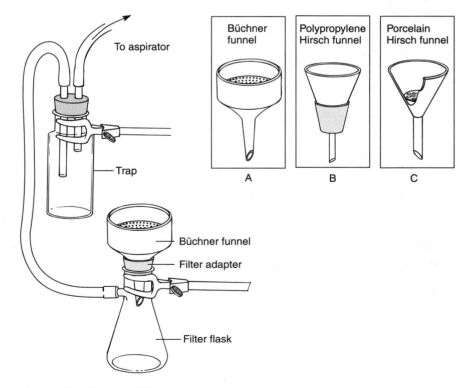

Figure 4.5   Vacuum filtration.

than vertical. The polypropylene Hirsch funnel (see Fig. 4.5B) is sealed to a 50-mL filter flask by a small section of Gooch tubing. This Hirsch funnel has a built-in adapter that forms a tight seal with some 25-mL filter flasks without the Gooch tubing. A polyethylene fritted disk fits into the bottom of the funnel. To prevent the holes in this disk from becoming clogged with solid material, the funnel should always be used with a circular filter paper that has the same diameter (1.27-cm) as the polyethylene disk. With a polypropylene Hirsch funnel, it is also important to moisten the paper with a small amount of solvent before beginning the filtration.

The porcelain Hirsch funnel is sealed to the filter flask with a rubber stopper or a Neoprene adapter. In this Hirsch funnel, the filter paper must also cover all the holes in the bottom but must not extend up the sides.

Because the filter flask is attached to a source of vacuum, a solution poured into a Büchner funnel or Hirsch funnel is literally "sucked" rapidly through the filter paper. For this reason, a vacuum filtration is generally not used to separate fine particles such as decolorizing charcoal, since the small particles would likely be pulled through the filter paper. However, this problem can be alleviated when desired by the use of specially prepared filter beds (see Section 4.4).

4.4 FILTERING MEDIA

It is occasionally necessary to use specially prepared filter beds to separate fine particles when using vacuum filtration. Often, very fine particles either pass right through a paper filter or they clog it so completely that the filtering stops. This is avoided by using a substance called Filter Aid, or Celite. This material is also called **diatomaceous earth** because of its source. It is a finely divided inert material derived from the microscopic shells of dead diatoms (a type of phytoplankton that grows in the sea).

> **Caution:** LUNG IRRITANT
> When using Filter Aid, take care not to breathe the dust.

Filter Aid will not clog the fiber pores of filter paper. It is **slurried,** mixed with a solvent to form a rather thin paste, and filtered through a Hirsch or Büchner funnel (with filter paper in place) until a layer of diatoms about 2–3 mm thick is formed on top of the filter paper. The solvent in which the diatoms were slurried is poured from the filter flask, and, if necessary, the filter flask is cleaned before the actual filtration is begun. Finely divided particles can now be suction-filtered through this layer and will be caught in the Filter Aid. This technique is used for removing impurities, not for collecting a product. The filtrate (filtered solution) is the desired material in this procedure. If the material caught in the filter were the desired material, you would have to try to separate the product from all those diatoms! Filtration with Filter Aid is not appropriate when the desired substance is likely to precipitate or crystallize from solution.

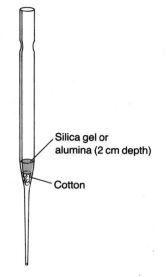

Figure 4.6   Pasteur pipet with filtering media.

In microscale work, it may sometimes be more convenient to use a column prepared with a Pasteur pipet to separate fine particles from a solution. The Pasteur pipet is packed with alumina or silica gel, as shown in Figure 4.6.

4.5 THE ASPIRATOR

The most common source of vacuum (approximately 10–20 mmHg) in the laboratory is the water aspirator, or "water pump," illustrated in Figure 4.7. This device passes water rapidly past a small hole to which a sidearm is attached. The water pulls air in

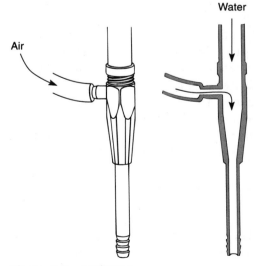

Figure 4.7   Aspirator.

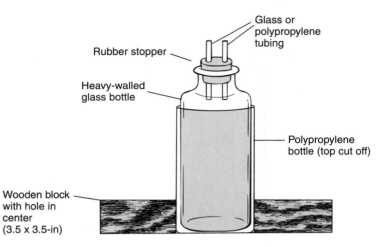

Figure 4.8   Simple aspirator trap and holder.

through the sidearm. This phenomenon, called the Bernoulli effect, causes a reduced pressure along the side of the rapidly moving water stream and creates a partial vacuum in the sidearm.

> The aspirator works most effectively when the water is turned on to the fullest extent.

A water aspirator can never lower the pressure beyond the vapor pressure of the water used to create the vacuum. Hence, there is a lower limit to the pressure (on cold days) of 9–10 mmHg. A water aspirator does not provide as high a vacuum in the summer as in the winter, due to this water-temperature effect.

A trap must be used with an aspirator. One type of trap is illustrated in Figure 4.5. Another method for securing this type of trap is shown in Figure 4.8. This simple holder can be constructed from readily available material and can be placed anywhere on the laboratory bench. Although not often needed, a trap can prevent water from contaminating your experiment. If the water pressure in the laboratory drops suddenly, the pressure in the filter flask may suddenly become lower than the pressure in the water aspirator. This would cause water to be drawn from the aspirator stream into the filter flask and contaminate the filtrate or even the material in the filter. The trap stops this reverse flow. A similar flow will occur if the water flow at the aspirator is stopped before the tubing connected to the aspirator sidearm is disconnected.

> Always disconnect the tubing before stopping the aspirator.

If a "back-up" begins, disconnect the tubing as rapidly as possible before the trap fills with water. Some chemists like to fit a stopcock into the stopper on top of the trap. A three-hole stopper is required for this purpose. With a stopcock in the trap, the system can be vented before the aspirator is shut off. Then, water cannot back up into the trap.

Aspirators do not work well if too many people use the water line at the same time because the water pressure is lowered. Also, the sinks at the ends of the lab benches or the lines that carry away the water flow may have a limited capacity for draining the resultant water flow from too many aspirators. Care must be taken to avoid floods.

4.6 FILTER-TIP PIPET

The filter-tip pipet, illustrated in Figure 4.9, has two common uses. The first is to remove a small amount of solid, such as dirt or filter paper fibers, from a small volume of liquid (1–2 mL). A filter-tip pipet is very useful for removing small particles from solutions of samples prepared for NMR analysis. It can also be helpful when using a Pasteur pipet to transfer a highly volatile liquid, especially during an extraction procedure (see Technique 7, Section 7.6, p. 693).

Preparing a filter-tip pipet is similar to preparing a filtering pipet, except that a much smaller amount of cotton is used. A very tiny piece of cotton is loosely shaped into a ball and placed into the large end of a Pasteur pipet. Using a wire with a diameter slightly smaller than the inside diameter of the narrow end of the pipet, the ball of cotton is pushed to the bottom of the pipet. If it becomes difficult to push the cotton, you have probably started with too much cotton; if the cotton slides through the narrow end with little resistance, you probably have not used enough.

To use a filter-tip pipet as a filter, the mixture is drawn up into the Pasteur pipet using a pipet bulb and then expelled. With this procedure, a small amount of solid will be captured by the cotton. However, very fine particles, such as activated charcoal, cannot

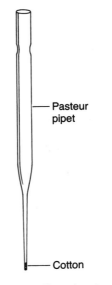

Pasteur
pipet

Cotton

Figure 4.9   Filter-tip pipet.

be removed efficiently with a filter-tip pipet, and this technique is not effective in removing more than a trace amount of solid from a liquid.

Transferring many organic liquids with a Pasteur pipet can be a somewhat difficult procedure for two reasons. First, the liquid may not adhere well to the glass. Second, as you handle the Pasteur pipet, the temperature of the liquid in the pipet increases slightly, and the increased vapor pressure may tend to "squirt" the liquid out the end of the pipet. This problem can be particularly troublesome when separating two liquids during an extraction procedure. The purpose of the cotton plug in this situation is to slow the rate of flow through the end of the pipet so that you can control the movement of liquid in the Pasteur pipet more easily.

4.7 CRAIG TUBES

The **Craig tube,** illustrated in Figure 4.10, is used primarily to separate crystals from a solution after a microscale crystallization procedure has been performed (Technique 5, Section 5.4, p. 658). Although it may not be a filtration procedure in the traditional sense, the outcome is similar. The outer part of the Craig tube is similar to a test tube, except that the diameter of the tube becomes wider part of the way up the tube, and the glass is ground at this point so that the inside surface is rough. The inner part (plug) of the Craig tube may be made of Teflon or glass. If this part is glass, the end of the plug is also ground. With either a glass or a Teflon inner plug, there is only a partial seal where the plug and the outer tube come together. Liquid may pass through, but solid will not. This is where the solution is separated from the crystals.

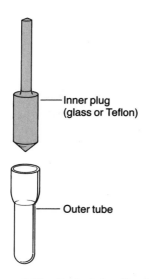

Inner plug
(glass or Teflon)

Outer tube

Figure 4.10   Craig tube (2 mL).

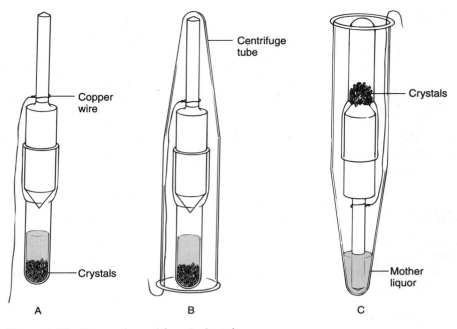

Figure 4.11 Separation with a Craig tube.

After crystallization has been completed in the outer Craig tube, replace the inner plug (if necessary) and connect a thin copper wire or strong thread to the narrow part of the inner plug, as indicated in Figure 4.11A. While holding the Craig tube in an upright position, a plastic centrifuge tube is placed over the Craig tube so that the bottom of the centrifuge tube rests on top of the inner plug, as shown in Figure 4.11B. The copper wire should extend just below the lip of the centrifuge tube and is now bent upward around the lip of the centrifuge tube. This apparatus is then turned over so that the centrifuge tube is in an upright position. The Craig tube is spun in a centrifuge (be sure it is balanced by placing another tube filled with water on the opposite side of the centrifuge) for several minutes until the **mother liquor** (solution from which the crystals grew) goes to the bottom of the centrifuge tube and the crystals collect on the end of the inner plug (see Fig. 4.11C). Depending on the consistency of the crystals and the speed of the centrifuge, the crystals may spin down to the inner plug, or (if you are unlucky) they may remain at the other end of the Craig tube. If the latter situation occurs, it may be helpful to centrifuge the Craig tube longer, or, if this problem is anticipated, to stir the crystal and solution mixture with a spatula or stirring rod before centrifugation.

Using the copper wire, the Craig tube is then pulled out of the centrifuge tube. If the crystals collected on the end of the inner plug, it is now a simple procedure to remove the plug and scrape the crystals with a spatula onto a watch glass, a clay plate, or a piece of smooth paper. Otherwise, it will be necessary to scrape the crystals from the inside surface of the outer part of the Craig tube.

4.8 CENTRIFUGATION

Sometimes centrifugation is more effective in removing solid impurities than are conventional filtration techniques. Centrifugation is particularly effective in removing suspended particles which are so small that the particles would pass through most filtering devices. Another situation in which centrifugation may be useful is when the mixture must be kept hot to prevent premature crystallization while the solid impurities are removed.

Centrifugation is performed by placing the mixture in one or two centrifuge tubes (be sure to balance the centrifuge) and centrifuging for several minutes. The supernatant liquid is then decanted (poured off) or removed with a Pasteur pipet.

PROBLEMS

1. In each of the following situations, what type of filtration device would you use?
   (a) Remove powdered decolorizing charcoal from 20 mL of solution.
   (b) Collect crystals obtained from crystallizing a substance from about 1 mL of solution.
   (c) Remove a very small amount of dirt from 1 mL of liquid.
   (d) Isolate 2.0 g of crystals from about 50 mL of solution after performing a crystallization.
   (e) Remove dissolved colored impurities from about 3 mL of solution.
   (f) Remove solid impurities from 5 mL of liquid at room temperature.

T E C H N I Q U E   5

Crystallization: Purification of Solids

Organic compounds that are solid at room temperature are usually purified by crystallization. The general technique involves dissolving the material to be crystallized in a *hot* solvent (or solvent mixture) and cooling the solution slowly. The dissolved material has a decreased solubility at lower temperatures and will separate from the solution as it is cooled. This phenomenon is called either **crystallization** if the crystal growth is relatively slow and selective or **precipitation** if the process is rapid and nonselective. Crystallization is an equilibrium process and produces very pure material. A small seed crystal is formed initially, and it then grows layer by layer in a reversible manner. In a sense, the crystal "selects" the correct molecules from the solution. In precipitation, the crystal lattice is formed so rapidly that impurities are trapped within the lattice. Therefore, any attempt at purification with too rapid a process should be avoided.

The method of crystallization described in detail in this chapter is called **standard-scale crystallization.** This technique, which is carried out with an Erlenmeyer flask to dissolve the material and a Büchner funnel to filter the crystals, is normally used when the weight of solid to be crystallized is more than 0.1 g. Another method, which is performed with a Craig tube, is used with smaller amounts of solid. Referred to as **microscale crystallization,** this technique is discussed briefly in Section 5.4.

5.1 SOLUBILITY

The first problem in performing a crystallization is selecting a solvent in which the material to be crystallized shows the desired solubility behavior. In the ideal case, the material should be sparingly soluble at room temperature and yet quite soluble at the boiling point of the solvent selected. The solubility curve should be steep, as can be seen in line A of Figure 5.1. A curve with a low slope (line B, Fig. 5.1) would not cause significant crystallization when the temperature of the solution was lowered. A solvent in which the material is very soluble at all temperatures (line C, Fig. 5.1) also would not be a suitable crystallization solvent. The basic problem in performing a crystallization is to select a solvent (or mixed solvent) that provides a steep solubility-vs.-temperature curve for the material to be crystallized. A solvent that allows the behavior shown in line A is an ideal crystallization solvent.

The solubility of organic compounds is a function of the polarities of both the solvent and the **solute** (dissolved material). A general rule is "like dissolves like." If the solute is very polar, a very polar solvent is needed to dissolve it; if it is nonpolar, a nonpolar solvent is needed. Usually, compounds having functional groups that can form hydrogen bonds (for example, —OH, —NH—, —COOH, —CONH—) will be more soluble in hydroxylic solvents such as water or methanol than in hydrocarbon solvents such as toluene or hexane. However, if the functional group is not a major part of the molecules, this solubility behavior may be reversed. For instance, dodecyl alcohol $CH_3(CH_2)_{10}CH_2OH$ is almost insoluble in water; its 12-carbon chain causes it to behave more like a hydrocarbon than an alcohol. The list in Table 5.1 gives an approximate order for decreasing polarity of organic functional groups.

The stability of the crystal lattice also affects solubility. Other things being equal, the higher the melting point (the more stable the crystal), the less soluble the compound. For instance, *p*-nitrobenzoic acid (mp 242°C) is, by a factor of 10, less soluble in a fixed amount of ethanol than the *ortho* (mp 147°C) and *meta* (mp 141°C) isomers.

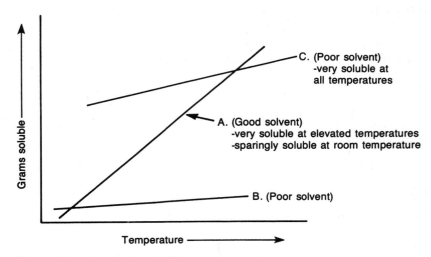

Figure 5.1 Graph of solubility vs. temperature.

TABLE 5.1. Solvents, in Decreasing Order of Polarity

	H_2O	Water
	RCOOH	Organic acids (acetic acid)
	$RCONH_2$	Amides (*N,N*-dimethylformamide)
	ROH	Alcohols (methanol, ethanol)
Decreasing Polarity (Approximate)	RNH_2	Amines (triethylamine, pyridine)
	RCOR	Aldehydes, ketones (acetone)
	RCOOR	Esters (ethyl acetate)
	RX	Halides ($CH_2Cl_2 > CHCl_3 > CCl_4$)
	ROR	Ethers (diethyl ether)
	ArH	Aromatics (benzene, toluene)
	RH	Alkanes (hexane, petroleum ether)

5.2 THEORY OF CRYSTALLIZATION

A successful crystallization depends on a large difference between the solubility of a material in a hot solvent and its solubility in the same solvent when it is cold. When the impurities in a substance are equally soluble in both the hot and the cold solvent, an effective purification is not easily achieved through crystallization. A material can be purified by crystallization when both the desired substance and the impurity have similar solubilities, but only when the impurity represents a small fraction of the total solid. The desired substance will crystallize on cooling, but the impurities will not.

For example, consider a case in which the solubilities of substance A and its impurity B are both 1 g/100 mL of solvent at 20°C and 10 g/100 mL of solvent at 100°C. In an impure sample of A, the composition is given to be 9 g of A and 2 g of B for this particular example. In the calculations for this example, it is assumed that the solubilities of both A and B are unaffected by the presence of the other substance. One hundred mL of solvent is used in each crystallization to make the calculations easier to understand. Normally, the minimum amount of solvent required to dissolve the solid would be used.

At 20°C, this total amount of material will not dissolve in 100 mL of solvent. However, if the solvent is heated to 100°C, all 11 g dissolve. The solvent has the capacity to dissolve 10 g of A *and* 10 g of B at this temperature. If the solution is cooled to 20°C, only 1 g of each solute can remain dissolved, so 8 g of A and 1 g of B crystallize, leaving 2 g of material in the solution. This crystallization is shown in Figure 5.2. The solution that remains after a crystallization is called the **mother liquor.** If the process is now repeated by treating the crystals with 100 mL of fresh solvent, 7 g of A will crystallize again, leaving 1 g of A and 1 g of B in the mother liquor. As a result of these operations, 7 g of pure A are obtained, but with the loss of 4 g of material (2 g of A plus 2 g of B). Again, this second crystallization step is illustrated in Figure 5.2. The final result illustrates an important aspect of crystallization—it is wasteful. Nothing can be done to prevent this waste; some A must be lost along with the impurity B for the method to be successful. Of course, if the impurity B were *more* soluble than A in the solvent, the

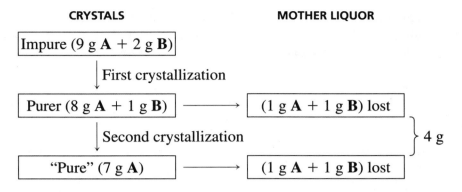

Figure 5.2 Purification of a mixture by crystallization.

losses would be reduced. Losses could also be reduced if the impurity were present in *much smaller* amounts than the desired material.

Note that in the preceding case, the method operated successfully because A was present in substantially larger quantity than its impurity B. If there had been a 50/50 mixture of A and B initially, no separation would have been achieved. In general, a crystallization is successful only if there is a *small* amount of impurity. As the amount of impurity increases, the loss of material must also increase. Two substances with nearly equal solubility behavior, present in equal amounts, cannot be separated. If the solubility behavior of two components present in equal amounts is different, however, a separation or purification is frequently possible.

In the preceding example, two crystallization procedures were performed. Normally this is not necessary; however, when it is, the second crystallization is more appropriately called **recrystallization.** As illustrated in this example, a second crystallization results in purer crystals, but the yield is lower.

In some experiments in this book, you will be instructed to cool the crystallizing mixture in an ice-water bath before collecting the crystals by filtration. Cooling the mixture increases the yield by decreasing the solubility of the substance; however, even at this reduced temperature, some of the product will be soluble in the solvent. It is not possible to recover all your product in a crystallization procedure.

5.3 STANDARD-SCALE CRYSTALLIZATION— BÜCHNER FUNNEL

The crystallization technique described in this section is used when the weight of solid to be crystallized is more than 0.1 g. The four main steps in a standard-scale crystallization are

1. Dissolving the solid
2. Removing insoluble impurities (when necessary)
3. Crystallization
4. Isolation of crystals

These steps are illustrated in Figure 5.3. It should be pointed out that a microscale crystallization with a Craig tube involves the same four steps, although the apparatus and procedures are somewhat different (see Section 5.4).

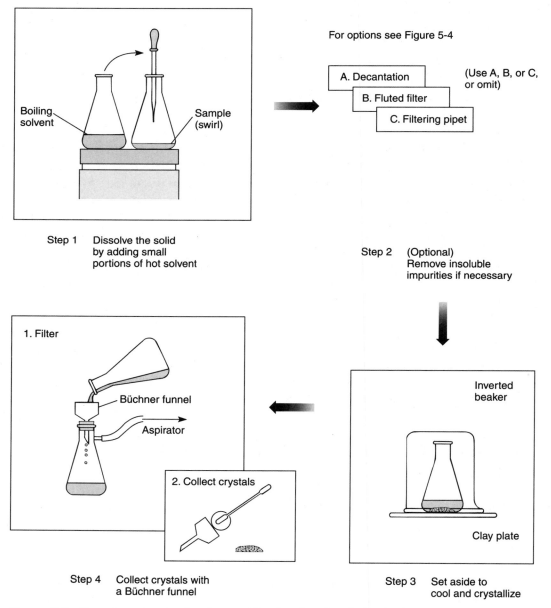

Figure 5.3 Steps in a standard-scale cystallization (no decolorization).

A.   DISSOLVING THE SOLID

To minimize losses of material to the mother liquor, it is desirable to *saturate* the boiling solvent with solute. This solution, when cooled, will return the maximum possible amount of solute as crystals. To achieve this high return, the solvent is brought to its boiling point, and the solute is dissolved in the *minimum amount (!) of boiling solvent.* For this procedure, it is advisable to maintain a container of boiling solvent (on a hotplate). From this container, a small portion (about 1–2 mL) of the solvent is added to the Erlenmeyer flask containing the solid to be crystallized, and this mixture is heated while swirling occasionally until it resumes boiling. Do not heat the flask containing the solid until after you have added the first portion of solvent. If the solid does not dissolve in the first portion of boiling solvent, then another small portion of boiling solvent is added to the flask. The mixture is heated again until it resumes boiling. If the solid dissolves, no more solvent is added. But if the solid has not dissolved, another portion of boiling solvent is added, as before, and the process is repeated until the solid dissolves. It is important to stress that the portions of solvent added each time are small, so that only the *minimum* amount of solvent necessary for dissolving the solid is added. It is also important to emphasize that the procedure requires the addition of solvent to solid. You must never add portions of solid to a fixed quantity of boiling solvent. By this latter method, it may be impossible to tell when saturation has been achieved.

In many of the experiments in this textbook, a specified amount of solvent for a given weight of solid will be recommended. In these cases, you should use the amount specified rather than the minimum amount of solvent necessary for dissolving the solid. The amount of solvent recommended has been selected to provide the optimum conditions for good crystal formation.

Occasionally, you can encounter an impure solid that contains small particles of insoluble impurities, pieces of dust, or paper fibers that will not dissolve in the hot crystallizing solvent. A common error is to add too much of the hot solvent in an attempt to dissolve these small particles, without realizing that they are not soluble. In such cases, you must be careful not to add too much solvent.

It is sometimes necessary to decolorize the solution by adding activated charcoal or by passing the solution through a column containing alumina or silica gel (see Section 5.6, Parts A and B, and Technique 12, Section 12.14, p. 782). (Note: Often, there may be a small amount of colored material that will remain in solution during the crystallization step. When you believe that this may be the case, omit the decolorizing step.)

B.   REMOVING INSOLUBLE IMPURITIES

It is necessary to use one of the following three methods only if insoluble material remains in the hot solution or if decolorizing charcoal has been used.

Caution: Indiscriminate use of the procedure can lead to needless loss of your product.

Decantation is the easiest method of removing solid impurities and should be considered first. If filtration is required, a filtering pipet is used when the volume of liquid to be filtered is less than 10 mL (see Technique 4, Section 4.1, Part C, p. 638), and you should use gravity filtration through a fluted filter when the volume is 10 mL or greater (see Technique 4, Section 4.1, Part B, p. 636). These three methods are illustrated in Figure 5.4, and each one is discussed below.

Decantation. If the solid particles are relatively large in size or they easily settle to the bottom of the flask, it may be possible to separate the hot solution from the impurities by carefully pouring off the liquid, leaving the solid behind. This is accomplished most easily by holding a glass stirring rod along the top of the flask and tilting the flask so that the liquid pours out along one end of the glass rod into another container. A technique similar in principle to decantation, which may be easier to perform with smaller amounts of liquid, is to use a **preheated Pasteur pipet** to remove the hot solution. With this method, it may be helpful to place the tip of the pipet against the bottom of the flask when removing the last portion of solution. The small space between the tip of the pipet and the inside surface of the flask prevents solid material from being drawn into the pipet. An easy way to preheat the pipet is to draw up a small portion of hot *solvent* (not the *solution* being transferred) into the pipet and expel the liquid. Repeat this process several times.

Fluted Filter. This method is the most effective way to remove solid impurities when the volume of liquid is greater than 10 mL or when decolorizing charcoal has been used (see Technique 4, Section 4.1, Part B, p. 636). You should first add a small amount of extra solvent to the hot mixture. This action helps to prevent crystal formation in the filter paper or the stem of the funnel during the filtration. The funnel is then fitted with a fluted filter and installed at the top of the Erlenmeyer flask to be used for the actual filtration. It is advisable to place a small piece of wire between the funnel and the mouth of the flask to relieve any increase in pressure caused by hot filtrate.

The Erlenmeyer flask containing the funnel and fluted paper is placed on top of a hotplate (low setting). The liquid to be filtered is brought to its boiling point and poured through the filter in portions. (If the volume of the mixture is less than 10 mL, it may be more convenient to transfer the mixture to the filter with a preheated Pasteur pipet.) It is necessary to keep the solutions in both flasks at their boiling temperatures to prevent premature crystallization. The refluxing action of the filtrate keeps the funnel warm and reduces the chance that the filter will clog with crystals that may have formed during the filtration. With low-boiling solvents, be aware that some solvent may be lost through evaporation. Consequently, extra solvent must be added to make up for this loss. If crystals begin to form in the filter during filtration, a minimum amount of boiling solvent is added to redissolve the crystals and to allow the solution to pass through the funnel. If the volume of liquid being filtered is less than 10 mL, a small amount of hot solvent should be used to rinse the filter after all the filtrate has been collected. The rinse solvent is then combined with the original filtrate.

After the filtration, it may be necessary to remove extra solvent by evaporation until the solution is once again saturated at the boiling point of the solvent (see Technique 3, Section 3.11, p. 630).

Filtering Pipet. If the volume of solution after dissolving the solid in hot solvent is less than 10 mL, gravity filtration with a filtering pipet may be used to remove solid

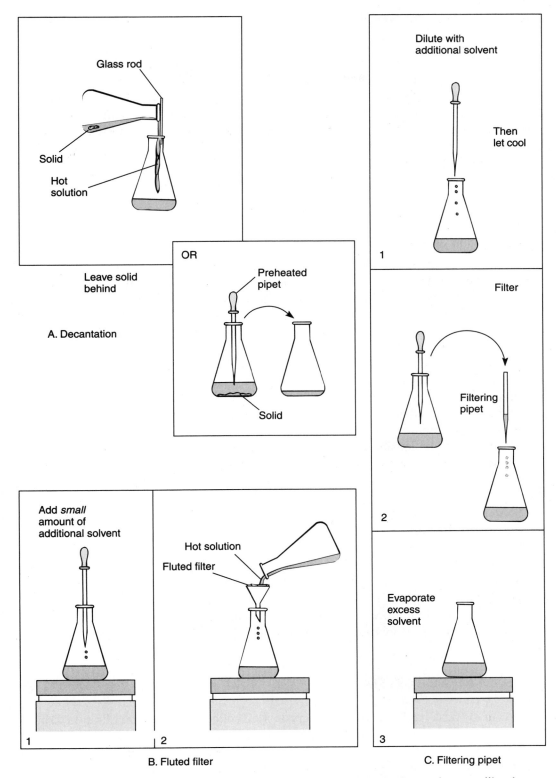

Glass rod

Solid

Hot solution

Leave solid behind

A. Decantation

OR

Preheated pipet

Solid

Dilute with additional solvent

Then let cool

1

Filter

Filtering pipet

2

Evaporate excess solvent

3

C. Filtering pipet

Add *small* amount of additional solvent

1

Hot solution

Fluted filter

2

B. Fluted filter

Figure 5.4 Methods for removing insoluble impurities in a semi-microscale crystallization.

impurities. However, using a filtering pipet to filter a hot solution saturated with solute can be difficult without premature crystallization. The best way to prevent this from occurring is to add enough solvent to dissolve the desired product at room temperature (be sure not to add too much solvent) and perform the filtration at room temperature, as described in Technique 4, Section 4.1, Part C, p. 638. After filtration, the excess solvent is evaporated by boiling until the solution is saturated at the boiling point of the mixture (see Technique 3, Section 3.11, p. 630). If powdered decolorizing charcoal was used, it will probably be necessary to perform two filtrations with a filtering pipet to remove all of the charcoal, or else a fluted filter can be used.

C. CRYSTALLIZATION

An Erlenmeyer flask, not a beaker, should be used for crystallization. The large open top of a beaker makes it an excellent dust catcher. The narrow opening of the Erlenmeyer flask reduces contamination by dust and allows the flask to be stoppered if it is to be set aside for a long period. Mixtures set aside for long periods must be stoppered after cooling to room temperature to prevent evaporation of solvent. If all the solvent evaporates, no purification is achieved, and the crystals originally formed become coated with the dried contents of the mother liquor. Even if the time required for crystallization to occur is relatively short, it is advisable to cover the top of the Erlenmeyer flask with a small watch glass or inverted beaker to prevent evaporation of solvent while the solution is cooling to room temperature.

The chances of obtaining pure crystals are improved if the solution cools to room temperature slowly. When the volume of solution is 10 mL or less, the solution is likely to cool more rapidly than is desired. This can be prevented by placing the flask on a surface that is a poor heat conductor and covering the flask with a beaker to provide a layer of insulating air. Appropriate surfaces include a clay plate or several pieces of filter paper on top of the laboratory bench. It may also be helpful to use a clay plate that has been warmed slightly on a hotplate or in an oven.

After crystallization has occurred, it is sometimes desirable to cool the flask in an ice-water bath. Because the solute is less soluble at lower temperatures, this will increase the yield of crystals.

If a cooled solution does not crystallize, it will be necessary to induce crystallization. Several techniques are described in Section 5.7, Part A.

D. ISOLATION OF CRYSTALS

After the flask has been cooled, the crystals are collected by vacuum filtration through a Büchner (or Hirsch) funnel (see Technique 4, Section 4.3, p. 640, and Fig. 4.5). The crystals should be washed with a small amount of *cold* solvent to remove any mother liquor adhering to their surface. Hot or warm solvent will dissolve some of the crystals. The crystals should then be left for a short time (usually 5–10 minutes) in the funnel, where air, as it passes, will dry them free of most of the solvent. It is often wise to cover the Büchner funnel with an oversize filter paper or towel during this air-drying. This precaution prevents accumulation of dust in the crystals. When the crystals are nearly dry, they should be gently scraped off (so paper fibers are not removed with the crystals) the filter paper onto a watch glass or clay plate for further drying (see Section 5.8).

The four steps in a standard scale crystallization are summarized in Figure 5.5

A. Dissolving the Solid

1. Find a solvent with a steep solubility-vs-temperature characteristic. (Done by trial and error using small amounts of material or by consulting a handbook.)
2. Heat the desired solvent to its boiling point.
3. Dissolve the solid in a **minimum** of boiling solvent in a flask.
4. If necessary, add decolorizing charcoal or decolorize the solution on a silica gel or alumina column.

B. Removing Insoluble Impurities

1. Decant or remove the solution with a Pasteur pipet, or
2. Filter the hot solution through a fluted filter, a filtering pipet, or a filter tip pipet to remove insoluble impurities or charcoal.

> **NOTE: If no decolorizing charcoal has been added or if there are no undissolved particles, Part B should be omitted.**

C. Crystallizing

1. Allow the solution to cool.
2. If crystals appear, cool the mixture in an ice-water bath (if desired) and go to Part D. If crystals do not appear, go to the next step.
3. Inducing crystallization.
    (a) Scratch the flask with a glass rod.
    (b) Seed the solution with original solid, if available.
    (c) Cool the solution in an ice-water bath.
    (d) Evaporate excess solvent and allow the solution to cool again.

D. Collecting and Drying

1. Collect crystals by vacuum filtration using a Büchner funnel.
2. Rinse crystals with a small portion of **cold** solvent.
3. Continue suction until crystals are nearly dry.
4. Drying.
    (a) Air-dry the crystals, or
    (b) Place the crystals in a drying oven, or
    (c) Dry the crystals *in vacuo.*

Figure 5.5 Steps in a crystallization.

5.4 MICROSCALE CRYSTALLIZATION—CRAIG TUBE

In most microscale experiments, the amount of solid to be crystallized is small enough (generally less than 0.1 g) that a **Craig tube** (see Technique 4, Fig. 4.10, p. 646) is the preferred method for crystallization. The main advantage of the Craig tube is that it minimizes the number of transfers of solid material, thus resulting in a greater yield of crystals. Also, the separation of the crystals from the mother liquor with the Craig tube is very efficient, and little time is required for drying the crystals. The steps involved are, in principle, the same as those performed when a crystallization is accomplished with an Erlenmeyer flask and a Büchner funnel.

The solid is transferred to the Craig tube, and small portions of hot solvent are added to the tube while the mixture is stirred with a spatula and heated. If there are any insoluble impurities present, they can be removed with a filter-tip pipet. The inner plug is then inserted into the Craig tube and the hot solution is cooled slowly to room temperature. When the crystals have formed, the Craig tube is placed into a centrifuge tube and the crystals are separated from the mother liquor by centrifugation (see Technique 4, Section 4.7, p. 646). The crystals are then scraped off the end of the inner plug or from inside the Craig tube onto a watch glass or piece of paper. Minimal drying will be necessary (see Section 5.8).

5.5 SELECTING A SOLVENT

A solvent that dissolves little of the material to be crystallized when it is cold but a great deal of the material when it is hot is a good solvent for the crystallization. With compounds that are well known, such as the compounds that are either isolated or prepared in this textbook, the correct crystallization solvent is already known through the experiments of earlier researchers. In such cases, the chemical literature can be consulted to determine which solvent should be used. Sources such as handbooks or tables frequently provide this information. Quite often, the correct crystallization solvents are indicated in the experimental procedures in this textbook.

When the appropriate solvent is not known, select a solvent for crystallization by experimenting with various solvents and a very small amount of the material to be crystallized. Experiments are conducted on a small test-tube scale before the entire quantity of material is committed to a particular solvent. Such trial-and-error methods are common when one is trying to purify a solid material that has not been previously studied.

When choosing a crystallization solvent, take care not to pick one whose boiling point is higher than the melting point of the substance to be crystallized. If the boiling point of the solvent is high, the solid may melt in the solvent rather than dissolve. In such a case, the solid may **oil out.** Oiling out occurs when the solid substance melts and forms a liquid that is not soluble in the solvent. On cooling, the liquid refuses to crystallize; rather, it becomes a supercooled liquid, or oil. Oils may solidify if the temperature is lowered, but often they will not crystallize. A solidified oil becomes an amorphous solid or a hardened mass—a condition that does not result in the purification of the substance. It can be very difficult to deal with oils when trying to obtain a pure substance. You must try to redissolve them and hope that they will precipitate as crystals with slow, careful cooling. During the cooling period, it may be helpful to scratch the glass container where the oil is present with a glass stirring rod that has not been fire-polished. Seeding the oil as it cools with a small sample of the original solid is another technique sometimes helpful in working with difficult oils. Other methods of inducing crystallization are discussed in Section 5.7.

One additional criterion for selecting the correct crystallization solvent is the **volatility** of that solvent. Volatile solvents are those that have low boiling points or evaporate easily. A solvent with a low boiling point may be removed from the crystals through evaporation without much difficulty. It will be difficult to remove a solvent with a high boiling point from the crystals without heating them under vacuum.

Table 5.2 lists common crystallization solvents. The solvents used most commonly are listed first in the table.

5.6 DECOLORIZATION

Small amounts of highly colored impurities may make the original crystallization solution appear colored; this color can often be removed by **decolorization,** either by using activated charcoal (often called Norit) or by passing the solution through a column packed with alumina or silica gel. A decolorizing step should be performed only if the color is due to impurities, not to the color of the desired product, and if the color is significant. Small amounts of colored impurities will remain in solution during crystalliza-

TABLE 5.2   Common Solvents for Crystallization

	Boils (°C)	Freezes (°C)	Soluble in H_2O	Flammability
Water	100	0	+	−
Methanol	65	*	+	+
95% Ethanol	78	*	+	+
Ligroin	60–90	*	−	+
Toluene	111	*	−	+
Chloroform†	61	*	−	−
Acetic acid	118	17	+	+
Dioxane†	101	11	+	+
Acetone	56	*	+	+
Diethyl ether	35	*	Slightly	+ +
Petroleum ether	30–60	*	−	+ +
Methylene chloride	41	*	−	−
Carbon tetrachloride†	77	*	−	−

*Lower than 0°C.

†Suspected carcinogen.

tion, making the decolorizing step unnecessary. The use of activated charcoal is described separately for standard scale and microscale crystallizations, and the column technique, which can be used with both crystallization techniques, is also described.

PART A.   STANDARD-SCALE—POWDERED CHARCOAL

As soon as the solute is dissolved in the minimum amount of boiling solvent, the solution is allowed to cool slightly and a small amount of Norit (powdered charcoal) is added to the mixture. The Norit adsorbs the impurities. When performing a crystallization in which the filtration is performed with a fluted filter, powdered Norit should be added because it has a larger surface area and can remove impurities more effectively. A reasonable amount of Norit would be what could be held on the end of a small spatula, or about 0.1–0.2 g. If too much Norit is used, it will adsorb product as well as impurities. A small amount of Norit should be used, and its use should be repeated if necessary. (It is difficult to determine if the initial amount added is sufficient until after the solution is filtered, because the suspended particles of black charcoal will obscure the color of the liquid.) Caution should be exercised so that the solution does not froth or erupt when the finely divided charcoal is added. The mixture is boiled for several minutes and then filtered by gravity, using a fluted filter (see Section 5.3 and Technique 4, Section 4.1, Part B, p. 636), and the crystallization is carried forward as described in Section 5.3.

The Norit preferentially absorbs the colored impurities and removes them from the solution. The technique seems to be most effective with hydroxylic solvents. In using Norit, be careful not to breathe the dust. Normally, small quantities are used so that little risk of lung irritation exists.

B.   DECOLORIZATION ON A COLUMN

Another method for decolorizing a solution is to pass the solution through a column containing alumina or silica gel. The adsorbent removes the colored impurities while allowing the desired material to pass through (see Technique 4, Fig. 4.6, p. 643, and Technique 12, Section 12.14, p. 782). If this technique is used, it will be necessary to dilute the solution with additional solvent to prevent crystallization from occurring during the process. The excess solvent must be evaporated after the solution is passed through the column (Technique 3, Section 3.11, p. 630), and the crystallization procedure is continued as described in Section 5.3.

C.   MICROSCALE—PELLETIZED NORIT

If the crystallization is being performed in a Craig tube, it is advisable to use pelletized Norit. Although this is not as effective in removing impurities as powdered Norit, it is easier to carry out the subsequent filtration, and the amount of pelletized Norit required is more easily determined because you can see the solution as it is being decolorized. Again, the Norit is added to the hot solution (the solution should not be boiling) after the solid has dissolved. This should be performed in a test tube rather than in a Craig tube. About 0.02 g is added, and the mixture is boiled for a minute or so to see if more Norit is required. More Norit is added, if necessary, and the liquid is boiled again. It is important not to add too much pelletized Norit because the Norit will also adsorb some of the desired material, and it is possible that not all the color can be removed no matter how much is added. The decolorized solution is then removed with a preheated filter-tip pipet (see Section 5.4 and Technique 4, Section 4.6, p. 645) to filter the mixture and transferred to a Craig tube for crystallization as described in Section 5.4.

5.7 INDUCING CRYSTALLIZATION

If a cooled solution does not crystallize, several techniques may be used to induce crystallization. Although identical in principle, the actual procedures vary slightly when performing standard scale and microscale crystallizations.

A.   STANDARD-SCALE

In the first technique, you should try scratching the inside surface of the flask vigorously with a glass rod that *has not been* fire-polished. The motion of the rod should be vertical (in and out of the solution) and should be vigorous enough to produce an audible scratching. Such scratching often induces crystallization, although the effect is not well understood. The high-frequency vibrations may have something to do with initiating crystallization; or perhaps—a more likely possibility—small amounts of solution dry by evaporation on the side of the flask, and the dried solute is pushed into the solution. These small amounts of material provide "seed crystals," or nuclei, on which crystallization may begin.

A second technique that can be used to induce crystallization is to cool the solution in an ice bath. This method decreases the solubility of the solute.

A third technique is useful when small amounts of the original material to be crystallized are saved. The saved material can be used to "seed" the cooled solution. A small crystal dropped into the cooled flask often will start the crystallization—this is called **seeding.**

If all these measures fail to induce crystallization, it is likely that too much solvent was added. The excess solvent must then be evaporated (Technique 3, Section 3.11, p. 630) and the solution allowed to cool.

B.   MICROSCALE

The strategy is basically the same as described for standard scale crystallizations. Scratching vigorously with a glass rod *should be avoided,* however, because the Craig tube is fragile and expensive. Scratching *gently* is allowed.

Another measure is to dip a spatula or glass stirring rod into the solution and allow the solvent to evaporate so that a small amount of solid will form on the surface of the spatula or glass rod. When placed back into the solution, the solid will seed the solution. A small amount of the original material, if some was saved, may also be used to seed the solution.

A third technique is to cool the Craig tube in an ice-water bath. This method may also be combined with either of the previous suggestions.

If none of these measures is successful, it is possible that too much solvent is present, and it may be necessary to evaporate some of the solvent (Technique 3, Section 3.11, p. 630) and allow the solution to cool again.

5.8 DRYING CRYSTALS

The most common method of drying crystals involves placing them on a watch glass, a clay plate, or a piece of paper and allowing them to dry in air. While the advantage of this method is that heat is not required, thus reducing the danger of decomposition or melting, exposure to atmospheric moisture may cause the hydration of strongly **hygroscopic** materials. A hygroscopic substance is one that absorbs moisture from the air. When crystals are allowed to air-dry, they should be covered with a watch glass or inverted beaker to protect them from air-borne dust particles.

Another method of drying crystals is to place the crystals on a watch glass, a clay plate, or a piece of absorbent paper in an oven. Although this method is simple, some possible difficulties deserve mention. Crystals that sublime readily should not be dried in an oven because they might vaporize and disappear. Care should be taken that the temperature of the oven does not exceed the melting point of the crystals. Remember that the melting point of crystals is lowered by the presence of solvent; allow for this melting-point depression when selecting a suitable oven temperature. Some materials decompose on exposure to heat, and they should not be dried in an oven. Finally, when many different samples are being dried in the same oven, crystals might be lost due to confusion or reaction with another person's sample. It is important to label the crystals when they are placed in the oven.

A third method, which requires neither heat nor exposure to atmospheric moisture, is drying *in vacuo*. Two procedures are illustrated in Figure 5.6.

Procedure A. In this method a desiccator is used. The sample is placed under vacuum in the presence of a drying agent. Two potential problems must be noted. The first deals with samples that sublime readily. Under vacuum, the likelihood of sublimation is increased. The second problem deals with the vacuum desiccator itself. Because the surface area of glass that is under vacuum is large, there is some danger that the desiccator could implode. A vacuum desiccator should never be used unless it has been placed within a protective metal container (cage). If a cage is not available, the desiccator can be wrapped with electrical or duct tape. If you use an aspirator as a source of vacuum, you should use a water trap (see Technique 4, Fig. 4.5, p. 641).

Procedure B. This method can be accomplished with a round-bottom flask and a thermometer adapter equipped with a short piece of glass tubing, as illustrated in Figure 5.6B. The glass tubing is connected by vacuum tubing to either an aspirator or a vacuum pump. A convenient alternative using a sidearm test tube is also shown in Figure 5.6B. With either apparatus, install a water trap when an aspirator is used.

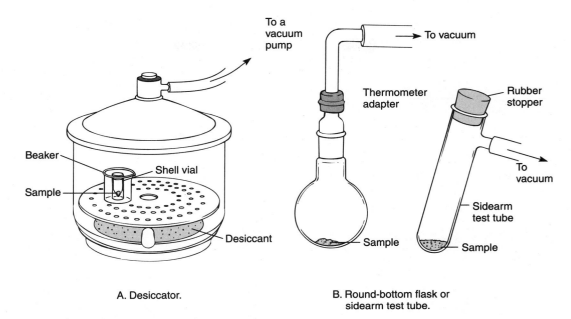

A. Desiccator.

B. Round-bottom flask or
   sidearm test tube.

Figure 5.6   Methods for drying crystals in a vacuum.

5.9 MIXED SOLVENTS

Often, the desired solubility characteristics for a particular compound are not found in a single solvent. In these cases, a mixed solvent may be used. You simply select a first solvent in which the solute is soluble and a second solvent, miscible with the first, in which the solute is relatively insoluble. The compound is dissolved in a minimum amount of the boiling solvent in which it is soluble. Following this, the second hot solvent is added to the boiling mixture, dropwise, until the mixture barely becomes cloudy. The cloudiness indicates precipitation. At this point, more of the first solvent should be added. Just enough is added to clear the cloudy mixture. At that point, the solution is saturated, and as it cools, crystals should separate. Common solvent mixtures are listed in Table 5.3.

TABLE 5.3.   Common Solvent Pairs for
                 Crystallization

Methanol–Water	Ether–Acetone
Ethanol–Water	Ether–Petroleum ether
Acetic acid–Water	Toluene–Ligroin
Acetone–Water	Methylene chloride–Methanol
Ether–Methanol	Dioxane*–Water

*Suspected carcinogen.

It is important not to add an excess of the second solvent or to cool the solution too rapidly. Either of these actions may cause the solute to oil out, or separate as a viscous liquid. If this happens, reheat the solution and add more of the first solvent.

PROBLEMS

1. Listed below are solubility-vs.-temperature data for an organic substance A dissolved in water.

Temperature (°C)	Solubility of A in 100 mL of Water
0	1.5 g
20	3.0 g
40	6.5 g
60	11.0 g
80	17.0 g

  (a) Graph the solubility of A vs. temperature. Use the data given in the table. Connect the data points with a smooth curve.

  (b) Suppose 0.1 g of A and 1.0 mL of water were mixed and heated to 80°C. Would all the substance A dissolve?

  (c) The solution prepared in (b) is cooled. At what temperature will crystals of A appear?

  (d) Suppose the cooling described in (c) were continued to 0°C. How many grams of A would come out of solution? Explain how you obtained your answer.

2. What would be likely to happen if a hot saturated solution were filtered by vacuum filtration using a Büchner funnel? (*Hint:* The mixture will cool as it comes in contact with the Büchner funnel.)

3. A compound you have prepared is reported in the literature to have a pale yellow color. When dissolving the substance in hot solvent to purify it by crystallization, the resulting solution is yellow. Should you use decolorizing charcoal before allowing the hot solution to cool? Explain your answer.

4. While performing a crystallization, you obtain a light tan solution after dissolving your crude product in hot solvent. A decolorizing step is determined to be unnecessary, and there are no solid impurities present. Should you perform a filtration to remove impurities before allowing the solution to cool? Why or why not?

5. (a) Draw a graph of a cooling curve (temperature vs. time) for a solution of a solid substance that shows no supercooling effects. Assume that the solvent does not freeze.

  (b) Repeat the instructions in (a) for a solution of a solid substance that shows some supercooling behavior but eventually yields crystals if the solution is cooled sufficiently.

6. A solid substance A is soluble in water to the extent of 1 g/100 mL of water at 25°C and 10 g/100 mL of water at 100°C. You have a sample that contains 10 g of A and an impurity B.

  (a) Assuming that 0.2 g of the impurity B is present along with 10 g of A, describe how you could purify A if B is completely insoluble in water.

  (b) Assuming that 0.2 g of the impurity B is present along with 10 g of A, describe how you could purify A if B had the same solubility behavior as A. Would one crystallization produce absolutely pure A?

(c) Assume that 3 g of the impurity B is present along with 10 g of A. Describe how you would purify A if B had the same solubility behavior as A. Each time, use the correct amount of water to just dissolve the solid. Would one crystallization produce absolutely pure A? How many crystallizations would be needed to produce pure A? How much A would have been recovered when the crystallizations had been completed?

T E C H N I Q U E   6

Physical Constants: Melting Points, Boiling Points, Density

6.1 PHYSICAL PROPERTIES

The physical properties of a compound are those properties that are intrinsic to a given compound when it is pure. Often, a compound may be identified simply by determining a number of its physical properties. The most commonly recognized physical properties of a compound include its color, melting point, boiling point, density, refractive index, molecular weight, and optical rotation. Modern chemists would include the various types of spectra (infrared, nuclear magnetic resonance, mass, and ultraviolet-visible) among the physical properties of a compound. A compound's spectra do not vary from one pure sample to another. In this chapter, we look at methods of determining the melting point, boiling point, and density of compounds. Refractive index, optical rotation, and spectra are considered separately in their own technique chapters.

Many reference books list the physical properties of substances. Useful works for finding lists of values for the nonspectroscopic physical properties include:

The Merck Index

The CRC Handbook of Chemistry and Physics

The Dictionary of Organic Compounds

Lange's Handbook of Chemistry

CRC Handbook of Tables for Organic Compound Identification

Complete citations for these references may be found in Technique 20 (Guide to the Chemical Literature, p. 861). Although the *CRC Handbook* has very good tables, it adheres strictly to IUPAC nomenclature. For this reason, it may be easier to use one of the other references, particularly *The Merck Index,* in your first attempt to locate information. *The Dictionary of Organic Compounds* is a multivolume work. A trip to the reference shelves of your library is required for you to use it, but it is a very complete source book.

PART A.   MELTING POINTS

6.2 THE MELTING POINT

The melting point of a compound is used by the organic chemist not only to iden-tify it, but also to establish its purity. A small amount of material is heated *slowly* in a special apparatus equipped with a thermometer or thermocouple, a heating bath or heat-ing coil, and a magnifying eyepiece for observing the sample. Two temperatures are noted. The first is the point at which the first drop of liquid forms among the crystals; the sec-ond is the point at which the whole mass of crystals turns to a *clear* liquid. The melting point is recorded by giving this range of melting. You might say, for example, that the melting point of a substance is 51–54°C. That is, the substance melted over a 3° range.

The melting point indicates purity in two ways. First, the purer the material, the higher its melting point. Second, the purer the material, the narrower the melting point range. Adding successive amounts of an impurity to a pure substance generally causes its melting point to decrease in proportion to the amount of impurity. Looking at it another way, adding impurities lowers the freezing point. The freezing point, a colligative prop-erty, is simply the melting point (solid $\longrightarrow$ liquid) approached from the opposite direc-tion (liquid $\longrightarrow$ solid).

Figure 6.1 is a graph of the usual melting-point behavior of mixtures of two sub-stances, A and B. The two extremes of the melting range (the low and high temperatures) are shown for various mixtures of the two. The upper curves indicate the temperature at which all the sample has melted. The lower curves indicate the temperature at which melt-ing is observed to begin. With pure compounds, melting is sharp and without any range. This is shown at the left- and right-hand edges of the graph. If you begin with pure A, the melting point decreases as impurity B is added. At some point, a minimum tempera-ture, or **eutectic,** is reached, and the melting point begins to increase to that of substance B. The vertical distance between the lower and upper curves represents the melting range. Notice that for mixtures that contain relatively small amounts of impurity (<15%) and

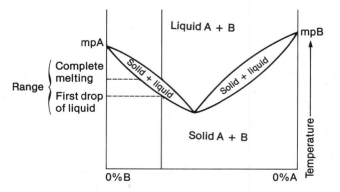

Figure 6.1   Melting point–composition curve.

are not close to the eutectic, the melting range increases as the sample becomes less pure. The range indicated by the lines in Figure 6.1 represents the typical behavior.

We can generalize the behavior shown in Figure 6.1. Pure substances melt with a narrow range of melting. With impure substances, the melting range becomes wider, and the entire melting range is lowered. Be careful to note, however, that at the minimum point of the melting point–composition curves, the mixture often forms a eutectic, which also melts sharply. Not all binary mixtures form eutectics, and some caution must be exercised in assuming that every binary mixture follows the previously described behavior. Some mixtures may form more than one eutectic, others might not form even one. In spite of these variations, both the melting point and its range are useful indications of purity, and they are easily determined by simple experimental methods.

6.3 MELTING POINT THEORY

Figure 6.2 is a phase diagram describing the usual behavior of a two-component mixture (A + B) on melting. The behavior on melting depends on the relative amounts of A and B in the mixture. If A is a pure substance (no B), then A melts sharply at its melting point t_A. This is represented by point A on the left side of the diagram. When B is a pure substance, it melts at t_B; its melting point is represented by point B on the right side of the diagram. At either point A or point B, the pure solid passes cleanly, with a narrow range, from solid to liquid.

In mixtures of A and B, the behavior is different. Using Figure 6.2, consider a mixture of 80% A and 20% B on a mole-per-mole basis (that is, mole percentage). The melting point of this mixture is given by t_M at point M on the diagram. That is, adding B to A has lowered the melting point of A from t_A to t_M. It has also expanded the melting range. The temperature t_M corresponds to the **upper limit** of the melting range.

Lowering the melting point of A by adding impurity B comes about in the following way. Substance A has the lower melting point in the phase diagram shown, and if

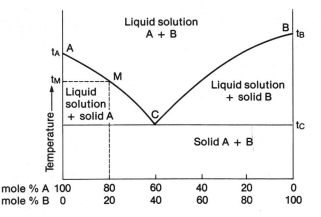

mole % A 100   80   60   40   20   0
mole % B   0   20   40   60   80   100

Figure 6.2 Phase diagram for melting in a two-component system.

heated, it begins to melt first. As A begins to melt, solid B begins to dissolve in the liquid A that is formed. When solid B dissolves in liquid A, the melting point is depressed. To understand this, consider the melting point from the opposite direction. When a liquid at a high temperature cools, it reaches a point at which it solidifies, or "freezes." The temperature at which a liquid freezes is identical to its melting point. Recall that the freezing point of a liquid can be lowered by adding an impurity. Because the freezing point and the melting point are identical, lowering the freezing point corresponds to lowering the melting point. Therefore, as more impurity is added to a solid, its melting point becomes lower. There is, however, a limit to how far the melting point can be depressed. You cannot dissolve an infinite amount of the impurity substance in the liquid. At some point, the liquid will become saturated with the impurity substance. The solubility of B in A has an upper limit. In Figure 6.2, the solubility limit of B in liquid A is reached at point C, the **eutectic point.** The melting point of the mixture cannot be lowered below t_C, the melting temperature of the eutectic.

Now consider what happens when the melting point of a mixture of 80% A and 20% B is approached. As the temperature is increased, A begins to "melt." This is not really a visible phenomenon in the beginning stages; it happens before liquid is visible. It is a softening of the compound to a point at which it can begin to mix with the impurity. As A begins to soften, it dissolves B. As it dissolves B, the melting point is lowered. The lowering continues until all B is dissolved or until the eutectic composition (saturation) is reached. When the maximum possible amount of B has been dissolved, actual melting begins, and one can observe the first appearance of liquid. The initial temperature of melting will be below t_A. The amount below t_A at which melting begins is determined by the amount of B dissolved in A, but will never be below t_C. Once all B has been dissolved, the melting point of the mixture begins to rise as more A begins to melt. As more A melts, the semisolid solution is diluted by more A, and its melting point rises. While all this is happening, you can observe *both* solid and liquid in the melting-point capillary. Once all A has begun to melt, the composition of the mixture M becomes uniform and will reach 80% A and 20% B. At this point, the mixture finally melts sharply, giving a clear solution. The maximum melting point will be t_M, because t_A is depressed by the impurity B that is present. The lower end of the melting range will always be t_C; however, melting will not always be observed at this temperature. An observable melting at t_C comes about only when a large amount of B is present. Otherwise, the amount of liquid formed at t_C will be too small to observe. Therefore, the melting behavior that is actually observed will have a smaller range, as shown in Figure 6.1.

6.4 MIXED MELTING POINTS

The melting point can be used as supporting evidence in identifying a compound in two different ways. Not only may the melting points of the two individual compounds be compared, but a special procedure called a **mixed melting point** may also be performed. The mixed melting point requires that an authentic sample of the same compound be available from another source. In this procedure, the two compounds (authentic and suspected) are finely pulverized and mixed together in equal quantities. Then the melting point of the mixture is determined. If there is a melting-point depression, or if the range of melting is expanded by a large amount, compared to the individual substances, you may conclude that one compound has acted as an impurity toward the other and that they are not the same compound. If there is no lowering of the melting point for the mixture (the melting point is identical with those of pure A and pure B), then A and B are almost certainly the same compound.

6.5 PACKING THE MELTING POINT TUBE

Melting points are usually determined by heating the sample in a piece of thin-walled capillary tubing (1 mm × 100 mm) that has been sealed at one end. To pack the tube, press the open end gently into a *pulverized* sample of the crystalline material. Crystals will stick in the open end of the tube. The amount of solid pressed into the tube should correspond to a column no more than 1–2 mm high. To transfer the crystals to the closed end of the tube, drop the capillary tube, closed end first, down a $\frac{2}{3}$-m length of glass tubing, which is held upright on the desk top. When the capillary tube hits the desk top, the crystals will pack down into the bottom of the tube. This procedure is repeated if necessary. Tapping the capillary on the desk top with fingers is not recommended, because it is easy to drive the small tubing into a finger if the tubing should break.

Some commercial melting-point instruments have a built-in vibrating device that is designed to pack capillary tubes. With these instruments, the sample is pressed into the open end of the capillary tube, and the tube is placed in the vibrator slot. The action of the vibrator will transfer the sample to the bottom of the tube and pack it tightly.

6.6 DETERMINING THE MELTING POINT— THE THIELE TUBE

There are two principal types of melting-point apparatus available: the Thiele tube and commercially available, electrically heated instruments. The Thiele tube, shown in Figure 6.3, is the simpler device. It is a glass tube designed to contain a heating oil (mineral oil or silicone oil) and a thermometer to which a capillary tube containing the sample is attached. The shape of the Thiele tube allows convection currents to form in the oil when it is heated. These currents maintain a uniform temperature distribution throughout the oil in the tube. The sidearm of the tube is designed to generate these convection currents and thus transfer the heat from the flame evenly and rapidly throughout the oil. The sample, which is in a capillary tube attached to the thermometer, is held by a rubber band or a thin slice of rubber tubing. It is important that this rubber band be above the level of the oil (allowing for expansion of the oil on heating), so that the oil does not soften the rubber and allow the capillary tubing to fall into the oil. If a cork or a rubber stopper is used to hold the thermometer, a triangular wedge should be sliced in it to allow pressure equalization.

The Thiele tube is usually heated by a microburner. During the heating, the rate of temperature increase should be regulated. Hold the burner by its cool base, and, using a low flame, move the burner slowly back and forth along the bottom of the arm of the

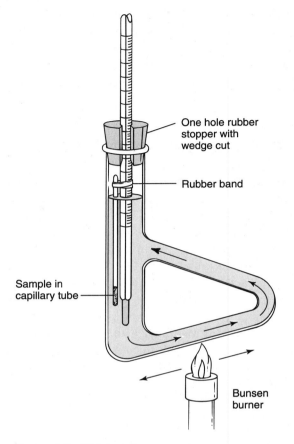

One hole rubber
stopper with
wedge cut

Rubber band

Sample in
capillary tube

Bunsen
burner

Figure 6.3   Thiele tube.

Thiele tube. If the heating is too fast, remove the burner for a few seconds, and then re-sume heating. The rate of heating should be *slow* near the melting point (about 1°C per min.) to ensure that the temperature increase is not faster than the rate at which heat can be transferred to the sample being observed. At the melting point, it is necessary that the mercury in the thermometer and the sample in the capillary tube be at temperature equi-librium.

6.7 DETERMINING THE MELTING POINT— ELECTRICAL INSTRUMENTS

Three types of electrically heated melting point instruments are illustrated in Figure 6.4. In each case, the melting-point tube is filled as described in Section 6.5 and placed in a holder located just behind the magnifying eyepiece. The apparatus is operated by moving the switch to the ON position, adjusting the potentiometric control dial for the

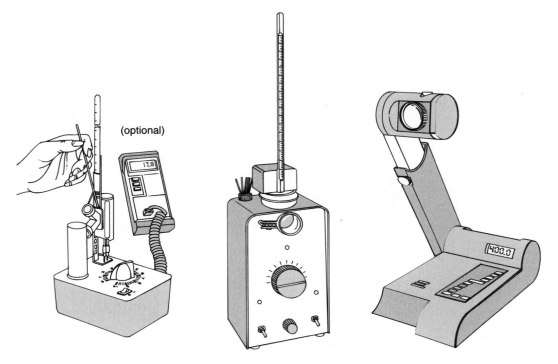

(optional)

Figure 6.4   Melting-point apparatus.

desired rate of heating, and observing the sample through the magnifying eyepiece. The temperature is read from a thermometer, or in the most modern instruments, from a digital display attached to a thermocouple. Your instructor will demonstrate and explain the type used in your laboratory.

Most electrically heated instruments do not heat or increase the temperature of the sample linearly. Although the rate of increase may be linear in the early stages of heating, it usually decreases and leads to a constant temperature at some upper limit. The upper-limit temperature is determined by the setting of the heating control. Thus, a family of heating curves is usually obtained for various control settings, as shown in Figure 6.5. The four hypothetical curves shown (1–4) might correspond to different control settings. For a compound melting at temperature t_1, the setting corresponding to curve 3 would be ideal. In the beginning of the curve, the temperature is increasing too rapidly to allow determination of an accurate melting point, but after the change in slope, the temperature increase will have slowed to a more usable rate.

If the melting point of the sample is unknown, you can often save time by preparing two samples for melting-point determination. With one sample, you can rapidly determine a crude melting-point value. Then, repeat the experiment more carefully using the second sample. For the second determination, you already have an approximate idea of what the melting-point temperature should be and a proper rate of heating can be chosen.

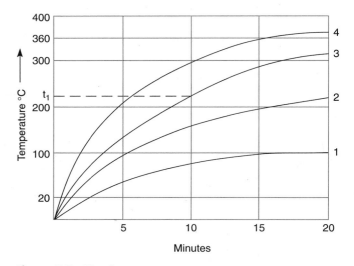

Figure 6.5   Heating-rate curves.

When measuring temperatures above 150°C, thermometer errors can become significant. For an accurate melting point with a high melting solid, you may wish to apply a **stem correction** to the thermometer as described in Section 6.13. An even better solution is to calibrate the thermometer as described in Section 6.12.

6.8 DECOMPOSITION, DISCOLORATION, SOFTENING, SHRINKAGE, AND SUBLIMATION

Many solid substances undergo some degree of unusual behavior before melting. At times it may be difficult to distinguish these types of behavior from actual melting. You should learn, through experience, how to recognize melting and how to distinguish it from decomposition, discoloration, and particularly softening and shrinkage.

Some compounds decompose on melting. This decomposition is usually evidenced by discoloration of the sample. Frequently, this decomposition point is a reliable physical property to be used in lieu of an actual melting point. Such decomposition points are indicated in tables of melting points by placing the symbol *d* immediately after the listed temperature. An example of a decomposition point is thiamine hydrochloride, whose melting point would be listed as 248°d, indicating that this substance melts with decomposition at 248°C. When decomposition is a result of reaction with the oxygen in air, it may be avoided by determining the melting point in a sealed, evacuated melting-point tube.

Figure 6.6 shows two simple methods of evacuating a packed tube. Method A uses an ordinary melting-point tube, while Method B constructs the melting-point tube from a disposable Pasteur pipet. Before using Method B, be sure to determine that the tip of the pipet will fit into the sample holder in your melting-point instrument.

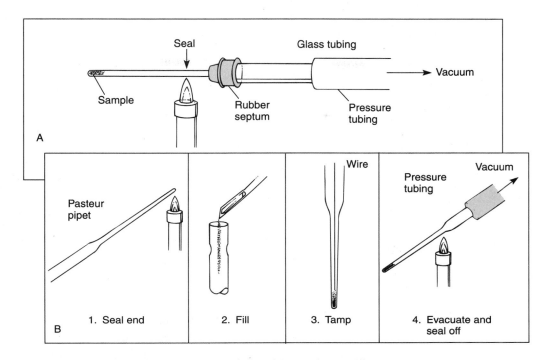

Figure 6.6   Evacuation and sealing of a melting-point capillary.

Method A.   In Method A, a hole is punched through a rubber septum using a large pin or a small nail, and the capillary tube is inserted from the inside, sealed end first. The septum is placed over a piece of glass tubing connected to a vacuum line. After evacuating the tube, the upper end of the tube may be sealed by heating and pulling it closed.

Method B.   In Method B, the thin section of a 9-inch Pasteur pipet is used to construct the melting-point tube. Carefully seal the tip of the pipet using a flame. Be sure to hold the tip *upward* as you seal it. This will prevent water vapor from condensing inside the pipet. When the sealed pipet has cooled, the sample may be added through the open end using a microspatula. A small wire may be used to compress the sample into the closed tip. (If your melting-point apparatus has a vibrator, it may be used in place of the wire to simplify the packing.) When the sample is in place, the pipet is connected to the vacuum line with tubing and evacuated. The evacuated sample tube is sealed by heating it with a flame and pulling it closed.

Some substances begin to decompose *below* their melting points. Thermally unstable substances may undergo elimination reactions or anhydride formation reactions during heating. The decomposition products formed represent impurities in the original sample, so the melting point of the substance may be lowered due to their presence.

It is normal for many compounds to soften or shrink immediately before melting. Such behavior represents not decomposition but a change in the crystal structure or a mixing with impurities. Some substances "sweat," or release solvent of crystallization, before melting. These changes do not indicate the beginning of melting. Actual melting begins

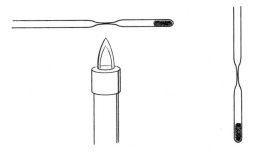

Figure 6.7   Sealing a tube for a substance that sublimes.

when the first drop of liquid becomes visible, and the melting range continues until the temperature at which all the solid has been converted to the liquid state. With experience, you soon learn to distinguish between softening, or "sweating," and actual melting. If you wish, the temperature of the onset of softening or sweating may be reported as a part of your melting point rage: 211°C (softens), 223–225°C (melts).

Some solid substances have such a high vapor pressure that they sublime at or below their melting points. In many handbooks, the sublimation temperature is listed along with the melting point. The symbols *sub, subl,* and sometimes *s* are used to designate a substance that sublimes. In such cases, the melting-point determination must be performed in a sealed capillary tube to avoid loss of the sample. The simplest way to accomplish sealing a packed tube is to heat the open end of the tube in a flame and pull it closed with a tweezers or forceps. A better way, although more difficult to master, is to heat the center of the tube in a small flame, rotating it about its axis, keeping the tube straight, until the center collapses. If this is not done quickly, the sample may melt or sublime while you are working. With the smaller chamber, the sample will not be able to migrate to the cool top of the tube that may be above the viewing area. Figure 6.7 illustrates the method.

PART B.   BOILING POINTS

6.9 THE BOILING POINT

As a liquid is heated, the vapor pressure of the liquid increases to the point where it just equals the applied pressure (usually atmospheric pressure). At this point, the liquid is observed to boil. The normal boiling point is measured at 760 mmHg (760 torr) or 1 atm. At a lower applied pressure, the vapor pressure needed for boiling is also lowered, and the liquid boils at a lower temperature. The relation between applied pressure and temperature of boiling for a liquid is determined by its vapor pressure–temperature behavior. Figure 6.8 is an idealization of the typical vapor pressure–temperature behavior of a liquid.

Because the boiling point is sensitive to pressure, it is important to record the barometric pressure when determining a boiling point if the determination is being conducted at an elevation significantly above or below sea level. Normal atmospheric variations may

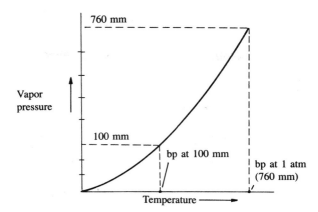

Figure 6.8   The vapor pressure–temperature curve for a typical liquid.

affect the boiling point, but they are usually of minor importance. However, if a boiling point is being monitored during the course of a vacuum distillation (Technique 9) that is being performed with an aspirator or a vacuum pump, the variation from the atmospheric value will be especially marked. In these cases, it is quite important to know the pressure as accurately as possible.

As a rule of thumb, the boiling point of many liquids drops about 0.5°C for a 10-mm decrease in pressure when in the vicinity of 760 mmHg. At lower pressures, a 10°C drop in boiling point is observed for each halving of the pressure. For example, if the observed boiling point of a liquid is 150°C at 10-mm pressure, then the boiling point would be about 140°C at 5 mmHg.

A more accurate estimate of the change in boiling point with a change of pressure can be made by using a **nomograph.** In Figure 6.9, a nomograph is given and a method is described for using it to obtain boiling points at various pressures when the boiling point is known at some other pressure.

6.10 DETERMINING BOILING POINTS— STANDARD-SCALE METHODS

Experimental methods of determining the boiling point of a liquid are easily available. When you have large quantities of material, you can simply record the boiling point (or boiling range) as indicated on a thermometer as the substance distills during a simple distillation (see Technique 8, Section 8.3, page 710). Alternatively, you may find it convenient to use a direct method, shown in Figure 6.10. With this method, the bulb of the thermometer can be immersed in vapor from the boiling liquid for a period long enough to allow it to equilibrate and give a good temperature reading.

The liquid can be brought to its boiling point very quickly when a test tube is used. Select a test tube that is long and narrow. You should avoid a large space between the sides of the test tube and the thermometer. Place the bulb of the thermometer as close as possible to the boiling liquid without actually touching it. You should use a small, inert

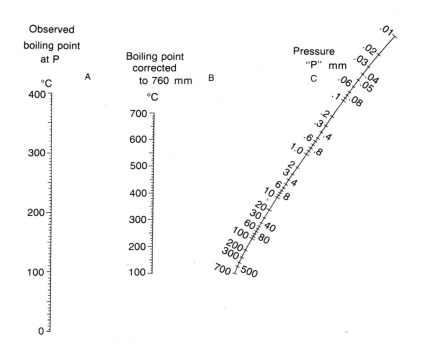

Figure 6.9 Pressure–temperature alignment nomograph. How to use the nomograph: Assume a reported boiling point of 100°C at 1 mm. To determine the boiling point at 18 mm, connect 100°C (column A) to 1 mm (column C) with a transparent plastic rule and observe where this line intersects column B (about 280°C). This value would correspond to the normal boiling point. Next, connect 280°C (column B) with 18 mm (column C) and observe where this intersects column A (151°C). The approximate boiling point will be 151°C at 18 mm. Reprinted by courtesy of MC/B Manufacturing Chemists, Inc.

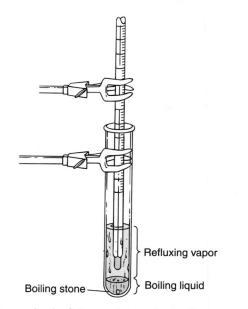

Figure 6.10 Standard-scale method of determining the boiling point.

carborundum (black) boiling stone. Do not use a marble or calcium carbonate (white) boiling chip. If safe operation permits, the best heating source is a small microburner. The flame will heat the liquid to boiling very quickly. The liquid must boil vigorously, such that you see a reflux ring and drops of liquid condensing on the sides of the test tube. You must, however, watch carefully so that you do not heat so much as to cause the liquid to boil out of the test tube, contact the flame, and cause a fire. The temperature reading on the thermometer must remain constant at its highest observed value. If the temperature continues to rise, the liquid has not yet reached its true boiling point.

6.11 DETERMINING BOILING POINTS— MICROSCALE METHODS

When you have smaller amounts of material, you can carry out a microscale or semi-microscale determination of the boiling point by using the apparatus shown in Figure 6.11.

Semi-Microscale Method. To carry out the semi-microscale determination, a piece of 5-mm glass tubing sealed at one end is attached to a thermometer with a rubber band

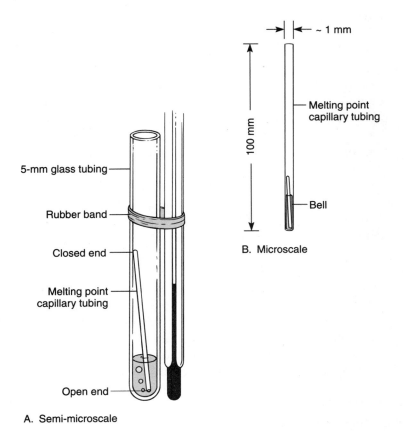

A. Semi-microscale

Figure 6.11 Boiling-point determinations.

or a thin slice of rubber tubing. The liquid whose boiling point is being determined is introduced with a Pasteur pipet into this piece of tubing and a short piece of melting-point capillary (sealed at one end) is dropped in with the open end down. The whole unit is then placed in a Thiele tube. The rubber band should be placed above the level of the oil in the Thiele tube; otherwise, the band may soften in the hot oil. When positioning the band, keep in mind that the oil will expand when heated. Next, the Thiele tube is heated in the same fashion as described in Section 6.6 for determining a melting point. Heating is continued until a rapid and continuous stream of bubbles emerges from the inverted capillary. At this point, you should stop heating. Soon, the stream of bubbles slows down and stops. When the bubbles stop, the liquid enters the capillary tube. The moment at which the liquid enters the capillary tube corresponds to the boiling point of the liquid, and the temperature is recorded.

Microscale Method. In microscale experiments, there is often too little product available to use the semi-microscale method described above. However, the method can be scaled down in the following manner. The liquid is placed in a 1-mm melting-point capillary tube to a depth of about 4–6 mm. Use a syringe or a Pasteur pipet that has had its tip drawn thinner to transfer the liquid into the capillary tube. It may be necessary to use a centrifuge to transfer the liquid to the bottom of the tube. Next, prepare an appropriately sized inverted capillary, or **bell.**

The easiest way to prepare a bell is to use a commercial micropipet, such as a 10-μL Drummond "microcap." These are available in vials of 50 or 100 microcaps and are very inexpensive. To prepare the bell, cut the microcap in half with a file or scorer and then seal one end by inserting it a small distance into a flame, turning it on its axis until the opening closes.

If microcaps are not available, a piece of 1-mm open-end capillary tubing (same size as a melting-point capillary) can be rotated along its axis in a flame while being held horizontally. Use your index finger and thumbs to rotate the tube; do not change the distance between your two hands while rotating. When the tubing is soft, remove it from the flame and pull it to a thinner diameter. When pulling, keep the tube straight by *moving both your hands and your elbows outward* by about 4 inches. Hold the pulled tube in place a few moments until it cools. Using the edge of a file or your fingernail, break out the thin center section. Seal one end of the thin section in the flame; then, break it to a length that is about one and one-half times the height of your sample liquid (6–9 mm). Be sure the break is done squarely. Invert the bell (open end down) and place it in the capillary tube containing the sample liquid. Push the bell to the bottom with a fine copper wire if it adheres to the side of the capillary tube. A centrifuge may be used if you prefer. Figure 6.12 shows the construction method for the bell and the final assembly.

Place the microscale assembly in a standard melting-point apparatus (or a Thiele tube if an electrical apparatus is not available) to determine the boiling point. Heating is continued until a rapid and continuous stream of bubbles emerges from the inverted capillary. At this point, stop heating. Soon, the stream of bubbles slows down and stops. When the bubbles stop, the liquid enters the capillary tube. The moment at which the liquid enters the capillary tube corresponds to the boiling point of the liquid, and the temperature is recorded.

Explanation of the Method. During the initial heating, the air trapped in the inverted bell expands and leaves the tube, giving rise to a stream of bubbles. When the liquid begins boiling, most of the air has been expelled; the bubbles of gas are due to the boiling action of the liquid. Once the heating is stopped, most of the vapor pressure left in the bell comes from the vapor of the heated liquid that seals its open end. There is always vapor in equilibrium with a heated liquid. If the temperature of the liquid is above its boiling point, the pressure of the trapped vapor will either exceed or equal the atmospheric pressure. As the liquid cools, its vapor pressure decreases. When the vapor pressure drops just below atmospheric pressure (just below the boiling point), the liquid is forced into the capillary tube.

Difficulties.   Three problems are common to this method. The first arises when the liquid is heated so strongly that it evaporates or boils away. The second arises when the liquid is not heated above its boiling point before heating is discontinued. If the heating is stopped at any point below the actual boiling point of the sample, the liquid enters the bell *immediately,* giving an apparent boiling point that is too low. Be sure that a continuous stream of bubbles, too fast for individual bubbles to be distinguished, is observed before lowering the temperature. Also, be sure that the bubbling action decreases slowly before the liquid enters the bell. If your melting-point apparatus has fine enough control and fast response, you can actually begin heating again and force the liquid out of the bell be-

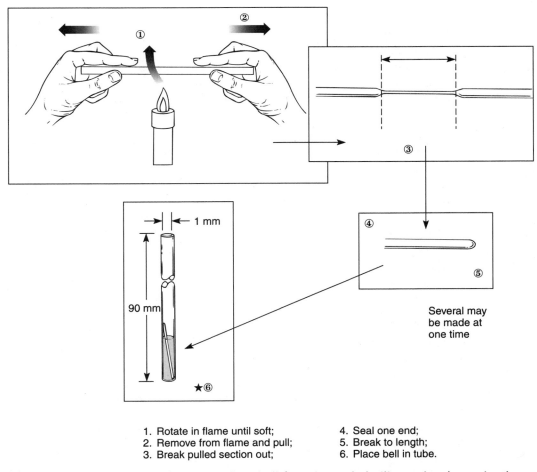

1. Rotate in flame until soft;
2. Remove from flame and pull;
3. Break pulled section out;
4. Seal one end;
5. Break to length;
6. Place bell in tube.

Several may be made at one time

Figure 6.12   Construction of microcapillary bell for microscale boiling-point determination.

fore it becomes completely filled with the liquid. This allows a second determination to be performed on the same sample. The third problem is that the bell may be so light that the bubbling action of the liquid causes the bell to move up the capillary tube. This problem can sometimes be solved by using a longer (heavier) bell or by sealing the bell so that a larger section of solid glass is formed at the sealed end of the bell.

When measuring temperatures above 150°C, thermometer errors can become significant. For an accurate boiling point with a high-boiling liquid, you may wish to apply a **stem correction** to the thermometer as described in Section 6.13, or to calibrate the thermometer as described in Section 6.12.

PART C.   THERMOMETER CALIBRATION AND CORRECTION

6.12 THERMOMETER CALIBRATION

When a melting-point or boiling-point determination has been completed, you expect to obtain a result that exactly duplicates the result recorded in a handbook or in the original literature. It is not unusual, however, to find a discrepancy of several degrees from the literature value. Such a discrepancy does not necessarily indicate that the experiment was incorrectly performed or that the material is impure; rather, it may indicate that the thermometer used for the determination was slightly in error. Most thermometers do not measure the temperature with perfect accuracy.

To determine accurate values, you must calibrate the thermometer that is used. This calibration is done by determining the melting points of a variety of standard substances with the thermometer. A plot is drawn of the observed temperature vs. the published value of each standard substance. A smooth line is drawn through the points to complete the chart. A correction chart prepared in this way is shown in Figure 6.13. This chart is used to correct any melting point determined with that particular thermometer. Each thermometer requires its own calibration curve. A list of suitable standard substances for calibrating thermometers is provided in Table 6.1. The standard substances, of course, must be pure in order for the corrections to be valid.

6.13 THERMOMETER STEM CORRECTIONS

Three types of thermometers are available: bulb immersion, stem immersion (partial immersion), and total immersion. **Bulb immersion** thermometers are calibrated by the manufacturer to give correct temperature readings when only the bulb (not the rest of the thermometer) is placed in the medium to be measured. **Stem immersion** thermometers are calibrated to give correct temperature readings when they are immersed to a specified depth in the medium to be measured. Stem immersion thermometers are easily rec-

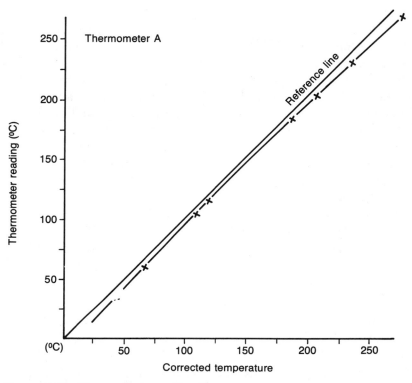

Figure 6.13   Thermometer calibration curve.

ognized because the manufacturer always scores a mark, or immersion ring, completely around the stem at the specified depth of immersion. The immersion ring is normally found below any of the temperature calibrations. **Total immersion** thermometers are calibrated when the entire thermometer is immersed in the medium to be measured. The three types of thermometer are often marked on the back (opposite side from the calibrations) by the words *bulb, immersion,* or *total,* but this may vary from one manufacturer to another. Because total immersion thermometers are less expensive, they are the type you are most likely to find in the laboratory.

TABLE 6.1   Melting-Point Standards

Compound	Melting Point (°C)
Ice (solid–liquid water)	0
Acetanilide	115
Benzamide	128
Urea	132
Succinic acid	189
3,5-Dinitrobenzoic acid	205

Manufacturers design total immersion thermometers to read correctly only when they are immersed totally in the medium to be measured. The entire mercury thread must be covered. Because this situation is rare, a **stem correction** should be added to the observed temperature. This correction, which is positive, can be fairly large when high temperatures are being measured. Keep in mind, however, that if your thermometer has been calibrated for its desired use (such as described in Section 6.12 for a melting-point apparatus), a stem correction should not be necessary for any temperature within the calibration limits. You are most likely to want a stem correction when you are performing a distillation. If you determine a melting point or boiling point using an uncalibrated, total-immersion thermometer, you will also want to use a stem correction.

When you wish to make a stem correction for a total immersion thermometer, the formula given below may be used. It is based on the fact that the portion of the mercury thread in the stem is cooler than the portion immersed in the vapor or the heated area around the thermometer. The mercury will not have expanded in the cool stem to the same extent as in the warmed section of the thermometer. The equation used is

$$(0.000154)(T - t_1)(T - t_2) = \text{correction to be added to } T \text{ observed.}$$

1. The factor 0.00154 is a constant, the coefficient of expansion for the mercury in the thermometer.
2. The term $T - t_1$ corresponds to the length of the mercury thread not immersed in the heated area. It is convenient to use the temperature scale on the thermometer itself for this measurement rather than an actual length unit. T is the observed temperature, and t_1 is the *approximate* place where the heated part of the stem ends and the cooler part begins.
3. The term $T - t_2$ corresponds to the difference between the temperature of the mercury in the vapor T and the temperature of the mercury in the air outside the heated area (room temperature). The term T is the observed temperature, and t_2 is measured by hanging another thermometer so that the bulb is close to the stem of the main thermometer.

Figure 6.14 shows how to apply this method for a distillation. By the formula given above, it can be shown that high temperatures are more likely to require a stem correction and that low temperatures need not be corrected. The calculations given below illustrate this point.

Example 1		Example 2	
$T = 200°C$		$T = 100°C$	
$t_1 = 0°C$		$t_1 = 0°C$	
$t_2 = 35°C$		$t_2 = 35°C$	
$(0.000154)(200)(165) = 5.1°$	stem correction	$(0.000154)(100)(65) = 1.0°C$	stem correction
$200°C + 5°C = 205°C$	corrected temp	$100°C + 1°C = 101°C$	corrected temp

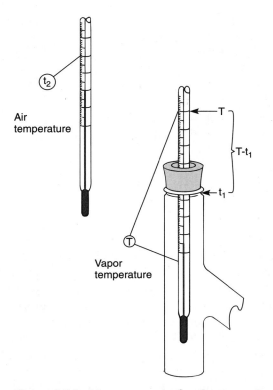

Figure 6.14   Measurement of a thermometer stem correction during distillation.

PART D.   DENSITY

6.14 DENSITY

Density is defined as mass per unit volume and is generally expressed in units of grams per milliliter (g/mL) for a liquid and grams per cubic centimeter (g/cm^3) for a solid.

$$\text{Density} = \frac{\text{Mass}}{\text{Volume}} \quad \text{or} \quad D = \frac{M}{V}$$

In organic chemistry, density is most commonly used in converting the weight of liquid to a corresponding volume, or vice versa. It is often easier to measure a volume of a liquid than to weigh it. As a physical property, density is also useful for identifying liquids in much the same way that boiling points are used.

Although precise methods that allow the measurement of the densities of liquids at the microscale level have been developed, they are often difficult to perform. An approx-

TABLE 6.2   Densities Determined by the Automatic
              Pipet Method (g/mL)

Substance	bp	lit	100 μL
Water	100	1.000	1.01
Hexane	69	0.660	0.66
Acetone	56	0.788	0.77
Dichloromethane	40	1.330	1.27
Diethyl ether	35	0.713	0.67

imate method for measuring densities can be found in using a 100-μL (0.100-mL) automatic pipet (Technique 1, Section 1.6, p. 597). Clean, dry, and pre-weigh one or more
small vials (including their caps and liners) and record their weights. You should handle
these vials with a tissue in order to avoid getting your fingerprints on them. Adjust the
automatic pipet to deliver 100 μL and fit it with a clean, new tip. Use the pipet to deliver
100 μL of the unknown liquid to each of your tared vials. Cap them so that the liquid
does not evaporate. Reweigh the vials and use the weight of the 100 μL of liquid delivered to calculate a density for each case. It is recommended that from three to five determinations be performed, that the calculations be performed to three significant figures,
and that all the calculations be averaged to obtain the final result. This determination of
the density will be accurate to within two significant figures. Table 6.2 compares some
literature values with those that could be obtained by this method.

PROBLEMS

1. Two substances, A and B, have the same melting point. How can you determine if they are
the same without using any form of spectroscopy? Explain in detail.

2. Using Figure 6.5, determine which heating curve would be most appropriate for a substance
with a melting point of about 150°C.

3. What steps can you take to determine the melting point of a substance that sublimes before it
melts?

4. Using the pressure–temperature alignment chart in Figure 6.9, answer the following questions.

   (a) What is the normal boiling point (at 760 mmHg) for a compound that boils at 150°C at
10 mmHg pressure?

   (b) At what temperature would the compound in (a) boil if the pressure were 40 mmHg?

   (c) A compound was distilled at atmospheric pressure and had a boiling point of 285°C. What
would be the approximate boiling range for this compound at 15 mmHg?

5. Calculate the corrected boiling point for nitrobenzene by using the method given in Section
6.13. The boiling point was determined using an apparatus similar to that shown in Figure 6.10.
The observed boiling point was 205°C. The reflux ring in the test tube just reached up to the 0°C
mark on the thermometer. A second thermometer suspended alongside the test tube, at a slightly
higher level than the one inside, gave a reading of 35°C.

6. Suppose that you had calibrated the thermometer in your melting-point apparatus against a series of melting-point standards. After reading the temperature and converting it using the calibration chart, should you also apply a stem correction? Explain.

7. The density of a liquid was determined by the automatic pipet method. A 100-μL automatic pipet was used. The liquid had a mass of 0.082 g. What was the density in grams per milliliter of the liquid?

8. A compound melting at 134°C was suspected to be either aspirin (mp 135°C) or urea (mp 133°C). Explain how you could determine whether one of these two suspected compounds was identical to the unknown compound without using any form of spectroscopy.

9. An unknown compound gave a melting point of 230°C. When the molten liquid solidified, the melting point was redetermined and found to be 131°C. Give a possible explanation for this discrepancy.

10. During the micro boiling-point determination of an unknown liquid, heating was discontinued at 154°C and the liquid immediately began to enter the inverted bell. Heating was begun again at once, and the liquid was forced out of the bell. Heating was again discontinued at 165°C, at which time a very rapid stream of bubbles emerged from the bell. On cooling, the rate of bubbling gradually diminished until the liquid reached a temperature of 161°C, and entered and filled the bell. Explain this sequence of events. What was the boiling point of the liquid?

T E C H N I Q U E   7

Extractions, Separations, and Drying Agents

7.1 EXTRACTION

Transferring a solute from one solvent into another is called **extraction,** or more precisely, liquid–liquid extraction. The solute is extracted from one solvent into the other because the solute is more soluble in the second solvent than in the first. The two solvents must not be **miscible** (mix freely), and they must form two separate **phases** or layers, in order for this procedure to work. Extraction is used in many ways in organic chemistry. Many **natural products** (organic chemicals that exist in nature) are present in animal and plant tissues having high water content. Extracting these tissues with a water-immiscible solvent is useful for isolating natural products. Often, diethyl ether (commonly referred to as "ether") is used for this purpose. Sometimes, alternative water-immiscible solvents such as hexane, petroleum ether, ligroin, and methylene chloride are used. For instance, caffeine, a natural product, can be extracted from an aqueous tea solution by shaking it successively with several portions of methylene chloride.

A generalized extraction process, using a specialized piece of glassware called a separatory funnel, is illustrated in Figure 7.1. The first solvent contains a mixture of black and white molecules (Fig. 7.1A). A second solvent that is not miscible with the first is added. After the separatory funnel is capped and shaken, the layers separate. In this ex-

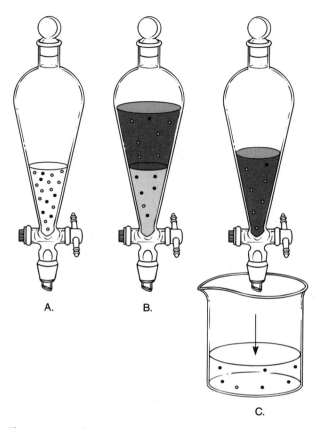

A. Solvent 1 contains a mixture of molecules (black and white).

B. After shaking with solvent 2 (shaded), most of the white molecules have been extracted into the new solvent. The white molecules are more soluble in the second solvent, while the black molecules are more soluble in the original solvent.

C. With removal of the lower phase, the black and white molecules have been partially separated.

Figure 7.1 The extraction process.

ample the second solvent (shaded) is less dense than the first, so it becomes the top layer (Fig. 7.1B). Because of differences in physical properties, the white molecules are more soluble in the second solvent, while the black molecules are more soluble in the first solvent. Most of the white molecules are in the upper layer, but there are some black molecules there, too. Likewise, most of the black molecules are in the lower layer. However, there are still a few white molecules in this lower phase. The lower phase may be separated from the upper phase by opening the stopcock at the bottom of the separatory funnel and allowing the lower layer to drain into a beaker (Fig. 7.1C). In this example, notice that it was not possible to effect a complete separation of the two types of molecules with a single extraction. This is a common occurrence in organic chemistry.

Many substances are soluble in both water and organic solvents. Water can be used to extract or "wash" water-soluble impurities from an organic reaction mixture. To carry out a "washing" operation, you add water and an immiscible organic solvent to the reaction mixture contained in a separatory funnel. After stoppering the funnel and shaking it, you allow the organic layer and the aqueous (water) layer to separate from each other. A water wash removes highly polar and water-soluble materials, such as sulfuric acid, hydrochloric acid, or sodium hydroxide from the organic layer. The washing operation helps to purify the desired organic compound present in the original reaction mixture.

7.2 DISTRIBUTION COEFFICIENT

When a solution (solute A in solvent 1) is shaken with a second solvent (solvent 2) with which it is not miscible, the solute distributes itself between the two liquid phases. When the two phases have separated again into two distinct solvent layers, an equilibrium will have been achieved such that the ratio of the concentrations of the solute in each layer defines a constant. The constant, called the **distribution coefficient** (or partition coefficient) K, is defined by

$$K = \frac{C_2}{C_1}$$

where C_1 and C_2 are the concentrations at equilibrium, in grams per liter or milligrams per milliliter of solute A in solvent 1 and in solvent 2, respectively. This relationship is a ratio of two concentrations and is independent of the actual amounts of the two solvents mixed. The distribution coefficient has a constant value for each solute considered and depends on the nature of the solvents used in each case.

Not all the solute will be transferred to solvent 2 in a single extraction unless K is very large. Usually, it takes several extractions to remove all the solute from solvent 1. In extracting a solute from a solution, it is always better to use several small portions of the second solvent than to make a single extraction with a large portion. Suppose, as an illustration, a particular extraction proceeds with a distribution coefficient of 10. The system consists of 5.0 g of organic compound dissolved in 100 mL of water (solvent 1). In this illustration, the effectiveness of three 50-mL extractions with ether (solvent 2) is compared with one 150-mL extraction with ether. In the first 50-mL extraction, the amount extracted into the ether layer is given by the following calculation. The amount of compound remaining in the aqueous phase is given by x.

$$K = 10 = \frac{C_2}{C_1} = \frac{\left(\dfrac{5.0 - x}{50} \dfrac{\text{g}}{\text{mL ether}}\right)}{\left(\dfrac{x}{100} \dfrac{\text{g}}{\text{mL H}_2\text{O}}\right)} ; \quad 10 = \frac{(5.0 - x)(100)}{50x}$$

$$500x = 500 - 100x$$
$$600x = 500$$
$$x = 0.83 \text{ g remaining in the aqueous phase}$$
$$5.0 - x = 4.17 \text{ g in the ether layer}$$

As a check on the calculation, it is possible to substitute the value 0.83 g for x in the original equation and demonstrate that the concentration in the ether layer divided by the concentration in the water layer equals the distribution coefficient.

$$\frac{\left(\dfrac{5.0 - x}{50} \dfrac{\text{g}}{\text{mL ether}}\right)}{\left(\dfrac{x}{100} \dfrac{\text{g}}{\text{mL H}_2\text{O}}\right)} = \frac{\dfrac{4.17}{50}}{\dfrac{0.83}{100}} = \frac{0.083 \text{ g/mL}}{0.0083 \text{ g/mL}} = 10 = K$$

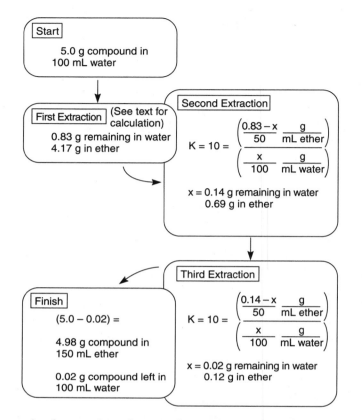

Figure 7.2   The result of extraction of 5.0 g of compound in 100 mL of water by three successive 50-mL portions of ether. Compare this result with that of Figure 7.3.

The second extraction with another 50-mL portion of fresh ether is performed on the aqueous phase which now contains 0.83 g of the solute. The amount of solute extracted is given by the calculation shown in Figure 7.2. Also shown in the figure is a calculation for a third extraction with another 50-mL portion of ether. This third extraction will transfer 0.12 g of solute into the ether layer, leaving 0.02 g of solute remaining in the water layer. A total of 4.98 g of solute will be extracted into the combined ether layers, and 0.02 g will remain in the aqueous phase.

Figure 7.3 shows the result of a *single* extraction with 150 mL of ether. As shown there, 4.69 g of solute was extracted into the ether layer, leaving 0.31 g of compound in the aqueous phase. One can see that three successive 50-mL ether extractions (Fig. 7.2) succeeded in removing 0.29 g more solute from the aqueous phase than using one 150-mL portion of either (Fig. 7.3). This differential represents 5.8% of the total material.

7.3 CHOOSING AN EXTRACTION METHOD AND A SOLVENT

Three types of apparatus are used for extractions: conical vials, centrifuge tubes, and separatory funnels. These are shown in Figure 7.4. Conical vials may be used with volumes

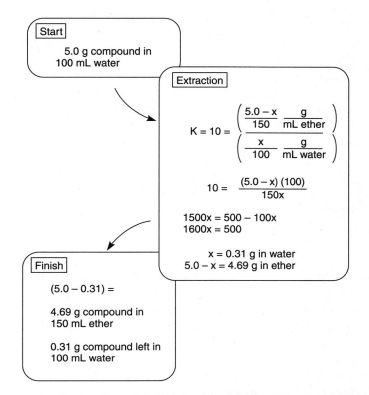

Figure 7.3 The result of extraction of 5.0 g of compound in 100 mL of water with one 150-mL portion of ether. Compare this result with that of Figure 7.2

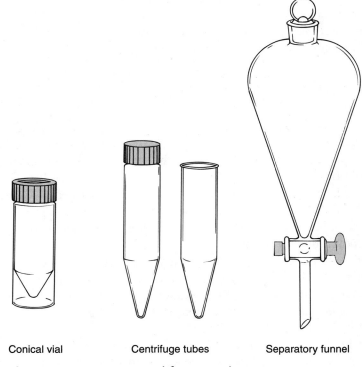

Conical vial   Centrifuge tubes   Separatory funnel

Figure 7.4 Apparatus used for extraction.

TABLE 7.1   Densities of Common
Extraction Solvents

Solvent	Density (g/mL)
Ligroin	0.67–0.69
Diethyl ether	0.71
Toluene	0.87
Water	1.00
Methylene chloride	1.330

of less than 4 mL; volumes of up to 10 mL may be handled in centrifuge tubes. A centrifuge tube equipped with a screwcap is particularly useful for extractions. The separatory funnel is used in large-scale reactions. Each type of equipment is discussed in a separate section.

Most extractions consist of an aqueous phase and an organic phase. In order to extract a substance from an aqueous phase, an organic solvent that is not miscible with water must be used. Table 7.1 lists a number of the common organic solvents that are not miscible with water and are used for extractions.

Those solvents that have a density less than that of water (1.00 g/mL) will separate as the top layer when shaken with water. Those solvents that have a greater density than water will separate into the lower layer. For instance, diethyl ether ($d = 0.71$ g/mL) when shaken with water will form the upper layer, whereas methylene chloride ($d = 1.33$ g/mL) will form the lower layer. When performing an extraction, slightly different methods are used when you wish to separate the lower layer (whether or not it is the aqueous layer or the organic layer) than when you wish to separate the upper layer.

7.4 THE SEPARATORY FUNNEL

A separatory funnel is illustrated in Figure 7.5. It is the piece of equipment used for carrying out extractions with medium to large quantities of material, and it will be used frequently in the procedures in this text. To fill the separatory funnel, support it in an iron ring attached to a ring stand. Since it is easy to break a separatory funnel by "clanking" it against the metal ring, pieces of rubber tubing are often attached to the ring to cushion the funnel as shown in Figure 7.5. These are short pieces of tubing, cut to a length of about 3 cm and slit open along their length. When slipped over the inside of the ring, they cushion the funnel in its resting place.

When beginning an extraction, the first step is to close the stopcock. (Don't forget!) Using a powder funnel (wide bore) placed in the top of the separatory funnel, fill it with both the solution to be extracted and the extraction solvent. Swirl the funnel gently by holding it by its upper neck, and then stopper it. Pick up the separatory funnel with two hands and hold it as shown in Figure 7.6. Hold then stopper in place firmly because the two immiscible liquids will build pressure when they mix, and this pressure may force the stopper out of the separatory funnel. To release this pressure, vent the funnel by hold-

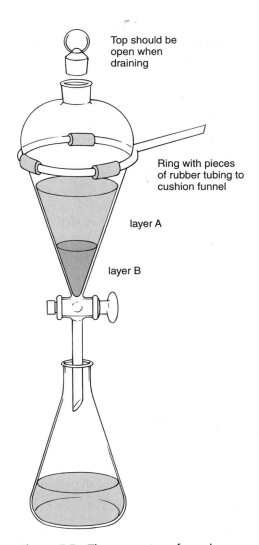

Top should be
open when
draining

Ring with pieces
of rubber tubing to
cushion funnel

layer A

layer B

Figure 7.5 The separatory funnel.

ing it upside down (hold the stopper securely) and slowly open the stopcock. Usually the rush of vapors out of the opening can be heard. Continue shaking and venting until the "whoosh" is no longer audible. Now continue shaking the mixture gently for about one minute. This can be done by inverting the funnel in a rocking motion repeatedly or, if the formation of an emulsion is not a problem (see Section 7.9, p. 699), by shaking the funnel more vigorously for less time.

There is an art to shaking and venting a separatory funnel correctly, and it usually seems awkward to the beginner. The technique is best learned by observing a person, such as your instructor, who is thoroughly familiar with the separatory funnel's use.

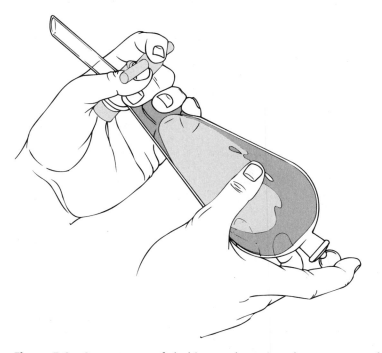

Figure 7.6 Correct way of shaking and venting the separatory funnel.

When you have finished mixing the liquids, place the separatory funnel in the iron ring and remove the top stopper immediately. The two immiscible solvents separate into two layers after a short time, and they can be separated from one another by draining most of the lower layer through the stopcock.[1] Allow a few minutes to pass so that any of the lower phase adhering to the inner glass surfaces of the separatory funnel can drain down. Open the stopcock again and allow the remainder of the lower layer to drain until the interface between the upper and lower phases just begins to enter the bore of the stopcock. At this moment, close the stopcock and remove the remaining upper layer by pouring it from the top opening of the separatory funnel.

To minimize contamination of the two layers, the lower layer should always be drained from the bottom of the separatory funnel and the upper layer poured out from the top of the funnel.

When methylene chloride is used as the extracting solvent with an aqueous phase, it will settle to the bottom and be removed through the stopcock. The aqueous layer remains in the funnel. A second extraction of the remaining aqueous layer with fresh methylene chloride may be needed.

With a diethyl ether (ether) extraction of an aqueous phase, the organic layer will form on top. Remove the lower aqueous layer through the stopcock and pour the upper

---

[1]A common error is to try to drain the separatory funnel without removing the top stopper. Under this circumstance, the funnel will not drain since a partial vacuum is created in the space above the liquid.

ether layer from the top of the separatory funnel. Pour the aqueous phase back into the separatory funnel and extract it a second time with fresh ether. The combined organic phases must be dried using a suitable drying agent (Section 7.8) before the solvent is removed.

7.5 HOW DO YOU DETERMINE WHICH ONE IS THE ORGANIC LAYER?

A common problem that you might encounter during an extraction is trying to determine which of the two layers is the organic layer and which is the aqueous (water) layer. The most common situation is when the aqueous layer is on the bottom in the presence of an upper organic layer consisting of ether, ligroin, petroleum ether, or hexane (see densities in Table 7.1). However, the aqueous layer will be on the top when you use methylene chloride as a solvent (again, see Table 7.1). Although the textbook frequently identifies the expected relative positions of the organic and aqueous layers, sometimes their actual positions are reversed. Surprises usually occur in situations where the aqueous layer contains a high concentration of sulfuric acid or a dissolved ionic compound, such as sodium chloride. Dissolved substances greatly increase the density of the aqueous layer, which may lead to the aqueous layer being found on the bottom even when coexisting with a relatively dense organic layer such as methylene chloride.

Always keep both layers until you have actually isolated the desired compound or until you are certain where your desired substance is located.

To determine if a particular layer is an aqueous one, add a few drops of water to the layer. Observe closely as you add the water to see where it goes. If the layer is water, then the drops of added water will dissolve in the aqueous layer and increase its volume. If the added water forms droplets or a new layer, however, you can assume that the suspected aqueous layer is actually organic. You can use a similar procedure to identify a suspected organic layer. This time, try adding more of the solvent, such as methylene chloride. The organic layer should increase in size, without separation of a new layer, if the tested layer is actually organic.

7.6 THE CONICAL VIAL

The 5-mL conical vial is the most useful piece of equipment for carrying out extractions with small amounts of material. Before you begin you should check the capped conical vial for leaks. Leaks are especially serious with a conical vial since you are dealing with a small volume of liquid from the very beginning. To check for leaks, place some water in the conical vial, place the Teflon liner in the cap, and screw the cap securely onto the conical vial. Shake the vial vigorously and check for leaks. Conical vials used for extractions must not be chipped on the top edge of the vial. The cap and liner seal against this top edge and, if there are chips, the seal may not be adequate. If there is a leak, try tightening the cap or replacing the liner with a new one. The liners are a sandwich with two sides, a hard Teflon surface (thin) on one side and a softer rubber surface (thick) on the other. Sometimes the rubber surface will provide a better seal to the top edge than the Teflon surface. If all else fails, replace the glass vial.

As an alternative to shaking, the contents of a conical vial can be mixed by spinning your microspatula in the vial for several minutes (twirl it between your thumb and your index finger). Another mixing technique involves repeatedly drawing the mixture into a disposable Pasteur pipet and squirting it rapidly back into the vial for a period of several minutes.

The conical vial, with a capacity of 5 mL of liquid, is commonly used in microscale procedures. In this section, we consider explicitly the method whereby the lower layer is removed by a disposable Pasteur pipet. A concrete example would be the extraction of the desired product from an aqueous layer using methylene chloride ($d = 1.33$ g/mL) as the extraction solvent.

Always remember to place a conical vial in a small beaker when it is not being held. Although the vials have a flat bottom, they can tip easily and roll off the bench.

Figure 7.7 shows the steps in using a conical vial for a separation. Figure 7.7A shows the solution to be extracted. The extraction solvent is added, and the vial is capped and

A. The aqueous solution contains the desired product.

B. Methylene chloride is used to extract the aqueous phase.

C. The Pasteur-filter tip pipet is placed in the vial.

D. The lower organic layer is removed from the aqueous phase.

E. The organic layer is transferred to a dry test tube or conical vial. The aqueous layer remains in the original extraction vial.

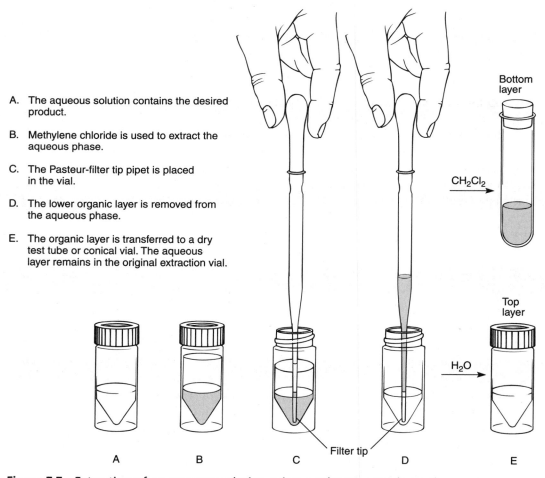

Figure 7.7 Extraction of an aqueous solution using a solvent more dense than water: Methylene chloride.

shaken (with occasional venting to relieve pressure). Figure 7.7B shows the mixture after the immiscible layers have reseparated. The lower layer is removed using a disposable Pasteur pipet (Fig. 7.7C) and transferred to a *dry* storage container such as a test tube. This operation requires a little practice. You must learn to depress the rubber bulb just the right amount to remove only the lower layer, and to leave the upper layer behind. To avoid spillage, it is best to have the storage test tube or another vial held in the same hand as the source vial as shown in Figure 7.8.

Many workers prefer to use a **filter tip pipet** when separating the layers. A filter tip pipet is a Pasteur pipet that has been modified to have a loose plug of cotton in the end. The filter tip pipet both fills and empties more slowly than an unmodified Pasteur pipet. This allows for better control when trying to separate the two layers in the conical vial. A *very small amount* of cotton or glass wool is used to make the filter plug, and it is pushed in place with a short length of copper wire. Care must be taken to avoid compressing the cotton or glass wool so tightly that liquid will not pass through the tip of the pipet.

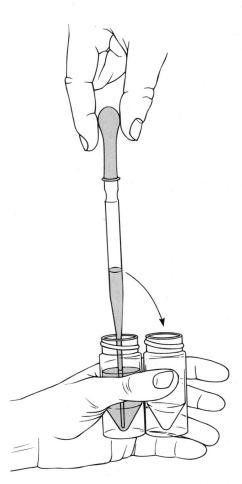

Figure 7.8 Method of holding vials while transferring liquids.

In Figure 7.7, methylene chloride, a solvent more dense than water, was illustrated. If you are extracting with a solvent less dense than water (for example, diethyl ether), you will want to save the upper layer. In that case, the extraction sequence shown in Figure 7.7 will have the positions of the water and the solvent reversed at stage E of the separation. The water layer will be in the test tube, and the ether in the conical vial.

7.7 THE SCREW-CAP CENTRIFUGE TUBE

If you require an extraction that uses a larger volume than a conical vial can accommodate (5 mL), a centrifuge tube can often be used. A commonly available size of centrifuge tube has a volume of about 15 mL, and is supplied with a screw cap. In performing an extraction with a screwcapped centrifuge tube, the same procedures outlined for the conical vial (Section 7.6) are used. As is the case for a conical vial, the tapered bottom of the centrifuge tube makes it easy to withdraw the lower layer with a Pasteur pipet.

A centrifuge tube has a large advantage over other methods of extraction. If an emulsion (Section 7.9) forms, you can use a centrifuge to aid in the separation of the layers.

You should check the capped centrifuge tube for leaks by filling it with water and shaking it vigorously. If it leaks, try replacing the cap with a different one. A **vortex mixer,** if available, provides an alternative to shaking the tube. In fact, a vortex mixer works well with a variety of containers, including small flasks, test tubes, conical vials, and centrifuge tubes. You start the mixing action on a vortex mixer by holding the test tube or other container on one of the neoprene pads. The unit mixes the sample by high-frequency vibration.

7.8 DRYING AGENTS

After an organic solvent has been shaken with an aqueous solution, it will be "wet"; that is, it will have dissolved some water even though its miscibility with water is not great. The amount of water dissolved varies from solvent to solvent: diethyl ether represents a solvent in which a fairly large amount of water dissolves (about 1.5% by weight). To remove water from a wet ether solution obtained from an extraction, a **drying agent** is used. A drying agent is an insoluble, *anhydrous* inorganic salt that acquires waters of hydration when exposed to moist air or a wet solution:

Insoluble   **Insoluble**

$$Na_2SO_4(s) + WET\ SOLUTION\ (nH_2O) \longrightarrow Na_2SO_4 \cdot nH_2O\ (s) + DRY\ SOLUTION$$

**Anhydrous   Hydrated
drying agent   drying agent**

The insoluble drying agent is placed directly into the solution, where it acquires water molecules and becomes hydrated. If enough drying agent is used, all of the water can be removed from a wet solution, making it "dry," or free of water. The following anhydrous salts are commonly used: sodium sulfate, magnesium sulfate, calcium chloride, calcium sulfate (Drierite), and potassium carbonate.

These drying agents differ in their properties and applications. Two major properties are defined for a drying agent: capacity and completeness. **Capacity** refers to the amount of water a drying agent absorbs per unit weight (usually per gram). **Completeness** refers to a compound's effectiveness in removing all of the water from a solution by the time equilibrium has been reached with the solution. Sodium and magnesium sulfates both absorb a large amount of water (capacity), but magnesium sulfate usually leaves less water in the solution (completeness). You can equalize this difference by drying with several smaller portions of sodium sulfate, where you might use one large portion of magnesium sulfate. A third sulfate compound, calcium sulfate, dries a solution completely, but has a low capacity (you would have to use a lot more of it than of magnesium sulfate, but a single drying might suffice).

Many magnesium compounds are good drying agents, but magnesium ion is a strong Lewis acid, and it can sometimes cause rearrangements of sensitive compounds such as epoxides. Calcium chloride is a good drying agent but cannot be used with many compounds containing oxygen or nitrogen because it forms complexes. Calcium chloride absorbs methanol and ethanol in addition to water, so it is also useful for removing these solvents when they are present as impurities or they make a solution "wet." Potassium carbonate is a base and is used for drying solutions of basic substances. Potassium hydroxide is often used to dry solutions of aromatic amines.

Anhydrous sodium sulfate is the most widely used drying agent. The granular variety is recommended, because it is easier to remove the dried solution from it than from the powdered variety. Sodium sulfate is mild and effective. It will remove water from most common solvents, with the possible exception of diethyl ether, in which case a prior drying with saturated salt solution may be advised (see p. 699). Sodium sulfate must be used at room temperature to be effective; it cannot be used with boiling solutions. Table 7.2 compares the various common drying agents.

Large-Scale Reactions.   To dry a large amount of solution, you should add enough granular anhydrous sodium sulfate to give a 1–3 mm layer on the bottom of the flask, depending on the volume of the solution. Dry the solution for at least 15 minutes, occasionally swirling the flask. The mixture is dry if it appears clear and shows the common signs of a dry solution given in Table 7.3.

If the solution remains cloudy after treatment with the first batch of drying agent, add more drying agent and repeat the drying procedure. If the drying agent clumps badly, with no drying agent that will flow freely when the flask is swirled, you should transfer the solution to a clean, dry flask and add a fresh portion of drying agent. If droplets of water are present, you should also transfer the solution to a new container and add a fresh quantity of drying agent. You should not add more drying agent if a puddle (water layer) forms. Instead, separate the layers, using a separatory funnel if necessary, and then add fresh drying agent to the organic layer, placed in a new container. If the first batch of drying agent does not completely dry the solution, it is not at all unusual to have to transfer (decant) the solution to a clean, dry flask and add a fresh portion of drying agent.

TABLE 7.2 Common Drying Agents

	Acidity	Hydrated	Capacity*	Complete-ness†	Rate‡	Use
Magnesium sulfate	Neutral	$MgSO_4 \cdot 7H_2O$	High	Medium	Rapid	General
Sodium sulfate	Neutral	$Na_2SO_4 \cdot 7H_2O$ $Na_2SO_4 \cdot 10H_2O$	High	Low	Medium	General
Calcium chloride	Neutral	$CaCl_2 \cdot 2H_2O$ $CaCl_2 \cdot 6H_2O$	Low	High	Rapid	Hydro-carbons Halides
Calcium sulfate (Drierite)	Neutral	$CaSO_4 \cdot \frac{1}{2}H_2O$ $CaSO_4 \cdot 2H_2O$	Low	High	Rapid	General
Potassium carbonate	Basic	$K_2CO_3 \cdot 1\frac{1}{2}H_2O$ $K_2CO_3 \cdot 2H_2O$	Medium	Medium	Medium	Amines, esters, bases, ketones
Potassium hydroxide	Basic	—	—	—	Rapid	Amines only
Molecular sieves (3 or 4 Å)	Neutral	—	High	Extremely high	—	General

*Amount of water removed per given weight of drying agent.
†Refers to amount of H_2O still in solution at equilibrium with drying agent.
‡Refers to rate of action (drying).

When the solution is dry, the drying agent should be removed by using decantation (pouring carefully to leave the drying agent behind). With granular sodium sulfate, decantation is quite easy to perform because of the size of the drying agent particles. If a powdered drying agent, such as magnesium sulfate, is used, it may be necessary to use a gravity filtration (Technique 4, Section 4.1B, p. 636) to remove the drying agent. The solvent is removed by distillation (Technique 8, Section 8.3, p. 710) or evaporation (Technique 3, Section 3.11, p. 630)

Microscale Reactions. Before attempting to dry an organic layer, check closely to see that there are no visible signs of water. If you see droplets of water in the organic layer or water droplets clinging to the sides of the conical vial or test tube, transfer the organic layer

---

TABLE 7.3 Common Signs That Indicate a Solution Is Dry

---

1. There are no visible water droplets on the side of flask or suspended in solution.

2. There is not a separate layer of liquid or a "puddle."

3. The solution is clear, not cloudy. Cloudiness indicates water is present.

4. The drying agent (or a portion of it) flows freely on the bottom of the container when stirred or swirled and does not "clump" together as a solid mass.

---

with a *dry* Pasteur pipet to a *dry* container before adding any drying agent. Now add one spatulaful of granular anhydrous sodium sulfate (or other drying agent) from the V-grooved end of a microspatula into a solution contained in a conical vial or test tube. If all the drying agent "clumps," add another spatulaful of sodium sulfate. Dry the solution for at least 15 minutes. Stir the mixture occasionally with a spatula during that period. The mixture is dry if there are no visible signs of water and the drying agent flows freely in the container when stirred with a microspatula. The solution should not be cloudy. Add more drying agent if necessary. You should not add more drying agent if a "puddle" (water layer) forms or if drops of water are visible. Instead, you should transfer the organic layer to a dry container before adding fresh drying agent. When dry, use a *dry* Pasteur pipet or a *dry* filter-tip pipet (Technique 4, Section 4.6, p. 645) to remove the solution from the drying agent and transfer the solution to a *dry* conical vial. Rinse the drying agent with a small amount of fresh solvent and transfer this solvent to the vial containing the solution. Remove the solvent by evaporation using heat and a stream of air or nitrogen (Technique 3, Section 3.11, p. 630).

An alternative method of drying an organic phase is to pass it through a filtering pipet (Technique 4, Section 4.1C, p. 638) that has been packed with a small amount (ca. 2 cm) of drying agent. Again, the solvent is removed by evaporation.

Saturated Salt Solution. At room temperature, diethyl ether (ether) dissolves 1.5% by weight of water, and water dissolves 7.5% of ether. Ether, however, dissolves a much smaller amount of water from a saturated aqueous sodium chloride solution. Hence, the bulk of water in ether, or ether in water, can be removed by shaking it with a saturated aqueous sodium chloride solution. A solution of high ionic strength is usually not compatible with an organic solvent and forces separation of it from the aqueous layer. The ether phase (organic layer) will be on top, and the saturated sodium chloride solution will be on the bottom ($d = 1.2$ g/mL). After removing the organic phase from the aqueous sodium chloride, dry the organic layer completely with sodium sulfate or with one of the other drying agents listed in Table 7.2.

7.9 EMULSIONS

An **emulsion** is a colloidal suspension of one liquid in another. Minute droplets of an organic solvent are often held in suspension in an aqueous solution when the two are mixed or shaken vigorously; these droplets form an emulsion. This is especially true if any gummy or viscous material was present in the solution. Emulsions are often encountered in performing extractions. Emulsions may require a long time to separate into two layers and are a nuisance to the organic chemist.

Fortunately, several techniques may be used to break a difficult emulsion once it has formed.

1. Often, an emulsion will break up if it is allowed to stand for some time. Patience is important here. Gently stirring with a stirring rod or spatula may also be useful.
2. If one of the solvents is water, adding a saturated aqueous sodium chloride solution will help destroy the emulsion. This makes the aqueous and organic layers less compatible, thereby forcing separation.

3. With microscale and many small-scale experiments, the mixture may be transferred to a centrifuge tube. The emulsion will often break during centrifugation. Remember to place another tube filled with water on the opposite of the centrifuge to balance it.

4. Adding a very small amount of a water-soluble detergent may also help. This method has been used in the past for combating oil spills. The detergent helps to solubilize the tightly bound oil droplets.

5. Gravity filtration (see Technique 4, Section 4.1, p. 634) may help to destroy an emulsion by removing a gummy polymeric substance. With large-scale reactions, you might try filtering the mixture through a fluted filter (Technique 4, Section 4.1B, p. 636) or a piece of cotton. With microscale reactions, a filtering pipet may work (Technique 4, Section 4.1C, p. 638). In many cases, once the gum is removed, the emulsion breaks up rapidly.

6. If you are using a separatory funnel, you might try to use a gentle swirling action in the funnel to help break an emulsion. Gently stirring with a stirring rod may also be useful.

When you know through prior experience that a mixture may form a difficult emulsion, you should avoid shaking the mixture vigorously. When using conical vials for extractions, it may be better to use a magnetic spin vane for mixing and not shake the mixture at all. When using separatory funnels, extractions should be performed with gentle swirling instead of shaking or with several gentle inversions of the separatory funnel. Do not shake the separatory funnel vigorously in these cases. It is important that you use a longer extraction period if the more gentle techniques described in this paragraph are being employed. Otherwise, you will not transfer all the material from the first phase to the second one.

7.10 PURIFICATION AND SEPARATION METHODS

In nearly all the synthetic experiments undertaken in this textbook, a series of operations involving extractions is used after the actual reaction has been concluded. These extractions form an important part of the purification. Using them, the desired product is separated from unreacted starting materials or from undesired side products in the reaction mixture. These extractions may be grouped into three categories, depending on the nature of the impurities they are designed to remove.

The first category involves extracting or "washing" an organic mixture with water. Water washes are designed to remove highly polar materials, such as inorganic salts, strong acids or bases, and low-molecular-weight, polar substances including alcohols, carboxylic acids, and amines. Many organic compounds containing fewer than five carbons are water-soluble. Water extractions are also used immediately following extractions of a mixture with either acid or base to ensure that all traces of acid or base have been removed.

The second category concerns extraction of an organic mixture with a dilute acid, usually 5 to 10% hydrochloric acid. Acid extractions are intended to remove basic impurities, especially such basic impurities as organic amines. The bases are converted to their corresponding cationic salts by the acid used in the extraction. If an amine is one of the reactants, or if pyridine or another amine is a solvent, such an extraction might be used to remove any excess amine present at the end of a reaction.

$$RNH_2 + HCl \longrightarrow RNH_3^+Cl^-$$
(Water-soluble salt)

Cationic salts are usually soluble in the aqueous solution, and they are thus extracted from the organic material. A water extraction may be used immediately following the acid extraction to ensure that all traces of the acid have been removed from the organic material.

The third category is extraction of an organic mixture with a dilute base, usually 5% sodium bicarbonate, although extractions with dilute sodium hydroxide can also be used. Such basic extractions are intended to convert acidic impurities, such as organic acids, to their corresponding anionic salts. For example, in the preparation of an ester, a sodium bicarbonate extraction might be used to remove any excess carboxylic acid that is present.

$$RCOOH + NaHCO_3 \longrightarrow RCOO^-Na^+ + H_2O + CO_2$$
(pKa ~ 5)   **(Water-soluble salt)**

Anionic salts, being highly polar, are soluble in the aqueous phase. As a result, these acidic impurities are extracted from the organic material into the basic solution. A water extraction may be used after the basic extraction to ensure that all the base has been removed from the organic material.

Occasionally, phenols may be present in a reaction mixture as impurities, and removing them by extraction may be desired. Because phenols, although they are acidic, are about 10^5 times less acidic than carboxylic acids, basic extractions may be used to separate phenols from carboxylic acids by a careful selection of the base. If sodium bicarbonate is used as a base, carboxylic acids are extracted into the aqueous base, but phenols are not. Phenols are not sufficiently acidic to be deprotonated by the weak base, bicarbonate. Extraction with sodium hydroxide, on the other hand, extracts both carboxylic acids and phenols into the aqueous basic solution, because hydroxide ion is a sufficiently strong base to deprotonate phenols.

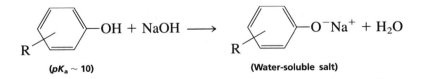

(pK_a ~ 10)   **(Water-soluble salt)**

It would be a useful exercise for you to examine the instructions for some of the preparative experiments given in the textbook. While you are examining these procedures, you should try to identify which impurities are being removed at each extraction step. Mixtures of acidic, basic, and neutral compounds are easily separated by extraction techniques. One such example is shown in Figure 7.9.

Materials that have been extracted can be regenerated by neutralizing the extraction reagent. If an acidic material has been extracted with aqueous base, the material can be

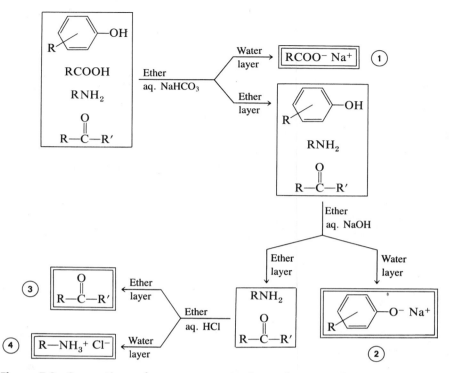

Figure 7.9   Separating a four-component mixture by extraction.

regenerated by acidifying the extract until the solution becomes acidic to blue litmus. The material will separate from the acidified solution. A basic material can be recovered from the acidic extract by adding base to the extract. These substances can then be removed from the neutralized aqueous solutions by extraction with an organic solvent such as ether. After the ether phase is dried with a drying agent, evaporation of the ether yields the isolated compounds.

7.11 CONTINUOUS SOLID–LIQUID EXTRACTION

The technique of liquid–liquid extraction was described in Sections 7.1–7.7. In this section, solid–liquid extraction is described. Solid–liquid extraction is often used to extract a solid natural product from a natural source, such as a plant. A solvent is chosen that selectively dissolves the desired compound but that leaves behind the undesired insoluble solid. A continuous solid–liquid extraction apparatus, called a Soxhlet extractor, is commonly used in a research laboratory (Fig. 7.10).

As shown in Figure 7.10, the solid to be extracted is placed in a thimble made from filter paper, and the thimble is inserted into the central chamber. A low boiling solvent, such as diethyl ether, is placed in the round-bottom distilling flask and is heated to reflux. The vapor rises through the left sidearm into the condenser where it liquefies. The con-

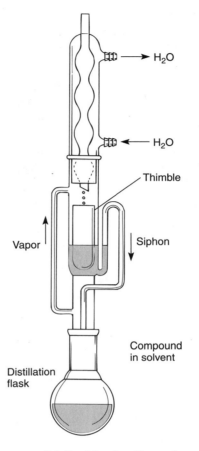

Figure 7.10   Continuous solid–liquid extraction using a Soxhlet extractor.

densate (liquid) drips into the thimble containing the solid. The hot solvent begins to fill the thimble and extracts the desired compound from the solid. Once the thimble is filled with solvent, the sidearm on the right acts as a siphon, and the solvent, which now contains the dissolved compound, drains back into the distillation flask. The vaporization, condensation, extraction, siphoning process is repeated hundreds of times, and the desired product is concentrated in the distillation flask. The product is concentrated in the flask, because it has a boiling point higher than that of the solvent or because it is a solid.

7.12 CONTINUOUS LIQUID–LIQUID EXTRACTION

When a product is very soluble in water, it is often difficult to extract using the techniques described in Sections 7.4–7.7, because of an unfavorable distribution coefficient. In this case, you need to extract the aqueous solution numerous times with fresh batches of an immiscible organic solvent to remove the desired product from water. A less labor-intensive technique involves the use of a continuous liquid–liquid extraction apparatus. One type of extractor, used with solvents that are less dense than water, is shown in Figure 7.11. Diethyl ether is usually the solvent of choice.

The aqueous phase is placed in the extractor, which is then filled with diethyl ether up to the sidearm. The round-bottom distillation flask is partially filled with ether. The ether is heated to reflux in the round-bottom flask, and the vapor is liquefied in the water-cooled condenser. The ether drips into the central tube, passes through the porous sintered glass tip, and flows through the aqueous layer. The solvent extracts the desired compound from the aqueous phase, and the ether is recycled back into the round-bottom flask. The product is concentrated in the flask. The extraction is rather inefficient and must be placed in operation for at least 24 hours to remove the compound from the aqueous phase.

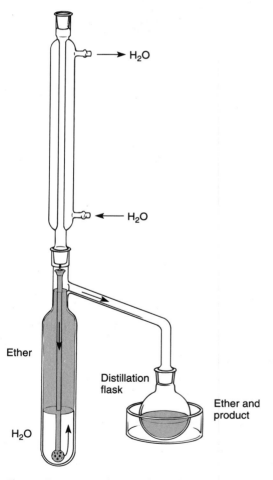

Figure 7.11 Continuous liquid–liquid extraction using a solvent less dense than water.

PROBLEMS

1. Suppose solute A has a distribution coefficient of 1.0 between water and diethyl ether. Demonstrate that if 100 mL of a solution of 5.0 g of A in water were extracted with two 25-mL portions of ether, a smaller amount of A would remain in the water than if the solution were extracted with one 50-mL portion of ether.

2. Write an equation to show how you could recover the parent compounds from their respective salts (1, 2, and 4) shown in Figure 7.9.

3. Aqueous hydrochloric acid was used *after* the sodium bicarbonate and sodium hydroxide extractions in the separation scheme shown in Figure 7.9. Is it possible to use this reagent earlier in the separation scheme so as to achieve the same overall result? If so, explain where you would perform this extraction.

4. Using aqueous hydrochloric acid, sodium bicarbonate, or sodium hydroxide solutions, devise a separation scheme using the style shown in Figure 7.9 to separate the following two-component mixtures. All the substances are soluble in ether. Also indicate how you would recover each of the compounds from their respective salts.

(**a**) Give two different methods for separating this mixture.

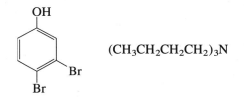

(**b**) Give two different methods for separating this mixture.

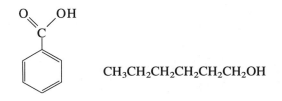

(**c**) Give one method for separating this mixture.

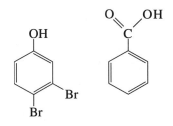

5. Solvents other than those in Table 7.1 may be used for extractions. Determine the relative positions of the organic layer and the aqueous layer in a conical vial or separatory funnel after shaking each of the following solvents with an aqueous phase. Find the densities for each of these solvents in a handbook (Technique 20. Section 20.1, p. 861).

(a) 1,1,1-Trichloroethane

(b) Hexane

6. A student prepares ethyl benzoate by the reaction of benzoic acid with ethanol using a sulfuric acid catalyst. The following compounds are found in the crude reaction mixture: ethyl benzoate (major component), benzoic acid, ethanol, and sulfuric acid. Using a handbook, obtain the solubility properties in water for each of these compounds (Technique 20, Section 20.1, p. 861). Indicate how you would remove benzoic acid, ethanol, and sulfuric acid from ethyl benzoate. At some point in the purification, you should also use an aqueous sodium bicarbonate solution.

7. Calculate the weight of water that could be removed from a wet organic phase using 50.0 mg of magnesium sulfate. Assume that it gives the hydrate listed in Table 7.2.

8. Explain the following laboratory instructions in a procedure:

(a) "Wash the organic layer with 5.0 mL of 5% aqueous sodium bicarbonate."

(b) "Extract the aqueous layer three times with 2-mL portions of methylene chloride."

9. Just prior to drying an organic layer with a drying agent, you notice water droplets in the organic layer. What should you do next?

10. What should you do if there is some question about which layer is the organic one during an extraction procedure?

11. Saturated aqueous sodium chloride ($d = 1.2$ g/mL) is added to the following mixtures in order to dry the organic layer. Which layer is likely to be on the bottom in each case?

(a) Sodium chloride layer or a layer containing a high-density organic compound dissolved in methylene chloride ($d = 1.4$ g/mL)

(b) Sodium chloride layer or a layer containing a low-density organic compound dissolved in methylene chloride ($d = 1.1$ g/mL)

T E C H N I Q U E   8

Simple Distillation

Distillation is the process of vaporizing a liquid, condensing the vapor, and collecting the condensate in another container. This technique is very useful for separating a liquid mixture when the components have different boiling points, or when one of the components will not distill. It is one of the principal methods of purifying a liquid. Four basic distillation methods are available to the chemist: simple distillation, vacuum distillation (distillation at reduced pressure), fractional distillation, and steam distillation. This technique chapter will discuss simple distillation. Vacuum distillation will be discussed in Technique 9. Fractional distillation will be discussed in Technique 10, and steam distillation will be discussed in Technique 11.

8.1 THE EVOLUTION OF DISTILLATION EQUIPMENT

There are probably more types and styles of distillation apparatus than exist for any other technique in chemistry. Over the centuries, chemists have devised just about every conceivable design. The earliest known types of distillation apparatus were the **alembic** and the **retort** (Fig. 8.1). They were used by alchemists in the Middle Ages and the

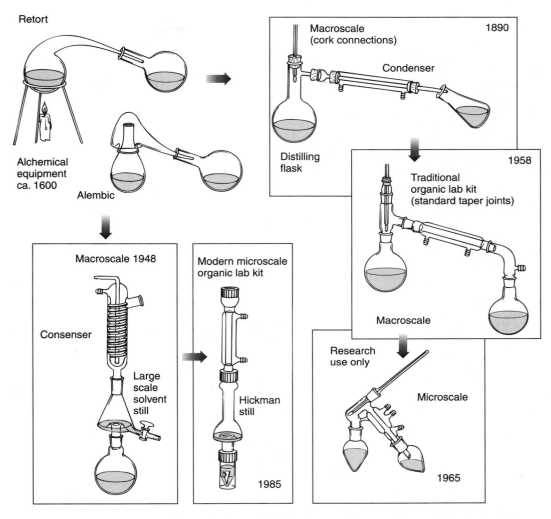

Figure 8.1   Some stages in the evolution of distillation equipment from alchemical equipment (dates represent approximate time of use).

Renaissance, and probably even earlier by Arabic chemists. Most other distillation equipment has evolved as variations on these designs.

Figure 8.1 shows several stages in the evolution of distillation equipment as it relates to the organic laboratory. It is not intended to be a complete history; rather, it is representative. Up until recent years, equipment based on the retort design was common in the laboratory. Although the retort itself was still in use early in this century, it had evolved by that time into the distillation flask and water-cooled condenser combination. This early equipment was connected with drilled corks. By 1958, most introductory laboratories were beginning to use "organic lab kits" that included glassware connected by standard-taper glass joints. The original lab kits contained large ͳ 24/40 joints. Within a short time, they became smaller with ͳ 19/22 and even ͳ 14/20 joints. These later kits are still being used today in many "macroscale" laboratory courses such as yours.

In the 1960s, researchers developed even smaller versions of these kits for working at the "microscale" level (in Fig. 8.1, see the box labeled 1965, research use only), but this glassware is generally too expensive to use in an introductory laboratory. However, in the mid 1980s, several groups developed a different style of microscale distillation equipment based on the alembic design (see the box labeled 1985, Hickman still). This new microscale equipment has ℑ 14/10 standard-taper joints, threaded outer joints with screwcap connectors, and an internal O-ring for a compression seal. Microscale equipment similar to this is now used in many introductory courses. The advantages of this glassware are that there is less material used (lower cost), lower personal exposure to chemicals, and less waste generated. Because both types of equipment are in use today, after we describe macroscale equipment, we will also show the equivalent microscale distillation apparatus.

8.2 DISTILLATION THEORY

In the traditional distillation of a pure substance, vapor rises from the distillation flask and comes into contact with a thermometer that records its temperature. The vapor then passes through a condenser, which reliquefies the vapor and passes it into the receiving flask. The temperature observed during the distillation of a **pure substance** remains constant throughout the distillation so long as both vapor *and* liquid are present in the system (see Fig. 8.2A). When a **liquid mixture** is distilled, often the temperature does not remain constant but increases throughout the distillation. The reason for this is that the composition of the vapor that is distilling varies continuously during the distillation (see Fig. 8.2B).

For a liquid mixture, the composition of the vapor in equilibrium with the heated solution is different from the composition of the solution itself. This is shown in Figure 8.3, which is a phase diagram of the typical vapor–liquid relation for a two-component system (A + B).

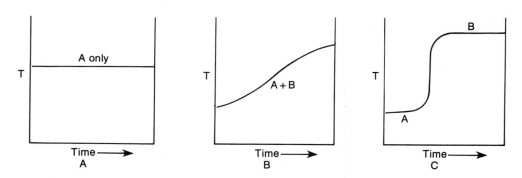

Figure 8.2   Three types of temperature behavior during a simple distillation. (A) A single pure component. (B) Two components of similar boiling points. (C) Two components with widely differing boiling points. Good separations are achieved in A and C.

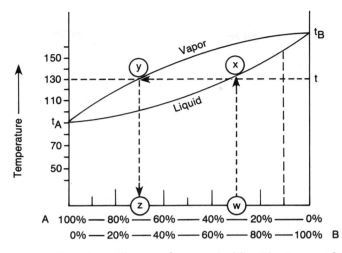

Figure 8.3   Phase diagram for a typical liquid mixture of two components.

On this diagram, horizontal lines represent constant temperatures. The upper curve represents vapor composition, and the lower curve represents liquid composition. For any horizontal line (constant temperature), like that shown at t, the intersections of the line with the curves give the compositions of the liquid and the vapor that are in equilibrium with each other at that temperature. In the diagram, at temperature t, the intersection of the curve at x indicates that liquid of composition w will be in equilibrium with vapor of composition z, which corresponds to the intersection at y. Composition is given as a mole percentage of A and B in the mixture. Pure A, which boils at temperature t_A, is represented at the left. Pure B, which boils at temperature t_B, is represented on the right. For either pure A or pure B, the vapor and liquid curves meet at the boiling point. Thus, either pure A or pure B will distill at a constant temperature (t_A or t_B). Both the vapor and the liquid must have the same composition in either of these cases. This is not the case for mixtures of A and B.

A mixture of A and B of composition w will have the following behavior when heated. The temperature of the liquid mixture will increase until the boiling point of the mixture is reached. This corresponds to following line wx from w to x, the boiling point of the mixture t. At temperature t the liquid begins to vaporize, which corresponds to line xy. The vapor has the composition corresponding to z. In other words, the first vapor obtained in distilling a mixture of A and B does not consist of pure A. It is richer in A than the original mixture but still contains a significant amount of the higher boiling component B, *even from the very beginning of the distillation.* The result is that it is never possible to separate a mixture completely by a simple distillation. However, in two cases it is possible to get an acceptable separation into relatively pure components. In the first case, if the boiling points of A and B differ by a large amount ($>100°$), and if the distillation is carried out carefully, it will be possible to get a fair separation of A and B. In the second case, if A contains a fairly small amount of B ($<10\%$), a reasonable separation of A from B can be achieved. When the boiling-point differences are not large, and when highly pure components are desired, it is necessary to perform a **fractional distil-**

lation. Fractional distillation is described in Technique 10, where the behavior during a simple distillation is also considered in detail. Note only that as vapor distills from the mixture of composition w (Fig. 8.3), it is richer in A than is the solution. Thus, the composition of the material left behind in the distillation becomes richer in B (moves to the right from w toward pure B in the graph). A mixture of 90% B (dotted line on the right side in Fig. 8.3) has a higher boiling point than at w. Hence, the temperature of the liquid in the distillation flask will increase during the distillation, and the composition of the distillate will change (as is shown in Fig. 8.2B).

When two components that have a large boiling-point difference are distilled, the temperature remains constant while the first component distills. If the temperature remains constant, a relatively pure substance is being distilled. After the first substance distills, the temperature of the vapors rises, and the second component distills, again at a constant temperature. This is shown in Figure 8.2C. A typical application of this type of distillation might be an instance of a reaction mixture containing the desired component A (bp 140°C) contaminated with a small amount of undesired component B (bp 250°C) and mixed with a solvent such as diethyl ether (bp 36°C). The ether is removed easily at low temperature. Pure A is removed at a higher temperature and collected in a separate receiver. Component B can then be distilled, but it usually is left as a residue and not distilled. This separation is not difficult and represents a case where simple distillation might be used to advantage.

8.3 SIMPLE DISTILLATION—STANDARD APPARATUS

For a simple distillation, the apparatus shown in Figure 8.4 is used. Six pieces of specialized glassware are used:

1. Distilling flask
2. Distillation head
3. Thermometer adapter
4. Water condenser
5. Vacuum takeoff adapter
6. Receiving flask

The apparatus is usually heated electrically, using a heating mantle. The distilling flask, condenser, and vacuum adapter should be clamped. Two different methods of clamping this apparatus were shown in Technique 3 (Fig. 3.4, page 616 and Fig. 3.6, page 617). The receiving flask should be supported by removable wooden blocks or a wire gauze on an iron ring attached to a ring stand. The various components are each discussed below, along with some other important points.

Distilling Flask.   The distilling flask should be a round-bottom flask. This type of flask is designed to withstand the required input of heat, and to accommodate the boiling action. It gives a maximized heating surface. The size of the distilling flask should be chosen so that it is never filled more than two-thirds full. When the flask is filled beyond this point, the neck constricts and "chokes" the boiling action, resulting in bumping. The sur-

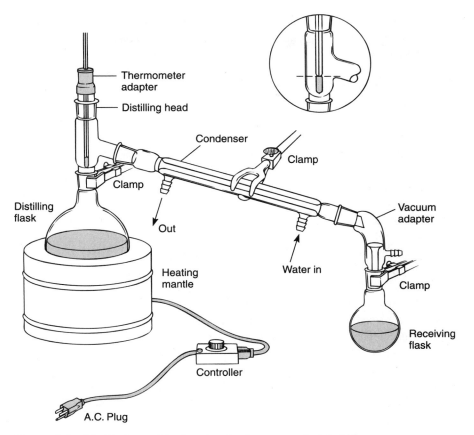

Figure 8.4 Distillation with the standard macroscale lab kit.

face area of the boiling liquid should be kept as large as possible. However, too large a distilling flask should also be avoided. With too large a flask, the **holdup** is excessive; the holdup is the amount of material that cannot distill since some vapor must fill the empty flask. When you cool the apparatus at the end, this material drops back into the distilling flask.

Boiling Stones. A boiling stone (Technique 3, Section 3.6, page 622) should be used during distillation to prevent bumping. As an alternative, the liquid being distilled may be rapidly stirred using a magnetic stirrer and stir bar (Technique 3, Section 3.5, page 621). If you forget a boiling stone, cool the mixture before adding it. If you add a boiling stone to a hot superheated liquid, it may "erupt" into vigorous boiling, breaking your apparatus and spilling hot solvent everywhere.

Grease. In most cases, it is unnecessary to grease standard-taper joints for a simple distillation. The grease makes cleanup more difficult, and it may contaminate your product.

Distillation Head. The distillation head directs the distilling vapors into the condenser and allows the connection of a thermometer via the thermometer adapter. The thermometer should be positioned in the distillation head so that it is placed directly in the stream of vapor that is distilling. This can be accomplished if the entire bulb of the thermometer is positioned *below* the side arm of the distilling head (see the circular inset in Fig. 8.4). The entire bulb must be immersed in the vapor to achieve an accurate temperature reading. When distilling, you should be able to see a reflux ring (Technique 3, Section 3.4, page 619) positioned well above both the thermometer bulb and the bottom of the side arm.

Thermometer Adapter. The thermometer adapter connects to the top of the distillation head (see Fig. 8.4). There are two parts to the thermometer adapter: a glass joint with an open rolled edge on the top, and a rubber adapter that fits over the rolled edge and holds the thermometer. The thermometer fits in a hole in the top of the rubber adapter and can be adjusted upward and downward by sliding it in the hole. Adjust the bulb to a point below the side arm.

Water Condenser. The joint between the distillation head and the water condenser is the joint most prone to leak in this entire apparatus. Since the distilling liquid is both hot and vaporized when it reaches this joint, it will leak out of any small opening between the two joint surfaces. The odd angle of the joint, neither vertical or horizontal, also makes a good connection more difficult. Be sure this joint is well sealed. If possible, use one of the plastic joint clips described in Technique 3, Figure 3.5, page 617. Otherwise, adjust your clamps to be sure that the joint surfaces are pressed together, and not pulled apart.

The condenser will remain full of cooling water only if the water flows *upward,* not downward. The water input hose should be connected to the lower opening in the jacket, and the exit hose should be attached to the upper opening. Place the other end of the exit hose in a sink. A moderate water flow will perform a good deal of cooling. A high rate of water flow may cause the tubing to pop off the joints and cause a flood. If you hold the exit hose horizontally and point the end into a sink, the flow rate is correct if the water stream continues horizontally for about two inches before bending downward.

If a distillation apparatus is to be left untended for a period of time, it is a good idea to wrap copper wire around the ends of the tubing and twist it tight. This will help to prevent the hoses from popping off of the connectors if there is an unexpected water pressure change.

Vacuum Adapter. In a simple distillation the vacuum adapter is not connected to a vacuum, but is left open. It is merely an opening to the outside air so that pressure does not build up in the distillation system. If you plug this opening, you will have a **closed system** (no outlet). It is always dangerous to heat a closed system. Enough pressure can build up in the closed system so that it can explode. The vacuum adapter, in this case, merely directs the distillate into the receiving, or collection, flask.

If the substance you are distilling is water sensitive, you can attach a calcium chloride drying tube to the vacuum connection to protect the freshly distilled liquid from atmospheric water vapor. Air that enters the apparatus will have to pass through the calcium chloride and be dried. Depending on the severity of the problem, drying agents other than calcium chloride may also be used.

The vacuum adapter has a disturbing tendency to obey the laws of Newtonian physics and fall off the slanted condenser onto the desk and break. If they are available, it is a good idea to use plastic joint clips on both ends of this piece. The top clip will secure the vacuum adapter to the condenser, and the bottom clip will secure the receiving flask, preventing it from falling.

Rate of Heating.   The rate of heating for the distillation can be adjusted to the proper rate of **takeoff,** the rate at which distillate leaves the condenser, by watching drops of liquid emerge from the bottom of the vacuum adapter. A rate of from one to three drops per second is considered a proper rate of takeoff for most applications. At a greater rate, equilibrium is not established within the distillation apparatus, and the separation may be poor. A slower rate of takeoff is also unsatisfactory since the temperature recorded on the thermometer is not maintained by a constant vapor stream, thus leading to an inaccurately low observation of the boiling point.

Receiving Flask.   The receiving flask collects the distilled liquid, and is usually a round-bottom flask. If the liquid you are distilling is extremely volatile, and there is danger of losing some of it to evaporation, it is sometimes advisable to cool the receiving flask in an ice-water bath.

Fractions.   The material being distilled is called the **distillate.** Frequently a distillate is collected in contiguous portions, called **fractions.** This is accomplished by replacing the collection flask with clean ones at regular intervals. If a small amount of liquid is collected at the beginning of a distillation and not saved or used further, it is called a **forerun.** Subsequent fractions will have higher boiling ranges, and each fraction should be labeled with its correct boiling range when the fraction is taken. For a simple distillation of a pure material, most of the material will be collected in a single large **midrun** fraction, with only a small forerun. In some small-scale distillations, the volume of the forerun will be so small that you will not be able to collect it separately from the midrun fraction. The material left behind is called the **residue.** It is usually advised that you discontinue a distillation before the distilling flask becomes empty. Typically, the residue becomes increasingly dark in color during distillation, and it frequently contains thermal decomposition products. In addition, a dry residue may explode on overheating, or the flask may melt or crack when it becomes dry. Don't distill until the distilling flask is completely dry!

8.4 MICROSCALE AND SEMI-MICROSCALE EQUIPMENT

When you wish to distill quantities that are smaller than 4–5 mL, different equipment is required. What you use depends on how small a quantity you wish to distill.

A.   SEMI-MICROSCALE

One possibility is to use equipment identical in style to that used with conventional macroscale procedures, but to "downsize" it using ⊤ 14/10 joints. The major manufacturers do make distillation heads and vacuum takeoff adapters with ⊤ 14/10 joints. This equipment will allow you to handle quantities of from 5 to 15 mL. An example of such a "semi-microscale" apparatus is given in Figure 8.5. Although the manufacturers make ⊤ 14/10 condensers, the condenser has been left out in this example. This can be done if the material to be distilled is not extremely volatile or is high boiling. It is also possible to omit the condenser if you not have a large amount of material and can cool the receiving flask in an ice-water bath as shown in the figure.

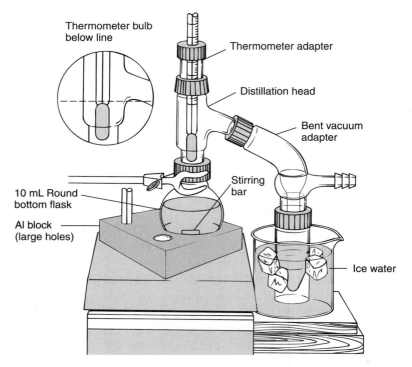

Thermometer bulb
below line

Thermometer adapter

Distillation head

Bent vacuum
adapter

10 mL Round
bottom flask

Stirring
bar

Al block
(large holes)

Ice water

Figure 8.5 Semi-microscale distillation.

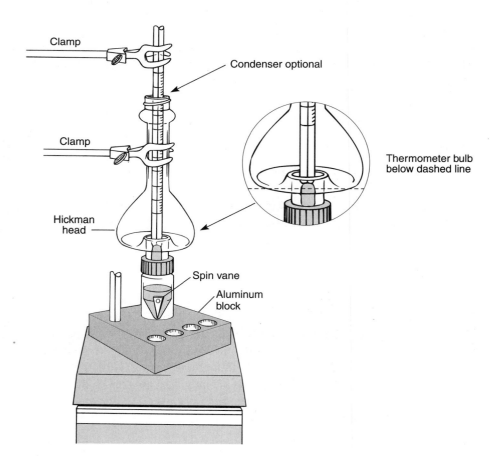

Clamp

Condenser optional

Clamp

Thermometer bulb
below dashed line

Hickman
head

Spin vane

Aluminum
block

Figure 8.6 Basic microscale distillation.

B.   MICROSCALE—STUDENT EQUIPMENT

Figure 8.6 shows the typical distillation setup for those students who are taking a microscale laboratory course. Instead of a distillation head, condenser, and vacuum takeoff, this equipment uses a single piece of glassware called a **Hickman head.** The Hickman head provides a "short path" for the distilled liquid to travel before it is collected. The liquid is boiled, moves upward through the central stem of the Hickman head, condenses on the walls of the "chimney," and then runs down the sides into the circular well surrounding the stem. With very volatile liquids, a condenser can be placed on top of the Hickman head to improve its efficiency. The apparatus shown uses a 5-mL conical vial as the distilling flask, meaning that this apparatus can distill from 1 to 3 mL of liquid. Unfortunately, the well in most Hickman heads holds only about 0.5–1.0 mL. That means the well must be emptied several times. This is done using a disposable Pasteur pipet as shown in Figure 8.7. The figure shows two different styles of Hickman head. The one with the side port makes removal of the distillate easier.

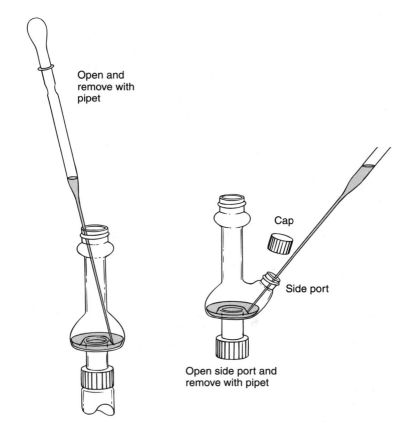

Open and
remove with
pipet

Cap

Side port

Open side port and
remove with pipet

Figure 8.7   Two styles of Hickman head.

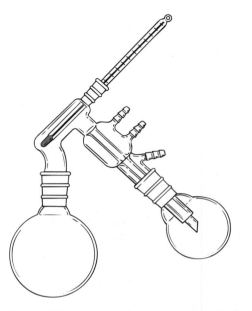

Figure 8.8   A research-style short-path distillation apparatus.

C.   MICROSCALE—RESEARCH EQUIPMENT

Figure 8.8 shows a very-well designed research-style, short-path distillation head. Note how the equipment has been "unitized," eliminating several joints, and decreasing the holdup.

PROBLEMS

1. Using Figure 8.3, answer the following questions.
(a) What is the molar composition of the vapor in equilibrium with a boiling liquid that has a composition of 60% A and 40% B?
(b) A sample of vapor has the composition 50% A and 50% B. What is the composition of the boiling liquid that produced this vapor?
2. Use an apparatus similar to that shown in Figure 8.4, and assume that the round-bottom flask holds 100 mL and the distilling head has an internal volume of 12 mL in the vertical section. At the end of a distillation, vapor would fill this volume, but it could not be forced through the system. No liquid would remain in the distillation flask. Assuming this holdup volume of 112 mL, use the ideal gas law and assume a boiling point of 100°C (760 mm Hg) to calculate the number of milliliters of liquid ($d = 0.9$ g/mL, $MW = 200$) that would recondense into the distillation flask upon cooling.
3. Explain the significance of a horizontal line connecting a point on the lower curve with a point on the upper curve (such as line xy) in Figure 8.3.
4. Using Figure 8.3, determine the boiling point of a liquid having a molar composition of 50% A and 50% B.
5. Where should the thermometer bulb be located in
(a) a microscale distillation apparatus using a Hickman head?
(b) a macroscale distillation apparatus using a distilling head, condenser, and vacuum takeoff adapter?
6. Under what conditions can a good separation be achieved with a simple distillation?

T E C H N I Q U E   9

Vacuum Distillation, Manometers

Vacuum distillation (distillation at reduced pressure) is used for compounds that have high boiling points (above 200°C). Such compounds often undergo thermal decomposition at the temperatures required for their distillation at atmospheric pressure. The boiling point of a compound is lowered substantially by reducing the applied pressure. Vacuum distillation is also used for compounds that, when heated, might react with the oxygen present in air. It is also used when it is more convenient to distill at a lower temperature because of experimental limitations. For instance, a heating device may have difficulty heating to a temperature in excess of 250°C.

The effect of pressure on the boiling point is discussed more thoroughly in Technique 6 (Section 6.9, p. 674). A nomograph is given (Fig. 6.9, p. 676) that allows you to estimate the boiling point of a liquid at a pressure different from the one at which it is reported. For example, a liquid reported to boil at 200°C at 760 mmHg would be expected to boil at 90°C at 20 mmHg. This is a significant decrease in temperature, and it would be advantageous to use a vacuum distillation if any problems were to be expected. Counterbalancing this advantage, however, is the fact that separations of liquids of different boiling points may not be as effective with a vacuum distillation as with a simple distillation.

9.1 STANDARD SCALE METHODS

When working with glassware that is to be evacuated, you should wear safety glasses at all times. There is always danger of an implosion.

> **Caution:** Safety glasses must be worn at all times during vacuum distillation.

It is a good idea to work in a hood when performing a vacuum distillation. If the experiment will involve high temperatures (>220°C) for distillation, or an extremely low pressure (<0.1 mmHg), for your own safety you should definitely work in a hood, behind a shield.

A basic apparatus similar to the one shown in Figure 9.1 may be used for vacuum distillations. The major differences to be found when comparing this assembly to one for simple distillation (Fig. 8.4, p. 711) are that a Claisen head has been inserted between the distillation flask and the distilling head, and that the opening to the atmosphere has been replaced by a connection **A** to a vacuum source. In addition, an air inlet tube **B** has been added to the top of the Claisen head. When connecting to a vacuum source, an aspirator (Technique 4, Section 4.5, p. 643), a mechanical vacuum pump (Section 9.6, p. 728), or a "house" vacuum system (one piped directly to the laboratory bench) may be used. The

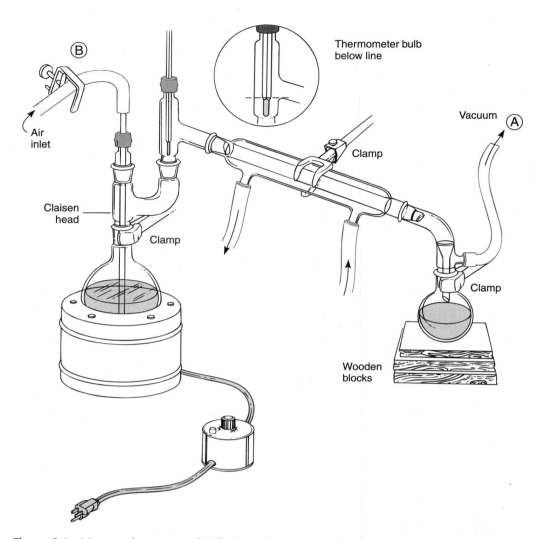

Figure 9.1 Macroscale vacuum distillation using the standard organic laboratory kit.

aspirator is probably the simplest of these sources and the vacuum source most likely to be available. However, if pressures below 10–20 mmHg are required, a mechanical vacuum pump must be used.

Assembling the Apparatus

When assembling an apparatus for vacuum distillation, it is important to check the following points.

Glassware.   Before assembly, check all glassware to be sure there are no cracks and that there are no chips in the standard-taper joints. Cracked glassware may break when evacuated. Joints that have chips may not be air-tight and they will leak.

Greasing Joints.   With standard scale equipment it is necessary to grease all standard-taper joints lightly. Take care not to use too much grease. Grease can become a very serious contaminant if it oozes out the bottom of the joints into your system. Apply a small amount of grease (thin film) completely around the top of the *inner* joint; then mate the joints and press or turn them slightly to spread the grease evenly. If you have used the correct amount of grease, it will not ooze out the bottom of the joint; rather the entire joint will appear clear and without striations or uncovered areas.

Claisen Head.   The Claisen head is placed between the distilling flask and the distilling head to help prevent material from "bumping over" into the condenser.

Ebulliator Tube.   The air inlet tube on top of the Claisen head is called an ebulliator (*ebb-u-lay-tor*) tube. Using the screw clamp **B** on the attached heavy-wall tubing (see pressure tubing below), the ebulliator is adjusted to admit a slow continuous stream of air bubbles into the distillation flask while you are distilling. Since boiling stones will not work in a vacuum, these bubbles keep the solution stirred and help to prevent bumping. The ebulliator tube is drawn to a point at its lower end. The end of the tube should be adjusted so that it is just above the bottom of the distilling flask.

Most standard ground-glass kits contain an ebulliator tube. If one is not available, an ebulliator can be prepared easily by heating a section of glass tubing and drawing it out about 3 cm. The glass is then scored in the middle of this drawn-out section and broken, making two tubes at once. In Figure 9.1, the ebulliator is inserted into a thermometer adapter. If you do not have a second thermometer adapter, a one-hole rubber stopper may be used, placing the stopper directly into the joint on top of the Claisen head.

Wooden Applicator Sticks.   An alternative to an ebulliator tube that is sometimes used is the wooden pine splint or wooden applicator stick. Air is trapped in the pores of the wood. Under vacuum, the stick will emit a slow stream of bubbles to stir the solution. The disadvantage is that, each time you open the system, you must use a new stick.

Thermometer Placement.   Be sure that the thermometer is positioned so that the entire mercury bulb is below the sidearm in the distilling head (see the circular insert in Fig. 9.1). If it is placed higher, it may not be surrounded by a constant stream of vapor from the material being distilled. If the thermometer is not exposed to a continuous stream of vapor, it may not reach temperature equilibrium. As a result, the temperature reading would be incorrect (low).

Joint Clips.   If plastic joint clips are available (Technique 3, Figure 3.5, p. 617), they should be used to secure the greased joints, particularly those on either side of the condenser, and the one at the bottom of the vacuum adapter where the receiving flask is attached.

Pressure Tubing.   The connection to the vacuum source **A** is made using pressure tubing. Pressure tubing (also called vacuum tubing), unlike the more common thin-walled tubing used to carry water or gas, has heavy walls and will not collapse inward when it is evacuated. A comparison of the two types of tubing is shown in Figure 9.2.

Make doubly sure that any connections to pressure tubing are tight. If a tight connection cannot be made, you may have the wrong size of tubing (either the rubber tubing or the glass tubing to which it is attached). Keep the lengths of pressure tubing relatively short. The pressure tubing should be relatively new and without cracks. If the tubing shows cracks when you stretch it or bend it, it may be old and leak air into the system. Replace any tubing that shows its age.

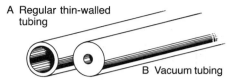

Figure 9.2   Comparison of tubing.

Rubber Stoppers.   Always use soft rubber stoppers in a vacuum apparatus; corks will not give an air-tight seal. Rubber stoppers harden with age and use. If a rubber stopper is not soft (will not squeeze), discard it. Glass tubing should fit securely into any rubber stoppers. If you can move the tubing up and down with only gentle force, it is too loose and you should obtain a larger size.

Receiving Flask.   When more than one fraction is expected from a vacuum distillation, it is considered good practice to have several pre-weighed receiver flasks, including the original, available before the distillation begins. Such preparation permits the rapid changing of receiving flasks during the distillation. The pre-weighing allows easy calculation of the weight of distillate in each fraction without the need to transfer the distillate to yet another flask.

To change receiving flasks, heating must be stopped, and the system vented at both ends, before replacing a flask. Complete directions for this procedure are given in the next section.

Vacuum Traps.   When performing a vacuum distillation, it is customary to place a "trap" in the line that connects to the vacuum source. Two common trap arrangements are shown in Figures 9.3 and 9.4. This type of trap is essential if an aspirator or a house vacuum is used as the source of vacuum. A mechanical vacuum pump requires a different type of trap (see Fig. 9.8, p. 728). Variations in pressure are to be expected when using an aspiratory or a house vacuum. With an aspirator, if the pressure drops low enough, the vacuum in the system will draw water from the aspirator into the connecting line. The trap allows you to see this happening and take corrective action (i.e., prevent water from entering the distillation apparatus). The correct action for anything but a small amount of water is to "vent" the system. This can be accomplished by opening the screw clamp **C** at the top of the trap to let air into the system. This is also the way air is admitted into the system at the end of the distillation.

> **Caution:** Note also that it is always necessary to vent the system *before* the aspirator is stopped. If you fail to vent the system, water may be drawn into it, contaminating your product. Be sure, however, that you vent *both ends* of the system. After venting the vacuum trap, you should immediately open the screw clamp on top of the ebulliator tube.

The trap, which contains a large volume, also acts as a buffer to pressure changes, evening out small variations in the line. In the house vacuum system, it prevents oil and water (often present in house lines) from entering your system.

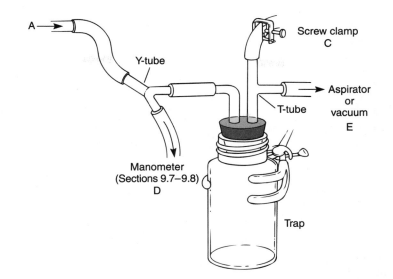

Figure 9.3 Vacuum trap using a gas bottle. The assembly connects to Figure 9.1 by joining the tubing at the point marked A. (The Y-tube connection to a manometer is optional.)

Manometer Connection.   A manometer allows measurement of pressure. A Y-tube (or T-tube) connection **D** is shown in the line from the apparatus to the trap. This branching connection is optional, but is required if you wish to monitor the actual pressure of your system when using a manometer. The operation of manometers is discussed in

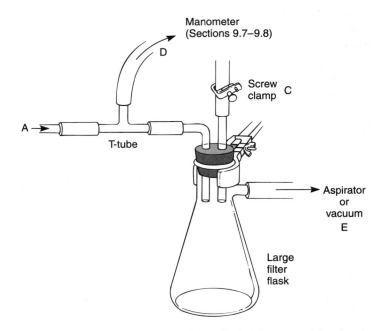

Figure 9.4 Vacuum trap using a heavy-wall filter flask. The assembly connects to Figure 9.1 by joining the tubing at the point A. (The T-tube connection to the manometer is optional.)

Sections 9.7 and 9.8. A suitable manometer should be included in the system at least part of the time during the distillation to measure the pressure at which the distillation is being conducted. A boiling point is of little value if the pressure is not known! After use, the manometer can be removed if a screw clamp is used to close the connection.

> **Caution:** The manometer must be vented very slowly to prevent a rush of mercury from breaking out the end of the tubing.

A manometer is also very useful in troubleshooting your system. It can be attached to the aspirator or house vacuum to determine the working pressure. In this way a defective aspirator (not uncommon) can be spotted and replaced. When you connect your apparatus, you can adjust all the joints and connections to obtain the best working pressure *before* you begin to distill. Generally a working pressure of from 25–50 mmHg is adequate for the procedures in this text.

Aspirators. In many labs the most convenient source of vacuum for a reduced-pressure distillation is the aspirator. The aspirator, or other vacuum source, is attached to the trap. The aspirator can theoretically pull a vacuum equal to the vapor pressure of the water flowing through it. The vapor pressure of flowing water depends on its temperature (24 mm Hg at 25°C; 18 mm Hg at 20°C; 9 mm Hg at 10°C). However, in the typical laboratory, the pressures attained are higher than expected due to reduced water pressure when many students are using their aspirators simultaneously. Good laboratory practice requires that only a few students on a given bench use the aspirator at the same time. It may be necessary to establish a schedule for aspirator use, or at least to have some students wait until others are finished.

House Vacuum. As stated for aspirators, depending on the capacity of the system, it may not be possible for everyone to use the vacuum system at once. Students may have to take turns or work in rotation. A typical house vacuum system will have a base pressure of about 35–100 mmHg when it is not overloaded.

9.2 VACUUM DISTILLATION: STEPWISE DIRECTIONS

The procedures in applying vacuum distillation are described in this section.

> **Caution:** Safety glasses must be worn at all times during vacuum distillation.

Evacuating the Apparatus

1. Assemble the apparatus shown in Figure 9.1 as discussed in Section 9.1, and attach a trap (either Fig. 9.3 or 9.4). The connection is made at the points labeled **A**. Next, attach the trap to either an aspirator or a house vacuum system at point **E**. Do not close any clamps at this time.

2. Weigh each empty receiving flask to be used in collecting the various fractions during the distillation.

3. Concentrate the material to be distilled in an Erlenmeyer flask or beaker by removing all volatile solvents, such as ether, using a steam bath or a water bath in the hood. Use boiling stones and a stream of air to help the solvent removal.

4. Remove the distilling flask from the vacuum distillation apparatus, remove the grease by wiping with a towel, and transfer the concentrate to the flask, using a funnel. Complete the transfer by rinsing with a *small* amount of solvent. Again, concentrate the material until no additional volatile solvent can be removed (boiling will cease). The flask should be no more than half-full after concentration. Regrease the joint and reattach the flask to the distilling apparatus. Make sure all joints are tight.

5. On the trap assembly (Figure 9.3 or 9.4) open the clamp at **C** and attach a manometer at point **D**.

6. Turn on the aspirator (or house vacuum) (Fig. 9.3) to the maximum extent.

7. Tighten the screw clamp at **B** (Fig. 9.1) until the tubing is nearly closed.

8. Going back to the trap (Fig. 9.3), slowly tighten the screw clamp at point **C**. Watch the bubbling action of the ebulliator tube to see that it is not too vigorous or too slow. Any volatile solvents you could not remove during concentration will be removed now. Once the loss of volatiles slows down, close screw clamp **C** to the fullest extent.

9. Adjust the ebulliator tube at **B** until a fine steady stream of bubbles is formed.

10. Wait a few minutes and then record the pressure obtained.

11. If the pressure is not satisfactory, check all connections to see that they are tight. Gently twist any hoses to snug them down. Press down on any rubber stoppers. Check the fit of all glass tubing. Press any joints together until they appear evenly greased and well joined. If you crimp the rubber tubing between the apparatus and the trap with your hand and the pressure decreases, you will know that there is a leak in the glassware assembly. If there is no change, the problem may be with the aspirator or the trap. Readjust the ebulliator screw clamp at **B** if necessary.

> Do not proceed until you have a good vacuum. Ask your instructor for help if necessary.

12. Once your vacuum has been established, record the pressure, and the manometer may be removed for use by another student if necessary. Place a screw clamp ahead of the manometer at **D** and tighten it. With careful venting, the manometer may now be removed.

Beginning Distillation

13. Raise the heat source into position with wooden blocks, or other means, and begin to heat.

14. Increase the temperature. Eventually a reflux ring will contact the thermometer bulb, and distillation will begin.

15. Record the temperature range and the pressure range (if the manometer is still connected) during the distillation. The distillate should be collected at a rate of about 1 drop per second.
16. If the reflux ring is in the Claisen head but will not rise into the distilling head, it may be necessary to insulate these pieces by wrapping them with glass wool and aluminum foil (shiny side in). The insulation should aid the distillate to pass into the condenser.
17. The boiling point should be relatively constant so long as the pressure is constant. A rapid increase in pressure may be due to increased use of the aspirators in the lab (or additional connections to the house vacuum). It could also be due to decomposition of the material being distilled. Decomposition will produce a dense white fog in the distilling flask. If this happens, reduce the temperature of the heat source, or remove it, and *stand back* until the system cools. When the fog subsides you can investigate the cause.

Changing Receiving Flasks

18. To change receiving flasks during distillation when a new component begins to distill (higher boiling point at the same pressure), carefully open the clamp on top of the trap assembly at **C** and immediately lower the heat source.

> Watch the ebulliator for excessive backup! It may also be necessary to open the clamp at **B**.

19. Remove the wooden blocks, or other support under the receiving flask, release the clamp, and replace the flask with a clean, pre-weighed receiver. Use a small amount of grease, if necessary, to reestablish a good seal.
20. Reclose the clamp at **C** and allow several minutes for the system to reestablish the reduced pressure. If you opened the ebulliator screw clamp at **B** you will have to close and readjust it. Bubbling will not recommence until any liquid is drawn back out of the ebulliator. This liquid may have been forced into the ebulliator when the vacuum was interrupted.
21. Raise the heating source back into position under the distilling flask and continue with the distillation.
22. When the temperature falls at the thermometer, this usually indicates that distillation is complete. If a significant amount of liquid remains, however, the bubbling may have stopped, the pressure may have risen, the heating source may not be hot enough, or perhaps insulation of the distillation head is required. Adjust accordingly.

Shutdown

23. At the end of the distillation, remove the heat source, and slowly open the screw clamps at **C** and **B**. When the system is vented you may shut off the aspirator or house vacuum and disconnect the tubing.

24. Remove the receiving flask and clean all glassware as soon as possible after disassembly (let it cool a bit) to keep the ground-glass joints from sticking.

> If you used grease, thoroughly clean all grease off the joints, or it will contaminate your samples in other procedures.

9.3 ROTARY FRACTION COLLECTORS

With the types of apparatus we have discussed previously, the vacuum must be stopped to remove fractions when a new substance (fraction) begins to distill. Quite a few steps are required to perform this change, and it is quite inconvenient when there are several fractions to be collected. Two pieces of semi-microscale apparatus that are designed to alleviate the difficulty of collecting fractions while working under vacuum are shown in Figure 9.5. The collector, which is shown to the right, is sometimes called a "cow" because of its appearance. With these rotary fraction collecting devices, all you need to do is rotate the device to collect fractions.

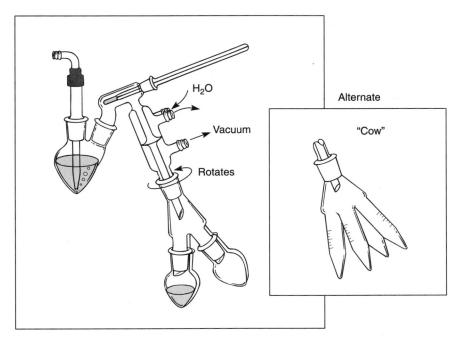

Figure 9.5   Rotary fraction collector.

9.4 MICROSCALE METHODS—STUDENT APPARATUS

Figure 9.6 shows the type of vacuum distillation equipment that would be used by a student enrolled in a microscale laboratory program. This apparatus, which uses a 5-mL conical vial as a distilling flask, can distill from 1 to 3 mL of liquid. The Hickman head replaces the Claisen head, distilling head, condenser, and receiving flask with a single piece of glassware.

9.5 BULB-TO-BULB DISTILLATION

The ultimate in microscale methods is to use a bulb-to-bulb distillation apparatus. This apparatus is shown in Figure 9.7. The sample to be distilled is placed in the

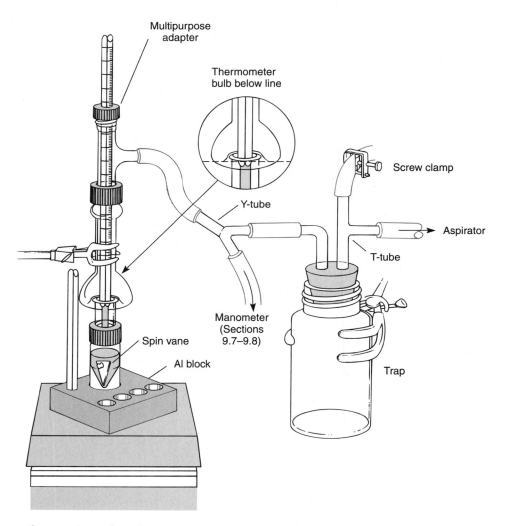

Figure 9.6 Reduced-pressure microscale distillation.

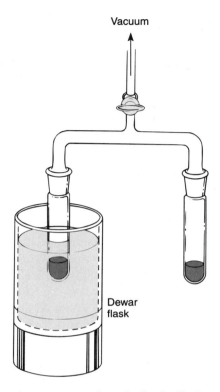

Vacuum

Dewar
flask

Figure 9.7   Bulb-to-bulb distillation.

glass container attached to one of the arms of the apparatus. The sample is frozen solid, usually by using liquid nitrogen, but dry ice in 2-propanol or an ice–salt-water mixture may also be used. The coolant container shown in the figure is a **Dewar flask.** The Dewar flask has a double wall with the space between the walls evacuated and sealed. A vacuum is a very good thermal insulator, and there is little heat loss from the cooling solution.

After freezing the sample, the entire apparatus is evacuated by opening the stopcock. When the evacuation is complete, the stopcock is closed, and the Dewar flask is removed. The sample is allowed to thaw and then it is frozen again. This freeze–thaw–freeze cycle removes any air or gases that were trapped in the frozen sample. Next, the stopcock is opened to evacuate the system again. When the second evacuation is complete, the stopcock is closed, and the Dewar flask is moved to the other arm to cool the empty container. As the sample warms, it will vaporize, travel to the other side, and be frozen or liquefied by the cooling solution. This transfer of the liquid from one arm to the other may take quite a while, but *no heating is required.*

The bulb-to-bulb distillation is most effective when liquid nitrogen is used as coolant and when the vacuum system can achieve a pressure of 10^{-3} mmHg or lower. This requires a vacuum pump; an aspirator cannot be used.

9.6 THE MECHANICAL VACUUM PUMP

The aspirator is not capable of yielding pressures below about 5 mmHg. This is the vapor pressure of water at 0°C, and water freezes at this temperature. A more realistic value of pressure for an aspirator is about 20 mmHg. When pressures below 20 mmHg are required, a vacuum pump will have to be employed. Figure 9.8 illustrates a mechanical vacuum pump and its associated glassware. The vacuum pump operates on a principle similar to that of the aspirator, but the vacuum pump uses a high-boiling oil, rather than water, to remove air from the attached system. The oil used in a vacuum pump, a silicone oil or a high molecular weight hydrocarbon-based oil, has a very low vapor pressure, and very low system pressures can be achieved. A good vacuum pump, with new oil, can achieve pressures of 10^{-3} or 10^{-4} mmHg. Instead of discarding the oil as it is used, it is recycled continuously through the system.

A cooled trap is required when using a vacuum pump. This trap protects the oil in the pump from any vapors that may be present in the system. If vapors from organic solvents, or from the organic compounds being distilled, dissolve in the oil, the oil's vapor pressure will increase, rendering it less effective. A special type of vacuum trap is illustrated in Figure 9.8. It is designed to fit into an insulated Dewar flask so that the coolant will last for a long period. At a minimum, this flask should be filled with ice water, but a dry ice–acetone mixture or liquid nitrogen is required to achieve lower temperatures and

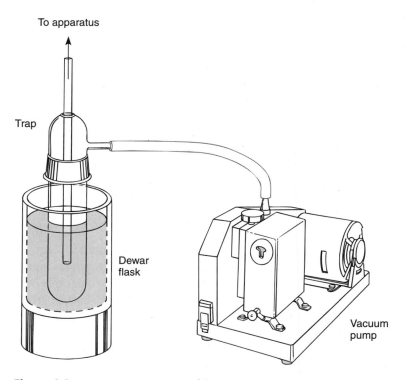

Figure 9.8   A vacuum pump and its trap.

better protect the oil. Often two traps are used; the first trap contains ice water and the second trap dry ice–acetone or liquid nitrogen. The first trap liquefies low-boiling vapors that might freeze or solidify in the second trap and block it.

9.7 THE CLOSED-END MANOMETER

The principal device used to measure pressures in a vacuum distillation is the **closed-end manometer.** Two basic types are shown in Figures 9.9 and 9.10. The manometer shown in Figure 9.9 is widely used because it is relatively easy to construct. It consists of a U-tube that is closed at one end and mounted on a wooden support. You can construct the manometer from 9-mm glass capillary tubing and fill it, as shown in Figure 9.11.

> **Caution:** Mercury is a very toxic metal with cumulative effects. Because mercury has a high vapor pressure, it must not be spilled in the laboratory. You must not touch it with your skin. Seek immediate help from an instructor in case of a spill or if you break a manometer. Spills must be cleaned immediately.

A small filling device is connected to the U-tube with pressure tubing. The U-tube is evacuated with a good vacuum pump; then the mercury is introduced by tilting the mer-

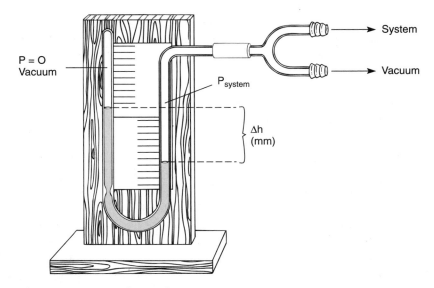

Figure 9.9   A simple U-tube manometer.

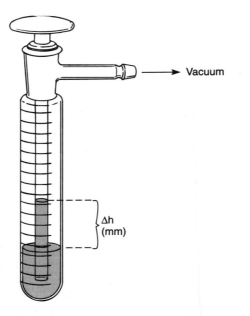

Δh
(mm)

Figure 9.10   Commercial "stick" manometer.

cury reservoir. The entire filling operation should be conducted in a shallow pan in order to contain any spills that might occur. Enough mercury should be added to form a column about 20 cm in total length. When the vacuum is interrupted by admitting air, the mercury is forced by atmospheric pressure to the end of the evacuated tube. The manometer is then ready for use. The constriction shown in Figure 9.11 helps to protect the manometer against breakage when the pressure is released. Be sure that the column of mercury is long enough to pass through this constriction.

When an aspirator or any other vacuum source is used, a manometer can be connected into the system. As the pressure is lowered, the mercury rises in the right tube and drops in the left tube until Δh corresponds to the approximate pressure of the system (see Fig. 9.9).

$$\Delta h = (P_{\text{system}} - P_{\text{reference arm}}) = (P_{\text{system}} - 10^{-3} \text{ mmHg}) \approx P_{\text{system}}$$

A short piece of metric ruler or a piece of graph paper ruled in millimeter squares is mounted on the support board to allow Δh to be read. No addition or subtraction is necessary, because the reference pressure (created by the initial evacuation when filling) is approximately zero (10^{-3} mmHg) when referred to readings in the 10–50 mmHg range. To determine the pressure, count the number of millimeter squares beginning at the top of the mercury column on the left and continuing downward to the top of the mercury column on the right. This is the height difference Δh, and it gives the pressure in the system directly.

A commercial counterpart to the U-tube manometer is shown in Figure 9.10. With this manometer, the pressure is given by the difference in the mercury levels in the inner and outer tubes.

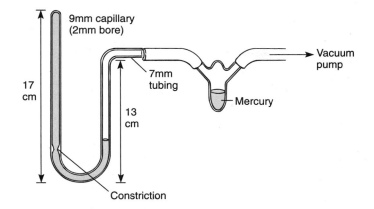

Figure 9.11   Filling a U-tube manometer.

The manometers described here have a range of about 1–150 mmHg in pressure. They are convenient to use when an aspirator is the source of vacuum. For high-vacuum systems (pressures below 1 mmHg), a more elaborate manometer or an electronic measuring device must be used. These devices will not be discussed here.

9.8 CONNECTING AND USING A MANOMETER

The most common use of a closed-end manometer is to monitor pressure during a reduced-pressure distillation. The manometer is placed in a vacuum distillation system, as shown in Figure 9.12. Generally, an aspirator is the source of vacuum. Both the manometer and the distillation apparatus should be protected by a trap from possible backups in the water line. Alternatives to the trap arrangements shown in Figure 9.12 appear in Figures 9.3 and 9.4. Notice in each case that the trap has a device (screw clamp or stopcock) for opening the system to the atmosphere. This is especially important in using a manometer, because you should always make pressure changes slowly. If this is not done, there is a danger of spraying mercury throughout the system, breaking the manometer, or spurting mercury into the room. In the closed-end manometer, if the system is opened suddenly, the mercury rushes to the closed end of the U-tube. The mercury rushes with such speed and force that the end will be broken out of the manometer. Air should be admitted *slowly* by opening the valve cautiously. In a similar fashion, the valve should be closed slowly when the vacuum is being started, or mercury may be forcefully drawn into the system through the open end of the manometer.

If the pressure in a reduced-pressure distillation is lower than desired, it is possible to adjust it by means of a **bleed valve.** The stopcock can serve this function in Figure 9.12 if it is opened only a small amount. In those systems with a screw clamp on the trap (Figs. 9.3 and 9.4), remove the screw clamp from the trap valve and attach the base of a Tirrill-style Bunsen burner. The needle valve in the base of the burner can be used to adjust precisely the amount of air that is admitted (bled) to the system and, hence, control the pressure.

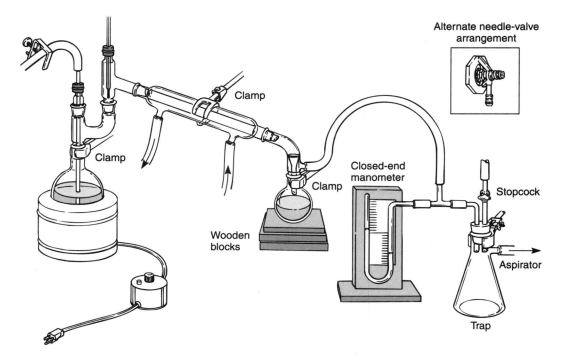

Figure 9.12   Connecting a manometer to the system. To construct a "bleed," the needle valve may replace the stopcock.

PROBLEMS

1. Give some reasons that would lead you to purify a liquid by using vacuum distillation rather than by using simple distillation.

2. When using an aspirator as a source of vacuum in a vacuum distillation, do you turn off the aspirator before venting the system? Explain.

3. A compound was distilled at atmospheric pressure and had a boiling range of 310–325°C. What would be the approximate boiling range of this liquid if it was distilled under vacuum at 20 mmHg?

4. Boiling stones generally do not work when performing a vacuum distillation. What substitutes may be used?

5. What is the purpose of the trap that is used during a vacuum distillation performed with an aspirator?

T E C H N I Q U E   1 0

Fractional Distillation, Azeotropes

Simple distillation, described in Technique 8, works well for most routine separation and purification procedures for organic compounds. When boiling-point differences of components to be separated are not large, however, **fractional distillation** must be used to achieve a good separation.

Part A.   Fractional Distillation

10.1 DIFFERENCES BETWEEN SIMPLE AND FRACTIONAL DISTILLATION

When an ideal solution of two liquids, such as benzene (bp 80°C) and toluene (bp 110°C), is distilled by simple distillation, the first vapor produced will be enriched in the lower-boiling component (benzene). However, when that initial vapor is condensed and analyzed, the distillate will not be pure benzene. The boiling point difference of benzene and toluene (30°C) is too small to achieve a complete separation by simple distillation. Following the principles outlined in Technique 8, Section 8.2 (pp. 708–710), and using the vapor-liquid composition curve given in Figure 10.1, you can see what would happen if you started with a equimolar mixture of benzene and toluene.

Following the dashed lines shows that an equimolar mixture (50 mole percent benzene) would begin to boil at about 91°C and, far from being 100% benzene, the distillate would contain about 74 mole percent benzene and 26 mole percent toluene. As the distillation continued, the composition of the undistilled liquid would move in the direction of A′ (there would be increased toluene, due to removal of more benzene than toluene), and the corresponding vapor would contain a progressively smaller amount of benzene. In effect, the temperature of the distillation would continue to increase throughout the distillation (as in Figure 8.2B, p. 708), and it would be impossible to obtain any fraction that consisted of pure benzene.

Suppose, however, that we are able to collect a small quantity of the first distillate that was 74 mole percent benzene and redistill it. Using Figure 10.1, we can see that this liquid would begin to boil at about 84°C and would give an initial distillate containing 90

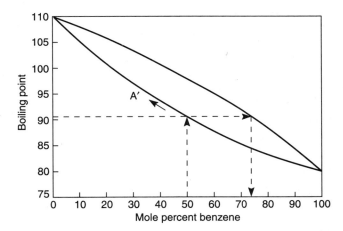

Figure 10.1 The vapor-liquid composition curve for mixtures of benzene and toluene.

mole percent of benzene. If we were experimentally able to continue taking small fractions at the beginning of each distillation, and redistill them, we would eventually reach a liquid with a composition of nearly 100 mole per cent benzene. However, since we took only a small amount of material at the beginning of each distillation, we would have lost most of the material we started with. To recapture a reasonable amount of benzene, we would have to process each of the fractions left behind in the same way as our early fractions. As each of them was partially distilled, the material advanced would become progressively richer in benzene, while that left behind would become progressively richer in toluene. It would require thousands (maybe millions) of such microdistillations to separate benzene from toluene.

Obviously, the procedure just described would be very tedious; fortunately, it need not be performed in usual laboratory practice. **Fractional distillation** accomplishes the same result. You simply have to use a column inserted between the distillation flask and the distilling head, as shown in Figure 10.2. This **fractionating column** is filled, or **packed,** with a suitable material such as a stainless steel sponge. This packing allows a mixture of benzene and toluene to be subjected continuously to many vaporization–condensation cycles as the material moves up the column. With each cycle within the column, the composition of the vapor is progressively enriched in the lower-boiling component (benzene).

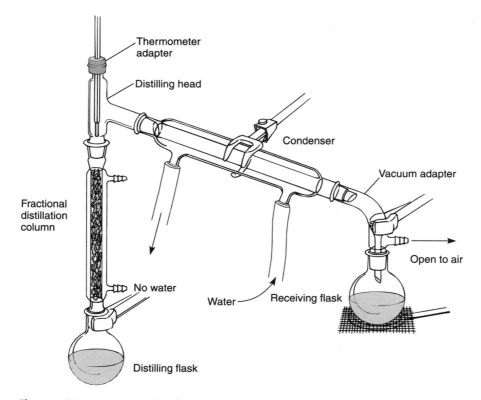

Figure 10.2 Fractional distillation apparatus.

Nearly pure benzene (bp 80°C) finally emerges from the top of the column, condenses, and passes into the receiving head or flask. This process continues until all the benzene is removed. The distillation must be carried out slowly to ensure that numerous vaporization–condensation cycles occur. When nearly all the benzene has been removed, the temperature begins to rise and a small amount of a second fraction, which contains some benzene and toluene, may be collected. When the temperature reaches 110°C, the boiling point of pure toluene, the vapor is condensed and collected as the third fraction. A plot of boiling-point versus volume of condensate (distillate) would resemble Figure 8.2C (p. 708). This separation would be much better than that achieved by simple distillation (Figure 8.2B (p. 708).

10.2 VAPOR–LIQUID COMPOSITION DIAGRAMS

A vapor–liquid composition phase diagram like the one in Figure 10.3 can be used to explain the operation of a fractionating column with an **ideal solution** of two liquids, A and B. An ideal solution is one in which the two liquids are chemically similar, miscible (mutually soluble) in all proportions, and do not interact. Ideal solutions obey **Raoult's Law.** Raoult's Law is explained in detail in Section 10.3.

The phase diagram relates the compositions of the boiling liquid (lower curve) and its vapor (upper curve) as a function of temperature. Any horizontal line drawn across the diagram (a constant-temperature line) intersects the diagram in two places. These intersections relate the vapor composition to the composition of the boiling liquid that pro-

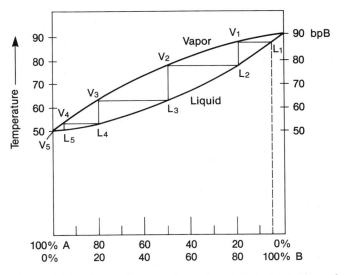

Figure 10.3 Phase diagram for a fractional distillation of an ideal two-component system.

duces that vapor. By convention, composition is expressed either in **mole fraction** or in **mole percentage.** The mole fraction is defined as follows:

$$\text{Mole fraction A} = N_A = \frac{\text{Moles A}}{\text{Moles A} + \text{Moles B}}$$

$$\text{Mole fraction B} = N_B = \frac{\text{Moles B}}{\text{Moles A} + \text{Moles B}}$$

$$N_A + N_B = 1$$

$$\text{Mole percentage A} = N_A \times 100$$

$$\text{Mole percentage B} = N_B \times 100$$

The horizontal and vertical lines shown in Figure 10.3 represents the processes that occur during a fractional distillation. Each of the **horizontal lines** (L_1V_1, L_2V_2, etc.) represents the **vaporization** step of a given vaporization–condensation cycle and represents the composition of the vapor in equilibrium with liquid at a given temperature. For example, at 63°C a liquid with a composition of 50% A (L_3 on the diagram) would yield vapor of composition 80% A (V_3 on diagram) at equilibrium. The vapor is richer in the lower-boiling component A than the original liquid was.

Each of the **vertical lines** (V_1L_2, V_2L_3, etc.) represents the **condensation** step of a given vaporization–condensation cycle. The composition does not change as the temperature drops on condensation. The vapor at V_3, for example, condenses to give a liquid (L_4 on the diagram) of composition 80% A with a drop in temperature from 63 to 53°C.

In the example shown in Figure 10.3, pure A boils at 50°C and pure B boils at 90°C. These two boiling points are represented at the left- and right-hand edges of the diagram, respectively. Now consider a solution that contains only 5% of A but 95% of B. (Remember that these are *mole* percentages.) This solution is heated (following the dashed line) until it is observed to boil at L_1 (87°C). The resulting vapor has composition V_1 (20% A, 80% B). The vapor is richer in A than the original liquid, but it is by no means pure A. In a simple distillation apparatus, this vapor would be condensed and passed into the receiver in a very impure state. However, with a fractionating column in place, the vapor is condensed in the **column** to give liquid L_2 (20% A, 80% B). Liquid L_2 is immediately revaporized (bp 78°C) to give a vapor of composition V_2 (50% A, 50% B), which is condensed to give liquid L_3. Liquid L_3 is revaporized (bp 63°C) to give vapor of composition V_3 (80% A, 20% B), which is condensed to give liquid L_4. Liquid L_4 is revaporized (bp 53°C) to give vapor of composition V_4 (95% A, 5% B). This process continues to V_5, which condenses to give nearly pure liquid A. The fractionating process follows the stepped lines in the figure downward and to the left.

As this process continues, all of liquid A is removed from the distillation flask or vial, leaving nearly pure B behind. If the temperature is raised, liquid B may be distilled as a nearly pure fraction. Fractional distillation will have achieved a separation of A and B, a separation that would have been nearly impossible with simple distillation. Notice that the boiling point of the liquid becomes lower each time it vaporizes. Because the temperature at the bottom of a column is normally higher than the temperature at the top, successive vaporizations occur higher and higher in the column as the composition of the distillate approaches that of pure A. This process is illustrated in Figure 10.4, where the composition of the liquids, their boiling points, and the composition of the vapors present are shown alongside the fractionating column.

10.3 RAOULT'S LAW

Two liquids (A and B) that are miscible and that do not interact form an **ideal solution** and follow Raoult's Law. The law states that the partial vapor pressure of component A in the solution (P_A) equals the vapor pressure of pure A (P_A^0) times its mole frac-

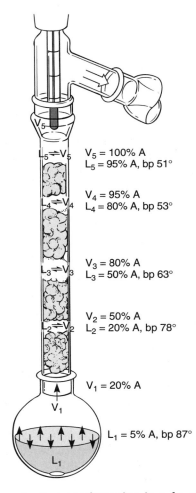

$V_5 = 100\%$ A
$L_5 = 95\%$ A, bp 51°

$V_4 = 95\%$ A
$L_4 = 80\%$ A, bp 53°

$V_3 = 80\%$ A
$L_3 = 50\%$ A, bp 63°

$V_2 = 50\%$ A
$L_2 = 20\%$ A, bp 78°

$V_1 = 20\%$ A

$L_1 = 5\%$ A, bp 87°

Figure 10.4 Vaporization-condensation in a fractionation column.

tion (N_A) (Eq.1). A similar expression can be written for component B (Eq. 2). The mole fractions N_A and N_B were defined in Section 10.2.

$$\text{Partial vapor pressure of A in solution} = P_A = (P_A^0)(N_A) \tag{1}$$

$$\text{Partial vapor pressure of B in solution} = P_B = (P_B^0)(N_B) \tag{2}$$

P_A^0 is the vapor pressure of pure A, independent of B. P_B^0 is the vapor pressure of B, independent of A. In a mixture of A and B, the partial vapor pressures are added to give the total vapor pressure above the solution (Eq. 3). When the total pressure (sum of the partial pressures) equals the applied pressure, the solution boils.

$$P_{\text{total}} = P_A + P_B = P_A^0 N_A + P_B^0 N_B \tag{3}$$

The composition of A and B in the vapor produced is given by Equations 4 and 5.

$$N_A \text{ (vapor)} = \frac{P_A}{P_{\text{total}}} \tag{4}$$

$$N_B \text{ (vapor)} = \frac{P_B}{P_{\text{total}}} \tag{5}$$

Several exercises involving applications of Raoult's Law are illustrated in Figure 10.5. Note, particularly in the result from Equation 4, that the vapor is richer ($N_A = 0.67$) in the lower-boiling (higher vapor pressure) component A than it was before vaporization ($N_A = 0.50$). This proves mathematically what was described in Section 10.2.

The consequences of Raoult's Law for distillations are shown schematically in Figure 10.6. In Part A the boiling points are identical (vapor pressures the same), and no separation is attained regardless of how the distillation is conducted. In Part B a fractional distillation is required, while in Part C a simple distillation provides an adequate separation.

When a solid B (rather than another liquid) is dissolved in a liquid A, the boiling point is increased. In this extreme case, the vapor pressure of B is negligible, and the vapor will be pure A no matter how much solid B is added. Consider a solution of salt in water.

$$P_{\text{total}} = P_{\text{water}}^0 N_{\text{water}} + P_{\text{salt}}^0 N_{\text{salt}}$$

$$P_{\text{salt}}^0 = 0$$

$$P_{\text{total}} = P_{\text{water}}^0 N_{\text{water}}$$

A solution whose mole fraction of water is 0.7 will not boil at 100°C, because $P_{\text{total}} = (760)(0.7) = 532$ mmHg and is less than atmospheric pressure. If the solution is heated to 110°C, it will boil because $P_{\text{total}} = (1085)(0.7) = 760$ mmHg. Although the solution must be heated at 110°C to boil it, the vapor is pure water and has a boiling-point temperature of 100°C. (The vapor pressure of water at 110°C can be looked up in a handbook; it is 1085 mmHg.)

Consider a solution at 100°C where $N_A = 0.5$ and $N_B = 0.5$.

1. What is the partial vapor pressure of A in the solution if the vapor pressure of pure A at 100°C is 1020 mmHg?

    Answer: $P_A = P°_A N_A = (1020)(0.5) = 510$ mmHg

2. What is the partial vapor pressure of B in the solution if the vapor pressure of pure B at 100°C is 500 mmHg?

    Answer: $P_B = P°_B N_B = (500)(0.5) = 250$ mmHg

3. Would the solution boil at 100°C if the applied pressure were 760 mmHg?

    Answer: Yes. $P_{total} = P_A + P_B = (510 + 250) = 760$ mmHg

4. What is the composition of the vapor at the boiling point?

    Answer: The boiling point is 100°C.

    $$N_A \text{ (vapor)} = \frac{P_A}{P_{total}} = 510/760 = 0.67$$

    $$N_B \text{ (vapor)} = \frac{P_B}{P_{total}} = 250/760 = 0.33$$

Figure 10.5   Sample calculations with Raoult's Law.

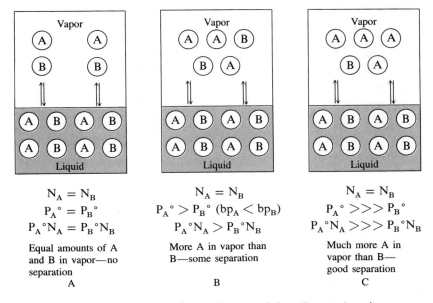

$N_A = N_B$	$N_A = N_B$	$N_A = N_B$
$P_A° = P_B°$	$P_A° > P_B°$ (bp$_A$ < bp$_B$)	$P_A° >>> P_B°$
$P_A°N_A = P_B°N_B$	$P_A°N_A > P_B°N_B$	$P_A°N_A >>> P_B°N_B$
Equal amounts of A and B in vapor—no separation	More A in vapor than B—some separation	Much more A in vapor than B— good separation
A	B	C

Figure 10.6   Consequences of Raoult's Law. (A) Boiling points (vapor pressures) are identical—no separation. (B) Boiling points somewhat less for A than for B— requires fractional distillation. (C) Boiling points much less for A than for B—simple distillation will suffice.

10.4 COLUMN EFFICIENCY

A common measure of the efficiency of a column is given by its number of **theoretical plates.** The number of theoretical plates in a column is related to the number of vaporization–condensation cycles that occur as a liquid mixture travels through it. Using the example mixture in Figure 10.3, if the first distillate (condensed vapor) had the composition at L_2 when starting with liquid of composition L_1, the column would be said to have *one theoretical plate.* This would correspond to a simple distillation, or one vaporization–condensation cycle. A column would have two theoretical plates if the first distillate had the composition at L_3. The two-theoretical-plate column essentially carries out "two simple distillations." According to Figure 10.3, *five theoretical plates* would be required to separate the mixture that started with composition L_1. Notice that this corresponds to the number of "steps" that need to be drawn in the figure to arrive at a composition of 100% A.

Most columns do not allow distillation in discrete steps, as indicated in Figure 10.3. Instead, the process is *continuous,* allowing the vapors to be continuously in contact with liquid of changing composition as they pass through the column. Any material can be used to pack the column as long as it can be wetted by the liquid and as long as it does not pack so tightly that vapor cannot pass.

The approximate relationship between the number of theoretical plates needed to separate an ideal two-component mixture and the difference in boiling points is given in Table 10.1. Notice that more theoretical plates are required as the boiling-point differences between the components decrease. For instance, a mixture of A (bp 130°C) and B (bp 166°C) with a boiling-point difference of 36°C would be expected to require a column with a minimum of five theoretical plates.

TABLE 10.1 Theoretical Plates Required to Separate Mixtures, Based on Boiling-Point Differences of Components

Boiling-Point Difference	Number of Theoretical Plates
108	1
72	2
54	3
43	4
36	5
20	10
10	20
7	30
4	50
2	100

10.5 TYPES OF FRACTIONATING COLUMNS AND PACKINGS

Several types of fractionating columns are shown in Figure 10.7. The Vigreux column, shown in Part A, has indentations that incline downward at angles of 45° and are in pairs on opposites sides of the column. The projections into the column provide increased possibilities for condensation and for the vapor to equilibrate with the liquid. Vigreux columns are popular in cases where only a small number of theoretical plates are required. They are not very efficient (a 20-cm column might have only 2.5 theoretical plates), but they allow for rapid distillation and have a small **holdup** (the amount of liquid retained by the column). A column packed with a stainless steel sponge is a more effective fractionating column than a Vigreux column, but not by a large margin. Glass beads, or glass helices, can also be used as a packing material, and they have a slightly greater efficiency yet. The air condenser or the water condenser can be used as an improvised column if an actual fractionating column is unavailable. If a condenser is packed with glass beads, glass helices, or sections of glass tubing, the packing must be held in place by inserting a small plug of stainless steel sponge into the bottom of the condenser.

The most effective type of column is the **spinning-band column.** In the most elegant form of this device, a tightly fitting, twisted platinum screen or a Teflon rod with helical threads is rotated rapidly inside the bore of the column (Fig. 10.8). A spinning-band column that is available for microscale work is shown in Figure 10.9. This spinning-band column has a band about 2–3 cm in length and provides 4–5 theoretical

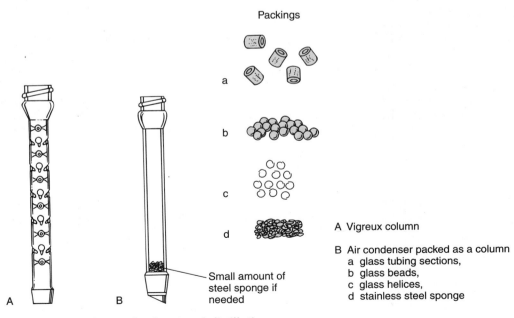

Figure 10.7   Columns for fractional distillation.

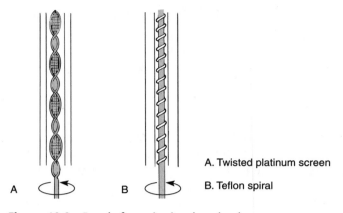

A. Twisted platinum screen

B. Teflon spiral

Figure 10.8   Bands for spinning-band columns.

Figure 10.9   A commercially available microscale spinning-band column.

plates. It can separate 1–2 mL of a mixture with a 30°C boiling-point difference. Larger research models of this spinning-band column can provide as many as 20 or 30 theoretical plates and can separate mixtures with a boiling-point difference of as little as 5–10°C.

Manufacturers of fractionating columns often offer them in a variety of lengths. Because the efficiency of a column is a function of its length, longer columns have more theoretical plates than shorter ones. It is common to express efficiency of a column in a unit called **HETP,** the **H**eight of a column that is **E**quivalent to one **T**heoretical **P**late. HETP is usually expressed in units of cm/plate. When the height of the column (in centimeters) is divided by this value, the total number of theoretical plates is specified.

10.6 FRACTIONAL DISTILLATION: METHODS AND PRACTICE

In performing a fractional distillation, the column should be clamped in a vertical position. The distillation should be conducted as slowly as possible, but the rate of distillation should be steady enough to produce a constant temperature reading at the thermometer.

Many fractionating columns must be insulated so that temperature equilibrium is maintained at all times. Additional insulation will not be required for columns that have an evacuated outer jacket, but those that do not can benefit from being wrapped in insulation.

Glass wool and aluminum foil (shiny side in) are often used for insulation. You can wrap the column with glass wool and then use a wrapping of the aluminum foil to keep it in place. An especially effective method is to make an insulation blanket by placing a layer of glass wool or cotton between two rectangles of aluminum foil, placed shiny side in. The sandwich is bound together with duct tape. This blanket, which is reusable, can be wrapped around the column and held in place with twist ties or tape.

The **reflux ratio** is defined as the ratio of the number of drops of distillate that return to the distillation flask compared to the number of drops of distillate collected. In an efficient column, the reflux ratio should equal or exceed the number of theoretical plates. A high reflux ratio ensures that the column will achieve temperature equilibrium and achieve its maximum efficiency. This ratio is not easy to determine; in fact, it is impossible to determine when using a Hickman head, and it should not concern a beginning student. In some cases, the **throughput,** or **rate of takeoff,** of a column may be specified. This is expressed as the number of milliliters of distillate that can be collected per unit of time, usually as mL/min.

Standard Scale Apparatus. Figure 10.2 illustrates a fractional distillation assembly that can be used for larger-scale distillations. It has a glass-jacketed column that is packed with a stainless steel sponge. This apparatus would be common in situations where quantities of liquid in excess of 10 mL were to be distilled.

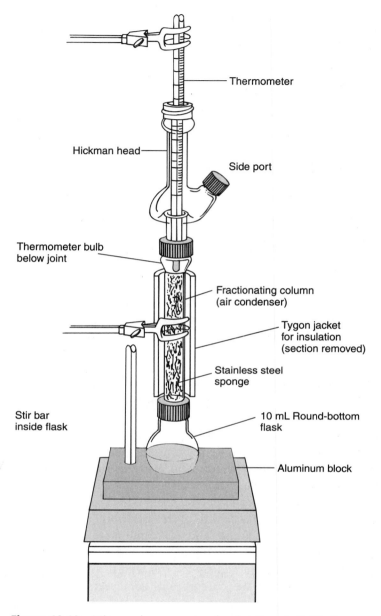

Figure 10.10 Microscale apparatus for fractional distillation.

Microscale Apparatus. The apparatus shown in Figure 10.10 is the one you are most likely to use in the microscale laboratory. If your laboratory is one of the better equipped ones, you may have access to spinning-band columns like those shown in Figure 10.9.

Part B.   Azeotropes

10.7 NONIDEAL SOLUTIONS: AZEOTROPES

Some mixtures of liquids, because of attractions or repulsions between the molecules, do not behave ideally; they do not follow Raoult's Law. There are two types of vapor–liquid composition diagrams that result from this nonideal behavior: **minimum-boiling-point** and **maximum-boiling-point** diagrams. The minimum or maximum points in these diagrams correspond to a constant-boiling mixture called an **azeotrope.** An azeotrope is a mixture with a fixed composition that cannot be altered by either simple or fractional distillation. An azeotrope behaves as if it were a pure compound, and it distills from the beginning to the end of its distillation at a constant temperature, giving a distillate of constant (azeotropic) composition. The vapor in equilibrium with an azeotropic liquid has the same composition as the azeotrope. Because of this, an azeotrope is represented as a *point* on a vapor–liquid composition diagram.

A.   MINIMUM-BOILING-POINT DIAGRAMS

A minimum-boiling-point azeotrope results from a slight incompatibility (repulsion) between the liquids being mixed. This incompatibility leads to a higher-than-expected combined vapor pressure from the solution. This higher combined vapor pressure brings about a lower boiling point for the mixture than is observed for the pure components. The most common two-component mixture that gives a minimum-boiling-point azeotrope is the ethanol–water system shown in Figure 10.11. The azeotrope at V_3 has a composition

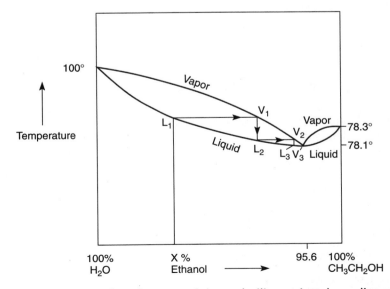

Figure 10.11 Ethanol–water minimum-boiling-point phase diagram.

of 96% ethanol–4% water and a boiling point of 78.1°C. This boiling point is not much lower than that of pure ethanol (78.3°C), but it means that it is impossible to obtain pure ethanol from the distillation of any ethanol–water mixture that contains more than 4% water. Even with the best fractionating column, you cannot obtain 100% ethanol. The remaining 4% of water can be removed by adding benzene and removing a different azeotrope, the ternary benzene–water–ethanol azeotrope (bp 65°C). Once the water is removed, the excess benzene is removed as an ethanol–benzene azeotrope (bp 68°C). The resulting material is free of water and is called "absolute" ethanol.

The fractional distillation of an ethanol–water mixture of composition X can be described as follows. The mixture is heated (follow line XL_1) until it is observed to boil at L_1. The resulting vapor at V_1 will be richer in the lower-boiling component, ethanol, than the original mixture.[1] The condensate at L_2 is vaporized to give V_2. The process continues, following the lines to the right, until the azeotrope is obtained at V_3. The liquid that distills is not pure ethanol, but it has the azeotropic composition of 96% ethanol and 4% water, and it distills at 78.1°C. The azeotrope, which is richer in ethanol than the original mixture, continues to distill. As it distills, the percentage of water left behind in the distillation flask continues to increase. When all the ethanol has been distilled (as the azeotrope), pure water remains behind in the distillation flask, and it distills at 100°C.

If the azeotrope obtained by the preceding procedure is redistilled, it distills from the beginning to the end of the distillation at a constant temperature of 78.1°C as if it were a pure substance. There is no change in the composition of the vapor during the distillation.

Some common minimum-boiling azeotropes are given in Table 10.2. Numerous other azeotropes are formed in two- and three-component systems; such azeotropes are common. Water forms azeotropes with many substances; therefore, water must be carefully removed with **drying agents** whenever possible before compounds are distilled. Extensive azeotropic data are available in references such as the *Handbook of Chemistry and Physics*.[2]

B. MAXIMUM-BOILING-POINT DIAGRAMS

A maximum-boiling point azeotrope results from a slight attraction between the component molecules. This attraction leads to lower combined vapor pressure than expected in the solution. The lower combined vapor pressures cause a higher boiling point than what would be characteristic for the components. A two-component maximum-boiling-point azeotrope is illustrated in Figure 10.12. Because the azeotrope has a higher boil-

---

[1]Keep in mind that this distillate is not pure ethanol but is an ethanol–water mixture.
[2]More examples of azeotropes, with their compositions and boiling points, can be found in the *CRC Handbook of Chemistry and Physics;* also in L.H. Horsley, ed., *Advances in Chemistry Series,* no. 116. Azeotropic Data, III (Washington: American Chemical Society, 1973).

TABLE 10.2   Common Minimum-Boiling Azeotropes

Azeotrope	Composition (Weight percentage)	Boiling Point (°C)
Ethanol–water	95.6% C_2H_5OH, 4.4% H_2O	78.17
Benzene–water	91.1% C_6H_6, 8.9% H_2O	69.4
Benzene–water–ethanol	74.1% C_6H_6, 7.4% H_2O, 18.5% C_2H_5OH	64.9
Methanol–carbon tetrachloride	20.6% CH_3OH, 79.4% CCl_4	55.7
Ethanol–benzene	32.4% C_2H_5OH, 67.6% C_6H_6	67.8
Methanol–toluene	72.4% CH_3OH, 27.6% $C_6H_5CH_3$	63.7
Methanol–benzene	39.5% CH_3OH, 60.5% C_6H_6	58.3
Cyclohexane–ethanol	69.5% C_6H_{12}, 30.5% C_2H_5OH	64.9
2-Propanol–water	87.8% $(CH_3)_2CHOH$, 12.2% H_2O	80.4
Butyl acetate–water	72.9% $CH_3COOC_4H_9$, 27.1% H_2O	90.7
Phenol–water	9.2% C_6H_5OH, 90.8% H_2O	99.5

ing point than any of the components, it will be concentrated in the distillation flask as the distillate (pure B) is removed. The distillation of a solution of composition X would follow to the right along the lines in Figure 10.12. Once the composition of the material remaining in the flask has reached that of the azeotrope, the temperature will rise, and the azeotrope will begin to distill. The azeotrope will continue to distill until all the material in the distillation flask has been exhausted.

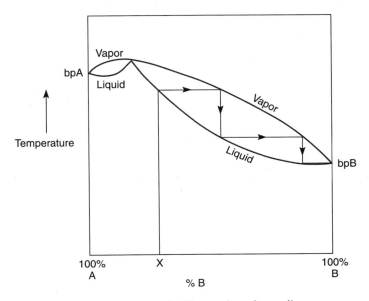

Figure 10.12   A maximum-boiling-point phase diagram.

TABLE 10.3 Maximum-Boiling Azeotropes

Azeotrope	Composition (Weight percentage)	Boiling Point (°C)
Acetone–chloroform	20.0% CH_3COCH_3, 80.0% $CHCl_3$	64.7
Chloroform–methyl ethyl ketone	17.0% $CHCl_3$, 83.0% $CH_3COCH_2CH_3$	79.9
Hydrochloric acid	20.2% HCl, 79.8% H_2O	108.6
Acetic acid–dioxane	77.0% CH_3COCH, 23.0% $C_4H_8O_2$	119.5
Benzaldehyde–phenol	49.0% C_6H_5CHO, 51.0% C_6H_5OH	185.6

Some maximum-boiling-point azeotropes are listed in Table 10.3. They are not nearly as common as minimum-boiling-point azeotropes.[3]

C. GENERALIZATIONS

There are some generalizations that can be made about azeotropic behavior. They are presented here without explanation, but you should be able to verify them by thinking through each case using the phase diagrams given. (Note that pure A is always to the left of the azeotrope in these diagrams, while pure B is to the right of the azeotrope.)

Minimum-Boiling-Point Azeotropes

Initial Composition	Experimental Result
to left of azeotrope	azeotrope distills first, pure A second
azeotrope	unseparable
to right of azeotrope	azeotrope distills first, pure B second

Maximum-Boiling Point Azeotropes

Initial Composition	Experimental Result
to left of azeotrope	pure A distills first, azeotrope second
azeotrope	unseparable
to right of azeotrope	pure B distills first, azeotrope second

[3]See Footnote 2.

10.8 AZEOTROPIC DISTILLATION: APPLICATIONS

There are numerous examples of chemical reactions in which the amount of product is low because of an unfavorable equilibrium. An example is the direct acid-catalyzed esterification of a carboxylic acid with an alcohol:

$$R-\overset{\overset{\displaystyle O}{\|}}{C}-OH + R-O-H \overset{H^+}{\rightleftharpoons} R-\overset{\overset{\displaystyle O}{\|}}{C}-OR + H_2O$$

Because the equilibrium does not favor formation of the ester, it must be shifted to the right, in favor of the product, by using an excess of one of the starting materials. In most cases, the alcohol is the least expensive reagent and is the material used in excess. Isopentyl acetate (Experiment 8) and methyl salicylate (Experiment 10) are examples of esters prepared by using one of the starting materials in excess.

Another way of shifting the equilibrium to the right is to remove one of the products from the reaction mixture as it is formed. In the above example, water can be removed as it is formed by **azeotropic distillation.** A common large-scale method is to use the Dean–Stark water separator shown in Figure 10.13A. In this technique, an in-

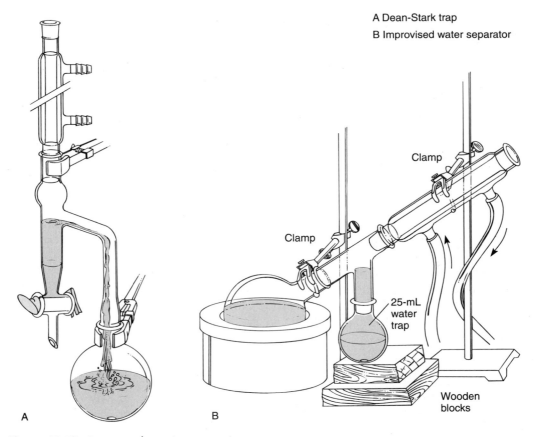

A Dean-Stark trap

B Improvised water separator

Clamp

Clamp

25-mL water trap

Wooden blocks

A   B

Figure 10.13   Large-scale water separators.

ert solvent, commonly benzene or toluene, is added to the reaction mixture contained in the round-bottom flask. The side-arm of the water separator is also filled with this solvent. If benzene is used, as the mixture is heated under reflux, the benzene–water azeotrope (bp 69.4°C, Table 10.2) distills out of the flask.[4] When the vapor condenses, it enters the side-arm directly below the condenser, and water separates from the benzene–water condensate; benzene and water mix as vapors, but they are not miscible as cooled liquids. Once the water (lower phase) separates from the benzene (upper phase), liquid benzene overflows from the side-arm back into the flask. The cycle is repeated continuously until no more water forms in the side-arm. You may calculate the weight of water that should theoretically be produced and compare this value with the amount of water collected in the side-arm. Because the density of water is 1.0, the volume of water collected can be compared directly with the calculated amount, assuming 100% yield.

An improvised water separator, constructed from the components found in the traditional organic kit, is shown in Figure 10.13B. Although this requires the condenser to be placed in a nonvertical position, it works quite well.

At the microscale level, water separation can be achieved using a standard distillation assembly with a water condenser and a Hickman head (Fig. 10.14). The side-ported variation of the Hickman head is the most convenient one to use for this purpose, but it is not essential. In this variation, you simply remove all the distillate (both solvent and water) several times during the course of the reaction. Use a Pasteur pipet to remove the distillate, as shown in Technique 8 (Fig. 8.7, p. 715). Because both the solvent and water are removed in this procedure, it may be desirable to add more solvent from time to time, adding it through the condenser with a Pasteur pipet.

The most important consideration in using azeotropic distillation to prepare an ester (described on p. 749) is that the azeotrope containing water must have a **lower boiling point** than the alcohol used. With ethanol, the benzene–water azeotrope boils at a much lower temperature (69.4°C) than ethanol (78.3°C), and the technique previously described works well. With higher-boiling-point alcohols, azeotropic distillation works well because of the large boiling-point difference between the azeotrope and the alcohol.

With methanol (bp 65°C), however, the boiling point of the benzene–water azeotrope is actually *higher* by about 5°C, and methanol distills first. Thus, in esterifications involving methanol, a totally different approach must be taken. For example, you can mix the carboxylic acid, methanol, the acid catalyst, and *1,2-dichloroethane* in a conventional reflux apparatus (Technique 3, Fig. 3.8, p. 619) without a water separator. During the reaction, water separates from the 1,2-dichloroethane because it is not miscible; however, the remainder of the components are soluble, so the reaction can continue. The equilibrium is shifted to the right by the "removal" of water from the reaction mixture.

---

[4]Actually, with ethanol, a lower-boiling-point, three-component azeotrope distills at 64.9°C (see Table 10.2). It consists of benzene–water–ethanol. Because some ethanol is lost in the azeotropic distillation, a large excess of ethanol is used in esterification reactions. The excess also helps to shift the equilibrium to the right.

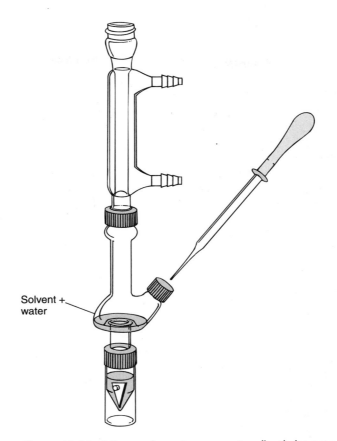

Figure 10.14   Microscale water separator (both layers are removed).

Solvent + water

Azeotropic distillation is also used in other types of reactions, such as ketal or acetal formation, and in enamine formation.

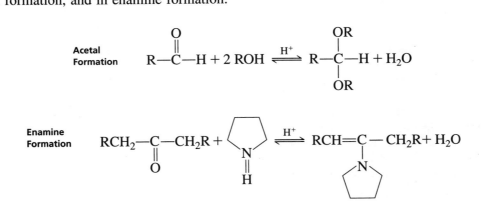

Acetal Formation

$$R-\overset{\overset{\displaystyle O}{\|}}{C}-H + 2\,ROH \;\rightleftharpoons\; R-\overset{\overset{\displaystyle OR}{|}}{\underset{\underset{\displaystyle OR}{|}}{C}}-H + H_2O$$

Enamine Formation

$$RCH_2-\overset{}{\underset{\underset{\displaystyle O}{\|}}{C}}-CH_2R + \underset{\underset{\displaystyle H}{|}}{\overset{}{N}} \;\rightleftharpoons\; RCH=C-CH_2R + H_2O$$

PROBLEMS

1. In the accompanying chart are approximate vapor pressures for benzene and toluene at various temperatures.

Temp (°C)	mmHg	Temp (°C)	mmHg
Benzene 30	120	Toluene 30	37
40	180	40	60
50	270	50	95
60	390	60	140
70	550	70	200
80	760	80	290
90	1010	90	405
100	1340	100	560
		110	760

   (a) What is the mole fraction of each component if 3.9 g of benzene C_6H_6 is dissolved in 4.6 g of toluene C_7H_8?

   (b) Assuming that this mixture is ideal, that is, it follows Raoult's Law, what is the partial vapor pressure of benzene in this mixture at 50°C?

   (c) Estimate to the nearest degree the temperature at which the vapor pressure of the solution equals 1 atm (bp of the solution).

   (d) Calculate the composition of the vapor (mole fraction of each component) that is in equilibrium in the solution at the boiling point of this solution.

   (e) Calculate the composition in weight percentage of the vapor that is in equilibrium with the solution.

   2. Estimate how many theoretical plates are needed to separate a mixture that has a mole fraction of B equal to 0.70 (70% B) in Figure 10.3.

   3. Two moles of sucrose are dissolved in 8 moles of water. Assume that the solution follows Raoult's Law and that the vapor pressure of sucrose is negligible. The boiling point of water is 100°C. The distillation is carried out at 1 atm (760 mmHg).

   (a) Calculate the vapor pressure of the solution when the temperature reaches 100°C.

   (b) What temperature would be observed during the entire distillation?

   (c) What would be the composition of the distillate?

   (d) If a thermometer were immersed below the surface of the liquid of the boiling flask, what temperature would be observed?

   4. Explain why the boiling point of a two-component mixture rises slowly throughout a simple distillation when the boiling-point differences are not large.

   5. Given the boiling points of several known mixtures of A and B (mole fractions are known) and the vapor pressures of A and B in the pure state (P_A^0 and P_B^0) at these same temperatures, how would you construct a boiling-point-composition phase diagram for A and B? Give a stepwise explanation.

6. Describe the behavior on distillation of a 98% ethanol solution through an efficient column. Refer to Figure 10.11.

7. Construct an approximate boiling-point-composition diagram for a benzene-methanol system. The mixture shows azeotropic behavior (see Table 10.2). Include on the graph the boiling points of pure benzene and pure methanol and the boiling point of the azeotrope. Describe the behavior for a mixture that is initially rich in benzene (90%) and then for a mixture that is initially rich in methanol (90%).

8. Construct an approximate boiling-point-composition diagram for an acetone–chloroform system, which forms a maximum boiling azeotrope (Table 10.3). Describe the behavior on distillation of a mixture that is initially rich in acetone (90%), then describe the behavior of a mixture that is initially rich in chloroform (90%).

9. Two components have boiling points of 130 and 150°C. Estimate the number of theoretical plates needed to separate these substances in a fractional distillation.

10. A spinning-band column has an HETP of 0.25 in./plate. If the column has 12 theoretical plates, how long is it?

T E C H N I Q U E   1 1

Steam Distillation

The simple, vacuum, and fractional distillations described in Techniques 8, 9, and 10 are applicable to completely soluble (miscible) mixtures only. When liquids are *not* mutually soluble (immiscible), they can also be distilled, but with a somewhat different result. A mixture of immiscible liquids will boil at a lower temperature than the boiling points of any of the separate components as pure compounds. When steam is used to provide one of the immiscible phases, the process is called **steam distillation.** The advantage of this technique is that the desired material distills at a temperature below 100°C. Thus, if unstable or very high-boiling substances are to be removed from a mixture, decomposition is avoided. Because all gases mix, the two substances can mix in the vapor and codistill. Once the distillate is cooled, the desired component, which is not miscible, separates from the water. Steam distillation is used widely in isolating liquids from natural sources. It is also used in removing a reaction product from a tarry reaction mixture.

11.1 DIFFERENCES BETWEEN DISTILLATION OF MISCIBLE AND IMMISCIBLE MIXTURES

$$\text{MISCIBLE LIQUIDS} \quad P_{\text{total}} = P_A^0 N_A + P_B^0 N_B \tag{1}$$

Two liquids A and B that are mutually soluble (miscible), and that do not interact, form an ideal solution and follow Raoult's Law, as shown in Equation 1. Note that the vapor pressures of pure liquids P_A^0 and P_B^0 are not added directly to give the total pressure P_{total} but are reduced by the respective mole fractions N_A and N_B. The total pressure above a miscible or homogeneous solution will depend on P_A^0 and P_B^0 and also N_A and N_B. Thus, the composition of the vapor will also depend on *both* the vapor pressures and the mole fractions of each component.

$$\text{IMMISCIBLE LIQUIDS} \quad P_{total} = P_A^0 + P_B^0 \tag{2}$$

In contrast, when two mutually insoluble (immiscible) liquids are "mixed" to give a heterogeneous mixture, each exerts its own vapor pressure, independently of the other, as shown in Equation 2. The mole fraction term does not appear in this equation, because the compounds are not miscible. You simply add the vapor pressures of the pure liquids P_A^0 and P_B^0 at a given temperature to obtain the total pressure above the mixture. When the total pressure equals 760 mmHg, the mixture boils. The composition of the vapor from an immiscible mixture, in contrast to the miscible mixture, is determined only by the vapor pressures of the two substances codistilling. Equation 3 defines the composition of the vapor from an immiscible mixture. Calculations involving this equation are given in Section 11.2.

$$\frac{\text{Moles A}}{\text{Moles B}} = \frac{P_A^0}{P_B^0} \tag{3}$$

A mixture of two immiscible liquids boils at a lower temperature than the boiling points of either component. The explanation for this behavior is like that given for minimum-boiling-point azeotropes (Technique 10, Section 10.7). Immiscible liquids behave as they do because an extreme incompatibility between the two liquids leads to higher combined vapor pressures than Raoult's Law would predict. The higher combined vapor pressures cause a lower boiling point for the mixture than for either single component. Thus, you may think of steam distillation as a special type of azeotropic distillation in which the substance is completely insoluble in water.

The differences in behavior of miscible and immiscible liquids, where it is assumed that P_A^0 equals P_B^0, are shown in Figure 11.1. Note that with miscible liquids, the composition of the vapor depends on the relative amounts of A and B present (Fig. 11.1A). Thus, the composition of the vapor must change during a distillation. In contrast, the composition of the vapor with immiscible liquids is independent of the amounts of A and B present (Fig. 11.1B). Hence, the vapor composition must remain *constant* during the distillation of such liquids, as predicted by Equation 3. Immiscible liquids act as if they were being distilled simultaneously from separate compartments, as shown in Figure 11.1B, even though in practice they are "mixed" during a steam distillation. Because all gases mix, they do give rise to a homogeneous vapor and codistill.

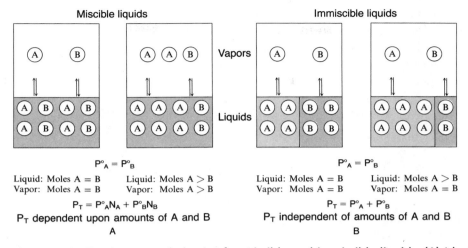

Miscible liquids   Immiscible liquids

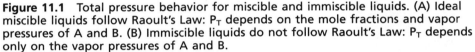

$P°_A = P°_B$ $P°_A = P°_B$

Liquid: Moles A = B   Liquid: Moles A > B   Liquid: Moles A = B   Liquid: Moles A > B
Vapor: Moles A = B   Vapor: Moles A > B   Vapor: Moles A = B   Vapor: Moles A = B

$P_T = P°_A N_A + P°_B N_B$ $P_T = P°_A + P°_B$

P_T dependent upon amounts of A and B P_T independent of amounts of A and B

A   B

Figure 11.1   Total pressure behavior for miscible and immiscible liquids. (A) Ideal miscible liquids follow Raoult's Law: P_T depends on the mole fractions and vapor pressures of A and B. (B) Immiscible liquids do not follow Raoult's Law: P_T depends only on the vapor pressures of A and B.

11.2 IMMISCIBLE MIXTURES: CALCULATIONS

The composition of the distillate is constant during a steam distillation, as is the boiling point of the mixture. The boiling points of steam-distilled mixtures will always be below the boiling point of water (bp 100°C) as well as the boiling point of any of the other substances distilled. Some representative boiling points and compositions of steam distillates are given in Table 11.1. Note that the higher the boiling point of a pure substance, the more closely the temperature of the steam distillate approaches, but does not exceed, 100°C. This is a reasonably low temperature, and it avoids the decomposition that might result at high temperatures with a simple distillation.

TABLE 11.1   Boiling Points and Compositions of Steam Distillates

Mixture	Boiling Point of Pure Substance (°C)	Boiling Point of Mixture (°C)	Composition (% Water)
Benzene–water	80.1	69.4	8.9%
Toluene–water	110.6	85.0	20.2%
Hexane–water	69.0	61.6	5.6%
Heptane–water	98.4	79.2	12.9%
Octane–water	125.7	89.6	25.5%
Nonane–water	150.8	95.0	39.8%
1-Octanol–water	195.0	99.4	90.0%

For immiscible liquids, the molar proportions of two components in a distillate equal the ratio of their vapor pressures in the boiling mixture, as given in Equation 3. When Equation 3 is rewritten for an immiscible mixture involving water, Equation 4 results. Equation 4 can be modified by substituting the relation moles = (weight/molecular weight) to give Equation 5.

$$\frac{\text{Moles substance}}{\text{Moles water}} = \frac{P^0_{\text{substance}}}{P^0_{\text{water}}} \tag{4}$$

$$\frac{\text{Wt substance}}{\text{Wt water}} = \frac{(P^0_{\text{substance}})(\text{Molecular weight}_{\text{substance}})}{(P^0_{\text{water}})(\text{Molecular weight}_{\text{water}})} \tag{5}$$

A sample calculation using this equation is given in Figure 11.2. Notice that the result of this calculation is very close to the experimental value given in Table 11.1.

11.3 STEAM DISTILLATION: METHODS

Two methods for steam distillation are in general use in the laboratory: the **direct method** and the **live steam method.** In the first method, steam is generated *in situ* (in place) by heating a distillation flask containing the compound and water. In the second method, steam is generated outside and is passed into the distillation flask using an inlet tube.

Problem   How many grams of water must be distilled to steam distill 1.55 g of 1-octanol from an aqueous solution? What will be the composition (wt %) of the distillate? The mixture distills at 99.4°C.

Answer   The vapor pressure of water at 99.4°C must be obtained from the CRC Handbook (= 744 mmHg).

(a) Obtain the partial pressure of 1-octanol.

$$P^\circ_{\text{1-octanol}} = P_{\text{total}} - P^\circ_{\text{water}}$$
$$P^\circ_{\text{1-octanol}} = (760 - 744) = 16 \text{ mmHg}$$

(b) Obtain the composition of the distillate.

$$\frac{\text{wt 1-octanol}}{\text{wt water}} = \frac{(16)(130)}{(744)(18)} = 0.155 \text{ g/g-water}$$

(c) Clearly, 10 g of water must be distilled.

$$(0.155 \text{ g/g-water})(10 \text{ g-water}) = 1.55 \text{ g 1-octanol}$$

(d) Calculate the **weight** percentages.

$$\text{1-octanol} = 1.55 \text{ g}/(10 \text{ g} + 1.55 \text{ g}) = 13.4\%$$
$$\text{water} = 10 \text{ g}/(10 \text{ g} + 1.55 \text{ g}) = 86.6\%$$

Figure 11.2 Sample calculations for a steam distillation.

A.   DIRECT METHOD

Standard Scale.   A large-scale direct method steam distillation is illustrated in Figure 11.3. Although a heating mantle may be used, it is probably best to use a flame with this method, because a large volume of water must be heated rapidly. A boiling stone must be used to prevent bumping. The separatory funnel allows more water to be added during the course of the distillation.

Distillate is collected as long as it is either cloudy or milky white in appearance. Cloudiness indicates that an immiscible liquid is separating. When the distillate runs clear in the distillation, it is usually a sign that only water is distilling. However, there are some steam distillations where the distillate is never cloudy, even though material has codistilled. You must observe carefully, and be sure to collect enough distillate that all of the organic material codistills.

Microscale.   The direct method of steam distillation is the only one suitable for microscale reactions. Steam is produced in the conical vial or distillation flask (*in situ*) by heating water to its boiling point in the presence of the compound to be distilled. This method works well for small amounts of materials. A microscale steam distillation apparatus is shown in Figure 11.4. Water and the compound to be distilled are placed in the flask and heated. A stirring bar or a boiling stone should be used to prevent bumping. The vapors of the water and the desired compound codistill when they are heated. They are

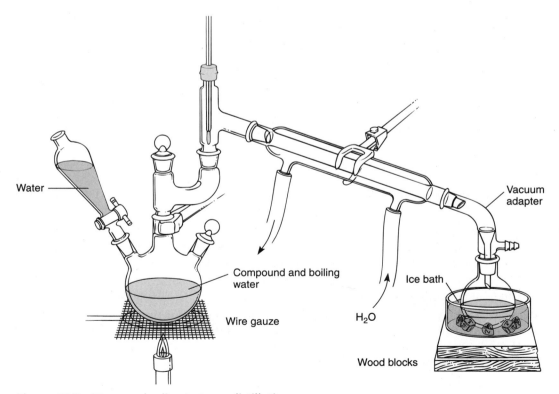

Figure 11.3   Macroscale direct steam distillation.

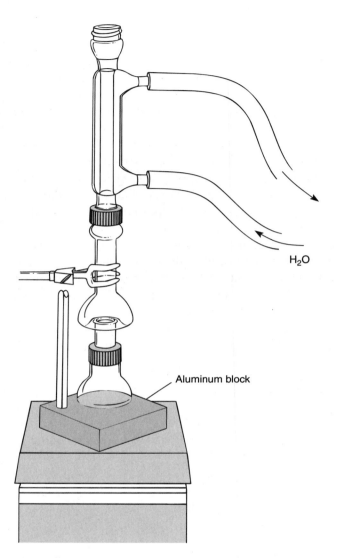

Aluminum block

H₂O

Figure 11.4 Microscale steam distillation.

condensed and collect in the Hickman head. When the Hickman head fills, the distillate is removed with a Pasteur pipet and placed in another vial for storage. For the typical microscale experiment, it will be necessary to fill the well and remove the distillate three or four times. All of these distillate fractions are placed in the same storage container. The efficiency in collecting the distillate can sometimes be improved if the inside walls of the Hickman head are rinsed several times into the well. A Pasteur pipet is used to perform the rinsing. Distillate is withdrawn from the well, and then it is used to wash the walls of the Hickman head all the way around the head. After the walls have been washed and when the well is full, the distillate can be withdrawn and transferred to the storage container. It may be necessary to add more water during the course of the distillation. More water is added (remove the condenser if used) through the center of the Hickman head by using a Pasteur pipet.

Semi-Microscale.   The apparatus shown in Figure 8.5, page 714, may also be used to perform a steam distillation at the microscale level or slightly above. This apparatus avoids the need to empty the collected distillate during the course of the distillation as is required when a Hickman head is used.

B.   LIVE STEAM METHOD

Standard Scale.   A large-scale steam distillation using the live steam method is shown in Figure 11.5. If steam lines are available in the laboratory, they may be attached directly to the steam trap (purge them first to drain water). If steam lines are not available, an external steam generator (see inset) must be prepared. The external generator usually will require a flame to produce steam at a rate fast enough for the distillation. When the distillation is first

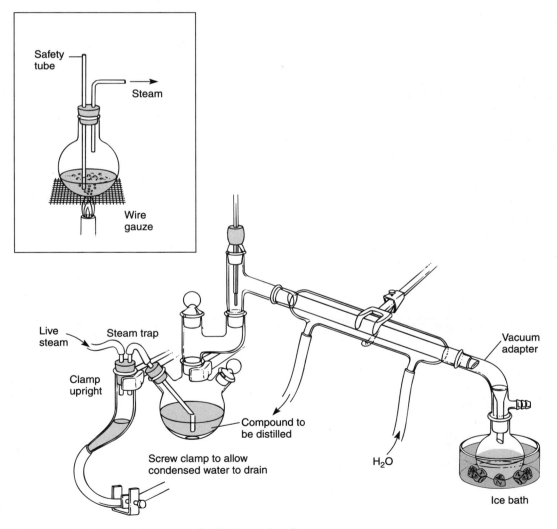

Figure 11.5   Macroscale steam distillation using live steam.

started, the clamp at the bottom of the steam trap is left open. The steam lines will have a large quantity of condensed water in them until they are well heated. When the lines become hot and condensation of steam ceases, the clamp may be closed. Occasionally, the clamp will have to be reopened to remove condensate. In this method, the steam agitates the mixture as it enters the bottom of the flask, and a stirrer or boiling stone is not required.

> **Caution.** Hot steam can produce very severe burns.

Sometimes it is helpful to heat the three-necked distilling flask with a heating mantle (or flame) to prevent excessive condensation at that point. Steam must be admitted at a fast enough rate for you to see the distillate condensing as a milky white fluid in the condenser. The vapors that codistill will separate on cooling to give this cloudiness. When the condensate becomes clear, the distillation is near the end. The flow of water through the condenser should be faster than in other types of distillation to help cool the vapors. Make sure the vacuum adapter remains cool to the touch. An ice bath may be used to cool the receiving flask if desired. When the distillation is to be stopped, the screw clamp on the steam trap should be opened, and the steam inlet tube must be removed from the three-necked flask. If this is not done, liquid will back up into the tube and steam trap.

PROBLEMS

1. Calculate the weight of benzene codistilled with each gram of water and the percentage composition of the vapor produced during a steam distillation. The boiling point of the mixture is 69.4°C. The vapor pressure of water at 69.4°C is 227.7 mmHg. Compare the result with the data in Table 11.1.

2. Calculate the approximate boiling point of a mixture of bromobenzene and water at atmospheric pressure. A table of vapor pressures of water and bromobenzene at various temperatures is given.

Temperature (°C)	Vapor Pressures (mmHg)	
	Water	**Bromobenzene**
93	588	110
94	611	114
95	634	118
96	657	122
97	682	127
98	707	131
99	733	136

3. Calculate the weight of nitrobenzene that codistills (bp 99°C of mixture) with each gram of water during a steam distillation. You may need the data given in Problem 2.

4. A mixture of *p*-nitrophenol and *o*-nitrophenol can be separated by steam distillation. The *o*-nitrophenol is steam volatile and the *para* isomer is not volatile. Explain. Base your answer on the ability of the isomers to form hydrogen bonds internally.

TECHNIQUE 12

Column Chromatography

The most modern and sophisticated methods of separating mixtures that the organic chemist has available all involve **chromatography.** Chromatography is defined as the separation of a mixture of two or more different compounds or ions by distribution between two phases, one of which is stationary and the other moving. Various types of chromatography are possible, depending on the nature of the two phases involved: **solid–liquid** (column, thin-layer, and paper), **liquid–liquid,** (high-performance liquid), and **gas–liquid** (vapor–phase) chromatographic methods are common.

All chromatography works on much the same principle as solvent extraction (Technique 7). Basically, the methods depend on the differential solubilities or adsorptivities of the substances to be separated relative to the two phases between which they are to be partitioned. In this chapter, column chromatography, a solid–liquid method, is considered. High-performance liquid chromatography is discussed in Technique 13. Thin-layer chromatography is examined in Technique 14; gas chromatography, a gas–liquid method, is discussed in Technique 15.

12.1 ADSORBENTS

Column chromatography is a technique based on both adsorptivity and solubility. It is a solid–liquid phase-partitioning technique. The solid may be almost any material that does not dissolve in the associated liquid phase; the solids used most commonly are silica gel $SiO_2 \cdot xH_2O$, also called silicic acid, and alumina $Al_2O_3 \cdot xH_2O$. These compounds are used in their powdered or finely ground forms (usually 200- to 400-mesh).

Most alumina used for chromatography is prepared from the impure ore bauxite $Al_2O_3 \cdot xH_2O + Fe_2O_3$. The bauxite is dissolved in hot sodium hydroxide and filtered to remove the insoluble iron oxides; the alumina in the ore forms the soluble amphoteric hydroxide $Al(OH)_4^-$. The hydroxide is precipitated by CO_2, which reduces the pH, as $Al(OH)_3$. When heated, the $Al(OH)_3$ loses water to form pure alumina Al_2O_3.

$$\text{Bauxite (crude)} \xrightarrow{\text{hot NaOH}} Al(OH)_4^- (aq) + Fe_2O_3 \text{ (insoluble)}$$

$$Al(OH)_4^- (aq) + CO_2 \longrightarrow Al(OH)_3 + HCO_3^-$$

$$2\ Al(OH)_3 \xrightarrow{\text{heat}} Al_2O_3(s) + 3\ H_2O$$

Alumnia prepared in this way is called **basic alumina** because it still contains some hydroxides. Basic alumina cannot be used for chromatography of compounds that are base-sensitive. Therefore, it is washed with acid to neutralize the base, giving **acid-washed alumina.** This material is unsatisfactory unless it has been washed with enough water to remove *all* the acid; on being so washed, it becomes the best chromatographic material, called **neutral alumina.** If a compound is acid-sensitive, either basic or neutral alumina must be used. You should be careful to ascertain what type of alumnia is being used for chromatography. Silica gel is not available in any form other than that suitable for chromatography.

12.2 INTERACTIONS

If powdered or finely ground alumina (or silica gel) is added to a solution containing an organic compound, some of the organic compound will *adsorb* onto or adhere to the fine particles of alumina. Many kinds of intermolecular forces cause organic molecules to bind to alumnia. These forces vary in strength according to their type. Nonpolar compounds bind to the alumina using only van der Waals forces. These are weak forces, and nonpolar molecules do not bind strongly unless they have extremely high molecular weights. The most important interactions are those typical of polar organic compounds. Either these forces are of the dipole–dipole type or they involve some direct interaction (coordination, hydrogen-bonding, or salt formation). These types of interactions are illustrated in Figure 12.1, which for convenience shows only a portion of the alumina structure. Similar interactions occur with silica gel. The strengths of such interactions vary in the approximate order:

Salt formation > Coordination > Hydrogen-bonding > Dipole–dipole > van der Waals

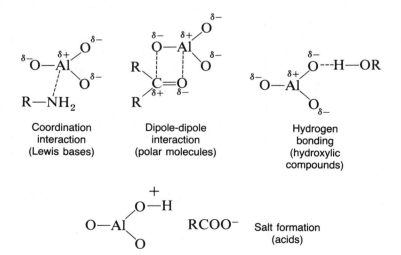

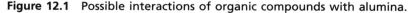

Figure 12.1 Possible interactions of organic compounds with alumina.

Strength of interaction varies among compounds. For instance, a strongly basic amine would bind more strongly than a weakly basic one (by coordination). In fact, strong bases and strong acids often interact so strongly that they *dissolve* alumina to some extent. You can use the following rule of thumb:

The more polar the functional group, the stronger the bond to alumina (or silica gel).

A similar rule holds for solubility. Polar solvents dissolve polar compounds more effectively than nonpolar solvents: nonpolar compounds are dissolved best by nonpolar solvents. Thus, the extent to which any given solvent can wash an adsorbed compound from alumina depends almost directly on the relative polarity of the solvent. For example, although a ketone adsorbed on alumina might not be removed by hexane, it might be removed completely by chloroform. For any adsorbed material, a kind of **distribution** equilibrium can be envisioned between the adsorbent material and the solvent. This is illustrated in Figure 12.2.

The distribution equilibrium is *dynamic,* with molecules constantly *adsorbing* from the solution and *desorbing* into it. The average number of molecules remaining adsorbed on the solid particles at equilibrium depends both on the particular molecule (RX) involved and the dissolving power of the solvent with which the adsorbent must compete.

12.3 PRINCIPLE OF COLUMN CHROMATOGRAPHIC SEPARATION

The dynamic equilibrium mentioned previously, and the variations in the extent to which different compounds adsorb on alumina or silica gel, underlie a versatile and ingenious method for *separating* mixtures of organic compounds. In this method, the mixture

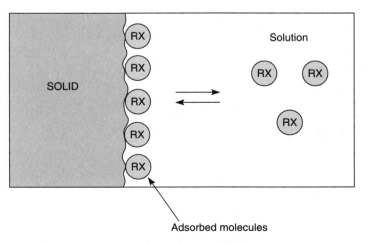

Figure 12.2   Dynamic adsorption equilibrium.

of compounds to be separated is introduced onto the top of a cylindrical glass column (Figure 12.3) **packed** or filled with fine alumina particles (stationary solid phase). The adsorbent is continuously washed by a flow of solvent (moving phase) passing through the column.

Initially, the components of the mixture adsorb onto the alumina particles at the top of the column. The continuous flow of solvent through the column **elutes,** or washes, the solutes off the alumina and sweeps them down the column. The solutes (or materials to be separated) are called **eluates** or **elutants;** and the solvents are called **eluents.** As the solutes pass down the column to fresh alumina, new equilibria are established between the adsorbent, the solutes, and the solvent. The constant equilibration means that different compounds will move down the column at differing rates depending on their relative affinity for the adsorbent on one hand, and for the solvent on the other. Because the number of alumina particles is large, because they are closely packed, and because fresh solvent is being added continuously, the number of equilibrations between adsorbent and solvent that the solutes experience is enormous.

As the components of the mixture are separated, they begin to form moving bands (or zones), each band containing a single component. If the column is long enough and the other parameters (column diameter, adsorbent, solvent, and flow rate) are correctly chosen, the bands separate from one another, leaving gaps or bands of pure solvent in between. As each band (solvent and solute) passes out the bottom of the column, it can be collected before the next band arrives. If the parameters mentioned are poorly chosen, the various bands either overlap or coincide, in which case either a poor separation or no separation at all is the result. A successful chromatographic separation is illustrated in Figure 12.4.

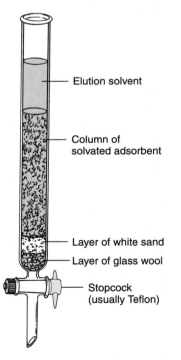

Elution solvent

Column of
solvated adsorbent

Layer of white sand
Layer of glass wool

Stopcock
(usually Teflon)

Figure 12.3 Chromatographic column.

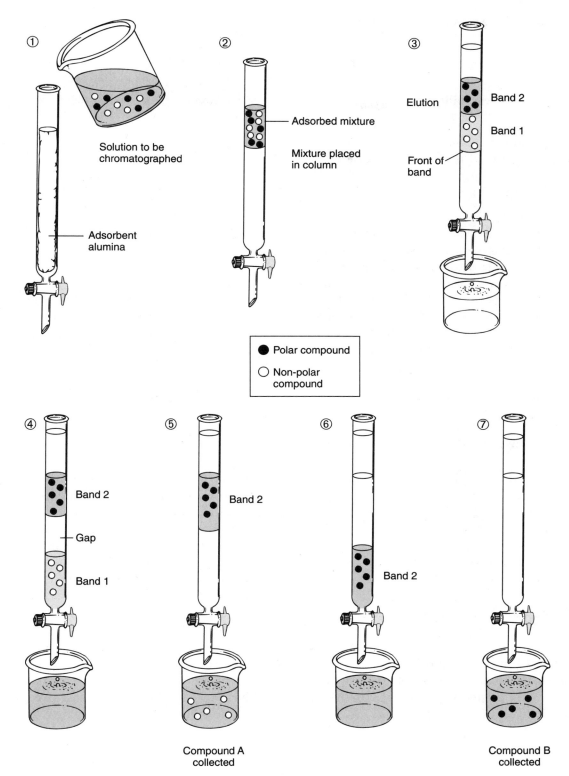

Figure 12.4 Sequence of steps in a chromatographic separation.

12.4 PARAMETERS AFFECTING SEPARATION

The versatility of column chromatography results from the many factors that can be adjusted. These include:

1. Adsorbent chosen
2. Polarity of the column or solvents chosen
3. Size of the column (both length and diameter) relative to the amount of material to be chromatographed
4. Rate of elution (or flow)

By carefully choosing the conditions, almost any mixture can be separated. This technique has even been used to separate optical isomers. An optically active solid-phase adsorbent was used to separate the enantiomers.

Two fundamental choices for anyone attempting a chromatographic separation are the kind of adsorbent and the solvent system. In general, nonpolar compounds pass through the column faster than polar compounds, because they have a smaller affinity for the adsorbent. If the adsorbent chosen binds all the solute molecules (both polar and nonpolar) strongly, they will not move down the column. On the other hand, if too polar a solvent is chosen, all the solutes (polar and nonpolar) may simply be washed through the column, with no separation taking place. The adsorbent and the solvent should be chosen so that neither is favored excessively in the equilibrium competition for solute molecules.[1]

A. ADSORBENTS

In Table 12.1, various kinds of adsorbents (solid phases) used in column chromatography are listed. The choice of adsorbent often depends on the types of compounds to be separated. Cellulose, starch, and sugars are used for polyfunctional plant and animal materials (natural products) that are very sensitive to acid–base interactions. Magnesium silicate is often used for separating acetylated sugars, steroids, and essential oils. Silica gel and Florisil are relatively mild toward most compounds and are widely used for a variety of functional groups—hydrocarbons, alcohols, ketones, esters, acids, azo compounds, amines. Alumina is the most widely used adsorbent and is obtained in the three forms mentioned in Section 12.1: acidic, basic, and neutral. The pH of acidic or acid-washed alumina is approximately 4. This adsorbent is particularly useful for separating acidic materials such as carboxylic acids and amino acids. Basic alumina has a pH of 10 and is useful in separating amines. Neutral alumina can be used to separate a variety of nonacidic and nonbasic materials.

The approximate strength of the various adsorbents listed in Table 12.1 is also given. The order is only approximate and therefore it may vary. For instance, the strength, or separating abilities, of alumina and silica gel largely depend on the amount of water pre-

---

[1]Often the chemist uses thin-layer chromatography (TLC), which is described in Technique 14, to arrive at the best choices of solvents and adsorbents for the best separation. The TLC experimentation can be performed quickly and with extremely small amounts (microgram quantities) of the mixture to be separated. This saves significant time and materials.

TABLE 12.1   Solid Adsorbents for Column Chromatography

Paper	
Cellulose	
Starch	
Sugars	
Magnesium silicate	
Calcium sulfate	**Increasing strength of**
Silicic acid	**binding interactions**
Silica gel	**toward polar compounds**
Florisil	
Magnesium oxide	
Aluminum oxide (Alumina)*	
Activated charcoal (Norit)	

*Basic, acid-washed, and neutral.

sent. Water binds very tightly to either adsorbent, taking up sites on the particles that could otherwise be used for equilibration with solute molecules. If one adds water to the adsorbent, it is said to have been **deactivated.** Anhydrous alumina or silica gel are said to be highly **activated.** High activity is usually avoided with these adsorbents. Use of the highly active forms of either alumina or silica gel, or of the acidic or basic forms of alumina, can often lead to molecular rearrangement or decomposition in certain types of solute compounds.

The chemist can select the degree of activity that is appropriate to carry out a particular separation. To accomplish this, highly activated alumina is mixed thoroughly with a precisely measured quantity of water. The water partially hydrates the alumina and thus reduces its activity. By carefully determining the amount of water required, the chemist can have available an entire spectrum of possible activities.

B.   SOLVENTS

In Table 12.2, some common chromatographic solvents are listed along with their relative ability to dissolve polar compounds. Sometimes, a single solvent can be found that will separate all the components of a mixture. Sometimes, a mixture of solvents can be found that will achieve separation. More often, you must start elution with a nonpolar solvent to remove relatively nonpolar compounds from the column and then gradually increase the solvent polarity to force compounds of greater polarity to come down the column, or elute. The approximate order in which various classes of compounds elute by this procedure is given in Table 12.3. In general, nonpolar compounds travel through the column faster (elute first), and polar compounds travel more slowly (elute last). However, molecular weight is also a factor in determining the order of elution. A nonpolar compound of high molecular weight travels more slowly than a nonpolar compound of low molecular weight, and it may even be passed by some polar compounds.

TABLE 12.2   Solvents (Eluents) for Chromatography

Petroleum ether	
Cyclohexane	
Carbon tetrachloride*	
Toluene	
Chloroform*	
Methylene chloride	**Increasing polarity**
Diethyl ether	**and "solvent power"**
Ethyl acetate	**toward polar functional**
Acetone	**groups**
Pyridine	
Ethanol	
Methanol	
Water	
Acetic acid	

*Suspected carcinogens.

When the polarity of the solvent has to be changed during a chromatographic separation, some precautions must be taken. Rapid changes from one solvent to another are to be avoided (especially when silica gel or alumina is involved). Usually, small percentages of a new solvent are mixed slowly into the one in use until the percentage reaches the desired level. If this is not done, the column packing often "cracks" as a result of the heat liberated when alumina or silica gel is mixed with a solvent. The solvent solvates the adsorbent, and the formation of a weak bond generates heat.

$$\text{Solvent} + \text{Alumina} \longrightarrow (\text{Alumina} \cdot \text{Solvent}) + \text{Heat}$$

TABLE 12.3   Elution Sequence for Compounds

Hydrocarbons	**Fastest (will elute with nonpolar solvent)**
Olefins	
Ethers	
Halocarbons	
Aromatics	
Ketones	**Order of elution**
Aldehydes	
Esters	
Alcohols	
Amines	
Acids, strong bases	**Slowest (need a polar solvent)**

Often, enough heat is generated locally to evaporate the solvent. The formation of vapor creates bubbles, which forces a separation of the column packing; this is called **cracking.** A cracked column does not produce a good separation, because it has discontinuities in the packing. The way in which a column is packed or filled is also very important in preventing cracking.

That the solvent itself has a tendency to adsorb on the alumina is an important factor in how compounds move down the column. The solvent can displace the adsorbed compound if it is more polar than the compound and, hence, can move it down the column. Thus, a more polar solvent not only dissolves more compound but also is effective in removing the compound from the alumina, because it displaces the compound from its site of adsorption.

Certain solvents should be avoided with alumina or silica gel, especially with the acidic, basic, and highly active forms. For instance, with any of these adsorbents, acetone dimerizes *via* an aldol condensation to give diacetone alcohol. Mixtures of esters *transesterify* (exchange their alcoholic portions) when ethyl acetate or an alcohol is the eluent. Finally, the most active solvents (pyridine, methanol, water, and acetic acid) dissolve and elute some of the adsorbent itself. Generally, try to avoid going to solvents more polar than diethyl ether or methylene chloride in the eluent series (Table 12.2).

C.   COLUMN SIZE AND ADSORBENT QUANTITY

The column size and the amount of adsorbent must also be selected correctly to separate a given amount of sample well. As a rule of thumb, the amount of adsorbent should be 25 to 30 times, by weight, the amount of material to be separated by chromatography. Furthermore, the column should have a height-to-diameter ratio of about 8:1. Some typical relations of this sort are given in Table 12.4.

Note, as a caution, that the difficulty of the separation is also a factor in determining the size and length of the column to be used and in the amount of adsorbent needed. Compounds that do not separate easily may require larger columns and more adsorbent than specified in Table 12.4. For easily separated compounds, a smaller column and less adsorbent may suffice.

TABLE 12.4   Size of Column and Amount of Adsorbent for Typical Sample Sizes

Amount of Sample (g)	Amount of Adsorbent (g)	Column Diameter (mm)	Column Height (mm)
0.01	0.3	3.5	30
0.10	3.0	7.5	60
1.00	30.0	16.0	130
10.00	300.0	35.0	280

D. FLOW RATE

The rate at which solvent flows through the column is also significant in the effectiveness of a separation. In general, the time the mixture to be separated remains on the column is directly proportional to the extent of equilibration between stationary and moving phases. Thus, similar compounds eventually separate if they remain on the column long enough. The time a material remains on the column depends on the flow rate of the solvent. If the flow is too slow, however, the dissolved substances in the mixture may diffuse faster than the rate at which they move down the column. Then the bands grow wider and more diffuse, and the separation becomes poor.

12.5 PACKING THE COLUMN: TYPICAL PROBLEMS

The most critical operation in column chromatography is packing (filling) the column with adsorbent. The **column packing** must be evenly packed and free of irregularities, air bubbles, and gaps. As a compound travels down the column, it moves in an advancing zone, or **band.** It is important that the leading edge, or **front,** of this band be horizontal, or perpendicular to the long axis of the column. If two bands are close together and do not have horizontal band fronts, it is impossible to collect one band while completely excluding the other. The leading edge of the second band begins to elute before the first band has finished eluting. This condition can be seen in Figure 12.5. There

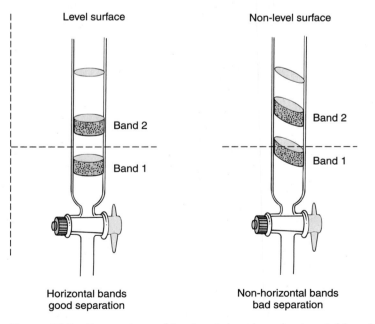

Figure 12.5 Comparison of horizontal and nonhorizontal band fronts.

are two main reasons for this problem. First, if the top surface edge of the adsorbent packing is not level, nonhorizontal bands result. Second, bands may be nonhorizontal if the column is not held in an exactly vertical position in both planes (front-to-back and side-to-side). When you are preparing a column, you must watch both these factors carefully.

Another phenomenon, called **streaming** or **channeling,** occurs when part of the band front advances ahead of the major part of the band. Channeling occurs if there are any cracks or irregularities in the adsorbent surface or any irregularities caused by air bubbles in the packing. A part of the advancing front moves ahead of the rest of the band by flowing through the channel. Two examples of channeling are shown in Figure 12.6.

12.6 PACKING THE COLUMN: STANDARD-SCALE AND SEMI-MICROSCALE METHODS

The following methods are used to avoid problems resulting from uneven packing and column irregularities. These procedures should be followed carefully in preparing a chromatography column. Failure to pay close attention to the preparation of the column may well affect the quality of the separation.

Preparation of a column involves two distinct stages. In the first stage, a support base on which the packing will rest is prepared. This must be done so that the packing, a finely divided material, does not wash out of the bottom of the column. In the second stage, the column of adsorbent is deposited on top of the supporting base.

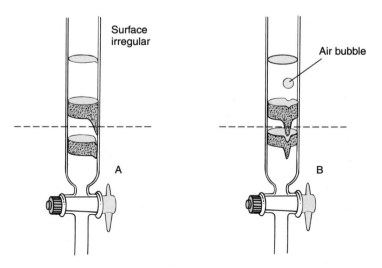

Figure 12.6 Channeling complications.

A. PREPARING THE SUPPORT BASE

Standard-Scale Columns.   For large-scale applications, clamp a chromatography column upright (vertically). The column (Fig. 12.3) is a piece of cylindrical glass tubing with a stopcock attached at one end. The stopcock usually has a Teflon plug, because stopcock grease (used on glass plugs) dissolves in many of the organic solvents used as eluents. Stopcock grease in the eluent will contaminate the eluates.

Instead of a stopcock, attach a piece of flexible tubing to the bottom of the column, with a screw clamp used to stop or regulate the flow (Fig. 12.7). When a screw clamp is used, care must be taken that the tubing is not dissolved by the solvents that will pass through the column during the experiment. Rubber, for instance, dissolves in chloroform, benzene, methylene chloride, toluene, or tetrahydrofuran (THF). Tygon tubing dissolves (actually, the plasticizer is removed) in many solvents, including benzene, methylene chloride, chloroform, ether, ethyl acetate, toluene, and THF. Polyethylene tubing is the best choice for use at the end of a column, because it is inert with most solvents.

Next, the column is partially filled with a quantity of solvent, usually a nonpolar solvent like hexane, and a support for the finely divided adsorbent is prepared in the following way. A loose plug of glass wool is tamped down into the bottom of the column with a long glass rod until all entrapped air is forced out as bubbles. Take care not to plug the column totally by tamping the glass wool too hard. A small layer of clean white sand is formed on top of the glass wool by pouring sand into the column. The column is tapped to level the surface of the sand. Any sand adhering to the side of the column is washed down with a small quantity of solvent. The sand forms a base that supports the column

Figure 12.7   Tubing with screw clamp to regulate solvent flow on a chromatography column.

of adsorbent and prevents it from washing through the stopcock. The column is packed in one of two ways: by the slurry method or by the dry pack method.

Semi-Microscale Columns.   An alternative apparatus for small-scale column chromatography is a commercial column, such as the one shown in Figure 12.8. This type of column is made of glass and has a solvent-resistant plastic stopcock at the bottom.[1] The stopcock assembly contains a filter disc to support the adsorbent column. An optional upper fitting, also made of solvent-resistant plastic, serves as a solvent reservoir. The column shown in Figure 12.8 is equipped with the solvent reservoir. This type of column is available in a variety of lengths, ranging from 100 to 300 mm. Because the column has a built-in filter disc, it is not necessary to prepare a support base before the adsorbent is added.

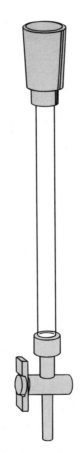

Figure 12.8 Commercial semi-microscale chromatography column. (The column is shown equipped with an optional solvent reservoir.)

---

[1]Note to the instructor: With certain organic solvents, we have found that the "solvent-resistant plastic" stopcock may tend to dissolve! We recommend that instructors test their equipment with the solvents that they intend to use before the beginning of the laboratory class.

B.   DEPOSITING THE ADSORBENT

Slurry Method

In the slurry method, the adsorbent is packed into the column as a mixture of a solvent and an undissolved solid. The slurry is prepared in a separate container by adding the solid adsorbent, a little at a time, to a quantity of the solvent. This order of addition (adsorbent added to solvent) should be followed strictly, because the adsorbent solvates and liberates heat. If the solvent is added to the adsorbent, it may boil away almost as fast as it is added due to heat evolved. This will be especially true if ether or another low-boiling solvent is used. When this happens, the final mixture will be uneven and lumpy. Enough adsorbent is added to the solvent, and mixed by swirling the container, to form a thick, but flowing, slurry. The container should be swirled until the mixture is homogenous and relatively free of entrapped air bubbles.

For standard-sized or semi-microscale columns, the procedure is as follows. When the slurry has been prepared, the column is filled about half-full with solvent, and the stopcock is opened to allow solvent to drain slowly into a large beaker. The slurry is mixed by swirling and is then poured in portions into the top of the draining column (a wide-necked funnel may be useful here). Be sure to swirl the slurry thoroughly before each addition to the column. The column is tapped constantly and *gently* on the side, during the pouring operation, with the fingers or with a pencil fitted with a rubber stopper. A short piece of large-diameter pressure tubing may also be used for tapping. The tapping promotes even settling and mixing and gives an evenly packed column free of air bubbles. Tapping is continued until all the material has settled, showing a well-defined level at the top of the column. Solvent from the collecting beaker may be re-added to the slurry if it becomes too thick to be poured into the column at one time. In fact, the collected solvent should be cycled through the column several times to ensure that settling is complete and that the column is firmly packed. The downward flow of solvent tends to compact the adsorbent. You should take care never to let the column run dry during packing.

Dry Pack Method 1

In the first of the dry pack methods introduced here, the column is filled with solvent and allowed to drain *slowly*. The dry adsorbent is added, a little at a time, while the column is tapped gently with a pencil, finger, or glass rod.

Standard-Scale Columns.   A plug of cotton is placed at the base of the column, and an even layer of sand is formed on top (see p. 764). The column is filled about half-full with solvent, and the solid adsorbent is added carefully from a beaker, while the solvent is allowed to flow slowly from the column. As the solid is added, the column is tapped as described for the slurry method in order to ensure that the column is packed evenly. When the column has the desired length, no more adsorbent is added. This method also produces an evenly packed column. Solvent should be cycled through this column (for standard scale applications) several times before each use. The same portion of solvent that has drained from the column during the packing is used to cycle through the column.

Semi-Microscale Columns. The procedure to fill a commercial semi-microscale column is essentially the same as that used to fill a Pasteur pipet (Section 12.7). The commercial column has the advantage that it is much easier to control the flow of solvent from the column during the filling process, because the stopcock can be adjusted appropriately. It is not necessary to use a cotton plug or to deposit a layer of sand before adding the adsorbent. The presence of the fritted disc at the base of the column acts to prevent adsorbent from escaping from the column.

Dry Pack Method 2

Standard-Scale Columns. Standard-scale columns can also be packed by a dry pack method that is similar to the microscale methods described in Section 12.7. The disadvantages described for the microscale method also apply to the macroscale method. This method is not recommended for use with silica gel or alumina, because the combination leads to uneven packing, air bubbles, and cracking, especially if a solvent that has a highly exothermic heat of solvation is used.

Semi-Microscale Columns. The Dry Pack Method 2 is similar to that described for Pasteur pipets (Section 12.7), except that the plug of cotton is not required. The flow rate of solvent through the column can be controlled using the stopcock, which is part of the column assembly (see Fig. 12.8).

12.7 PACKING THE COLUMN: MICROSCALE METHODS

As with standard-scale or semi-microscale columns, the procedures described in this section should be followed carefully in preparing a microscale chromatography column. Failure to pay close attention to the details of these procedures may adversely affect the quality of the separation.

Again, preparation of a column involves two distinct stages: preparation of the support base and filling the column with adsorbent.

A.   PREPARING THE SUPPORT BASE

For microscale applications, select a Pasteur pipet ($5\frac{3}{4}$-inch) and clamp it upright (vertically). In order to reduce the amount of solvent needed to fill the column, break off most of the tip of the pipet. Place a small ball of cotton in the pipet and tamp it into position using a glass rod or a piece of wire. Take care not to plug the column totally by tamping the cotton too hard. The correct position of the cotton is shown in Figure 12.9. A microscale chromatography column is packed by one of the dry pack methods described in Part B of this section.

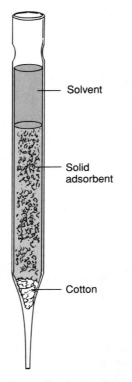

Solvent

Solid
adsorbent

Cotton

Figure 12.9   Microscale chromatography column.

B. DEPOSITING THE ADSORBENT

The slurry method is not recommended as a microscale method for use with Pasteur pipets. On a very small scale, it is too difficult to pack the column with the slurry without losing the solvent before the packing has been completed. Microscale columns should be packed by the one of the dry pack methods, as described in the following discussion.

Dry Pack Method 1

To fill a microscale column, fill the Pasteur pipet (with the cotton plug, prepared as described in Section A) about half-full with solvent. Using a microspatula, add the solid adsorbent slowly to the solvent in the column. As you add the solid, tap the column *gently* with a pencil, a finger, or a glass rod. The tapping promotes even settling and mixing and gives an evenly packed column free of air bubbles. As the adsorbent is added, solvent flows out of the Pasteur pipet. Because the adsorbent must not be allowed to dry during the packing process, you must use a means of controlling the solvent flow. If a piece of small-diameter plastic tubing is available, it can be fitted over the narrow tip of the Pasteur pipet. The flow rate can then be controlled using a screw clamp. A simple approach to controlling the flow rate is to use a finger over the top of the Pasteur pipet, much as you control the flow of liquid in a volumetric pipet. Continue adding the adsorbent slowly, with constant tapping, until the level of the adsorbent has reached the desired level. As you pack the column, be careful not to let the column run dry. The final column should appear as shown in Figure 12.9.

Dry Pack Method 2

An alternative dry pack method for microscale columns is to fill the Pasteur pipet with *dry* adsorbent, without any solvent. A plug of cotton is positioned in the bottom of the Pasteur pipet. The desired amount of adsorbent is added slowly, and the pipet tapped constantly, until the level of adsorbent has reached the desired height. Figure 12.9 can be used as a guide to judge the correct height of the column of adsorbent. When the column is packed, added solvent is allowed to percolate through the adsorbent until the entire column is moistened. The solvent is not added until just before the column is to be used.

This method is not recommended for use with silica gel nor for experiments where a very careful separation is required.

This method is useful when the adsorbent is alumina, but it does not produce satisfactory results with silica gel. Even with alumina, poor separations can arise due to uneven packing, air bubbles, and cracking, especially if a solvent that has a highly exothermic heat of solvation is used.

12.8 APPLYING THE SAMPLE TO THE COLUMN

The solvent (or solvent mixture) used to pack the column is normally the least polar elution solvent that can be used during chromatography. The compounds to be chromatographed are not highly soluble in the solvent. If they were, they would probably have a greater affinity for the solvent than for the adsorbent and would pass right through the column without equilibrating with the stationary phase.

The first elution solvent, however, is generally not a good solvent to use in preparing the sample to be placed on the column. Because the compounds are not highly soluble in nonpolar solvents, it takes a large amount of the initial solvent to dissolve the compounds, and it is difficult to get the mixture to form a narrow band on top of the column. A narrow band is ideal for an optimum separation of components. For the best separation, therefore, the compound is applied to the top of the column undiluted if it is a liquid, or in a *very small* amount of highly polar solvent if it is a solid. Water must not be used to dissolve the initial sample being chromatographed, because it reacts with the column packing.

In adding the sample to the column, use the following procedure. Lower the solvent level to the top of the adsorbent column by draining the solvent from the column. Add the sample (either a pure liquid or a solution) to form a small layer on top of the adsorbent. A Pasteur pipet is convenient for adding the sample to the column. Take care not to disturb the surface of the adsorbent. This is best accomplished by touching the pipet to the inside of the glass column and slowly draining it so as to allow the sample to spread into a thin film, which slowly descends to cover the entire adsorbent surface. Drain the pipet close to the surface of the adsorbent. When all the sample has been added, drain this small layer of liquid into the column until the top surface of the column *just begins* to dry. Then add a small layer of the chromatographic solvent carefully with a Pasteur pipet, again being careful not to disturb the surface. Drain this small layer of solvent into the column until the top surface of the column just dries. Add another small layer of fresh solvent, if necessary, and repeat the process until it is clear that the sample is strongly adsorbed on the top of the column. If the sample is colored and the fresh layer of solvent acquires some of this color, the sample has not been properly adsorbed. Once the sample has been properly applied, you can protect the level surface of the adsorbent by carefully filling the top of the column with solvent and sprinkling clean, white sand into the column so as to form a small protective layer on top of the adsorbent. For microscale applications, this layer of sand is not required.

Separations are often better if the sample is allowed to stand a short time on the column before elution. This allows a true equilibrium to be established. In columns that stand for too long, however, the adsorbent often compacts or even swells, and the flow can become annoyingly slow. Diffusion of the sample to widen the bands also becomes a problem if a column is allowed to stand over an extended period. For microscale chromatography, using Pasteur pipets, there is no stopcock, and it is not possible to stop the flow. In this case, it is not considered necessary to allow the column to stand.

12.9 ELUTION TECHNIQUES

Solvents for analytical and preparative chromatography should be pure reagents. Commercial-grade solvents often contain small amounts of residue, which remains when the solvent is evaporated. For normal work, and for relatively easy separations that take only small amounts of solvent, the residue usually presents few problems. For large-scale work, commercial-grade solvents may have to be redistilled before use. This is especially true for hydrocarbon solvents, which tend to have more residue than other solvent types. Most of the experiments in this laboratory manual have been designed to avoid this particular problem.

One usually begins elution of the products with a nonpolar solvent, like hexane or petroleum ether. The polarity of the elution solvent can be increased gradually by adding successively greater percentages of either ether or toluene (for instance, 1, 2, 5, 10, 15, 25, 50, 100%) or some other solvent of greater solvent power (polarity) than hexane. The transition from one solvent to another should not be too rapid in most solvent changes. If the two solvents to be changed differ greatly in their heats of solvation in binding to the adsorbent, enough heat can be generated to crack the column. Ether is especially troublesome in this respect, as it has both a low boiling point and a relatively high heat of solvation. Most organic compounds can be separated on silica gel or alumina using hexane–ether or hexane–toluene combinations for elution, and following these by pure methylene chloride. Solvents of greater polarity are usually avoided for the various reasons mentioned previously. In microscale work, the usual procedure is to use only one solvent for the chromatography.

The flow of solvent through the column should not be too rapid, or the solutes will not have time to equilibrate with the adsorbent as they pass down the column. If the rate of flow is too low, or stopped for a period, diffusion can become a problem—the solute band will diffuse, or spread out, in all directions. In either of these cases, separation will be poor. As a general rule (and only an approximate one), most standard scale columns are run with flow rates ranging from 5 to 50 drops of effluent per minute; a steady flow of solvent is usually avoided. Microscale columns made from Pasteur pipets do not have a means of controlling the solvent flow rate, but commercial semimicroscale columns are equipped with stopcocks. The solvent flow rate in this type of column can be adjusted in a manner similar to that used with larger columns. To avoid diffusion of the bands, do not stop the column and do not set it aside overnight.

12.10 RESERVOIRS

When large quantities of solvent are used in a chromatographic separation, it is often convenient to use a solvent reservoir to forestall having to add small portions of fresh solvent continually. The simplest type of reservoir, a feature of many columns, is created by fusing the top of the column to a round-bottom flask (Figure 12.10A). If the column has a standard-taper joint at its top, a reservoir can be created by joining a standard-taper separatory funnel to the column (Figure 12.10B). In this arrangement, the stopcock is left open, and no stopper is placed in the top of the separatory funnel. A third common arrangement is shown in Figure 12.10C. A separatory funnel is filled with solvent; its stopper is wetted with solvent and put *firmly* in place. The funnel is inserted into the empty filling space at the top of the chromatographic column, and the stopcock is opened. Solvent flows out of the funnel, filling the space at the top of the column until the solvent level is well above the outlet of the separatory funnel. As solvent drains from the column, this arrangement automatically refills the space at the top of the column by allowing air to enter through the stem of the separatory funnel. Some semi-microscale columns, such as that shown in Figure 12.8, are equipped with a solvent reservoir that fits onto the top of the column. It functions just like the reservoirs described in this section.

For a microscale chromatography, the portion of the Pasteur pipet above the adsorbent is used as a reservoir of solvent. Fresh solvent, as needed, is added by means of another Pasteur pipet. When it is necessary to change solvent, the new solvent is also added in this manner. In some cases, the chromatography may proceed too slowly; the rate of solvent flow can be accelerated by attaching a rubber dropper bulb to the top of the Pasteur pipet column and squeezing *gently*. The additional air pressure forces the solvent through the column more rapidly. If this technique is used, however, care must be taken to remove the rubber bulb from the column before releasing it. Otherwise, air may be drawn up through the bottom of the column, destroying the column packing.

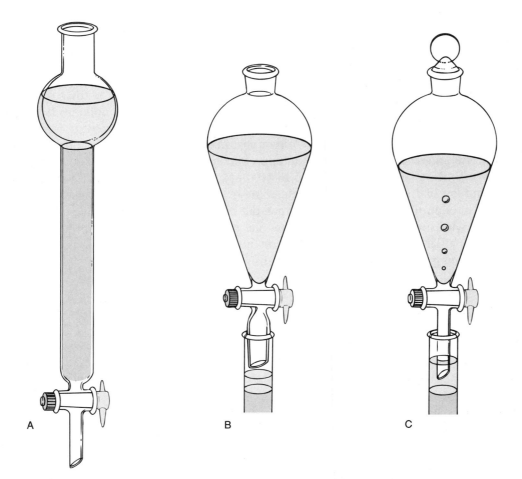

Figure 12.10   Various types of solvent-reservoir arrangements for chromatographic columns.

12.11 MONITORING THE COLUMN

It is a happy instance when the compounds to be separated are colored. The separation can then be followed visually and the various bands collected separately as they elute from the column. For the majority of organic compounds, however, this lucky circumstance does not exist, and other methods must be used to determine the positions of the bands. The most common method of following a separation of colorless compounds is to collect **fractions** of constant volume in pre-weighed flasks, to evaporate the solvent from each fraction, and to reweigh the flask plus any residue. A plot of fraction number *versus* the weight of the residues after evaporation of solvent gives a plot like that in Figure 12.11. Clearly, fractions 2 through 7 (Peak 1) may be combined as a single compound, and so can fractions 8 through 11 (Peak 2) and 12 through 15 (Peak 3). The size of the fractions collected (1, 10, 100, or 500 mL) depends on the size of the column and the ease of separation.

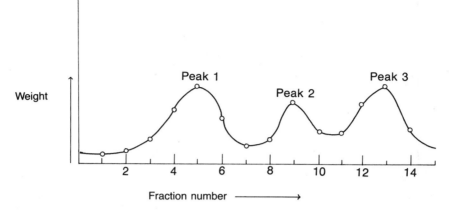

Figure 12.11 Typical elution graph.

Another common method of monitoring the column is to mix an inorganic phosphor into the adsorbent used to pack the column. When the column is illuminated with an ultraviolet light, the adsorbent treated in this way fluoresces. However, many solutes have the ability to **quench** the fluorescence of the indicator phosphor. In areas in which solutes are present, the adsorbent does not fluoresce, and a dark band is visible. In this type of column, the separation can also be followed visually.

Thin-layer chromatography is often used to monitor a column. This method is described in Technique 14 (Section 14.10, p. 804). Several sophisticated instrumental and spectroscopic methods, which we shall not detail, can also monitor a chromatographic separation.

12.12 TAILING

Often, when a single solvent is used for elution, an elution curve (weight versus fraction) like that shown as a solid line in Figure 12.12 is observed. An ideal elution curve is shown by dashed lines. In the nonideal curve, the compound is said to be **tailing.** Tailing can interfere with the beginning of a curve or a peak of a second component and lead to a poor separation. One way to avoid this is to increase the polarity of the solvent constantly while eluting. In this way, at the tail of the peak, where the solvent polarity is increasing, the compound will move slightly faster than at the front and allow the tail to squeeze forward, forming a more nearly ideal band.

12.13 RECOVERING THE SEPARATED COMPOUNDS

In recovering each of the separated compounds of a chromatographic separation when they are solids, the various correct fractions are combined and evaporated. If the combined fractions contain sufficient material, they may be purified by recrystallization. If the

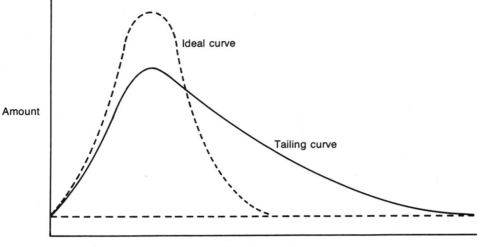

Figure 12.12   Elution curves: One ideal and one that "tails."

compounds are liquids, the correct fractions are combined, and the solvent is evaporated. If sufficient material has been collected, liquid samples can be purified by distillation. The combination of chromatography–crystallization or chromatography–distillation usually yields very pure compounds. For microscale applications, the amount of sample collected is too small to allow a purification by crystallization or distillation. The samples that are obtained after the solvent has been evaporated are considered to be sufficiently pure, and no additional purification is attempted.

12.14 DECOLORIZATION BY COLUMN CHROMATOGRAPHY

A common outcome of organic reactions is the formation of a product that is contaminated by highly colored impurities. Very often these impurities are highly polar, and they have a high molecular weight, as well as being colored. The purification of the desired product requires that these impurities be removed. Section 5.6 of Technique 5 (pp. 659–661) details methods of decolorizing an organic product. In most cases, these methods involve the use of a form of activated charcoal, or Norit.

An alternative, which is applied conveniently in microscale experiments, is to remove the colored impurity by column chromatography. Because of the polarity of the impurities, the colored components are strongly adsorbed on the stationary phase of the column, and the less polar desired product passes through the column and is collected.

Microscale decolorization of a solution on a chromatography column requires that a column be prepared in a Pasteur pipet, using either alumina or silica gel as the adsorbent (Section 12.6). The sample to be decolorized is diluted to the point where crystallization within the column will not take place, and it is then passed through the column in the usual manner. The desired compound is collected as it exits the column, and the excess solvent is removed by evaporation (Technique 3, Section 3.11, p. 630).

12.15 GEL CHROMATOGRAPHY

The stationary phase in gel chromatography consists of a cross-linked polymeric material. Molecules are separated according to their *size* by their ability to penetrate a sieve-like structure. Molecules permeate the porous stationary phase as they move down the column. Small molecules penetrate the porous structure more easily than large ones. Thus, the large molecules move through the column faster than the smaller ones and elute first. The separation of molecules by gel chromatography is depicted in Figure 12.13. With adsorption chromatography using materials such as alumina or silica, the order is usually the reverse. Small molecules (of low molecular weight) pass through the column *faster* than large molecules (of high molecular weight) because large molecules are more strongly attracted to the polar stationary phase.

Equivalent terms used by chemists for the gel-chromatography technique are **gel filtration** (biochemistry term), **gel-permeation chromatography** (polymer chemistry term),

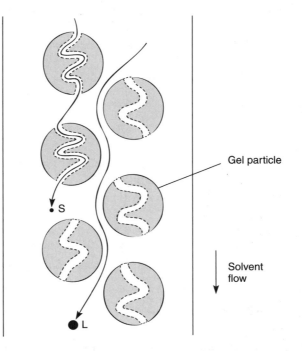

Gel particle

S

Solvent flow

L

Figure 12.13   Gel chromatography: Comparison of the paths of large (L) and small (S) molecules through the column during the same interval of time.

and **molecular sieve chromatography. Size-exclusion chromatography** is a general term for the technique, and it is perhaps the most descriptive term for what occurs on a molecular level.

Sephadex is one of the most popular materials for gel chromatography. It is widely used by biochemists for separating proteins, nucleic acids, enzymes, and carbohydrates. Most often, water or aqueous solutions of buffers are used as the moving phase. Chemically, Sephadex is a polymeric carbohydrate that has been cross-linked. The degree of cross-linking determines the size of the "holes" in the polymer matrix. In addition, the hydroxyl groups on the polymer can adsorb water, which causes the material to swell. As it expands, "holes" are created in the matrix. Several different gels are available from manufacturers, each with its own set of characteristics. For example, a typical Sephadex gel, such as G-75, can separate molecules in the molecular-weight (MW) range 3000 to 70,000. Assume for the moment that one has a four-component mixture containing compounds with molecular weights of 10,000, 20,000, 50,000, and 100,000. The 100,000-MW compound would pass through the column first, because it cannot penetrate the polymer matrix. The 50,000-, 20,000-, and 10,000-MW compounds penetrate the matrix to varying degrees and would be separated. The molecules would elute in the order given (decreasing order of molecular weights). The gel separates on the basis of molecular size and configuration, rather than molecular weight.

Sephadex LH-20 has been developed for nonaqueous solvents. Some of the hydroxyl groups have been alkylated, and thus the material can swell under both aqueous and nonaqueous conditions (it now has "organic" character). This material can be used with several organic solvents, such as alcohol, acetone, methylene chloride, and aromatic hydrocarbons.

Another type of gel is based on a polyacrylamide structure (Bio-Gel P and Poly-Sep AA). A portion of a polyacrylamide chain is shown below.

$$-CH_2-CH-CH_2-CH-CH_2-CH-$$

$$\begin{array}{ccc} C{=}O & C{=}O & C{=}O \\ | & | & | \\ NH_2 & NH_2 & NH_2 \end{array}$$

Gels of this type can also be used in water and some polar organic solvents. They tend to be more stable than Sephadex, especially under acidic conditions. Polyacrylamides can be used for many biochemical applications involving macromolecules. For separating synthetic polymers, cross-linked polystyrene beads (copolymer of styrene and divinyl benzene) find common application. Again, the beads are swollen before use. Common organic solvents can be used to elute the polymers. As with other gels, the higher-molecular-weight compounds elute before the lower-molecular-weight compounds.

12.16 FLASH CHROMATOGRAPHY

One of the drawbacks to column chromatography is that, for large-scale preparative separations, the time required to complete a separation may be very long. Furthermore, the resolution that is possible for a particular experiment tends to deteriorate as the time for the experiment grows longer. This latter effect arises because the bands of compounds that move very slowly through a column tend to "tail."

A technique that can be useful in overcoming these problems has been developed. This technique, called **flash chromatography,** is actually a very simple modification of an ordinary column chromatography. In flash chromatography, the adsorbent is packed into a relatively short glass column, and air pressure is used to force the solvent through the adsorbent.

The apparatus used for flash chromatography is shown in Figure 12.14. The glass column is fitted with a Teflon stopcock at the bottom to control the flow rate of solvent. A plug of glass wool is placed in the bottom of the column to act as a support for the adsorbent. A layer of sand may also be added on top of the glass wool. The column is filled with adsorbent using the dry pack method. When the column has been filled, a fitting is attached to the top of the column, and the entire apparatus is connected to a source of high-pressure air or nitrogen. The fitting is designed so that the pressure applied to the top of the column can be adjusted precisely. The source of the high-pressure air is often a specially adapted air pump.

A typical column would use silica gel adsorbent (particle size = 40 to 63 μm) packed to a height of 5 inches in a glass column of 20-mm diameter. The pressure applied to the column would be adjusted to achieve a solvent flow rate such that the solvent level in the column would decrease by about 2 in/min. This system would be appropriate to separate the components of a 250-mg sample.

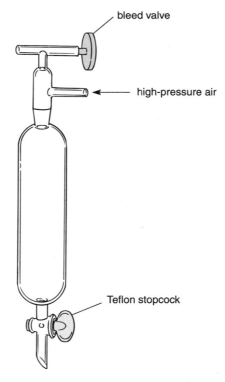

bleed valve

high-pressure air

Teflon stopcock

Figure 12.14   Apparatus for flash chromatography.

The high-pressure air forces the solvent through the column of adsorbent at a rate that is much greater than what would be achieved if the solvent flowed through the column under the force of gravity. Because the solvent is caused to flow faster, the time required for substances to pass through the column is reduced. By itself, simply applying air pressure to the column might reduce the clarity of the separation, because the components of the mixture would not have time to establish themselves into distinctly separate bands. However, in flash chromatography, you can use a much finer adsorbent than would be used in ordinary chromatography. With a much smaller particle size for the adsorbent, the surface area is increased, and the resolution possible thereby improves.

A simple variation on this idea does not use air pressure. Instead, the lower end of the column is inserted into a stopper, which is fitted into the top of a suction flask. Vacuum is applied to the system, and the vacuum acts to draw the solvent through the adsorbent column. The overall effect of this variation is similar to that obtained when air pressure is applied to the top of the column.

REFERENCES

Deyl, Z., Macek, K., and Janák, J. *Liquid Column Chromatography.* Amsterdam: Elsevier, 1975.

Heftmann, E. *Chromatography.* 3rd ed. New York: Van Nostrand Reinhold, 1975.

Jacobson, B.M. "An Inexpensive Way to Do Flash Chromatography." *Journal of Chemical Education, 65* (May 1988): 459.

Still, W. C., Kahn, M., and Mitra, A. "Rapid Chromatographic Technique for Preparative Separations with Moderate Resolution." *Journal of Organic Chemistry, 43* (1978): 2923.

PROBLEMS

1. A sample was placed on a chromatography column. Methylene chloride was used as the eluting solvent. All of the components eluted off the column, but no separation was observed. What must have been happening during this experiment? How would you change the experiment in order to overcome this problem?

2. You are about to purify an impure sample of naphthalene by column chromatography. What solvent should you use to elute the sample?

3. Consider a sample that is a mixture composed of biphenyl, benzoic acid, and benzyl alcohol. Predict the order of elution of the components in this mixture. Assume that the chromatography uses a silica column, and the solvent system is based on cyclohexane, with an increasing proportion of methylene chloride being added as a function of time.

4. An orange compound was added to the top of a chromatography column. Solvent was added immediately, with the result that the entire volume of solvent in the solvent reservoir turned orange. No separation could be obtained from the chromatography experiment. What went wrong?

5. A yellow compound, dissolved in methylene chloride, is added to a chromatography column. The elution is begun using petroleum ether as the solvent. After 6 L of solvent had passed through the column, the yellow band still had not traveled down the column appreciably. What should be done to make this experiment work better?

6. You have 0.50 g of a mixture that you wish to purify by column chromatography. How much adsorbent should you use to pack the column? Estimate the appropriate column diameter and height.

7. In a particular sample, you wish to collect the component with the *highest* molecular weight as the *first* fraction. What chromatographic technique should you use?

8. A colored band shows an excessive amount of tailing as it passes through the column. What can you do to rectify this problem?

9. How would you monitor the progress of a column chromatography when the sample is colorless? Describe at least two methods.

T E C H N I Q U E   1 3

High-Performance Liquid Chromatography (HPLC)

The separation that can be achieved is greater if the column packing used in column chromatography is made more dense by using an adsorbent that has a small particle size. The solute molecules encounter a much larger surface area on which they can be adsorbed as they pass through the column packing. At the same time, the solvent spaces between the particles are reduced in size. As a result of this tight packing, equilibrium between the liquid and solid phases can be established very rapidly with a fairly short column, and the degree of separation is markedly improved. The disadvantage of making the column packing more dense is that the solvent flow rate becomes very slow or even stops. Gravity is not strong enough to pull the solvent through a tightly packed column.

A modern technique can be applied to obtain much better separations with tightly packed columns. A pump forces the solvent through the column packing. As a result, solvent flow rate is increased and the advantage of better separation is retained. This technique, called **high-performance liquid chromatography (HPLC),** is becoming widely applied to problems where separations by ordinary column chromatography are unsatisfactory. Because the pump often provides pressures in excess of 1000 pounds per square inch (psi), this method is also known as **high-pressure liquid chromatography.** High pressures are not required, however, and satisfactory separations can be achieved with pressures as low as 100 psi.

The basic design of an HPLC instrument is shown in Figure 13.1. The instrument contains the following essential components:

1. Solvent reservoir
2. Solvent filter and degasser
3. Pump
4. Pressure gauge
5. Sample injection system
6. Column
7. Detector
8. Amplifier and electronic controls
9. Chart recorder

There may be other variations on this simple design. Some instruments have heated ovens in order to maintain the column at a specified temperature, fraction collectors, and microprocessor-controlled data-handling systems. Additional filters for the solvent and

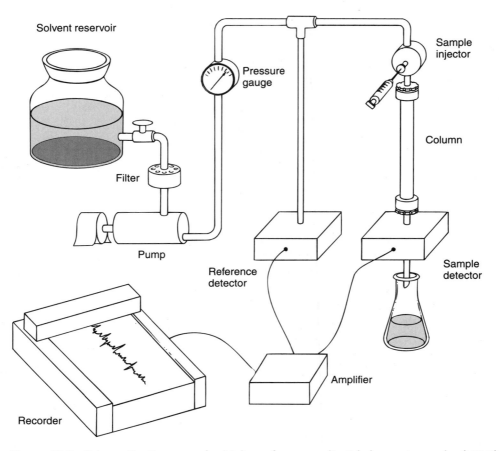

Figure 13.1   Schematic diagram of a high-performance liquid chromatography (HPLC).

sample may also be included. You may find it interesting to compare this schematic diagram with that shown in Technique 15 (Fig. 15.2, p. 809) for a gas chromatography instrument. Many of the essential components are common to both types of instruments.

13.1 ADSORBENTS AND COLUMNS

The most important factor to consider when choosing a set of experimental conditions is the nature of the material packed into the column. You must also consider the size of the column that will be selected. The chromatography column is generally packed with silica or alumina adsorbents. Unlike column chromatography, however, the adsorbents used for HPLC have a much smaller particle size. Typically, particle size ranges from 5 to 20 μm in diameter for HPLC; it is on the order of 100 μm for column chromatography.

The adsorbent is packed into a column that can withstand the elevated pressures typical of this type of experiment. Generally, the column is constructed of stainless steel, although some columns that are constructed of a rigid polymeric material ("PEEK"—Poly Ether Ether Ketone) are available commercially. A strong column is required to withstand the high pressures that may be used. The columns are fitted with stainless steel connectors, which ensure a pressure-tight fit between the column and the tubing that connects the column to the other components of the instrument.

Columns that fulfill a large number of specialized purposes are available. In this chapter, we consider only the four most important types of columns:

1. Normal-phase chromatography
2. Reversed-phase chromatography
3. Ion-exchange chromatography
4. Size exclusion chromatography

In most types of chromatography, the adsorbent is more polar than the mobile phase. For example, the solid packing material, which may be either silica or alumina, has a stronger affinity for polar molecules than does the solvent. As a result, the molecules in the sample adhere strongly to the solid phase, and their progress down the column is much slower than the rate at which solvent moves through the column. The time required for a substance to move through the column can be altered by changing the polarity of the solvent. In general, as the solvent becomes more polar, substances move faster through the column. This type of behavior is known as **normal-phase chromatography.** In HPLC, you inject a sample onto a normal-phase column and elute it by varying the polarity of the solvent, much as you do with ordinary column chromatography. Disadvantages of normal-phase chromatography are that retention times tend to be long, and bands have a tendency to "tail."

These disadvantages can be ameliorated by selecting a column in which the solid support is *less polar* than the moving solvent phase. In this type of chromatography, known as **reversed-phase chromatography,** the silica column packing is treated with alkylating agents. As a result, nonpolar alkyl groups are bonded to the silica surface, making the adsorbent nonpolar. The alkylating agents that are used most commonly can attach methyl ($-CH_3$), octyl ($-C_8H_{17}$), or octadecyl ($-C_{18}H_{37}$) groups to the silica surface. The latter variation, where an 18-carbon chain is attached to the silica, is the most popular. This type of column is known as a **C-18 column.** The bonded alkyl groups have an effect similar to that which would be produced by an extremely thin organic solvent layer coating the surface of the silica particles. The interactions that take place between the substances dissolved in the solvent and the stationary phase thus become more like those observed in a liquid–liquid extraction. The solute particles distribute themselves between the two "solvents"—that is, between the moving solvent and the organic coating on the silica. The longer the alkyl groups that are bonded to the silica, the more effective the alkyl groups are as they interact with solute molecules.

Reversed-phase chromatography is widely used because the rate at which solute molecules exchange between moving phase and stationary phase is very rapid, which means that substances pass through the column relatively quickly. Furthermore, problems arising from the "tailing" of peaks are reduced. A disadvantage of this type of column, however, is that the chemically bonded solid phases tend to decompose. The organic groups are slowly hydrolyzed from the surface of the silica, which leaves a normal silica surface exposed. Thus, the chromatographic process that takes place on the column slowly shifts from a reversed-phase to a normal-phase separation mechanism.

Another type of solid support that is sometimes used in reversed-phase chromatography is organic polymer beads. These beads present a surface to the moving phase that is largely organic in nature.

For solutions of ions, a column that is packed with an ion-exchange resin is used. This type of chromatography is known as **ion-exchange chromatography.** The ion-exchange resin that is chosen can be either an anion-exchange resin or a cation-exchange resin, depending upon the nature of the sample being examined.

A fourth type of column is known as a **size exclusion column** or a **gel filtration column.** The interaction that takes place on this type of column is similar to that described in Technique 12, Section 12.15, page 783.

13.2 COLUMN DIMENSIONS

The dimensions of the column that you use depend upon the application. For analytical applications, a typical column is constructed of tubing that has an inside diameter of between 4 and 5 mm, although analytical columns with inside diameters of 1 or 2 mm are also available. A typical analytical column has a length of about 7.5 to 30 cm. This type of column is suitable for the separation of a 0.1- to 5-mg sample. With columns of smaller diameter, it is possible to perform an analysis with samples smaller than 1 *micro*gram.

High-performance liquid chromatography is an excellent analytical technique, but the separated compounds may also be isolated. The technique can be used for preparative experiments. Just as in column chromatography, the fractions can be collected into individual receiving containers as they pass through the column. The solvents can be evaporated from these fractions, allowing you to isolate separated components of the original mixture. Samples that range in size from 5 to 100 mg can be separated on a **semiprep column.** The dimensions of a semiprep column are typically 8 mm inside diameter and 10 cm in length. A semiprep column is a practical choice when you wish to use the same column for both analytical and preparative separations. A semiprep column is small enough to provide reasonable sensitivity in analyses, but it is also capable of handling moderate-size samples when you need to isolate the components of a mixture. Even larger samples can be separated using a **preparative column.** This type of column is useful when you wish to collect the components of a mixture and then use the pure samples for additional study (e.g., for a subsequent chemical reaction or for spectroscopic analysis). A preparative column may be as large as 20 mm in inside diameter and 30 cm in length. A preparative column can handle samples as large as 1 g per injection.

13.3 SOLVENTS

The choice of solvent used for an HPLC run depends on the type of chromatographic separation selected. For a normal-phase separation, the solvent is selected based on its polarity. The criteria described in Technique 12, Section 12.4B, page 767, are used. A solvent of very low polarity might be pentane, petroleum ether, hexane, or carbon tetrachloride; a solvent of very high polarity might be water, acetic acid, methanol, or 1-propanol.

For a reversed-phase experiment, a less polar solvent causes solutes to migrate *faster*. For example, for a mixed methanol–water solvent, as the percentage of methanol in the solvent increases (solvent becomes less polar), the time required to elute the components of a mixture from a column decreases. The behavior of solvents as eluents in a reversed-phase chromatography would be the reverse of the order shown in Table 12.2 on page 768.

If a single solvent (or solvent mixture) is used for the entire separation, the chromatogram is said to be **isochratic.** Special electronic devices are available with HPLC instruments that allow you to program changes in the solvent composition from the beginning to the end of the chromatography. These are called **gradient elution systems.** With gradient elution, the time required for a separation may be shortened considerably.

The need for pure solvents is especially acute with HPLC. The narrow bore of the column and the very small particle size of the column packing require that solvents be particularly pure and free of insoluble residue. In most cases, the solvents must be filtered through ultrafine filters and **degassed** (have dissolved gases removed) before they can be used.

The solvent gradient is chosen so that the eluting power of the solvent increases over the duration of the experiment. The result is that components of the mixture that tend to move very slowly through the column are caused to move faster as the eluting power of the solvent gradually increases. The instrument can be programmed to change the composition of the solvent following a linear gradient or a nonlinear gradient, depending upon the specific requirements of the separation.

13.4 DETECTORS

A flow-through **detector** must be provided to determine when a substance has passed through the column. In most applications, the detector detects either the change in index of refraction of the liquid as its composition changes or the presence of solute by its absorption of ultraviolet or visible light. The signal generated by the detector is amplified and treated electronically in a manner similar to that found in gas chromatography (Technique 15, Section 15.6, p. 814).

A detector that responds to changes in the index of refraction of the solution may be considered the most universal of the HPLC detectors. The refractive index of the liquid passing through the detector changes slightly, but significantly, as the liquid changes from pure solvent to a liquid where the solvent contains some type of organic solute. This change in refractive index can be detected and compared to the refractive index of pure solvent. The difference in index values is then recorded as a peak on a chart. A disadvantage of this type of detector is that it must respond to very small changes in refractive index. As a result, the detector tends to be unstable and difficult to balance.

When the components of the mixture have some type of absorption in the ultraviolet or visible regions of the spectrum, a detector that is adjusted to detect absorption at a particular wavelength of light can be used. This type of detector is much more stable, and the readings tend to be more reliable. Unfortunately, many organic compounds do not absorb ultraviolet light, and this type of detector cannot be used.

REFERENCES

Karger, Barry L. "HPLC: Early and Recent Perspectives." *Journal of Chemical Education, 74* (January 1997): 45.

Meyer, Veronika R. *Practical High-performance Liquid Chromatography.* Chichester: John Wiley & Sons, Ltd., 1988.

Rubinson, K. A. "Liquid Chromatography." Chapter 14 in *Chemical Analysis.* Boston: Little, Brown and Co., 1987.

Snyder, Lloyd R. "Modern Practice of Liquid Chromatography." *Journal of Chemical Education, 74* (January 1997): 37.

Van Arman, Scott A., and Thomsen, Marcus W. "HPLC for Undergraduate Introductory Laboratories." *Journal of Chemical Education, 74* (January 1997): 49.

PROBLEMS

1. For a mixture of biphenyl, benzoic acid, and benzyl alcohol, predict the order of elution and describe any differences that you would expect for a normal-phase HPLC experiment (in hexane solvent) compared with a reversed-phase experiment (in tetrahydrofuran–water solvent).

2. How would the *gradient elution program* differ between normal-phase and reversed-phase chromatography?

T E C H N I Q U E   1 4

Thin-Layer Chromatography

Thin-layer chromatography (TLC) is a very important technique for the rapid separation and qualitative analysis of small amounts of material. It is ideally suited for the analysis of mixtures and reaction products in small-scale and microscale experiments. The technique is closely related to column chromatography. In fact, TLC can be considered simply column chromatography *in reverse,* with the solvent ascending the adsorbent, rather than descending. Because of this close relationship to column chromatography, and because the principles governing the two techniques are similar, Technique 12, on column chromatography, should be read before reading this one.

14.1 PRINCIPLES OF THIN-LAYER CHROMATOGRAPHY

Like column chromatography, TLC is a solid–liquid partitioning technique. However, the moving liquid phase is not allowed to percolate down the adsorbent; it is caused to *ascend* a thin layer of adsorbent coated onto a backing support. The most typical backing is a glass plate, but other materials are also used. A thin layer of the adsorbent is spread onto the plate and allowed to dry. A coated and dried plate of glass is called a **thin-layer plate** or a **thin-layer slide.** (The reference to *slide* comes about because microscope slides are often used to prepare small thin-layer plates.) When a thin-layer plate is placed upright in a vessel that contains a shallow layer of solvent, the solvent ascends the layer of adsorbent on the plate by capillary action.

In TLC, the sample is applied to the plate before the solvent is allowed to ascend the adsorbent layer. The sample is usually applied as a small spot near the base of the plate; this technique is often referred to as **spotting.** The plate is spotted by repeated applications of a sample solution from a small capillary pipet. When the filled pipet touches the plate, capillary action delivers its contents to the plate, and a small spot is formed.

As the solvent ascends the plate, the sample is partitioned between the moving liquid phase and the stationary solid phase. During this process, you are **developing,** or **running,** the thin-layer plate. In development, the various components in the applied mixture are separated. The separation is based on the many equilibrations the solutes experience between the moving and the stationary phases. (The nature of these equilibrations was thoroughly discussed in Technique 12, Sections 12.2 and 12.3, pp. 762–765.) As in column chromatography, the least polar substances advance faster than the most polar substances. A separation results from the differences in the rates at which the individual components of the mixture advance upward on the plate. When many substances are present in a mixture, each has its own characteristic solubility and adsorptivity properties, depending on the functional groups in its structure. In general, the stationary phase is strongly polar and strongly binds polar substances. The moving liquid phase is usually less polar than the adsorbent and most easily dissolves substances that are less polar or even nonpolar. Thus, substances that are the most polar travel slowly upward, or not at all, and nonpolar substances travel more rapidly if the solvent is sufficiently nonpolar.

When the thin-layer plate has been developed, it is removed from the developing tank and allowed to dry until it is free of solvent. If the mixture that was originally spotted on the plate was separated, there will be a vertical series of spots on the plate. Each spot corresponds to a separate component or compound from the original mixture. If the components of the mixture are colored substances, the various spots will be clearly visible after development. More often, however, the "spots" will not be visible because they correspond to colorless substances. If spots are not apparent, they can be made visible only if a **visualization method** is used. Often, spots can be seen when the thin-layer plate is held under ultraviolet light; the ultraviolet lamp is a common visualization method. Also common is the use of iodine vapor. The plates are placed in a chamber containing iodine crystals and left to stand for a short time. The iodine reacts with the various compounds adsorbed on the plate to give colored complexes that are clearly visible. Because iodine often changes the compounds by reaction, the components of the mixture cannot be recovered from the plate when the iodine method is used. (Other methods of visualization are discussed in Section 14.7.)

14.2 COMMERCIALLY PREPARED TLC PLATES

The most convenient type of TLC plate is prepared commercially and sold in a ready-to-use form. Many manufacturers supply glass plates precoated with a durable layer of silica gel or alumina. More conveniently, plates are also available that have either a flexible plastic backing or an aluminum backing. The most common types of commercial TLC plates are composed of plastic sheets that are coated with silica gel and polyacrylic acid, which serves as a binder. A fluorescent indicator may be mixed with the silica gel. The indicator renders the spots due to the presence of compounds in the sample visible under ultraviolet light (see Section 14.7). Although these plates are relatively expensive when compared with plates prepared in the laboratory, they are far more convenient to use, and they provide more consistent results. The plates are manufactured quite uniformly. Because the plastic backing is flexible, they have the additional advantage that the coating does not flake off the plates easily. The plastic sheets can also be cut with a pair of scissors to whatever size may be required.

14.3 PREPARATION OF THIN-LAYER SLIDES AND PLATES

The two adsorbent materials used most often for TLC are alumina G (aluminum oxide) and silica gel G (silicic acid). The G designation stands for gypsum (calcium sulfate). Calcined gypsum $CaSO_4 \cdot \frac{1}{2}H_2O$ is better known as plaster of Paris. When exposed to water or moisture, gypsum sets in a rigid mass $CaSO_4 \cdot 2H_2O$, which binds the adsorbent together and to the glass plates used as a backing support. In the adsorbents used for TLC, about 10%–13% by weight of gypsum is added as a binder. The adsorbent materials are otherwise like those used in column chromatography; the adsorbents used in column chromatography have a larger particle size, however. The material for thin-layer work is a fine powder. The small particle size, along with the added gypsum, makes it impossible to use silica gel G or alumina G for column work. In a column, these adsorbents generally set so rigidly that solvent virtually stops flowing through the column.

A. MICROSCOPE SLIDE TLC PLATES

For qualitative work such as identifying the number of components in a mixture or trying to establish that two compounds are identical, small TLC plates made from microscope slides are especially convenient. Coated microscope slides are easily made by dipping the slides into a container holding a slurry of the adsorbent material. Although numerous solvents can be used to prepare a slurry, methylene chloride is probably the best choice. It has the two advantages of low boiling point (40°C) and inability to cause the adsorbent to set or form lumps. The low boiling point means that it is not necessary to dry the coated slides in an oven. Its inability to cause the gypsum binder to set means that slurries made with it are stable for several days. The layer of adsorbent formed is fragile, however, and must be treated carefully. For this reason, some persons prefer to add a small amount of methanol to the methylene chloride to enable the gypsum to set more firmly.

The methanol solvates the calcium sulfate much as water does. More durable plates can be made by dipping plates into a slurry prepared from water. These plates must be oven-dried before use. Also, a slurry prepared from water must be used soon after its preparation. If it is not, it begins to set and form lumps. Thus, an aqueous slurry must be prepared immediately before use; it cannot be used after it has stood for any length of time. For microscope slides, a slurry of silica gel G in methylene chloride is not only convenient but also adequate for most purposes.

Preparing the Slurry. The slurry is most conveniently prepared in a 4-oz wide-mouthed screwcap jar. About 3 mL of methylene chloride is required for each gram of silica gel G. For a smooth slurry without lumps, the silica gel should be added to the solvent while the mixture is being either stirred or swirled. Adding solvent to the adsorbent usually causes lumps to form in the mixture. When the addition is complete, the cap should be placed on the jar tightly and the jar shaken vigorously to ensure thorough mixing (use protective gloves). The slurry may be stored in the tightly capped jar until it is to be used. More methylene chloride may have to be added to replace evaporation losses.

> **Caution:** Avoid breathing silica dust or methylene chloride, prepare and use the slurry in a hood, and avoid getting methylene chloride or the slurry mixture on your skin. Use protective gloves.

Preparing the Slides. If new microscope slides are available, you can use them without any special treatment. However, it is more economical to reuse or recycle microscope slides. Wash the slides with soap and water, rinse with water, and then rinse with 50% aqueous methanol. Allow the plates to dry thoroughly on paper towels. They should be handled by the edges, because fingerprints on the plate surface will make it difficult for the adsorbent to bind to the glass.

Coating the Slides. The slides are coated with adsorbent by dipping them into the container of slurry. You can coat two slides simultaneously by sandwiching them together before dipping them in the slurry.

> Perform the coating operation under a hood. Use gloves.

Shake the slurry vigorously just before dipping the slides. Because the slurry settles on standing, it should be mixed in this way before each set of slides is dipped. The depth of the slurry in the jar should be about 3 inches, and the plates should be dipped into the slurry until only about 0.25 inches at the top remains uncoated. The dipping operation should be performed smoothly. The plates may be held at the top (see Fig. 14.1), where they will not be coated. They are dipped into the slurry and withdrawn with a slow and steady motion. The dipping operation takes about 2 seconds. Some practice may be required to get the correct timing. After dipping, replace the cap on the jar, and hold the plates for a minute until most of the solvent has evaporated. Separate the plates and place them on paper towels to complete the drying.

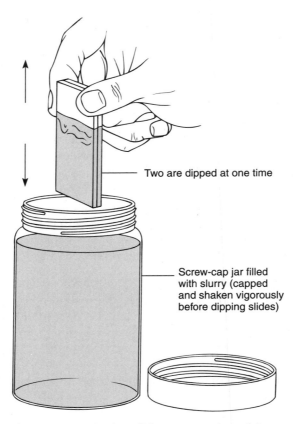

Two are dipped at one time

Screw-cap jar filled
with slurry (capped
and shaken vigorously
before dipping slides)

Figure 14.1 Dipping slides to coat them (gloves recommended).

The plates should have an even coating; there should be no streaks and no thin spots where glass shows through the adsorbent. The plates should not have a thick and lumpy coating.

Two conditions cause thin and streaked plates. First, the slurry may not have been mixed thoroughly before the dipping operation; the adsorbent might then have settled to the bottom of the jar, and the thin slurry at the top would not have coated the slides properly. Second, the slurry simply may not have been thick enough; more silica gel G must then be added to the slurry until the consistency is correct. If the slurry is too thick, the coating on the plates will be thick, uneven, and lumpy. To correct this, dilute the slurry with enough solvent to achieve the proper consistency.

Plates with an unsatisfactory coating may be wiped clean with a paper towel and redipped. Take care to handle the plates only from the top or by the sides, to avoid getting fingerprints on the glass surface.

B. LARGER THIN-LAYER PLATES

For separations involving large amounts of material, or for difficult separations, it may be necessary to use larger thin-layer plates. Plates with dimensions up to 20–25 cm^2

are common. With larger plates, it is desirable to have a more durable coating, and a water slurry of the adsorbent should be used to prepare them. If silica gel is used, the slurry should be prepared in the ratio about 1 g silica gel G to each 2 mL of water. The glass plate used for the thin-layer plate should be washed, dried, and placed on a sheet of newspaper. Place two strips of masking tape along two edges of the plate. Use more than one layer of masking tape if a thicker coating is desired on the plate. A slurry is prepared, shaken well, and poured along one of the untaped edges of the plate.

Observe the precautions stated on page 795.

A heavy piece of glass rod, long enough to span the taped edges, is used to level and spread the slurry over the plate. While the rod is resting on the tape, it is pushed along the plate from the end at which the slurry was poured toward the opposite end of the plate. This is illustrated in Figure 14.2. After the slurry is spread, the masking tape strips are removed, and the plates are dried in a 110°C oven for about 1 hour. Plates of 10–25 cm^2 are easily prepared by this method. Larger plates present more difficulties. Many laboratories have a commercially manufactured spreading machine that makes the entire operation simpler.

14.4 SAMPLE APPLICATION: SPOTTING THE PLATES

Preparing a Micropipet

To apply the sample that is to be separated to the thin-layer plate, one uses a micropipet. A micropipet is easily made from a short length of thin-walled capillary tubing like that used for melting-point determinations. The capillary tubing is heated at its midpoint with a microburner and rotated until it is soft. When the tubing is soft, the heated

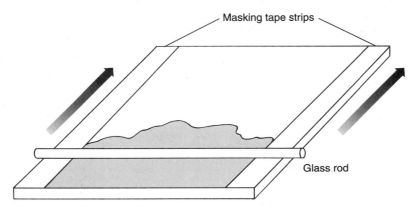

Figure 14.2 Preparing a large plate.

portion of the tubing is drawn out until a constricted portion of tubing 4–5 cm long is formed. After cooling, the constricted portion of tubing is scored at its center with a file or scorer and broken. The two halves yield two capillary micropipets. Figure 14.3 shows how to make such pipets.

Spotting the Plate

To apply a sample to the plate, begin by placing about 1 mg of a solid test substance, or one drop of a liquid test substance, in a small container like a watch glass or a test tube. Dissolve the sample in a few drops of a volatile solvent. Acetone or methylene chloride is usually a suitable solvent. If a solution is to be tested, it can often be used directly. The small capillary pipet, prepared as described, is filled by dipping the pulled end into the solution to be examined. Capillary action fills the pipet. Empty the pipet by touching it *lightly* to the thin-layer plate at a point about 1 cm from the bottom (Fig. 14.4). The spot must be high enough that it does not dissolve in the developing solvent. It is important to touch the plate very lightly and not to gouge a hole in the adsorbent. When the pipet touches the plate, the solution is transferred to the plate as a small spot. The pipet should be touched to the plate *very briefly* and then removed. If the pipet is held to the plate, its entire contents will be delivered to the plate. Only a small amount of material is needed. It is often helpful to blow gently on the plate as the sample is applied. This helps to keep the spot small by evaporating the solvent before it can spread out on the plate. The smaller the spot formed, the better the separation obtainable. If needed, additional material can be applied to the plate by repeating the spotting procedure. You should repeat the procedure with several small amounts, rather than apply one large amount. The solvent should be allowed to evaporate between applications. If the spot is not small (about 2 mm in diameter), a new plate should be prepared. The capillary pipet may be used several times if it is rinsed between uses. It

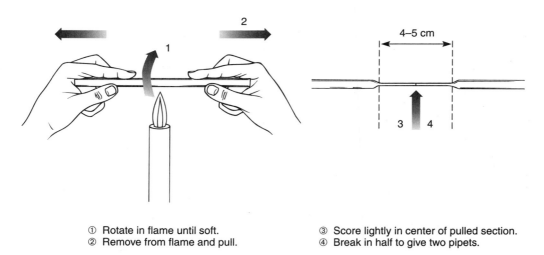

① Rotate in flame until soft.   ③ Score lightly in center of pulled section.
② Remove from flame and pull.   ④ Break in half to give two pipets.

Figure 14.3 Construction of two capillary micropipets.

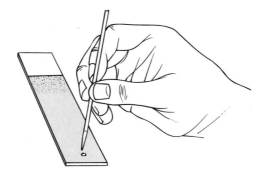

Figure 14.4   Spotting the plate with a drawn capillary pipet.

is repeatedly dipped into a small portion of solvent to rinse it and touched to a paper towel to empty it.

As many as three different spots may be applied to a microscope-slide TLC plate. Each spot should be about 1 cm from the bottom of the plate, and all spots should be evenly spaced, with one spot in the center of the plate. Due to diffusion, spots often increase in diameter as the plate is developed. To keep spots containing different materials from merging, and to avoid confusing the samples, do not place more than three spots on a single plate. Larger plates can accommodate many more samples.

14.5 DEVELOPING (RUNNING) TLC PLATES

Preparing a Development Chamber

A convenient development chamber for microscope-slide TLC plates can be made from a 4-oz wide-mouthed jar. An alternative development chamber can be constructed from a beaker, using aluminum foil to cover the opening. The inside of the jar or beaker should be lined with a piece of filter paper, cut so that it does not quite extend around the inside of the jar. A small vertical opening (2–3 cm) should be left in the filter paper for observing the development. Before development, the filter paper inside the jar or beaker should be thoroughly moistened with the development solvent. The solvent-saturated liner helps to keep the chamber saturated with solvent vapors, thereby speeding the development. Once the liner is saturated, the level of solvent in the bottom of the development chamber is adjusted to a depth of about 5 mm, and the chamber is capped (or covered with aluminum foil) and set aside until it is to be used. A correctly prepared development chamber (with slide in place) is shown in Figure 14.5.

Developing the TLC Plate

Once the spot has been applied to the thin-layer plate and the solvent has been selected (see Sections 14.4–14.6), the plate is placed in the chamber for development. The

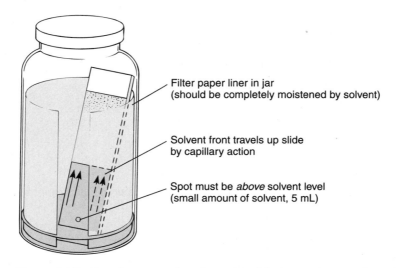

Filter paper liner in jar
(should be completely moistened by solvent)

Solvent front travels up slide
by capillary action

Spot must be *above* solvent level
(small amount of solvent, 5 mL)

Figure 14.5 Development chamber with thin-layer plate undergoing development.

plate must be placed in the chamber carefully so that none of the coated portion touches the filter paper liner. In addition, the solvent level in the bottom of the chamber must not be above the spot that was applied to the plate, or the spotted material will dissolve in the pool of solvent instead of undergoing chromatography. Once the plate has been placed correctly, replace the cap on the developing chamber and wait for the solvent to advance up the plate by capillary action. This generally occurs rapidly, and you should watch carefully. As the solvent rises, the plate becomes visibly moist. When the solvent has advanced to within 5 mm of the end of the coated surface, the plate should be removed, and the position of the solvent front should be marked *immediately* by scoring the plate along the solvent line with a *pencil.* The solvent front must not be allowed to travel beyond the end of the coated surface. The plate should be removed before this happens. The solvent will not actually advance beyond the end of the plate, but spots allowed to stand on a completely moistened plate on which the solvent is not in motion expand by diffusion. Once the plate has dried, any visible spots should be outlined on the plate with a pencil. If no spots are apparent, a visualization method (Section 14.7) may be needed.

14.6 CHOOSING A SOLVENT FOR DEVELOPMENT

The development solvent used depends on the materials to be separated. You may have to try several solvents before a satisfactory separation is achieved. Because microscope slides can be prepared and developed rapidly, an empirical choice is usually not hard to make. A solvent that causes all the spotted material to move with the solvent front is too polar. One that does not cause any of the material in the spot to move is not polar enough. As a guide to the relative polarity of solvents, consult Table 12.2 in Technique 12 (p. 768).

Methylene chloride and toluene are solvents of intermediate polarity and good choices for a wide variety of functional groups to be separated. For hydrocarbon materials, good first choices are hexane, petroleum ether (ligroin), or toluene. Hexane or petroleum ether with varying proportions of toluene or ether gives solvent mixtures of moderate polarity that are useful for many common functional groups. Polar materials may require ethyl acetate, acetone, or methanol.

A rapid way to determine a good solvent is to apply several sample spots to a single plate. The spots should be placed a minimum of 1 cm apart. A capillary pipet is filled with a solvent and gently touched to one of the spots. The solvent expands outward in a circle. The solvent front should be marked with a pencil. A different solvent is applied to each spot. As the solvents expand outward, the spots expand as concentric rings. From the appearance of the rings, you can judge approximately the suitability of the solvent. Several types of behavior experienced with this method of testing are shown in Figure 14.6.

14.7 VISUALIZATION METHODS

It is fortunate when the compounds separated by TLC are colored because the separation can be followed visually. More often than not, however, the compounds are colorless. In that case, the separated materials must be made visible by some reagent or some method that makes the separated compounds visible. Reagents that give rise to colored spots are called **visualization reagents.** Methods of viewing that make the spots apparent are **visualization methods.**

The visualization reagent used most often is iodine. Iodine reacts with many organic materials to form complexes that are either brown or yellow. In this visualization method, the developed and dried TLC plate is placed in a 4-oz wide-mouthed screwcap jar along with a few crystals of iodine. The jar is capped and gently warmed on a steam bath or a hot plate at low heat. The jar fills with iodine vapors and the spots begin to appear. When the spots are sufficiently intense, the plate is removed from the jar, and the spots are outlined with a pencil. The spots are not permanent. Their appearance results from the formation of complexes the iodine makes with the organic substances. As the iodine sublimes off the plate, the spots fade. Hence, they should be marked immediately. Nearly all

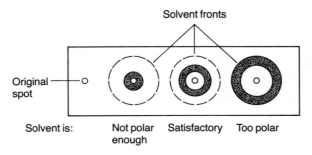

Figure 14.6 Concentric ring method of testing solvents.

compounds except saturated hydrocarbons and alkyl halides form complexes with iodine. The intensities of the spots do not accurately indicate the amount of material present, except in the crudest way.

The second most common method of visualization is by an ultraviolet (UV) lamp. Under UV light, compounds often look like bright spots on the plate. This often suggests the structure of the compound because certain types of compounds shine very brightly under UV light because they fluoresce.

Another method that provides good results involves adding a fluorescent indicator to the adsorbent used to coat the plates. A mixture of zinc and cadmium sulfides is often used. When treated in this way and held under UV light, the entire plate fluoresces. However, dark spots appear on the plate where the separated compounds are seen to quench this fluorescence.

In addition to the preceding methods, several chemical methods are available that either destroy or permanently alter the separated compounds through reaction. Many of these methods are specific for particular functional groups.

Alkyl halides can be visualized if a dilute solution of silver nitrate is sprayed on the plates. Silver halides are formed. These halides decompose if exposed to light, giving rise to dark spots (free silver) on the TLC plate.

Most organic functional groups can be made visible if they are charred with sulfuric acid. Concentrated sulfuric acid is sprayed on the plate, which is then heated in an oven at 110°C to complete the charring. Permanent spots are thus created.

Colored compounds can be prepared from colorless compounds by making derivatives before spotting them on the plate. An example of this is the preparation of 2,4-dinitrophenylhydrazones from aldehydes, and ketones to produce yellow and orange compounds. You may also spray the 2,4-dinitrophenylhydrazine reagent on the plate after the ketones or aldehydes have separated. Red and yellow spots form where the compounds are located. Other examples of this method are using ferric chloride for visualizing phenols and using bromocresol green for detecting carboxylic acids. Chromium trioxide, potassium dichromate, and potassium permanganate can be used for visualizing compounds that are easily oxidized. *p*-Dimethylaminobenzaldehyde easily detects amines. Ninhydrin reacts with amino acids to make them visible. Numerous other methods and reagents available from various supply outlets are specific for certain types of functional groups. These visualize only the class of compounds of interest.

14.8 PREPARATIVE PLATES

If you use large plates (Section 14.3B), materials can be separated and the separated components can be recovered individually from the plates. Plates used in this way are called **preparative plates.** For preparative plates, a thick layer of adsorbent is generally used. Instead of being applied as a spot or a series of spots, the mixture to be separated is applied as a line of material about 1 cm from the bottom of the plate. As the plate is developed, the separated materials form bands. After development, you can observe the separated bands, usually by UV light, and outline the zones in pencil. If the method of visualization is destructive, most of the plate is covered with paper to protect it, and the reagent is applied only at the extreme edge of the plate.

Once the zones have been identified, the adsorbent in those bands is scraped from the plate and extracted with solvent to remove the adsorbed material. Filtration removes the adsorbent, and evaporation of the solvent gives the recovered component from the mixture.

14.9 THE R_f VALUE

Thin layer chromatography conditions include:

1. Solvent system
2. Adsorbent
3. Thickness of the adsorbent layer
4. Relative amount of material spotted

Under an established set of such conditions, a given compound always travels a fixed distance relative to the distance the solvent front travels. This ratio of the distance the compound travels to the distance the solvent travels is called the **R_f value.** The symbol R_f stands for "retardation factor," or "ratio-to-front," and it is expressed as a decimal fraction:

$$R_f = \frac{Distance\ traveled\ by\ substance}{Distance\ traveled\ by\ solvent\ front}$$

When the conditions of measurement are completely specified, the R_f value is constant for any given compound, and it corresponds to a physical property of that compound.

The R_f value can be used to identify an unknown compound, but like any other identification based on a single piece of data, the R_f value is best confirmed with some additional data. Many compounds can have the same R_f value, just as many compounds have the same melting point.

It is not always possible, in measuring an R_f value, to duplicate exactly the conditions of measurement another researcher has used. Therefore, R_f values tend to be of more use to a single researcher in one laboratory than they are to researchers in different laboratories. The only exception to this is when two researchers use TLC plates from the same source, as in commercial plates, or know the *exact* details of how the plates were prepared. Nevertheless, the R_f value can be a useful guide. If exact values cannot be relied on, the relative values can provide another researcher with useful information about what to expect. Anyone using published R_f values will find it a good idea to check them by comparing them with standard substances whose identity and R_f values are known.

To calculate the R_f value for a given compound, measure the distance that the compound has traveled from the point at which it was originally spotted. For spots that are not too large, measure to the center of the migrated spot. For large spots, the measurement should be repeated on a new plate, using less material. For spots that show tailing, the measurement is made to the "center of gravity" of the spot. This first distance measurement is then divided by the distance the solvent front has traveled from the same original spot. A sample calculation of the R_f values of two compounds is illustrated in Figure 14.7.

14.10 THIN-LAYER CHROMATOGRAPHY APPLIED IN ORGANIC CHEMISTRY

Thin-layer chromatography has several important uses in organic chemistry. It can be used in the following applications:

1. To establish that two compounds are identical
2. To determine the number of components in a mixture
3. To determine the appropriate solvent for a column chromatographic separation
4. To monitor a column chromatographic separation
5. To check the effectiveness of a separation achieved on a column, by crystallization or by extraction
6. To monitor the progress of a reaction

In all these applications, TLC has the advantage that only small amounts of material are necessary. Material is not wasted. With many of the visualization methods, less than a tenth of a microgram (10^{-7} g) of material can be detected. On the other hand, samples as large as a milligram may be used. With preparative plates that are large (about 9 in. on a side) and have a relatively thick coating of adsorbent (>500 μm), it is often possible to separate from 0.2 to 0.5 g of material at one time. The main disadvantage of TLC is that volatile materials cannot be used, because they evaporate from the plates.

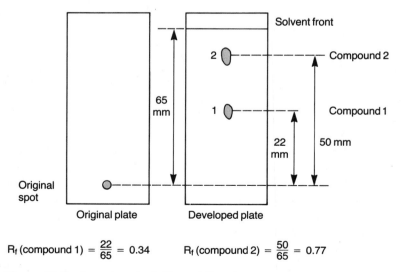

$$R_f \text{ (compound 1)} = \frac{22}{65} = 0.34 \qquad R_f \text{ (compound 2)} = \frac{50}{65} = 0.77$$

Figure 14.7 Sample calculation of R_f values.

Thin-layer chromatography can establish that two compounds suspected to be identical are in fact identical. Simply spot both compounds side by side on a single plate and develop the plate. If both compounds travel the same distance on the plate (have the same R_f), they are probably identical. If the spot positions are not the same, the compounds are definitely not identical. It is important to spot both compounds *on the same plate*. This is especially important with hand-dipped microscope slides. Because they vary widely from plate to plate, no two plates have exactly the same thickness of adsorbent. If you use commercial plates, this precaution is not necessary, although it is nevertheless a good idea.

Thin-layer chromatography can establish whether a sample is a single substance or a mixture. A single substance gives a single spot no matter what solvent is used to develop the plate. On the other hand, the number of components in a mixture can be established by trying various solvents on a mixture. A word of caution should be given. It may be difficult, in dealing with compounds of very similar properties, isomers for example, to find a solvent that will separate the mixture. Inability to achieve a separation is not absolute proof that a sample is a single pure substance. Many compounds can be separated only by *multiple developments* of the TLC slide with a fairly nonpolar solvent. In this method, you remove the plate after the first development and allow it to dry. After being dried, it is placed in the chamber again and developed once more. This effectively doubles the length of the slide. At times, several developments may be necessary.

When a mixture is to be separated, you can use TLC to choose the best solvent to separate it if column chromatography is contemplated. You can try various solvents on a plate coated with the same adsorbent as will be used in the column. The solvent that resolves the components best will probably work well on the column. These small-scale experiments are quick, use very little material, and save time that would be wasted by attempting to separate the entire mixture on the column. Similarly, TLC plates can **monitor** a column. A hypothetical situation is shown in Figure 14.8. A solvent was found that

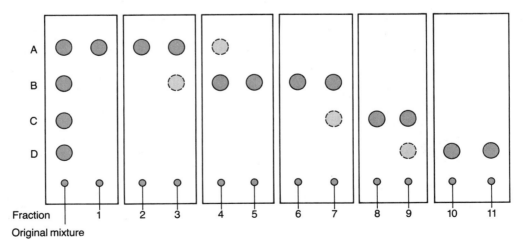

Figure 14.8   Monitoring a column.

would separate the mixture into four components (A–D). A column was run using this solvent, and 11 fractions of 15 mL each were collected. Thin-layer analysis of the various fractions showed that Fractions 1–3 contained Component A; Fractions 4–7, Component B; Fractions 8–9, Component C; and Fractions 10–11, Component D. A small amount of cross-contamination was observed in Fractions 3, 4, 7, and 9.

In another TLC example, a researcher found a product from a reaction to be a mixture. It gave two spots, A and B, on a TLC slide. After the product was crystallized, the crystals were found by TLC to be pure A, whereas the mother liquor was found to have a mixture of A and B. The crystallization was judged to have purified A satisfactorily.

Finally, it is often possible to monitor the progress of a reaction by TLC. At various points during a reaction, samples of the reaction mixture are taken and subjected to TLC analysis. An example is given in Figure 14.9. In this case, the desired reaction was the conversion of A to B. At the beginning of the reaction (0 hr), a TLC slide was prepared that was spotted with pure A, pure B, and the reaction mixture. Similar slides were prepared at 0.5, 1, 2, and 3 hours after the start of the reaction. The slides showed that the reaction was complete in 2 hours. When the reaction was run longer than 2 hours, a new compound, side-product C, began to appear. Thus, the optimum reaction time was judged to be 2 hours.

14.11 PAPER CHROMATOGRAPHY

Paper chromatography is often considered to be related to thin-layer chromatography. The experimental techniques are somewhat like those of TLC, but the principles are more closely related to those of extraction. Paper chromatography is actually a liquid–

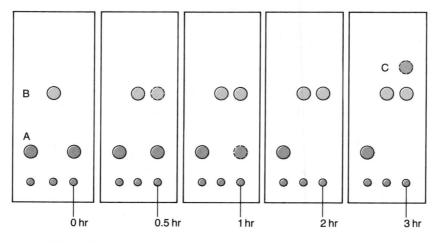

Figure 14.9 Monitoring a reaction.

liquid partitioning technique, rather than a solid–liquid technique. For paper chromatography, a spot is placed near the bottom of a piece of high-grade filter paper (Whatman No. 1 is often used). Then the paper is placed in a developing chamber. The development solvent ascends the paper by capillary action and moves the components of the spotted mixture upward at differing rates. Although paper consists mainly of pure cellulose, the cellulose itself does not function as the stationary phase. Rather, the cellulose absorbs water from the atmosphere, especially from an atmosphere saturated with water vapor. Cellulose can absorb up to about 22% of water. It is this water adsorbed on the cellulose that functions as the stationary phase. To ensure that the cellulose is kept saturated with water, many development solvents used in paper chromatography contain water as a component. As the solvent ascends the paper, the compounds are partitioned between the stationary water phase and the moving solvent. Because the water phase is stationary, the components in a mixture that are most highly water-soluble, or those that have the greatest hydrogen-bonding capacity, are the ones that are held back and move most slowly. Paper chromatography applies mostly to highly polar compounds or to those that are polyfunctional. The most common use of paper chromatography is for sugars, amino acids, and natural pigments. Because filter paper is manufactured consistently, R_f values can often be relied on in paper chromatographic work. However, R_f values are customarily measured from the leading edge (top) of the spot—not from its center, as is customary in TLC.

PROBLEMS

1. A student spots an unknown sample on a TLC plate and develops it in dichloromethane solvent. Only one spot, for which the R_f value is 0.95, is observed. Does this indicate that the unknown material is a pure compound? What can be done to verify the purity of the sample?

2. You and another student were each given an unknown compound. Both samples contained colorless material. You each prepared your own hand-dipped TLC plates and developed the plate using the same solvent. Each of you obtained a single spot of $R_f = 0.75$. Were the samples that you and the other student were assigned necessarily the same substance? How could you prove unambiguously that they were identical, using TLC?

3. Consider a sample that is a mixture composed of biphenyl, benzoic acid, and benzyl alcohol. The sample is spotted on a TLC plate and developed in a dichloromethane–cyclohexane solvent mixture. Predict the *relative* R_f values for the three components in the sample. *Hint:* See Table 12.3.

4. Calculate the R_f value of a spot that travels 5.7 cm, with a solvent front that travels 13 cm.

5. A student spots an unknown sample on a TLC plate and develops it in pentane solvent. Only one spot, for which the R_f value is 0.05, is observed. Is the unknown material a pure compound? What can be done to verify the purity of the sample?

6. A *colorless* unknown substance is spotted on a TLC plate and developed in the correct solvent. The spots do not appear when visualization with a UV lamp or iodine vapors is attempted. What could you do in order to visualize the spots if the compound is

  (a) an alkyl halide
  (b) a ketone
  (c) an amino acid
  (d) a sugar

TECHNIQUE 15

Gas Chromatography

Gas chromatography is one of the most useful instrumental tools for separating and analyzing organic compounds that can be vaporized without decomposition. Common uses include testing the purity of a substance and separating the components of a mixture. The relative amounts of the components in a mixture may also be determined. In some cases, gas chromatography can be used to identify a compound. In microscale work, it can also be used as a preparative method to isolate pure compounds from a small amount of a mixture.

Gas chromatography resembles column chromatography in principle, but it differs in three respects. First, the partitioning processes for the compounds to be separated are carried out between a **moving gas phase** and a **stationary liquid phase.** (Recall that in column chromatography the moving phase is a liquid and the stationary phase is a solid adsorbent.) A second difference is that the temperature of the gas system can be controlled, because the column is contained in an insulated oven. And third, the concentration of any given compound in the gas phase is a function of its vapor pressure only. Because gas chromatography separates the components of a mixture primarily on the basis of their vapor pressures (or boiling points), this technique is also similar in principle to fractional distillation. In microscale work, it is sometimes used to separate and isolate compounds from a mixture; fractional distillation would normally be used with larger amounts of material.

Gas chromatography (GC) is also known as vapor-phase chromatography (VPC) and as gas–liquid partition chromatography (GLPC). All three names, as well as their indicated abbreviations, are often found in the literature of organic chemistry. In reference to the technique, the last term, GLPC, is the most strictly correct and is preferred by most authors, but GC is used most often.

15.1 THE GAS CHROMATOGRAPH

The apparatus used to carry out a gas–liquid chromatographic separation is generally called a **gas chromatograph.** A typical student-model gas chromatograph, the GOW-MAC model 69–350, is illustrated in Figure 15.1. A schematic block diagram of a basic gas chromatograph is shown in Figure 15.2. The basic elements of the apparatus are apparent. In short, the sample is injected into the chromatograph, and it is immediately vaporized in a heated injection chamber and introduced into a moving stream of gas, called the **carrier gas.** The vaporized sample is then swept into a column filled with particles coated with a liquid adsorbent. The column is contained in a temperature-controlled oven. As the sample passes through the column, it is subjected to many gas–liquid partitioning processes, and the components are separated. As each component leaves the column, its presence is detected by an electrical detector that generates a signal that is recorded on a strip chart recorder.

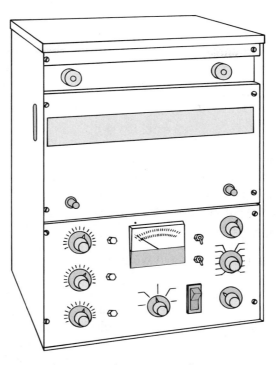

Figure 15.1   Gas chromatograph.

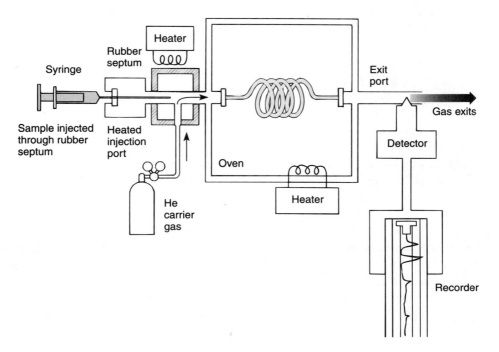
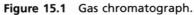

Figure 15.2   Schematic diagram of gas chromatograph.

Many modern instruments are also equipped with a microprocessor, which can be programmed to change parameters, such as the temperature of the oven, while a mixture is being separated on a column. With this capability, it is possible to optimize the separation of components and to complete a run in a relatively short time.

15.2 THE COLUMN

The heart of the gas chromatograph is the packed column. This column is usually made of copper or stainless steel tubing, but sometimes glass is used. The most common diameters of tubing are $\frac{1}{8}$ in. (3 mm) and $\frac{1}{4}$ in. (6 mm). To construct a column, cut a piece of tubing to the desired length and attach the proper fittings on each of the two ends to connect it to the apparatus. The most common length is 4–12 ft, but some columns may be up to 50 ft in length.

The tubing (column) is then packed with the **stationary phase.** The material chosen for the stationary phase is usually a liquid, a wax, or a low-melting solid. This material should be relatively nonvolatile, that is, it should have a low vapor pressure and a high boiling point. Liquids commonly used are high-boiling hydrocarbons, silicone oils, waxes, and polymeric esters, ethers, and amides. Some typical substances are listed in Table 15.1.

The liquid phase is usually coated onto a **support material.** A common support material is crushed firebrick. Many methods exist for coating the high-boiling liquid phase onto the support particles. The easiest is to dissolve the liquid (or low-melting wax or solid) in a volatile solvent like methylene chloride (bp 40°C). The firebrick (or other support) is added to this solution, which is then slowly evaporated (rotary evaporator) so as to leave each particle of support material evenly coated. Other support materials are listed in Table 15.2.

In the final step, the liquid-phase-coated support material is packed into the tubing as evenly as possible. The tubing is bent or coiled so that it fits into the oven of the gas chromatograph with its two ends connected to the gas entrance and exit ports.

Selection of a liquid phase usually revolves about two factors. First, most of them have an upper temperature limit above which they cannot be used. Above the specified limit of temperature, the liquid phase itself will begin to "bleed" off the column. Second, the materials to be separated must be considered, For polar samples, it is usually best to use a polar liquid phase; for nonpolar samples, a nonpolar liquid phase is indicated. The liquid phase performs best when the substances to be separated *dissolve* in it.

Most researchers today buy packed columns from commercial sources, rather than pack their own. A wide variety of types and lengths are available.

Alternatives to packed columns are Golay or glass capillary columns of diameters 0.1–0.2 mm. With these columns, no solid support is required, and the liquid is coated directly on the inner walls of the tubing. Liquid phases commonly used in glass capillary columns are similar in composition to those used in packed columns. They include DB-1 (similar to SE-30), DB-17 (similar to DC-710), and DB-WAX (similar to Carbowax 20M).

TABLE 15.1 Typical Liquid Phases

Increasing polarity →

Type	Composition	Maximum Temperature (°C)	Typical Use
Apiezons (L, M, N, etc.) Hydrocarbon greases (varying MW)	Hydrocarbon mixtures	250–300	Hydrocarbons
SE–30 Methyl silicone rubber	Like silicone oil, but cross-linked	350	General applications
DC–200 Silicone oil (R = CH$_3$)	$R_3Si-O-\left[\begin{array}{c} R \\ \vert \\ Si-O \\ \vert \\ R \end{array}\right]_n -SiR_3$	225	Aldehydes, ketones, halocarbons
DC–710 Silicone oil (R = CH$_3$) (R' = C$_6$H$_5$)	$\left[\begin{array}{c} R' \\ \vert \\ Si-O \\ \vert \\ R \end{array}\right]_n -$	300	General applications
Carbowaxes (400–20M) Polyethylene glycols (varying chain lengths)	Polyether $HO-(CH_2CH_2-O)_n-CH_2CH_2OH$	Up to 250	Alcohols, ethers, halocarbons
DEGS Diethylene glycol succinate	Polyester $\left(-CH_2CH_2-O-\underset{\underset{O}{\parallel}}{C}-(CH_2)_2-\underset{\underset{O}{\parallel}}{C}-O-\right)_n$	200	General applications

---

TABLE 15.2   Typical Solid Supports

Crushed firebrick	Chromosorb T (Teflon beads)
Nylon beads	Chromosorb P (Pink diatomaceous earth, highly absorptive, pH 6–7)
Glass beads	
Silica	
Alumina	Chromosorb W (White diatomaceous earth, medium absorptivity, pH 8–10)
Charcoal	
Molecular sieves	
	Chromosorb G (like the above, low absorptivity, pH 8.5)

The length of a capillary column is usually very long, typically 50–100 ft. Because of the length and small diameter, there is increased interaction between the sample and the stationary phase. Gas chromatographs equipped with these small-diameter columns are able to separate components more effectively than instruments using packed columns.

15.3 PRINCIPLES OF SEPARATION

After a column is selected, packed, and installed, the **carrier gas** (usually helium, argon, or nitrogen) is allowed to flow through the column supporting the liquid phase. The mixture of compounds to be separated is introduced into the carrier gas stream, where its components are equilibrated (or partitioned) between the moving gas phase and the stationary liquid phase (Figure 15.3). The latter is held stationary because it is adsorbed onto the surfaces of the support material.

The sample is introduced into the gas chromatograph by a microliter syringe. It is injected as a liquid or as a solution through a rubber septum into a heated chamber, called the **injection port,** where it is vaporized and mixed with the carrier gas. As this mixture reaches the column, which is heated in a controlled oven, it begins to equilibrate between the liquid and gas phases. The length of time required for a sample to move through the column is a function of how much time it spends in the vapor phase and how much time it spends in the liquid phase. The more time it spends in the vapor phase, the faster it gets to the end of the column. In most separations, the components of a sample have similar solubilities in the liquid phase. Therefore, the time the different compounds spend in the vapor phase is primarily a function of their vapor pressure, and the more volatile component arrives at the end of the column first, as illustrated in Figure 15.3. By selecting the correct temperature of the oven and the correct liquid phase, the compounds in the injected mixture travel through the column at different rates and are separated.

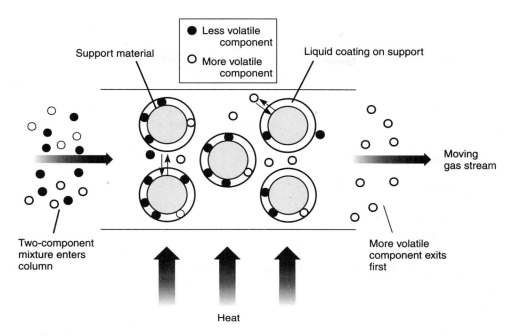

Figure 15.3   The separation process.

15.4 FACTORS AFFECTING SEPARATION

Several factors determine the rate at which a given compound travels through a gas chromatograph. First of all, compounds of low boiling point will generally travel through the gas chromatograph faster than compounds of higher boiling point. This is because the column is heated, and low-boiling compounds always have higher vapor pressures than compounds of higher boiling point. In general, therefore, for compounds with the same functional group, the higher the molecular weight the longer the retention time. For most molecules, the boiling point increases as the molecular weight increases. If the column is heated to a temperature that is too high, however, the entire mixture to be separated is flushed through the column at the same rate as the carrier gas, and no equilibration takes place with the liquid phase. On the other hand, at too low a temperature, the mixture dissolves in the liquid phase and never revaporizes. Thus, it is retained on the column.

Second, the rate of flow of the carrier gas is important. The carrier gas must not move so rapidly that molecules of the sample in the vapor phase cannot equilibrate with those dissolved in the liquid phase. This may result in poor separation between components in the injected mixture. If the rate of flow is too slow, however, the bands broaden significantly, leading to poor resolution (see Section 15.8).

The third factor is the choice of liquid phase used in the column. The molecular weights, functional groups, and polarities of the component molecules in the mixture to be separated must be considered when a liquid phase is being chosen. One generally uses a different type of material for hydrocarbons, for instance, than for esters. The materials to be separated should *dissolve* in the liquid. The useful temperature limit of the liquid phase selected must also be considered.

Fourth, the length of the column is important. Compounds that resemble one another closely, in general, require longer columns than dissimilar compounds. Many kinds of isomeric mixtures fit into the "difficult" category. The components of isomeric mixtures are so much alike that they travel through the column at very similar rates. You need a longer column, therefore, to take advantage of any differences that may exist.

15.5 ADVANTAGES OF GAS CHROMATOGRAPHY

All factors that have been mentioned must be adjusted by the chemist for any mixture to be separated. Considerable preliminary investigation is often required before a mixture can be separated successfully into its components by gas chromatography. Nevertheless, the advantages of the technique are many.

First, many mixtures can be separated by this technique when no other method is adequate. Second, as little as 1–10 μL (1 μL = 10^{-6} L) of a mixture can be separated by this technique. This advantage is particularly important when working at the microscale level. Third, when gas chromatography is coupled with an electronic recording device (see following discussion), the amount of each component present in the separated mixture can be estimated quantitatively.

The range of compounds that can be separated by gas chromatography extends from gases, such as oxygen (bp −183°C) and nitrogen (bp −196°C), to organic compounds with boiling points over 400°C. The only requirement for the compounds to be separated is that they have an appreciable vapor pressure at a temperature at which they can be separated and that they be thermally stable at this temperature.

15.6 MONITORING THE COLUMN (THE DETECTOR)

To follow the separation of the mixture injected into the gas chromatograph, it is necessary to use an electrical device called a **detector.** Two types of detectors in common use are the **thermal conductivity detector (TCD)** and the **flame ionization detector (FID).**

The thermal conductivity detector is simply a hot wire placed in the gas stream at the column exit. The wire is heated by constant electrical voltage. When a steady stream of carrier gas passes over this wire, the rate at which it loses heat and its electrical conductance have constant values. When the composition of the vapor stream changes, the rate of heat flow from the wire, and hence its resistance, changes. Helium, which has a higher thermal conductivity than most organic substances, is a common carrier gas. Thus, when a substance elutes in the vapor stream, the thermal conductivity of the moving gases will be lower than with helium alone. The wire then heats up, and its resistance decreases.

A typical TCD operates by difference. Two detectors are used: one exposed to the actual effluent gas and the other exposed to a reference flow of carrier gas only. To achieve this situation, a portion of the carrier gas stream is diverted *before* it enters the injection port. The diverted gas is routed through a reference column into which no sample has been admitted. The detectors mounted in the sample and reference columns are arranged so as to form the arms of a Wheatstone bridge circuit, as shown in Figure 15.4. As long as the carrier gas alone flows over both detectors, the circuit is in balance. However, when a sample elutes from the sample column, the bridge circuit becomes unbalanced, creating an electrical signal. This signal can be amplified and used to activate a strip chart recorder. The recorder is an instrument that plots, by means of a moving pen, the unbalanced bridge current versus time on a continuously moving roll of chart paper. This record of detector response (current) versus time is called a **chromatogram.** A typical gas chromatogram is illustrated in Figure 15.5. Deflections of the pen are called **peaks.**

When a sample is injected, some air (CO_2, H_2O, N_2, and O_2) is introduced along with the sample. The air travels through the column almost as rapidly as the carrier gas; as it passes the detector, it causes a small pen response, thereby giving a peak, called the **air peak.** At later times (t_1, t_2, t_3), the components also give rise to peaks on the chromatogram as they pass out of the column and past the detector.

In a flame ionization detector, the effluent from the column is directed into a flame produced by the combustion of hydrogen, as illustrated in Figure 15.6. As organic compounds burn in the fame, ion fragments are produced which collect on the ring above the flame. The resulting electrical signal is amplified and sent to a recorder in a similar manner to a TCD, except that a FID does not produce an air peak. The main advantage of the FID is that it is more sensitive and can be used to analyze smaller quantities of sample. Also, because a FID does not respond to water, a gas chromatograph with this detector can be used to analyze aqueous solutions. Two disadvantages are that it is more difficult

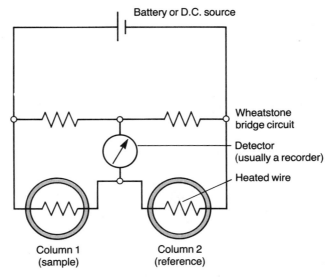

Figure 15.4 Typical thermal conductivity detector.

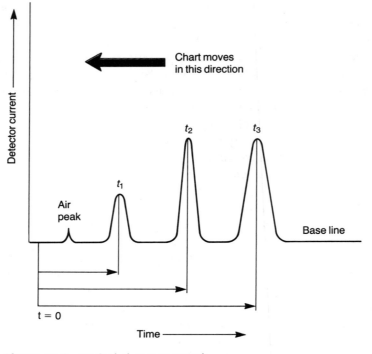

Figure 15.5 Typical chromatograph.

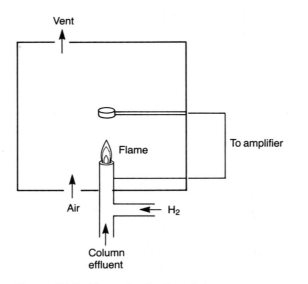

Figure 15.6 Flame ionization detector.

to operate and the detection process destroys the sample. Therefore, an FID gas chromatograph cannot be used to do preparative work, which is often desired in the microscale laboratory.

15.7 RETENTION TIME

The period following injection that is required for a compound to pass through the column is called the **retention time** of that compound. For a given set of constant conditions (flow rate of carrier gas, column temperature, column length, liquid phase, injection port temperature, carrier), the retention time of any compound is always constant (much like the R_f value in thin-layer chromatography, as described in Technique 14, Section 14.9, p. 803). The retention time is measured from the time of injection to the time of maximum pen deflection (detector current) for the component being observed. This value, when obtained under controlled conditions, can identify a compound by a direct comparison of it with values for known compounds determined under the same conditions. For easier measurement of retention times, most strip chart recorders are adjusted to move the paper at a rate that corresponds to time divisions calibrated on the chart paper. The retention times (t_1, t_2, t_3) are indicated in Figure 15.5 for the three peaks illustrated.

15.8 POOR RESOLUTION AND TAILING

The peaks in Figure 15.5 are well **resolved.** That is, the peaks are separated from one another, and between each pair of adjacent peaks the tracing returns to the **baseline.** In Figure 15.7, the peaks overlap and the resolution is not good. Poor resolution is often caused by using too much sample, too high a column temperature, too short a column, a liquid phase that does not discriminate well between the two components, a column with too large a diameter, or, in short, almost any wrongly adjusted parameter. When peaks are poorly resolved, it is more difficult to determine the relative amount of each component. Methods for determining the relative percentages of each component are given in Section 15.11.

Another desirable feature illustrated by the chromatogram in Figure 15.5 is that each peak is symmetrical. A common example of an unsymmetrical peak is one in which **tailing** has occurred, as shown in Figure 15.8. Tailing usually results from injecting too much sample into the gas chromatograph. Another cause of tailing occurs with polar compounds,

Figure 15.7   Poor resolution or peaks overlap.

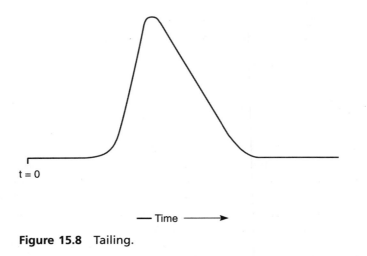

Figure 15.8   Tailing.

such as alcohols and aldehydes. These compounds may be temporarily adsorbed on column walls or areas of the support material that are not adequately coated by the liquid phase. Therefore, they do not leave in a band, and tailing results.

15.9 QUALITATIVE ANALYSIS

A disadvantage of the gas chromatograph is that it gives no information whatever about the identities of the substances it has separated. The little information it does provide is given by the retention time. It is hard to reproduce this quantity from day to day, however, and exact duplications of separations performed last month may be difficult to make this month. It is usually necessary to **calibrate** the column each time it is used. That is, you must run pure samples of all known and suspected components of a mixture individually, just before chromatographing the mixture, to obtain the retention time of each known compound. As an alternative, each suspected component can be added, one by one, to the unknown mixture while the operator looks to see which peak has its intensity increased relative to the unmodified mixture. Another solution is to *collect* the components individually as they emerge from the gas chromatograph. Each component can then be identified by other means, such as by infrared or nuclear magnetic resonance spectroscopy or by mass spectrometry.

15.10 COLLECTING THE SAMPLE

For gas chromatographs with a thermal conductivity detector, it is possible to collect samples that have passed through the column. One uses a form of cooled trap connected to the outlet of the column as a collecting device. The simplest form of trap is just a U-shaped tube dipped into a bath of coolant (ice water, liquid nitrogen, or dry ice–acetone), as shown in Figure 15.9.

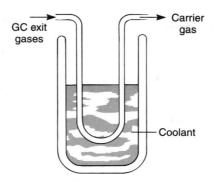

Figure 15.9 Collection trap.

For instance, if the coolant is liquid nitrogen (bp −196°C) and the carrier gas is helium (bp −269°C), compounds boiling above the temperature of liquid nitrogen generally are condensed or trapped at the bottom of the U-shaped tube, while the carrier gas continues to flow. However, if the compound leaving is frozen in the form of a fine mist, it may be carried through the trap even though it is frozen. When this happens, special trapping arrangements are required. To collect each component of the mixture, one must change the trap each time a new peak begins to appear on the chromatogram. Since even a gas chromatograph with large-diameter columns can rarely handle more than about 0.5 mL of a mixture at a time, a chemist or student having 50 mL of a mixture would have to make 50 to 100 separate injections, changing the trap for each component as it comes off the column. Fortunately, there are accessories that can be attached to a gas chromatograph to do this task automatically.

A microscale method uses a gas collection tube (see Fig. 15.10), which is included in most microscale glassware kits. A collection tube is joined to the exit port of the col-

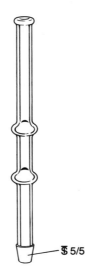

Figure 15.10 Gas chromatograph collection tube.

umn by inserting the ℑ 5/5 inner joint into a metal adapter, which is connected to the exit port. When a sample is eluted from the column in the vapor state, it is cooled by the connecting adapter and the gas collection tube and condenses in the collection tube. The gas collection tube is removed from the adapter when the recorder indicates that the desired sample has completely passed through the column. After the first sample has been collected, the process can be repeated with another gas collection tube.

To isolate the liquid, the tapered joint of the collection tube is inserted into a 0.1-mL conical vial, which has a ℑ 5/5 outer joint. The assembly is placed into a test tube, as illustrated in Figure 15.11. During centrifugation, the sample is forced into the bottom of the conical vial. After disassembling the apparatus, the liquid can be removed from the vial with a syringe for a boiling-point determination or analysis by infrared spectroscopy. If a determination of the sample weight is desired, the empty conical vial and cap should be tared and reweighed after the liquid has been collected. It is advisable to dry the gas collection tube and the conical vial in an oven before use in order to prevent contamination by water or other solvents used in cleaning this glassware.

15.11 QUANTITATIVE ANALYSIS

The area under a gas chromatograph peak is proportional to the amount (moles) of compound eluted. Hence, the molar percentage composition of a mixture can be approximated by comparing relative peak areas. This method of analysis assumes that the de-

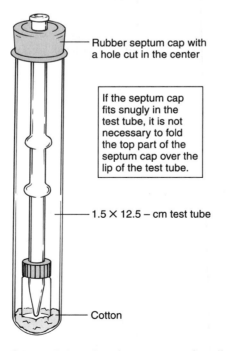

Rubber septum cap with a hole cut in the center

If the septum cap fits snugly in the test tube, it is not necessary to fold the top part of the septum cap over the lip of the test tube.

1.5 × 12.5 – cm test tube

Cotton

Figure 15.11   Gas chromatograph collection tube and 0.1-mL conical vial.

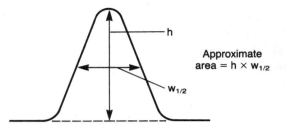

Figure 15.12 Triangulation of a peak.

tector is equally sensitive to all compounds eluted and that it gives a linear response with respect to amount. Nevertheless, it gives reasonably accurate results.

The simplest method of measuring the area of a peak is by geometrical approximation, or triangulation. In this method, you multiply the height h of the peak above the baseline of the chromatogram by the width of the peak at half of its height $w_{1/2}$. This is illustrated in Figure 15.12. The baseline is approximated by drawing a line between the two side-arms of the peak. This method works well only if the peak is symmetrical. If the peak has tailed or is unsymmetrical, it is best to cut out the peaks with scissors and weigh the pieces of paper on an **analytical balance.** Because the weight per area of a piece of good chart paper is reasonably constant from place to place, the ratio of the areas is the same as the ratio of the weights. To obtain a percentage composition for the mixture, first add all the peak areas (weights). Then, to calculate the percentage of any component in the mixture, divide its individual area by the total area and multiply the result by 100. A sample calculation is illustrated in Figure 15.13. If peaks overlap (see Fig. 15.7), either

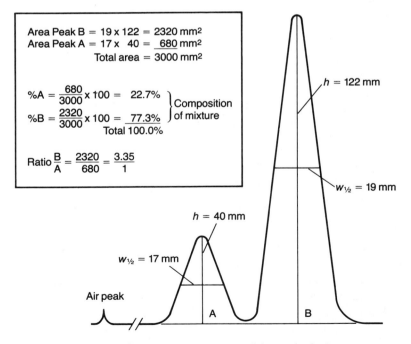

Figure 15.13 Sample percentage composition calculation.

the gas chromatographic conditions must be readjusted to achieve better resolution of the peaks or the peak shape must be estimated.

There are various instrumental means, which are built into recorders, of detecting the amounts of each sample automatically. One method uses a separate pen that produces a trace that integrates the area under each peak. Another method employs an electronic device that automatically prints out the area under each peak and the percentage composition of the sample.

For the experiments in this textbook, we have assumed that the detector is equally sensitive to all compounds eluted. Compounds with different functional groups or with widely varying molecular weights, however, produce different responses with both TCD and FID gas chromatographs. With a TCD, the responses are different because not all compounds have the same thermal conductivity. Different compounds analyzed with a FID gas chromatograph also give different responses because the detector response varies with the type of ions produced. For both types of detectors, it is possible to calculate a **response factor** for each compound in a mixture. Response factors are usually determined by making up an equimolar mixture of two compounds, one of which is considered to be the reference. The mixture is separated on a gas chromatograph, and the relative percentages are calculated using one of the methods described previously. From these percentages you can determine a response factor for the compound being compared to the reference. If you do this for all the components in a mixture, you can then use these correction factors to make more accurate calculations of the relative percentages for the compounds in the mixture.

Consider the following example, which illustrates how response factors are determined. In this example, an equimolar mixture of benzene, hexane, and ethyl acetate is prepared and analyzed using a flame-ionization gas chromatograph. The peak areas obtained are:

Benzene	966463
Hexane	831158
Ethyl Acetate	1449695

In most cases, benzene is taken as the standard, and its response factor is defined to be equal to 1.00. Calculation of the response factors for the other components of the test mixture proceeds as follows:

Benzene	966463/966463 = 1.00 (by definition)
Hexane	831158/966463 = 0.86
Ethyl Acetate	1449695/966463 = 1.50

Notice that the response factors calculated in this example are *molar* response factors. It is necessary to correct these values by the relative molecular weights of each substance to obtain *weight* response factors.

When you use a flame-ionization gas chromatograph for quantitative analysis, it is first necessary to determine the response factors for each component of the mixture being analyzed. For a quantitative analysis, it is likely that you will have to convert molar response factors into *weight* response factors. Next, the chromatography experiment using the unknown samples is performed. The observed peak areas for each component are corrected using the response factors in order to arrive at the correct weight percentage of each component in the sample.

PROBLEMS

1. (a) A sample consisting of 1-bromopropane and 1-chloropropane is injected into a gas chromatograph equipped with a nonpolar column. Which compound has the shorter retention time? Explain your answer.

    (b) If the same sample were run several days later with the conditions as nearly the same as possible, would you expect the retention times to be identical to those obtained the first time? Explain.

2. Using triangulation, calculate the percentage of each component in a mixture composed of two substances, A and B. The chromatogram is shown in Figure 15.14.

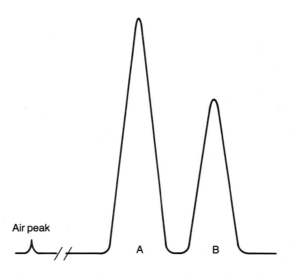

Figure 15.14   Chromatogram for Problem 2.

3. Make a photocopy of the chromatogram in Figure 15.14. Cut out the peaks and weigh them on an analytical balance. Use the weights to calculate the percentage of each component in the mixture. Compare your answer to what you calculated in Problem 2.

4. What would happen to the retention time of a compound if the following changes were made?

    (a) Decrease the flow rate of the carrier gas

    (b) Increase the temperature of the column

    (c) Increase the length of the column

T E C H N I Q U E   1 6

Sublimation

In Technique 6, the influence of temperature on the change in vapor pressure of a liquid was considered (see Fig. 6.8, p. 675). It was shown that the vapor pressure of a liquid increases with temperature. Because the boiling point of a liquid occurs when its vapor pressure is equal to the applied pressure (normally atmospheric pressure), the vapor pressure of a liquid equals 760 mmHg at its boiling point. The vapor pressure of a solid also varies with temperature. Because of this behavior, some solids can pass directly into the vapor phase without going through a liquid phase. This process is called **sublimation.** Because the vapor can be resolidified, the overall vaporization–solidification cycle can be used as a purification method. The purification can be successful only if the impurities have significantly lower vapor pressures than the material being sublimed.

16.1 VAPOR PRESSURE BEHAVIOR OF SOLIDS AND LIQUIDS

In Figure 16.1, vapor pressure curves for solid and liquid phases for two different substances are shown. Along lines *AB* and *DF,* the sublimation curves, the solid and vapor are at equilibrium. To the left of these lines, the solid phase exists, and to the right of these lines, the vapor phase is present. Along lines *BC* and *FG,* the liquid and vapor are at equilibrium. To the left of these lines, the liquid phase exists, and to the right, the vapor is present. The two substances vary greatly in their physical properties, as shown in Figure 16.1.

In the first case (Fig. 16.1A), the substance shows normal change-of-state behavior on being heated, going from solid to liquid to gas. The dashed line, which represents an atmospheric pressure of 760 mmHg, is located *above* the melting point *B* in Figure 16.1A. Thus, the applied pressure (760 mmHg) is *greater* than the vapor pressure of the solid–liquid phase at the melting point. Starting at *A,* as the temperature of the solid is raised, the vapor pressure increases along *AB* until the solid is observed to melt at *B*. At *B* the vapor pressures of *both* the solid and liquid are identical. As the temperature continues to rise, the vapor pressure will increase along *BC* until the liquid is observed to boil at *C*. The description given is for the "normal" behavior expected for a solid substance. All three states (solid, liquid, and gas) are observed sequentially during the change in temperature.

In the second case (Fig. 16.1B), the substance develops enough vapor pressure to vaporize completely at a temperature below its melting point. The substance shows a solid-to-gas transition only. The dashed line is now located *below* the melting point *F* of this substance. Thus, the applied pressure (760 mmHg) is *less* than the vapor pressure of the solid–liquid phase at the melting point. Starting at *D,* the vapor pressure of the solid rises as the temperature increases along line *DF.* However, the vapor pressure of the solid reaches atmospheric pressure (point *E*) *before* the melting point at *F* is attained. Therefore, sublimation occurs at *E*. No melting behavior will be observed at atmospheric pressure for

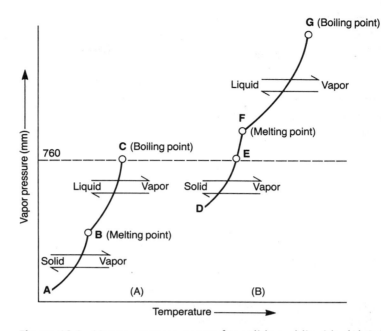

Figure 16.1 Vapor pressure curves for solids and liquids. (A) Substance shows normal solid to liquid to gas transitions at 760-mmHg pressure. (B) Substance shows a solid to gas transition at 760-mmHg pressure.

this substance. For a melting point to be reached and the behavior along line *FG* to be observed, an applied pressure greater than the vapor pressure of the substance at point *F* would be required. This could be achieved by using a sealed pressure apparatus.

The sublimation behavior just described is relatively rare for substances at atmospheric pressure. Several compounds exhibiting this behavior—carbon dioxide, perfluorocyclohexane, and hexachloroethane—are listed in Table 16.1. Notice that these compounds have vapor pressures *above* 760 mmHg at their melting points. In other words, their vapor pressures reach 760 mmHg below their melting points and they sublime rather than melt. Anyone trying to determine the melting point of hexachloroethane at atmospheric pressure will see vapor pouring from the end of the melting point tube! With a sealed capillary tube, the melting point of 186°C is observed.

16.2 SUBLIMATION BEHAVIOR OF SOLIDS

Sublimation is usually a property of relatively nonpolar substances that also have highly symmetrical structures. Symmetrical compounds have relatively high melting points and high vapor pressures. The ease with which a substance can escape from the solid state is determined by the strength of intermolecular forces. Symmetrical molecular structures have a relatively uniform distribution of electron density and a small dipole moment. A smaller dipole moment means a higher vapor pressure because of lower electrostatic attractive forces in the crystal.

TABLE 16.1   Vapor Pressures of Solids at Their Melting Points

Compound	Vapor Pressure of Solid at MP (mmHg)	Melting Point (°C)
Carbon dioxide	3876 (5.1 atm)	−57
Perfluorocyclohexane	950	59
Hexachloroethane	780	186
Camphor	370	179
Iodine	90	114
Naphthalene	7	80
Benzoic acid	6	122
p-Nitrobenzaldehyde	0.009	106

Solids sublime if their vapor pressures are greater than atmospheric pressure at their melting points. Some compounds with the vapor pressures at their melting points are listed in Table 16.1. The first three entries in the table were discussed in Section 16.1. At atmospheric pressure they would sublime rather than melt, as shown in Figure 16.1B.

The next four entries in Table 16.1 (camphor, iodine, naphthalene, and benzoic acid) exhibit typical change-of-state behavior (solid, liquid, and gas) at atmospheric pressure, as shown in Figure 16.1A. These compounds sublime readily under reduced pressure, however. Vacuum sublimation is discussed in Section 16.3.

Compared with many other organic compounds, camphor, iodine, and naphthalene have relatively high vapor pressures at relatively low temperatures. For example, they have a vapor pressure of 1 mmHg at 42, 39, and 53°C, respectively. Although this vapor pressure does not seem very large, it is high enough to lead, after a time, to **evaporation** of the solid from an open container. Mothballs (naphthalene and 1,4-dichlorobenzene) show this behavior. When iodine stands in a closed container over a period of time, you can observe movement of crystals from one part of the container to another.

Although chemists often refer to any solid–vapor transition as sublimation, the process described for camphor, iodine, and naphthalene is really an **evaporation** of a solid. Strictly speaking, a sublimation point is like a melting point or a boiling point. It is defined as the point at which the vapor pressure of the solid *equals* the applied pressure. Many liquids readily evaporate at temperatures far below their boiling points. It is, however, much less common for solids to evaporate. Solids that readily sublime (evaporate) must be stored in sealed containers. When the melting point of such a solid is being determined, some of the solid may sublime and collect toward the open end of the melting-point tube while the rest of the sample melts. To solve the sublimation problem, one seals the capillary tube or rapidly determines the melting point. It is possible to use the sublimation behavior to purify a substance. For example, at atmospheric pressure, camphor can be readily sublimed, just below its melting point at 175°C. At 175°C the vapor pressure of camphor is 320 mmHg. The vapor solidifies on a cool surface.

16.3 VACUUM SUBLIMATION

Many organic compounds sublime readily under reduced pressure. When the vapor pressure of the solid equals the applied pressure, sublimation occurs, and the behavior is identical to that shown in Figure 16.1B. The solid phase passes directly into the vapor phase. From the data given in Table 16.1, you should expect camphor, naphthalene, and benzoic acid to sublime at or below the respective applied pressures of 370, 7, and 6 mmHg. In principle, you can sublime *p*-nitrobenzaldehyde (last entry in the table), but it would not be practical because of the low applied pressure required.

16.4 SUBLIMATION METHODS

Sublimation can be used to purify solids. The solid is warmed until its vapor pressure becomes high enough for it to vaporize and condense as a solid on a cooled surface placed closely above. Three types of apparatus are illustrated in Figure 16.2. Since all of the parts fit securely, they are all capable of holding a vacuum. Chemists usually perform vacuum sublimations because most solids undergo the solid to gas transition only at low pressures. Reduction of pressure also helps to prevent thermal decomposition of substances that would require high temperatures to sublime at ordinary pressures. One end of a piece of rubber pressure tubing is attached to the apparatus and the other end is attached to an aspirator or to the house vacuum system or to a vacuum pump.

A sublimation is probably best carried out using one of the pieces of microscale equipment shown in Figures 16.2A and B. It is recommended that the laboratory instructor make available either one type or the other to be used on a communal basis. Each of the apparatus shown employs a central tube (closed on one end) filled with ice-cold water that serves as a condensing surface. The tube is filled with ice chips and a minimum of water. If the cooling water becomes warm before the sublimation is completed, a Pasteur pipet can be used to remove the warm water. The tube is then refilled with more ice-cold water. Warm water is undesirable because the vapor will not condense efficiently to form a solid as readily on a warm surface as it would on a cold surface. A poor recovery of solid results.

The apparatus shown in Figure 16.2C can be constructed from a side-arm test tube, a neoprene adapter, and a piece of glass tubing sealed at one end. Alternatively, a 15 × 125-mm test tube may be used instead of the piece of glass tubing. The test tube is inserted into a #2 neoprene adapter using a little water as a lubricant. All pieces must fit securely in order to be able to obtain a good vacuum and to avoid water being drawn into the side-arm test tube around the rubber adapter. In order to achieve an adequate seal, the side-arm test tube may need to be flared somewhat. A complete description of the assembly of the apparatus, similar to that shown in Figure 16.2C, is given in Experiment 5A, p. 73. Detailed procedures for using this apparatus are also given in that experiment.

A flame is the preferred heating device because the sublimation will occur more quickly than with other heating devices. The sublimation will be finished before the ice water warms significantly. The burner can be held by its cool base (not the hot barrel!) and moved up and down the sides of the outer tube to "chase" any solid that has formed

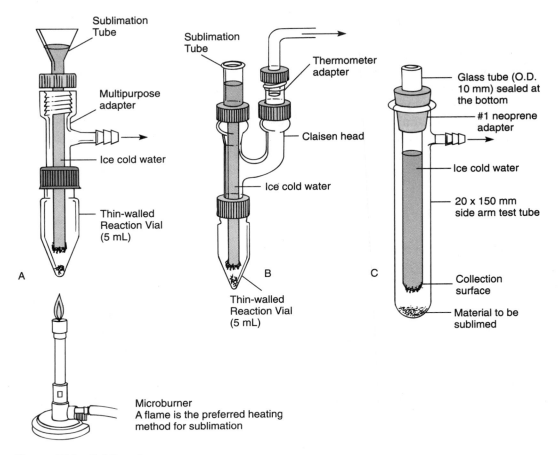

Figure 16.2   Sublimation apparatus.

on the sides toward the cold tube in the center. When using the apparatus shown in Figures 16.2A and B with a flame, you will need to use a thin-wall vial. Thicker glass can shatter when heated with a flame.

Remember that while performing a sublimation, it is important to keep the temperature below the melting point of the solid. After sublimation, the material that has collected on the cooled surface is recovered by removing the central tube (cold-finger) from the apparatus. Take care in removing this tube to avoid dislodging the crystals that have collected. The deposit of crystals is scraped from the inner tube with a spatula. If reduced pressure has been used, the pressure must be released carefully to keep a blast of air from dislodging the crystals.

16.5 ADVANTAGES OF SUBLIMATION

One advantage of sublimation is that no solvent is used and therefore none needs to be removed later. Sublimation also removes occluded material, like molecules of solvent, from the sublimed substance. For instance, caffeine (sublimes at 178°C, melts at 236°C) absorbs water gradually from the atmosphere to form a hydrate. During sublimation, this water is lost, and anhydrous caffeine is obtained. If too much solvent is present in a sample to be sublimed, however, instead of becoming lost, it condenses on the cooled surface and thus interferes with the sublimation.

Sublimation is a faster method of purification than crystallization but not as selective. Similar vapor pressures are often a factor in dealing with solids that sublime; consequently, little separation can be achieved. For this reason, solids are far more often purified by crystallization. Sublimation is most effective in removing a volatile substance from a nonvolatile compound, particularly a salt or other inorganic material. Sublimation is also effective in removing highly volatile bicyclic or other symmetrical molecules from less volatile reaction products. Examples of volatile bicyclic compounds are borneol, isoborneol, and camphor.

PROBLEMS

1. Why is solid carbon dioxide called dry ice? How does it differ from solid water in behavior?
2. Under what conditions can you have *liquid* carbon dioxide?
3. A solid substance has a vapor pressure of 800 mmHg at its melting point (80°C). Describe how the solid behaves as the temperature is raised from room temperature to 80°C, while the atmospheric pressure is held constant at 760 mmHg.
4. A solid substance has a vapor pressure of 100 mmHg at the melting point (100°C). Assuming an atmospheric pressure of 760 mmHg, describe the behavior of this solid as the temperature is raised from room temperature to its melting point.
5. A substance has a vapor pressure of 50 mmHg at the melting point (100°C). Describe how you would experimentally sublime this substance.

TECHNIQUE 17

Polarimetry

17.1 NATURE OF POLARIZED LIGHT

Light has a dual nature because it shows properties of both waves and particles. The wave nature of light can be demonstrated by two experiments: polarization and interference. Of the two, polarization is the more interesting to organic chemists, because they can take advantage of polarization experiments to learn something about the structure of an unknown molecule.

Ordinary white light consists of wave motion in which the waves have a variety of wavelengths and vibrate in all possible planes perpendicular to the direction of propagation. Light can be made to be **monochromatic** (of one wavelength or color) by using fil-

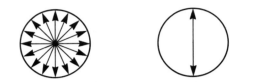

Figure 17.1   Ordinary versus plane-polarized light.

ters or special light sources. Frequently, a sodium lamp (sodium D line = 5893 Å) is used. Although the light from this lamp consists of waves of only one wavelength, the individual light waves still vibrate in all possible planes perpendicular to the beam. If we imagine that the beam of light is aimed directly at the viewer, ordinary light can be represented by showing the edges of the planes oriented randomly around the path of the beam, as in the left part of Figure 17.1.

A Nicol prism, which consists of a specially prepared crystal of Iceland spar (or calcite), has the property of serving as a screen that can restrict the passage of light waves. Waves that are vibrating in one plane are transmitted; those in all other planes are rejected (either refracted in another direction or absorbed). The light that passes through the prism is called **plane-polarized light,** and it consists of waves that vibrate in only one plane. A beam of plane-polarized light aimed directly at the viewer can be represented by showing the edges of the plane oriented in one particular direction, as in the right portion of Figure 17.1.

Iceland spar has the property of **double refraction,** that is, it can split, or doubly refract, an entering beam of ordinary light into two separate emerging beams of light. Each of the two emerging beams (labeled A and B in Fig. 17.2) has only a single plane of vibration, and the plane of vibration in Beam A is perpendicular to the plane of Beam B. In other words, the crystal has separated the incident beam of ordinary light into two beams of plane-polarized light, with the plane of polarization of Beam A perpendicular to the plane of Beam B.

To generate a single beam of plane-polarized light, one can take advantage of the double-refracting property of Iceland spar. A Nicol prism, invented by the Scottish physicist William Nicol, consists of two crystals of Iceland spar cut to specified angles and ce-

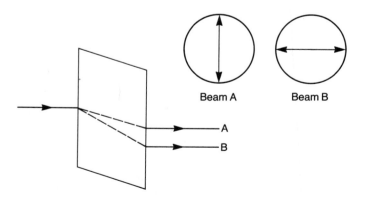

Beam A   Beam B

Figure 17.2   Double refraction.

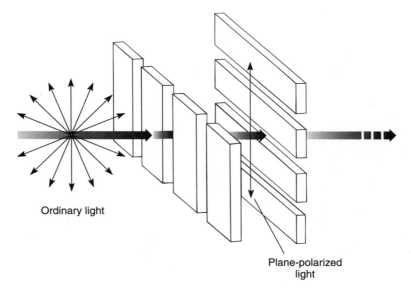

Figure 17.3 The picket-fence analogy.

mented by Canada balsam. This prism transmits one of the two beams of plane-polarized light while reflecting the other at a sharp angle so that it does not interfere with the transmitted beam. Plane-polarized light can also be generated by a Polaroid filter, a device invented by E. H. Land, an American physicist. Polaroid filters consist of certain types of crystals, embedded in transparent plastic and capable of producing plane-polarized light.

After passing through a first Nicol prism, plane-polarized light can pass through a second Nicol prism, but only if the second prism has its axis oriented so that it is *parallel* to the incident light's plane of polarization. Plane-polarized light is *absorbed* by a Nicol prism that is oriented so that its axis is *perpendicular* to the incident light's plane of polarization. These situations can be illustrated by the picket-fence analogy, as shown in Figure 17.3. Plane-polarized light can pass through a fence whose slats are oriented in the proper direction but is blocked out by a fence whose slats are oriented perpendicularly.

An **optically active substance** is one that interacts with polarized light to rotate the plane of polarization through some angle α. Figure 17.4 illustrates this phenomenon.

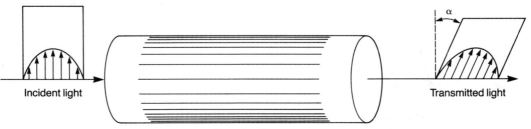

Figure 17.4 Optical activity.

17.2 THE POLARIMETER

An instrument called a **polarimeter** is used to measure the extent to which a substance interacts with polarized light. A schematic diagram of a polarimeter is shown in Figure 17.5. The light from the source lamp is polarized by being passed through a fixed Nicol prism, called a polarizer. This light passes through the sample, with which it may or may not interact to have its plane of polarization rotated in one direction or the other. A second, rotatable Nicol prism, called the analyzer, is adjusted to allow the maximum amount of light to pass through. The number of degrees and the direction of rotation required for this adjustment are measured to give the **observed rotation** α.

So that data determined by several persons under different conditions can be compared, a standardized means of presenting optical rotation data is necessary. The most common way of presenting such data is by recording the **specific rotation** $[\alpha]_\lambda^t$, which has been corrected for differences in concentration, cell path length, temperature, solvent, and wavelength of the light source. The equation defining the specific rotation of a compound in solution is

$$[\alpha]_\lambda^t = \frac{\alpha}{cl}$$

where α = observed rotation in degrees, c = concentration in grams per milliliter of solution, l = length of sample tube in decimeters, λ = wavelength of light (usually indicated as "D," for the sodium D line), and t = temperature in degrees Celsius. For pure liquids, the density d of the liquid in grams per milliliter replaces c in the preceding formula. You may occasionally want to compare compounds of different molecular weights, so a **molecular rotation,** based on moles instead of grams, is more convenient than a specific rotation. The molecular rotation M_λ^t is derived from the specific rotation $[\alpha]_\lambda^t$ by

$$M_\lambda^t = \frac{[\alpha_\lambda^t] \times \text{Molecular weight}}{100}$$

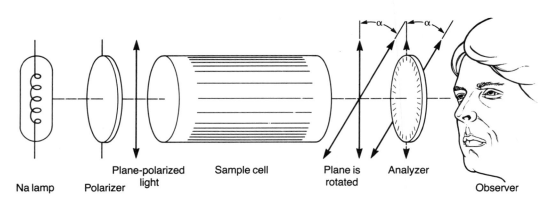

Plane-polarized light   Sample cell   Plane is rotated   Analyzer

Na lamp   Polarizer   Observer

Figure 17.5 Schematic diagram of a polarimeter.

Usually, measurements are made at 25°C with the sodium D line as a light source; consequently, specific rotations are reported as $[\alpha]_D^{25}$.

Polarimeters which are now available incorporate electronics to determine the angle of rotation of chiral molecules. These instruments are essentially automatic. The only real difference between an automatic polarimeter and a manual one is that a light detector replaces the eye. No visual observation of any kind is made with an automatic instrument. A microprocessor adjusts the analyzer until the light reaching the detector is at a minimum. The angle of rotation is displayed digitally in an LCD window, including the sign of rotation. The simplest instrument is equipped with a sodium lamp that gives rotations based on the D line of sodium (589 nm). More expensive instruments use a tungsten lamp and filters so that wavelengths can be varied over a range of values. Using the latter instrument, a chemist can observe rotations at different wavelengths.

17.3 THE SAMPLE CELLS

It is important for the solution whose optical rotation is to be determined to contain no suspended particles of dust or dirt that might disperse the incident polarized light. Therefore, you must clean the sample cell carefully and make certain that there are no air bubbles trapped in the path of the light. The sample cells contain an enlarged ring near one end, in which the air bubbles may be trapped. Sample cells are available in various lengths, with 0.5 and 1.0 dm being the most common.

The sample is generally prepared by dissolving 0.1–0.5 g of the substance to be studied in 25 mL of solvent, usually water, ethanol, or methylene chloride (chloroform was used in the past). If the specific rotation of the substance is very high or very low, you may need to make the concentration of the solution respectively lower or higher, but usually this is determined after first trying a concentration range such as that suggested previously. The sample cell shown in Figure 17.6 is filled with solution. It is then tilted upward and tapped until the air bubbles move into the enlarged ring. It is important not to get fingerprints on the glass endplate in reassembling the cell.

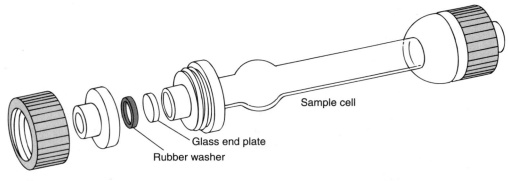

Screw cap

Rubber washer

Glass end plate

Sample cell

Figure 17.6 Polarimeter cell assembly.

17.4 OPERATION OF THE POLARIMETER

The procedures given here for preparing the cells and for operating the instrument are appropriate for the Zeiss polarimeter with the circular scale; other models of polarimeter are operated similarly. It is necessary before beginning the experiments to turn the power switch to the ON position and wait 5–10 minutes until the sodium lamp is properly warmed.

The instrument should be checked initially by making a zero reading with a sample cell filled only with solvent. If the zero reading does not correspond with the zero-degree calibration mark, then the difference in readings must be used to correct all subsequent readings. The reading is determined by laying the sample tube in the cradle, enlarged end up (making sure that there are no air bubbles in the light path), closing the cover, and turning the knob until the proper angle of the analyzer is reached. Most instruments, including the Zeiss polarimeter, are of the double-field type, in which the eye sees a split field whose sections must be matched in light intensity. The value of the angle through which the plane of polarized light has been rotated (if any) is read directly from the scale that can be seen through the eyepiece directly below the split-field image. Figure 17.7 shows how this split field might appear.

The cell containing the solution of the sample is then placed in the polarimeter, and the observed angle of rotation is measured in the same way. Be sure to record not only the numerical value of the angle of rotation in degrees but also the direction of rotation. Rotations clockwise are due to **dextrorotatory** substances and are indicated by the sign "+." Rotations counterclockwise are due to **levorotatory** substances and are indicated by the sign "−." It is best, in making a determination, to take several readings, including readings for which the actual value was approached from both sides. In other words, where the actual reading might be +75°, first approach this reading upward from a reading near zero; on the next measurement approach this reading downward from an angle greater than +75°. Duplicating readings and approaching the observed value from both sides reduce errors. The readings are then averaged to get the observed rotation α. This rotation is then corrected by the appropriate factors, according to the formulas in Section 17.2, to provide the specific rotation. The specific rotation is always reported as a function of temperature, indicating the wavelength by "D" if a sodium lamp is used and reporting the concentration and solvent used. For example $[\alpha]_D^{20} = +43.8°$ ($c = 7.5$ g/100 mL in absolute ethanol).

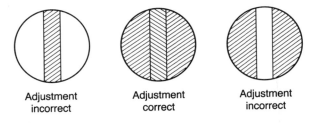

Adjustment   Adjustment   Adjustment
incorrect   correct   incorrect

Figure 17.7 Split-field image in the polarimeter.

17.5 OPTICAL PURITY

When you prepare a sample of an enantiomer by a resolution method, the sample is not always 100% of a single enantiomer. It frequently is contaminated by residual amounts of the opposite stereoisomer. If you know the amount of each enantiomer in a mixture, you can calculate the **optical purity.** Some chemists prefer to use the term **enantiomeric excess (ee)** rather than optical purity. The two terms can be used interchangeably. The percentage enantiomeric excess or optical purity is calculated as follows:

$$\% \text{ Optical purity} = \frac{\text{moles one enantiomer} - \text{moles of other enantiomer}}{\text{total moles of both enantiomers}} \times 100$$

$$\% \text{ Optical purity} = \% \text{ Enantiomeric excess (ee)}$$

Often, it is difficult to apply the equation shown above because you do not know the exact amount of each enantiomer present in a mixture. It is far easier to calculate the optical purity (enantiomeric excess) by using the observed specific rotation of the mixture and dividing it by the specific rotation of the pure enantiomer. Values for the pure enantiomers can sometimes be found in literature sources.

$$\% \text{ Optical purity} = \% \text{ Enantiomeric excess} = \frac{\text{Observed specific rotation}}{\text{Specific rotation of pure enantiomer}} \times 100$$

This latter equation only holds true for mixtures of two chiral molecules that are mirror images of each other (enantiomers). If some other chiral substance is present in the mixture as an impurity, then the actual optical purity will deviate from the value calculated.

In a racemic ($\pm$) mixture, there is no excess enantiomer and the optical purity (enantiomeric excess) is zero; in a completely resolved material, the optical purity (enantiomeric excess) is 100%. A compound that is $x\%$ optically pure contains $x\%$ of one enantiomer and $(100 - x)\%$ of a racemic mixture.

Once the optical purity (enantiomeric excess) is known, the relative percentages of the each of the enantiomers can be calculated easily. If the predominant form in the impure, optically active mixture is assumed to be the ($+$) enantiomer, the percentage of the ($+$) enantiomer is

$$\left[x + \left(\frac{100 - x}{2} \right) \right] \%$$

and the percentage of the ($-$) enantiomer is $[(100 - x)/2]\%$. The relative percentages of ($+$) and ($-$) forms in a partially resolved mixture of enantiomers can be calculated as shown next. Consider a partially resolved mixture of camphor enantiomers. The specific rotation for pure (+)-camphor is +43.8° in absolute ethanol, but the mixture shows a specific rotation of +26.3°.

$$\text{Optical purity} = \frac{+26.3°}{+43.8°} \times 100 = 60\% \text{ optically pure}$$

$$\% \text{ (+) enantiomer} = 60 + \left(\frac{100 - 60}{2}\right) = 80\%$$

$$\% \text{ (−) enantiomer} = \left(\frac{100 - 60}{2}\right) = 20\%$$

Notice that the difference between these two calculated values equals the optical purity or enantiomeric excess (ee).

PROBLEMS

1. Calculate the specific rotation of a substance that is dissolved in a solvent (0.4 g/mL) and that has an observed rotation of −10° as determined with a 0.5-dm cell.
2. Calculate the observed rotation for a solution of a substance (2.0 g/mL) that is 80% optically pure. A 2-dm cell is used. The specific rotation for the optically pure substance is +20°.
3. What is the optical purity of a partially racemized product if the calculated specific rotation is −8° and the pure enantiomer has a specific rotation of −10°? Calculate the percentage of each of the enantiomers in the partially racemized product.

T E C H N I Q U E   1 8

Refractometry

The **refractive index** is a useful physical property of liquids. Often, a liquid can be identified from a measurement of its refractive index. The refractive index can also provide a measure of the purity of the sample being examined. This is accomplished by comparing the experimentally measured refractive index with the value reported in the literature for an ultrapure sample of the compound. The closer the measured sample's value to the literature value, the purer the sample.

18.1 THE REFRACTIVE INDEX

The refractive index has as its basis the fact that light travels at a different velocity in condensed phases (liquids, solids) than in air. The refractive index n is defined as the ratio of the velocity of light in air to the velocity of light in the medium being measured:

$$n = \frac{V_{\text{air}}}{V_{\text{liquid}}} = \frac{\sin \theta}{\sin \phi}$$

It is not difficult to measure the ratio of the velocities experimentally. It corresponds to $(\sin \theta/\sin \phi)$, where θ is the angle of incidence for a beam of light striking the surface of the medium and ϕ is the angle of refraction of the beam of light *within* the medium. This is illustrated in Figure 18.1.

The refractive index for a given medium depends on two variable factors. First, it is *temperature*-dependent. The density of the medium changes with temperature; hence, the speed of light in the medium also changes. Second, the refractive index is *wavelength*-dependent. Beams of light with different wavelengths are refracted to different extents in the same medium and give different refractive indices for that medium. It is usual to report refractive indices measured at 20°C, with a sodium discharge lamp as the source of illumination. The sodium lamp gives off yellow light of 589-nm wavelength, the so-called sodium D line. Under these conditions, the refractive index is reported in the following form

$$n_{\text{D}}^{20} = 1.4892$$

The superscript indicates the temperature, and the subscript indicates that the sodium D line was used for the measurement. If another wavelength is used for the determination, the D is replaced by the appropriate value, usually in nanometers ($1 \text{ nm} = 10^{-9} \text{ m}$).

Notice that the hypothetical value reported above has four decimal places. It is easy to determine the refractive index to within several parts in 10,000. Therefore, n_{D} is a very

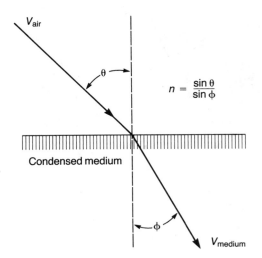

Figure 18.1   The refractive index.

accurate physical constant for a given substance and can be used for identification. However, it is sensitive to even small amounts of impurity in the substance measured. Unless the substance is purified *extensively,* you will not usually be able to reproduce the last two decimal places given in a handbook or other literature source. Typical organic liquids have refractive index values between 1.3400 and 1.5600.

18.2 THE ABBÉ REFRACTOMETER

The instrument used to measure the refractive index is called a **refractometer.** Although many styles of refractometer are available, by far the most common instrument is the Abbé refractometer. This style of refractometer has the following advantages:

1. White light may be used for illumination; the instrument is compensated, however, so that the index of refraction obtained is actually that for the sodium D line.
2. The prisms can be temperature-controlled.
3. Only a small sample is required (a few drops of liquid using the standard method, or about 5 μL using a modified technique).

A common type of Abbé refractometer is shown in Figure 18.2.

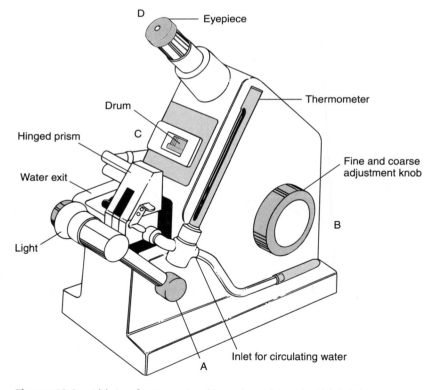

Figure 18.2   Abbé refractometer (Bausch and Lomb Abbé 3L).

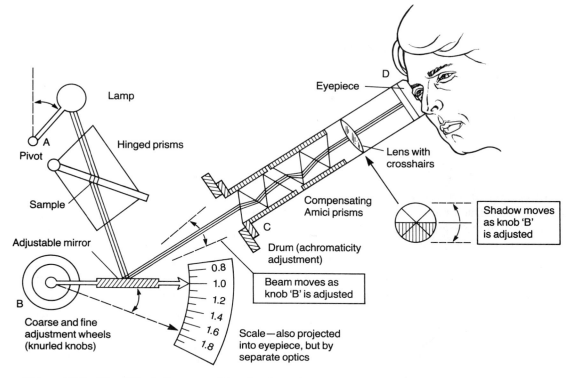

Figure 18.3 Simplified diagram of a refractometer.

The optical arrangement of the refractometer is very complex; a simplified diagram of the internal workings is given in Figure 18.3. The letters *A, B, C,* and *D* label corresponding parts in both Figures 18.2 and 18.3. A complete description of refractometer optics is too difficult to attempt here, but Figure 18.3 gives a simplified diagram of the essential operating principles.

Using the standard method, the sample to be measured is introduced between the two prisms. If it is a free-flowing liquid, it may be introduced into a channel along the side of the prisms, injected from a Pasteur pipet. If it is a viscous sample, the prisms must be opened (they are hinged) by lifting the upper one; a few drops of liquid are applied to the lower prism with a Pasteur pipet or a wooden applicator. If a Pasteur pipet is used, take care not to touch the prisms, because they become scratched easily. When the prisms are closed, the liquid should spread evenly to make a thin film. With highly volatile samples, the remaining operations must be performed rapidly. Even when the prisms are closed, evaporation of volatile liquids can readily occur.

Next, you turn on the light and look into the eyepiece *D*. The hinged lamp is adjusted to give the maximum illumination to the visible field in the eyepiece. The light rotates at pivot *A*.

Rotate the coarse and fine adjustment knobs at *B* until the dividing line between the light and dark halves of the visual field coincide with the center of the cross hairs

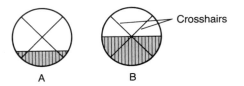

Figure 18.4   (A) Refractometer incorrectly adjusted. (B) Correct adjustment.

(Fig. 18.4). If the cross hairs are not in sharp focus, adjust the eyepiece to focus them. If the horizontal line dividing the light and dark areas appears as a colored band, as in Figure 18.5, the refractometer shows **chromatic aberration** (color dispersion). This can be adjusted with the knob labeled *C* drum. This knurled knob rotates a series of prisms, called Amici prisms, that color-compensate the refractometer and cancel out dispersion. Adjust the knob to give a sharp, uncolored division between the light and dark segments. When you have adjusted everything correctly (as in Fig. 18.4B), read the refractive index. In the instrument described here, press a small button on the left side of the housing to make the scale visible in the eyepiece. In other refractometers, the scale is visible at all times, frequently through a separate eyepiece.

Occasionally, the refractometer will be so far out of adjustment that it may be difficult to measure the refractive index of an unknown. When this happens, it is wise to place a pure sample of known refractive index in the instrument, set the scale to the correct value of refractive index, and adjust the controls for the sharpest line possible. Once this is done, it is easier to measure an unknown sample. It is especially helpful to perform this procedure prior to measuring the refractive index of a highly volatile sample.

There are many styles of refractometer, but most have adjustments similar to those described here.

In the procedure described above, several drops of liquid are required to obtain the refractive index. In some experiments, you may not have enough sample to use this standard method. It is possible to modify the procedure so that a reasonably accurate refractive index can be obtained on about 5 μL of liquid. Instead of placing the sample directly onto the prism, the sample is applied to a small piece of lens paper. The lens paper can be

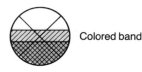

Figure 18.5   Refractometer showing chromatic aberration (color dispersion). The dispersion is incorrectly adjusted.

conveniently cut with a hand-held paper punch,[1] and the paper disc (0.6 cm diameter) is placed in the center of the bottom prism of the refractometer. To avoid scratching the prism, a forceps or tweezers with plastic tips should be used to handle the disc. About 5 μL of liquid is carefully placed on the lens paper using a microliter syringe. After closing the prisms, the refractometer is adjusted as described above and the refractive index is read. With this method, the horizontal line dividing the light and dark areas may not be as sharp as it is in the absence of the lens paper. It may also be impossible to eliminate color dispersion completely. Nonetheless, the refractive index values determined by this method are usually within 10 parts in 10,000 of the values determined by the standard procedure.

18.3 CLEANING THE REFRACTOMETER

In using the refractometer, you should always remember that if the prisms are scratched, the instrument will be ruined.

Do not touch the prisms with any hard object.

This admonition includes Pasteur pipets and glass rods.

When measurements are completed, the prisms should be cleaned with ethanol or petroleum ether. *Soft* tissues are moistened with the solvent, and the prisms are wiped *gently*. When the solvent has evaporated from the prism surfaces, the prisms should be locked together. The refractometer should be left with the prisms closed to avoid collection of dust in the space between them. The instrument should also be turned off when it is no longer in use.

18.4 TEMPERATURE CORRECTIONS

Most refractometers are designed so that circulating water at a constant temperature can maintain the prisms at 20°C. If this temperature-control system is not used, or if the water is not at 20°C, a temperature correction must be made. Although the magnitude of the temperature correction may vary from one class of compound to another, a value of 0.00045 per degree Celsius is a useful approximation for most substances. The index of refraction of a substance *decreases* with *increasing* temperature. Therefore, add the correction to the observed n_D value for temperatures higher than 20°C and subtract it for temperatures lower than 20°C. For example, the reported n_D value for nitrobenzene is 1.5529. One would observe a value at 25°C of 1.5506. The temperature correction would be made as follows:

$$n_D^{20} = 1.5506 + 5(0.00045) = 1.5529$$

---

[1]In order to cut the lens paper more easily, place several sheets between two pieces of heavier paper, such as that used for file folders.

PROBLEMS

1. A solution consisting of isobutyl bromide and isobutyl chloride is found to have a refractive index of 1.3931 at 20°C. The refractive indices at 20°C of isobutyl bromide and isobutyl chloride are 1.4368 and 1.3785, respectively. Determine the molar composition (in percent) of the mixture by assuming a linear relation between the refractive index and the molar composition of the mixture.

2. The refractive index of a compound at 16°C is found to be 1.3982. Correct this refractive index to 20°C.

T E C H N I Q U E   1 9

Preparation of Samples for Spectroscopy

Modern organic chemistry requires sophisticated scientific instruments. Most important among these instruments are the two spectroscopic instruments: the infrared (IR) and nuclear magnetic resonance (NMR) spectrometers. These instruments are indispensable to the modern organic chemist in proving the structures of unknown substances, in verifying that reaction products are indeed the predicted ones, and in characterizing organic compounds. The theory underlying these instruments can be found in most standard lecture textbooks in organic chemistry. Additional information, including correlation charts, to help in interpreting spectra are found in this textbook in Appendix 3 (Infrared Spectroscopy), Appendix 4 (Nuclear Magnetic Resonance Spectroscopy) and Appendix 5 (Carbon-13 Nuclear Magnetic Resonance Spectroscopy). This technique chapter concentrates on the preparation of samples for these spectroscopic methods. Part A covers techniques used in infrared spectroscopy, and Part B describes sample preparation for nuclear magnetic resonance spectroscopy.

Part A.   Infrared Spectroscopy

19.1 INTRODUCTION

To determine the infrared spectrum of a compound, one must place it in a sample holder or cell. In infrared spectroscopy this immediately poses a problem. Glass, quartz, and plastics absorb strongly throughout the infrared region of the spectrum (any compound with covalent bonds usually absorbs) and cannot be used to construct sample cells. Ionic substances must be used in cell construction. Metal halides (sodium chloride, potassium bromide, silver chloride) are commonly used for this purpose.

Sodium Chloride Cells.   Single crystals of sodium chloride are cut and polished to give plates that are transparent throughout the infrared region. These plates are then used to fabricate cells that can be used to hold *liquid* samples. Because sodium chloride is water-soluble, samples must be *dry* before a spectrum can be obtained. In general, sodium chloride plates are preferred for most applications involving liquid samples.

Silver Chloride Cells.   Cells may be constructed of silver chloride. These plates may be used for *liquid* samples that contain small amounts of water, because silver chloride is water-insoluble. However, because water absorbs in the infrared region, as much water as possible should be removed, even when using silver chloride. Silver chloride plates must be stored in the dark, and they cannot be used with compounds that have an amino functional group. Amines react with silver chloride.

Solid Samples.   A *solid* sample is usually held in place by making a potassium bromide pellet that contains a small amount of dispersed compound. A solid sample may also be suspended in mineral oil, which absorbs only in specific regions of the infrared spectrum. Another method is to dissolve the solid compound in an appropriate solvent and place the solution between two sodium chloride or silver chloride plates.

19.2 LIQUID SAMPLES—NaCl PLATES

The simplest method of preparing the sample, if it is a liquid, is to place a thin layer of the liquid between two sodium chloride plates that have been ground flat and polished. This is the method of choice when you need to determine the infrared spectrum of a pure liquid. A spectrum determined by this method is referred to as a **neat** spectrum. No solvent is used. The polished plates are expensive because they are cut from a large, single crystal of sodium chloride. Salt plates break easily, and they are water-soluble.

Preparing the Sample.   Obtain two sodium chloride plates and a holder from the desiccator where they are stored. Moisture from fingers will mar and occlude the polished surfaces. Samples that contain water will destroy the plates.

> The plates should be touched only on their edges. Be certain to use a sample that is dry or free from water.

Add a drop of the liquid to the surface of one plate, then place the second plate on top. The pressure of this second plate causes the liquid to spread out and form a thin, capillary film between the two plates. As shown in Figure 19.1, set the plates between the bolts in a holder and place the metal ring carefully on the salt plates. Use the hex nuts to hold the salt plates in place.

> Do not overtighten the nuts or the salt plates will cleave or split.

Tighten the nuts firmly, but do not use any force at all to turn them. Spin them with the fingers until they stop; then turn them just another fraction of a full turn, and they will be tight enough. If the nuts have been tightened carefully, you should observe a transparent film of sample (a uniform wetting of the surface). If a thin film has not been obtained, either loosen one or more of the hex nuts and adjust them so that a uniform film is obtained or add more sample.

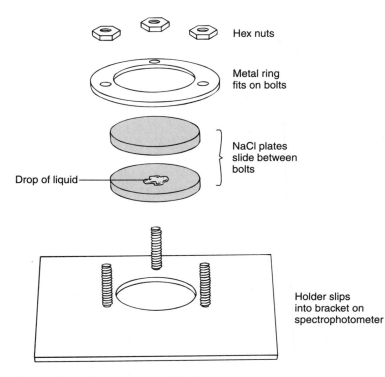

Hex nuts

Metal ring
fits on bolts

NaCl plates
slide between
bolts

Drop of liquid

Holder slips
into bracket on
spectrophotometer

Figure 19.1   Salt plates and holder.

The thickness of the film obtained between the two plates is a function of two factors: (1) the amount of liquid placed on the first plate (one drop, two drops, etc.) and (2) the pressure used to hold the plates together. If more than one or two drops of liquid have been used, it will probably be too much, and the resulting spectrum will show strong absorptions that are off the scale of the chart paper. Only enough liquid to wet both surfaces is needed.

If the sample has a very low viscosity, you may find that the capillary film is too thin to produce a good spectrum. Another problem you may find is that the liquid is so volatile that the sample evaporates before the spectrum can be determined. In these cases, you may need to use the silver chloride plates discussed in Section 19.3, or a solution cell described in Section 19.5. Often, you can obtain a reasonable spectrum by assembling the cell quickly and running the spectrum before the sample runs out of the salt plates or evaporates.

Determining the Infrared Spectrum.   Slide the holder into the slot in the sample beam of the spectrophotometer. Determine the spectrum according to the instructions provided by your instructor. In some cases, your instructor may ask you to calibrate your spectrum. If this is the case, refer to Section 19.8.

Cleaning and Storing the Salt Plates.   Once the spectrum has been determined, demount the holder and rinse the salt plates with methylene chloride (or *dry* acetone). (Keep the plates away from water!) Use a soft tissue, moistened with the solvent, to wipe the plates. If some of your compound remains on the plates, you may observe a shiny surface. Continue to clean the plates with solvent until no more compound remains on the surfaces of the plates.

> **Caution:** Avoid direct contact with methylene chloride. Return the salt plates and holder to the desiccator for storage. Use gloves.

19.3 LIQUID SAMPLES—AgCl PLATES

The mini-cell[1] shown in Figure 19.2 may also be used with liquids. The cell assembly consists of a two-piece threaded body, an O-ring, and two silver chloride plates. The plates are flat on one side, and there is a circular depression (0.025 mm or 0.10 mm deep) on the other side of the plate. The advantages of using silver chloride plates are that they may be used with wet samples or solutions. A disadvantage is that silver chloride darkens when exposed to light for extended periods. They also scratch more easily than salt plates and react with amines.

Preparing the Sample.   Silver chloride plates should be handled in the same way as salt plates. Unfortunately, they are smaller and thinner (about like a contact lens) than salt plates, and care must be taken to not lose them! Remove them from the light-tight container with care. It is difficult to tell which side of the plate has the slight circular depression. Your instructor may have etched a letter on each plate to indicate which side is the flat one. If you are going to determine the infrared spectrum of a pure liquid (neat

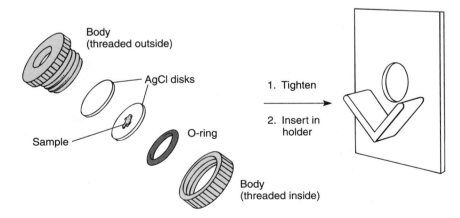

Figure 19.2   AgCl mini-liquid cell and V-mount holder.

---

[1]The Wilks Mini-Cell liquid sample holder is available from the Foxboro Company, 151 Woodward Avenue, South Norwalk, CT 06856. We recommend the AgCl cell windows with a 0.10-mm depression, rather than the 0.025-mm depression.

spectrum), you should select the flat side of each silver chloride plate. Insert the O-ring into the cell body as shown in Figure 19.2, place the plate into the cell body with the flat surface up, and add one drop or less of liquid to the plate.

Do not use amines with AgCl plates.

Place the second plate on top of the first with the flat side down. The orientation of the silver chloride plates is shown in Figure 19.3A. This arrangement is used to obtain a capillary film of your sample. Screw the top of the mini-cell into the body of the cell so that the silver chloride plates are held firmly together. A tight seal forms because AgCl deforms under pressure.

Other combinations may be used with these plates. For example, you may vary the sample path length by using the orientations shown in Figures 19.3B and 19.3C. If you add your sample to the 0.10-mm depression of one plate and cover it with the flat side of the other one, you obtain a path length of 0.10 mm (Fig. 19.3B). This arrangement is useful for analyzing volatile or low-viscosity liquids. Placement of the two plates with their depressions toward each other gives a path length of 0.20 mm (Fig. 19.3C). This orientation may be used for a solution of a solid (or liquid) in carbon tetrachloride (Section 19.5B).

Determining the Spectrum.   Slide the V-mount holder shown in Figure 19.2 into the slot on the infrared spectrophotometer. Set the cell assembly in the V-mount holder and determine the infrared spectrum of the liquid.

Cleaning and Storing the AgCl Plates.   Once the spectrum has been determined, the cell assembly holder should be demounted and the AgCl plates rinsed with methylene chloride or acetone. Do not use tissue to wipe the plates as they scratch easily. AgCl plates are light-sensitive. Store the plates in a light-tight container.

19.4 SOLID SAMPLES—KBr PELLETS

The easiest method of preparing a solid sample is to make a potassium bromide (KBr) pellet. When KBr is placed under pressure, it melts, flows, and seals the sample into a solid solution, or matrix. Because potassium bromide does not absorb in the infrared spectrum, a spectrum can be obtained on a sample without interference.

Preparing the Sample.   Remove the agate mortar and pestle from the desiccator for use in preparing the sample. (Take care of them, they are expensive.) Grind 1 mg (0.001 g) of the solid sample for 1 minute in the agate mortar. At this point, the particle

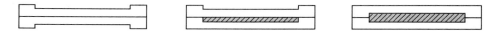

A. Capillary film   B. 0.10–mm path length   C. 0.20–mm path length

Figure 19.3 Path length variations for AgCl plates.

size will become so small that the surface of the solid appears shiny. Add 80 mg (0.080 g) of *powdered* potassium bromide and grind the mixture for about 30 seconds with the pestle. Scrape the mixture into the middle with a spatula and grind the mixture again for about 15 seconds. This grinding operation helps to mix the sample thoroughly with the KBr. You should work as rapidly as possible, because KBr absorbs water. The sample and KBr must be finely ground or the mixture will scatter the infrared radiation excessively. Using your spatula, heap the mixture in the center of the mortar. Return the bottle of potassium bromide to the desiccator where it is stored when it is not in use.

The sample and potassium bromide should be weighed on an analytical balance the first few times that a pellet is prepared. After some experience, you can estimate these quantities quite accurately by eye.

Making a Pellet Using a KBr Hand Press.   Two methods are commonly used to prepare KBr pellets. The first method uses the hand press apparatus shown in Figure 19.4.[2]

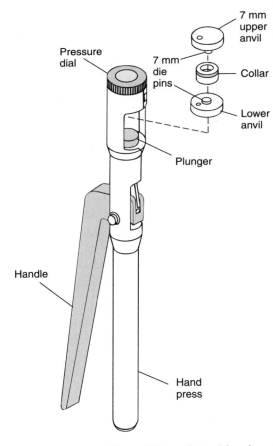

Figure 19.4 Making a KBr pellet with a hand press.

[2]KBr Quick Press unit is available from Wilmad Glass Company, Inc., Route 40 and Oak Road, Buena, NJ 08310.

Remove the die set from the storage container. Take extreme care to avoid scratching the polished surfaces of the die set. Place the anvil with the shorter die pin (lower anvil in Figure 19.4) on a bench. Slip the collar over the pin. Remove about one-fourth of your KBr mixture with a spatula and transfer it into the collar. The powder may not cover the head of the pin completely but do not be concerned about this. Place the anvil with the longer die pin into the collar so that the die pin comes into contact with the sample. Never press the die set unless it contains a sample.

Lift the die set carefully by holding onto the lower anvil so that the collar stays in place. If you are careless with this operation, the collar may move enough to allow the powder to escape. Open the handle of the hand press slightly, tilt the press back a bit, and insert the die set into the press. Make sure that the die set is seated against the side wall of the chamber. Close the handle. It is imperative that the die set be seated against the side wall of the chamber so that the die is centered in the chamber. Pressing the die in an off-centered position can bend the anvil pins.

With the handle in the closed position, rotate the pressure dial so that the upper ram of the hand press just touches the upper anvil of the die assembly. Tilt the unit back so that the die set does not fall out of the press. Open the handle and rotate the pressure dial clockwise about one-half turn. Slowly compress the KBr mixture by closing the handle. The pressure should be no greater than that exerted by a very firm handshake. Do not apply excessive pressure or the dies may be damaged. If in doubt, rotate the pressure dial counterclockwise to lower the pressure. If the handle closes too easily, open the handle, rotate the pressure dial clockwise, and compress the sample again. Compress the sample for about 60 seconds.

After this time, tilt the unit back so that the die set does not fall out of the hand press. Open the handle and carefully remove the die set from the unit. Turn the pressure dial counterclockwise about one full turn. Pull the die set apart and inspect the KBr pellet. Ideally, the pellet should appear clear like a piece of glass, but usually it will be translucent or somewhat opaque. There may be some cracks or holes in the pellet. The pellet will produce a good spectrum, even with imperfections, as long as light can travel through the pellet.

Making a Pellet with a KBr Minipress. The second method of preparing a pellet uses the minipress apparatus shown in Figure 19.5. Obtain a ground KBr mixture as

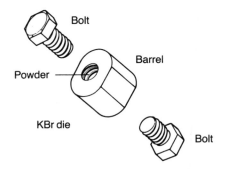

Figure 19.5 Making a KBr pellet with a minipress.

described above in "Preparing the Sample," and transfer a portion of the finely ground powder (usually not more than half) into a die that compresses it into a translucent pellet. As shown in Figure 19.5, the die consists of two stainless steel bolts and a threaded barrel. The bolts have their ends ground flat. To use this die, screw one of the bolts into the barrel, but not all the way; leave one or two turns. Carefully add the powder with a spatula into the open end of the partly assembled die and tap it lightly on the bench top to give an even layer on the face of the bolt. While keeping the barrel upright, carefully screw the second bolt into the barrel until it is finger-tight. Insert the head of the bottom bolt into the hexagonal hole in a plate bolted to the bench top. This plate keeps the head of one bolt from turning. The top bolt is tightened with a torque wrench to compress the KBr mixture. Continue to turn the torque wrench until you hear a loud click (the ratchet mechanism makes softer clicks) or until you reach the appropriate torque value (20 ft-lb). If you tighten the bolt beyond this point, you may twist the head off one of the bolts. Leave the die under pressure for about 60 seconds; then reverse the ratchet on the torque wrench or pull the torque wrench in the opposite direction to open the assembly. When the two bolts are loose, hold the barrel horizontally and carefully remove the two bolts. You should observe a clear or translucent KBr pellet in the center of the barrel. Even if the pellet is not totally transparent, you should be able to obtain a satisfactory spectrum as long as light passes through the pellet.

Determining the Infrared Spectrum.   To obtain the spectrum, slide the holder appropriate for the type of die that you are using into the slot on the infrared spectrophotometer. Set the die containing the pellet in the holder so that the sample is centered in the optical path. Obtain the infrared spectrum. If you are using a double-beam instrument, you may be able to compensate (at least partially) for a marginal pellet by placing a wire screen or attenuator in the reference beam, thereby balancing the lowered transmittance of the pellet.

Problems with an Unsatisfactory Pellet.   If the pellet is unsatisfactory (too cloudy to pass light), one of several things may have been wrong:

1. The KBr mixture may not have been ground finely enough, and the particle size may be too big. The large particle size creates too much light scattering.
2. The sample may not be dry.
3. Too much sample may have been used for the amount of KBr taken.
4. The pellet may be too thick; that is, too much of the powdered mixture was put into the die.
5. The KBr may have been "wet" or have acquired moisture from the air while the mixture was being ground in the mortar.
6. The sample may have a low melting point. Low-melting solids not only are difficult to dry but also melt under pressure. You may need to dissolve the compound in a solvent and run the spectrum in solution (Section 19.5).

Cleaning and Storing the Equipment.   After you have determined the spectrum, punch the pellet out of the die with a wooden applicator stick (a spatula should not be used as it may scratch the dies). Remember that the polished faces of the die set must not be scratched or they become useless. Pull a piece of Kimwipe through the die unit to remove all the sample. Also wipe any surfaces with a Kimwipe. *Do not wash the dies with* *water.* Check with your instructor to see if there are additional instructions for cleaning the die set. Return the dies to the storage container. Wash the mortar and pestle with water and acetone, dry them carefully with paper towels, and return them to the desiccator. Return the KBr powder to its desiccator.

19.5 SOLID SAMPLES—SOLUTION SPECTRA

Method A—Solution Between Salt (NaCl) Plates

For substances that are soluble in carbon tetrachloride, a quick and easy method for determining the spectra of solids is available. Dissolve as much solid as possible in 0.1 mL of carbon tetrachloride. Place one or two drops of the solution between sodium chloride plates in precisely the same manner as that used for pure liquids (Section 19.2). The spectrum is determined as described for pure liquids using salt plates (Section 19.2). You should work as quickly as possible. If there is a delay, the solvent will evaporate from between the plates before the spectrum is recorded. Because the spectrum contains the absorptions of the solute superimposed on the absorptions of carbon tetrachloride, it is important to remember that any absorption that appears near 800 cm^{-1} may be due to the stretching of the C—Cl bond of the solvent. Information contained to the right of about 900 cm^{-1} is not usable in this method. There are no other interfering bands for this solvent (see Fig. 19.6), and any other absorptions can be attributed to your sample. Chloroform solutions should not be studied by this method, because the solvent has too many interfering absorptions (see Fig. 19.7).

> **Caution:** Carbon tetrachloride is a hazardous solvent. Work under the hood! Use gloves.

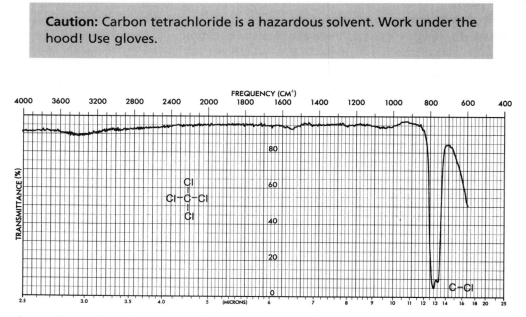

Figure 19.6   Infrared spectrum of carbon tetrachloride.

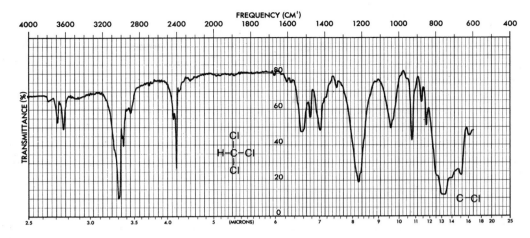

Figure 19.7 Infrared spectrum of chloroform.

Carbon tetrachloride, besides being toxic, is suspected of being a carcinogen. In spite of the health problems associated with its use, there is no suitable alternative solvent for infrared spectroscopy. Other solvents have too many interfering infrared absorption bands. Handle carbon tetrachloride very carefully to minimize the adverse health effects. The spectroscopic-grade carbon tetrachloride should be stored in a glass stoppered bottle in a hood. A Pasteur pipet should be attached to the bottle, possibly by storing it in a test tube taped to the side of the bottle. All sample preparation should be conducted in a hood. Rubber or plastic gloves should be worn. The cells should also be cleaned in the hood. All carbon tetrachloride used in preparing samples should be disposed of in an appropriately marked waste container.

Method B—AgCl Mini-Cell

The AgCl mini-cell described in Section 19.3 may be used to determine the infrared spectrum of a solid dissolved in carbon tetrachloride. Prepare a 5–10% solution (5–10 mg in 0.1 mL) in carbon tetrachloride. If it is not possible to prepare a solution of this concentration because of low solubility, dissolve as much solid as possible in the solvent. Following the instructions given in Section 19.3, position the AgCl plates as shown in Figure 19.3C to obtain the maximum possible path length of 0.20 mm. When the cell is tightened firmly, it will not leak.

As indicated in Method A, the spectrum will contain the absorptions of the dissolved solid superimposed on the absorptions of carbon tetrachloride. A strong absorption appears near 800 cm^{-1} for C—Cl stretch in the solvent. No useful information may be obtained for the sample to the right of about 900 cm^{-1}, but other bands that appear in the spectrum will belong to your sample. Read the safety material provided in Method A. Carbon tetrachloride is toxic, and it should be used under a hood.

Care should be taken in cleaning the AgCl plates. Because AgCl plates scratch easily, they should not be wiped with tissue. Rinse them with methylene chloride and keep them in a dark place. Amines will destroy the plates.

Method C—Solution Cells (NaCl)

The spectra of solids may also be determined in a type of permanent sample cell called a **solution cell.** (The infrared spectra of liquids may also be determined in this cell.) The solution cell, shown in Figure 19.8, is made from two salt plates, mounted with a Teflon spacer between them to control the thickness of the sample. The top sodium chloride plate has two holes drilled in it so that the sample can be introduced into the cavity between the two plates. These holes are extended through the face plate by two tubular extensions designed to hold Teflon plugs, which seal the internal chamber and prevent evaporation. The tubular extensions are tapered so that a syringe body (Luer lock without a needle) will fit snugly into them from the outside. The cells are thus filled from a syringe; usually, they are held upright and filled from the bottom entrance port.

These cells are very expensive, and you should try either Method A or B before using solution cells. If you do need them, obtain your instructor's permission and receive instruction before using the cells. The cells are purchased in matched pairs, with identical path lengths. Dissolve a solid in a suitable solvent, usually carbon tetrachloride, and add the solution to one of the cells (**sample cell**) as described in the previous paragraph. The pure solvent, identical to that used to dissolve the solid, is placed in the other cell (**reference cell).** The spectrum of the solvent is subtracted from the spectrum of the so-

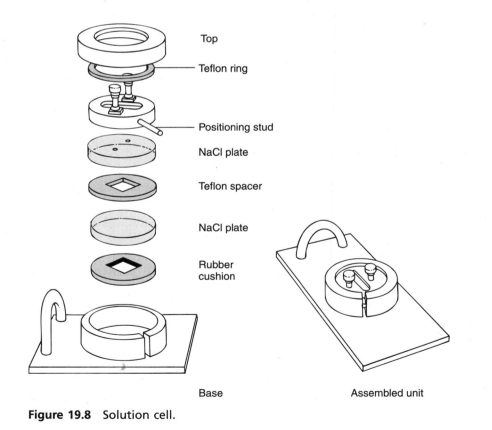

Figure 19.8   Solution cell.

lution (not always completely), and a spectrum of the solute is thus provided. For the solvent compensation to be as exact as possible and to avoid contamination of the reference cell, it is essential that one cell be used as a reference and that the other cell be used as a sample cell without ever being interchanged. After the spectrum is determined, it is important to clean the cells by flushing them with clean solvent. They should be dried by passing dry air through the cell.

Solvents most often used in determining infrared spectra are carbon tetrachloride, chloroform, and carbon disulfide. The spectra of these substances are shown in Figures 19.6, 19.7, and 19.9. A 5–10% solution of solid in one of these solvents usually gives a good spectrum. Carbon tetrachloride and chloroform are suspected carcinogens. However, because there are no suitable alternative solvents, these compounds must be used in infrared spectroscopy. The procedure outlined on page 851 for carbon tetrachloride should be followed. This procedure serves equally well for chloroform.

Before you use the solution cells, you must obtain the instructor's permission and instruction on how to fill and clean the cells.

19.6 SOLID SAMPLES—NUJOL MULLS

If an adequate KBr pellet cannot be obtained, or if the solid is insoluble in a suitable solvent, the spectrum of a solid may be determined as a Nujol mull. In this method, finely grind about 5 mg of the solid sample in an agate mortar with a pestle. Then add one or two drops of Nujol mineral oil (white) and grind the mixture to a very fine dispersion. The solid is not dissolved in the Nujol; it is actually a suspension. This mull is then placed between two salt plates using a rubber policeman. Mount the salt plates in the holder in the same way as for liquid samples (Section 19.2).

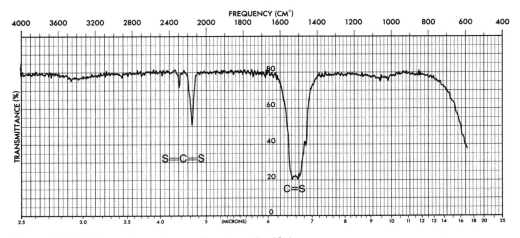

Figure 19.9   Infrared spectrum of carbon disulfide.

Nujol is a mixture of high-molecular-weight hydrocarbons. Hence, it has absorptions in the C—H stretch and CH_2 and CH_3 bending regions of the spectrum (Fig. 19.10). Clearly, if Nujol is used, no information can be obtained in these portions of the spectrum. In interpreting the spectrum, you must ignore these Nujol peaks. It is important to label the spectrum immediately after it was determined, noting that it was determined as a Nujol mull. Otherwise, you might forget that the C—H peaks belong to Nujol and not to the dispersed solid.

19.7 RECORDING THE SPECTRUM

The instructor will describe how to operate the infrared spectrophotometer, because the controls vary considerably, depending on the manufacturer, model of the instrument, and type. For example, some instruments involve pushing only a few buttons, while others use a more complicated computer interface system.

In all cases, it is important that the sample, the solvent, the type of cell or method used, and any other pertinent information be written on the spectrum immediately after the determination. This information may be important, and it is easily forgotten if not recorded. You may also need to calibrate the instrument (Section 19.8).

19.8 CALIBRATION

For some instruments, the frequency scale of the spectrum must be calibrated so that you know the position of each absorption peak precisely. You can recalibrate by recording a very small portion of the spectrum of polystyrene over the spectrum of your sample. The complete spectrum of polystyrene is shown in Figure 19.11. The most important

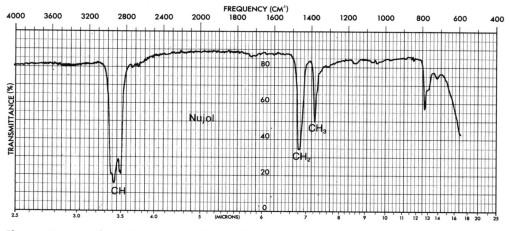

Figure 19.10   Infrared spectrum of Nujol (mineral oil).

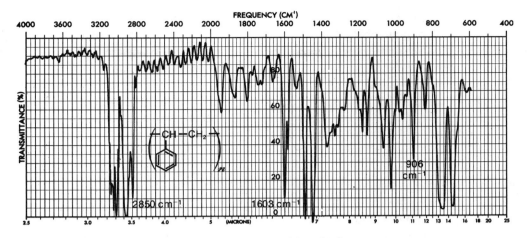

Figure 19.11 Infrared spectrum of polystyrene (thin film).

of these peaks is at 1603 cm^{-1}; other useful peaks are at 2850 and 906 cm^{-1}. After you record the spectrum of your sample, substitute a thin film of polystyrene for the sample cell and record the **tips** (not the entire spectrum) of the most important peaks over the sample spectrum.

It is always a good idea to calibrate a spectrum when the instrument uses chart paper with a preprinted scale. It is difficult to align the paper properly so that the scale matches the absorption lines precisely. You often need to know the precise values for certain functional groups (for example, the carbonyl group). Calibration is essential in these cases.

With computer-interfaced instruments, the instrument does not need to be calibrated. With this type of instrument, the spectrum and scale are printed on blank paper at the same time. The instrument has an internal calibration that ensures that the positions of the absorptions are known precisely and that they are placed at the proper positions on the scale. With this type of instrument, it is often possible to print a list of the locations of the major peaks as well as to obtain the complete spectrum of your compound.

Part B.   Nuclear Magnetic Resonance (NMR)

19.9 PREPARING A SAMPLE FOR PROTON NMR

The NMR sample tubes used in most instruments are approximately 0.5 cm $\times$ 18 cm in overall dimension and are fabricated of uniformly thin glass tubing. These tubes are very fragile and expensive, so care must be taken to avoid breaking the tubes.

To prepare the solution you must first choose the appropriate solvent. The solvent should not have NMR absorption peaks of its own, that is, no protons. Carbon tetrachloride CCl_4 fits this requirement and can be used in some instruments. However, because

Fourier transform (FT) NMR spectrometers require deuterium to stabilize (lock) the field (Section 19.10), organic chemists usually use deuterated chloroform $CDCl_3$ as a solvent. This solvent dissolves most organic compounds and is relatively inexpensive. You can use this solvent with any NMR instrument. You should not use normal chloroform $CHCl_3$, because the solvent contains a proton. Deuterium ^2H does not absorb in the proton region and is thus "invisible," or not seen, in the proton NMR spectrum. Use deuterated chloroform to dissolve your sample unless you are instructed by your instructor to use another solvent, such as carbon tetrachloride CCl_4.

Routine Sample Preparation Using Deuterated Chloroform

1. Most organic liquids and low-melting solids will dissolve in deuterated chloroform. However, you should first determine if your sample will dissolve in ordinary $CHCl_3$ before using the deuterated solvent. If your sample does not dissolve in chloroform, consult your instructor about a possible alternate solvent, or consult the section entitled "Nonroutine Sample Preparation" later in this section.

> **Caution:** Chloroform, deuterated chloroform, and carbon tetrachloride are all toxic solvents. In addition, they may be carcinogenic substances (see p. 21). Use protective gloves.

2. If you are using an FT-NMR spectrometer, add 30 mg (0.030 g) of your liquid or solid sample to a tared conical vial or test tube. Use a Pasteur pipet to transfer a liquid or a spatula to transfer a solid. Non-FT instruments usually require a more concentrated solution in order to obtain an adequate spectrum. Typically, a 10–30% sample concentration (weight/weight) is used.
3. With the help of your instructor, transfer about 0.5 mL of the deuterated chloroform with a *clean and dry Pasteur pipet* to your sample. Swirl the test tube or conical vial to help dissolve the sample. At this point, the sample should have completely dissolved. Add a little more solvent, if necessary, to dissolve the sample fully.
4. Transfer the solution to the NMR tube using a clean and dry Pasteur pipet. Be careful when transferring the solution into the NMR tube so that you avoid breaking the edge of the fragile NMR tube. It is best to hold the NMR tube and the container with the solution in the same hand when making the transfer.
5. Once the solution has been transferred to the NMR tube, use a clean pipet to add enough deuterated chloroform to bring the total solution height[1] to about 35 mm total (Fig. 19.12). In some cases, you will need to add a small amount of tetramethylsilane (TMS) as a reference substance (Section 19.11). Check with your instructor to see if you need to add TMS to your sample. Deuterated chloroform has a small amount of $CHCl_3$ impurity, which gives rise to a low-intensity peak in the NMR spectrum at 7.27 ppm. This impurity may also help you to "reference" your spectrum.

---

[1] Your instructor may recommend a different height than the one specified here (35 mm).

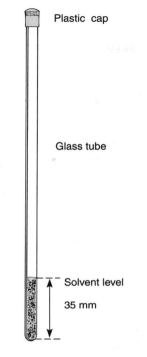

Plastic cap

Glass tube

Solvent level

35 mm

Figure 19.12   NMR sample tube.

6. Cap the NMR tube. Do this firmly, but not too tightly. If you jam the cap on, you may have trouble removing it later without breaking the end off of the very thin glass tube. Make sure that the cap is on straight. Invert the NMR tube several times to mix the contents.

7. You are now ready to record the NMR spectrum of your sample. Insert the NMR tube into its holder and adjust its depth by using the gauge provided to you. See Section 19.12.

Cleaning the NMR Tube

1. Carefully uncap the tube so that you do not break it. Turn the tube upside-down and hold it vertically over a beaker. Shake the tube up and down gently so that the contents of the tube empties into the beaker.

2. Partially refill the NMR tube with acetone using a Pasteur pipet. Carefully replace the cap and invert the tube several times to rinse it.

3. Remove the cap and drain the tube as before. Place the open tube upside-down in a beaker with a Kimwipe or paper towel placed in the bottom of the beaker. Leave the tube standing in this position for at least one laboratory period so that the acetone completely evaporates. Alternatively, you may place the beaker and NMR tube in an oven for at least 2 hours. If you need to use the NMR tube before the acetone has fully evaporated, attach a piece of pressure tubing to the tube, and pull a vacuum with an aspirator. After several

minutes, the acetone should have fully evaporated. Because acetone contains protons, you must not use the NMR tube until the acetone has evaporated completely.

4. Once the acetone is evaporated, place the clean tube and its cap (do not cap the tube) in its storage container and place it in your desk. The storage container will prevent the tube from being crushed.

Nonroutine Sample Preparation. With highly polar substances you may find that your sample will not dissolve in deuterated chloroform. If this is the case, you may be able to dissolve the sample in deuterium oxide, D_2O. Spectra determined in D_2O often show a small peak at about 5 ppm because of OH impurity. If the sample compound has acidic hydrogens, they may *exchange* with D_2O, leading to the appearance of an OH peak in the spectrum and the *loss* of the original absorption from the acidic proton, owing to the exchanged hydrogen. In many cases, this will also alter the splitting patterns of a compound.

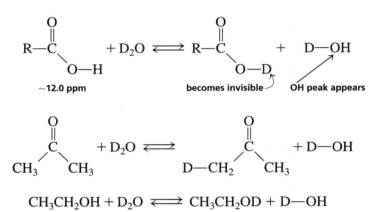

Most solid carboxylic acids do not dissolve in CCl_4, $CDCl_3$, or even D_2O. In such cases, a small piece of sodium metal is added to about 1 mL of D_2O. The acid is then dissolved in this solution. The resulting basic solution enhances the solubility of the carboxylic acid. In such a case, the hydroxyl proton of the carboxylic acid cannot be observed in the NMR spectrum, because it exchanges with the solvent. A large DOH peak is observed, however, due to the exchange and the H_2O impurity in the D_2O solvent.

When the above solvents fail, other special solvents can be used. Acetone, acetonitrile, dimethylsulfoxide, pyridine, benzene, and dimethylformamide can be used if you are not interested in the region or regions of the NMR spectrum in which they give rise to absorption. The deuterated (but expensive) analogues of these compounds are also used in special instances (for example, acetone-d_6, dimethylsulfoxide-d_6, dimethylformamide-d_7, and benzene-d_6).[2] If the sample is not sensitive to acid, trifluoroacetic acid (which has no protons with $\delta < 12$) can be used. You must be aware that these solvents often lead to different chemical shift values from those determined in CCl_4 or $CDCl_3$. Variations of as much as 0.5–1.0 ppm have been observed. In fact, it is sometimes possible, by switching

---

[2]Unisol, a commercial mixture of dimethylsulfoxide-d_6 and $CDCl_3$, dissolves most carboxylic acids.

to pyridine, benzene, acetone, or dimethylsulfoxide as solvents, to separate peaks that overlap when CCl_4 or $CDCl_3$ solutions are used.

Health Hazards Associated with NMR Solvents.   Carbon tetrachloride, chloroform (and chloroform-d), and benzene (and benzene-d_6) are hazardous solvents. Besides being highly toxic, they are also suspected carcinogens. In spite of these health problems, these solvents are commonly used in NMR spectroscopy, because there are no suitable alternatives. These solvents are used because they contain no protons and because they are excellent solvents for most organic compounds. Therefore, you must learn to handle these solvents with great care to minimize the hazard. These solvents should be stored either under a hood or in septum-capped bottles. If the bottles have screwcaps, a pipet should be attached to each bottle. A recommended way of attaching the pipet is to store it in a test tube taped to the side of the bottle. Septum-capped bottles can be used only by withdrawing the solvent with a hypodermic syringe that has been designated solely for this use. All samples should be prepared under a hood and solutions should be disposed of in an appropriately designated waste container that is stored under the hood. Wear rubber or plastic gloves when preparing or discarding samples.

19.10 PREPARING A SAMPLE FOR CARBON-13 NMR

Section 19.9 describes the technique for preparing samples for proton NMR. Much of what is described there also applies to carbon NMR. There are some differences, however, in determining a carbon spectrum. Fourier transform instruments require a deuterium signal to stabilize (lock) the field. Therefore, the solvents must contain deuterium. Deuterated chloroform $CDCl_3$ is used most commonly for this purpose because of its relatively low cost. Other deuterated solvents may also be used.

Because of the low natural abundance of carbon-13 in a sample, you often need to acquire multiple scans over a long period (Appendix 5, Section CMR.1, p. 931). You can save considerable time by using a relatively concentrated sample.

Modern FT-NMR spectrometers allow chemists to obtain both the proton and carbon NMR spectra of the same sample in the same NMR tube. After changing several parameters in the program operating the spectrometer, you can obtain both spectra without removing the sample from the probe. The only real difference is that a proton spectrum may be obtained after a few scans, whereas the carbon spectrum may require several thousand scans to obtain a suitable spectrum.

19.11 REFERENCE SUBSTANCES

Proton NMR.   To provide the internal reference standard, tetramethylsilane (TMS) must be added to the sample solution. This substance has the formula $(CH_3)_4Si$. By universal convention, the chemical shifts of the protons in this substance are defined as 0.00 ppm ($0.00\ \delta$). The spectrum should be shifted so that the TMS signal appears at this position on precalibrated paper.

The concentration of TMS in the sample should range from 1 to 3%. Some people prefer to add one to two drops of TMS to the sample just before determining the spectrum. Because TMS has 12 equivalent protons, not much of it needs to be added. A Pasteur pipet or a syringe may be used for the addition. It is far easier to have available in the laboratory a prepared solvent that already contains TMS. Deuterated chloroform and carbon tetrachloride often have TMS added to them. Because TMS is highly volatile (bp 26.5°C), such solutions should be stored, tightly stoppered, in a refrigerator. Tetramethylsilane itself is best stored in a refrigerator as well.

Tetramethylsilane does not dissolve in D_2O. For spectra determined in D_2O, a different internal standard, sodium 2,2-dimethyl-2-silapentane-5-sulfonate, must be used. This standard is water-soluble and gives a resonance peak at 0.00 ppm (0.00 δ).

$$CH_3 - \underset{\underset{CH_3}{|}}{\overset{\overset{CH_3}{|}}{Si}} - CH_2 - CH_2 - CH_2 - SO_3^- Na^+$$

Sodium 2,2-dimethyl-2-silapentane-5-sulfonate (DSS)

Carbon NMR.   TMS may be added as an internal reference standard where the chemical shift of the methyl carbon is defined as 0.00 ppm. Alternatively, you may use the center peak of the $CDCl_3$ pattern, which is found at 77.0 ppm. This pattern can be observed as a small "triplet" near 77.0 ppm in a number of the spectra given in Appendix 5. (For example, see Fig. CMR.3 on p. 933).

In most instances, the instructor or some qualified laboratory assistant will actually record your NMR spectrum. If you are permitted to operate the NMR spectrometer, the instructor will provide instructions. Because the controls of NMR spectrometers vary, depending on the make or model of the instrument, we shall not try to describe these controls.

Do not operate the NMR spectrometer unless you have been properly instructed.

PROBLEMS

1. Comment on the suitability of running the infrared spectrum under each of the following conditions. If there is a problem with the conditions given, provide a suitable alternative method.

  (a) A neat spectrum of a liquid with a boiling point of 150°C is determined using salt plates.

  (b) A neat spectrum of a liquid with a boiling point of 35°C is determined using salt plates.

  (c) A KBr pellet is prepared with a compound that melts at 200°C.

  (d) A KBr pellet is prepared with a compound that melts at 30°C.

  (e) A solid aliphatic hydrocarbon compound is determined as a Nujol mull.

  (f) Silver chloride plates are used to determine the spectrum of aniline.

  (g) Sodium chloride plates are selected to run the spectrum of a compound that contains some water.

2. Describe the method that you should employ to determine the proton NMR spectrum of a carboxylic acid, which is insoluble in *all* the common organic solvents that your instructor is likely to make available.

3. In order to save money, a student uses chloroform instead of deuterated chloroform to run a carbon-13 NMR spectrum. Is this a good idea?

4. Look up the solubilities for the following compounds and decide whether you would select deuterated chloroform or deuterated water to dissolve the substances for NMR spectroscopy.
  (a) Glycerol (1,2,3-propanetriol)
  (b) 1,4-Diethoxybenzene
  (c) Propyl pentanoate (propyl ester of pentanoic acid)

5. What would happen if you ran a proton NMR spectrum without any TMS in the sample?

T E C H N I Q U E   2 0

Guide to the Chemical Literature

Often, it may be necessary to go beyond the information contained in the typical organic chemistry textbook and to use reference material in the library. At first glance, using library materials may seem formidable because of the numerous sources the library contains. If, however, you adopt a systematic approach, the task can prove rather useful. This description of various popular sources and an outline of logical steps to follow in the typical literature search should be helpful.

20.1 LOCATING PHYSICAL CONSTANTS: HANDBOOKS

To find information on routine physical constants, such as melting points, boiling points, indices of refraction, and densities, you should first consider a handbook. Examples of suitable handbooks are

S. Budavari, ed. *The Merck Index.* 12th ed. Whitehouse Station, NJ: Merck & Co., 1996.

J. A. Dean, ed. *Lange's Handbook of Chemistry.* 14th ed. New York: McGraw-Hill, 1992.

D. R. Lide, ed. *CRC Handbook of Chemistry and Physics.* 76th ed. Boca Raton, FL: CRC Press, 1993.

The *Handbook of Chemistry and Physics* is the handbook consulted most often. For organic chemistry, however, *The Merck Index* is probably better suited. *The Merck Index* also contains literature references on the isolation, structure determination, and synthesis of a substance, along with its molecular formula, elemental analysis, and certain properties of medicinal interest (e.g., toxicity and medicinal and veterinary uses).

A more complete handbook is

J. Buckingham, ed., *Dictionary of Organic Compounds.* New York: Chapman & Hall/Methuen, 1982–1992.

This is a revised version of an earlier four-volume handbook edited by I. M. Heilbron and H. M. Bunbury. In its present form, it consists of seven volumes with 10 supplements.

20.2 GENERAL SYNTHETIC METHODS

Many standard introductory textbooks in organic chemistry provide tables that summarize most of the common reactions, including side reactions, for a given class of compounds. These books also describe alternative methods of preparing compounds.

W. H. Brown. *Organic Chemistry,* 2nd ed. Philadelphia: Saunders College Publishing, 1997.

F. A. Carey. *Organic Chemistry.* 3rd ed. New York: McGraw-Hill, 1996.

S. Ege. *Organic Chemistry.* 3rd ed. Lexington, MA: D. C. Heath, 1994.

R. J. Fessenden and J. S. Fessenden. *Organic Chemistry.* 5th ed. Pacific Grove, CA: Brooks/Cole, 1994.

M. A. Fox and J. K. Whitesell. *Organic Chemistry,* 2nd ed. Boston: Jones and Bartlett, 1997.

G. M. Loudon. *Organic Chemistry.* 3rd ed. Menlo Park, CA: Benjamin/Cummings, 1995.

J. March. *Advanced Organic Chemistry: Reactions, Mechanisms, and Structure.* 4th ed. New York: John Wiley, 1992. Appendix B.

J. McMurry. *Organic Chemistry.* 4th ed. Pacific Grove, CA: Brooks/Cole, 1996.

R. T. Morrison and R. N. Boyd. *Organic Chemistry.* 6th ed. Englewood Cliffs, NJ: Prentice Hall, 1992.

S. H. Pine. *Organic Chemistry.* 5th ed. New York: McGraw-Hill, 1987.

T. W. G. Solomons. *Organic Chemistry.* 6th ed. New York: John Wiley, 1996.

A. Streitwieser, C. H. Heathcock, and E. M. Kosower. *Introduction to Organic Chemistry.* 4th ed. New York: Macmillan, 1992.

K. P. C. Vollhardt and N. E. Schore. *Organic Chemistry.* 2nd ed. New York: W. H. Freeman, 1994.

L. G. Wade, Jr. *Organic Chemistry.* 3rd ed. Englewood Cliffs, NJ: Prentice Hall, 1995.

20.3 SEARCHING THE CHEMICAL LITERATURE

If the information you are seeking is not available in any of the handbooks mentioned in Section 20.1 or if you are searching for more detailed information than they can provide, then a proper literature search is in order. Although an examination of standard textbooks can provide some help, you often must use all the resources of the library, including journals, reference collections, and abstracts. The following sections of this chapter outline how the various types of sources should be used and what sort of information can be obtained from them.

The methods for searching the literature discussed in this chapter use mainly printed materials. Modern search methods also make use of computerized databases and are discussed in Section 20.11. These are vast collections of data and bibliographic materials that can be scanned very rapidly from remote computer terminals. Although computerized searching is widely available, its use may not be readily accessible to undergraduate students. The following references provide excellent introductions to the literature of organic chemistry:

C. Carr. "Teaching and Using Chemical Information." *Journal of Chemical Education, 70* (September 1993): 719.

J. March. *Advanced Organic Chemistry: Reactions, Mechanisms, and Structure.* 4th ed. New York: John Wiley, 1992. Appendix A.

A. N. Somerville. "Information Sources for Organic Chemistry, 1: Searching by Name Reaction and Reaction Type." *Journal of Chemical Education, 68* (July 1991): 553.

A. N. Somerville. "Information Sources for Organic Chemistry, 2: Searching by Functional Group." *Journal of Chemical Education, 68* (October 1991): 842.

A. N. Somerville. "Information Sources for Organic Chemistry, 3: Searching by Reagent." *Journal of Chemical Education, 69* (May 1992): 379.

G. Wiggins. *Chemical Information Sources.* New York: McGraw Hill, 1990. Integrates printed materials and computer sources of information.

20.4 COLLECTIONS OF SPECTRA

Collections of infrared, nuclear magnetic resonance, and mass spectra can be found in the following catalogs of spectra:

A. Cornu and R. Massot. *Compilation of Mass Spectral Data.* 2nd ed. London: Heyden and Sons, Ltd., 1975.

High-Resolution NMR Spectra Catalog. Palo Alto, CA: Varian Associates. Volume 1, 1962; Volume 2, 1963.

L. F. Johnson and W. C. Jankowski. *Carbon-13 NMR Spectra.* New York: John Wiley, 1972.

C. J. Pouchert. *Aldrich Library of Infrared Spectra.* 3rd ed. Milwaukee: Aldrich Chemical Co., 1981.

C. J. Pouchert. *Aldrich Library of FT-IR Spectra.* Milwaukee: Aldrich Chemical Co., 1985.

C. J. Pouchert. *Aldrich Library of NMR Spectra,* 2nd ed. Milwaukee: Aldrich Chemical Co., 1983.

C. J. Pouchert and J. Behnke. *Aldrich Library of ^{13}C and ^{1}H FT NMR Spectra.* Milwaukee: Aldrich Chemical Co., 1993.

Sadtler Standard Spectra. Philadelphia: Sadtler Research Laboratories. Continuing collection.

E. Stenhagen, S. Abrahamsson, and F. W. McLafferty. *Registry of Mass Spectral Data.* New York: John Wiley-Interscience, 1974. Four-volume set.

The American Petroleum Institute has also published collections of infrared, nuclear magnetic resonance, and mass spectra.

20.5 ADVANCED TEXTBOOKS

Much information about synthetic methods, reaction mechanisms, and reactions of organic compounds is available in any of the many current advanced textbooks in organic chemistry. Examples of such books are

F. A. Carey and R. J. Sundberg. *Advanced Organic Chemistry. Part A. Structure and Mechanisms; Part B. Reactions and Synthesis.* 3rd ed. New York: Plenum Press, 1990.

W. Carruthers. *Some Modern Methods of Organic Synthesis.* 3rd ed. Cambridge, U.K.: Cambridge
     University Press, 1986.
E. J. Corey and Xue-Min Cheng. *The Logic of Chemical Synthesis.* New York: John Wiley, 1989.
L. F. Fieser and M. Fieser. *Advanced Organic Chemistry.* New York: Reinhold, 1961.
I. L. Finar. *Organic Chemistry.* 6th ed. London: Longman Group, Ltd., 1986.
H. O. House. *Modern Synthetic Reactions.* 2nd ed. Menlo Park, CA: W. H. Benjamin, 1972.
J. March. *Advanced Organic Chemistry: Reactions, Mechanisms, and Structure.* 4th ed. New York:
     John Wiley, 1992.
C. R. Noller. *Chemistry of Organic Compounds.* 3rd ed. Philadelphia: W. B. Saunders, 1965.
M. B. Smith. *Organic Synthesis.* New York: McGraw-Hill, 1994.
J. C. Stowell. *Intermediate Organic Chemistry.* 2nd ed. New York: John Wiley, 1993.
S. Warren. *Organic Synthesis: The Disconnection Approach.* New York: John Wiley, 1982.

These books often contain references to original papers in the literature for students wanting to follow the subject further. Consequently, you obtain not only a review of the subject from such a textbook but also a key reference that is helpful toward a more extensive
literature search. The textbook by March is particularly useful for this purpose.

20.6 SPECIFIC SYNTHETIC METHODS

Anyone interested in locating information about a particular method of synthesizing
a compound should first consult one of the many general textbooks on the subject. Useful
ones are

N. Anand, J. S. Bindra, and S. Ranganathan. *Art in Organic Synthesis.* 2nd ed. New York: John
     Wiley, 1988.
D. Barton and W. D. Ollis, eds. *Comprehensive Organic Chemistry.* Oxford: Pergamon Press, 1979.
     Six-volume set.
C. A. Buehler and D. E. Pearson. *Survey of Organic Syntheses.* New York: John Wiley-Interscience,
     1970 and 1977. Two-volume set.
F. A. Carey and R. J. Sundberg. *Advanced Organic Chemistry. Part B. Reactions and Synthesis.*
     3rd ed. New York: Plenum Press, 1990.
Compendium of Organic Synthetic Methods. New York: John Wiley-Interscience, 1971–1995. This
     is a continuing series, now in eight volumes.
L. F. Fieser and M. Fieser. *Reagents for Organic Synthesis.* New York: John Wiley-Interscience,
     1967–1994. This is a continuing series, now in 17 volumes.
T. W. Greene and P. G. M. Wuts. *Protective Groups in Organic Synthesis.* 2nd ed. New York: John
     Wiley, 1991.
H. O. House. *Modern Synthetic Reactions.* 2nd ed. Menlo Park, CA: W. H. Benjamin, 1972.
R. C. Larock. *Comprehensive Organic Transformations.* New York: VCH Press, 1989.
J. March. *Advanced Organic Chemistry: Reactions, Mechanisms, and Structure.* 4th ed. New York:
     John Wiley, 1992.
B. P. Mundy and M. G. Ellerd. *Name Reactions and Reagents in Organic Synthesis.* New York:
     John Wiley, 1988.
S. Patai, ed. *The Chemistry of the Functional Groups.* London: Interscience, 1964–present. This
     series consists of many volumes, each one specializing in a particular functional group.

A. I. Vogel. Revised by members of the School of Chemistry, Thames Polytechnic. *Vogel's Textbook of Practical Organic Chemistry, Including Qualitative Organic Analysis.* 5th ed. London: Longman Group, Ltd., 1989.

R. B. Wagner and H. D. Zook. *Synthetic Organic Chemistry.* New York: John Wiley, 1956.

More specific information, including actual reaction conditions, exists in collections specializing in organic synthetic methods. The most important of these is

Organic Syntheses. New York: John Wiley, 1921–present. Published annually.
Organic Syntheses, Collective Volumes. New York: John Wiley, 1941–1993.
   Vol. 1, 1941, Annual Volumes 1–9
   Vol. 2, 1943, Annual Volumes 10–19
   Vol. 3, 1955, Annual Volumes 20–29
   Vol. 4, 1963, Annual Volumes 30–39
   Vol. 5, 1973, Annual Volumes 40–49
   Vol. 6, 1988, Annual Volumes 50–59
   Vol. 7, 1990, Annual Volumes 60–64
   Vol. 8, 1993, Annual Volumes 65–69

It is much more convenient to use the collective volumes where the earlier annual volumes of *Organic Syntheses* are combined in groups of nine or ten in the first six collective volumes (Vol. 1–6), and then in groups of five for the next two volumes (Vol. 7 and 8). Useful indices are included at the end of each of the collective volumes that classify methods according to the type of reaction, type of compound prepared, formula of compound prepared, preparation or purification of solvents and reagents, and use of various types of specialized apparatus.

The main advantage of using one of the *Organic Syntheses* procedures is that they have been tested to make sure that they work as written. Often, an organic chemist will adapt one of these tested procedures to the preparation of another compound. One of the features of the advanced organic textbook by March is that it includes references to specific preparative methods contained in *Organic Syntheses.*

More advanced material on organic chemical reactions and synthetic methods may be found in any one of a number of annual publications that review the original literature and summarize it. Examples include

Advances in Organic Chemistry: Methods and Results. New York: John Wiley, 1960–present.
Annual Reports of the Chemical Society, Section B. London: Chemical Society, 1905–present.
   Specifically, the section on *Synthetic Methods.*
Annual Reports in Organic Synthesis. Orlando, FL: Academic Press, 1985–1995.
Progress in Organic Chemistry. New York: John Wiley, 1952–1973.
Organic Reactions. New York: John Wiley, 1942–present.

Each of these publications contains a great many citations to the appropriate articles in the original literature.

20.7 ADVANCED LABORATORY TECHNIQUES

The student who is interested in reading about more advanced techniques than those described in this textbook, or in more complete descriptions of techniques, should consult one of the advanced textbooks specializing in organic laboratory techniques. Besides focusing on apparatus construction and the performance of complex reactions, these books also provide advice on purifying reagents and solvents. Useful sources of information on organic laboratory techniques include

R. B. Bates and J. P. Schaefer. *Research Techniques in Organic Chemistry*. Englewood Cliffs, NJ: Prentice-Hall, 1971.

A. J. Krubsack. *Experimental Organic Chemistry*. Boston: Allyn and Bacon, 1973.

J. Leonard, B. Lygo, and G. Procter. *Advanced Practical Organic Chemistry*, 2nd ed. London: Chapman and Hall, 1995.

R. S. Monson, *Advanced Organic Synthesis: Methods and Techniques*. New York: Academic Press, 1971.

Techniques of Chemistry. New York: John Wiley, 1970–present. Currently 21 volumes. The successor to *Technique of Organic Chemistry,* this series covers experimental methods of chemistry, such as purification of solvents, spectral methods, and kinetic methods.

A. Weissberger, et al., eds. *Technique of Organic Chemistry*. 3rd ed. New York: John Wiley-Interscience, 1959–1969. This work is in 14 volumes.

K. B. Wiberg. *Laboratory Technique in Organic Chemistry*. New York: McGraw-Hill, 1960.

J. W. Zubrick. *The Organic Chem Lab Survival Manual: A Student's Guide to Techniques*. 4th ed. New York: John Wiley, 1997.

Numerous works and some general textbooks specialize in particular techniques. The above list is only representative of the most common books in this category. The following books deal specifically with micro and semi-microscale techniques.

N. D. Cheronis. "Micro and Semimicro Methods." In: A Weissberger, ed. *Technique of Organic Chemistry,* Volume 6. New York: John Wiley-Interscience, 1954.

N. D. Cheronis and T. S. Ma. *Organic Functional Group Analysis by Micro and Semimicro Methods*. New York: John Wiley-Interscience, 1964.

T. S. Ma and V. Horak. *Microscale Manipulations in Chemistry*. New York: John Wiley-Interscience, 1976.

20.8 REACTION MECHANISMS

As with the case of locating information on synthetic methods, you can obtain a great deal of information about reaction mechanisms by consulting one of the common textbooks on physical organic chemistry. The textbooks listed here provide a general description of mechanisms, but they do not contain specific literature citations. Very general textbooks include

A. Miller. *Writing Reaction Mechanisms in Organic Chemistry*. San Diego: Academic Press, 1992.

P. Sykes. *A Guidebook to Mechanism in Organic Chemistry*. 6th ed. London: Longman Group, Ltd., 1986.

More advanced textbooks include

F. A. Carey and R. J. Sundberg. *Advanced Organic Chemistry. Part A. Structure and Mechanisms.* 3rd ed. New York: Plenum Press, 1990.

L. P. Hammett. *Physical Organic Chemistry: Reaction Rates, Equilibria, and Mechanisms.* 2nd ed. New York: McGraw-Hill, 1970.

J. Hine. *Physical Organic Chemistry.* 2nd ed. New York: McGraw-Hill, 1962.

C. K. Ingold. *Structure and Mechanism in Organic Chemistry.* 2nd ed. Ithaca, NY: Cornell University Press, 1969.

N. S. Isaacs. *Physical Organic Chemistry.* 2nd ed. New York: John Wiley, 1995.

R. A. Y. Jones. *Physical and Mechanistic Organic Chemistry.* 2nd ed. Cambridge, U.K.: Cambridge University Press, 1984.

T. H. Lowry and K. S. Richardson. *Mechanism and Theory in Organic Chemistry.* 3rd ed. New York: Harper & Row, 1987.

J. March. *Advanced Organic Chemistry: Reactions, Mechanisms, and Structure.* 4th ed. New York: John Wiley, 1992.

J. W. Moore and R. G. Pearson. *Kinetics and Mechanism.* 3rd ed. New York: John Wiley, 1981.

These books include extensive bibliographies that permit the reader to delve more deeply into the subject.

Most libraries also subscribe to annual series of publications that specialize in articles dealing with reaction mechanisms. Among these are

Advances in Physical Organic Chemistry. London: Academic Press, 1963–present.
Annual Reports of the Chemical Society. Section B. London: Chemical Society, 1905–present. Specifically, the section on *Reaction Mechanisms.*
Organic Reaction Mechanisms. Chichester, U.K.: John Wiley, 1965–present.
Progress in Physical Organic Chemistry. New York: Interscience, 1963–present.

These publications provide the reader with citations from the original literature that can be very useful in an extensive literature search.

20.9 ORGANIC QUALITATIVE ANALYSIS

Experiment 57 contains a procedure for identifying organic compounds through a series of chemical tests and reactions. Occasionally, you might require a more complete description of analytical methods or a more complete set of tables of derivatives. Textbooks specializing in organic qualitative analysis should fill this need. Examples of sources for such information include

N. D. Cheronis and J. B. Entriken. *Identification of Organic Compounds: A Student's Text Using Semimicro Techniques.* New York: Interscience, 1963.

D. J. Pasto and C. R. Johnson. *Laboratory Text for Organic Chemistry: A Source Book of Chemical and Physical Techniques.* Englewood Cliffs, NJ: Prentice-Hall, 1979.

Z. Rappoport, ed. *Handbook of Tables for Organic Compound Identification.* 3rd ed. Cleveland: Chemical Rubber Co., 1967.

R. L. Shriner, R. C. Fuson, D. Y. Curtin, and T. C. Morrill. *The Systematic Identification of Organic Compounds: A Laboratory Manual.* 6th ed. New York: John Wiley, 1980.

A. I. Vogel, *Elementary Practical Organic Chemistry. Part 2. Qualitative Organic Analysis.* 2nd ed. New York: John Wiley, 1966.

A. I. Vogel. Revised by members of the School of Chemistry, Thames Polytechnic. *Vogel's Textbook of Practical Organic Chemistry, Including Qualitative Organic Analysis.* 5th ed. London: Longman Group, Ltd., 1989.

20.10 BEILSTEIN AND CHEMICAL ABSTRACTS

One of the most useful sources of information about the physical properties, synthesis, and reactions of organic compounds is *Beilsteins Handbuch der Organischen Chemie.* This is a monumental work, initially edited by Friedrich Konrad Beilstein, and updated through several revisions by the Beilstein Institute in Frankfurt am Main, Germany. The original edition (the *Hauptwerk,* abbreviated H) was published in 1918 and covers completely the literature to 1909. Five supplementary series (*Ergänzungswerken*) have been published since that time. The first supplement (*Erstes Ergänzungswerk,* abbreviated E I) covers the literature from 1910–1919; the second supplement (*Zweites Ergänzungswerk,* E II) covers 1920–1929; the third supplement (*Drittes Ergänzungswerk,* E III) covers 1930–1949; the fourth supplement (*Viertes Ergänzungswerk,* E IV) covers 1950–1959; and the fifth supplement (in English) covers 1960–1979. Volumes 17–27 of supplementary series III and IV, covering heterocyclic compounds, are combined in a joint issue, E III/IV. Supplementary series III, IV, and V are not complete, so the coverage of *Handbuch der Organischen Chemie* can be considered complete to 1929, with partial coverage to 1979.

Beilsteins Handbuch der Organischen Chemie, usually referred to simply as *Beilstein,* also contains two types of cumulative indices. The first of these is a name index (*Sachregister*) and the second is a formula index (*Formelregister*). These indices are particularly useful for a person wishing to locate a compound in *Beilstein.*

The principal difficulty in using *Beilstein* is that it is written in German through the fourth supplement. The fifth supplement is in English. Although some reading knowledge of German is useful, you can obtain information from the work by learning a few key phrases. For example, *Bildung* is "formation" or "structure." *Darst* or *Darstellung* is "preparation," K_P or *Siedepunkt* is "boiling point," and *F* or *Schmelzpunkt* is "melting point." Furthermore, the names of some compounds in German are not cognates of the English names. Some examples are *Apfelsäure* for "malic acid" (*säure* means "acid"), *Harnstoff* for "urea," *Jod* for "iodine," and *Zimtsäure* for "cinnamic acid." If you have access to a German-English dictionary for chemists, many of these difficulties can be overcome. The best such dictionary is

A. M. Patterson. *German–English Dictionary for Chemists.* 3rd ed. New York: John Wiley, 1959.

Beilstein is organized according to a very sophisticated and complicated system. To locate a compound in *Beilstein,* you can learn all the intricacies of this system. However, most students do not wish to become experts on *Beilstein* to this extent. A simpler, though slightly less reliable, method is to look for the compound in the formula index that accompanies the second supplement. Under the molecular formula, you will find the names of compounds that have that formula. After that name will be a series of numbers that indicate the pages and volume in which that compound is listed. Suppose, as an example, that you are searching for information on *p*-nitroaniline. This compound has the molecular formula $C_6H_6N_2O_2$. Searching for this formula in the formula index to the second supplement, you find

4-Nitro-anilin 12 711, I 349, II 383

This information tells us that *p*-nitroaniline is listed in the main edition (*Hauptwerk*) in volume 12, page 711. Locating this particular volume, which is devoted to isocyclic monoamines, we turn to page 711 and find the beginning of the section on *p*-nitroaniline. At the left side of the top of this page, we find "Syst. No. 1671." This is the system number given to compounds in this part of Volume 12. The system number is useful, as it can help you find entries for this compound in subsequent supplements. The organization of *Beilstein* is such that all entries on *p*-nitroaniline in each of the supplements will be found in Volume 12. The entry in the formula index also indicates that material on this compound may be found in the first supplement on page 349 and in the second supplement on page 383. On page 349 of volume 12 of the first supplement, there is a heading, **"XII, 710–712,"** and on the left is "Syst. No. 1671." Material on *p*-nitroaniline is found in each supplement on a page that is headed with the volume and page of the *Hauptwerk* in which the same compound is found. On page 383 of volume 12 of the second supplement, the heading in the center of the top of the page is **"H12, 710–712."** On the left you find "Syst. No. 1671." Again, because *p*-nitroaniline appeared in volume 12, page 711, of the main edition, you can locate it by searching through volume 12 of any supplement until you find a page with the heading corresponding to volume 12, page 711. Because the third and fourth supplements are not complete, there is no comprehensive formula index for these supplements. However, you can still find material on *p*-nitroaniline by using the system number and the volume and page in the main work. In the third supplement, because the amount of information available has grown so much since the early days of Beilstein's work, volume 12 has now expanded so that it is found in several bound parts. However, you select the part that includes system number 1671. In this part of volume 12, you look through the pages until you find a page headed "Syst. No. 1671/H 711." The information on *p*-nitroaniline is found on this page (p. 1580). If volume 12 of the fourth supplement were available, you would go on in the same way to locate more recent data on *p*-nitroaniline. This example is meant to illustrate how you can locate information on particular compounds without having to learn the Beilstein system of classification. You might do well to test your ability at finding compounds in *Beilstein* as we have described here.

Guidebooks to using *Beilstein,* which include a description of the Beilstein system, are recommended for anyone who wants to work extensively with *Beilstein.* Among such sources are

E. H. Huntress. *A Brief Introduction to the Use of Beilsteins Handbuch der Organischen Chemie.* 2nd ed. New York: John Wiley, 1938.

How to Use Beilstein. Beilstein Institute, Frankfurt am Main. Berlin: Springer-Verlag.
O. Weissbach. *The Beilstein Guide: A Manual for the Use of Beilsteins Handbuch der Organischen Chemie.* New York: Springer-Verlag, 1976.

Beilstein reference numbers are listed in such handbooks as *CRC Handbook of Chemistry and Physics* and *Lange's Handbook of Chemistry.* Additionally, Beilstein numbers are included in the *Aldrich Catalog Handbook of Fine Chemicals,* issued by the Aldrich Chemical Company. If the compound you are seeking is listed in one of these handbooks, you will find that using *Beilstein* is simplified.

Another very useful publication for finding references for research on a particular topic is *Chemical Abstracts,* published by the Chemical Abstracts Service of the American Chemical Society. *Chemical Abstracts* contains abstracts of articles appearing in more than 10,000 journals from virtually every country conducting scientific research. These abstracts list the authors, the journal in which the article appeared, the title of the paper, and a short summary of the contents of the article. Abstracts of articles that appeared originally in a foreign language are provided in English, with a notation indicating the original language.

To use *Chemical Abstracts,* you must know how to use the various indices that accompany it. At the end of each volume there appears a set of indices, including a formula index, a general subject index, a chemical substances index, an author index, and a patent index. The listings in each index refer the reader to the appropriate abstract according to the number assigned to it. There are also collective indices that combine all the indexed material appearing in a 5-year period (10-year period before 1956). In the collective indices, the listings include the volume number as well as the abstract number.

For material after 1929, *Chemical Abstracts* provides the most complete coverage of the literature. For material before 1929, use *Beilstein* before consulting *Chemical Abstracts.* *Chemical Abstracts* has the advantage that it is written entirely in English. Nevertheless, most students perform a literature search to find a relatively simple compound. Finding the desired entry for a simple compound is much easier in *Beilstein* than in *Chemical Abstracts.* For simple compounds, the indices in *Chemical Abstracts* are likely to contain very many entries. To locate the desired information, you must comb through this multitude of listings—potentially a very time-consuming task.

The opening pages of each index in *Chemical Abstracts* contain a brief set of instructions on using that index. If you want a more complete guide to *Chemical Abstracts,* consult a textbook designed to familiarize you with these abstracts and indices. Two such books are

CAS Printed Access Tools: A Workbook. Washington, D.C.: Chemical Abstracts Service, American Chemical Society, 1977.
How to Search Printed CA. Washington, D.C.: Chemical Abstracts Service, American Chemical Society, 1989.

Chemical Abstracts Service maintains a computerized database that permits users to search through *Chemical Abstracts* very rapidly and thoroughly. This service, which is called *CA Online,* is described in Section 20.11. *Beilstein* is also available for online searching by computer.

20.11 COMPUTER ONLINE SEARCHING

You can search a number of chemistry databases online by using a computer and modem. Many academic and industrial libraries can access these databases through their computers. One organization that maintains a large number of databases is the Scientific and Technical Information Network (STN International). The fee charged to the library for this service depends on the total time used in making the search, the type of information being asked for, the time of day when the search is being conducted, and the type of database being searched.

The Chemical Abstracts Service database (*CA Online*) is one of many databases available on STN. It is particularly useful to chemists. Unfortunately, this database extends back only to about 1967, although some earlier references are available. Searches earlier than 1967 must be made with printed abstracts (Section 20.10). An online search is much faster than searching in the printed abstracts. In addition, you can tailor the search in a number of ways by using keywords and the Chemical Abstracts Substance Registry Number (CAS number) as part of the search routine. For the more common organic compounds, you can easily obtain CAS numbers from the catalogues of most of the companies that supply chemicals. Another advantage of performing an online search is that the *Chemical Abstracts* files are updated much more quickly than the printed versions of abstracts. This means that your search is more likely to reveal the most current information available.

Other useful databases available from STN include *Beilstein* and *CASREACTS*. As described in Section 20.10, *Beilstein* is very useful to organic chemists. Currently, there are over 3.5 million compounds listed in the database. You can use the CAS Registry Numbers to help in a search that has the potential of going back to 1830. *CASREACTS* is a chemical reactions database derived from over 100 journals covered by *Chemical Abstracts,* starting in 1985. With this database, you can specify a starting material and a product using the CAS Registry Numbers. Further information on *CA Online, Beilstein, CASREACTS,* and other databases can be obtained from the following references:

J. March. *Advanced Organic Chemistry: Reactions, Mechanisms, and Structure.* 4th ed. New York: John Wiley, 1992. Appendix A contains a summary.

A. N. Somerville. "Information Sources for Organic Chemistry, 2: Searching by Functional Group." *Journal of Chemical Education, 68* (October 1991): 842.

A. N. Somerville. "Subject Searching of Chemical Abstracts Online." *Journal of Chemical Education, 70* (March 1993): 200.

G. Wiggins. *Chemical Information Sources.* New York: McGraw Hill, 1990. Integrates printed materials and computer sources of information.

20.12 SCIENTIFIC JOURNALS

Ultimately, someone wanting information about a particular area of research will be required to read articles from the scientific journals. These journals are of two basic types: review journals and primary scientific journals. Journals that specialize in review articles summarize all the work that bears on the particular topic. These articles may focus on the contributions of one particular researcher but often consider the contributions of many researchers to the subject. These articles also contain extensive bibliographies, which refer you to the original research articles. Among the important journals devoted, at least partly, to review articles are

Accounts of Chemical Research

Angewandte Chemie (International Edition, in English)

Chemical Reviews

Chemical Society Reviews (formerly known as *Quarterly Reviews*)

Nature

Science

The details of the research of interest appear in the primary scientific journals. Although there are thousands of journals published in the world, a few important journals specializing in articles dealing with organic chemistry might be mentioned here. These are

Canadian Journal of Chemistry

Chemische Berichte

Journal of the American Chemical Society

Journal of the Chemical Society, Chemical Communications

Journal of the Chemical Society, Perkin Transactions (Parts I and II)

Journal of Organic Chemistry

Journal of Organometallic Chemistry

Liebigs Annalen der Chemie

Organometallics

Synthesis

Tetrahedron

Tetrahedron Letters

20.13 TOPICS OF CURRENT INTEREST

The following journals and magazines are good sources for topics of educational and current interest. They specialize in news articles and focus on current events in chemistry or in science in general. Articles in these journals (magazines) can be useful in keeping you abreast of developments in science that are not part of your normal specialized scientific reading.

American Scientist

Chemical and Engineering News

Chemistry in Britain

Chemistry and Industry

Chemtech

Discover

Journal of Chemical Education

Nature

New Scientist

Omni

Science

Science Digest

Scientific American

SciQuest (formerly *Chemistry*)

20.14 HOW TO CONDUCT A LITERATURE SEARCH

The easiest method to follow in searching the literature is to begin with secondary sources and then go to the primary sources. In other words, you would try to locate material in a textbook, *Beilstein,* or *Chemical Abstracts.* From the results of that search, you would then consult one of the primary scientific journals.

A literature search that ultimately requires you to read one or more papers in the scientific journals is best conducted if you can identify a particular paper central to the study. Often, you can obtain this reference from a textbook or a review article on the subject. If this is not available, a search through *Beilstein* is required. A search through one of the handbooks that provides *Beilstein* reference numbers (see Section 20.10) may be helpful. Searching through *Chemical Abstracts* would be considered the next logical step. From these sources, you should be able to identify citations from the original literature on the subject.

Additional citations may be found in the references cited in the journal article. In this way, the background leading to the research can be examined. It is also possible to conduct a search forward in time from the date of the journal article through the *Science Citation Index.* This publication provides the service of listing articles and the papers in which these articles were cited. Although the *Science Citation Index* consists of several types of indices, the *Citation Index* is most useful for the purposes described here. A person who knows of a particular key reference on a subject can examine the *Science Citation Index* to obtain a list of papers that have used that seminal reference in support of the work described. The *Citation Index* lists papers by their senior author, journal, volume, page, and date, followed by citations of papers that have referred to that article, author, journal, volume, page, and date of each. The *Citation Index* is published in annual volumes, with quarterly supplements issued during the current year. Each volume contains a complete list of the citations of the key articles made during that year. A disadvantage is that *Science Citation Index* has been available only since 1961. An additional disadvantage is that you may miss journal articles on the subject of interest if they failed to cite that particular key reference in their bibliographies—a reasonably likely possibility.

You can, of course, conduct a literature search by a "brute force" method, by beginning the search with *Beilstein* or even with the indices in *Chemical Abstracts*. However, the task can be made much easier by performing a computer search (Section 20.11) or by starting with a book or an article of general and broad coverage, which can provide a few citations for starting points in the search.

The following guides to using the chemical literature are provided for the reader who is interested in going further into this subject.

R. T. Bottle and J. F. B. Rowland, eds. *Information Sources in Chemistry,* 4th ed. New York: Bowker-Saur, 1992.

M. G. Mellon. *Chemical Publications.* 5th ed. New York: McGraw-Hill, 1982.

R. E. Maizell. *How to Find Chemical Information: A Guide for Practicing Chemists, Teachers, and Students,* 2nd ed. New York: John Wiley-Interscience, 1987.

G. Wiggins. *Chemical Information Sources.* New York: McGraw Hill, 1990. Integrates printed materials and computer sources of information.

PROBLEMS

1. Using the *Merck Index* discussed in Section 20.1, find and draw structures for the following compounds:

    (a) atropine

    (b) quinine

    (c) saccharin

    (d) benzo[a]pyrene (benzpyrene)

    (e) itaconic acid

    (f) adrenosterone

    (g) chrysanthemic acid (chrysanthemumic acid)

    (h) cholesterol

    (i) vitamin C (ascorbic acid)

2. Find the melting points for the following compounds in the *Handbook of Chemistry and Physics* or *Lange's Handbook of Chemistry* (Section 20.1):

    (a) biphenyl

    (b) 4-bromobenzoic acid

    (c) 3-nitrophenol

3. Find the boiling point for each compound in the references listed in Problem 2:

    (a) octanoic acid at reduced pressure

    (b) 4-chloroacetophenone at atmosphere and reduced pressure

    (c) 2-methyl-2-heptanol

4. Find the index of refraction n_D and density for the liquids listed in Problem 3.

5. Using the references given in Problem 2, give the specific rotations $[\alpha]_D$ for the enantiomers of camphor.

6. Read the section on carbon tetrachloride in the *Merck Index* and list some of the health hazards for this compound.

7. Find the following compounds in the formula index for the *Second Supplement of Beilstein* (Section (20.10). (i) List the page numbers from the Main Work and the Supplements (First and Second). (ii) Using these page numbers, look up the System Number (Syst. No.) and the Main Work number (Hauptwerk number, H) for each compound in the Main Work, and the First and

Second supplements. In some cases, a compound may not be found in all three places. (iii) Now use the System Number and Main Work number to find each of these compounds in the Third and Fourth Supplements. List the page numbers where these compounds are found.

    (a) 2,5-hexanedione (acetonylacetone)

    (b) 3-nitroacetophenone

    (c) 4-*tert*-butylcyclohexanone

    (d) 4-phenylbutanoic acid (4-phenylbutyric acid, γ-phenylbuttersäure)

  8. Using the *Science Citation Index* (Section 20.14), list five research papers by complete title and journal citation for each of the following chemists who have been awarded the Nobel Prize. Use the *Five-Year Cumulation Source Index* for the years 1980–1984 as your source.

    (a) H. C. Brown

    (b) R. B. Woodward

    (c) D. J. Cram

    (d) G. Olah

  9. The reference book by J. March is listed in Section 20.2. Using Appendix B in this book, give two methods for preparing the following functional groups. You will need to provide equations.

    (a) carboxylic acids

    (b) aldehydes

    (c) esters (Carboxylic esters)

10. *Organic Synthesis* is described in Section 20.6. There are currently eight collective volumes in the series, each with its own index. Find the compounds listed below and provide the equations for preparing each compound.

    (a) 2-methylcyclopentane-1,3-dione

    (b) *cis*-Δ^4-tetrahydrophthalic anhydride (listed as tetrahydrophthalic anhydride)

11. Provide four ways that may be used to oxidize an alcohol to an aldehyde. Give complete literature references for each method as well as equations. Use the *Compendium of Organic Synthetic Methods* or *Survey of Organic Syntheses* by Buehler and Pearson (Section 20.6).

Appendices

APPENDIX 1

Tables of Unknowns and Derivatives

More extensive tables of unknowns may be found in Z. Rappoport, ed. *Handbook of Tables for Organic Compound Identification,* 3rd ed. Cleveland: Chemical Rubber Co., 1967

ALDEHYDES

Compound	BP	MP	Semi-carbazone*	2,4-Dinitro-phenyl-hydrazone*
Ethanal (acetaldehyde)	21	—	162	168
Propanal (propionaldehyde)	48	—	89	148
Propenal (acrolein)	52	—	171	165
2-Methylpropanal (isobutyraldehyde)	64	—	125	187
Butanal (butyraldehyde)	75	—	95	123
3-Methylbutanal (isovaleraldehyde)	92	—	107	123
Pentanal (valeraldehyde)	102	—	—	106
2-Butenal (crotonaldehyde)	104	—	199	190
2-Ethylbutanal (diethylacetaldehyde)	117	—	99	95
Hexanal (caproaldehyde)	130	—	106	104
Heptanal (heptaldehyde)	153	—	109	108
2-Furaldehyde (furfural)	162	—	202	212
2-Ethylhexanal	163	—	254	114
Octanal (caprylaldehyde)	171	—	101	106
Benzaldehyde	179	—	222	237
Phenylethanal (phenylacetaldehyde)	195	33	153	121
2-Hydroxybenzaldehyde (salicylaldehyde)	197	—	231	248
4-Methylbenzaldehyde (*p*-tolualdehyde)	204	—	234	234
3,7-Dimethyl-6-octenal (citronellal)	207	—	82	77
2-Chlorobenzaldehyde	213	11	229	213
4-Methoxybenzaldehyde (*p*-anisaldehyde)	248	2.5	210	253
trans-Cinnamaldehyde	250 d.	—	215	255
3,4-Methylenedioxybenzaldehyde (piperonal)	263	37	230	266 d.
2-Methoxybenzaldehyde (*o*-anisaldehyde)	245	38	215 d.	254
4-Chlorobenzaldehyde	214	48	230	254
3-Nitrobenzaldehyde	—	58	246	293
4-Dimethylaminobenzaldehyde	—	74	222	325
Vanillin	285 d.	82	230	271

ALDEHYDES *(Cont.)*

Compound	BP	MP	Semi-carbazone*	2,4-Dinitro-phenyl-hydrazone*
4-Nitrobenzaldehyde	—	106	221	320 d.
4-Hydroxybenzaldehyde	—	116	224	280 d.
(±)-Glyceraldehyde	—	142	160 d.	167

NOTE: "*d*" indicates "*decomposition*."
*See Appendix 2, "Procedures for Preparing Derivatives."

KETONES

Compound	BP	MP	Semi-carbazone*	2,4-Dinitro-phenyl hydrazone*
2-Propanone (acetone)	56	—	187	126
2-Butanone (methyl ethyl ketone)	80	—	146	117
3-Methyl-2-butanone (isopropyl methyl ketone)	94	—	112	120
2-Pentanone (methyl propyl ketone)	101	—	112	143
3-Pentanone (diethyl ketone)	102	—	138	156
Pinacolone	106	—	157	125
4-Methyl-2-pentanone (isobutyl methyl ketone)	117	—	132	95
2,4-Dimethyl-3-pentanone (diisopropyl ketone)	124	—	160	95
2-Hexanone (methyl butyl ketone)	128	—	125	106
4-Methyl-3-penten-2-one (mesityl oxide)	130	—	164	205
Cyclopentanone	131	—	210	146
2,3-Pentanedione	134	—	122 (mono) 209 (di)	209
2,4-Pentanedione (acetylacetone)	139	—	—	122 (mono) 209 (di)
4-Heptanone (dipropyl ketone)	144	—	132	75
2-Heptanone (methyl amyl ketone)	151	—	123	89
Cyclohexanone	156	—	166	162
2,6-Dimethyl-4-heptanone (diisobutyl ketone)	168	—	122	92
2-Octanone	173	—	122	58
Cycloheptanone	181	—	163	148
2,5-Hexanedione (acetonylacetone)	191	−9	185 (mono) 224 (di)	257 (di)
Acetophenone (methyl phenyl ketone)	202	20	198	238

KETONES *(Cont.)*

Compound	BP	MP	Semi-carbazone*	2,4-Dinitro-phenyl hydrazone*
Phenyl-2-propanone (phenylacetone)	216	27	198	156
Propiophenone (ethyl phenyl ketone)	218	21	182	191
4-Methylacetophenone	226	—	205	258
2-Undecanone	231	12	122	63
4-Chloroacetophenone	232	12	204	236
4-Phenyl-2-butanone (benzylacetone)	235	—	142	127
4-Chloropropiophenone	—	36	176	223
4-Phenyl-3-buten-2-one	—	37	187	227
4-Methoxyacetophenone	258	38	198	228
Benzophenone	305	48	167	238
4-Bromoacetophenone	225	51	208	230
2-Acetonaphthone	—	54	235	262
Desoxybenzoin	320	60	148	204
3-Nitroacetophenone	202	80	257	228
9-Fluorenone	345	83	234	283
Benzoin	344	136	206	245
4-Hydroxypropiophenone	—	148	—	229
($\pm$)-Camphor	205	179	237	177

*See Appendix 2, "Procedures for Preparing Derivatives."

CARBOXYLIC ACIDS

Compound	BP	MP	*p*-Toluidide*	Anilide*	Amide*
Formic acid	101	8	53	47	43
Acetic acid	118	17	148	114	82
Propenoic acid (acrylic acid)	139	13	141	104	85
Propanoic acid (propionic acid)	141	—	124	103	81
2-Methylpropanoic acid (isobutyric acid)	154	—	104	105	128
Butanoic acid (butyric acid)	162	—	72	95	115
2-Methylpropenoic acid (methacrylic acid)	163	16	—	87	102
Trimethylacetic acid (pivalic acid)	164	35	—	127	178
Pyruvic acid	165 d.	14	109	104	124
3-Methylbutanoic acid (isovaleric acid)	176	—	109	109	135

CARBOXYLIC ACIDS *(Cont.)*

Compound	BP	MP	*p*-Toluidide*	Anilide*	Amide*
Pentanoic acid (valeric acid)	186	—	70	63	106
2-Methylpentanoic acid	186	—	80	95	79
2-Chloropropanoic acid	186	—	124	92	80
Dichloroacetic acid	194	6	153	118	98
Hexanoic acid (caproic acid)	205	—	75	95	101
2-Bromopropanoic acid	205	24	125	99	123
Octanoic acid (caprylic acid)	237	16	70	57	107
Nonanoic acid	254	12	84	57	99
Decanoic acid (capric acid)	268	32	78	70	108
4-Oxopentanoic acid (levulinic acid)	246	33	108	102	108 d.
Dodecanoic acid (lauric acid)	299	43	87	78	100
3-Phenylpropanoic acid (hydrocinnamic acid)	279	48	135	98	105
Bromoacetic acid	208	50	—	131	91
Tetradecanoic acid (myristic acid)	—	54	93	84	103
Trichloroacetic acid	198	57	113	97	141
Hexadecanoic acid (palmitic acid)	—	62	98	90	106
Chloroacetic acid	189	63	162	137	121
Octadecanoic acid (stearic acid)	—	69	102	95	109
trans-2-Butenoic acid (crotonic acid)	—	72	132	118	158
Phenylacetic acid	—	77	136	118	156
2-Methoxybenzoic acid (*o*-anisic acid)	200	101	—	131	129
2-Methylbenzoic acid (*o*-toluic acid)	—	104	144	125	142
Nonanedioic acid (azelaic acid)	—	106	201 (di)	107 (mono) 186 (di)	93 (mono) 175 (di)
3-Methylbenzoic acid (*m*-toluic acid)	263 s.	110	118	126	94
(±)-Phenylhydroxyacetic acid (mandelic acid)	—	118	172	151	133
Benzoic acid	249	122	158	163	130
2-Benzoylbenzoic acid	—	127	—	195	165
Maleic acid	—	130	142 (di)	198 (mono) 187 (di)	172 (mono) 260 (di)
Decanedioic acid (sebacic acid)	—	133	201 (di)	122 (mono) 200 (di)	170 (mono) 210 (di)
Cinnamic acid	300	133	168	153	147

NOTE: "*s*" indicates "*sublimination*"; "*d*" indicates "*decomposition.*"
*See Appendix 2, "Procedures for Preparing Derivatives."

CARBOXYLIC ACIDS (*Cont.*)

Compound	BP	MP	*p*-Toluidide*	Anilide*	Amide*
2-Chlorobenzoic acid	—	140	131	118	139
3-Nitrobenzoic acid	—	140	162	155	143
2-Aminobenzoic acid (anthranilic acid)	—	146	151	131	109
Diphenylacetic acid	—	148	172	180	167
2-Bromobenzoic acid	—	150	—	141	155
Benzilic acid	—	150	190	175	154
Hexanedioic acid (adipic acid)	—	152	239	151 (mono) 241 (di)	125 (mono) 220 (di)
Citric acid	—	153	189 (tri)	199 (tri)	210 (tri)
4-Chlorophenoxyacetic acid	—	158	—	125	133
2-Hydroxybenzoic acid (salicylic acid)	—	158	156	136	142
5-Bromo-2-hydroxybenzoic acid (5-bromosalicylic acid)	—	165	—	222	232
Methylenesuccinic acid (itaconic acid)	—	166 d.	—	152 (mono)	191 (di)
(+)-Tartaric acid	—	169	—	180 (mono) 264 (di)	171 (mono) 196 (di)
4-Chloro-3-nitrobenzoic acid	—	180	—	131	156
4-Methylbenzoic acid (*p*-toluic acid)	—	180	160	145	160
4-Methoxybenzoic acid (*p*-anisic acid)	280	184	186	169	167
Butanedioic acid (succinic acid)	235 d.	188	180 (mono) 255 (di)	143 (mono) 230 (di)	157 (mono) 260 (di)
3-Hydroxybenzoic acid	—	201	163	157	170
3,5-Dinitrobenzoic acid	—	202	—	234	183
Phthalic acid	—	210 d.	150 (mono) 201 (di)	169 (mono) 253 (di)	144 (mono) 220 (di)
4-Hydroxybenzoic acid	—	214	204	197	162
Pyridine-3-carboxylic acid (nicotinic acid)	—	236	150	132	128
4-Nitrobenzoic acid	—	240	204	211	201
4-Chlorobenzoic acid	—	242	—	194	179
Fumaric acid	—	300	—	233 (mono) 314 (di)	270 (mono) 266 (di)

NOTE: "*d*" indicates "*decomposition.*"

*See Appendix 2, "Procedures for Preparing Derivatives."

PHENOLS[†]

Compound	BP	MP	α-Naphthylurethane*	Bromo Derivative*			
				Mono	*Di*	*Tri*	*Tetra*
2-Chlorophenol	176	7	120	48	76	—	—
3-Methylphenol (*m*-cresol)	203	12	128	—	—	84	—
2-Methylphenol (*o*-cresol)	191	32	142	—	56	—	—
2-Methoxyphenol (guaiacol)	204	32	118	—	—	116	—
4-Methylphenol (*p*-cresol)	202	34	146	—	49	—	198
Phenol	181	42	133	—	—	95	—
4-Chlorophenol	217	43	166	33	90	—	—
2,4-Dichlorophenol	210	45	—	68	—	—	—
4-Ethylphenol	219	45	128	—	—	—	—
2-Nitrophenol	216	45	113	—	117	—	—
2-Isopropyl-5-methylphenol (thymol)	234	51	160	55	—	—	—
3,4-Dimethylphenol	225	64	141	—	—	171	—
4-Bromophenol	238	64	169	—	—	95	—
3,5-Dimethylphenol	220	68	109	—	—	166	—
2,5-Dimethylphenol	212	75	173	—	—	178	—
1-Naphthol (α-naphthol)	278	96	152	—	105	—	—
2-Hydroxyphenol (catechol)	245	104	175	—	—	—	192
3-Hydroxyphenol (resorcinol)	281	109	275	—	—	112	—
4-Nitrophenol	—	112	150	—	142	—	—
2-Naphthol (β-naphthol)	286	121	157	84	—	—	—
1,2,3-Trihydroxybenzene (pyrogallol)	309	133	—	—	158	—	—
4-Phenylphenol	305	164	—	—	—	—	—

*See Appendix 2, "Procedures for Preparing Derivatives."
†Also check:
   Salicylic acid (2-hydroxybenzoic acid)
   Esters of salicylic acid (salicylates)
   Salicylaldehyde (2-hydroxybenzaldehyde)
   4-Hydroxybenzaldehyde
   4-Hydroxypropiophenone
   3-Hydroxybenzoic acid
   4-Hydroxybenzoic acid
   4-Hydroxybenzophenone

PRIMARY AMINES[†]

Compound	BP	MP	Benzamide*	Picrate*	Acetamide*
t-Butylamine	46	—	134	198	101
Propylamine	48	—	84	135	—

PRIMARY AMINES† *(Cont.)*

Compound	BP	MP	Benzamide*	Picrate*	Acetamide*
Allylamine	56	—	—	140	—
sec-Butylamine	63	—	76	139	—
Isobutylamine	69	—	57	150	—
Butylamine	78	—	42	151	—
Cyclohexylamine	135	—	149	—	104
Furfurylamine	145	—	—	150	—
Benzylamine	184	—	105	194	60
Aniline	184	—	163	198	114
2-Methylaniline (*o*-toluidine)	200	—	144	213	110
3-Methylaniline (*m*-toluidine)	203	—	125	200	65
2-Chloroaniline	208	—	99	134	87
2,6-Dimethylaniline	216	11	168	180	177
2-Methoxyaniline (*o*-anisidine)	225	6	60	200	85
3-Chloroaniline	230	—	120	177	74
2-Ethoxyaniline (*o*-phenetidine)	231	—	104	—	79
4-Chloro-2-methylaniline	241	29	142	—	140
4-Ethoxyaniline (*p*-phenetidine)	250	2	173	69	137
4-Methylaniline (*p*-toluidine)	200	43	158	182	147
2-Ethylaniline	210	47	147	194	111
2,5-Dichloroaniline	251	50	120	86	132
4-Methoxyaniline (*p*-anisidine)	—	58	154	170	130
4-Bromoaniline	245	64	204	180	168
2,4,5-Trimethylaniline	—	64	167	—	162
4-Chloroaniline	—	70	192	178	179
2-Nitroaniline	—	72	110	73	92
Ethyl *p*-aminobenzoate	—	89	148	—	110
o-Phenylenediamine	258	102	301 (di)	208	185 (di)
2-Methyl-5-nitroaniline	—	106	186	—	151
2-Chloro-4-nitroaniline	—	108	161	—	139
3-Nitroaniline	—	114	157	143	155
4-Chloro-2-nitroaniline	—	118	—	—	104
2,4,6-Tribromoaniline	300	120	200	—	232 (mono) 127 (di)
2-Methyl-4-nitroaniline	—	130	—	—	202
2-Methoxy-4-nitroaniline	—	138	149	—	153
p-Phenylenediamine	267	140	128 (mono) 300 (di)	—	162 (mono) 304 (di)
4-Nitroaniline	—	148	199	100	215
4-Aminoacetanilide	—	162	—	—	304
2,4-Dinitroaniline	—	180	202	—	120

*See Appendix 2, "Procedures for Preparing Derivatives."

†Also check: 4-Aminobenzoic acid and its esters.

SECONDARY AMINES

Compound	BP	MP	Benzamide*	Picrate*	Acetamide*
Diethylamine	56	—	42	155	—
Diisopropylamine	84	—	—	140	—
Pyrrolidine	88	—	Oil	112	—
Piperidine	106	—	48	152	—
Dipropylamine	110	—	—	75	—
Morpholine	129	—	75	146	—
Diisobutylamine	139	—	—	121	86
N-Methylcyclohexylamine	148	—	85	170	—
Dibutylamine	159	—	—	59	—
Benzylmethylamine	184	—	—	117	—
N-Methylaniline	196	—	63	145	102
N-Ethylaniline	205	—	60	132	54
N-Ethyl-m-toluidine	221	—	72	—	—
Dicyclohexylamine	256	—	153	173	103
N-Benzylaniline	298	37	107	48	58
Indole	254	52	68	—	157
Diphenylamine	302	52	180	182	101
N-Phenyl-1-naphthylamine	335	62	152	—	115

*See Appendix 2, "Procedures for Preparing Derivatives."

TERTIARY AMINES†

Compound	BP	MP	Picrate*	Methiodide*
Triethylamine	89	—	173	280
Pyridine	115	—	167	117
2-Methylpyridine (α-picoline)	129	—	169	230
3-Methylpyridine (β-picoline)	144	—	150	92
Tripropylamine	157	—	116	207
N,N-Dimethylbenzylamine	183	—	93	179
N,N-Dimethylaniline	193	—	163	228 d.
Tributylamine	216	—	105	186
N,N-Diethylaniline	217	—	142	102
Quinoline	237	—	203	133

NOTE: "*d*" indicates "*decomposition.*"

*See Appendix 2, "Procedures for Preparing Derivatives."

†Also check: Nicotinic acid and its esters.

ALCOHOLS

Compound	BP	MP	3,5-Dinitrobenzoate*	Phenylurethane*
Methanol	65	—	108	47
Ethanol	78	—	93	52
2-Propanol (isopropyl alcohol)	82	—	123	88
2-Methyl-2-propanol (*t*-butyl alcohol)	83	26	142	136
2-Propen-1-ol (allyl alcohol)	97	—	49	70
1-Propanol	97	—	74	57
2-Butanol (*sec*-butyl alcohol)	99	—	76	65
2-Methyl-2-butanol (*t*-pentyl alcohol)	102	−8.5	116	42
2-Methyl-3-butyn-2-ol	104	—	112	—
2-Methyl-1-propanol (isobutyl alcohol)	108	—	87	86
2-Propyn-1-ol (propargyl alcohol)	114	—	—	—
3-Pentanol	115	—	101	48
1-Butanol	118	—	64	61
2-Pentanol	119	—	62	—
3-Methyl-3-pentanol	123	—	96	43
2-Methoxyethanol	124	—	—	(113)†
2-Chloroethanol	129	—	95	51
3-Methyl-1-butanol (isoamyl alcohol)	130	—	70	31
4-Methyl-2-pentanol	132	—	65	143
1-Pentanol	138	—	46	46
Cyclopentanol	140	—	115	132
2-Ethyl-1-butanol	146	—	51	—
2,2,2-Trichloroethanol	151	—	142	87
1-Hexanol	157	—	58	42
Cyclohexanol	160	—	113	82
(2-Furyl)-methanol (furfuryl alcohol)	170	—	80	45
1-Heptanol	176	—	47	60
2-Octanol	179	—	32	114
1-Octanol	195	—	61	74
3,7-Dimethyl-1,6-octadien-3-ol (linalool)	196	—	—	66
Benzyl alcohol	204	—	113	77
1-Phenylethanol	204	20	92	95
2-Phenylethanol	219	—	108	78
1-Decanol	231	7	57	59
3-Phenylpropanol	236	—	45	92
1-Dodecanol (lauryl alcohol)	—	24	60	74
3-Phenyl-2-propen-1-ol (cinnamyl alcohol)	250	34	121	90
1-Tetradecanol (myristyl alcohol)	—	39	67	74
(−)-Menthol	212	41	158	111
1-Hexadecanol (cetyl alcohol)	—	49	66	73

ALCOHOLS *(Cont.)*

Compound	BP	MP	3,5-Dinitrobenzoate*	Phenylurethane*
1-Octadecanol (stearyl alcohol)	—	59	77	79
Diphenylmethanol (benzhydrol)	288	68	141	139
Benzoin	—	133	—	165
Cholesterol	—	147	—	168
(+)-Borneol	—	208	154	138

*See Appendix 2, "Procedures for Preparing Derivatives."

†α-Naphthylurethane.

ESTERS

Compound	BP	MP	Compound	BP	MP
Methyl formate	34	—	Ethyl lactate	154	—
Ethyl formate	54	—	Ethyl hexanoate (ethyl caproate)	168	—
Vinyl acetate	72	—	Methyl acetoacetate	170	—
Ethyl acetate	77	—	Dimethyl malonate	180	—
Methyl propanoate (methyl propionate)	80	—	Ethyl acetoacetate	181	—
Methyl acrylate	80	—	Diethyl oxalate	185	—
2-Propyl acetate (isopropyl acetate)	85	—	Methyl benzoate	199	—
Ethyl chloroformate	93	—	Ethyl octanoate (ethyl caprylate)	207	—
Methyl 2-methylpropanoate (methyl isobutyrate)	93	—	Ethyl cyanoacetate	210	—
			Ethyl benzoate	212	—
2-Propenyl acetate (isopropenyl acetate)	94	—	Diethyl succinate	217	—
2-(2-Methylpropyl acetate (*t*-butyl acetate)	98	—	Methyl phenylacetate	218	—
			Diethyl fumarate	219	—
Ethyl acrylate	99	—	Methyl salicylate	222	—
Ethyl propanoate (ethyl propionate)	99	—	Diethyl maleate	225	—
Methyl methacrylate	100	—	Ethyl phenylacetate	229	—
Methyl trimethylacetate (methyl pivalate)	101	—	Ethyl salicylate	234	—
Propyl acetate	102	—	Dimethyl suberate	268	—
Methyl butanoate (methyl butyrate)	102	—	Ethyl cinnamate	271	—
2-Butyl acetate (*sec*-butyl acetate)	111	—	Diethyl phthalate	298	—
Methyl 3-methylbutanoate (methyl isovalerate)	117	—	Dibutyl phthalate	340	—
			Methyl cinnamate	—	36
Ethyl butanoate (ethyl butyrate)	120	—	Phenyl salicylate	—	42
Butyl acetate	127	—	Methyl *p*-chlorobenzoate	—	44
Methyl pentanoate (methyl valerate)	128	—	Ethyl *p*-nitrobenzoate	—	56
Methyl chloroacetate	130	—	Phenylbenzoate	314	69

ESTERS *(Cont.)*

Compound	BP	MP	Compound	BP	MP
Ethyl 3-methylbutanoate (ethyl isovalerate)	132	—	Methyl *m*-nitrobenzoate	—	78
Pentyl acetate (*n*-amyl acetate)	142	—	Methyl *p*-bromobenzoate	—	81
3-Methylbutyl acetate (isoamyl acetate)	142	—	Ethyl *p*-aminobenzoate	—	90
Ethyl chloroacetate	143	—	Methyl *p*-nitrobenzoate	—	94

A P P E N D I X   2

Procedures for Preparing Derivatives

Caution: Some of the chemicals used in preparing derivatives are suspected carcinogens. The list of suspected carcinogens on page 21 should be consulted before beginning any of these procedures. Care should be exercised in handling these substances.

ALDEHYDES AND KETONES

Semicarbazones.   Place 0.5 mL of a 2*M* stock solution of semicarbazide hydrochloride (or 0.5 mL of a solution prepared by dissolving 1.11 g of semicarbazide hydrochloride [MW = 111.5] in 5 mL of water) in a small test tube. Add an estimated 1 millimole (mmol) of the unknown compound to the test tube. If the unknown does not dissolve in the solution, or if the solution becomes cloudy, add enough methanol (maximum of 10 mL) to dissolve the solid and produce a clear solution. If a solid or cloudiness remains after adding 10 mL of methanol, do not add any more methanol and continue this procedure with the solid present. Using a Pasteur pipet, add 10 drops of pyridine and heat the mixture in a hot-water bath (about 60°C) for about 10–15 minutes. By that time, the product should have begun to crystallize. Collect the product by vacuum filtration. The product can be recrystallized from ethanol if necessary.

Semicarbazones (Alternative Method).   Dissolve 0.25 g of semicarbazide hydrochloride and 0.38 g of sodium acetate in 1.3 mL of water. Then dissolve 0.25 g of the unknown in 2.5 mL of ethanol. Mix the two solutions together in a 25-mL Erlenmeyer flask and heat the mixture to boiling for about 5 minutes. After heating, place the reaction flask in a beaker of ice and scratch the sides of the flask with a glass rod to induce crystallization of the derivative. Collect the derivative by vacuum filtration and recrystallize it from ethanol.

2,4-Dinitrophenylhydrazones.   Place 10 mL of a solution of 2,4-dinitrophenyl-hydrazine (prepared as described for the classification test in Experiment 57D) in a test tube and add an estimated 1 mmol of the unknown compound. If the unknown is a solid, it should be dissolved in the minimum amount of 95% ethanol or 1,2-dimethoxyethane before it is added. If crystallization is not immediate, gently warm the solution for a minute in a hot-water bath (90°C) and then set it aside to crystallize. Collect the product by vacuum filtration.

CARBOXYLIC ACIDS

Working in a hood, place 0.5 g of the acid and 2 mL of thionyl chloride into a small round-bottom flask. Add a magnetic stir bar and attach a condenser and a drying tube packed with calcium chloride. While stirring, heat the reaction mixture almost to boiling with a heating mantle with a magnetic stirrer or a hotplate/stirrer. *If the mixture does not turn color,* heat the solution for 30 minutes. Allow the mixture to cool to room temperature. *If the mixture begins to turn color,* remove the heat source and stir the mixture for 45 minutes at room temperature. Use the mixture for one of the following three procedures.

Amides.   Working in a hood, pour the reaction mixture into a beaker containing 10 mL of ice cold concentrated ammonium hydroxide and stir it vigorously. When the reaction is complete, collect the product by vacuum filtration and recrystallize it from water or from water–ethanol, using the mixed-solvents method (Technique 5, Section 5.9).

Anilides.   Dissolve 1 g of aniline in 25 mL of methylene chloride in a 125-mL Erlenmeyer flask. Using a Pasteur pipet, carefully add the reaction mixture to this solution. Warm the mixture for an additional 5 minutes on a hotplate, unless a significant color change occurs. *If a color change occurs,* discontinue heating, add a magnetic stirring bar, and stir the mixture for 20 minutes at room temperature. Then transfer the methylene chloride solution to a small separatory funnel and wash it sequentially with 5 mL of water, 5 mL of 5% hydrochloric acid, 5 mL of 5% sodium hydroxide, and a second 5-mL portion of water (the methylene chloride solution should be the bottom layer). Dry the methylene chloride layer over a small amount of anhydrous sodium sulfate. Decant the methylene chloride layer away from the drying agent into a small flask and evaporate the methylene chloride on a warm hotplate in the hood. Use a stream of air or nitrogen to speed up the evaporation. Recrystallize the product from water or from ethanol–water, using the mixed-solvents method (Technique 5, Section 5.9).

p-Toluidides.   Use the same procedure as that described for the anilide but substitute *p*-toluidine for aniline.

PHENOLS

α-Naphthylurethanes.   Follow the procedure given later for preparing phenylurethanes from alcohols but substitute α-naphthylisocyanate for phenylisocyanate.

Bromo Derivatives. First, if a stock brominating solution is not available, prepare one by dissolving 0.75 g of potassium bromide in 5 mL of water and adding 0.5 g of bromine. Dissolve 0.1 g of the phenol in 1 mL of methanol or 1,2-dimethoxyethane; then add 1 mL of water. Add 1 mL of the brominating mixture to the phenol solution and swirl the mixture vigorously. Then, continue adding the brominating solution dropwise while swirling, until the color of the bromine reagent persists. Finally, add 3–5 mL of water and shake the mixture vigorously. Collect the precipitated product by vacuum filtration and wash it well with water. Recrystallize the derivative from methanol–water, using the mixed-solvents method (Technique 5, Section 5.9).

AMINES

Acetamides. Place an estimated 1 mmol of the amine and 0.5 mL of acetic anhydride in a small Erlenmeyer flask. Heat the mixture for about 5 minutes; then add 5 mL of water and stir the solution vigorously to precipitate the product and hydrolyze the excess acetic anhydride. If the product does not crystallize, it may be necessary to scratch the walls of the flask with a glass rod. Collect the crystals by vacuum filtration and wash them with several portions of cold 5% hydrochloric acid. Recrystallize the derivative from methanol–water, using the mixed-solvents method (Technique 5, Section 5.9).

Aromatic amines, or those amines that are not very basic, may require pyridine (2 mL) as a solvent and a catalyst for the reaction. If pyridine is used, a longer period of heating is required (up to 1 hour), and the reaction should be carried out in an apparatus equipped with a reflux condenser. After reflux, the reaction mixture must be extracted with 5–10 mL of 5% sulfuric acid to remove the pyridine.

Benzamides. Using a centrifuge tube suspend an estimated 1 mmol of the amine in 1 mL of 10% sodium hydroxide solution and add 0.5 g of benzoyl chloride. Cap the tube and shake the mixture vigorously for about 10 minutes. After shaking, add enough dilute hydrochloric acid to bring the pH of the solution to pH 7 or 8. Collect the precipitate by vacuum filtration, wash it thoroughly with cold water, and recrystallize it from ethanol–water, using the mixed-solvents method (Technique 5, Section 5.9).

Benzamides (Alternative Method). Dissolve 0.5 g of the amine in a solution of 2.5 mL of pyridine and 5 mL of toluene. Add 0.5 mL of benzoyl chloride to the solution and heat the mixture under reflux about 30 minutes. Pour the cooled reaction mixture into 50 mL of water and stir the mixture vigorously to hydrolyze the excess benzoyl chloride. Separate the toluene layer and wash it, first with 3 mL of water and then with 3 mL of 5% sodium carbonate. Dry the toluene over anhydrous sodium sulfate, decant the toluene into a small flask, and remove the toluene by evaporation on a hotplate in the hood. Use a stream of air or nitrogen to speed up the evaporation. Recrystallize the benzamide from ethanol or ethanol–water, using the mixed-solvents method (Technique 5, Section 5.9).

Picrates. Dissolve 0.2 g of the unknown in about 5 mL of ethanol and add 5 mL of a saturated solution of picric acid in ethanol. Heat the solution to boiling and then allow it to cool slowly. Collect the product by vacuum filtration and rinse it with a small amount of cold ethanol.

Methiodides.   Mix equal-volume quantities of the amine and methyl iodide in a test tube (about 0.25 mL is sufficient) and allow the mixture to stand for several minutes. Then heat the mixture gently under reflux for about 5 minutes. The methiodide should crystallize on cooling. If it does not, you can induce crystallization by scratching the walls of the test tube with a glass rod. Collect the product by vacuum filtration and recrystallize it from ethanol or ethyl acetate.

ALCOHOLS

3,5-Dinitrobenzoates.   *Liquid Alcohols.* Dissolve 0.25 g of 3,5-dinitrobenzoyl chloride[1] in 0.25 mL of the alcohol and heat the mixture for about 5 minutes. Allow the mixture to cool and add 1.5 mL of a 5% sodium carbonate solution and 1 mL of water. Stir the mixture vigorously and crush any solid that forms. Collect the product by vacuum filtration and wash it with cold water. Recrystallize the derivative from ethanol–water, using the mixed-solvent method (Technique 5, Section 5.9).

Solid Alcohols. Dissolve 0.25 g of the alcohol in 1.5 mL of dry pyridine and add 0.25 g of 3,5-dinitrobenzoyl chloride.[2] Heat the mixture under reflux for 15 minutes. Pour the cooled reaction mixture into a cold mixture of 2.5 mL of 5% sodium carbonate and 2.5 mL of water. Keep the solution cooled in an ice bath until the product crystallizes and stir it vigorously during the entire period. Collect the product by vacuum filtration, wash it with cold water, and recrystallize it from ethanol–water, using the mixed-solvents method (Technique 5, Section 5.9).

Phenylurethanes.   Place 0.25 g of the *anhydrous* alcohol in a dry test tube and add 0.25 mL of phenylisocyanate (α-naphthylisocyanate for a phenol). If the compound is a phenol, add one drop of pyridine to catalyze the reaction. If the reaction is not spontaneous, heat the mixture in a hot-water bath (90°C) for 5–10 minutes. Cool the test tube in a beaker of ice and scratch the tube with a glass rod to induce crystallization. Decant the liquid from the solid product or, if necessary, collect the product by vacuum filtration. Dissolve the product in 2.5–3 mL of hot ligroin or hexane and filter the mixture by gravity (preheat funnel) to remove any unwanted and insoluble diphenylurea present. Cool the filtrate to induce crystallization of the urethane. Collect the product by vacuum filtration.

ESTERS

We recommend that esters be characterized by spectroscopic methods whenever possible. A derivative of the alcohol part of an ester can be prepared with the procedure described below. For other derivatives, consult a comprehensive textbook. Several are listed in Experiment 57 (p. 485).

---

[1]This is an acid chloride and undergoes hydrolysis readily. The purity of this reagent should be checked before its use by determining its melting point. When the carboxylic acid is present, the melting point will be high.
[2]See Footnote 1.

3,5-Dinitrobenzoates.   Place 2.0 mL of the ester and 1.5 g of 3,5-dinitrobenzoic acid in a small round-bottom flask. Add 4 drops of concentrated sulfuric acid and a boiling stone to the flask and attach a condenser. If the boiling point of the ester is below 150°C, heat under reflux while stirring for 30–45 minutes. If the boiling point of the ester is above 150°C, heat the mixture at about 150°C for 30–45 minutes. Cool the mixture and transfer it to a small separatory funnel. Add 20 mL of ether. Extract the ether layer two times with 10 mL of 5% aqueous sodium carbonate (save the ether layer). Wash the organic layer with 10 mL of water and dry the ether solution over magnesium sulfate. Evaporate the ether in a hot-water bath in the hood. Use a stream of air or nitrogen to speed the evaporation. Dissolve the residue, usually an oil, in 4 mL of boiling ethanol and add water dropwise until the mixture becomes cloudy. Cool the solution to induce crystallization of the derivative.

A P P E N D I X   3

Infrared Spectroscopy

Almost any compound having covalent bonds, whether organic or inorganic, will be found to absorb frequencies of electromagnetic radiation in the infrared region of the spectrum. The infrared region of the electromagnetic spectrum lies at wavelengths longer than those associated with visible light, which includes wavelengths from approximately 400 to 800 nm (1 nm = 10^{-9} m), but at wavelengths shorter than those associated with radio waves, which have wavelengths longer than 1 cm. For chemical purposes, we are interested in the *vibrational* portion of the infrared region. This portion is defined as that which includes radiations with wavelengths (λ) between 2.5 and 15 μm (1 μm = 10^{-6} m). The relation of the infrared region to others included in the electromagnetic spectrum is illustrated in Figure IR.1.

As with other types of energy absorption, molecules are excited to a higher energy state when they absorb infrared radiation. The absorption of infrared radiation is, like other

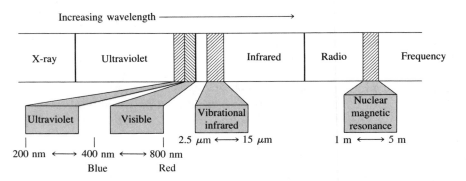

Figure IR.1 Portion of electromagnetic spectrum, showing relation of vibrational infrared to other types of radiation.

absorption processes, a quantized process. Only selected frequencies (energies) of infrared radiation are absorbed by a molecule. The absorption of infrared radiation corresponds to energy changes on the order of 8–40 kJ/mole (2–10 kcal/mole). Radiation in this energy range corresponds to the range encompassing the stretching and bending vibrational frequencies of the bonds in most covalent molecules. In the absorption process, those frequencies of infrared radiation that match the natural vibrational frequencies of the molecule in question are absorbed, and the energy absorbed increases the *amplitude* of the vibrational motions of the bonds in the molecule.

Most chemists refer to the radiation in the vibrational infrared region of the electromagnetic spectrum by units called **wavenumbers** *(v)*. Wavenumbers are expressed in reciprocal centimeters (cm^{-1}) and are easily computed by taking the reciprocal of the wavelength (λ) expressed in centimeters. This unit has the advantage, for those performing calculations, of being directly proportional to energy. Thus, the vibrational infrared region of the spectrum extends from about 4000 to 650 cm^{-1} (or wavenumbers).

Wavelengths (μm) and wavenumbers (cm^{-1}) can be interconverted by the following relationships:

$$cm^{-1} = \frac{1}{(\mu m)} \times 10,000$$

$$\mu m = \frac{1}{(cm^{-1})} \times 10,000$$

IR.1 USES OF THE INFRARED SPECTRUM

Because every type of bond has a different natural frequency of vibration, and because the same type of bond in two different compounds is in a slightly different environment, no two molecules of different structure have exactly the same infrared absorption pattern, or **infrared spectrum.** Although some of the frequencies absorbed in the two cases might be the same, in no case of two different molecules will their infrared spectra (the patterns of absorption) be identical. Thus, the infrared spectrum can be used for molecules much as a fingerprint can be used for people. By comparing the infrared spectra of two substances thought to be identical, you can establish whether or not they are in fact identical. If the infrared spectra of two substances coincide peak for peak (absorption for absorption), in most cases, the substances are identical.

A second and more important use of the infrared spectrum is that it gives structural information about a molecule. The absorptions of each type of bond (N—H, C—H, O—H, C—X, C=O, C—O, C—C, C=C, C≡C, C≡N, and so on) are regularly found only in certain small portions of the vibrational infrared region. A small range of absorption can be defined for each type of bond. Outside this range, absorptions will normally be due to some other type of bond. Thus, for instance, any absorption in the range 3000 ± 150 cm^{-1} will almost always be due to the presence of a CH bond in the molecule; an absorption in the range 1700 ± 100 cm^{-1} will normally be due to the presence of a C=O bond (carbonyl group) in the molecule. The same type of range applies to each type of bond. The

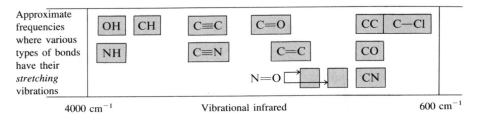

Figure IR.2 Approximate regions in which various common types of bonds absorb. (Bending and twisting and other types of bond vibration have been omitted for clarity).

way these are spread out over the vibrational infrared is illustrated schematically in Figure IR.2. It is a good idea to remember this general scheme for future convenience.

IR.2 MODES OF VIBRATION

The simplest types, or **modes,** of vibrational motion in a molecule that are **infrared-active,** that is, give rise to absorptions, are the stretching and bending modes.

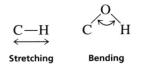

Other, more complex types of stretching and bending are also active, however. To introduce several words of terminology, the normal modes of vibration for a methylene group are shown on page 895.

In any group of three or more atoms—at least two of which are identical—there are *two* modes of stretching or bending: the symmetric mode and the asymmetric mode. Examples of such groups are $-CH_3$, $-CH_2-$, $-NO_2$, $-NH_2$, and anhydrides $(CO)_2O$. For the anhydride, owing to asymmetric and symmetric modes of stretch, this functional group gives *two* absorptions in the $C=O$ region. A similar phenomenon is seen for amino groups, where primary amines usually have *two* absorptions in the NH stretch region, whereas secondary amines R_2NH have only one absorption peak. Amides show similar bands. There are two strong $N=O$ stretch peaks for a nitro group, which are caused by asymmetric and symmetric stretching modes.

IR.3 WHAT TO LOOK FOR IN EXAMINING INFRARED SPECTRA

The instrument that determines the absorption spectrum for a compound is called an **infrared spectrophotometer.** The spectrophotometer determines the relative strengths and positions of all the absorptions in the infrared region and plots this information on a piece of calibrated chart paper. This plot of absorption intensity versus wavenumber or wavelength is referred to as the **infrared spectrum** of the compound. A typical infrared spectrum, that of methyl isopropyl ketone, is shown in Figure IR.3.

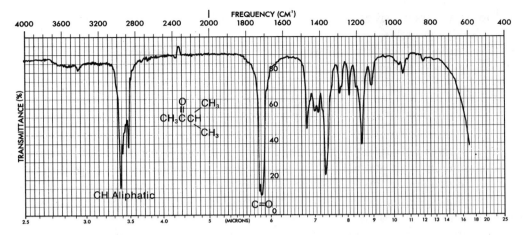

Figure IR.3 Infrared spectrum of methyl isopropyl ketone (neat liquid, salt plates.)

The strong absorption in the middle of the spectrum corresponds to $C{=}O$, the carbonyl group. Note that the $C{=}O$ peak is quite intense. In addition to the characteristic position of absorption, the **shape** and **intensity** of this peak are also unique to the $C{=}O$ bond. This is true for almost every type of absorption peak; both shape and intensity characteristics can be described, and these characteristics often make it possible to distinguish the peak in a confusing situation. For instance, to some extent both $C{=}O$ and $C{=}C$ bonds absorb in the same region of the infrared spectrum:

$$C{=}O \quad 1850\text{--}1630 \text{ cm}^{-1}$$
$$C{=}C \quad 1680\text{--}1620 \text{ cm}^{-1}$$

However, the $C{=}O$ bond is a strong absorber, whereas the $C{=}C$ bond generally absorbs only weakly. Hence, a trained observer would not normally interpret a strong peak at 1670

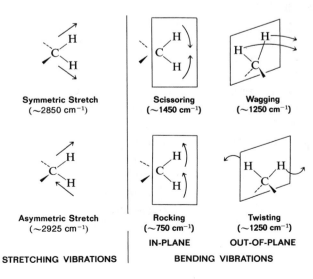

cm^{-1} to be a carbon–carbon double bond nor a weak absorption at this frequency to be due to a carbonyl group.

The shape of a peak often gives a clue to its identity as well. Thus, although the NH and OH regions of the infrared overlap,

$$\text{OH} \quad 3650\text{--}3200 \text{ cm}^{-1}$$
$$\text{NH} \quad 3500\text{--}3300 \text{ cm}^{-1}$$

NH usually gives a **sharp** absorption peak (absorbs a very narrow range of frequencies), and OH, when it is in the NH region, usually gives a **broad** absorption peak. Primary amines give *two* absorptions in this region, whereas alcohols give only one.

Therefore, while you are studying the sample spectra in the pages that follow, you should also notice shapes and intensities. They are as important as the frequency at which an absorption occurs, and you must train your eye to recognize these features. Often, in the literature of organic chemistry, you will find absorptions referred to as strong (s), medium (m), weak (w), broad, or sharp. The author is trying to convey some idea of what the peak looks like without actually drawing the spectrum. Although the intensity of an absorption often provides useful information about the identity of a peak, you should be aware that the relative intensities of all the peaks in the spectrum are dependent on the amount of sample that is used and the sensitivity setting of the instrument. Therefore, the *actual* intensity of a particular peak may vary from spectrum to spectrum, and you must pay attention to *relative* intensities.

IR.4 CORRELATION CHARTS AND TABLES

To extract structural information from infrared spectra, you must know the frequencies or wavelengths at which various functional groups absorb. Infrared **correlation tables** present as much information as is known about where the various functional groups absorb. The books listed at the end of this appendix present extensive lists of correlation tables (see page 911). Sometimes the absorption information is given in a chart, called a **correlation chart.** A simplified correlation table is given in Table IR.1.

Although you may think assimilating the mass of data in Table IR.1 will be difficult, it is not if you make a modest start and then gradually increase your familiarity with the data. An ability to interpret the fine details of an infrared spectrum will follow. This is most easily accomplished by first establishing the broad visual patterns of Figure IR.2 firmly in mind. Then, as a second step, a "typical absorption value" can be memorized for each of the functional groups in this pattern. This value will be a single number that can be used as a pivot value for the memory. For instance, start with a simple aliphatic ketone as a model for all typical carbonyl compounds. The typical aliphatic ketone has a carbonyl absorption of $1715 \pm 10 \text{ cm}^{-1}$. Without worrying about the variation, memorize 1715 cm^{-1} as the base value for carbonyl absorption. Then, learn the extent of the carbonyl range and the visual pattern of how the different kinds of carbonyl groups are arranged throughout this region. See, for instance, Figure IR.15, which gives typical val-

TABLE IR.1   A Simplified Correlation Table

Type of Vibration			Frequency (cm^{-1})	Intensity
C—H	Alkanes	(stretch)	3000–2850	s
	—CH$_3$	(bend)	1450 and 1375	m
	—CH$_2$—	(bend)	1465	m
	Alkenes	(stretch)	3100–3000	m
		(bend)	1700–1000	s
	Aromatics	(stretch)	3150–3050	s
		(out-of-plane bend)	1000–700	s
	Alkyne	(stretch)	ca. 3300	s
	Aldehyde		2900–2800	w
			2800–2700	w
C—C	Alkane	Not interpretatively useful		
C=C	Alkene		1680–1600	m–w
	Aromatic		1600–1400	m–w
C≡C	Alkyne		2250–2100	m–w
C=O	Aldehyde		1740–1720	s
	Ketone (acyclic)		1725–1705	s
	Carboxylic acid		1725–1700	s
	Ester		1750–1730	s
	Amide		1700–1640	s
	Anhydride		ca. 1810	s
			ca. 1760	s
C—O	Alcohols, ethers, esters, carboxylic acids		1300–1000	s
O—H	Alcohol, phenols			
	Free		3650–3600	m
	H-Bonded		3400–3200	m
	Carboxylic acids		3300–2500	m
N—H	Primary and secondary amines		ca. 3500	m
C≡N	Nitriles		2260–2240	m
N=O	Nitro (R—NO$_2$)		1600–1500	s
			1400–1300	s
C—X	Fluoride		1400–1000	s
	Chloride		800–600	s
	Bromide, iodide		< 600	s

NOTE: s, strong; m, medium; w, weak.

TABLE IR.2 Base Values for Absorptions
of Bonds

OH	3600 cm^{-1}	C≡C	2150 cm^{-1}
NH	3500 cm^{-1}	C=O	1715 cm^{-1}
CH	3000 cm^{-1}	C=C	1650 cm^{-1}
C≡N	2250 cm^{-1}	C—O	1100 cm^{-1}

---

ues for carbonyl compounds. Also learn how factors like ring size (when the functional group is contained in a ring) and conjugation affect the base values (that is, in which direction the values are shifted). Learn the trends—always remembering the base value (1715 cm^{-1}). It might prove useful as a beginning to memorize the base values in Table IR.2 for this approach. Notice that there are only eight values.

IR.5 ANALYZING A SPECTRUM (OR WHAT YOU CAN TELL AT A GLANCE)

In trying to analyze the spectrum of an unknown, you should concentrate first on trying to establish the presence (or absence) of a few major functional groups. The most conspicuous peaks are C=O, C—H, N—H, C—O, C=C, C≡C, C≡N, and NO$_2$. If they are present, they give immediate structural information. Do not try to analyze in detail the CH absorptions near 3000 cm^{-1}; almost all compounds have these absorptions. Do not worry about subtleties of the exact type of environment in which the functional group is found. A checklist of the important gross features follows:

1. Is a carbonyl group present?
   The C=O group gives rise to a strong absorption in the region 1820–1600 cm^{-1}. The peak is often the strongest in the spectrum and of medium width. You can't miss it.

2. If C=O is present, check the following types. (If it is absent, go to 3.)

   | ACIDS | Is OH also present? |
   | | **Broad** absorption near 3300–2500 cm^{-1} (usually overlaps C—H). |

   | AMIDES | Is NH also present? |
   | | Medium absorption near 3500 cm^{-1}, sometimes a double peak, equivalent halves. |

   | ESTERS | Is C—O also present? |
   | | Medium intensity absorptions near 1300–1000 cm^{-1}. |

   | ANHYDRIDES | Have *two* C=O absorptions near 1810 and 1760 cm^{-1}. |

ALDEHYDES	Is aldehyde CH present?
	Two weak absorptions near 2850 and 2750 cm^{-1} on the right side of CH absorptions.
KETONES	The above five choices have been eliminated.

3. If C=O is absent

ALCOHOLS or PHENOLS	Check for OH.
	Broad absorption near 3600–3300 cm^{-1}.
	Confirm this by finding C—O near 1300–1000 cm^{-1}.
AMINES	Check for NH.
	Medium absorption(s) near 3500 cm^{-1}.
ETHERS	Check for C—O (and absence of OH) near 1300–1000 cm^{-1}.

4. Double Bonds or Aromatic Rings or Both

C=C is a **weak** absorption near 1650 cm^{-1}.

Medium to strong absorptions in the region 1650–1450 cm^{-1} often imply an aromatic ring.

Confirm the above by consulting the CH region.

Aromatic and vinyl CH occur to the left of 3000 cm^{-1} (aliphatic CH occurs to the right of this value).

5. Triple Bonds

C≡N is a medium, sharp absorption near 2250 cm^{-1}.

C≡C is a weak but sharp absorption near 2150 cm^{-1}.

Check also for acetylenic CH near 3300 cm^{-1}.

6. Nitro Groups

Two strong absorptions 1600–1500 cm^{-1} and 1390–1300 cm^{-1}.

7. Hydrocarbons

None of the above is found.

Main absorptions are in CH region near 3000 cm^{-1}.

Very simple spectrum, only other absorptions near 1450 cm^{-1} and 1375 cm^{-1}.

The beginning student should resist the idea of trying to assign or interpret *every* peak in the spectrum. You simply will not be able to do this. Concentrate first on learning the principal peaks and recognizing their presence or absence. This is best done by carefully studying the illustrative spectra in the section that follows.

> **Note:** In describing the shifts of absorption peaks or their relative positions, we have used the phrases "to the left" and "to the right." This was done to simplify descriptions of peak positions. The meaning is clear, because all spectra are conventionally presented left to right from 4000 to 600 cm^{-1}.

IR.6 SURVEY OF THE IMPORTANT FUNCTIONAL GROUPS

Alkanes

Spectrum is usually simple with few peaks.

C—H   Stretch occurs around 3000 cm^{-1}.

(a) In alkanes (except strained ring compounds), absorption always occurs to the right of 3000 cm^{-1}.

(b) If a compound has vinylic, aromatic, acetylenic, or cyclopropyl hydrogens, the CH absorption is to the left of 3000 cm^{-1}.

CH$_2$ Methylene groups have a characteristic absorption at approximately 1450 cm^{-1}.

CH$_3$ Methyl groups have a characteristic absorption at approximately 1375 cm^{-1}.

C—C   Stretch—not interpretatively useful—has many peaks.

The spectrum of decane is shown in Figure IR.4.

Alkenes

=C—H   Stretch occurs to the left of 3000 cm^{-1}.

=C—H   Out-of-plane ("oop") bending at 1000–650 cm^{-1}.

The C—H out-of-plane absorptions often allow you to determine the type of substitution pattern on the double bond, according to the number of absorptions and their positions. The correlation chart in Figure IR.5 shows the positions of these bands.

C=C   Stretch 1675–1600 cm^{-1}, often weak.

Conjugation moves C=C stretch to the right.

Symmetrically substituted bonds, as in 2,3-dimethyl-2-butene, do not absorb in the infrared region (no dipole change). Highly substituted double bonds are often vanishingly weak in absorption.

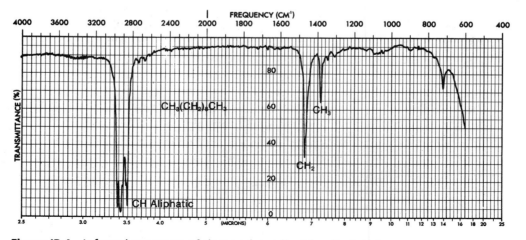

Figure IR.4   Infrared spectrum of decane (neat liquid, salt plates).

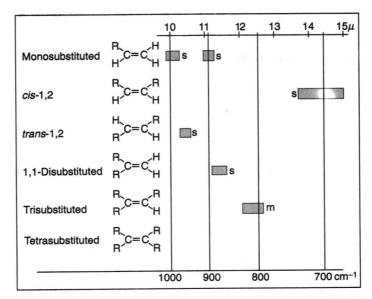

Figure IR.5 The C—H out-of-plane bending vibrations for substituted alkenes.

The spectrum of styrene is shown in Figure IR.6. The spectrum of 4-methylcyclohexene in shown in Experiment 27.

Aromatic Rings

=C—H   Stretch is always to the left of 3000 cm^{-1}.

=C—H   Out-of-plane (oop) bending at 900 to 690 cm^{-1}.

The CH out-of-plane absorptions often allow you to determine the type of ring substitution by their numbers, intensities, and positions. The correlation chart in Figure IR.7A indicates the positions of these bands.

The patterns are generally reliable—most particularly reliable for rings with alkyl substituents, least for polar substituents.

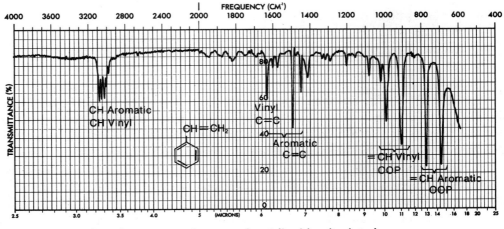

Figure IR.6 Infrared spectrum of styrene (neat liquid, salt plates).

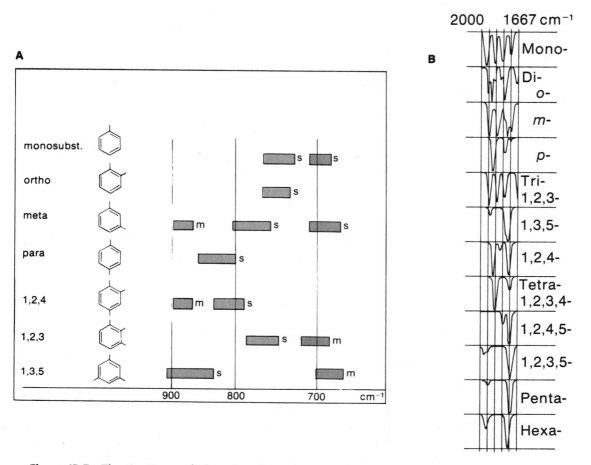

Figure IR.7 The C—H out-of-plane bending vibrations for substituted benzenoid compounds. (B) The 2000–1667 cm^{-1} region for substituted benzenoid compounds. (From John R. Dyer, *Applications of Absorption Spectroscopy of Organic Compounds.* Englewood Cliffs, NJ: Prentice-Hall, 1965.)

Ring Absorptions (C=C). There are often four sharp absorptions that occur in pairs at 1600 and 1450 cm^{-1} and are characteristic of an aromatic ring. See, for example, the spectra of anisole (Fig. IR.11), benzonitrile (Fig. IR.14), and methyl benzoate (Fig. IR.18).

There are many weak combination and overtone absorptions that appear between 2000 and 1667 cm^{-1}. The relative shapes and numbers of these peaks can be used to tell whether an aromatic ring is monosubstituted or di-, tri-, tetra-, penta-, or hexasubstituted. Positional isomers can also be distinguished. Because the absorptions are weak, these bands are best observed by using neat liquids or concentrated solutions. If the compound has a high-frequency carbonyl group, this absorption overlaps the weak overtone bands, so that no useful information can be obtained from analyzing this region. The various patterns that are obtained in this region are shown in Figure IR.7B.

The spectra of styrene and *o*-dichlorobenzene are shown in Figures IR.6 and IR.8.

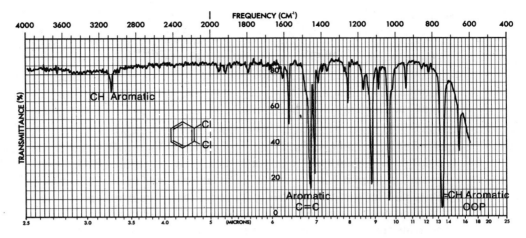

Figure IR.8   Infrared spectrum of o-dichlorobenzene (neat liquid, salt plates).

Alkynes

≡C—H   Stretch is usually near 3300 cm^{-1}.

C≡C   Stretch is near 2150 cm^{-1}.

Conjugation moves C≡C stretch to the right.

Disubstituted or symmetrically substituted triple bonds give either no absorption or weak absorption.

The spectrum of propargyl alcohol is shown in Figure IR.9.

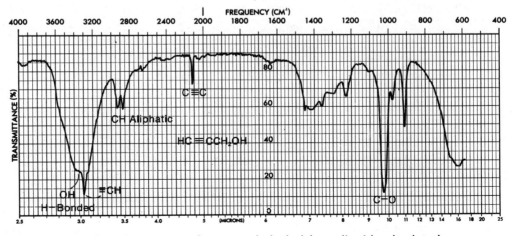

Figure IR.9   Infrared spectrum of propargyl alcohol (neat liquid, salt plates).

Alcohols and Phenols

O—H   Stretch is a sharp peak at 3650–3600 cm^{-1} if no hydrogen bonding takes place. (This is usually observed only in dilute solutions.)

If there is hydrogen bonding (usually in neat or concentrated solutions), the absorption is **broad** and occurs more to the right at 3500–3200 cm^{-1}, sometimes overlapping C—H stretch absorptions.

C—O   Stretch is usually in the range 1300–1000 cm^{-1}.

Phenols are like alcohols. The 2-naphthol shown in Figure IR.10 has some molecules hydrogen-bonded and some free. The spectrum of 4-methylcyclohexanol is given in Experiment 27. This alcohol, which was determined neat, would also have had a free OH spike to the left of its hydrogen-bonded band if it had been determined in dilute solution. The solution spectra of borneol and isoborneol are shown in Experiment 20.

Ethers

C—O   The most prominent band is due to C—O stretch at 1300–1000 cm^{-1}. Absence of C=O and O—H bands is required to be sure C—O stretch is not due to an alcohol or ester. Phenyl and vinyl ethers are found in the left portion of the range, aliphatic ethers in the right. (Conjugation with the oxygen moves the absorption to the left.)

The spectrum of anisole is shown in Figure IR.11.

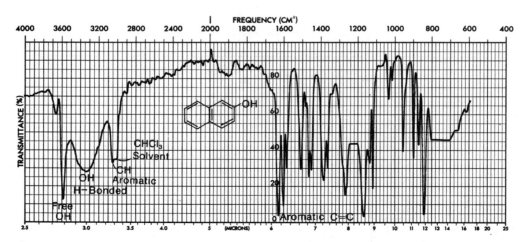

Figure IR.10 Infrared spectrum of 2-naphthol, showing both free and hydrogen-bonded OH (CHCl$_3$ solution).

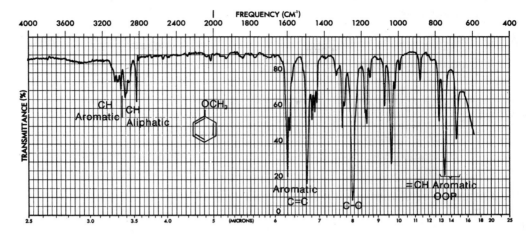

Figure IR.11   Infrared spectrum of anisole (neat liquid, salt plates).

Amines

N—H   Stretch occurs in the range of 3500–3300 cm^{-1}.
Primary amines have *two* bands typically 30 cm^{-1} apart.
Secondary amines have one band, often vanishingly weak.
Tertiary amines have no NH stretch.

C—N   Stretch is weak and occurs in the range of 1350–1000 cm^{-1}.

N—H   Scissoring mode occurs in the range of 1640–1560 cm^{-1} (broad).
An out-of-plane bending absorption can sometimes be observed at about 800 cm^{-1}.

The spectrum of *n*-butylamine is shown in Figure IR.12.

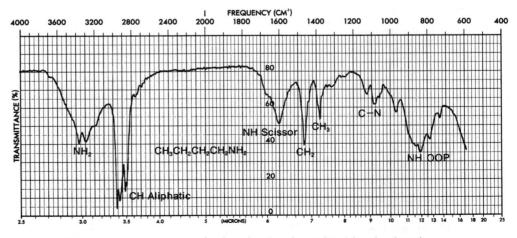

Figure IR.12   Infrared spectrum of *n*-butylamine (neat liquid, salt plates).

Nitro Compounds

N=O　Stretch is usually two strong bonds at 1600–1500 cm^{-1} and 1390–1300 cm^{-1}.

The spectrum of nitrobenzene is shown in Figure IR.13.

Nitriles

C≡N　Stretch is a sharp absorption near 2250 cm^{-1}.
　　　Conjugation with double bonds or aromatic rings moves the absorption to the right.

The spectrum of benzonitrile is shown in Figure IR.14.

Carbonyl Compounds

The carbonyl group is one of the most strongly absorbing groups in the infrared region of the spectrum. This is mainly due to its large dipole moment. It absorbs in a variety of compounds (aldehydes, ketones, acids, esters, amides, anhydrides, and so on) in the range of 1850–1650 cm^{-1}. In Figure IR.15 the normal values for the various types of carbonyl groups are compared. In the sections that follow, each type is examined separately.

Aldehydes

C=O　Stretch at approximately 1725 cm^{-1} is normal.
　　　Aldehydes *seldom* absorb to the left of this value.
　　　Conjugation moves the absorption to the right.

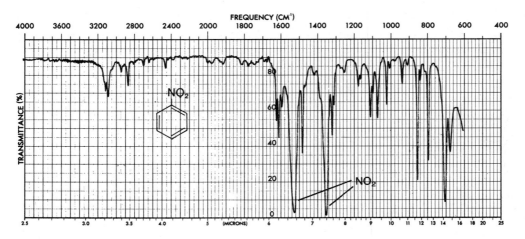

Figure IR.13 Infrared spectrum of nitrobenzene, neat.

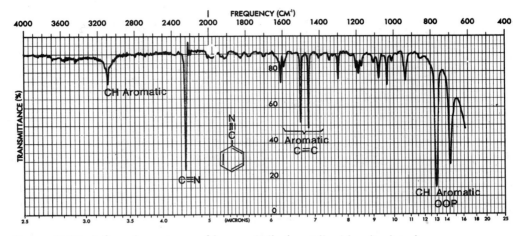

Figure IR.14   Infrared spectrum of benzonitrile (neat liquid, salt plates).

C—H   Stretch, aldehyde hydrogen (—CHO), consists of *weak* bands at about 2750 and 2850 cm^{-1}. Note that the CH stretch in alkyl chains does not usually extend this far to the right.

The spectrum of nonanal is shown in Figure IR.16. In addition, the spectrum of benzaldehyde is shown in Experiment 40.

Ketones

C=O   Stretch at approximately 1715 cm^{-1} is normal.
Conjugation moves the absorption to the right.
Ring strain moves the absorption to the left in cyclic ketones.

The spectra of methyl isopropyl ketone and mesityl oxide are shown in Figure IR.3 and IR.17. The spectrum of camphor is shown in Experiment 20.

1810	1760	1735	1725	1715	1710	1690	cm^{-1}
Anhydride (Band 1)		Esters		Ketones		Amides	
	Anhydride (Band 2)		Aldehydes		Carboxylic acids		

Figure IR.15   Normal values (±10 cm^{-1}) for various types of carbonyl groups.

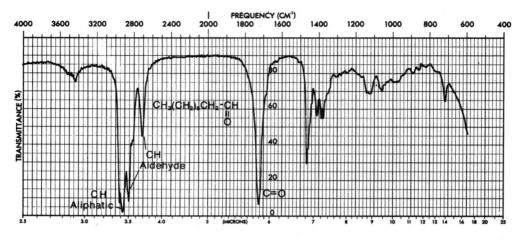

Figure IR.16 Infrared spectrum of nonanal (neat liquid, salt plates).

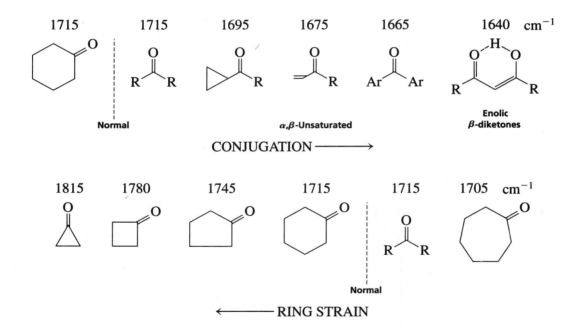

Acids

O—H   Stretch, usually **very broad** (strongly hydrogen-bonded) at 3300–2500 cm^{-1}, often interferes with C—H absorptions.

C=O   Stretch, broad, 1730–1700 cm^{-1}.
Conjugation moves the absorption to the right.

C—O   Stretch, in range of 1320–1210 cm^{-1}, strong.

The spectrum of benzoic acid is shown in Experiment 30B.

REFERENCES

Bellamy, L. J. *The Infra-red Spectra of Complex Molecules.* 3rd ed. New York: Methuen, 1975.

Colthup, N. B., Daly, L. H., and Wiberly, S. E. *Introduction to Infrared and Raman Spectroscopy.* 3rd ed. San Diego, CA: Academic Press, 1990.

Dyer, J. R. *Applications of Absorption Spectroscopy of Organic Compounds.* Englewood Cliffs, NJ: Prentice-Hall, 1965.

Lin-Vien, D., Colthup, N. B., Fateley, W. G., and Grasselli, J. G. *Infrared and Raman Characteristic Frequencies of Organic Molecules.* San Diego, CA: Academic Press, 1991.

Nakanishi, K., and Soloman, P. H. *Infrared Absorption Spectroscopy.* 2nd ed. San Francisco: Holden-Day, 1977.

Pavia, D. L., Lampman, G. M., and Kriz, G. S. *Introduction to Spectroscopy: A Guide for Students of Organic Chemistry.* 2nd ed. Philadelphia: W. B. Saunders, 1996.

Silverstein, R. M., Bassler, G. C., and Morrill, T. C. *Spectrometric Identification of Organic Compounds.* 5th ed. New York: Wiley, 1991.

APPENDIX 4 ■

Nuclear Magnetic Resonance Spectroscopy

NMR.1 THE RESONANCE PHENOMENON

Nuclear magnetic resonance (NMR) spectroscopy is an instrumental technique that allows the number, type, and relative positions of certain atoms in a molecule to be determined. This type of spectroscopy applies only to those atoms that have nuclear magnetic moments because of their nuclear spin properties. Although many atoms meet this requirement, hydrogen atoms ($_1^1H$) and carbon atoms ($_6^{13}C$) are of the greatest interest to the organic chemist. Atoms of the ordinary isotopes of carbon ($_6^{12}C$) and oxygen ($_8^{16}O$) do not have nuclear magnetic moments, and ordinary nitrogen atoms ($_7^{14}N$), although they do have magnetic moments, generally fail to show typical NMR behavior for other reasons. The same is true of the halogen atoms, except for fluorine ($_9^{19}F$), which does show active NMR behavior. The hydrogen nucleus is discussed in this appendix and carbon-13 NMR is described in Appendix 5.

Nuclei of NMR-active atoms placed in a magnetic field can be thought of as tiny bar magnets. In hydrogen, which has two allowed nuclear spin states ($+\frac{1}{2}$ and $-\frac{1}{2}$), either the nuclear magnets of individual atoms can be aligned with the magnetic field (spin $+\frac{1}{2}$), or they can be opposed to it (spin $-\frac{1}{2}$). A slight majority of the nuclei are aligned with the field, as this spin orientation constitutes a slightly lower-energy spin state. If radio-frequency waves of the appropriate energy are supplied, nuclei aligned with the field can absorb this radiation and reverse their direction of spin or become reoriented so that the nuclear magnet opposes the applied magnetic field (Fig. NMR.1).

The frequency of radiation required to induce spin conversion is a direct function of the strength of the applied magnetic field. When a spinning hydrogen nucleus is placed

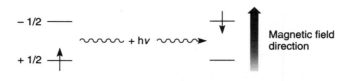

Figure NMR.1 The NMR absorption process.

in a magnetic field, the nucleus begins to precess with angular frequency ω, much like a child's toy top. This precessional motion is depicted in Figure NMR.2. The angular frequency of nuclear precession ω increases as the strength of the applied magnetic field is increased. The radiation that must be supplied to induce spin conversion in a hydrogen nucleus of spin $+\frac{1}{2}$ must have a frequency that just matches the angular precessional frequency ω. This is called the resonance condition, and spin conversion is said to be a resonance process.

For the average proton (hydrogen atom), if a magnetic field of approximately 14,000 gauss is applied, radiofrequency radiation of 60 MHz (60,000,000 cycles per second) is required to induce a spin transition. Fortunately, the magnetic-field strength required to induce the various protons in a molecule to absorb 60-MHz radiation varies from proton to proton within the molecule and is a sensitive function of the immediate *electronic* environment of each proton. The typical proton nuclear magnetic resonance spectrometer supplies a basic radiofrequency radiation of 60 MHz to the sample being measured and *increases* the strength of the applied magnetic field over a range of several parts per million from the basic field strength. As the field increases, various protons come into resonance (absorb 60-MHz energy), and a resonance signal is generated for each proton. An NMR spectrum is a plot of the strength of the magnetic field versus the intensity of the absorptions. A typical NMR spectrum is shown in Figure NMR.3.

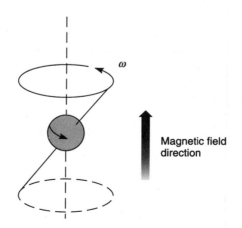

Figure NMR.2 Precessional motion of a spinning nucleus in an applied magnetic field.

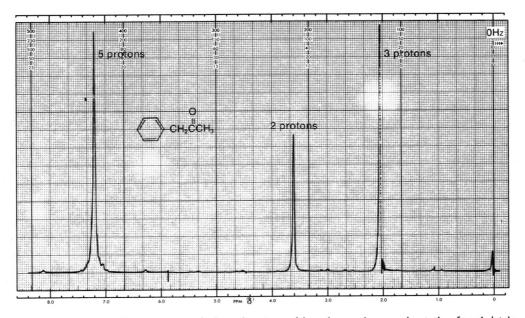

Figure NMR.3   NMR spectrum of phenylacetone (the absorption peak at the far right is caused by the added reference substance, TMS).

NMR.2 THE CHEMICAL SHIFT

The differences in the applied field strengths at which the various protons in a molecule absorb 60-MHz radiation are extremely small. The different absorption positions amount to a difference of only a few parts per million (ppm) in the magnetic field strength. Because it is experimentally difficult to measure the precise field strength at which each proton absorbs to less than one part in a million, a technique has been developed whereby the *difference* between two absorption positions is measured directly. To achieve this measurement, a standard reference substance is used, and the positions of the absorptions of all other protons are measured relative to the values for the reference substance. The reference substance that has been universally accepted is tetramethylsilane $(CH_3)_4Si$, which is also called TMS. The proton resonances in this molecule appear at a higher field strength than the proton resonances in most all other molecules do, and all the protons of TMS have resonance at the same field strength.

To give the position of absorption of a proton a quantitative measurement, a parameter called the **chemical shift** (δ) has been defined. One δ unit corresponds to a 1-ppm change in the magnetic field strength. To determine the chemical shift value for the various protons in a molecule, the operator determines an NMR spectrum of the molecule with a small quantity of **TMS** added directly to the sample. That is, both spectra are determined *simultaneously.* The TMS absorption is adjusted to correspond to the $\delta = 0$ ppm position on the recording chart, which is calibrated in δ units, and the δ values of the absorption peaks for all other protons can be read directly from the chart.

Because the NMR spectrometer increases the magnetic field as the pen moves from left to right on the chart, the TMS absorption appears at the extreme right edge of the spectrum ($\delta = 0$ ppm) or at the *upfield* end of the spectrum. The chart is calibrated in δ units (or ppm), and most other protons absorb at a lower field strength (or *downfield*) from TMS.

Because the frequency at which a proton precesses, and hence the frequency at which it absorbs radiation, is directly proportional to the strength of the applied magnetic field, a second method of measuring an NMR spectrum is possible. You could hold the magnetic field strength constant and vary the frequency of the radiofrequency radiation supplied. Thus, a given proton could be induced to absorb *either* by increasing the field strength, as described earlier, or alternatively, by decreasing the frequency of the radiofrequency oscillator. A 1-ppm decrease in the frequency of the oscillator would have the same effect as a 1-ppm increase in the magnetic field strength. For reasons of instrumental design, it is simpler to vary the strength of the magnetic field than to vary the frequency of the oscillator. Most instruments operate on the former principle. Nevertheless, the recording chart is calibrated not only in δ units but in Hertz (Hz) as well (1 ppm = 60 Hz when the frequency is 60 MHz), and the chemical shift is customarily defined and computed using Hertz rather than gauss:

$$\delta = \text{Chemical shift} = \frac{\text{Observed shift from TMS (in Hz)}}{60 \text{ MHz}} = \frac{\text{Hz}}{\text{MHz}} = \text{ppm}$$

Although the equation defines the chemical shift for a spectrometer operating at 14,100 gauss and 60 MHz, the chemical shift value that is calculated is *independent* of the field strength. For instance, at 23,500 gauss the oscillator frequency would have to be 100 MHz. Although the observed shifts from TMS (in Hertz) would be larger at this field strength, the divisor of the equation would be 100 MHz, instead of 60 MHz, and δ would turn out to be identical under either set of conditions.

NMR.3 CHEMICAL EQUIVALENCE—INTEGRALS

All the protons in a molecule that are in chemically identical environments often exhibit the same chemical shift. Thus, all the protons in tetramethylsilane (TMS) or all the protons in benzene, cyclopentane, or acetone have their own respective resonance values all at the same δ value. Each compound gives rise to a single absorption peak in its NMR spectrum. The protons are said to be **chemically equivalent.** On the other hand, molecules that have sets of protons that are chemically distinct from one another may give rise to an absorption peak from each set.

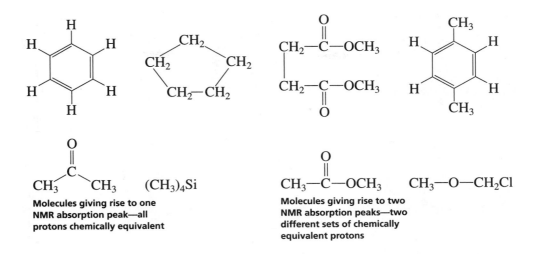

Molecules giving rise to one NMR absorption peak—all protons chemically equivalent

Molecules giving rise to two NMR absorption peaks—two different sets of chemically equivalent protons

The NMR spectrum given in Figure NMR.3 is that of phenylacetone, a compound having *three* chemically distinct types of protons:

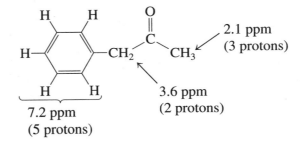

2.1 ppm (3 protons)

3.6 ppm (2 protons)

7.2 ppm (5 protons)

You can immediately see that the NMR spectrum furnishes valuable information on this basis alone. In fact, the NMR spectrum can not only distinguish how many types of protons a molecule has but also can reveal *how many* of each type are contained within the molecule.

In the NMR spectrum, the area under each peak is proportional to the number of hydrogens generating that peak. Hence, in the case of phenylacetone, the area ratio of the three peaks is 5:2:3, the same as the ratio of the numbers of each type of hydrogen. The NMR spectrometer can electronically "integrate" the area under each peak. It does this by tracing over each peak a vertically rising line, which rises in height by an amount proportional to the area under the peak. Shown in Figure NMR.4 is an NMR spectrum of benzyl acetate, with each of the peaks integrated in this way.

It is important to note that the height of the integral line does not give the absolute number of hydrogens; it gives the *relative* numbers of each type of hydrogen. For a given integral to be of any use, there must be a second integral to which it is referred. The benzyl acetate case gives a good example of this. The first integral rises for 55.5 divisions on the chart paper, the second for 22.0 divisions, and the third for 32.5 divisions. These num-

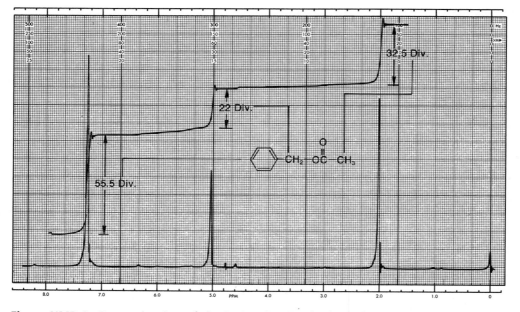

Figure NMR.4 Determination of the integral ratios for benzyl acetate.

bers are relative and give the *ratios* of the various types of protons. You can find these ratios by dividing each of the larger numbers by the smallest number:

$$\frac{55.5 \text{ div}}{22.0 \text{ div}} = 2.52 \qquad \frac{22.0 \text{ div}}{22.0 \text{ div}} = 1.00 \qquad \frac{32.5 \text{ div}}{22.0 \text{ div}} = 1.48$$

Thus, the number ratio of the protons of each type is $2.52:1.00:1.48$. If you assume that the peak at 5.1 ppm is really caused by two hydrogens, and if you assume that the integrals are slightly in error (this can be as much as 10%), then you can arrive at the true ratios by multiplying each figure by two and rounding off; we then get $5:2:3$. Clearly the peak at 7.3 ppm, which integrates for 5, arises from the resonance of the aromatic ring protons, and the peak at 2.0 ppm, which integrates for 3, is caused by the methyl protons. The two-proton resonance at 5.1 ppm arises from the benzyl protons. Notice then that the integrals give the simplest ratios, but not necessarily the true ratios, of the number of protons of each type.

NMR.4 CHEMICAL ENVIRONMENT AND CHEMICAL SHIFT

If the resonance frequencies of all protons in a molecule were the same, NMR would be of little use to the organic chemist. However, not only do different types of protons have different chemical shifts, they also have a value of chemical shift that characterizes the type of proton they represent. Every type of proton has only a limited range of δ values over which it gives resonance. Hence, the numerical value of the chemical shift for a proton indicates the *type of proton* originating the signal, just as the infrared frequency suggests the type of bond or functional group. Notice, for instance, that the aromatic protons of both phenylacetone (Fig. NMR.3) and benzyl acetate (Fig. NMR.4) have resonance near 7.3 ppm and that both methyl groups attached directly to a carbonyl group have a resonance of approximately 2.1 ppm. Aromatic protons characteristically have resonance near 7–8 ppm, and acetyl groups (the methyl protons) have their resonance near 2 ppm. These values of chemical shift are diagnostic. Notice also how the resonance of the benzyl (—CH_2—) protons comes at a higher value of chemical shift (5.1 ppm) in benzyl acetate than in phenylacetone (3.6 ppm). Being attached to the electronegative element, oxygen, these protons are more deshielded (see Section NMR.5) than the protons in phenylacetone. A trained chemist would have readily recognized the probable presence of the oxygen by the chemical shift shown by these protons.

It is important to learn the ranges of chemical shifts over which the most common types of protons have resonance. Figure NMR.5 is a correlation chart that contains the most essential and frequently encountered types of protons. For the beginner, it is often difficult to memorize a large body of numbers relating to chemical shifts and proton types. You need actually do this only crudely. It is more important to "get a feel" for the regions and the types of protons than to know a string of factual numbers.

The values of chemical shift given in Figure NMR.5 can be easily understood in terms of two factors: local diamagnetic shielding and anisotropy. These two factors are discussed in Sections NMR.5 and NMR.6.

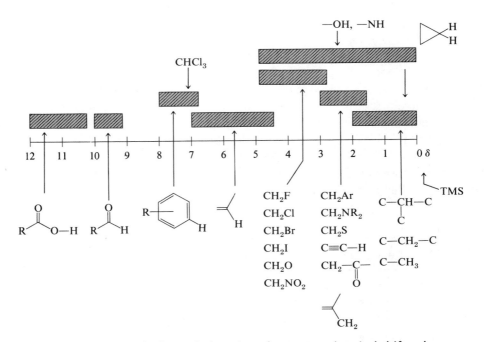

Figure NMR.5 Simplified correlation chart for proton chemical shift values.

NMR.5 LOCAL DIAMAGNETIC SHIELDING

The trend of chemical shifts that is easiest to explain is that involving electronegative elements substituted on the same carbon to which the protons of interest are attached. The chemical shift simply increases as the electronegativity of the attached element increases. This is illustrated in Table NMR.1 for several compounds of the type CH_3X.

Multiple substituents have a stronger effect than a single substituent. The influence of the substituent, an electronegative element having little effect on protons that are more than three carbons away, drops off rapidly with distance. These effects are illustrated in Table NMR.2.

Electronegative substituents attached to a carbon atom, because of their electron-withdrawing effects, reduce the valence electron density around the protons attached to that carbon. These electrons *shield* the proton from the applied magnetic field. This effect, called **local diamagnetic shielding,** occurs because the applied magnetic field induces the valence electrons to circulate and thus to generate an induced magnetic field, which *opposes* the applied field. This is illustrated in Figure NMR.6. Electronegative substituents on carbon reduce the local diamagnetic shielding in the vicinity of the attached protons because they reduce the electron density around those protons. Substituents that produce this effect are said to *deshield* the proton. The greater the electronegativity of the substituent, the more the deshielding of the protons and, hence, the greater the chemical shift of those protons.

NMR.6 ANISOTROPY

Figure NMR.5 clearly shows that several types of protons have chemical shifts not easily explained by simple consideration of the electronegativity of the attached groups. Consider, for instance, the protons of benzene or other aromatic systems. Aryl protons generally have a chemical shift that is as large as that for the proton of chloroform. Alkenes, alkynes, and aldehydes also have protons whose resonance values are not in line with the expected magnitude of any electron-withdrawing effects. In each of these cases, the effect is due to the presence of an unsaturated system (π electrons) in the vicinity of the proton in question. In benzene, for example, when the π electrons in the aromatic ring system are placed in a magnetic field, they are induced to circulate around the ring. This circulation is called a **ring current.** Moving electrons (the ring current) generate a mag-

TABLE NMR.1   Dependence of Chemical Shift of CH_3X on the Element X

Compound CH_3X	CH_3F	CH_3OH	CH_3Cl	CH_3Br	CH_3I	CH_4	$(CH_3)_4Si$
Element X	F	O	Cl	Br	I	H	Si
Electronegativity of X	4.0	3.5	3.1	2.8	2.5	2.1	1.8
Chemical shift (ppm)	4.26	3.40	3.05	2.68	2.16	0.23	0

TABLE NMR.2   Substitution Effects

	CHCl$_3$	CH$_2$Cl$_2$	CH$_3$Cl	—C$\underline{H}_2$Br	—C$\underline{H}_2$—CH$_2$Br	—C$\underline{H}_2$—CH$_2$CH$_2$Br
δ (ppm)	7.27	5.30	3.05	3.30	1.69	1.25

netic field much like that generated in a loop of wire through which a current is induced to flow. The magnetic field covers a spatial volume large enough to influence the shielding of the benzene hydrogens. This is illustrated in Figure NMR.7. The benzene hydrogens are deshielded by the **diamagnetic anisotropy** of the ring. An applied magnetic field is non-uniform (anisotropic) in the vicinity of a benzene molecule because of the labile electrons in the ring that interact with the applied field. Thus, a proton attached to a benzene ring is influenced by *three* magnetic fields: the strong magnetic field applied by the electromagnets of the NMR spectrometer and two weaker fields, one due to the usual shielding by the valence electrons around the proton and the other due to the anisotropy generated by the ring system electrons. It is this anisotropic effect that gives the benzene protons a greater chemical shift than is expected. These protons just happen to lie in a **deshielding** region of this anisotropic field. If a proton were placed in the center of the ring rather than on its periphery, the proton would be shielded, because the field lines would have the opposite direction.

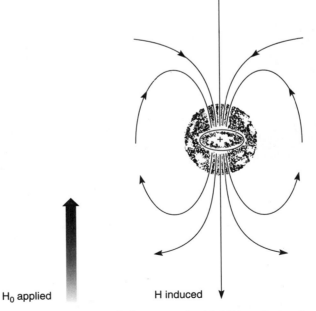

H$_0$ applied   H induced

Figure NMR.6   Local diamagnetic shielding of a proton due to its valence electrons.

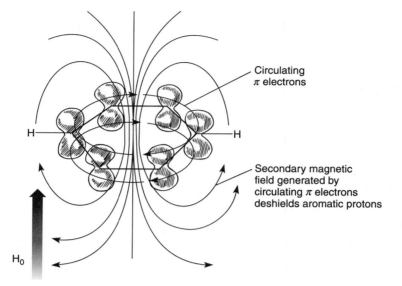

Figure NMR.7   Diamagnetic anisotropy in benzene.

All groups in a molecule that have π electrons generate secondary anisotropic fields. In acetylene, the magnetic field generated by induced circulation of π electrons has a geometry such that the acetylene hydrogens are **shielded.** Hence, acetylenic hydrogens come at a higher field than expected. The shielding and deshielding regions due to the various π electron functional groups have characteristic shapes and directions; they are illustrated in Figure NMR.8. Protons falling within the cones are shielded, and those falling outside the conical areas are deshielded. Because the magnitude of the anisotropic field diminishes with distance, beyond a certain distance anisotropy has essentially no effect.

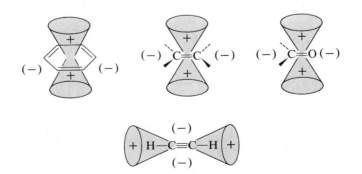

Figure NMR.8   Anisotropy caused by the presence of π electrons in some common multiple-bond systems.

NMR.7 SPIN–SPIN SPLITTING (*N* + 1 RULE)

We have already considered how the chemical shift and the integral (peak area) can give information about the number and type of hydrogens contained in a molecule. A third type of information available from the NMR spectrum is derived from spin–spin splitting. Even in simple molecules, each type of proton rarely gives a single resonance peak. For instance, in 1,1,2-trichloroethane there are two chemically distinct types of hydrogen:

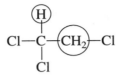

From information given this far, you would predict *two* resonance peaks in the NMR spectrum of 1,1,2-trichloroethane with an area ratio (integral ratio) of 2:1. In fact, the NMR spectrum of this compound has *five* peaks. A group of three peaks (called a **triplet**) exists at 5.77 ppm and a group of two peaks (called a **doublet**) is found at 3.95 ppm. The spectrum is shown in Figure NMR.9. The methine (CH) resonance (5.77 ppm) is split into a triplet, and the methylene resonance (3.95 ppm) is split into a doublet. The area under the three triplet peaks is *one,* relative to an area of *two* under the two doublet peaks.

This phenomenon is called **spin–spin splitting.** Empirically, spin–spin splitting can be explained by the "*n* + 1 rule." Each type of proton "senses" the number of equivalent protons (*n*) on the carbon atom or atoms next to the one to which it is bonded, and its resonance peak is split into *n* + 1 components.

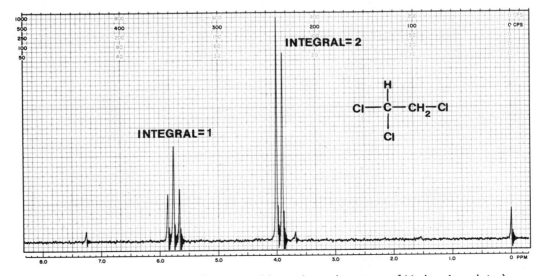

Figure NMR.9   NMR spectrum of 1,1,2-trochloroethane (courtesy of Varian Associates).

Let's examine the case at hand, 1,1,2-trichloroethane, using the $n + 1$ rule. First, the lone methine hydrogen is situated next to a carbon bearing two methylene protons. According to the rule, it has two equivalent neighbors ($n = 2$) and is split into $n + 1 = 3$ peaks (a triplet). The methylene protons are situated next to a carbon bearing only one methine hydrogen. According to the rule, they have one neighbor ($n = 1$) and are split into $n + 1 = 2$ peaks (a doublet).

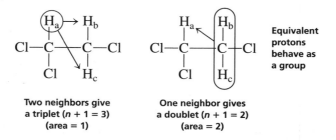

Two neighbors give a triplet ($n + 1 = 3$) (area = 1)

One neighbor gives a doublet ($n + 1 = 2$) (area = 2)

Equivalent protons behave as a group

The spectrum of 1,1,2-trichloroethane can be explained easily by the interaction, or coupling, of the spins of protons on adjacent carbon atoms. The position of absorption of proton H_a is affected by the spins of protons H_b and H_c attached to the neighboring (adjacent) carbon atom. If the spins of these protons are aligned with the applied magnetic field, the small magnetic field generated by their nuclear spin properties will augment the strength of the field experienced by the first-mentioned proton H_a. The proton H_a will thus be *deshielded*. If the spins of H_b and H_c are opposed to the applied field, they will decrease the field experienced by proton H_a. It will then be *shielded*. In each of these situations, the absorption position of H_a will be altered. Among the many molecules in the solution, you will find all the various possible spin combinations for H_b and H_c; hence, the NMR spectrum of the molecular solution will give *three* absorption peaks (a triplet) for H_a because H_b and H_c have three different possible spin combinations (Fig. NMR.10). By a similar analysis, it can be seen that protons H_b and H_c should appear as a doublet.

Some common splitting patterns that can be predicted by the $n + 1$ rule and that are frequently observed in a number of molecules are shown in Figure NMR.11. Notice particularly the last entry, where *both* methyl groups (six protons in all) function as a unit and split the methine proton into a septet ($6 + 1 = 7$).

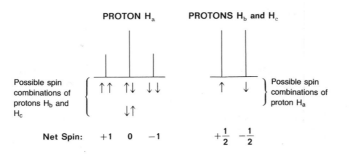

Figure NMR.10   Analysis of the spin–spin splitting pattern of 1,1,2-trichloroethane.

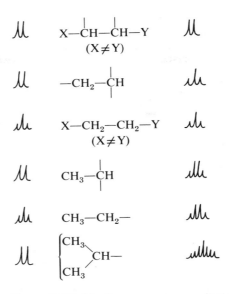

Figure NMR.11   Some common splitting patterns.

NMR.8 THE COUPLING CONSTANT

The quantitative amount of spin–spin interaction between two protons can be defined by the **coupling constant.** The spacing between the component peaks in a simple multiplet is called the coupling constant *J*. This distance is measured on the same scale as the chemical shift and is expressed in Hertz (Hz).

For the interaction of most aliphatic protons in acyclic systems, the magnitudes of the coupling constants are always near 7.5 Hz. See, for instance, the NMR spectrum of 1,1,2-trichloroethane in Figure NMR.9, where the coupling constant is approximately 6 Hz. Different magnitudes of *J* are found for different types of protons. For instance, the *cis* and *trans* protons substituted on a double bond commonly have values where $J_{trans} \cong$ 17 Hz and $J_{cis} \cong$ 10 Hz are typical coupling constants. In ordinary compounds, coupling constants may range anywhere from 0 to 18 Hz. The magnitude of *J* often provides structural clues. You can usually distinguish, for example, between a *cis* olefin and a *trans* olefin on the basis of the observed coupling constants for the vinyl protons. The approximate values of some representative coupling constants are given in Table NMR.3.

NMR.9 MAGNETIC EQUIVALENCE

In the example of spin–spin splitting in 1,1,2-trichloroethane, notice that the two protons H_b and H_c, which are attached to the same carbon atom, do not split one another.

TABLE NMR.3 Representative Coupling Constants and Approximate Values (Hz)

Structure	J (Hz)	Structure	J (Hz)	Structure	J (Hz)
H H | | C—C	6–8	benzene ortho H	ortho 6–10	cyclohexane a,e H	a,a 8–14 a,e 0–7 e,e 0–5
H₂C=CH (trans)	11–18	benzene meta H	meta 1–4	cyclopropane	cis 6–12 trans 4–8
H₂C=CH (cis)	6–15	benzene para H	para 0–2	epoxide (oxirane)	cis 2–5 trans 1–3
H₂C= geminal	0–5				
H₂C=CH—CH (allylic)	4–10	cyclohexene	8–11	cyclopentene	5–7
H—C≡C—CH	0–3				

They behave as an integral group. Actually the two protons H_b and H_c *are* coupled to one another; however, for reasons we cannot explain fully here, protons that are attached to the same carbon and both of which have the *same chemical shift* do not show spin–spin splitting. Another way of stating this is that protons coupled to the same extent to *all* other protons in a molecule do not show spin–spin splitting. Protons that have the same chemical shift and are coupled equivalently to all other protons are *magnetically equivalent* and do not show spin–spin splitting. Thus, in 1,1,2-trichloroethane, protons H_b and H_c have the same value of δ and are coupled by the same value of J to proton H_a. They are magnetically equivalent.

It is important to differentiate magnetic equivalence and chemical equivalence. Note the following two compounds:

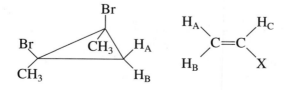

In the cyclopropane compound, the two geminal hydrogens are chemically equivalent; however, they are not magnetically equivalent. Proton H_A is on the same side of the ring as the two halogens. Proton H_B is on the same side of the ring as the two methyl groups. Protons H_A and H_B will have different chemical shifts, will couple to one another, and will show spin-spin splitting. Two doublets will be seen for H_A and H_B. For cyclopropane rings, $J_{geminal}$ is usually around 5 Hz.

Another situation in which protons are chemically equivalent but not magnetically equivalent exists in the vinyl compound. In this example, protons A and B are chemically equivalent but not magnetically equivalent. H_A and H_B have different chemical shifts. In addition, a second distinction can be made between H_A and H_B in this type of compound. Each has a different coupling constant with H_C. The constant J_{AC} is a *cis* coupling constant, and J_{BC} is a *trans* coupling constant. Whenever two protons have different coupling constants relative to a third proton, they are not magnetically equivalent. In the vinyl compound, H_A and H_B do not act as a group to split proton H_C. Each proton acts independently. Thus, H_B splits H_C with coupling constant J_{BC} into a doublet, and then H_A splits each of the components of the doublet into doublets with coupling constant J_{AC}. In such a case, the NMR spectrum must be analyzed graphically, splitting by splitting. An NMR spectrum of a vinyl compound is shown in Figure NMR.12. The graphical analysis of the vinyl portion of the NMR spectrum is in Figure NMR.13.

NMR.10 AROMATIC COMPOUNDS

The NMR spectra of protons on aromatic rings are often too complex to explain by simple theories. However, some simple generalizations can be made that are useful in an-

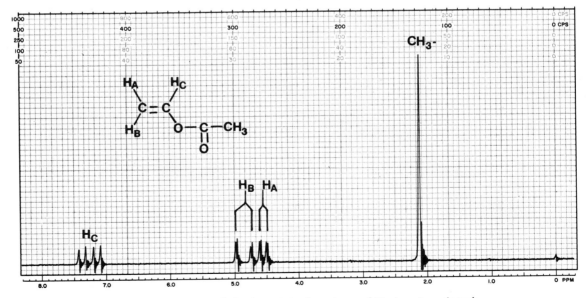

Figure NMR.12 NMR spectrum of vinyl acetate (courtesy of Varian Associates).

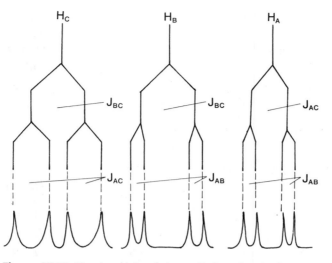

Figure NMR.13   Analysis of the splittings in vinyl acetate.

alyzing the aromatic region of the NMR spectrum. First of all, most aromatic protons have resonance near 7.0 ppm. In monosubstituted rings in which the ring substituent is an alkyl group, all the ring protons often have chemical shifts that are very nearly identical, and the five ring protons may appear as if they gave rise to an overly broad singlet (Fig. NMR.14A). If an electronegative group is attached to the ring, all the ring protons are

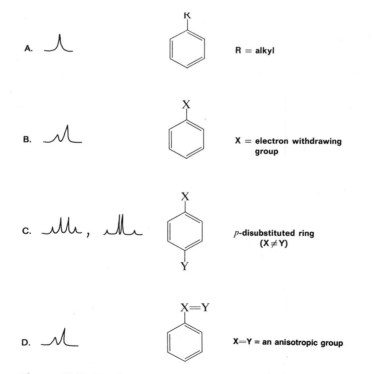

Figure NMR.14   Some common aromatic patterns.

shifted downfield from where they would appear in benzene. However, often the *ortho* protons are shifted more than the others, as they are more affected by the group. This often gives rise to an absorption pattern like that in Figure NMR.14B. In a *para*-disubstituted ring with two substituents X and Y that are identical, all the protons in the ring are chemically and magnetically equivalent, and a singlet is observed. If X is different from Y in electronegativity, however, a pattern like that shown in the left side of Figure NMR.14C is often observed, clearly identifying a *p*-disubstituted ring. If X and Y are more nearly similar, a pattern more like the one on the right is observed. In monosubstituted rings that have a carbonyl group or a double bond attached directly to the ring, a pattern like that in Figure NMR.14D is not uncommon. In this case, the *ortho* protons of the ring are influenced by the anisotropy of the π systems that make up the CO and CC double bonds and are deshielded by them. In other types of substitution, such as *ortho* or *meta,* or polysubstituted ring systems, the patterns may be much more complicated and require an advanced analysis.

NMR.11 PROTONS ATTACHED TO ATOMS OTHER THAN CARBON

Protons attached to atoms other than carbon often have a widely variable range of absorptions. Several of these groups are tabulated in Table NMR.4. In addition, under the usual conditions of determining an NMR spectrum, protons on heteroelements normally do not couple with protons on adjacent carbon atoms to give spin–spin splitting. This is primarily because such protons often exchange very rapidly with those of the solvent medium. The absorption position is variable because these groups also undergo varying degrees of hydrogen bonding in solutions of different concentrations. The amount of hydrogen bonding that occurs with a proton radically affects the valence electron density around that proton and produces correspondingly large changes in the chemical shift. The absorption peaks for protons that have hydrogen bonding or are undergoing exchange are frequently broad relative to other singlets and can often be recognized on that basis. For a different reason, called quadrupole broadening, protons attached to nitrogen atoms often show an extremely broad resonance peak, often almost indistinguishable from the baseline.

TABLE NMR.4   *Typical Ranges for Groups with Variable Chemical Shift*

Acids	RCOOH	10.5–12.0 ppm
Phenols	ArOH	4.0–7.0
Alcohols	ROH	0.5–5.0
Amines	RNH_2	0.5–5.0
Amides	$RCONH_2$	5.0–8.0
Enols	CH=CH—OH	≥ 15

NMR.12 SPECTRA AT HIGHER FIELD STRENGTH

Occasionally, the 60-MHz spectrum of an organic compound, or a portion of it, is almost undecipherable because the chemical shifts of several groups of protons are all very similar. In these cases, all the proton resonances occur in the same area of the spectrum, and peaks often overlap so extensively that individual peaks and splittings cannot be extracted. One of the ways in which such a situation can be simplified is by using a spectrometer that operates at a higher frequency. Although both 60- and 90-MHz instruments are quite common, it is not unusual to find instruments with operating frequencies of 100, 220, 300 MHz, or even higher.

Although NMR coupling constants are not dependent on the frequency or the field strength of operation of the NMR spectrometer, chemical shifts in Hertz are dependent on these parameters. This circumstance can often be used to simplify an otherwise undecipherable spectrum. Suppose, for instance, that a compound contained three multiplets: a quartet and two triplets derived from groups of protons with very similar chemical shifts. At 60 MHz these peaks might overlap, as illustrated in Figure NMR.15, and simply give an unresolved envelope of absorption.

Figure NMR.15 also shows the spectrum of the same compound at two higher field strengths (frequencies). In redetermining the spectrum at higher field strengths, the coupling constants do not change, but the chemical shifts in Hertz (not ppm) of the proton groups (H_A, H_B, H_C) responsible for the multiplets do increase. It should be noted that at 220 MHz the individual multiplets are cleanly separated and resolved.

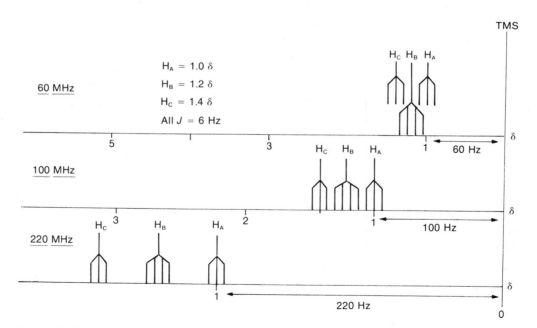

Figure NMR.15 A comparison of the spectrum of a compound with overlapping multiplets at 60 MHz with spectra of the same compound also determined at 100 MHz and 220 MHz. The drawing is to scale.

NMR.13 CHEMICAL SHIFT REAGENTS

Researchers have known for some time that interactions between molecules and solvents, such as those due to hydrogen bonding, can cause large changes in the resonance positions of certain types of protons (e.g., hydroxyl and amino). They have also known that the resonance positions of some groups of protons can be greatly affected by changing from the usual NMR solvents such as CCl_4 and $CDCl_3$ to solvents like benzene, which impose local anisotropic effects on surrounding molecules. In many cases, it was possible to resolve partially overlapping multiplets by such a solvent change. However, the use of chemical shift reagents for this purpose dates from about 1969. Most of these chemical shift reagents are organic complexes of paramagnetic rare earth metals from the lanthanide series of elements. When these metal complexes are added to the compound whose spectrum is being determined, profound shifts in the resonance positions of the various groups of protons are observed. The direction of the shift (upfield or downfield) depends primarily on which metal is being used. Complexes of europium, erbium, thulium, and ytterbium shift resonances to lower field; complexes of cerium, praseodymium, neodymium, samarium, terbium, and holmium generally shift resonances to higher field. The advantage of using such reagents is that shifts similar to those observed at higher field can be induced without the purchase of an expensive higher field instrument.

Of the lanthanides, europium is probably the most commonly used metal. Two of its widely used complexes are *tris*-(dipivalomethanato)europium and *tris*-(6,6,7,7,8,8,8-heptafluoro-2,2-dimethyl-3,5-octanedionato)europium. These are frequently abbreviated $Eu(dpm)_3$ and $Eu(fod)_3$, respectively.

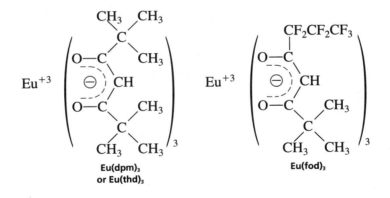

These lanthanide complexes produce spectral simplifications in the NMR spectrum of any compound that has a relatively basic pair of electrons (unshared pair) that can coordinate with Eu^{3+}. Typically, aldehydes, ketones, alcohols, thiols, ethers, and amines will all interact:

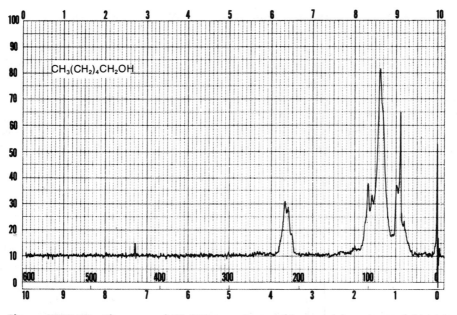

Figure NMR.16   The normal 60-MHz spectrum of hexanol (courtesy of Aldrich Chemical Co.).

The amount of shift that a given group of protons will experience depends (1) on the distance separating the metal (Eu^{3+}) and that group of protons, and (2) on the concentration of the shift reagent in the solution. Because of the latter dependence, it is necessary when reporting a lanthanide-shifted spectrum to report the number of mole equivalents of shift reagent used or its molar concentration.

The distance factor is illustrated in the spectra of hexanol, which are given in Figures NMR.16 and NMR.17. In the absence of shift reagent, the normal spectrum is obtained

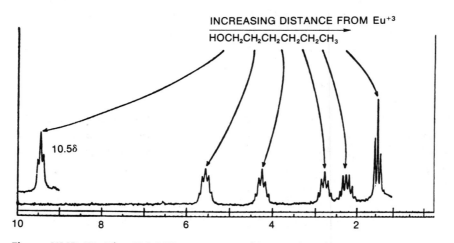

Figure NMR.17   The 100-MHz spectrum of hexanol with 0.29 mole equivalents of $Eu(dpm)_3$. (Reprinted with permission from J.K.M. Sanders and D.H. Williams, *Chem. Commun.*, 422[1970].)

(Figure NMR.16). Only the triplet of the terminal methyl group and the triplet of the methylene group next to the hydroxyl are resolved in the spectrum. The other protons (aside from OH) are found together in a broad unresolved group. With shift reagent added (Figure NMR.17), each of the methylene groups is clearly separated and resolved into the proper multiplet structure. The spectrum is first order and simplified; all the splittings are explained by the $n + 1$ rule.

One final consequence of using a shift reagent should be noted. Notice in Figure NMR.17 that the multiplets are not as nicely resolved into sharp peaks as you might expect. This is due to the fact that shift reagents cause a small amount of peak broadening. At high shift reagent concentrations this problem becomes serious, but at most useful concentrations the amount of broadening experienced is tolerable.

REFERENCES

Friebolin, H. *Basic One- and Two-Dimensional NMR Spectroscopy.* New York: VCH Publishers, 1991.

Jackman, L. M., and Sternhell, S. *Applications of Nuclear Magnetic Resonance in Organic Chemistry.* 2nd ed. New York: Pergamon Press, 1969.

Paudler, W. W. *Nuclear Magnetic Resonance.* New York: John Wiley, 1987.

Pavia, D. L., Lampman, G. M., and Kriz, G. S. *Introduction to Spectroscopy.* 2nd ed. Philadelphia: W. B. Saunders, 1996.

Sanders, J. K. M., and Hunter, B. K. *Modern NMR Spectroscopy.* 2nd ed. Oxford: Oxford University Press, 1993.

Silverstein, R. M., Bassler, G. C., and Morrill, T. C. *Spectrometric Identification of Organic Compounds.* 5th ed. New York: John Wiley 1991.

A P P E N D I X   5

Carbon-13 Nuclear Magnetic Resonance Spectroscopy

CMR.1 CARBON-13 NUCLEAR MAGNETIC RESONANCE

Carbon-12, the most abundant isotope of carbon, does not possess spin ($I = 0$); it has both an even atomic number and an even atomic weight. The second principal isotope of carbon, ^{13}C, however, does have the nuclear spin property ($I = \frac{1}{2}$). ^{13}C atom resonances are not easy to observe, due to a combination of two factors. First, the natural abundance of ^{13}C is low; only 1.08% of all carbon atoms are ^{13}C. Second, the magnetic moment μ of ^{13}C is low. For these two reasons, the resonances of ^{13}C are about *6000 times weaker* than those of hydrogen. With special Fourier transform instrumental techniques, which are not discussed here, it is possible to observe ^{13}C nuclear magnetic resonance (carbon-13) spectra on samples that contain only the natural abundance of ^{13}C.

The most useful parameter derived from carbon-13 spectra is the chemical shift. Integrals are unreliable and are not necessarily related to the relative numbers of ^{13}C atoms present in the sample. Hydrogens that are attached to ^{13}C atoms cause spin–spin splitting, but spin–spin interaction between adjacent carbon atoms is rare. With its low natural abundance (0.0108), the probability of finding two ^{13}C atoms adjacent to one another is extremely low.

CMR.2 COMPLETELY COUPLED ^{13}C SPECTRA

Figure CMR.1 shows the carbon-13 spectrum of ethyl phenylacetate. Consider first the upper trace shown in the figure. Chemical shifts, just as in proton NMR, are reported by the number of ppm (δ units) that the peak is shifted downfield from TMS. Keep in mind, however, that it is a ^{13}C atom of the methyl group of TMS that is being observed, not the 12 methyl hydrogens. Notice the extent of the scale. Although the chemical shifts of protons encompass a range of only about 20 ppm, ^{13}C chemical shifts cover an extremely wide range of up to 200 ppm! Under these circumstances, even adjacent —CH_2— carbons in a long hydrocarbon chain generally have their own distinct resonance peaks, and these peaks are clearly resolved. It is unusual to find any two carbon atoms in a molecule having resonance at the same chemical shift unless these two carbon atoms are equivalent by symmetry.

Returning to the upper spectrum in Figure CMR.1, you can see that the first quartet downfield from TMS (14.2 ppm) corresponds to the carbon of the methyl group. It is

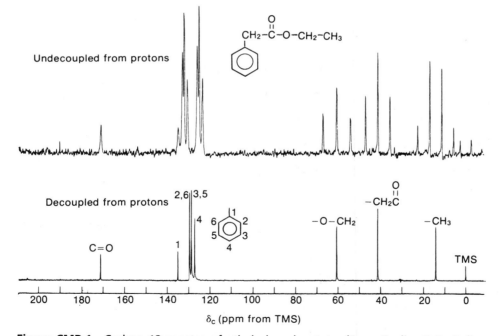

Figure CMR.1   Carbon-13 spectra of ethyl phenylacetate. (From Rodig, O.R., Bell, C.E., Jr., and Clark, A.K., *Organic Chemistry Laboratory; Standard and Microscale Experiments,* 1990 by Saunders College Publishing, Philadelphia, PA. Reprinted by permission of the publisher.)

split into a quartet ($J = 127$ Hz) by the three attached hydrogen atoms. In addition, although it cannot be seen on the scale of this spectrum, each of the quartet lines is split into a *closely spaced* triplet ($J = $ ca 1 Hz). This additional fine splitting is caused by the two protons on the adjacent —CH_2— group. These are geminal couplings (H—C—^{13}C) of a type that commonly occurs in carbon-13 spectra, with coupling constants that are generally small ($J = 0-2$ Hz). The quartet is caused by **direct coupling** (^{13}C—H). Direct coupling constants are larger, usually about 100–200 Hz and are more obvious on the scale in which the spectrum is presented.

There are two —CH_2— groups in ethyl phenylacetate. The one corresponding to the ethyl —CH_2— group is found further downfield (60.6 ppm), as this carbon is deshielded by the attached oxygen. It is a triplet because of the two attached hydrogens. Again, although it is not seen in this unexpanded spectrum, each of the triplet peaks is finely split into a quartet by the three hydrogens on the adjacent methyl group. The benzyl —CH_2— carbon is the intermediate triplet (41.4 ppm). Furthest downfield is the carbonyl group carbon (171.1 ppm). On the scale of presentation, it is a singlet (no directly attached hydrogens), but because of the adjacent benzyl —CH_2— group, it is actually split finely into a triplet. The aromatic ring carbons also appear in the spectrum, and they have resonances over the range from 127 ppm to 136 ppm.

CMR.3 BROAD-BAND DECOUPLED ^{13}C SPECTRA

Although the splittings in a simple molecule such as ethyl phenylacetate yield interesting structural information, namely the number of hydrogens attached to each carbon (as well as those adjacent if the spectrum is expanded), for large molecules the carbon-13 spectrum becomes very complex due to these splittings, and the splitting patterns often overlap. It is customary, therefore, to decouple *all* the protons in the molecule by irradiating them simultaneously with a broad spectrum of frequencies in the proper range. This type of spectrum is said to be **completely decoupled.** The completely decoupled spectrum is much simpler and, for larger molecules, much easier to interpret. The decoupled spectrum of ethyl phenylacetate is presented in the lower trace of Figure CMR.1.

In the completely decoupled carbon-13 spectrum, each peak represents a different carbon atom. If two carbons are represented by a single peak, they must be equivalent by symmetry. Thus, the carbons at positions 2 and 6 of the aromatic ring of ethyl phenylacetate give a single peak, and the carbons at positions 3 and 5 also give a single peak in the lower spectrum of Figure CMR.1.

CMR.4 CHEMICAL SHIFTS

Just as is the case for proton spectra, the chemical shift of each carbon indicates both its type and its structural environment. In fact, a correlation chart can be presented for ^{13}C chemical shift ranges, similar to the correlation chart for proton resonances shown in Figure NMR.5. Figure CMR.2 gives typical chemical shift ranges for the types of carbon resonances.

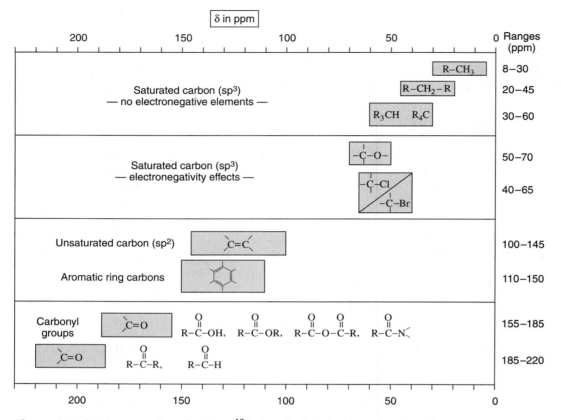

Figure CMR.2   A correlation chart for ^{13}C chemical shifts (ppm from TMS).

Electronegativity, hybridization, and anisotropy effects all influence ^{13}C chemical shifts, just as they do for protons, but in a more complex fashion. These factors are not discussed in any detail here, but note that the —CH_2— group carbon attached to oxygen in ethyl phenylacetate has a larger chemical shift than the —CH_2— carbon of the benzyl group. Note also that the carbonyl carbon appears relatively far downfield, probably due to an anisotropy effect.

CMR.5 SOME SAMPLE SPECTRA

The following spectra illustrate some of the effects that can be observed in carbon-13 spectra. The spectrum of 2,2-dimethylbutane is presented in Figure CMR.3. Notice that, although this compound has six total carbon atoms, 2,2-dimethylbutane shows only four peaks in the carbon-13 spectrum along with the solvent peaks ($CDCl_3$) and TMS. The carbon-13 atoms that are equivalent appear at the same chemical shift value. Thus, a single methyl carbon **a** appears at highest field (8.8 ppm), while the three equivalent methyl carbons **b** appear at 28.9 ppm. The quaternary carbon **c** gives rise to the small peak of 30.4 ppm, whereas the methylene carbon **d** appears at 36.5 ppm. The relative sizes of

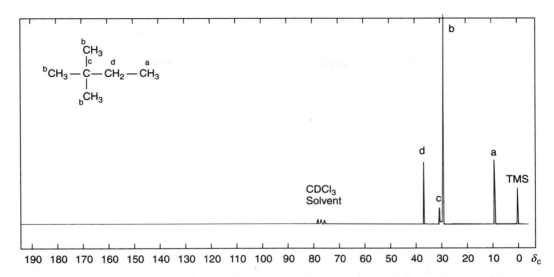

Figure CMR.3 Carbon-13 spectrum of 2,2-dimethylbutane. (From Johnson, L.F., and Jankowski, W.C., *Carbon-13 NMR Spectra: A Collection of Assigned, Coded, and Indexed Spectra,* 1972 by John Wiley and Sons, New York. Reprinted by permission of the publisher.)

peaks gives some idea of the numbers of each type of carbon atom present in the molecule. For example, in Figure CMR.3 notice that the peak at 28.9 ppm (**b**) is much larger than the other peaks. A characteristic of proton-decoupled ^{13}C NMR spectra is that carbon atoms that do not have hydrogens attached to them generally appear as weak peaks. Thus, the quaternary carbon at 30.4 ppm (**c**) is very weak (see Section CMR.6).

The presence of an electronegative element should deshield a carbon atom closest to it, as is illustrated in the cases of bromocyclohexane (Fig. CMR.4) and cyclohexanol

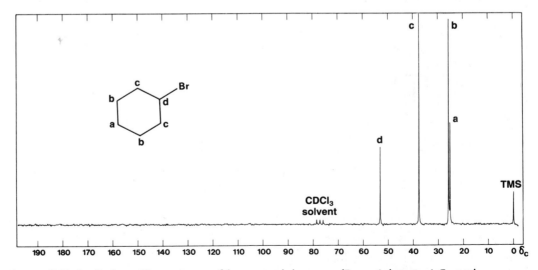

Figure CMR.4 Carbon-13 spectrum of bromocyclohexane. (From Johnson, L.F., and Jankowski, W.C., *Carbon-13 NMR Spectra: A Collection of Assigned, Coded, and Indexed Spectra,* 1972 by John Wiley and Sons, New York. Reprinted by permission of the publisher.)

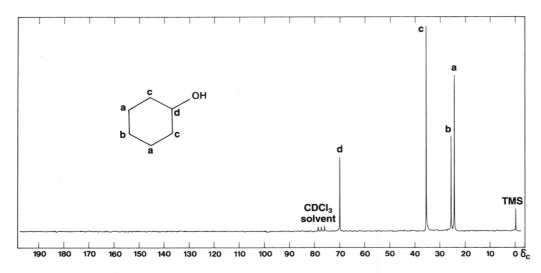

Figure CMR.5 Carbon-13 spectrum of cyclohexanol. (From Johnson, L.F., and Jankowski, W.C., *Carbon-13 NMR Spectra: A Collection of Assigned, Coded, and Indexed Spectra,* 1972 by John Wiley and Sons, New York. Reprinted by permission of the publisher.)

(Fig. CMR.5). The carbon bearing the bromine in bromocyclohexane appears at 53.0 ppm; the carbon bearing the hydroxyl group of cyclohexanol appears at 70.0 ppm. In each of these cases, note that as the ring carbons are located farther away from the electronegative element, their resonances appear at higher field. A carbon attached to a double bond appears deshielded, due to diamagnetic anisotropy. This effect can be seen in the spectrum of cyclohexene (Fig. CMR.6). The carbon atoms of the double bond appear at 127.2

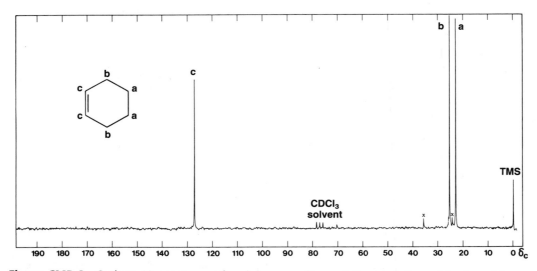

Figure CMR.6 Carbon-13 spectrum of cyclohexene. (From Johnson, L.F., and Jankowski, W.C., *Carbon-13 NMR Spectra: A Collection of Assigned, Coded, and Indexed Spectra,* 1972 by John Wiley and Sons, New York. Reprinted by permission of the publisher.)

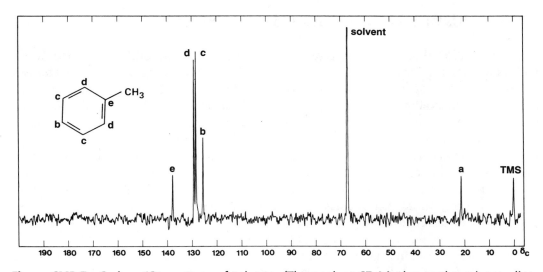

Figure CMR.7 Carbon-13 spectrum of toluene. (The peak at 67.4 is due to the solvent, dioxane.) (From Johnson, L.F., and Jankowski, W.C., *Carbon-13 NMR Spectra: A Collection of Assigned, Coded, and Indexed Spectra,* 1972 by John Wiley and Sons, New York. Reprinted by permission of the publisher.)

ppm. Again, it can be seen that as carbon atoms are located farther from the double bond, their resonances appear at higher field. The effect of diamagnetic anisotropy can be seen in the spectrum of toluene (Fig. CMR.7), where the carbon atoms of the aromatic ring appear at low field (125.5–137.7 ppm). Finally, the strong deshielding experienced by the carbon atom of a carbonyl group can be seen in the carbon-13 spectrum of cyclohexanone (Fig. CMR.8). The carbon atom appears at a chemical shift of 211.3 ppm.

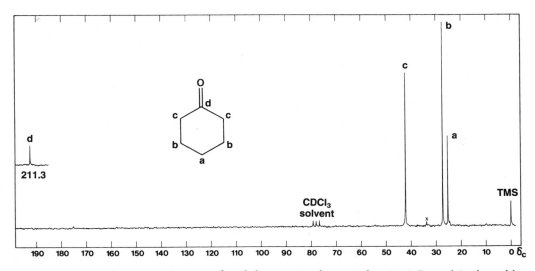

Figure CMR.8 Carbon-13 spectrum of cyclohexanone. (From Johnson, L.F., and Jankowski, W.C., *Carbon-13 NMR Spectra: A Collection of Assigned, Coded, and Indexed Spectra,* 1972 by John Wiley and Sons, New York. Reprinted by permission of the publisher.)

CMR.6 NUCLEAR OVERHAUSER EFFECT

As mentioned previously, integrals (areas under peaks) are not as reliable for carbon spectra as they are for hydrogen spectra. This is due in part to the **nuclear Overhauser effect.** This effect operates when two dissimilar adjacent atoms (in this case carbon and hydrogen) both exhibit spins and are NMR active. The atoms can influence the NMR absorption intensities of each other. The effect can be either positive or negative, but in the case of carbon-13 interacting with hydrogen, the effect is positive. As a result, carbon-13 NMR absorptions vary in intensity with respect to the number of hydrogen atoms that are directly attached to the carbon atom being observed. In general, the more hydrogens that are attached to a given carbon, the stronger its NMR absorption. Other factors also influence the absorption intensities (they are related to molecular relaxation phenomena), so the number of attached hydrogens can only be taken as a single factor influencing absorption intensity; this is often a very helpful factor in deciding which carbon to assign to a given absorption. In Figure CMR.1 note the low intensity of the carbonyl carbon (172 ppm), and in Figure CMR.7 note the low intensity of the ring carbon to which the methyl group is attached (138 ppm). The carbonyl peak in cyclohexanone (Fig. CMR.8) is also weak. None of these carbons has attached hydrogens.

CMR.7 AN EXAMPLE OF SYMMETRY

As one example of the utility of carbon-13 experiments, consider the cases of the isomers 1,2- and 1,3-dichlorobenzene. Although these isomers could be difficult to dis-

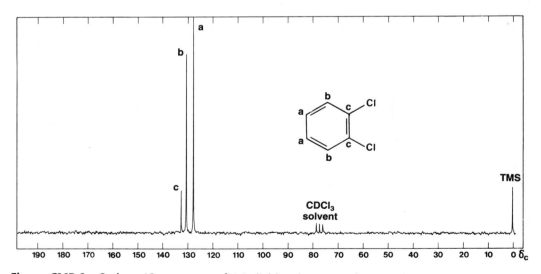

Figure CMR.9 Carbon-13 spectrum of 1,2-dichlorobenzene. (From Johnson, L.F., and Jankowski, W.C., *Carbon-13 NMR Spectra: A Collection of Assigned, Coded, and Indexed Spectra,* 1972 by John Wiley and Sons, New York. Reprinted by permission of the publisher.)

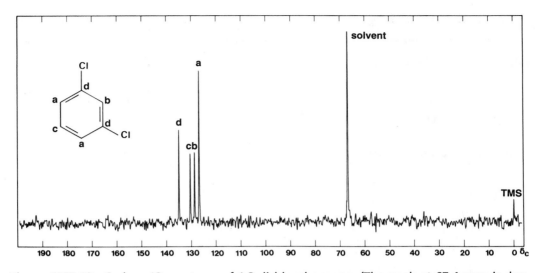

Figure CMR.10 Carbon-13 spectrum of 1,3-dichlorobenzene. (The peak at 67.4 ppm is due to the solvent, dioxane.) (From Johnson, L.F., and Jankowski, W.C., *Carbon-13 NMR Spectra: A Collection of Assigned, Coded, and Indexed Spectra,* 1972 by John Wiley and Sons, New York. Reprinted by permission of the publisher.)

tinguish from one another on the basis of their boiling points or their infrared spectra, each can be identified clearly by their carbon-13 spectra. 1,2-Dichlorobenzene has a plane of symmetry that gives it only three different types of carbon atoms. 1,3-Dichlorobenzene has a plane of symmetry that gives it four different types of carbon atoms. The proton-decoupled carbon-13 spectra of these two compounds are shown in Figures CMR.9 and CMR.10, respectively. It is easy to see the differences in the carbon-13 spectra of these two isomers.

REFERENCES

Breitmaier, E., and Voelter, W. *Carbon-13 NMR Spectroscopy.* 3rd ed. New York: VCH Publishers, 1987.

Friebolin, H. *Basic One- and Two-Dimensional NMR Spectroscopy.* New York: VCH Publishers, 1991.

Levy, G. C., Lichter, R. L., and Nelson, G. L. *Carbon-13 Nuclear Magnetic Resonance for Organic Chemists.* 2nd ed. New York: John Wiley, 1980.

Paudler, W. W. *Nuclear Magnetic Resonance.* New York: John Wiley, 1987.

Pavia, D. L., Lampman, G. M., and Kriz, G. S. *Introduction to Spectroscopy.* 2nd ed. Philadelphia: W. B. Saunders, 1996.

Sanders, J. K. M., and Hunter, B. K. *Modern NMR Spectroscopy.* 2nd ed. Oxford: Oxford University Press, 1993.

Silverstein, R. M., Bassler, G. C., and Morrill, T. C. *Spectrometric Identification of Organic Compounds.* 5th ed. New York: John Wiley, 1991.

A P P E N D I X   6

Index of Spectra

Infrared Spectra

^{1}H NMR Spectra

^{13}C NMR Spectra

Ultraviolet-Visible Spectra

Mixtures

INDEX